Representative Elements (*p* Series)

Transition Elements

	IB	IIB	IIIA	IVA	VA	VIA	VIIA	VIIIA
								2 Helium **He** 4.0026
			5 Boron **B** 10.811	6 Carbon **C** 12.0112	7 Nitrogen **N** 14.0067	8 Oxygen **O** 15.9994	9 Fluorine **F** 18.9984	10 Neon **Ne** 20.179
			13 Aluminum **Al** 26.9815	14 Silicon **Si** 28.086	15 Phosphorous **P** 30.9738	16 Sulfur **S** 32.064	17 Chlorine **Cl** 35.453	18 Argon **Ar** 39.948
28 Nickel **Ni** 58.71	29 Copper **Cu** 63.546	30 Zinc **Zn** 65.38	31 Gallium **Ga** 69.723	32 Germanium **Ge** 72.59	33 Arsenic **As** 74.922	34 Selenium **Se** 78.96	35 Bromine **Br** 79.904	36 Krypton **Kr** 83.80
46 Palladium **Pd** 106.4	47 Silver **Ag** 107.868	48 Cadmium **Cd** 112.40	49 Indium **In** 114.82	50 Tin **Sn** 118.69	51 Antimony **Sb** 121.75	52 Tellurium **Te** 127.60	53 Iodine **I** 126.904	54 Xenon **Xe** 131.30
78 Platinum **Pt** 195.09	79 Gold **Au** 196.967	80 Mercury **Hg** 200.59	81 Thalium **Tl** 204.37	82 Lead **Pb** 207.19	83 Bismuth **Bi** 208.980	84 Polonium **Po** (209)	85 Astatine **At** (210)	86 Radon **Rn** (222)

Transition Elements (*f* Series)

63 Europium **Eu** 151.96	64 Gadolinium **Gd** 157.25	65 Terbium **Tb** 158.925	66 Dysprosium **Dy** 162.50	67 Holmium **Ho** 164.930	68 Erbium **Er** 167.26	69 Thulium **Tm** 168.934	70 Ytterbium **Yb** 173.04	71 Lutetium **Lu** 174.97
95 Americium **Am** (243)	96 Curium **Cm** (247)	97 Berkelium **Bk** (247)	98 Californium **Cf** 242.058	99 Einsteinium **Es** (254)	100 Fermium **Fm** 257.095	101 Mendelevium **Md** 258.10	102 Nobelium **No** 259.101	103 Lawrencium **Lr** 260.105

third edition

CHEMISTRY

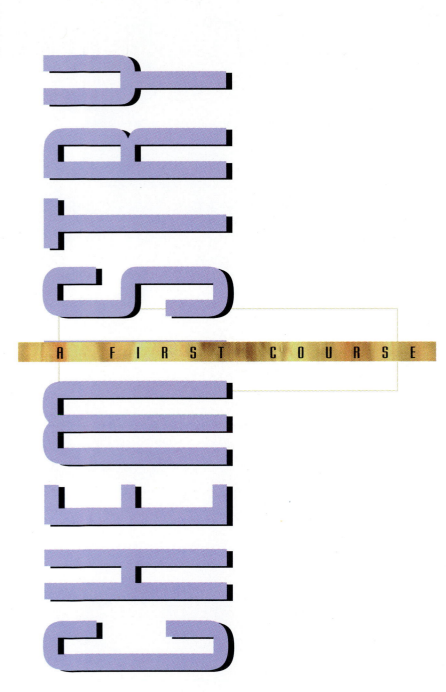

A FIRST COURSE

third edition

CHEMISTRY

A FIRST COURSE

jacqueline i. kroschwitz
•
melvin winokur
•
a. bryan lees
kean college of new jersey

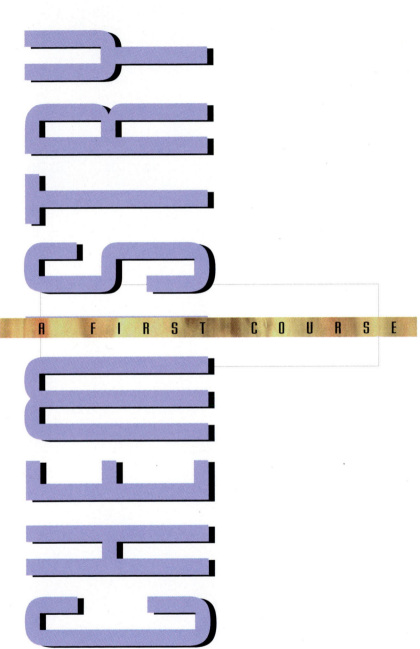

WCB Wm. C. Brown Publishers
Dubuque, IA Bogota Boston Buenos Aires Caracas Chicago
Guilford, CT London Madrid Mexico City Sydney Toronto

Book Team

Editor *Megan Johnson*
Developmental Editor *John Berns*
Production Editor *Karen L. Nickolas*
Art Editor *Rachel Imsland*
Photo Editor *Lori Hancock*
Permissions Coordinator *Mavis M. Oeth*

Wm. C. Brown Publishers
A Division of Wm. C. Brown Communications, Inc.

Vice President and General Manager *Beverly Kolz*
Vice President, Publisher *Jeffrey L. Hahn*
Vice President, Director of Sales and Marketing *Virginia S. Moffat*
Vice President, Director of Production *Colleen A. Yonda*
National Sales Manager *Douglas J. DiNardo*
Marketing Manager *Jane Ducham*
Advertising Manager *Janelle Keeffer*
Production Editorial Manager *Renée Menne*
Publishing Services Manager *Karen J. Slaght*
Permission/Records Manager *Connie Allendorf*

Wm. C. Brown Communications, Inc.

President and Chief Executive Officer *G. Franklin Lewis*
Senior Vice President, Operations *James H. Higby*
Corporate Senior Vice President, President of WCB Manufacturing *Roger Meyer*
Corporate Senior Vice President and Chief Financial Officer *Robert Chesterman*

Copyedited by Janet Jessup

Cover credit: Cardinal in the snow. © L. West
Photo Research by Susan Kaprov

The credits section for this book begins on page 617 and
is considered an extension of the copyright page

A Times Mirror Company

Library of Congress Catalog Card Number: 94–70438

ISBN 0–697–27282–6

Printed in the United States of America by Wm. C. Brown Communications, Inc.,
2460 Kerper Boulevard, Dubuque, IA 52001

10 9 8 7 6 5 4 3 2 1

brief contents

contents

chapter 5

electronic structure of the atom 117

chapter 6

chemical bonding 147

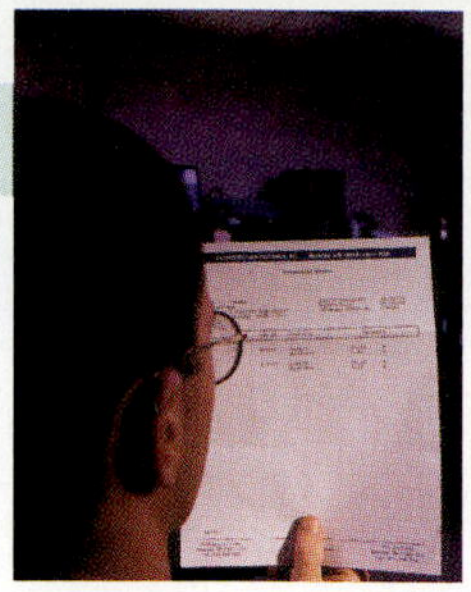

chapter 7

nomenclature, weight, and percentage composition 193

chapter 8

the mole concept 217

chapter 9

chemical reactions 243

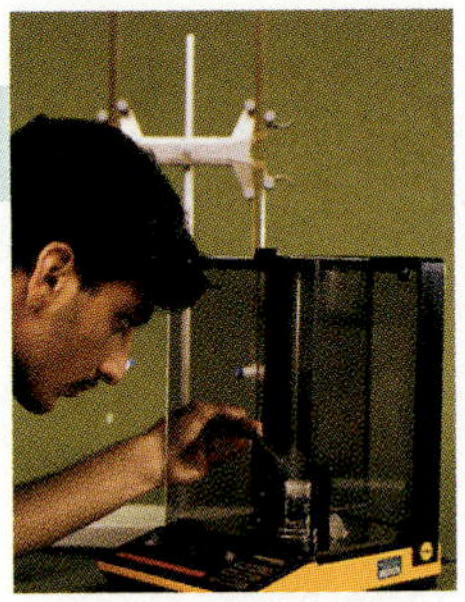

C10000

chapter 14

chemical equilibrium 427

chapter 15

acids and bases 453

chapter 16

oxidation-reduction 491

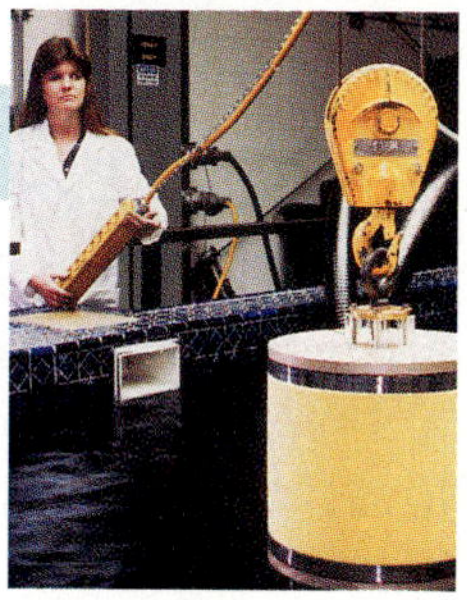

chapter 17

nuclear chemistry 525

appendix a
basic arithmetic review 556

appendix b
chemical arithmetic using a hand calculator 563

appendix c
arithmetic with signed numbers 566

appendix d
vapor pressure of water at various temperatures 570

appendix e
answers to odd-numbered problems 571

preface

Chemistry is the science that deals with matter and the transformations that it undergoes. To the student who has not previously taken a chemistry course this may be an intimidating topic. Therefore, *Chemistry: A First Course* was developed as a one-semester, preparatory textbook for students with a non-chemistry background. This book will be a foundation on which to build a successful progression into future two-semester General Chemistry courses. We will present chemistry to the student as an interdisciplinary subject with broad points of interest for people in all walks of life, integrating the function and uses of chemicals and chemistry in the everyday world.

It is for the students who seek a broad base of knowledge that we have written the present text. As you will later note, the text has undergone a great deal of scrutiny to assure the accuracy and versatility of this book. New supplements and pedagogical features have been added to aid the students and to utilize the wide range of benefits that other technologies have to offer.

new features in this edition

Highlighted Regions: To help the student in organizing material, the authors have included shaded summary areas. These shaded summaries include rules, tables, charts, etc.

ChemLab: Chemistry progress in the real world is accomplished through curiosity and experimentation. We have included in each chapter a laboratory exercise that introduces the student to a potentially hands-on experience. These laboratory exercises are closely tied to the material in the chapter, making them relevant to understanding the lecture material.

ChemQuest: These essays are located throughout the text to help the student make connections between the text and everyday applications. These essays lead the student to a series of questions that may form the basis of a term paper or group project.

Full-Color Art Program: Color is another way to reinforce learning and we have used color consistently throughout the text. Lavender is used for electrons, green for nuclei, blue for bases, red for acids, and so on. This consistent use of color not only improves the asthetics of the textbook, but also supports student comprehension.

Design: Essential to this edition is a new design. With a large trim size, we were better able to use white space and larger, clearer visuals effectively, providing the student with a textbook that is clear, inviting, and well-organized.

Features retained and improved from earlier editions include:

In-Chapter Exercises, Solutions, and Problems: Because problem-solving is most easily learned by example and practice, we have included throughout the text a number of examples that show the student, step-by-step, precisely how to determine the correct answer. Whenever possible, exercises are included with each example to build student self-confidence. The student will find the answers to the odd-numbered problems in the back of the book.

End-of-Chapter Accomplishments: This feature is a combination of chapter outline, student comprehension checklist, and chapter summary. This review of chapter content is an elemental tool used to help the student organize and categorize material that often appears foreign and overwhelming. By simply going through this detailed checklist the students will be able to assess their own understanding.

Key Terms: Key Terms appear in boldface when they are introduced within the text and are immediately defined in context. All key terms are also defined in the glossary.

The End-of-Chapter Problems: There are a wide variety and number of problems located at the end of each chapter. There are two different types of problems that require the students to think a little harder, testing their understanding. Furthermore, since students learn by various means, the problems at the end of the chapter contain not only the traditional mathematically based problems but problems that ask the student to explain, identify, predict, arrange, state evidence, etc. The students are encouraged to explore a new dimension of their understanding of the chapter material. All of this leads to students who are better prepared for any chemistry that they must take in the future.

End of Book Glossary of Key Terms: All new terms are in boldface and each term is defined in the alphabetized glossary at the end of the textbook.

ancillaries

An extensive supplemental package has been designed to support this text. It includes the following elements.

1. **Instructor's Manual** The Instructor's Manual contains the printed test item file, answers to the text's even-numbered problems and exercises, detailed solutions to the text's odd-numbered problems and exercises, and a list of the transparencies. Written by the authors, this unique ancillary also contains suggestions for organizing lectures, additional "Perspectives," and a list of each chapter's key problems and concepts.
2. **Student Study Guide/ Solutions Manual** A separate Student Study Guide/Solutions Manual is available for the students. It contains the answers and solutions for the half of the chapter problems found in the Instructor's Manual. It also offers students a variety of exercises and keys for testing their comprehension of basic, as well as difficult, concepts.
3. **Transparencies** A set of over 100 transparencies is available to help the instructor coordinate the lecture with the key illustrations from the text.
4. **Customized Transparency Service** For those adopters interested in receiving acetates of text figures not included in the standard transparency package, a select number of acetates will be custom-made upon request. Contact your local Wm. C. Brown Publishers sales representative for more information.
5. **Microtest** This computerized classroom management system/service includes a database of test questions, reproducible student self-quizzes, and a grade-recording program. Disks are available for IBM and Macintosh computers, and require no programming experience.
6. **Laboratory Manual** Written by Kathy Dodds Tyner of Southwest College. *Lab Exercises for Preparatory Chemistry* features 63 class-tested experiments. The manual can easily be customized to suit an instructor's individual needs. The instructor can delete experiments, add his or her own experiments, or change the arrangement to create a custom manual to fit specific class needs.
7. ***Is Your Math Ready for Chemistry?*** Developed by Walter Gleason of Bridgewater State College, this unique booklet provides a diagnostic test that measures your students' math ability. Part II of the booklet provides helpful hints on the necessary math skills needed to successfully complete a chemistry course.
8. ***Problem Solving Guide to General Chemistry*** Written by Ronald DeLorenzo of Middle Georgia College, this exceptional supplement provides your students with over 2,500 problems and questions. The guide holds students' interests by integrating the solution of chemistry problems with real-life applications, analogies, and anecdotes.
9. ***How to Study Science*** Written by Fred Drewes of Suffolk County Community College, this excellent workbook offers students helpful suggestions for meeting the considerable challenges of a science course. It offers tips on how to take notes and how to overcome science anxiety. The book's unique design helps to stir critical thinking skills, while facilitating careful note taking on the part of the student.
10. **Exploring Chemistry Videotapes** Narrated by Ken Hughes of the University of Wisconsin-Oshkosh, the tapes provide six hours of laboratory demonstrations. Many of the demonstrations are of high-interest experiments, too expensive or dangerous to be performed in the typical freshman laboratory. Contact your local Wm. C. Brown Publishers sales representative for more details.
11. **Doing Chemistry Videodisc** This critically acclaimed image database contains 136 experiments and demonstrations. It can be used as a prelab demonstration of equipment setups, laboratory techniques, and safety precautions. It may also be used as a substitute for lab experiences for which time or equipment is not available. Contact your local Wm. C. Brown Publishers sales representative for more details.
12. **ChemTALK Lecture Presentation Software** This unique presentation software contains lecture outlines and numerous color illustrations and animations designed to bring concepts of chemistry to life. Three software programs have been developed to be used in a variety of introductory courses including general chemistry, preparatory

chemistry, and allied health chemistry. The programs are available on both Windows and Macintosh formats.

Less cumbersome than writing on a blackboard, this software will allow you to present material to your students at the pace that you choose, while integrating full-color illustrations and animations into your lecture.

Students may purchase the corresponding lecture notebook which contains all the material from the software. Using the notebook to follow the lecture, students will spend more time listening and absorbing information.

13. **Student Study Art Notebook** Free with each new text, the Student Study Art Notebook contains all of the full-color art included in the transparency set. The notebook allows the student to focus on the lecture rather than trying to recopy art being displayed in class.

acknowledgments

We would like to acknowledge the may individuals who helped and encouraged us as we developed this text, including CUNY, Bloomfield College, and Kean College students who provided the original and sustaining inspiration and the many students who used the first two editions and provided us with their comments.

We are very grateful to our colleagues at Wm. C. Brown Publishers, especially our Developmental Editor, John Berns, and our Acquisitions Editor, Megan Johnson. We also thank Molly Kelchen, Editorial Secretary, Laurie Janssen, Electronic Paging Specialist, and Karen Nickolas, Production Editor. We also want to give a very special thank you to Carl Binz, of Loras College whose consistent diligence on this project helped to ensure the accuracy of this text. A special note of thanks goes to James Funston who provided many suggestions for improvements in the second edition. We also appreciate the special efforts of our reviewers whose comments on this edition helped us produce an excellent text. Throughout the publication process of this textbook, many steps were taken to ensure accuracy. Although one cannot guarantee that a text will be completely errorless, the following steps were taken to make this text as clean as possible. The check for accuracy began when thirty-one reviewers evaluated the manuscript's narrative, examples, and artwork. When the manuscript was in final form, Carl Binz carefully checked the text problems and all of the art. Two professional proofreaders and the author team then checked the typeset text and illustrations. Once final pages were produced, another check was made by Carl Binz, the author team, and another professional proofreader. As a result, every word, example, figure, and answer has been independently checked and rechecked by many individuals at four crucial stages in the publication process.

reviewers

Ronald Albrecht
Dodge City Community College

Joseph Barnes
Pasadena City College

Warren Bosch
Elgin Community College

Juliet Bryson
Chabot College

Gordon Ewing
New Mexico State University

Robert Farina
Western Kentucky University

Joseph Friederichs
Dawson Community College

Fred Gant
Jacksonville State University

Arthur Hayes
Rancho Santiago College

Richard Hoffman
Illinois Central College

Lindell Holtzmeier
Hocking Technical College

Rita Hoots
Yuba College

Norman Hunter
Western Kentucky University

T.G. Jackson
University of Southern Alabama

Catherine Keenan
Chaffey College

Floyd Kelly
Casper College

Robert Kelly
Ventura College

Jerome Maas
Oakton Community College

John McLean
University of Detroit - Mercy

David Moss
Hiram College

Barbara Murray
California State University

Judy Okamura
San Bernardino Valley

Gordon Parker
University of Michigan

Sandor Reichman
California State University

Steve Ruis
American River College

David Schroder
Lincoln College

Margot Schumm
Montgomery College

Bob Smith

Trudie Jo Wagner
Vincennes University

Gary Wenberg
Sudmi College

David Williamson
Cal Poly

chapter 1

classification of matter

A modern chemistry student is using a desktop computer to access a chemical database containing the properties of more than 10 million known compounds. Any chemist may use this information to help identify an unknown material or to verify that he or she may have prepared a new material with properties different from materials previously registered.

1.1 introduction

We invite you to explore with us some of the questions, beauty, and helpful results that have arisen from the study of chemistry and the efforts of chemists throughout the ages. As we proceed, you will discover some of the concerns of chemists, how they think, how they design experiments to answer their questions, and how they attempt to make new materials that are useful in improving our lives.

The basic aim of chemists is to understand the behavior of the 109 known elements and how those elements combine or may be combined in the laboratory to make new materials and to prepare known materials in a more useful and efficient manner.

Studying how and why atoms from different elements combine into new materials, how quickly they combine, and how much material is formed increases the chemist's ability to predict the behavior of elements without further experiment. It is amazing also to consider how the modern chemist is able to build, or synthesize, chemically complex materials from simpler or more readily available materials. Sometimes the new materials may become a useful drug, fiber, or fuel.

The number of materials catalogued in the Chemical Abstracts Service (CAS) Registry at present is more than 10 million, and a large number of those have been synthesized, that is, built to order in the laboratory. With so many materials sufficiently understood to the extent of being "laboratory prepared," you can appreciate the powerful scientific principles that chemists must have developed to guide them in their work. Those principles will form the basis of each of the chapters in this text.

Fortunately, chemists have organized materials into relatively few groups, so that we can all work with such enormous diversity in a systematic way. That organization or classification of materials will be our starting point in Chapter 1.

Chemistry is the study of matter and changes in matter, which includes everything you see around you. This book, your hand, a pencil, water, a tree, and invisible things such as air are all examples of matter. Two characteristics define **matter**: matter occupies space and has mass.

As you begin reading this text, your view of matter resembles that of the earliest chemists who set out to study matter. You probably perceive the world around you as boasting an unlimited number of different forms of matter (Figure 1.1). As it turns out, you can actually classify matter into a surprisingly small number of categories, which are, in fact, the focus of this chapter. By the time you finish reading Chapter 1, you will be familiar with the most fundamental categories for classifying matter and you will be able to use them to help you identify any of the millions of known materials. You will have begun to function as a modern chemist.

Classification of matter is the process of arranging materials systematically into a useful number of groups according to their **properties**, those characteristics that are distinct for each material and that help you to identify them. Classifying information is a process you do every day without consciously realizing it. You classify automobiles as those that you particularly like and those you don't, as those that have pizzazz and those that don't, and as those that are economical and dependable and those that aren't.

••• problem 1

Make a list of the categories you use when you think about food, clothes, friends, jobs, and courses you are taking (including this chemistry course).

Figure 1.1

All matter occupies space and has mass, including these hot air balloons, the gas within the balloons, the air around the balloons, the gondolas, and the grains of soil in the earth below.

If you compare the lists that you made for Problem 1 with those of a fellow student, you will find that some of your categories agree, but others do not. Each of you has made a list that is useful for your own needs. When scientists construct a list of categories for classifying matter, they are looking for categories that are as simple and as general as possible. Then these categories are likely to include all materials, yet serve the needs of everyone studying those materials or trying to identify them.

Discovering properties useful for the classification of matter will be our first task in this text, as it was for the earliest investigators of matter. By grouping materials according to similar properties, you will begin to understand the tremendous power and increased capability you attain by simply organizing the information you have on hand. In everyday language, classifying matter is just a way for you to "take charge." But "taking charge" of scientific information means that you will proceed in a critical manner so that your efforts will be as useful as possible to anyone who wants to build on your results.

1.2 classification of matter by physical state

As citizens of the late twentieth century, you have a decided advantage over early investigators because you have absorbed some of the compiled scientific knowledge of the last 200 years in your everyday experience. For example, consider the following list of samples and think about how you might group them into three categories based on properties of the samples you have observed.

Samples of matter (Figure 1.2)

Gasoline	Ice	Distilled water	Mercury
Table salt	Oxygen	Helium	
Iron	Carbon dioxide	Rubbing alcohol	

When you think about each of these materials, you usually visualize a particular form or **physical state** that each takes (Figure 1.2). For example, ice, iron, and table salt are **solids**; distilled water, gasoline, mercury, and rubbing alcohol are **liquids**; and carbon dioxide, helium, and oxygen are **gases**. So, a useful classification system that you may readily choose involves identifying these samples as solids, liquids, or gases. The grouping of these materials according to their physical state is shown below.

Solids	Liquids	Gases
Ice	Distilled water	Carbon dioxide
Iron	Gasoline	Helium
Table salt	Mercury	Oxygen
	Rubbing alcohol	

Because you have long been aware of the characteristics of the three physical states, you recognize solids, liquids, and gases without realizing that you are

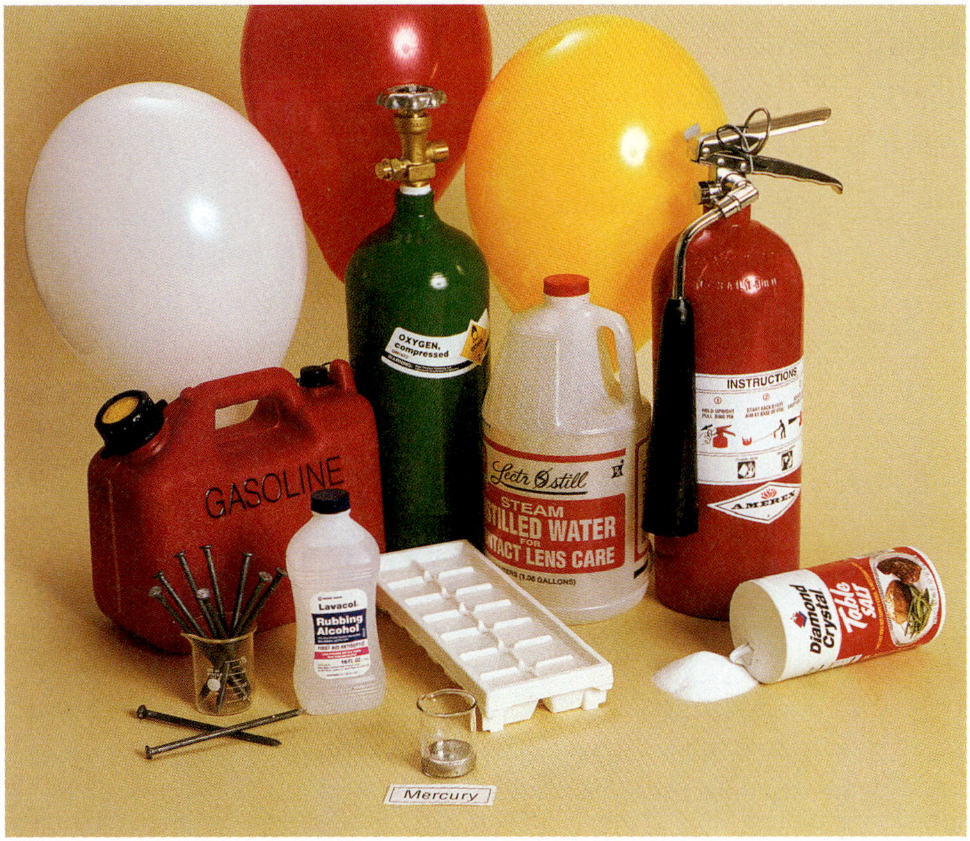

Figure 1.2

Common examples of solids, liquids, and gases. Solids include ice (in the ice cubes), iron (in the nails), and table salt. Liquids include distilled water, gasoline, mercury, and rubbing alcohol. Gases include carbon dioxide (in the fire extinguisher), helium (in the balloons), and oxygen. Water has been shown in two of its three states—solid ice and liquid.

table 1.1 shape and volume characteristics of the three physical states

Physical state	Shape	Volume
Solid	Definite shape. The shape does not depend on the container; e.g., the shape of an iron bar is the same in a box or on the table.	Definite volume. The volume is fixed; e.g., an iron bar occupies a clearly defined amount of space.
Liquid	Indefinite shape. The liquid takes on the shape of its container; e.g., the same amount of water changes shape in a cup, glass, pan, or puddle.	Definite volume. The volume is fixed; e.g., a cup of water has the same volume in a glass, a pan, or spilled on the floor.
Gas	Indefinite shape. A gas takes the shape of its container; e.g., the same amount of natural gas can be contained in a pipe or in a room.	Indefinite volume. The volume is not fixed but changes with the size of the container; e.g., a small amount of natural gas that leaked from a pipe in a large auditorium spreads to fill the entire room.

examining whether the shape and volume of each sample are definite when you make your classification. Table 1.1 summarizes the shape and volume characteristics of the three physical states. Note that only solids have *both* a definite shape and a definite volume. Liquids have a definite volume only, not a definite shape, because their shape takes on the shape of their container. Gases have neither a definite shape nor a definite volume; they take on the shape of the container and spread throughout the container to fill it no matter how big it is.

Usually you refer to the physical state of all matter at room temperature as a standard for comparison. This is necessary because the physical state may change as temperature changes. While you visualized one physical state for each of the

Figure 1.3
Classification of changes in physical states.

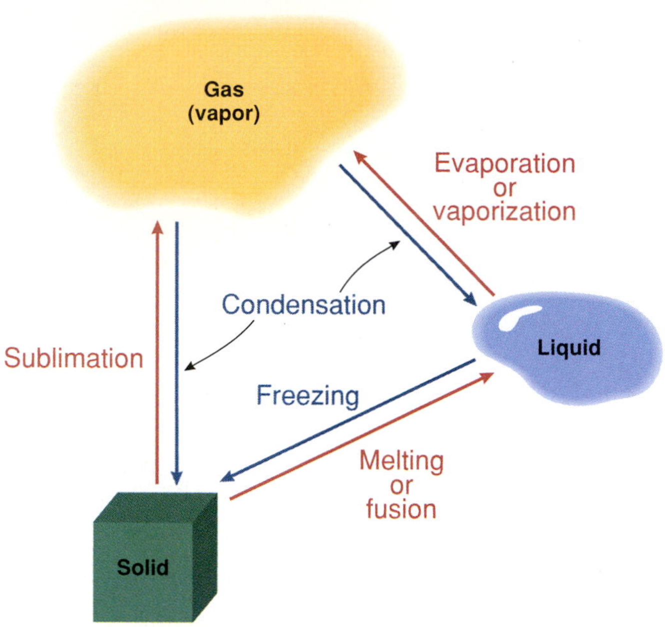

materials in the classification example, each material may appear as either a solid, liquid, or gas depending on its temperature. For example, water, which is a liquid at room temperature (25°C), changes its state to a solid (ice) as the temperature drops to freezing (0°C); it also changes its state to a gas (steam) as the temperature rises to boiling (100°C).

1.3 changes in the physical state of matter

The processes for changing physical states are given in Figure 1.3. The arrows are labeled with the names of each process, and the arrowheads indicate the direction of the change between states. Here you can quickly see how the changes between physical states may be more easily remembered because the changes have been classified into three groups: between solid and liquid, between liquid and gas, and between solid and gas. Each of these groups of changes occurs at its own specific temperature for a given material and each may occur in either direction, depending on whether heat is being added or taken away.

changes between solid and liquid

Melting is the term applied to the change as solid turns into liquid (Figure 1.4). **Freezing** corresponds to the reverse process in which liquid turns into solid. Both processes occur at the same temperature, called the **melting point, freezing point**, or **temperature of fusion.** The melting point is measured when the solid and liquid states are both present, as shown in Figure 1.5 with a combination of solid ice and liquid water.

Figure 1.4
Icicles melting on a sunny day.

changes between liquid and gas

Evaporation and **vaporization** are terms applied to the change as liquid turns into gas. **Condensation** corresponds to the reverse process in which gas turns into liquid. The temperature at which bubbles form throughout a liquid and the liquid becomes a gas is called the **boiling point**. The boiling point is measured when liquid and gas states are both present, as shown in Figure 1.6 with a combination of water and steam. Bubbles forming within the body of the liquid distinguish boiling from evaporation. Below the boiling point, liquids still evaporate, but more slowly than at the boiling point. You can smell the "fumes" (vapor) of gasoline or the aroma of perfume at temperatures below their boiling points (Figure 1.7).

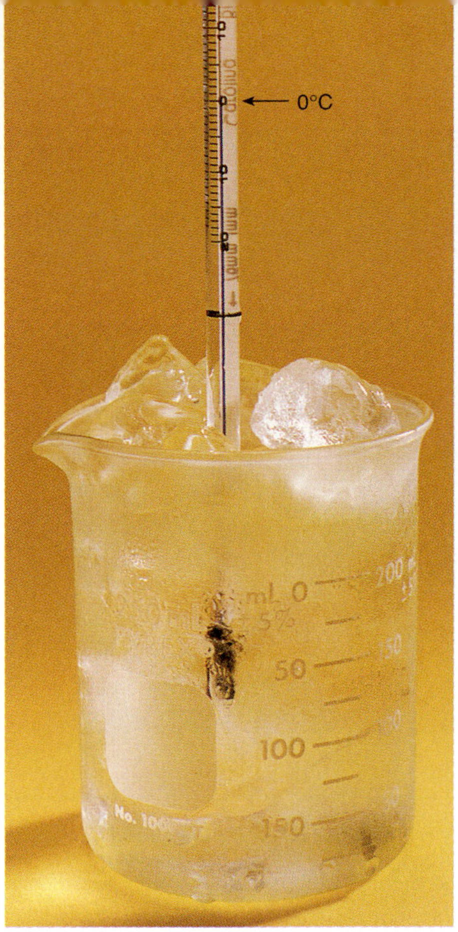

Figure 1.5
The melting point of water is measured when solid and liquid states are both present. The melting point of water is measured here as 0°C.

Figure 1.6
The boiling point of water is measured when liquid and gas (steam) are both present. Here you can see a fog of tiny liquid droplets that were formed from steam that cooled in the air above the boiling water. Bubbles forming within the body of the liquid distinguish boiling from evaporation. Note that a small "boiling chip" has been added to the flask for safety. The boiling chip reduces the size of the bubbles that form so that the liquid does not "bump" (collide violently) with the flask.

changes between solid and gas

Sublimation is the term applied to the change as solid turns directly into gas. *Condensation* once again refers to the reverse process in which a gas turns directly into a solid. Both processes also occur at the same temperature, called the **sublimation point** or **temperature of sublimation**. Also note that condensation is used as the name for two different processes, each process starting as a gas but changing into either the liquid or the solid state directly. Because of this ambiguity, the use of the term condensation may be confusing unless you specify the final state of the process you are describing.

Sublimation is usually the least familiar of the above processes, yet it is not uncommon to our everyday experience. The water in wet laundry that is hung outside on a freezing winter day turns to ice. The clothes become dry when the ice sublimes into the gaseous state and goes off into the air. Dry ice (solid carbon dioxide) is used to keep things cold. It is useful because its temperature is lower than that of ice and because it does not melt, but changes from solid directly into gas. Another application of subliming dry ice can be seen in the creation of mist or fog during a stage production or movie as seen in Figure 1.8. In this case, blocks of dry ice have been strategically placed beneath vents in the stage. The solid carbon dioxide blocks sublime to produce very cold carbon dioxide gas, which then freezes the water vapor in moist air that is blown over the blocks. The very fine grains of ice that are formed in the cold air then appear as fog on the stage.

Figure 1.7
Gasoline pump nozzles have been equipped with vapor restricting seals so that gasoline fumes do not escape into the atmosphere and contribute to the formation of smog.

Figure 1.8
The sublimation of carbon dioxide is used to create the fog that covers the stage in this production of the Wizard of Oz.

sample exercise 1

If you were to heat each of the following until they were at the temperature of boiling water (100°C), what would their physical states be?

a. Oxygen
b. Aluminum foil
c. Gasoline
d. Ice cream

solution

a. Gas. Oxygen is a gas at room temperature. Heating a gas only raises the temperature of the gas; it cannot change it into another physical state.
b. Solid. Aluminum foil remains a solid at 100°C. It melts at 660°C. You could look up its melting point in a reference book such as the *CRC Handbook of Chemistry & Physics* or *Lange's Handbook of Chemistry.*
c. Gas. Most gasolines boil at temperatures just slightly lower than the boiling point of water. You would not necessarily know this from your experience, however, and a reasonable answer might be that gasoline would still be a liquid at 100°C. What is important when "guessing" at some answers is not necessarily the correctness of the answer, but *the reasoning you use to arrive at your answer.* For example, if you were uncertain about the physical state of the gasoline and did not have a reference manual to look up the boiling point, the best answer would be: Gasoline is still a liquid if the gasoline boils above 100°C, but the gasoline would be a gas if its boiling point is below 100°C.
d. Liquid. Ice cream melts whenever the temperature is above 0°C.

••• problem 2

a. At room temperature, what is the physical state of a nail, of water, of cooking oil, and of air?
b. On a very cold winter day (below freezing), what is the physical state of each of these substances?

1.4 scientific method

In Sample Exercise 1, you used the physical states—solid, liquid, and gas—to classify specific samples of materials with which you are familiar. In classifying the substances in Problem 2, you used a line of thinking called the **scientific method.** Intuitively, you followed four basic steps: You used (1) *observations* of the physical

Figure 1.9
This surfer is applying the scientific method to find the best waves for surfing.

states of the substances to formulate a (2) *hypothesis* or suggestion that all of the substances were either a solid, liquid, or gas. Your hypothesis was then tested by (3) *experiment* or further observation, verifying that the substances did indeed fit into these categories. Once all of the substances were observed to be solids, liquids, or gases, you drew a (4) *conclusion* that physical state provided a general scheme for classifying matter. These steps are explicitly summarized below.

Scientific method	Illustration
1. Observation	The physical state of each sample was observed and compared.
2. Hypothesis (suggestion)	It was suggested or hypothesized that all matter can be classified as either a solid, a liquid, or a gas.
3. Experiment (further observation)	Observations of many other samples of matter were made to test the hypothesis.
4. Conclusion	Because the new observations confirm the hypothesis, we now have a means of classifying matter.

As you can see in Figure 1.9, the surfer is doing more than enjoying the view. He is *observing* the location of breaking waves to find a place where the waves break repeatedly. If waves continue to break in one particular place, he will *hypothesize* that the waves can be depended on to break at that point, probably because of shallow water above an underlying reef. Soon, he will *experiment* and test that hypothesis by paddling out to that spot to catch a ride. The *conclusion* follows from his success after numerous tries. The scientific method may be exhilarating when applied to sports as well as laboratory activities; it reduces uncertainty in your efforts and greatly improves your likelihood for enjoyment.

You may appreciate the scientific method as a useful procedure in your everyday life. The primary difference between your everyday usage and that of a scientist in the laboratory is the care with which a scientist constructs experiments and the caution he or she uses in not jumping to conclusions. To avoid jumping to conclusions, a scientist strives to limit the variables in an experiment so that only one parameter (or factor) is changed at a time. It is difficult for the surfer to sort out the effects of underwater depth, wind, and tide on the breaking of the waves unless a more careful study is done to check on each of these parameters individually.

A scientist welcomes and invites anyone to repeat the same experiment to make sure that the same conclusions can be reached independently and the same hypothesis verified. Accepted hypotheses are continuously tested by new experiments. Hypotheses not agreeing with experiments are modified, expanded, or rejected entirely.

An example of the need to modify a hypothesis might be seen by thinking back to your classification of matter in Sample Exercise 1. Some substances are not so easily classified as solids, liquids, and gases simply on the basis of definite shape and definite volume. Ice cream, gelatin, pudding, wax, bread dough, bubble gum, and numerous other substances, such as plastics, can flow—a property that removes their definite shape. Even though they flow, they flow *extremely slowly,* and they do not pour like a conventional liquid. While the classification of these special substances is beyond the scope of this text, it is important that you understand the need to expand scientific hypotheses. Scientists called rheologists have developed additional categories for classifying matter that include substances that flow very slowly and do not behave strictly as solids or liquids. For the classification of all materials discussed in this text, however, the basic hypothesis that all matter can be classified as solid, liquid, or gas will be adequate.

As you proceed with the classification of matter, the scientific method is the logical framework within which you hope to fit your observations.

1.5 pure substances and mixtures of pure substances

Any solid, liquid, or gas can be further classified into two broad categories, **pure substances** and **mixtures.** All matter can be separated into these categories, based on the **constancy of composition** of the sample, which is reflected in the properties of the sample. A pure substance, whether it be a solid, liquid, or gas, has a constant composition or makeup, and this is demonstrated by the fact that it always has the same properties for each physical state regardless of its origin. For example, solid copper metal, a pure substance, has the same melting point whether it is found in ore in the United States or South America or is reclaimed from copper tubing in an old building. Distilled water has the same freezing point and the same boiling point no matter where the water came from. Other examples of pure substances are any pure metal, un-iodized table salt, and refined sugar. The melting and boiling points of some pure substances with which you are familiar are listed in Table 1.2. Knowledge about the temperatures at which physical states change enables you to predict the physical state of a pure substance at any temperature, as you saw in Sample Exercise 1.

Just as pure substances can be solids, liquids, or gases, so too can mixtures. Mixtures, on the other hand, display changing properties as the proportions of the components of the mixture change. For example, bronze is a solid, metallic mixture of copper and tin; stainless steel is a mixture of iron, nickel, and chromium (Figure 1.10). Both bronze and stainless steel can be made with different proportions of their constituents, and different mixtures will have somewhat different colors, different melting points, and other different properties as a result of their different proportions.

table 1.2 physical properties of a few common pure substances

Substance	Familiarity	Color	Odor	Melting point (°C)*	Boiling point (°C)*
Acetic acid	Vinegar is a 5% solution of acetic acid	Colorless	Pungent	16.6	118.1
Carbon tetrachloride	Dry-cleaning solvent	Colorless	Like cleaning fluid	–23.0	76.5
Chlorine	Swimming pool disinfectant	Yellow-green	Irritating	–101.0	–34.6
Iron	Nails	Gray	Odorless	1,530	3,000
Methane	Natural gas	Colorless	Odorless†	–182.5	–161.5
Oxygen	Take a deep breath	Colorless	Odorless	–218.8	–183.0
Sodium chloride	Table salt	White	Odorless	808	1,465
Sucrose	Table sugar	White	Odorless	186; decomposes	Decomposes

*This unit of measurement is described in Chapter 3.

†A substance is added to natural gas to give it a detectable odor for reasons of safety. The methane itself has no odor.

(a)

(b)

Figure 1.10

(a) Steel is nearly pure iron, has its own color, and combines with oxygen in the air to form rust on its surface as seen on the girders of the bridge. (b) Stainless steel is a mixture of iron with nickel and chromium. The addition of nickel and chromium in the mixture gives it a more shiny, silvery appearance as seen in these bearings and prevents it from rusting as readily as pure iron.

A cup of coffee is a mixture, too, although in the liquid state. The properties of a cup of coffee are described as strong or weak, depending on how much material has been removed from the various coffee beans by water. The coffee may also be described as bitter or sweet depending on how many lumps of sugar have been added (Figure 1.11).

Similarly, a piece of steak is a complex mixture. It can be juicy and tasty or dry and tasteless, depending on its particular composition. Paint is a mixture. There is no definite composition and hence no definite properties corresponding to paint. It can contain water or no water; it can contain yellow pigment or blue pigment in varying amounts. These different compositions result in different solubility and color properties. Similarly, a solution of table salt mixed with water has a freezing point and a boiling point that changes with the amount of salt added. The more salt you put into the water, the lower the temperature at which the saltwater solution will freeze and the higher the temperature at which it will boil.

Gases also form mixtures, but because of their invisibility, they are often difficult to observe. However, the air we breathe is a mixture of mostly nitrogen and oxygen gases with small amounts of argon gas, carbon dioxide, water vapor, and trace amounts of several other gases. Carbonated soft drinks are mixtures of carbon dioxide gas, a sweetening, and flavored water. In soft drinks, carbon dioxide gas at a pressure higher than atmospheric pressure has been mixed with flavored water and sealed. Once opened, the bubbles indicate that the carbon dioxide gas is escaping from the water.

A second characteristic of mixtures is that they may be separated into pure materials by *physical* means; that is, the original components of the mixture can be obtained without any change in their original properties. For example, a mixture of sand and water can be separated by filtration—a process of pouring the mixture through a fine piece of paper called filter paper, which strains out the solid sand particles and allows the liquid water to pass through (Figure 1.12). After separation, the sand, which has been collected on the filter paper, has the same properties as the pure sand before it was mixed. And the water collected in the beaker beneath the funnel holding the filter paper has the same properties as pure water. Similarly, a mixture of sand and iron filings could be readily separated by using a magnet to attract all of the iron filings out of the mixture. Both the sand and the iron filings retain their original properties.

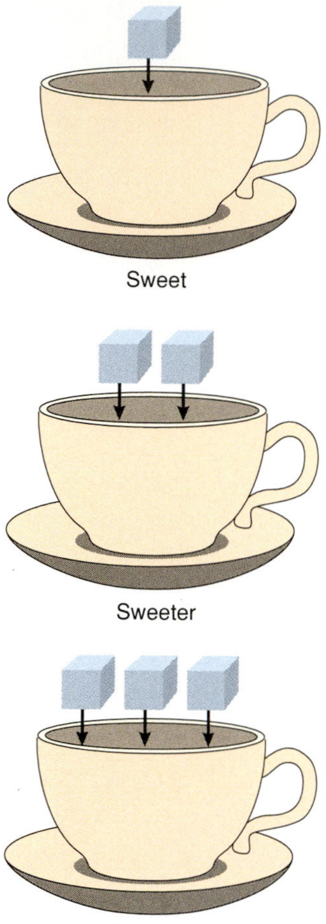

Figure 1.11

The properties of mixtures vary with the proportions of the ingredients. Three lumps of sugar give a sweeter taste to the same volume of coffee than do two lumps of sugar; two lumps of sugar give a sweeter taste than one lump.

Figure 1.12

Filtration of a sand-water mixture separates the solid sand on the filter paper and the liquid water in the beaker below the funnel.

A solution (liquid mixture) of salt and water can be separated by allowing the water to evaporate (turn into a gas), leaving the solid salt behind as a residue. If the gaseous water needs to be retained as a pure form of water, it can be collected on a cold surface or in a cold container so that it condenses back to pure liquid water. In general, this process is called distillation. When used specifically to remove salt from sea water to obtain pure water for drinking, this process is called desalinization. Desalinization is used extensively by countries that do not have an adequate supply of fresh water, such as those in the Middle East (Figure 1.13).

The test of the separability of components is a good experiment that distinguishes between mixtures and pure substances. In Sample Exercise 2 and in Problem 3, you are asked to make the classification without doing any experimental testing, and this is possible because, in each case, the constancy of composition is discernible through observation or through your previous everyday experiences.

sample exercise 2

Are the following mixtures or pure substances?

a. Urine
b. Sea water
c. Gasoline

solution

a. Urine is a mixture of water and various bodily waste products. Its composition varies depending on what you eat and your state of health. The standard test for diabetes measures the amount of sugar mixed in the urine. Urine is also tested for the presence of anabolic steroids in athletes or illegal drugs in ship captains, train engineers, and airplane pilots.
b. Sea water is a mixture of water, salt, seaweed, and various debris. The amount and type of debris in sea water has become hazardous to seashore recreation and has led to increased legislation that restricts ocean dumping.
c. Gasoline is a mixture. The composition is variable. It can be leaded or unleaded and contain additives specifically developed by each company.

Most matter you encounter is a complex mixture.

Figure 1.13
This desalinization plant uses the physical process of distillation to separate pure drinking water from sea water. Salt, a component of sea water, has collected on this pipe.

• • • problem 3

Are the following mixtures or pure substances? Give the reason for your answer.

a. A page in this textbook

b. Skin

c. Pure iron metal

d. A rusty nail

1.6 physical and chemical changes of matter

You have already observed physical change in Sections 1.4 and 1.5. Changes in state are examples of physical change. A **physical change** causes no change in the basic nature or composition of pure substances in a sample of matter. The freezing, melting, boiling, sublimation, or condensation of water are all physical changes since no change in the composition of water occurs in these processes.

Another common example of physical change is a change in the size or shape of matter. For example, tearing aluminum foil into pieces represents a physical change. The chemical composition is still aluminum, but you have several smaller pieces rather than one large one. The mixing together of two pure substances, such as iron filings and salt, without changing the composition of either is a physical change. Similarly, the separation of two or more pure substances from a mixture by filtration or a change in physical state is also a physical change.

Chemical change, by contrast, forms one or more new materials—pure substances with distinctly different properties. Old materials are converted into new ones, and this transformation is called a **chemical reaction**. In Figure 1.10, you saw that pure iron will combine, or react, with oxygen in the air to produce rust, a new, reddish-brown material with a different color and a different melting point from the original iron. Burning coal is an example of a process that produces chemical change. The solid, black coal burns (reacts with oxygen in the air), gives off heat and light, and produces colorless carbon dioxide gas. This new gaseous material has a composition and a set of properties that differ from those of coal; it also has a physical state different from that of the old material. However, we do not call this a physical change just because a new physical state has appeared. The solid coal is not simply becoming liquid coal or gaseous coal. Rather, the different physical state arises because the new carbon dioxide gas that has been produced in the reaction happens to exist in a physical state different from the old material.

Another example of a chemical reaction is the electrolysis of (passing an electric current through) table salt, the results of which are shown in Figure 1.14. The

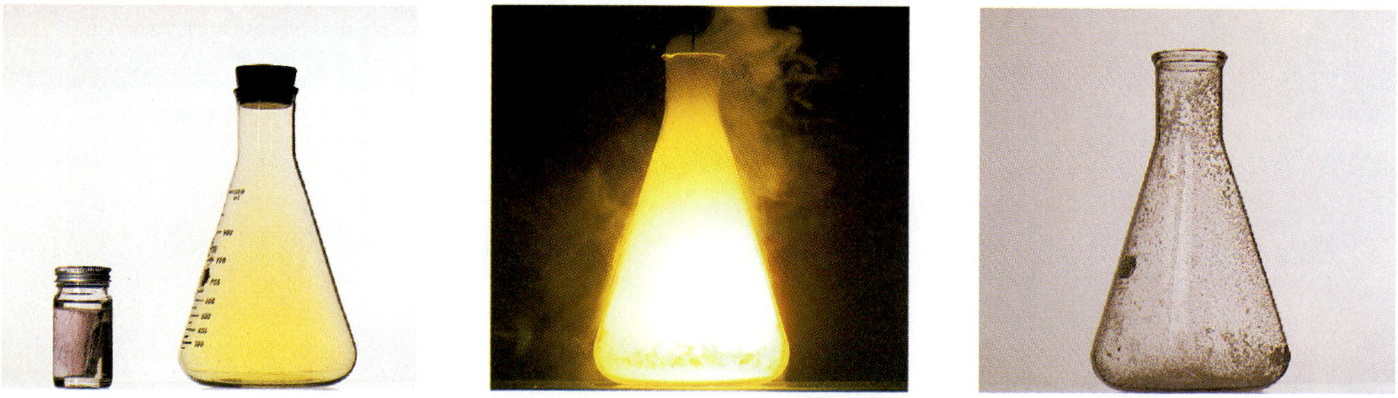

(a) (b) (c)

Figure 1.14

(a) Sodium metal and chlorine gas have obviously different physical properties. Chemically, sodium metal reacts violently with moist air so it must be stored under oil, as seen in the bottle on the left, to prevent contact with air. Yellow-green chlorine gas is a severe irritant that reacts rapidly with body tissue, so it must be tightly sealed. (b) When combined, sodium metal reacts violently with chlorine gas to release heat and light. (c) The product of reaction, sodium chloride, is also known as table salt. It is a white solid that does not react with air and is harmless to humans if not consumed in excess.

familiar white crystals of table salt are changed into shiny sodium metal and yellow-green chlorine gas. *Chemical reactions always affect the composition of pure substances.* In this example, two new, pure substances with distinctly different properties are produced during the electrolysis reaction.

Any change in the physical properties that characterized the pure reactants (the pure substances with which you started) gives evidence of a chemical change in which new materials are formed. New materials (products) in a reaction may be formed in any physical state. For example, in some reactions that occur in a liquid mixture, a new solid may be formed. If the new solid does not dissolve (is insoluble) in the liquid, it separates out of the liquid reaction mixture and is called a **precipitate** (Figure 1.15). In some reactions, a new liquid may be formed. If water is the liquid that is formed, it will not be observed directly since it will mix with the solution already present. Only if the new liquid does not dissolve (is insoluble) in the original solution will it separate into a distinct layer of its own. When a new gas is formed, however, the bubbles of the gas released from the solution will be observed easily. The formation of a precipitate, the appearance of a new liquid layer, and the evolution of a gas are means of observing chemical change.

Figure 1.15

A new compound, lead iodide (PbI_2), forms as a yellow precipitate when a clear solution of lead nitrate ($Pb(NO_3)_2$) is mixed with a clear solution of sodium iodide (NaI). The formation of a precipitate is an experimental means of observing a chemical change.

Additional evidence for *either a physical or chemical change* can be observed by any change in temperature of the substances studied. A temperature change can be measured with a thermometer, but it may often be determined by simply feeling the outside of the test tube or flask holding the materials to see if it is warmer, colder, or the same as room temperature. The amount of *heat absorbed or released* during a chemical or physical change is a property that is extremely useful to chemists. Some chemical changes, such as the burning of natural gas (methane), produce a tremendous amount of heat. You use the heat produced in the burning of natural gas to cook your food and heat your home. The explosive power of TNT is also related to how much heat is released in a chemical process. Some reactions do not release heat, but absorb it from the environment. As a result of heat being absorbed by the substances reacting, the temperature of the reaction mixture may drop considerably. The relation of heat to a change in energy will be discussed in Chapter 3; the relation of heat to a chemical reaction will be discussed further in Chapter 10.

Heat may be either released or absorbed in a physical change as well as a chemical change. You have seen in Section 1.5 that dissolving a solid in water is a physical process because the solid and the water may be recovered with no change in their properties. When sodium hydroxide (lye) is dissolved in cold water in your sink drain, a great deal of heat is released, which you can feel by carefully touching the trap below the sink. When other solids, such as ammonium nitrate, dissolve in water, they absorb heat from the surroundings and, as a result, lower the temperature of the mixture. Instant ice packs such as Kold-Kompress, which are available in drugstores, contain two separate containers, one with water and one with ammonium nitrate. When the contents of the containers are mixed, the packs become very cold.

sample exercise 3

Are the following examples of physical or chemical changes?

a. A ham is sliced.
b. The sugar in grain ferments into alcohol. Small bubbles are seen in the liquid.
c. Sugar dissolves in hot water.
d. Sugar is heated to a brownish coloration and a noncrystalline texture.
e. Oxygen gas is compressed and liquefied.

solution

a. Physical. It is still ham, but in smaller pieces, with each piece having the same physical properties.
b. Chemical. A new substance, alcohol, with new properties has been produced.
c. Physical. No new material is produced. The sugar and water have become intimately mixed but are still sugar and water and can be recovered as such. If the water evaporates, the sugar is left behind as a solid residue.
d. Chemical. Upon heating dry sugar (sucrose), a mixture of products called caramel is formed. That this is a new material can be seen by the changes in color and texture.
e. Physical. Gaseous oxygen becomes liquid oxygen. Each physical state has the same properties as pure oxygen obtained by any means. No new materials with different properties are produced.

••• problem 4

Are the following physical or chemical changes?

a. Food is digested.
b. Grass is cut.
c. Perspiration evaporates.
d. Hydrogen peroxide, the antiseptic and hair bleach, decomposes to form water and oxygen.
e. A tomato plant grows.

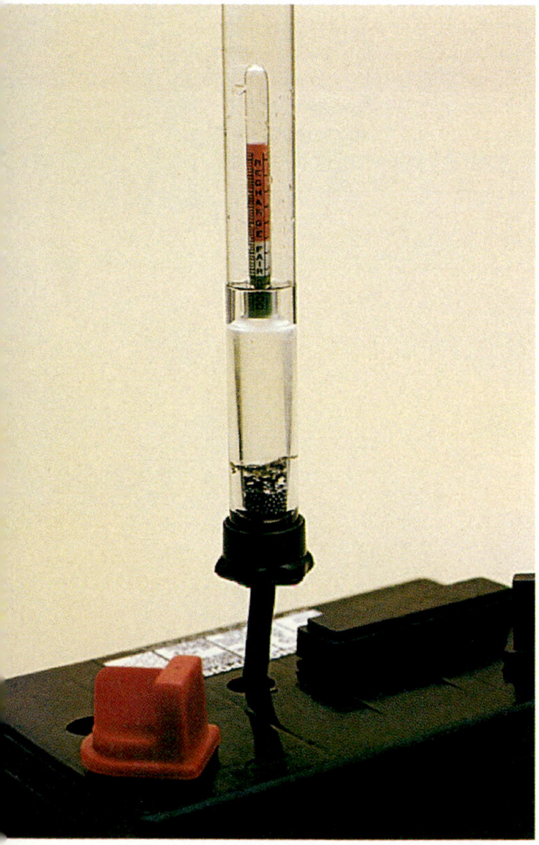

Figure 1.16
The density of battery fluid is a physical property that you may use to determine the charge on your car battery. The level of the float in the hydrometer indicates the charge on the battery.

1.7 physical and chemical properties identify pure substances

We have used the "properties" of matter throughout the previous sections in discussing its classification and as a basis for identifying pure materials. However, not all properties of materials are equally useful. While size, shape, and volume are properties of the particular sample you are studying, they are not characterizing properties. For example, many different materials may contain the same amount of matter. You could have one liter of water, milk, coffee, or antifreeze, and the amount would not help you to distinguish the antifreeze from the other liquids. To identify a pure substance, you need to focus on those properties that are *characteristic of that material* and *independent of the amount of the material.*

The first group of properties for pure substances that you need to consider are the **physical properties**—those properties that can be observed and measured without changing the composition of the substance being observed. So far, you have used the *boiling point, melting point, color,* and *physical state (at room temperature)* to help you identify some pure substances. Materials can also be identified by their *odor, electrical conductivity, ability to dissolve in a liquid (solubility),* and *crystalline form (if they are crystalline solids).* You can always identify the odor of a ripe banana or a particular perfume; you would never place an aluminum ladder, a good conductor of electricity, near electric power lines; you add "dry gas" to the gasoline in your car to make any water that may be present dissolve in the gasoline so the mixture will burn in your engine; and, without tasting, you can identify salt from sugar by observing the cubic shape characteristic of salt, but not of sugar crystals. In Chapter 3, we will define the *density* (or the mass per unit volume) as a particularly useful physical property that you can measure accurately in the laboratory or approximately with a hydrometer. When you test the fluid in your car battery to see if the battery is properly charged (Figure 1.16), you are measuring the density of the fluid with a hydrometer.

Taste is a physical property often used to identify a special group of pure substances called foods. Taste should never be observed in the laboratory and should never be used with chemical substances. Please reserve your tasting *only* for foods and beverages that you know are perfectly safe and are outside the chemical laboratory.

Notice how the experimentally determined physical properties in Table 1.2 give you additional means of distinguishing between look-alikes. For example, salt and sugar are both white solids at room temperature that have different crystalline forms. Observing these crystalline differences usually requires a good magnifying glass or microscope. Another simple experiment of heating the materials also readily identifies them. Sugar melts easily (186°C), but salt will not melt in a match or Bunsen burner flame. If you heat very vigorously, your identification will be made on the basis of a chemical property; i.e., sugar "burns" and salt does not. This simple example illustrates how experimentation can offer more information than observation alone. Obviously, to distinguish between sugar and salt, you could taste them the way you would in your kitchen. However, in the chemistry laboratory you must remember to refrain from tasting since we do not know ahead of time that the substances we are testing are foods. Even if you knew you were testing food substances, you would still not taste them in the laboratory since they may be contaminated with other toxic materials.

The second group of properties used to identify a material are its **chemical properties**, which are observed when a substance undergoes reaction. As you saw in the previous section, at least one new material must appear in a chemical change.

The chemical properties of a pure substance are much harder to describe at this stage than its physical properties because chemical properties are usually related to how that substance reacts with another. For example, *flammability* is an example of a chemical property, the property that some substances burn while others do not. Flammability depends on the ability of the substance to react rapidly with oxygen gas in the air to produce heat, light, and combustion products. When it burns, the original material disappears and carbon dioxide gas and water vapor are usually formed. Therefore, flammability is a chemical property that may serve to distinguish some substances from others. Iron is an example of a substance that does not burn, but it does combine slowly with oxygen gas in the air for the chemical property of *rusting,* as we have seen in Figure 1.10. This chemical property uniquely identifies iron. Stainless steel does not rust, however, and this property helps to distinguish it from pure iron.

Later in this book, you will see other chemical properties that are defined in terms of the reactivity of a material with other substances. A particularly common example is the reactivity with acids, which will be discussed in Chapter 15.

Chemical properties will provide additional means for classifying matter and for helping you to identify pure, unknown substances.

••• problem 5

List some of the physical properties of a glass test tube, shown in Figure 1.17.

sample exercise 4

What is the physical state of table salt at room temperature (25°C), at 1,000°C, and at 1,500°C?

solution

Because you are familiar with table salt at room temperature, you do not need Table 1.2 to tell you that it is a solid at that temperature. If it had been unfamiliar, you would have concluded from the table that it is solid because 25°C is well below its melting point. On the other hand, 1,000°C is above its melting point, so at that temperature salt would be in the liquid state. It would not be converted to a gas until the temperature reached 1,465°C. At any temperature above 1,465°C (for example, 1,500°C), salt would be a gas.

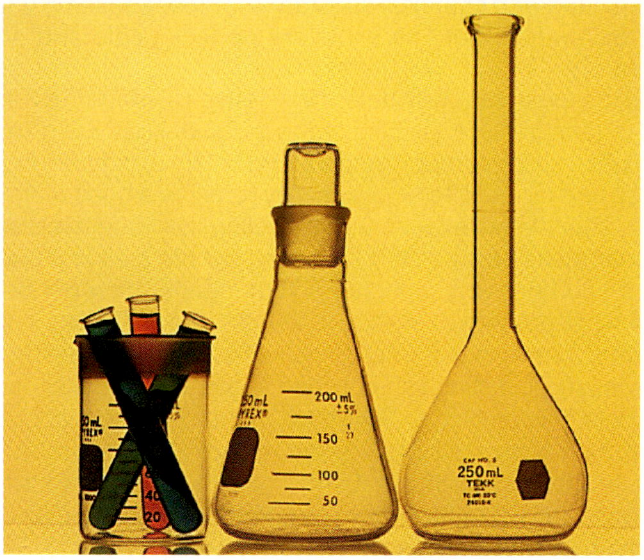

Figure 1.17
Glass test tubes have been placed in the 150 mL beaker shown on the left. In the center is a 250 mL Erlenmeyer flask used to dissolve solids or mix liquids. Its wider base permits swirling of liquid contents. On the right is a 250 mL volumetric flask used to prepare solutions with known concentrations.

••• problem 6

Chloroform looks and smells like carbon tetrachloride. Its boiling point is 61°C. How might you identify two unlabeled liquids if you already know that one is chloroform and the other carbon tetrachloride?

••• problem 7

Both methane and oxygen are colorless, odorless gases at room temperature. How might you distinguish a test tube of methane from one containing oxygen?

1.8 subdivisions of pure substances

Now that you have explored the distinctions between physical and chemical changes and physical and chemical properties, you are ready to complete the final classification of matter just like the early chemical investigators. Pure substances can be further divided into *elements* and *compounds,* as shown in Figure 1.18. Note, however, that the division between elements and compounds classifies the composition of matter in terms of its fundamental chemical building blocks and is essential to your understanding of all further chemical processes.

Some pure substances can be changed into distinctly different materials by certain reactions, whereas other pure substances defy attempts to alter them. The substances that *can* be broken down into simpler substances are called compounds; the substances that are so elementary that they *cannot* be broken down by chemical change are called elements. Table 1.3 lists the results of two experimental tests (heating and passage of an electric current) on pure substances and shows how you may use the results of these tests to classify materials as elements or compounds. All of the tests in Table 1.3 were conducted inside a vacuum, an environment in which no other matter except the sample is present. It is necessary to exclude other matter so that it does not react with the sample or interfere with your observations of the sample you are studying.

Notice that there are three possible outcomes for each experiment in Table 1.3. One possible outcome is no change, as you see for helium, iodine, and mercury. A second possible outcome is only a change in physical state. This frequently happens when the sample is heated; for example, water boils, producing steam, mercury vaporizes, and tin melts. You recall that a physical change results in no new substance formation. You also recall that physical changes are reversible; for example, when steam and mercury vapor are cooled, liquid re-forms, and

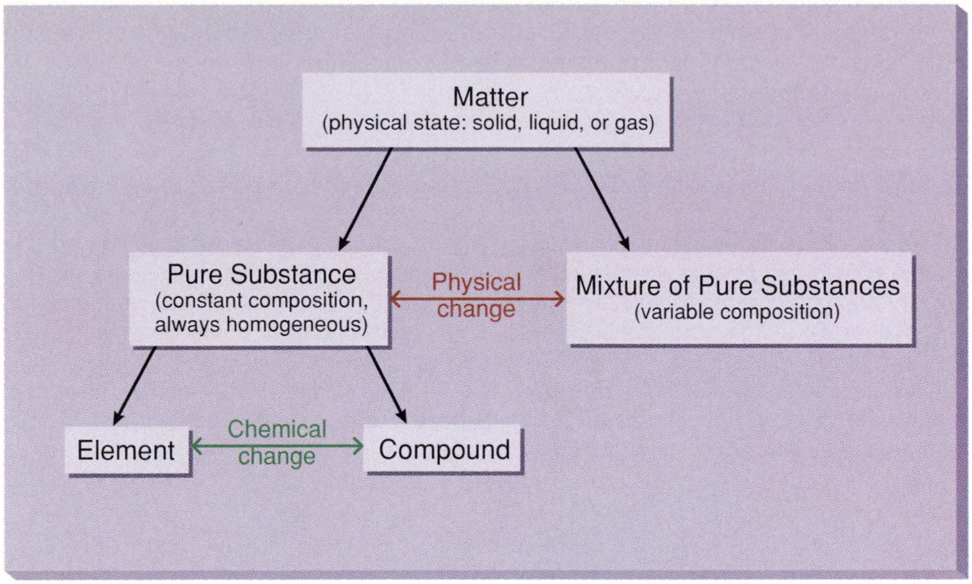

Figure 1.18

All matter is divided into two large categories: pure substances and mixtures of pure substances. Pure substances are composed of either elements or compounds.

table 1.3 effects of experiments on a selection of pure substances

Pure substance	Description	Effect of heating in a vacuum	Effect of an electrical current	Conclusion and classification
Helium	Colorless gas	Remains the same.	Does not conduct; no change.	No change; helium is an *element.*
Iodine	Purple solid	Sublimes and the solid re-forms upon cooling.	Does not conduct; no change.	No change; iodine is an *element.*
Malonic acid	White solid	A gas is given off and a white solid with different properties is formed.	Does not conduct; no change.	New materials are formed; malonic acid is a *compound.*
Mercuric oxide	Red powder	Turns black and a gas is given off.	Does not conduct; no change.	New materials are formed; mercuric oxide is a *compound.*
Mercury	Silvery liquid	Vaporizes at high temperature, but the liquid re-forms upon cooling.	Conducts electricity, but is not changed by it.	No change; mercury is an *element.*
Table salt	White solid	Melts at a very high temperature and resolidifies unchanged.	Passing an electric current through melted table salt produces shiny sodium metal and yellow-green chlorine gas.	New materials are formed; table salt is a *compound.*
Tin	Shiny solid	Can be melted and resolidified unchanged.	Conducts electricity, but is not changed by it.	No change; tin is an *element.*
Water	Colorless liquid	Heating produces steam; cooling restores the water.	Bubbles of gas with properties other than those of steam are released.	New materials are formed; water is a *compound.*

when the melted tin is cooled, it resolidifies. The third possible outcome is the appearance of a new substance, as observed for malonic acid, mercuric oxide, table salt, and water. These are all chemical changes that indicate that the pure substance tested is a compound. The chemical changes of malonic acid and mercuric oxide occur by heating in the absence of air; the chemical changes for table salt and water occur by electrolysis.

Not all substances can be correctly classified by these two simple experiments of heating and passage of an electric current. However, experimentation can subdivide all pure substances into elements and compounds, and these simple tests give you an idea of how it's done.

sample exercise 5

Classify potassium chlorate and silver as elements or compounds based on the following data:

Silver is a valuable metal that is an extremely good conductor of electricity and is used in cables for high-end audio equipment. It is also used for plating tableware and making jewelry and photographic chemicals. Silver melts at 961°C and resolidifies when cooled. Upon reheating, it again melts at 961°C. Silver conducts electricity, but it is not changed by the current.

Potassium chlorate is used in the manufacture of explosives, fireworks, matches, bleaches, and disinfectants. When potassium chlorate is heated, a gas is released and a white solid with properties different from potassium chlorate forms.

solution

Silver is an element according to the experimental tests described because it remains unaltered, as shown by the fact that the melting point remains the same.

Potassium chlorate must be a compound because it is broken down by heat.

1.9 elements and compounds

Chemically, the most important distinction of matter is your identification of pure substances as elements or compounds. From the foregoing discussions, you can distinguish between these two categories by the following definitions:

element:

A pure substance that cannot be broken down into simpler substances by ordinary chemical changes.

compound:

A pure substance that can be broken down into simpler substances (elements) by ordinary chemical changes. Compounds are composed of two or more elements combined in a fixed proportion.

elements

Elements are the simplest substances, the building blocks from which all other matter is made. You will see in the next section that compounds are made by chemically combining elements in various proportions, and the different proportions of elements account for the new properties of each compound.

Whereas there are unlimited numbers of compounds, there are only 109 known elements. Look at Figure 1.19 and inside the front cover of this text and you will see the names and symbols of all elements arranged in what is called the periodic table, which we will discuss more fully in Chapter 4. For understanding the chemistry developed in this text, you will rarely need to recognize more than 35 of the most common elements. Table 1.4 lists the names and symbols of the elements with which you need to be familiar.

Notice in Table 1.4 that each element has a symbol consisting of one or two letters related to the name of the element. Also notice that the elements are classified as either metals or nonmetals. The classification of elements as metals or nonmetals arose naturally when chemists first began studying the properties of elements. It was observed that there are two distinct sets of elements, each of

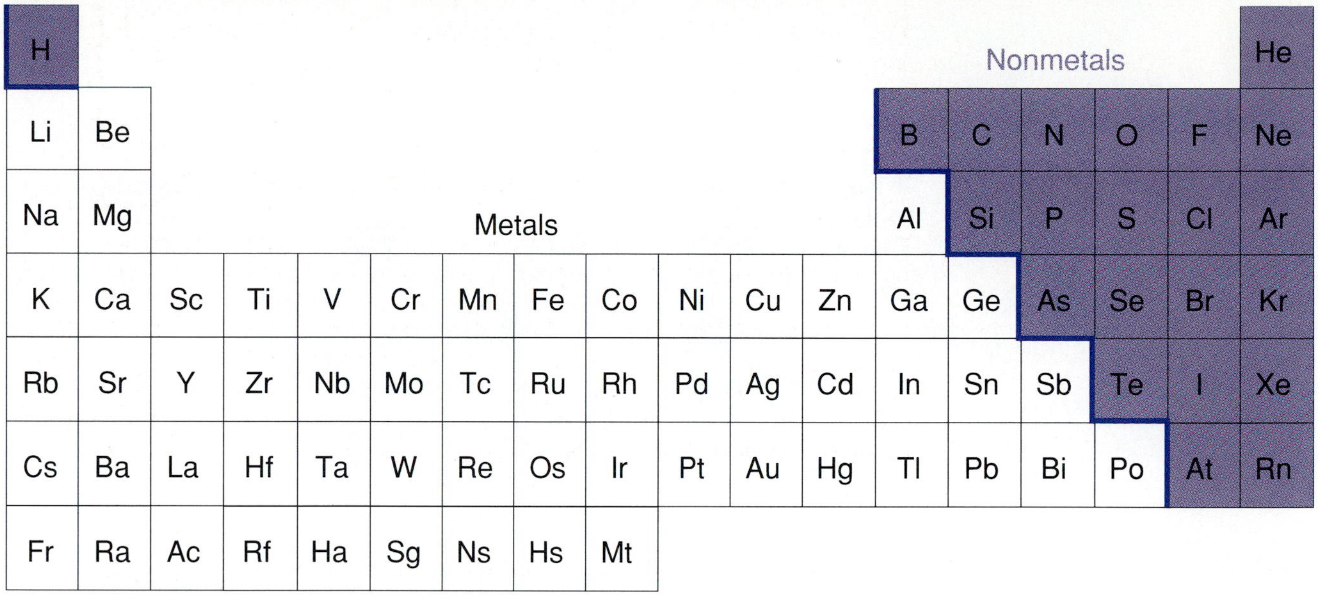
Nonmetals
Metals
H He
Li Be B C N O F Ne
Na Mg Al Si P S Cl Ar
K Ca Sc Ti V Cr Mn Fe Co Ni Cu Zn Ga Ge As Se Br Kr
Rb Sr Y Zr Nb Mo Tc Ru Rh Pd Ag Cd In Sn Sb Te I Xe
Cs Ba La Hf Ta W Re Os Ir Pt Au Hg Tl Pb Bi Po At Rn
Fr Ra Ac Rf Ha Sg Ns Hs Mt

Figure 1.19

The periodic table groups elements according to similar chemical properties and makes a division into metals and nonmetals with the stepped diagonal line. Hydrogen is an exception because it is shown with the metals even though it is usually a nonmetal in its chemical properties.

table 1.4 common elements classified as metals and nonmetals

Metal element	Symbol	Nonmetal element	Symbol
Aluminum	Al	Argon	Ar
Barium	Ba	Boron	B
Beryllium	Be	Bromine	Br
Calcium	Ca	Carbon	C
Chromium	Cr	Chlorine	Cl
Cobalt	Co	Fluorine	F
Copper	Cu	Helium	He
Gold	Au	Hydrogen	H
Iron	Fe	Iodine	I
Lead	Pb	Neon	Ne
Lithium	Li	Nitrogen	N
Magnesium	Mg	Oxygen	O
Manganese	Mn	Phosphorus	P
Mercury	Hg	Silicon*	Si
Nickel	Ni	Sulfur	S
Potassium	K		
Silver	Ag		
Sodium	Na		
Tin	Sn		
Zinc	Zn		

*While silicon is a nonmetal, it may also be classified as a metalloid (or semimetal) with a few other elements. Metalloids have both metallic and nonmetallic properties.

(a)

(b)

Figure 1.20

Elements are classified as metals or nonmetals because of their different properties. (a) In copper ore, the shiny metal can be seen embedded in earth and rock. (b) The nonmetal, sulfur, is a dull-yellow material.

which displays certain physical properties. One set contains 84 elements that are shiny, conduct electricity, and are malleable (i.e., they bend without breaking). These substances are called **metals.** You are already familiar with metals from everyday life, and you can probably describe the properties of such metals as aluminum, copper, and gold. The set of 22 **nonmetals,** on the other hand, has the opposite characteristics. Nonmetals generally lack luster, do not conduct electricity, are brittle, and many are gases at room temperature (Figure 1.20).

As you become familiar with the elements, it is a good idea to distinguish metals from nonmetals from the very start. It is not necessary to memorize which elements are metals and which are nonmetals, because the periodic table contains this information. Look at the periodic table and note the stepped diagonal line toward the right side. This line divides metals to the left from nonmetals to the right. The only exception is hydrogen, which is usually treated as a nonmetal despite its position on the left side of the periodic table. Figure 1.19 shows a periodic table that emphasizes these distinctions.

Most of the human body mass is composed of six nonmetallic elements: carbon, hydrogen, oxygen, nitrogen, phosphorus, and sulfur. These elements are not found free, that is, in their elemental state, in the body; rather, they are combined in intricate ways in various compounds. Only the elements oxygen and nitrogen are found in their elemental state in the body. Other nonmetals such as chlorine and iodine are required in the body in lesser amounts and are found combined, not as free elements.

Many of the metallic elements are also required for good health, and their absence or insufficiency in the body leads to a deficiency symptom. Metals are always found in their combined form as compounds or dissolved compounds in the body, however, never in their free, elemental state.

compounds

Compounds are formed when two or more elements are chemically combined (or react) to form a new material with a fixed composition. Composition is constant in elements because there is only one component; composition is constant in compounds because the ratio of component elements is fixed throughout the compound. For example, a fixed amount of nitrogen gas will combine with twice the amount of oxygen gas to form nitrogen dioxide, NO_2, a new material that is a highly toxic, reddish-brown gas. Nitrogen dioxide is used in the manufacture of acids; it is a major air pollutant formed from automobile exhaust and is responsible for the yellow-brown color of smog. Because the elements are chemically combined to form the compound, they no longer retain their original properties, but have become part of a totally new substance that has its own properties. The elements in nitrogen dioxide will not separate unless a chemical change is used to separate them. If, however, nitrogen dioxide *were* chemically decomposed into its elements, you would find a fixed amount of nitrogen gas and twice that amount of oxygen gas. Nitrogen dioxide was formed from nitrogen and oxygen in a fixed ratio: 1 part nitrogen to 2 parts oxygen.

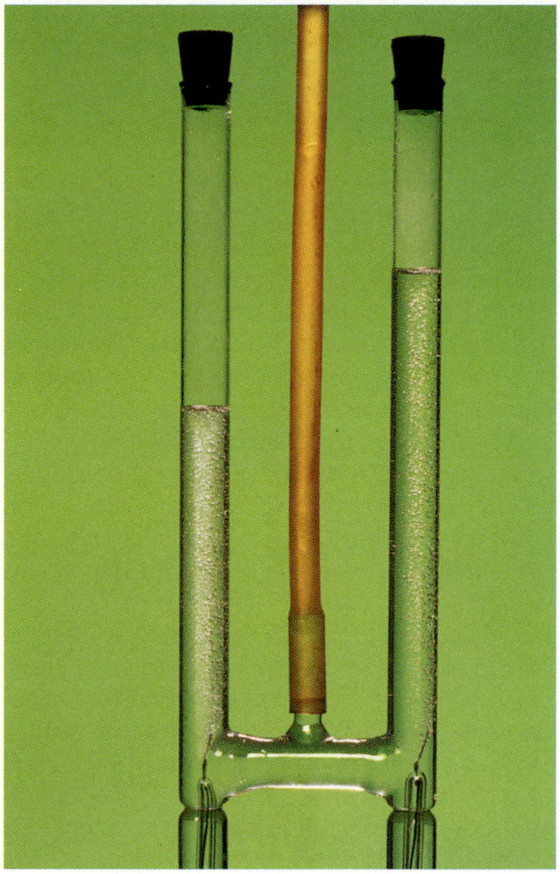

Figure 1.21

When an electric current is passed through water, the water breaks down into hydrogen gas and oxygen gas. The gases collect in the volume ratio of 2 parts of hydrogen (left side) to 1 part of oxygen (right side).

Under different conditions, a fixed amount of nitrogen gas will combine with half the amount of oxygen to form nitrous oxide, N_2O, another new material with totally different properties. Nitrous oxide is a colorless, sweet-smelling gas that produces a feeling of exhilaration when inhaled as an anesthetic called laughing gas. Again, it is a stable, new substance with its own unique properties. If laughing gas were chemically decomposed into its elements, however, you would find a fixed amount of nitrogen gas and half that amount of oxygen gas. Nitrous oxide was formed from nitrogen and oxygen in a fixed ratio: 2 parts nitrogen to 1 part oxygen.

You have seen that the same elements may react to form different materials, but that each new material has its own unique properties because of the different, fixed ratio of elements in each substance. This property—that a compound always contains its component elements in a fixed ratio *by weight*—was first noticed and formalized as the **law of definite composition** (or law of definite proportions) in 1797. The fixed-ratio notion is similar to the idea that, to make a certain cake turn out the same way every time, you always follow the same recipe, with specified amounts of each ingredient. This *constant,* fixed ratio of elements in compounds gives compounds the *constancy of composition* that is a characteristic of all pure substances. In Chapter 6 you will see how the weight ratios of elements are determined. For now, however, it will be more straightforward to discuss volumes of gases and how these volumes are related to chemical formulas of compounds as discussed in Section 1.11.

Water is another common example of a compound. The elements hydrogen, a flammable gas, and oxygen, the gas needed for life, combine in a fixed proportion to give the compound water, which has properties with which we are familiar. Only a chemical change such as that produced by an electric current can separate water into its components, as described in Table 1.3 and shown in Figure 1.21. The volume of hydrogen gas produced by the electrolysis of water is twice the volume of oxygen gas produced. This ratio of elements is reflected in the formula for water, H_2O, which, with the subscript 2, indicates twice as many hydrogen atoms as oxygen atoms in water.

1.10 mixtures

Mixtures are physical combinations of two or more pure substances, two or more elements, elements and compounds, or two or more compounds, in no fixed proportion. Because the combination is merely a physical mixing, the components of a mixture can be separated by physical means. For example, consider a mixture of iron filings (an element) and salt (a compound). You know this combination is a mixture because you can separate the components by a physical process. You can use a magnet to attract the iron away from the salt and the separation is complete. Or you can place the mixture in water to dissolve the salt and filter the solid iron filings from the salt solution, which will pass through the filter paper (refer to Figure 1.12 for the example of filtering sand from water). The salt (saline) solution is still a mixture, however, containing the compounds salt and water; it can be separated by the physical change of boiling the water away and leaving the salt behind as we discussed with regard to distillation (desalinization) of water in Figure 1.13.

In addition to physical separability, the other property of mixtures is the lack of a fixed proportion of components; that is, the composition of a mixture is *not constant.* Consider the preceding mixture. You could have a 50:50 mix of iron filings and salt, or you could have more iron and less salt, or much salt and little iron. Similarly, in making a salt solution, you could dissolve just a pinch of salt in a large amount of water or you could dissolve several tablespoons. This is what is meant by no fixed proportion or nonconstant composition.

The lack of constant composition of a mixture (Figure 1.18) shows up in its physical properties. For example, the boiling point and freezing point of a solution vary with its composition. Pure water always boils at 100°C (at 1 atmosphere pressure); however, the boiling point of a saltwater solution increases above 100°C as more salt is mixed with the water. As another example, pure water always freezes at 0°C (at 1 atmosphere pressure); however, the freezing point of a saltwater solution decreases below 0°C as more salt is mixed with the water.

sample exercise 6

Classify the following as elements, compounds, or mixtures.

a. Air
b. Gold
c. Iron filings and iodine crystals stirred together in a beaker
d. Iron filings and iodine combined into a new material in the proportion 56:254 by weight
e. Toothpaste

solution

a. Mixture. The principal components are the elements oxygen and nitrogen. Air also contains trace amounts of other gases and dust and other pollutants.
b. Element. See element 79 in the periodic table.
c. Mixture. A physical combination is described.
d. Compound. A chemical combination in a fixed proportion is described.
e. Mixture. If you read the ingredients on your toothpaste tube, the list of materials indicates a mixture.

•••problem 8

Classify the following as elements, compounds, or mixtures.

a. Wine
b. Chalk, a 4:12:48 (by weight) combination of calcium, carbon, and oxygen
c. Magnesium
d. Vitamin C, a 9:1:12 (by weight) combination of carbon, hydrogen, and oxygen

•••problem 9

Which of the materials listed in Sample Exercise 6 and in Problem 8 are pure substances?

sample exercise 7

Classify the following pure substances as elements or compounds.

a. Sodium bicarbonate
b. Carbon tetrachloride
c. Chlorine
d. Potassium

solution

Refer to Table 1.4 or the periodic table to identify the elements. The other pure substances must be compounds. Thus, a and b are compounds; c and d are elements.

1.11 using chemical language

Now that you have completed classifying matter, you are ready to combine words that represent matter into sentences that describe how different types of matter react with each other. The words that represent elements are called **atomic symbols**, and the words that represent compounds are called **chemical formulas**. The chemical sentence that represents a reaction is called a **chemical equation**. Once you understand accurately the symbolism in a chemical equation, you are ready to begin functioning as a responsible chemist.

For the sake of convenience and precise communication, chemists have devised a system of abbreviations for elements and compounds with which you must become familiar. To a large extent, your success in chemistry courses will depend on how accurately you understand the symbolic representations of matter. Knowing the names and symbols for the most common elements is the starting point for mastering chemistry. If you have not already memorized the names and symbols for the elements in Table 1.4, begin to do so now. Make your own list of these symbols that you can keep in front of you as you read formulas throughout the book. As you read, verify that you are using the correct name for each symbol as it appears.

Now that you have memorized the symbols for the most common elements, it will be helpful for you to think in terms of symbols. For example, when you hear the word *hydrogen, H* should appear in your mind's eye. The sound of the word *iron* should conjure up *Fe* rather than the letters *i-r-o-n.* Notice that the symbols always consist of either one capital letter or one capital letter followed by one small letter. These are the only correct forms. Carbon is always C, never c; calcium is always Ca, never CA, ca, or cA. Sticking to these correct forms for symbols is essential if you are to write correct representations of compounds.

In addition to visualizing the symbol for each element, it is now important for you to visualize the basic building blocks of elements called atoms. **Atoms** are very small, spherical units that fit together to make up each element. All atoms of each element look identical,[1] so that all atoms of the element copper, Cu, appear the same in size and structure. Each element has its own unique type of atoms, however. While all the atoms of hydrogen, H, appear identical, they are different from the atoms of Cu (Figure 1.22). It is the difference in size and structure between atoms of different elements that accounts for the different properties of each element, all of which you will learn in greater detail in Chapter 5.

Figure 1.22
All copper (Cu) atoms look identical to each other and hydrogen (H) atoms look identical to each other, but H atoms are different from Cu atoms in size and internal structure.

In the language of chemists, *compounds* are represented by *formulas.* The symbols of the combined elements are written without any space between them in a chemical formula. The formula tells not only the elements that have chemically combined to form the compound, but also the fixed proportion in which the chemical combination occurs. For example, the compound table salt has the formula NaCl, which immediately tells you sodium and chlorine have combined chemically. In Chapter 6, you will see how the formula tells you that the fixed proportion is 23:35.5 by weight. For now, it is far more useful to consider how the formula represents the combination of atoms of each element. The formula NaCl represents one atom of Na and one atom of Cl, which have combined. The formula for water is H_2O, which represents two atoms of H and one atom of O, which have combined into one new, basic building block called a **molecule.** Water molecules are all identical, all with the same size and same fixed geometry. Enormous numbers of these molecules are present in a sample of the compound water (Figure 1.23).

Other compounds also have atoms of elements chemically combined in different, fixed proportions. For example, sucrose (table sugar) with the formula $C_{12}H_{22}O_{11}$ has 12 C atoms, 22 H atoms, and 11 O atoms chemically combined into one sucrose molecule. All sucrose molecules are identical; each molecule has the same size and the same, fixed geometry. Enormous numbers of these molecules are present in a sample of the compound sucrose.

A more complete meaning of chemical formulas will unfold in Chapters 6 and 7. For now you should realize that the formula tells you the elements that have chemically combined. Also recognize that a correctly written formula employs

1. In Chapter 5, you will learn that some atoms of the same element, called isotopes, may have a different mass even though they look alike.

Figure 1.23
All water molecules are identical, with two H atoms and one O atom combined in a fixed proportion and with fixed geometry. All table sugar molecules are identical with 12 C atoms, 22 H atoms, and 11 O atoms combined in a fixed proportion and with fixed geometry.

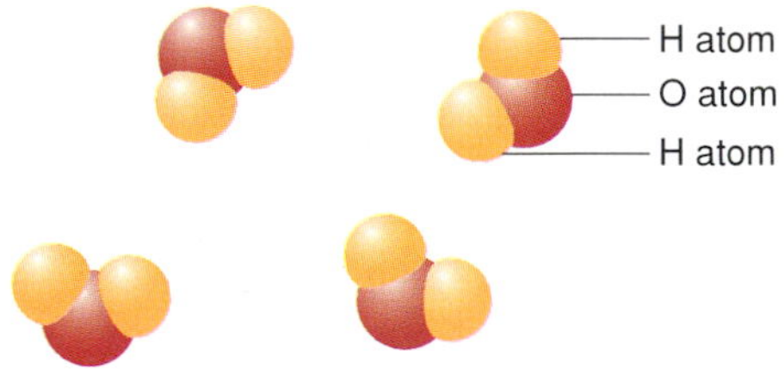

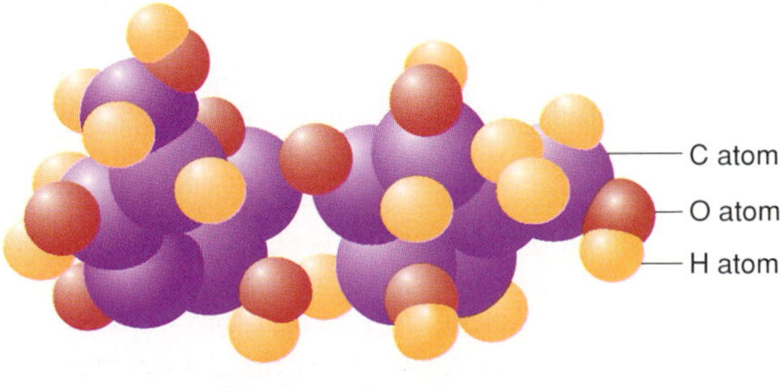

correct symbols, and the numbers of atoms of each element appear as subscripts (i.e., they appear below the line on which the symbols are written). It is the difference in the fixed proportion of the atoms of each element in compounds that accounts for the properties of each compound.

Chemical equations represent chemical reactions in the language of chemists. You have already seen that elements or compounds that react are called **reactants** and that the new material or materials formed are called the **products** of the reaction. The change of reactants into products is represented by an arrow in chemical language and is read as "yield." For example, the following reaction,

$$C(solid) + O_2(gas) \rightarrow CO_2(gas)$$

would be read as "carbon plus oxygen (react to) *yield* carbon dioxide." The reactants are the elements carbon, which is a solid at room temperature, and oxygen, which is a gas. The product is the new material, carbon dioxide, which is a gas. This reaction also teaches you new information about elemental oxygen at room temperature. It is not composed of separate atoms like many elements; rather, two atoms of oxygen gas have chemically combined to form oxygen as a **diatomic** molecule. *Di-* simply means two, and *diatomic* simply means two atoms. A diatomic molecule is a molecule made up of two atoms. Diatomic, *molecular oxygen* is the stable form of the element. Other elements, such as hydrogen, nitrogen, fluorine, chlorine, bromine, and iodine, also form diatomic molecules as their stable form.

The above reaction also uses everything you have learned about formulas for compounds. The product, carbon dioxide, is made of molecules in a fixed proportion, containing one atom of C and two atoms of O.

Chemical language is summarized in Figure 1.24. Using symbols, formulas, and equations, is a short way of illustrating chemical properties and representing much of what you know about the classification of matter. By simply looking at the reaction at the end of Figure 1.24, you now know that HBr, KOH, H_2O, and KBr are all compounds and you know the fixed ratio of elements in each compound. You also know that because they are compounds, they are pure substances. In addition, you know that the reaction starts with a mixture of two pure reactants, and when the reaction is finished, you end up with a mixture of two products and possibly some leftover reactants.

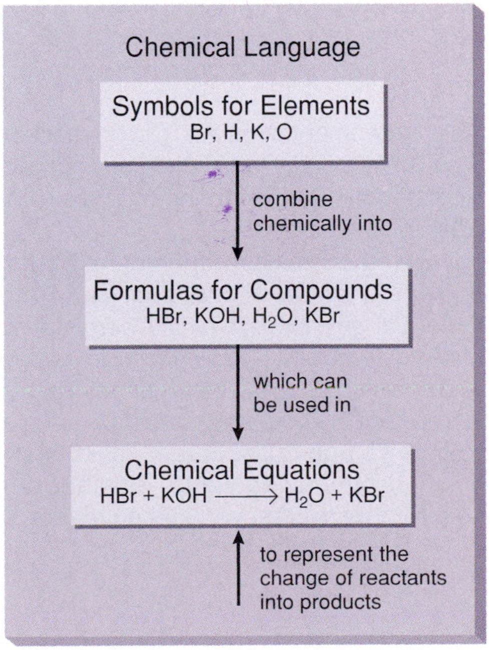

Figure 1.24
Summary of chemical language with symbols and formulas and its use in representing chemical changes.

sample exercise 8

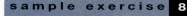

Classify the following as elements, compounds, or mixtures.

a. Al_2O_3
b. Al and S
c. CO_2 and H_2O
d. AgCl
e. Ca

solution

a. Compound, represented by a formula in which two atoms of Al are present for every three atoms of O. Aluminum oxide is a hard, naturally occurring mineral found as the gemstones ruby and sapphire and used in the manufacture of abrasives. The colors of ruby and sapphire are the result of small amounts of different impurities.
b. Mixture, the physical combination of the elements Al and S in any proportion. Aluminum is a silver-white metal, light in weight and not readily corroded or tarnished. It is used in lightweight utensils, airplane parts, and beverage containers. Sulfur is a nonmetal, appearing most commonly as a yellow, crystalline solid. It is used in making gunpowder and matches, in medicine, and in vulcanizing (imparting durability and elasticity to) rubber.
c. Mixture, the physical combination of the compounds CO_2, a gas, and H_2O, a liquid. Carbon dioxide is a colorless, odorless, inflammable gas used in carbonated beverages and in fire extinguishers.
d. Compound, silver chloride, in which the elements silver and chlorine occur in the fixed ratio of one atom of silver to one atom of chlorine. Silver chloride is a white, crystalline solid that is insoluble in water and darkens on exposure to light. It is used in photographic emulsions and in antiseptic silver preparations.
e. Element, calcium. Calcium is a silver-white metal found combined with other elements in limestone and chalk. It also occurs in animals and humans combined with other elements as a necessary component of bones, shells, and physiological functions such as nerve conduction and muscle contraction (including your beating heart).

1.12 classifying and identifying matter

You have begun to classify matter from its atomic symbol or from its chemical formula. Without conducting any experiment, you can determine immediately whether that substance is a metal or a nonmetal, an element or a compound, a pure substance or a mixture. From its formula or symbol, you can look up its properties in handbooks or databases.

Conversely, by using experimental results, you have also classified changes in substances as being physical or chemical, and you can recognize physical and chemical properties of materials. By measuring the properties of substances, you can uniquely identify unknown materials.

Many of the physical and chemical properties of materials are tabulated in handbooks, such as the *CRC Handbook of Chemistry & Physics* published by the CRC Press, Incorporated, *Lange's Handbook of Chemistry* published by McGraw-Hill, Incorporated, and *Merck Index of Chemicals & Biologicals* published by Merck & Company, Incorporated, which specializes in listing drugs and biological chemicals. All of these major reference works are in most libraries and should be readily available to you. Next time you go to the library, make a special point of finding these handbooks and perusing their contents. Whether or not you plan to continue studying science after this course, you can never predict when knowing the

CHEMLAB

Laboratory exercise 1: identifying a pure substance

Figure 1.25 shows photos of a few common vitamins and Table 1.5 gives their properties as listed in *Merck Index of Chemicals & Biologicals*. Use this table to answer the following questions.

Questions:

1. What property could be *immediately* used to identify a white solid as vitamin C or D?
2. What properties do vitamins A and D have in common?
3. How would you separate a mixture of vitamin C and vitamin A?
4. Which of these solids would be a liquid at 80°C?

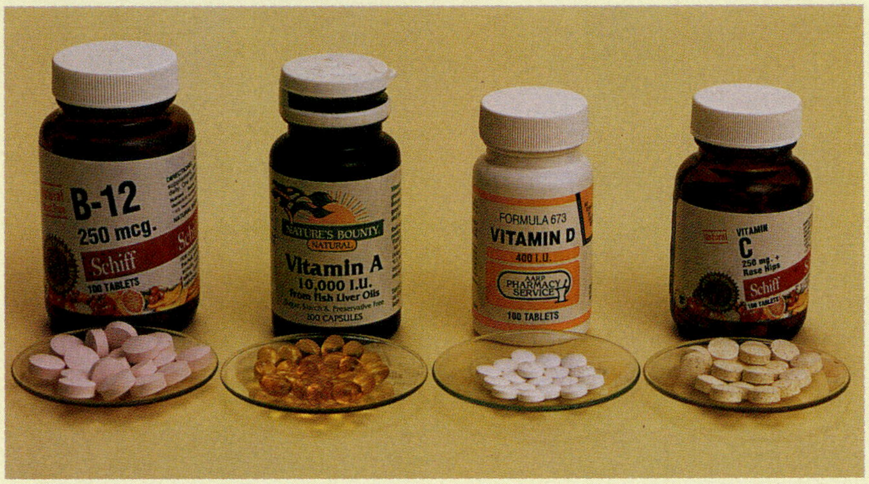

Figure 1.25
Vitamins B_{12}, A, D, and C, whose properties are listed in Table 1.5.

table 1.5 properties of some common vitamins

Vitamin	Appearance	Melting point (°C)	Solubility in water (ability to dissolve in water)
Vitamin A	Yellow solid	62–64	Insoluble
Vitamin B_{12}	Red solid	Decomposes without melting above 300°C	Slightly soluble
Vitamin C	White solid	190–192	Soluble
Vitamin D	White solid	84–85	Insoluble

properties of specific materials might be to your advantage. For example, the flammability of paint solvents, the toxicity of wood preservatives, the allergenic properties of cosmetics, and the hazards of individual pollutants in your drinking water raise potential questions that you should now be able to answer without having to rely on "experts" or feel the knowledge is beyond your grasp. Remember, studying chemistry and applying the scientific method in your life is just a part of your "taking charge." They enable you to be responsible for your personal safety, the safety of your environment, and your economic growth with safe chemical products, safe chemical processes, and life-improving pharmaceuticals.

As a chemist, you are now ready to use these handbooks or the computerized databases such as the CAS Registry referred to in the opening photograph of this chapter. For practice in classifying substances and identifying material, consider the hypothetical laboratory exercise at the top of this page.

CHEMQUEST

Separating two or more substances using a physical process is accomplished not only in the laboratory, but in everyday life as well. For example, the water you drink has been recycled innumerable times. Naturally, it has been separated from solid waste contaminants by being filtered through soil. The small pores between soil particles trap the solid contaminants while allowing the liquid water to pass through, just as the filter paper in a funnel traps solid iron filings.

Given enough soil acreage, the solid contaminants may be effectively removed from the water by the time the waste water returns to an underground well or a reservoir's watershed. However, in areas of dense population, municipal waste water treatment plants must be used to purify the sixty gallons of waste water produced daily by each person in the United States.

In spite of the fact that these plants provide three stages of treatment, many harmful contaminants, particularly those that dissolve in water, may not necessarily be removed from the water. Even though drinking water is commonly tested for toxic heavy metals, pesticides, herbicides, petroleum products, and cleaning agents, as well as harmful bacteria, measurable amounts of these items are frequently reported. You may have heard of tap water foaming due to dissolved detergents, or of concerns about the toxic lead levels in a city's water supply. Because of these types of problems, legislation restricting the disposal of harmful substances has been passed in an attempt to reduce pollution levels in drinking water.

Do you know what substances occur with greatest abundance in your drinking water? Do you know how these substances are introduced into your water supply? If these substances are potentially harmful, do you know what precautions are taken to keep them at a safe level? Separating pure water from other substances can be a long and expensive process, so which do you think is more cost effective: separating pure water from other substances or preventing these substances from entering the water supply in the first place?

chapter accomplishments

After completing this chapter, you should be able to define all key terms and do the following.

1.1 Introduction

- ❑ Define chemistry.
- ❑ Define matter.

1.2 Classification of matter by physical state

- ❑ State the shape and volume characteristics of the three physical states.
- ❑ Classify common samples of matter according to physical state.

1.3 Changes in the physical state of matter

- ❑ Name the processes by which matter changes physical state.

1.4 Scientific method

- ❑ State the four basic steps of the scientific method.

1.5 Pure substances and mixtures of pure substances

- ❑ Distinguish between the characteristics of pure substances and mixtures.
- ❑ Classify common samples of matter as pure substances or mixtures.

1.6 Physical and chemical changes of matter

- ❑ Distinguish between physical and chemical changes.

1.7 Physical and chemical properties identify pure substances

- ❑ Distinguish between physical and chemical properties.
- ❑ Given a list of samples and their melting and boiling points, classify the substances according to their physical state at a given temperature.
- ❑ Given a description of physical and chemical properties, distinguish between samples of matter.

1.8 Subdivisions of pure substances

- ❑ Given experimental data, classify pure substances as elements or compounds.

1.9 Elements and compounds

- ❑ Given a periodic table, classify elements as metals or nonmetals.

1.10 Mixtures

- ❑ Classify common samples of matter as elements, compounds, or mixtures.

1.11 Using chemical language

- ❑ Given the name, provide the symbol of each element in Table 1.4.
- ❑ Given the symbol, provide the name of each element in Table 1.4.
- ❑ Recognize a correctly written formula.
- ❑ Given the formula(s) of the substance(s) present, classify samples of matter as elements, compounds, or mixtures.

1.12 Classifying and identifying matter

- ❑ Given a table of physical and chemical properties, identify substances from their measured properties.

key terms

chemistry
matter
properties
physical state
solids
liquids
gases
melting
freezing
melting point
freezing point
temperature of fusion
evaporation
vaporization
condensation
boiling point
sublimation
sublimation point
temperature of sublimation
scientific method
pure substances
mixtures
constancy of composition
physical change
chemical change
chemical reaction
precipitate
physical properties
chemical properties
element
compound
metals
nonmetals
law of definite composition
atomic symbols
chemical formulas
chemical equation
atoms
molecule
reactants
products
diatomic

problems

1.1 Introduction

10. What does the study of chemistry involve?
11. What is matter?

1.2 Classification of matter by physical state

12. Describe experimental tests that will allow you to distinguish between the solid, liquid, and gaseous states.
13. What is meant by the statement "a gas has indefinite volume"?
14. Which of the three states of matter possess the property of fluidity?
15. Which of the three states of matter can be poured from one container into another?
16. For each of the following common substances, indicate whether it exists as a solid, liquid, or gas at room temperature and give the reason for your classification.
    a. Tap water
    b. Silver
    c. Skin
    d. Mercury
    e. Carbon dioxide
    f. Molasses
    g. Chocolate pudding
17. A sample of matter is in the liquid state. How might you convert it to a solid? How might you convert it to a gas?

1.3 Changes in the physical state of matter

18. Classify each of the materials below as a solid, liquid, or gas at room temperature and at "dry ice" temperature (–80°C).
    a. Ice cream
    b. Iodized table salt
    c. Cola
    d. A nickel
    e. Dry ice (frozen CO_2)
    f. Snow
    g. Methane
19. What name do we give to each of the following processes?
    a. Ice cubes turn to water.
    b. Steam forms water droplets on a mirror.
    c. Milk is made into ice milk.
    d. Frozen laundry "dries."
    e. Perspiration "dries."

20. What change in physical state occurs during the formation of the following?
    a. Rain
    b. Frost
    c. Snow
    d. Steam
21. A sealed glass bulb is half-filled with water, on which some ice and wood are floating. The remainder of the bulb is filled with air. How many physical states are present? Identify them.

1.4 Scientific method

22. Describe the steps of the scientific method as applied to a medical diagnosis by a physician.

1.5 Pure substances and mixtures of pure substances

23. Classify each of the materials in Problem 18 as a pure substance or a mixture.
24. Based on your everyday experience, how would you separate the following mixtures into their components?
    a. A mixture of sand and water
    b. A mixture of sugar and sand
    c. A dissolved mixture of sugar and water
25. Design and describe experiments that would prove that your classifications of cola and dry ice in Problem 23 were correct.

1.6 Physical and chemical changes of matter

26. Classify each of the following as a physical or chemical change.
    a. Water evaporates from a glass.
    b. The process of metabolism, that is, complex foodstuffs are broken down in the body to smaller substances, carbon dioxide and water.
    c. Wood is sanded.
    d. Meat is ground up and formed into hamburger patties.
    e. A firecracker explodes.
    f. A leaf turns color in autumn.
    g. Gasoline is burned in a car engine.
27. Identify the chemical and physical changes in the following sequences.
    a. A lump of sugar is ground to a powder and then heated in air. It melts, then darkens, and finally bursts into flame and burns.
    b. Gasoline is sprayed into the carburetor, mixed with air, converted to vapor, and burned, and the combustion products expand in the cylinder.
28. The electrolysis of water produces bubbles (Figure 1.21). Boiling water produces bubbles (Figure 1.6). How does the composition of the bubbles tell us that one process is chemical and the other is physical?
29. Which of the following changes are physical and which are chemical?
    a. Melting butter
    b. Iron rusting
    c. Flash cube going off
    d. Banging on a drum

1.7 Physical and chemical properties identify pure substances

30. Use the data in Table 1.2 to classify the following substances according to their physical state at the given temperature.
    a. Sucrose at 78°C
    b. Iron at 3,200°C
    c. Iron at 1,700°C
    d. Chlorine at –121°C
    e. Oxygen at –195°C
31. State three properties that may be used to identify specimens of pure substances.
32. What properties distinguish water from other colorless liquids, such as alcohol, benzene, and acetone?
33. What properties distinguish white solids, such as table salt and table sugar, from one another?
34. Are the following properties of a certain metal physical or chemical?
    a. It is a solid at room temperature.
    b. It is easily bent into shapes.
    c. It combines with water violently, giving off a gas.
    d. It melts at a low temperature.
35. Identify the physical and chemical properties indicated in the following description. Chlorine is a yellow-green gas that is toxic to humans. It combines with sodium to form sodium chloride, a solid that melts at 808°C.
36. Compare the similarities and differences between the processes of melting sugar and dissolving sugar in water.

1.8 Subdivisions of pure substances

37. A pure blue powder, when heated in a vacuum, releases a greenish-colored gas and leaves behind a white solid. Is the original blue powder a compound or an element?
38. A shiny, metallic substance conducts an electric current without a change in its properties. The substance is heated until it liquefies and then an electric current is passed through the liquid, again without a change in its properties. Is the substance likely to be an element or a compound?
39. Pure substances can be classified into how many categories? Identify them.
40. Classify the pure substances in Problem 18 as elements or compounds.
41. Are there more elements or compounds in the world? Explain.

1.9 Elements and compounds

42. Classify the following pure substances as elements or compounds.
    a. Copper
    b. Iron oxide
    c. Sulfur tetrafluoride
    d. Fluorine
43. Elements *A* and *B* combine to form *AB* in a fixed proportion of 3:1 by weight. What will happen if we try to combine 4 g of *A* with 1 g of *B*?
44. Describe the basic steps of the scientific method. Apply these steps to the problem of distinguishing elements as metals and nonmetals.

1.10 Mixtures

45. Classify the following as elements, compounds, or mixtures.
    a. Sodium
    b. Spaghetti with meatballs
    c. Carbon particles in a hydrogen gas atmosphere
    d. A cup of tea
    e. Laughing gas (14 parts N to 8 parts O by weight)
    f. Methane (75% carbon, 25% hydrogen by weight)
46. Identify the components of the following mixtures as elements or compounds.
    a. Club soda (CO_2 gas in water with a pinch of salt)
    b. "Clean" air
    c. Aerated water

1.11 Using chemical language

47. Name the elements whose symbols are H, Ca, Si, C, Cl, P, and K.
48. Give symbols for the elements bromine, nitrogen, mercury, silver, and gold.
49. Classify the elements in Problems 47 and 48 as metals or nonmetals.
50. Tell whether the following symbols and formulas are written correctly. If not, tell what is wrong and write them correctly.
    a. CL
    b. C_3H_6
    c. Zn
    d. f
    e. NaOH

1.12 Classifying and identifying matter

51. Classify the following as elements, compounds, or mixtures.
    a. C_2H_2
    b. O_2 and N_2
    c. Mg
    d. K_2SO_4
    e. H_2O and NaCl
52. Match each item on the left with as many descriptions on the right as apply.

a. C and P	
b. $C_6H_{12}O_6$ (blood sugar)	1. Pure compound
c. C_2H_5OH and H_2O (alcohol and water)	2. A mixture of compounds
	3. Elements
d. A chocolate chip cookie	4. A solution
e. Oxygen dissolved in water	5. A mixture of solids

53. Classify each of the following as a chemical or physical property.
    a. Aluminum melts at 660°C.
    b. Aluminum forms a thin oxide coating in air.
    c. Naphthalene (moth balls) sublimes at room temperature.
    d. Mercury (II) oxide forms mercury and oxygen when heated.
    e. Butter turns rancid when left unrefrigerated in open air.
    f. Water evaporates from an open container.
    g. Sugar dissolves in water.
    h. Oil burns, giving off carbon dioxide, water, and heat.
54. Classify each of the following as a physical or chemical change.
    a. The process of photosynthesis, that is, CO_2 and H_2O are converted to carbohydrates in plants.
    b. Antifreeze boils out of a radiator.
    c. A dish of cherries jubilee is flamed with brandy.
    d. A firefly emits light.
    e. A nail is magnetized.
    f. A nail rusts.
    g. Food spoils.

Additional problems

55. Name the elements whose symbols are Cu, Hg, Pb, Sn, S, and P.
56. Give symbols for the elements potassium, sodium, chlorine, iron, hydrogen, silver, and gold.
57. Tell whether the following symbols and formulas are written correctly. If not, tell what is wrong and write them correctly.
    a. CH_4
    b. AL
    c. cL
    d. Zn
    e. h
58. Classify the following as elements, compounds, or mixtures.
    a. KNO_3
    b. Ca and C
    c. B
    d. CaC_2
    e. CO and NO
59. A newly discovered substance is soft and shiny and conducts electricity well. Is it more likely to be a metal or nonmetal?
60. The symbol Th stands for thorium. Without looking at a periodic table, how can you be sure this symbol represents an element and not a compound?
61. Does each of the following describe a physical change or a chemical change? Explain your answers.
    a. Tea is sweetened by adding sugar to it.
    b. Tobacco is smoked in a pipe.
    c. Frozen lemonade is reconstituted by adding water to it.
    d. A pond's surface freezes in winter.
    e. A candle's light slowly diminishes and finally goes out.
62. Describe a chemical property exhibited by each of the following.
    a. Iron
    b. Alcohol
    c. Air
    d. Water
    e. Pizza

chapter 2

measurement

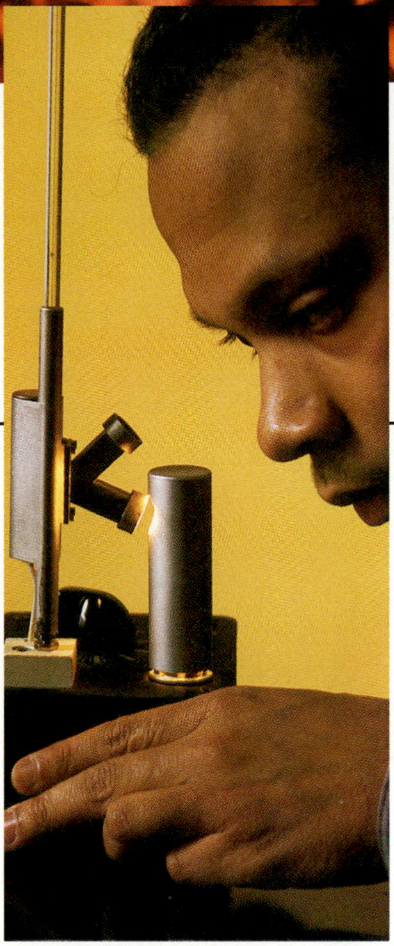

A student is using a Mel-Temp apparatus to measure the melting point of a solid sample. The melting point is a characteristic physical property that can be used to identify the sample. The Mel-Temp apparatus is a block of metal that is electrically heated at a slow rate and holds a thermometer and sample. The sample whose melting point is being measured has been ground into a fine powder and placed in a very thin glass tube. A magnifying lens and light allow you to observe the sample clearly as it melts and to record the temperature on the thermometer.

2.1 what is measurement?

Now that you can classify matter, use reference handbooks and databases, identify matter from its properties, and observe the properties of matter with experiments, you should be ready to measure matter and to understand how closely two or more measurements of the same quantity agree with one another. It is not possible to function successfully in daily life without some knowledge of weights and measures. How could you buy proper amounts of potatoes or liverwurst, decide how much fabric is needed to make a blouse, or mark out bases on a baseball diamond without understanding measurement? Because the understanding of scientific concepts is often based on measurements, it is very important that you consider exactly what the process of measurement is.

When you measure something, you are comparing it to some reference standard. For example, when you step on a bathroom scale calibrated or marked off in pounds, you are comparing your weight to the reference standard, the pound. The scale tells you how heavy you are relative to 1 pound. All systems of measurement involve this idea of comparison to a reference standard. All measurements are relative to a standard. This is an important concept that comes up repeatedly in chemistry. It is a good idea to think it through for some familiar measurements before tackling unfamiliar ones. The metric system is the set of reference standards used by scientists, and it is rapidly being introduced into everyday measurements. As a convenience in converting to metric units, many commodities are labeled with the more familiar British units of inches, feet, yards, ounces, and pounds, as well as with metric units of centimeters, meters, grams, and kilograms. Once everyone becomes familiar with metric units, there will no longer be a need for double labeling, and only metric units will be used (Figure 2.1).

In addition to becoming familiar with the metric system, you will come to understand that how a scientific measurement is written reflects how precise that measurement is, that is, how much uncertainty will occur if the same measurement is repeated. When you pay for a 3-pound steak at $6.99 per pound, you would be disturbed when you got home if you measured the same steak as 2.5 pounds. How reproducible a measurement is becomes important. Most packages of meat are now labeled with the weight given to two decimal places, as shown in Figure 2.2, so you know the weight to one-hundredth of a pound and you are charged accordingly. All scientific measurements are automatically written to reflect how reproducible they are.

Scientific measurements may vary over a tremendous range in size. Some measurements, such as the size or mass of atoms and molecules, are extremely small,

Figure 2.1
Many commodities are conveniently labeled with both British and metric units, allowing you to become familiar with the metric system.

Figure 2.2
This meat has been weighed to two decimal places so you know its weight to one-hundredth of a pound.

while other measurements, such as the number of atoms and molecules in a sample of matter, are extremely large. Scientists have developed a very helpful notation for writing numbers of any size. Familiarity with scientific notation will be useful throughout your scientific studies and will be our starting point in this chapter.

2.2 scientific notation

In scientific work, it is often necessary to use extremely large and extremely small numbers. The number of atoms and molecules in a sample of matter is extremely large. In Chapter 8, you will learn how to calculate the number of atoms or molecules in a sample using Avogadro's number, named in honor of the Italian physicist who lived in the early nineteenth century. Avogadro's number equals 602,200,000,000,000,000,000,000, which is unduly cumbersome to write. Similarly, the mass of an electron, 0.000000000000000000000000000091095 grams, is also tedious to write and even more difficult to read because of the lack of commas in decimal numbers. Consequently, you need a more convenient way to express such numbers. A way that eliminates the need to write all of the zeros is desirable if for no other reason than to save ink and elbow grease. The method used is known as **scientific notation**. As you review some of its basic principles, keep in mind the advantages it offers not only for writing numbers, but also for using those numbers in further calculations.

Exponents and bases. You are probably already familiar with the idea of raising a number to a power. For example, 10^2 is read as 10 to the second power and it is equal to 10×10, or 100. The 10 is known as the base and the 2 is called the exponent.

$$10^2$$
(2 = Exponent; 10 = Base)

The exponent tells the number of times the base is a factor in the multiplication as reviewed in the following examples.

$$10^6 = \underbrace{10 \times 10 \times 10 \times 10 \times 10 \times 10}_{\text{Six factors of 10}} = 1{,}000{,}000$$
(6 = Exponent; 10 = Base)

$$10^{10} = \underbrace{10 \times 10 \times 10 \times 10 \times 10 \times 10 \times 10 \times 10 \times 10 \times 10}_{\text{Ten factors of 10}} = 10{,}000{,}000{,}000$$

$$10^1 = \underbrace{10}_{\text{One factor of 10}}$$

$$10^0 = 1$$
No factors of 10

$$10^{-1} = \frac{1}{10^1} = \frac{1}{10} = 0.1$$
Negative exponent; $\frac{1}{10^1}$: Reciprocal of original; 10^1: Same exponent with a positive sign

$$10^{-3} = \frac{1}{10^3} = \frac{1}{1{,}000} = 0.001$$

Since $10^1 = 10$ and $10^{-1} = 0.1$, it is reasonable that $10^0 = 1$. This will be proved later in this section under "multiplying numbers in scientific notation."

A base may also have a negative exponent, for example, 10^{-1} or 10^{-3}. You may always write a base with a negative exponent as the reciprocal, with the sign of the exponent changed to positive. The reciprocal of a number A is defined as $1/A$.

Writing numbers in scientific notation. Any ordinary decimal number can be expressed as a decimal number between 1 and 10 multiplied by some power of 10. For example, the number 392,000, when written in scientific notation, looks like

$$392{,}000 = \underbrace{3.92}_{\text{Decimal number between 1 and 10}} \times \underbrace{10^5}_{\text{Power of 10}}$$

When a number is less than 1, negative exponents are used.

$$0.00432 = \underbrace{4.32}_{\text{Decimal number between 1 and 10}} \times \underbrace{10^{-3}}_{\text{Power of 10}}$$

The rules for converting ordinary numbers to scientific notation are subdivided into three cases, depending on the size of the number.

Case 1: if the number is equal to or greater than 10

1. Move the decimal point to the left, counting the number of places you must move it until you have a decimal number between 1 and 10.
2. Multiply this decimal number by 10 raised to a positive power equal to the number of places you moved the decimal point.

---

sample exercise 1

illustrating case 1

Write 138.34 in scientific notation.

solution

Step 1. You recognize that 138.34 is greater than 10. Therefore, move the decimal point to the left until you obtain a number between 1 and 10. Then, 138.34 becomes 1.3834. This requires a movement of two places.

Step 2. Now write down the new number and multiply it by 10^2, because the decimal was moved two places. (The sign is positive because the number is greater than 10.) Note that $10^2 = 100$, so multiplying by 10^2 is the same as multiplying by 100.

• **Answer:** $138.34 = 1.3834 \times 10^2$

Case 2: if the number is less than 1

1. Move the decimal point to the right, counting the number of places you must move it until you have a decimal number between 1 and 10.
2. Multiply this decimal number by 10 raised to a negative power equal to the number of places you moved the decimal point.

---

sample exercise 2

illustrating case 2

Write 0.000108 in scientific notation.

solution

Step 1. You recognize that 0.000108 is less than 1. Therefore, move the decimal point to the right until you obtain a number between 1 and 10. Then, 0.000108 becomes 1.08. This requires a movement of four places.

Step 2. Write down the new number and multiply it by 10^{-4} because the decimal was moved four places. The sign is negative because the number is less than 1.

• **Answer:** $0.000108 = 1.08 \times 10^{-4}$

Case 3: if the number is equal to or greater than 1.0 and less than 10.0

Simply write down the number and multiply it by 10^0, because no (zero) movement of the decimal point is required.

---

sample exercise 3

illustrating case 3

Write 3.85 in scientific notation.

solution

You recognize that the number is between 1.0 and 10.0. Therefore, write the number down and multiply it by 10^0.

• **Answer:** $3.85 = 3.85 \times 10^0$ Normally 10^0 is omitted, just as the number 1 often is.

••• problem 1

Write the following in scientific notation.

a. 173
b. 0.0029
c. 131,982
d. 0.00000401
e. 16.4

Converting scientific notation into ordinary decimal numbers. To convert a number in scientific notation, such as 4.923×10^2, into an ordinary decimal number, start by examining the exponent.

1. If the exponent is positive, move the decimal to the right a number of places equal to the value of the exponent. You may have to fill in zeros. Remember that positive exponents are associated with numbers greater than 10.
2. If the exponent is negative, move the decimal to the left a number of places equal to the value of the exponent. Again, you may have to fill in zeros. Remember that negative exponents are associated with numbers less than 1.

---

sample exercise 4

Write 4.923×10^2 as an ordinary decimal number.

solution

The exponent is positive 2, so we move the decimal point two places to the right.

$$4.923 \times 10^2 = 492.3$$

---

sample exercise 5

Write 9.23×10^{-5} as an ordinary decimal number.

solution

The exponent is negative 5, so we move the decimal point five places to the left, filling in zeros.

$$00009.23 \times 10^{-5} = 0.0000923$$

problem 2

Write the following as decimal numbers.

a. 7.31×10^3 c. 6.38×10^5

b. 1.92×10^{-4} d. 8.36×10^{-5}

Multiplying numbers in scientific notation. To multiply two numbers written in scientific notation, multiply the two decimal numbers together in the usual manner and then *add* the two exponents, being careful to treat the exponents as signed numbers. The sum is the correct power of 10 in the product.

$$(a \times 10^m)(b \times 10^n) = ab \times 10^{(m+n)}$$

Note that the exponents may be either positive or negative when multiplying numbers. Obtaining the sum of the exponents may require adding signed numbers. For a review of the addition of signed numbers, see Appendix C. Four examples of multiplying in scientific notation follow.

$$(2 \times 10^4)(3 \times 10^7) = 6 \times 10^{(4+7)} = 6 \times 10^{11}$$

$$(4.11 \times 10^{-1})(1.38 \times 10^3) = 5.67 \times 10^{(-1+3)} = 5.67 \times 10^2$$

$$(3.39 \times 10^{-8})(2.25 \times 10^{-4}) = 7.63 \times 10^{(-8+(-4))} = 7.63 \times 10^{-12}$$

$$(9.11 \times 10^5)(2.18 \times 10^3) = 19.9 \times 10^{(5+3)} = 19.9 \times 10^8$$

Notice in the last example that the decimal number of the answer turns out to be greater than 10, that is, 19.9. This answer, 19.9×10^8, is not in the correct form and must be adjusted so that the decimal number is between 1 and 10. You do this by recognizing that $19.9 = 1.99 \times 10^1$. Therefore,

Sum of exponents

$$19.9 \times 10^8 = 1.99 \times 10^{\textcircled{1}} \times 10^{\textcircled{8}} = 1.99 \times 10^9$$

Summary of the method for multiplication:

1. Multiply the decimal numbers in the usual manner.
2. Add the exponents.
3. Adjust the answer to the correct form of scientific notation.

sample exercise 6

Carry out the following multiplication.

$$(6.42 \times 10^{-9})(2.58 \times 10^2) =$$

solution

Follow the preceding summarized steps.

Step 1. $6.42 \times 2.58 = 16.6$

Step 2. $-9 + 2 = -7$

Step 3. But 16.6×10^{-7} must be adjusted because 16.6 is not between 1 and 10. Putting 16.6 in scientific notation, you get 1.66×10^1. Substituting this for 16.6 yields

Sum of exponents

$$16.6 \times 10^{-7} = 1.66 \times 10^{\textcircled{1}} \times 10^{\textcircled{-7}} = 1.66 \times 10^{-6}$$

••• problem 3

Multiply each of the following, expressing your answer in scientific notation.

a. $(2.76 \times 10^{2})(8.32 \times 10^{4}) =$

b. $(3.37 \times 10^{1})(4.89 \times 10^{0}) =$

c. $(2.51 \times 10^{-8})(1.53 \times 10^{-6}) =$

d. $(8.13 \times 10^{-5})(3.89 \times 10^{6}) =$

Now that you know how to multiply exponential numbers, you can accept the proof that $10^0 = 1$.

$$10^1 \times 10^{-1} = 10^0$$

$$10 \times 0.1 = 1$$

Since the left sides of both equations are equal, the right sides must also be equal. Therefore,

$$10^0 = 1$$

Dividing two numbers in scientific notation. The process of division in scientific notation is similar to that of multiplication. To divide two numbers written in scientific notation, you divide one decimal number by the other in the usual manner and then *subtract* the exponent in the denominator (divisor) from the exponent in the numerator (dividend), being careful to treat this as the subtraction of signed numbers. The difference is the correct power of 10 in the quotient.

$$\frac{a \times 10^m}{b \times 10^n} = \frac{a}{b} \times 10^{m-n}$$

Note that the exponents may again be either positive or negative. Obtaining the difference of the exponents may require you to subtract signed numbers. For a review of the subtraction of signed numbers, see Appendix C. Four examples of dividing in scientific notation follow.

$$\frac{8 \times 10^4}{4 \times 10^7} = 2 \times 10^{4-7} = 2 \times 10^{-3}$$

$$\frac{4.11 \times 10^{-1}}{1.38 \times 10^3} = 2.98 \times 10^{-1-3} = 2.98 \times 10^{-4}$$

$$\frac{3.39 \times 10^{-8}}{2.25 \times 10^{-4}} = 1.51 \times 10^{-8-(-4)} = 1.51 \times 10^{-4}$$

$$\frac{2.18 \times 10^3}{9.11 \times 10^5} = 0.239 \times 10^{3-5} = 0.239 \times 10^{-2}$$

Notice in the last example that the decimal number is not between 1 and 10, so you must rewrite this answer in the correct form. You do this by recognizing that $0.239 = 2.39 \times 10^{-1}$. Therefore,

$$0.239 \times 10^{-2} = 2.39 \times 10^{(-1)} \times 10^{(-2)} = 2.39 \times 10^{-3}$$

↑ Sum

Summary of the method for division:

1. Divide the decimal numbers in the usual manner.
2. Subtract the exponents (numerator minus denominator).
3. Adjust the answer to the correct form of scientific notation.

sample exercise 7

Carry out the following division.

$$\frac{1.25 \times 10^8}{6.12 \times 10^4} =$$

solution

Follow the preceding summarized steps.

Step 1. $1.25 \div 6.12 = 0.204$

Step 2. $8 - 4 = 4$

Step 3. But, 0.204×10^4 must be adjusted because 0.204 is not between 1 and 10. Putting 0.204 in scientific notation, you get 2.04×10^{-1}. Substituting this for 0.204, you obtain

$$0.204 \times 10^4 = 2.04 \times 10^{(-1)} \times 10^{(4)} = 2.04 \times 10^3$$

(3 ← Sum)

problem 4

Divide each of the following, expressing your answer in scientific notation.

a. $\dfrac{8.0 \times 10^7}{4.0 \times 10^3} =$ c. $\dfrac{4.19 \times 10^5}{2.08 \times 10^{-3}} =$

b. $\dfrac{8.53 \times 10^{-7}}{3.12 \times 10^3} =$ d. $\dfrac{3.97 \times 10^{-8}}{4.19 \times 10^{-7}} =$

Adding and subtracting numbers written in scientific notation. To add or subtract exponential numbers directly, the exponents of all numbers must be the same. If you were asked to add 4.5×10^4 plus 3.2×10^4, simply add the decimal numbers and maintain the same base and exponent. The sum would be 7.7×10^4. When the numbers to be added or subtracted have different exponents, the numbers must be rewritten in equivalent forms so that the exponents are the same.

sample exercise 8

$$(5.93 \times 10^7) - (4.0 \times 10^5) =$$

solution

The exponents are different, so before subtracting, you must change one of the exponents so that it matches the other exponent. It is generally easier to make the larger exponent smaller by "taking out tens." In this case, take out 10^2 from 5.93×10^7.

$$5.93 \times 10^7 = \underbrace{5.93 \times 10^2}_{} \times 10^5 = 593 \times 10^5$$

Now subtract:

$$\begin{array}{r} 593 \times 10^5 \\ -\ 4 \times 10^5 \\ \hline 589 \times 10^5 \end{array}$$

Rewrite the answer in proper form:

$$589 \times 10^5 = 5.89 \times 10^2 \times 10^5 = 5.89 \times 10^7$$

To adjust negative exponents, it is easiest to "take out tenths." For example, to add $(6.94 \times 10^{-3}) + (1.920 \times 10^{-2})$, begin by adjusting the –3 exponent to –2. This is done by "taking out" 10^{-1}:

$$6.94 \times 10^{-3} = \underbrace{6.94 \times 10^{-1}}_{} \times 10^{-2} = 0.694 \times 10^{-2}$$

Now the addition can be done:

$$\begin{array}{r} 0.694 \times 10^{-2} \\ 1.920 \times 10^{-2} \\ \hline 2.614 \times 10^{-2} \end{array}$$

Figure 2.3

Scientific calculators allow you to enter numbers in scientific notation directly and to use them in calculations.

2.3 using scientific notation on a hand calculator

You are probably familiar with using a hand calculator for arithmetic with decimal numbers. Scientific hand calculators are particularly convenient because they allow you to enter numbers directly in scientific notation. The procedure for arithmetical calculations on a scientific calculator is the same as for calculations on an ordinary calculator. It simply has the added advantage of accepting numbers in scientific notation and giving answers in scientific notation as well. For a review of any of these procedures, please refer to Appendix B.

With such tremendous advantages at low cost, it is strongly advised that you have your own scientific calculator for all further chemical calculations (Figure 2.3). Using the rules for scientific notation as outlined in Section 2.2 will aid you in checking the reasonableness of your answers and helping you to understand the size of your results.

2.4 metric system

The reference standards scientists use are those of the **metric system**, a highly logical, easy-to-use system based on multiples of 10. This system was originally developed during the French Revolution and was adopted throughout the world except for English-speaking countries. The United States is slowly converting to the metric system; new road signs in kilometers and food package weights in grams are just the beginning. Not only is the metric system the standard for scientists, but it is the standard for international commerce.

Since your children and grandchildren will grow up with the metric system, they will be able to skip this section of the text. Because you have been brought up with the British system of units using pounds and inches, you must now learn the metric system. Around 1960, scientists slightly revised and updated the metric system and called it **SI** after the French name, *Système International d'Unités.* The "base units" in the SI system of units are meters for length, kilograms for mass, and liters for volume.

Length, the distance between two points. For **length**, the distance between two points, the reference standard is represented by a meterstick, which has a length of 1 meter[1] (meter is abbreviated m). A yardstick, at 36 inches (inches is abbreviated in.), is somewhat shorter than a meterstick, which is 39.37 in. long. Look at Figure 2.4. The man is larger than the meterstick. His measurement is 1.8 m (1.8 times a meter). The child is smaller than the meterstick, 0.9 m (0.9 times a meter).

The meter is the base unit of length, and all length measurements could be expressed in meters. However, it is more convenient to have some larger and smaller units (just as in the British system we have inches, feet, yards, and miles). The beauty of the metric (SI) system is that all units are related by multiples (or divisions) of the base unit, and the multiple (or division) of the base

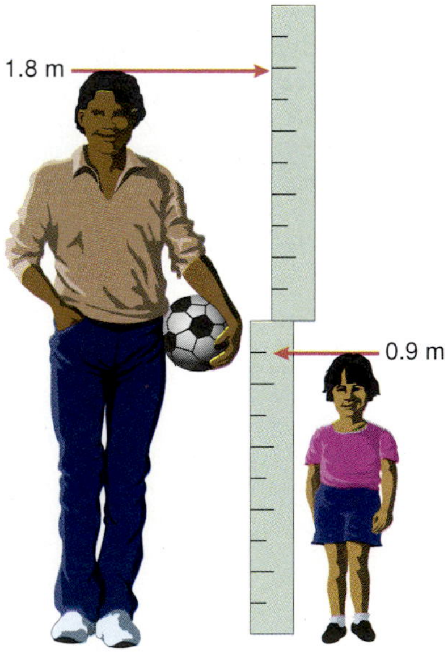

Figure 2.4

The reference standard for length in the metric system is the meter. The man is 1.8 times the length of the reference standard; the child is 0.9 times the length of the reference standard.

1. The meter is defined as the distance light will travel in 1/299,792,458 of a second. This distance is unchangeable and is readily measured with great accuracy by modern instruments. Metersticks are made the length of this distance as accurately as possible.

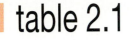

table 2.1 commonly encountered prefixes of the metric system, their abbreviations, and their mathematical meanings

		Multiply base unit by	
Prefix	**Abbreviation**	**Common number**	**Exponential number**
kilo-	k	1,000	10^3
deci-	d	1/10, or 0.1	10^{-1}
centi-	c	1/100, or 0.01	10^{-2}
milli-	m	1/1,000, or 0.001	10^{-3}
micro-	μ	1/1,000,000, or 0.000001	10^{-6}

table 2.2 metric prefixes applied to the base units of length and mass

Length		**Mass**	
1 kilometer (km)	= 1,000 meters (m)	1 kilogram (kg)	= 1,000 grams (g)
1 decimeter (dm)	= 0.1 meter (m)	1 decigram* (dg)	= 0.1 gram (g)
1 centimeter (cm)	= 0.01 m	1 centigram* (cg)	= 0.01 g
1 millimeter (mm)	= 0.001 m	1 milligram (mg)	= 0.001 g
1 micrometer (μm)	= 0.000001 m	1 microgram (μg)	= 0.000001 g

* This unit is not commonly used by scientists.

unit is a power of 10. The names of the units are constructed from the base unit name preceded by a prefix (kilo-, deci-, centi-, milli-, micro-), which tells which multiple is involved.

$$\underbrace{\text{Prefix}}_{\substack{\text{Tells the multiple}\\ \text{(power of 10)}}} \quad + \quad \text{Base unit}$$

For example,

$$\underbrace{\text{kilo-}}_{\substack{\text{Multiple is 1,000}\\ (10^3)}} \quad + \quad \underbrace{\text{meter}}_{\substack{\text{Base unit of}\\ \text{length}}}$$

Thus, *kilometer* (km) means "1,000 meters" or 1×10^3 m.

Because the abbreviation for kilo- is k and for meter is m, the abbreviation for kilometer is km. Prefixes and their abbreviations appear in Table 2.1. These prefixes and abbreviations always have mathematical meaning whenever they are encountered (Table 2.2).

One thousand metersticks laid end to end equal 1 km. Dividing the meterstick into 10 equal parts (dividing by 10 and multiplying by 1/10 are the same operation) results in 10 smaller units, each of which is called a decimeter (dm), so 10 dm = 1 m.

table 2.3 equalities among common metric units

Length	Mass	Volume
1 km = 1,000 m	1 kg = 1,000 g	
1 m = 10 dm		
1 m = 100 cm		
1 m = 1,000 mm	1 g = 1,000 mg	1,000 cm^3 = 1 liter (L) = 1,000 milliliters (mL)
1 m = 1,000,000 μm	1 g = 1,000,000 μg	

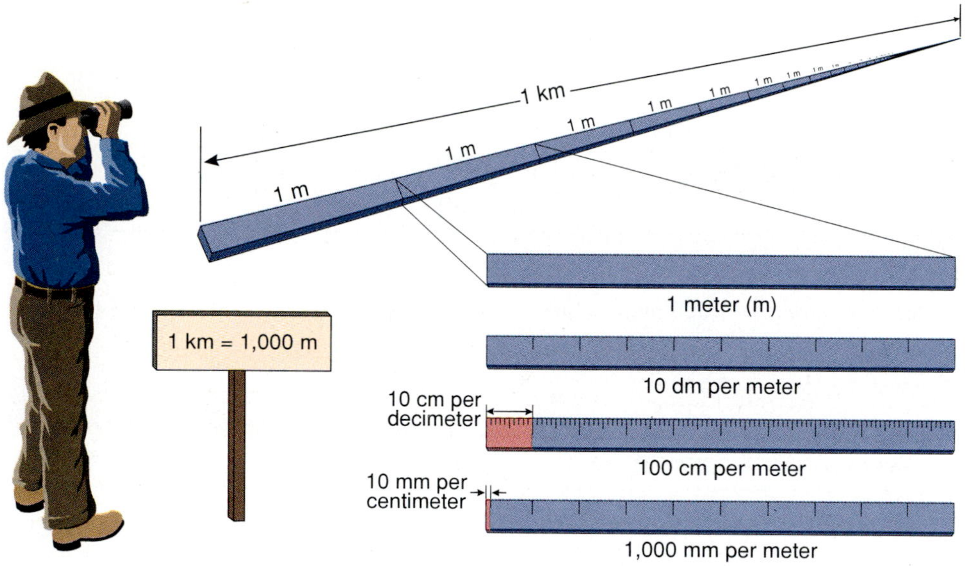

Figure 2.5
Factor-of-ten relationships between the common units of length in the metric system.

Chopping each decimeter into 10 parts results in 10 still smaller units, each of which is called a centimeter (cm), so 100 cm = 1 m. Further dividing each centimeter into 10 parts results in 10 smaller units, each of which is called a millimeter (mm), so 1,000 mm = 1 m. This is summarized in Table 2.3 (under "Length") and in Figure 2.5.

Mass, the amount of matter in an object. The metric reference standard for mass is a metal cylinder made of a platinum-iridium alloy stored in a vault near Paris, France. The mass of this block of metal is defined as 1 kilogram (abbreviated kg). In the British system, this is about 2.2 lb. The base unit of mass is the gram (g), and from the foregoing discussion about prefixes, you can see that 1 kilogram (1 kg) must equal 1,000 g. Since metric prefixes have the same mathematical meaning whenever they are used, they may be applied to other base units equally well. For example, we have seen that 1 km = 1,000 m. Just as 1 millimeter (1 mm) equals 1/1,000 of a meter (m), 1 milliliter (1 mL) equals 1/1,000 of a liter (L). Other relationships among the units of mass are summarized in Tables 2.2 and 2.3 (under "Mass").

Scientists prefer to measure **mass,** the fixed amount of matter in an object, rather than **weight,** the amount of gravitational attraction pulling on an object. Mass is always the same no matter where it is measured, whereas weight varies because it depends on the size of gravitational attraction. On earth, the measured value for weight depends on the mass of the earth and the distance the object being weighed is from the center of the earth. The weight of an object measured in Death Valley (low altitude) is slightly greater than the weight of the same object measured on top of Mount Everest (high altitude). On the moon, the measured value for weight is

Figure 2.6
Because they weigh less on the moon than on earth, astronauts must wear boots with extra mass to keep them from bouncing too high as they walk. Their mass is the same as it is on earth. On the moon, however, their mass is attracted less strongly toward the center of the moon than it would be if they were on earth and being attracted toward the earth's center. Astronaut John Young is shown approximately a meter above the moon's surface.

considerably smaller because of the smaller mass of the moon (Figure 2.6). Because they weigh less, astronauts can jump much higher on the moon. Their mass, however, is the same in all locations, whether they are on the earth or the moon.

Mass is measured by using a double-pan balance, as shown in Figure 2.7. The crucial feature of a device for measuring mass is the knife edge on which the weighing pan is balanced. Standard known masses (commonly called "weights") are used to balance the object being measured. Because gravity works on each pan equally at any altitude, on any planet or satellite, you are assured that you are actually measuring mass by comparing the unknown object with known masses. Modern single-pan laboratory balances compensate for gravity in a less obvious way than does the double-pan balance in Figure 2.8, but the idea is the same.

Although mass and weight are not the same thing, it is common practice to use the words interchangeably. We use the verb *weigh* to describe the measurement of either mass or weight. Because chemists always use balances to weigh things, chemists always measure mass, whether it is called "mass" or incorrectly called "weight."

Volume, the amount of space occupied by a three-dimensional object. The reference standard for **volume,** the amount of space occupied by a three-dimensional object, in the metric system is the liter,[2] which is just slightly larger than a quart. The *liter* (L) is the volume occupied by a perfect cube, 10 cm on each edge. Picture the cube as an empty box. You can calculate the volume of any rectangular solid in the unit of cubic centimeters (cm^3, also abbreviated cc) by multiplying its length by its width by its height. So, given that volume = $l \times w \times h$, we have in this case:

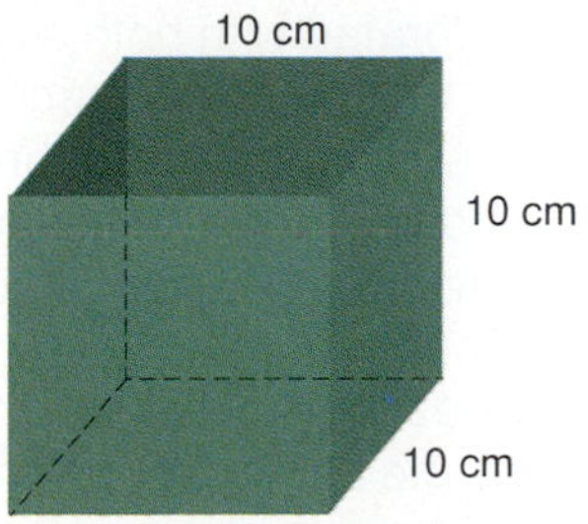

Volume = 10 cm x 10 cm x 10 cm = 1,000 cm^3

1 L = 1,000 cm^3. (An alternative abbreviation for cubic centimeter is cc.)

2. The metric system is the precursor of the SI system. The actual SI reference standard for volume is the cubic meter, but this unit is impractical in many settings, for example, in medical laboratories. Therefore, the older unit, the liter, is maintained and discussed in this text.

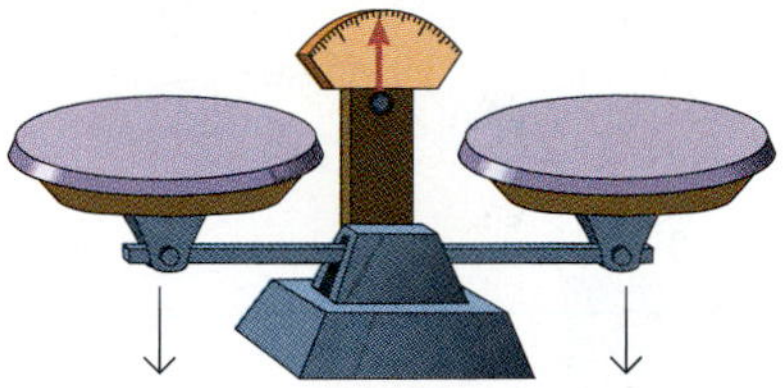

The balance is adjusted so that the pointer reads zero. This ensures that the pans have the same mass.

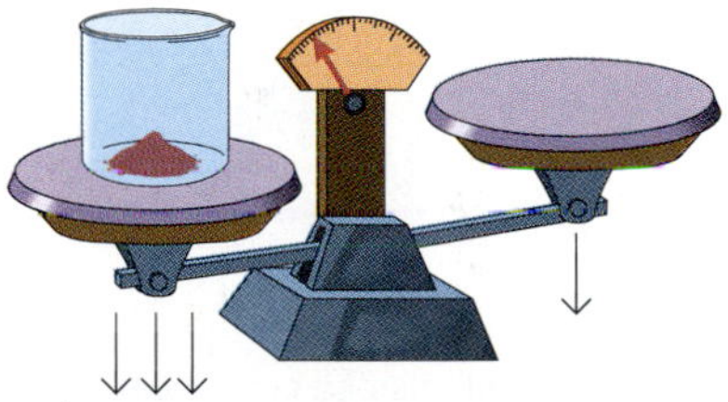

The object to be "weighed" is placed on the left pan.

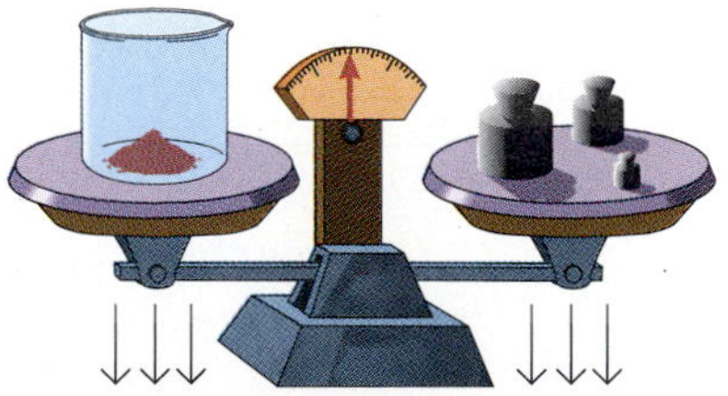

"Weights" of known mass are placed on the right pan until the pointer again reads zero. The object on the left then has the same mass as the sum of the known masses on the right.

Figure 2.7
Use of the two-pan balance. Notice that the force of gravity (↓) is pulling downward on both pans at all times. When the masses on both pans are equal, gravity affects the pans equally. Thus, you are assured that you are truly measuring mass by comparison to some reference standard.

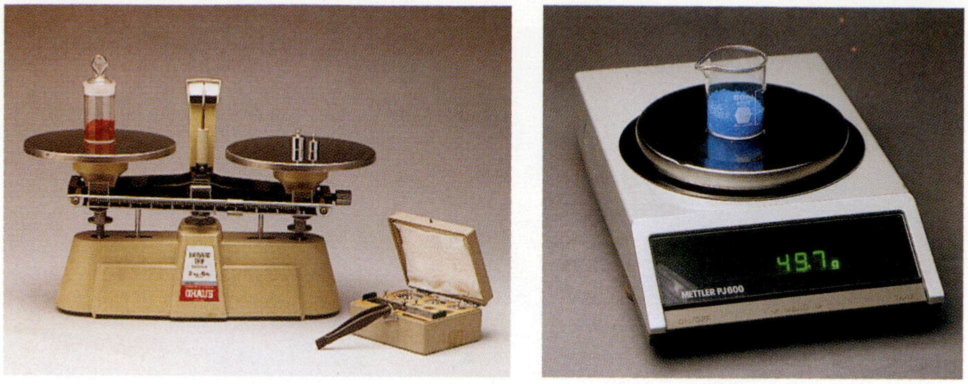

(a) (b)

Figure 2.8
The double-pan balance (a) is rapidly being replaced by the single-pan balance (b) in the laboratory because of the convenience of not needing additional standard masses to make a "weighing."

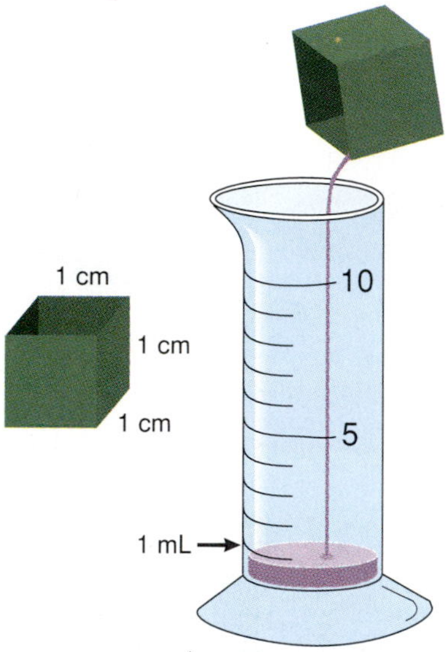

Figure 2.9
The liquid contents of a cube 1 cm on a side are equivalent to 1 mL in the graduated cylinder. One mL of water always occupies the same volume, but its shape may change as the shape of the container changes.

Now imagine filling up the box with water. The volume of the water is 1,000 cm^3, or 1 L. If you pour the 1 L of water into different containers, it is still 1 L, despite the fact that its shape changes. Volume is the amount of space occupied and does not depend on shape.

Because 1 L = 1,000 cm^3 and 1 L = 1,000 mL, it must be true that

$$1{,}000 \text{ cm}^3 = 1{,}000 \text{ mL}$$

When each side of the equation is divided by 1,000, it becomes apparent that

$$1 \text{ cm}^3 = 1 \text{ mL}$$

Thus, the liquid contents of a cube 1 cm on a side will come up to the 1-mL mark of a graduated cylinder, as you can see in Figure 2.9.

Most often you will be concerned with measuring the volumes of liquids. Chemists have invented several kinds of equipment to do this conveniently (Figure 2.10). You will learn the special advantages of each type of measuring device as we apply them in later chapters. These devices are all calibrated in milliliters. As the prefix tells you, 1 mL is a tiny part, namely, one-thousandth, of a liter. There are 1,000 mL in 1 L. Notice in Table 2.3 under "Volume" that this is the only equality listed. This is because scientists do not commonly use other units. However, the wine industry does. For example, a "fifth" of wine is exactly 75 centiliters (cL).

2.5 temperature

Temperature is a measure of the hotness or coldness of an object, and from Chapter 1, you recall that it is a property that is independent of the size and shape of the sample. It is convenient to measure temperature using a thermometer. The reference standard for measuring temperature in British units is **degrees Fahrenheit** (°F), and the metric equivalent is **degrees Celsius** (°C). Weather reports now commonly give values on both scales for the daily temperature. Even home thermometers are calibrated with both scales for your convenience.

Specific temperatures, such as the boiling point and melting point, are characteristic of a substance and can be used to identify that substance. By observing the boiling point and the freezing point of water, you know that you are measuring two unique temperatures. Measuring each temperature on the Celsius and Fahrenheit thermometers enables you to determine a mathematical relationship between degrees Fahrenheit and degrees Celsius (Figure 2.11). Table 2.4 summarizes the data gained from the Fahrenheit and Celsius thermometers used in this experiment.

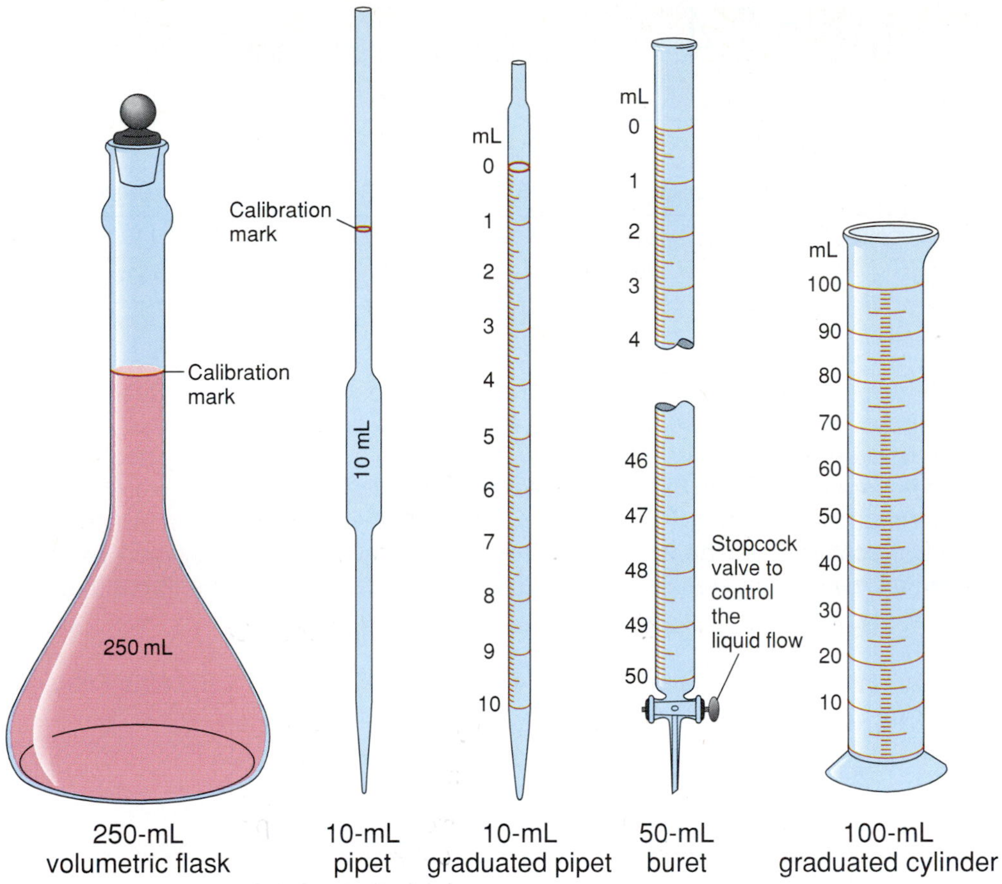

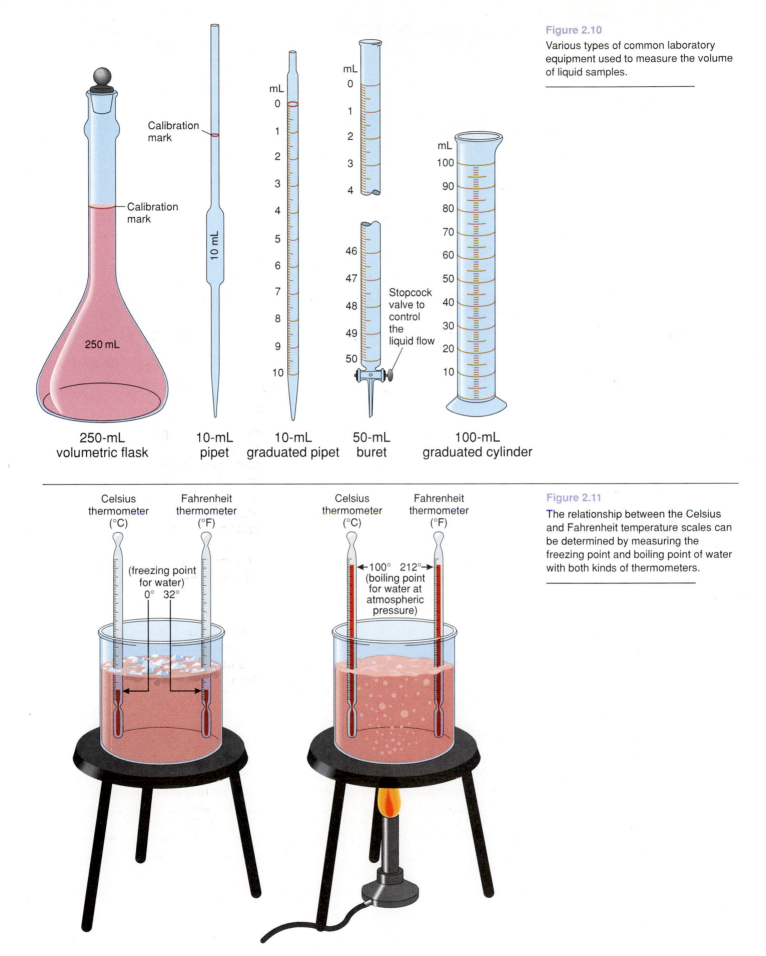

Figure 2.10
Various types of common laboratory equipment used to measure the volume of liquid samples.

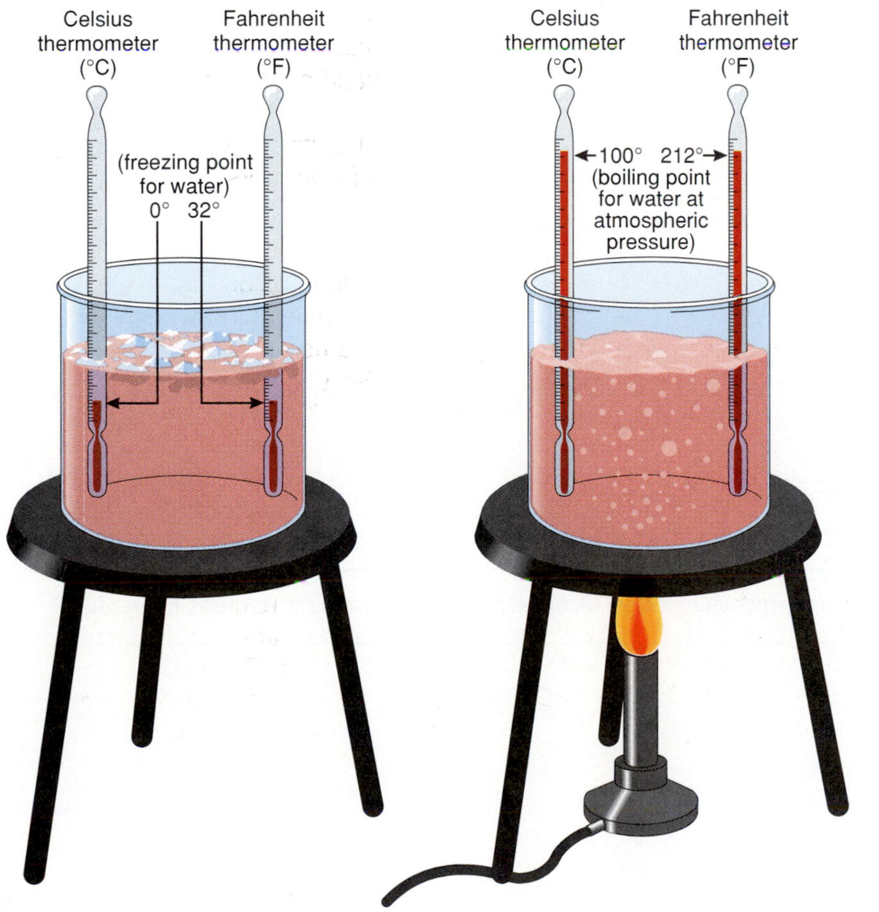

Figure 2.11
The relationship between the Celsius and Fahrenheit temperature scales can be determined by measuring the freezing point and boiling point of water with both kinds of thermometers.

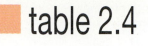

table 2.4 comparison of Fahrenheit and Celsius temperature scales

	Normal boiling point (BP) of water	Normal freezing point (FP) of water	BP – FP	Degree ratio
Fahrenheit	212°F	32°F	212° – 32° = 180°	$\frac{180}{100} = 1.8$
Celsius	100°C	0°C	100° – 0° = 100°	

From these data, you see that there are 180° between water's freezing point and its boiling point on the Fahrenheit temperature scale, but that there are only 100° between these points on the Celsius scale. As a result, for every degree that the temperature rises on the Celsius thermometer, it will rise 180/100 = 1.8° on the Fahrenheit thermometer. This leads to Equation 2.1, which relates temperature on the two scales. Equation 2.1 has a form that is convenient to use when you are given a Celsius temperature and asked to calculate a Fahrenheit temperature.

$$°F = 1.8\ (°C) + 32 \tag{2.1}$$

Similarly, the Celsius temperature is found to be (100/180) or (1/1.8) of the Fahrenheit temperature minus 32, as shown in Equation 2.2.

$$°C = \frac{(°F - 32)}{1.8} \tag{2.2}$$

The subtraction of 32 takes into account the difference in displacement between the freezing point of water on the Fahrenheit scale (32°F) and the Celsius scale (0°C).

sample exercise 9

A scientist measures the temperature of her laboratory as 25°C. What is the temperature in degrees Fahrenheit?

solution

You are given the number of degrees Celsius; therefore, substitute this number in Equation 2.1.

$$°F = 1.8\ (°C) + 32$$

Substituting gives

$$°F = 1.8\ (25) + 32$$

To complete this calculation, multiply before adding because the plus sign is outside the parentheses. Thus,

$$°F = 45 + 32 = 77$$

• **Answer:** 25°C = 77°F

• • • problem 5

A home thermostat reads 20°C. What is this temperature in degrees Fahrenheit?

sample exercise 10

One day last winter, the temperature was reported as −10°C. What was the Fahrenheit reading?

solution

Use the preceding method, being careful to treat the minus sign correctly.

The given temperature is in degrees Celsius. Therefore, we select Equation 2.1 and substitute for degrees Celsius.

$$°F = 1.8\,(°C) + 32 \tag{2.1}$$

Substituting gives

$$°F = 1.8\,(-10) + 32$$

Multiply first:

$$°F = -18 + 32$$

Then add:

$$°F = +14$$

• **Answer:** −10°C = 14°F

sample exercise 11

Body temperature is 98.6°F. What is body temperature in degrees Celsius?

solution

You are given the number of degrees Fahrenheit; therefore, substitute this number into Equation 2.2.

$$°C = \frac{(°F - 32)}{1.8} \tag{2.2}$$

Substitution yields

$$°C = \frac{(98.6 - 32)}{1.8}$$

To complete this calculation, you must subtract before multiplying by the fraction because the minus is inside the parentheses.

$$°C = \frac{(66.6)}{1.8} = 37$$

• **Answer:** 98.6°F = 37.0°C

• • • problem 6

A house thermostat reads 72°F. What is this temperature in degrees Celsius?

The general method of problem solving for temperature conversions is as follows.

1. Identify the kind of degrees given.
2. Select a correct equation relating degrees Celsius and degrees Fahrenheit, and substitute for the degrees given.
3. Complete the calculation, being careful to multiply or add, subtract or divide, in the proper order.

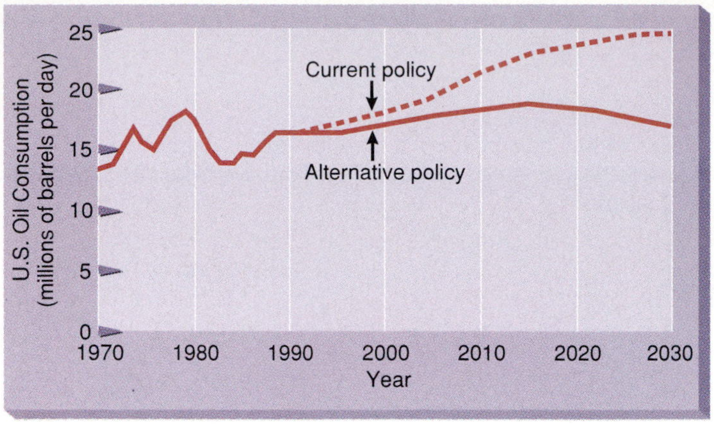

Figure 2.12

Graphs visually show the variation of one quantity with the change in another. In this case, you see how the consumption of oil changed from 1970 to 1990 and how it will continue to change under two policies: (1) if the current energy policy continues and (2) if conservation and alternative energy measures are instituted.
Source: National Energy Strategy.

2.6 graphing

Graphing helps to make measurements more understandable and useful. A **graph** is a visual display from which the viewer can see the change in one quantity compared to a change in another quantity. For example, in Figure 2.12, the plotted data show the variation of oil consumption (*y* axis) in the United States with time (*x* axis). By the end of this section, you should be able to use Figure 2.12 to see that instituting conservation measures could drastically reduce the consumption of oil in the United States over the next 30 years. Since the world supply of oil is limited, the slower it is used, the longer the supply will last.

Most common graphs and the one we will examine in this section are constructed on a grid bordered by perpendicular lines called the horizontal (*x*) and vertical (*y*) axes. Some of the guidelines below may be familiar to you, but are presented here for completeness. Constructing and reading graphs are basic skills you will find useful in many endeavors.

Guidelines for constructing a graph given *x, y* data.

1. Title the graph.
2. Draw in or darken the lines that will represent the horizontal and vertical axes and label each axis with the varying quantity, including the units of the quantity.
3. Establish appropriate and convenient scales for the axes based on the data given; that is, the data plotted on each axis must fit within the borders of the axis and the edge of the paper. The appropriate scale depends on the range of the given data (lowest to highest value) and the number of subdivisions on your graph paper. The ratio of these two quantities yields an appropriate scale value.

$$\frac{\text{Range of data}}{\text{Number of subdivisions}} = \text{Appropriate scale value}$$

For example, if the range of data for the *x* axis is 500 cm and the number of boxes (subdivisions) from the vertical axis to approximately the edge of the paper is 50, an appropriate scale would be

$$\frac{500\text{ cm}}{50\text{ boxes}} = 10\ \frac{\text{cm}}{\text{box}}$$

In some cases, use of this formula leads to an inconvenient scale value. For example, if the range of data had been 170 cm with 50 subdivisions, the appropriate scale calculated would be 3.4 cm/box. Such a scale is less convenient than an integral value for each subdivision, and it is preferable to choose the next larger integral value. In this example, this means 4 cm/box. It is not necessary for each axis to have the same numerical scale, but once a scale for a particular axis is chosen, it must be used uniformly throughout that axis.

4. Locate data points by moving along the *x* axis until the *x* value is reached and then moving vertically to the *y* value. Mark each point with a dot or an X.
5. Draw the best-fitting curve through the points. Even when the graph is a straight line, the line will not generally pass through all the points. The best-fitting line will be simultaneously closest to all points; it may be that it actually goes through none of them. Although made-up data may fit a line exactly, in experimental data, there is always some error in the measurement process, and thus a perfect fit is not obtained.

sample exercise 12

Plot distance on the *y* axis against time on the *x* axis for the values in the following table.

Distance, meter (m)	Time, seconds (s)
0	0
3	1
6	2
8	3
13	4
15	5

solution

Following the guidelines above.

Step 1. Title the graph "Distance vs. Time."

Step 2. Label the vertical axis "Distance (m)" and the horizontal axis "Time (s)."

Step 3. The range over the distance data is 15 meters (m), and the number of marked boxes on the vertical axis below is 20. An appropriate scale would be 15 m/20 boxes = 0.75 m/box. Since all the distance data are integers, however, it is more convenient to choose the higher scale value of 1.0 m/box. The range of time data is 5 seconds (s), and the number of subdivisions on the horizontal axis is 20. An appropriate scale is 5 s/20 boxes = 0.25 s/box. Although the time data are given in integers so that a higher scale value of 1.0 s/box could be used, this would compress the graph because the time range is small. A better display is obtained by using the 0.25-s/box scale and marking every fourth box with the appropriate integer.

Step 4. Mark the data points in the graph. Dotted lines are drawn from the axes to the second data point to show that it is at $x = 1$ and $y = 3$.

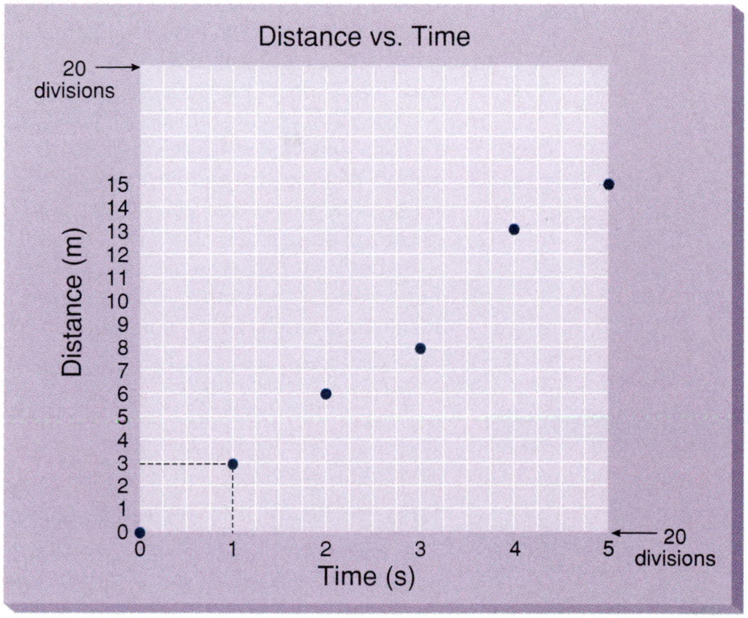

Step 5. Draw a line so that the distance of points not touching the line on one side is balanced by the distance of points not touching the line on the other side.

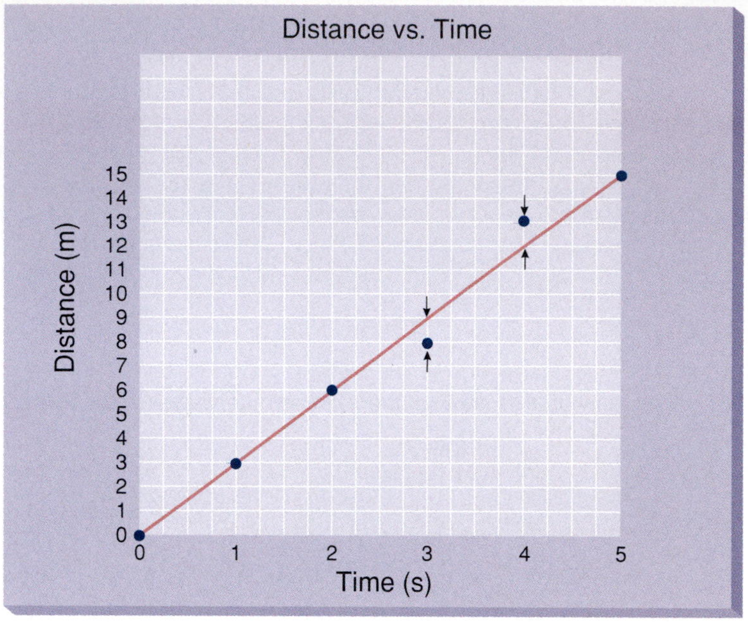

The purpose of constructing a graph is to make use of the information it relates. A graph can be used to determine the *y* (or *x*) value corresponding to an *x* (or *y*) value of a point not in the data table. Move along the *x* axis to the chosen *x* value and then up to the drawn curve. From this point in the curve, move horizontally across to the vertical axis and obtain the *y* value.

sample exercise 13

Determine the distance corresponding to a time of 2.5 s for the graph plotted in Sample Exercise 12.

solution

Move along the time axis to 2.5 s, then straight up to the drawn line. Moving horizontally across from the drawn line, you intersect the vertical axis at a point between 7 and 8 m. The estimated *y* value is 7.5 m. Note that the graph enabled us to determine a time-distance point that is not in the data table.

Now that you are familiar with using a graph to see how one value changes with another, go back and look carefully at Figure 2.12. For any particular year, you can determine the number of millions of barrels of oil used per day. Find the number of millions of barrels of oil that will be used in the year 2010 if our present energy policy continues; also find the number of millions of barrels of oil that will be used in the year 2010 if conservation is instituted. As noted at the beginning of the section, you will find that the millions of barrels of oil used per day will be much higher in the year 2010 if our current energy policy continues (found on the upper curve), but it will be much lower if conservation is introduced (found on the lower curve).

••• problem 7

Make a graph from the data given in the following table, and then determine the mass corresponding to a volume of 6.5 mL.

Mass, grams (g)	Volume, milliliters (mL)
2.5	2.0
4.0	3.0
5.2	4.0
6.4	5.0
7.6	6.0
10.3	8.0

CHEMLAB

Laboratory exercise 2: graphing

The use of measurements in scientific notation will be repeated continually throughout this textbook. On the other hand, you will not have to construct a great number of graphs; but being able to do so and being able to understand the information they convey are essential for developing your growing knowledge of chemistry. The following exercise will continue to build your skills of constructing and reading graphs.

A beaker of water containing two thermometers, one measuring degrees Celsius and the other measuring degrees Fahrenheit, is heated slowly with a Bunsen burner (Figure 2.13). The following data were recorded.

Temperature, degrees Fahrenheit (°F)	Temperature, degrees Celsius (°C)
32	0
52	10
68	20
86	30
212	100

Establish convenient scales for Celsius degrees (°C) on the horizontal (*x*) axis and Fahrenheit degrees (°F) on the vertical (*y*) axis. Let the range of Celsius degrees extend below the data measured so that it goes from –50°C to 100°C. Let the range of Fahrenheit degrees also extend below the data measured so that it goes from –50°C to 220°F.

Plot your data on a full sheet of graph paper and circle each data point so that it stands out. Draw the best straight line through your data and answer the following questions.

Questions:

1. When the temperature is 37°C, what is the Fahrenheit temperature?
2. When the temperature is 0°F, what is the Celsius temperature?
3. On a cold day in Juneau, Alaska, when the temperature measures –40°C, what is the Fahrenheit temperature?

Figure 2.13
This apparatus is used to measure the Celsius and Fahrenheit temperatures of the same sample of water. The boiling chips in the beaker prevent the water from "bumping" when it boils. The thermometers are suspended so they do not touch the sides of the beaker, but measure the temperature of the water only. Each thermometer is clamped in place with a slit rubber stopper to protect the glass from breaking.

2.7 what are significant figures?

You have used measurements from experiments in the previous sections, and you have visually represented those measurements with graphs. Yet, so far, we have not discussed the error associated with those measurements. There is always some uncertainty associated with any measurement because you must estimate the value of the last digit you record. The last digit is always one place beyond the smallest division of your measuring instrument. For example, at the beginning of this chapter,

Figure 2.14

Uncertainty in measurement: The nature of the measured object introduces uncertainty, and the measuring device has limitations. (a) With the decimeter ruler, you can say with certainty that the pencil measures between 0.1 and 0.2 m. You can estimate that it is 0.14 m. (b) With the centimeter ruler, you can say with certainty that the pencil measures between 14 and 15 cm (between 0.14 and 0.15 m). You can estimate that it is 14.5 cm or 0.145 m.

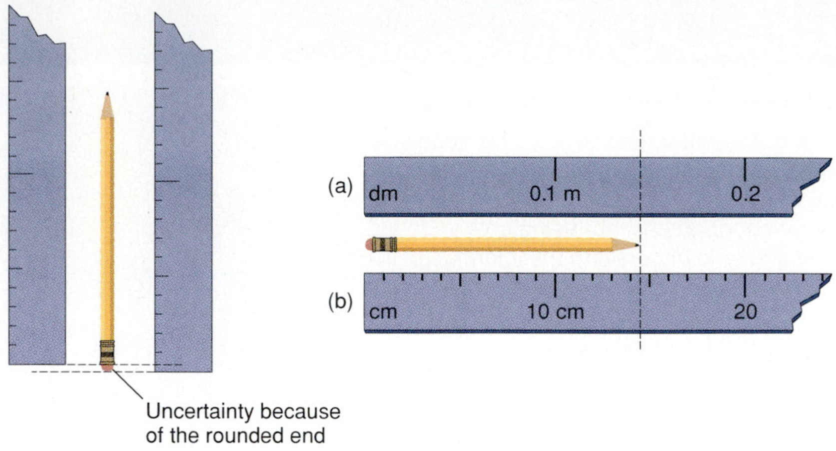

you saw how important it is for the butcher to weigh your meat to two decimal places, that is, to the nearest hundredth of a pound, so that your uncertainty in the amount of meat you paid for is small. Significant figures indicate the uncertainty in the numbers you measure. When you write a measurement as 2.97 lb, it means that the uncertainty is in the last digit, that is, in the hundredths of pounds. When you write a measurement as 3.0 lb, it means the uncertainty is in the tenths of pounds. All scientific measurements are written with the correct number of significant figures, so that you know automatically how much error is in that measurement.

The uncertainty in any measurement comes about because of the nature of the object measured and the limitations of the measuring device. For example, consider the attempts at measuring the length of a common pencil shown in Figure 2.14. There is an uncertainty because the shape of the pencil makes it difficult to line up the ends for accurate measurement. This is an unavoidable uncertainty, but you can see that the measuring device also limits our knowledge. In Figure 2.14a, the use of a meterstick calibrated only in decimeters allows you to say with certainty that the pencil measures between 0.1 and 0.2 m. You can estimate that it is about 0.14 m.

In Figure 2.14b, using a meterstick calibrated in centimeters allows you to say with certainty that the pencil measures between 14 cm (0.14 m) and 15 cm (0.15 m). You can estimate that it is about 14.5 cm (0.145 m).

In Figure 2.14a, you say that the measurement is good to two significant figures (2 sig figs for short), one you know for certain and the one you guessed.

0.14 m

Certain — Good guess

2 sig figs

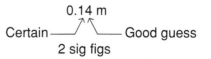

In Figure 2.14b, using a more sensitive measuring device, the measurement is good to 3 sig figs:

0.145 m

Certain — Good guess

3 sig figs

For any measurement, the number of ***significant figures*** *that can be reported is the number of figures that can be read accurately from the measuring device plus one more figure that must be estimated.*

sample exercise 14

How many sig figs can be reported in each of the following measurements?

(a) Bathroom scale calibrated in pounds

(b) Speedometer calibrated in kilometers per hour

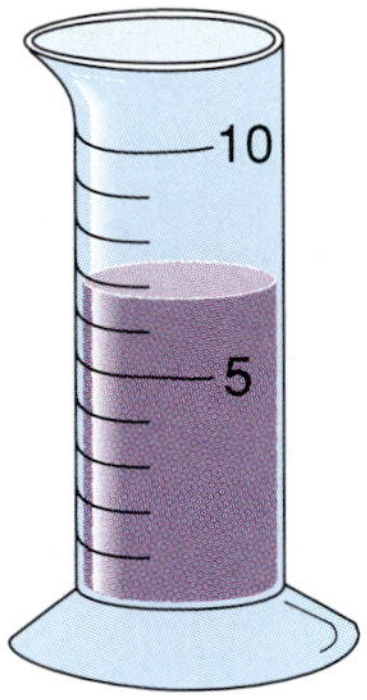

(c) Graduated cylinder calibrated in milliliters

solution

a. You can tell with accuracy that the weight is between 136 and 137 lb. Then you can estimate that the pointer is midway between 136 and 137 and report 4 sig figs:

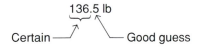

136.5 lb

Certain (136) — Good guess (5)

b. The speed is definitely between 70 and 75 km/hour (km/h). Because there are no calibration lines between 70 and 75, you must estimate the nearest kilometer. A guess is 72 km/h. Therefore, you are allowed 2 sig figs: the 7 is certain, the 2 is a good guess.

c. The volume is between 6 and 7 mL. You estimate that it is 6.8 mL. Two sig figs are allowed.

When you actually do or see the measurement, as in Sample Exercise 14, it is easy to determine how many sig figs are allowed and are represented by the reported number. More often, you are given a measured number and asked to tell how many sig figs it contains. To do this requires a set of rules.

Rules for counting sig figs in a given number.

1. All *nonzero* figures are significant.
2. All *zeros between nonzero* figures are significant.
3. When a decimal point is shown, *zeros to the right of nonzero* figures are significant. (When a decimal point is not shown, the number is ambiguous and the number of sig figs cannot be determined.)
4. *Zeros to the left* of the first nonzero figure are not significant.

An example demonstrating all of these rules follows.

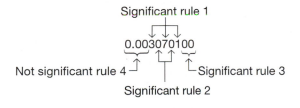

Significant rule 1

0.003070100

Not significant rule 4 — Significant rule 2 — Significant rule 3

Thus, this decimal number contains a total of 7 sig figs.

sample exercise 15

For each of the following measurements, tell how many sig figs are represented.

a. 4.065 m	d. 20.00 s
b. 0.32 g	e. 0.00040 km
c. 57.98 mL	f. 604.0820 kg

solution

a. 4 sig figs (rules 1 and 2)
b. 2 sig figs (rules 1 and 4)
c. 4 sig figs (rule 1)
d. 4 sig figs (rules 1 and 3)
e. 2 sig figs (rules 1, 3, and 4) 0.00040

Not significant (0.000) Significant (40)

f. 7 sig figs (rules 1, 2, and 3)

problem 8

Tell the number of sig figs in each of the following.

a. 21.94 mL	c. 0.08040 cm
b. 3.90 g	d. 0.9000 g

sample exercise 16

How many sig figs are represented by each of the following measurements?

a. 3.24×10^{-8} cm	c. 7.0300×10^{2} mL
b. 5.0×10^{3} g	d. 4.705×10^{-4} km

solution

To determine the number of sig figs in a number written in scientific notation, apply the rules to the number preceding the times sign. You will find that if the number is in the correct notation form, then all digits are significant.

a. 3	c. 5
b. 2	d. 4

2.8 rounding off

Now that you understand which digits are significant in your measurements, you are prepared to use your measurements in calculations such as those shown in the examples found in the next section. You will find that the answers to your calculations will also be limited to a definite number of significant figures. Even though the results on your hand calculator may have many digits, only those digits that are significant represent physical reality and are meaningful to a scientist. As a result, you will frequently need to round off answers to fewer digits than given by your calculator.

In Section 2.9, you will learn to determine the number of figures that are significant in the results of your calculations. Once you have determined the number of significant figures, you will apply the following rules to **round off** an answer, that is, to reduce the number of digits in your answer to only those that are significant. For example, if you calculated a distance as 3.5677 cm, but know that your answer is only good to three significant figures, you will report the correct answer as 3.57 cm. To round off correctly, you must look at the three significant figures, which are the first three digits in your answer. In this case, the significant figures are 3.56. Then you must look at the next (fourth) digit in your answer, which is 7. If this next digit, beyond those that are significant, is 5 or more (here it is 7), you increase the last significant digit by 1. Thus, 3.56 becomes 3.57 because the 6 was followed by 7, a number greater than 5. If the next digit, beyond those that are significant, had been less than 5, no change would have been made to the three significant digits; your answer would have been 3.56.

Rules for rounding off.

1. If the first figure to be dropped is less than 5, the preceding digit is not changed. Therefore, 1.234 is rounded to 1.23 (3 sig figs) or 1.2 (2 sig figs) or 1. (1 sig fig).
2. If the first figure to be dropped is 5 or more, the preceding digit is increased by 1. Therefore, 1.756 is rounded to 1.76 (3 sig figs) or 1.8 (2 sig figs) or 2. (1 sig fig).

2.9 sig figs in calculations

In Section 2.7, you learned that significant figures indicate the size of the error in each measurement. After using those measurements in a calculation, you would still like to know how much error is in the final result. By writing the answer to your calculation with the correct number of significant figures, you will still be indicating the size of the error you expect in the answer to that calculation, just as you did with the measurements themselves. However, you must learn rules to help you determine the correct number of significant figures in different calculations. These are outlined below.

Multiplication and division. In the course of problem solving and making calculations, you will be given measurements such as 2.3 and 5.69 that require multiplication. Your hand calculator will give 13.087 as the product. But this is not a proper answer to the multiplication because, in multiplying or dividing measurements, the answer should contain no more sig figs than the measurement with the fewest number of sig figs. Therefore, in multiplying 2.3 (2 sig figs) × 5.69 (3 sig figs), we are allowed only 2 sig figs in the answer. So we must round off to 2 sig figs:

$$2.3 \times 5.69 = 13.087$$
$$= 13$$

sample exercise 17

The mass and volume of a sample of blood are measured and found to be 12.193 g and 11.5 mL, respectively. In the next chapter, you will learn to calculate the density of a sample by dividing the mass of the sample by its volume. Carry out this division and express the answer in the proper number of sig figs.

solution

$$\text{Density} = \frac{\text{Mass}}{\text{Volume}}$$

Using a calculator,

$$\text{Density} = \frac{12.193\text{ g}}{11.5\text{ mL}} = 1.0602609\ \frac{\text{g}}{\text{mL}}$$

But, because you know the volume to only 3 sig figs, you are allowed only 3 sig figs in the final answer. Therefore, you round off and report the density as 1.06 g/mL.

Even when there is a series of multiplications and divisions, the measurement with the smallest number of sig figs governs the number of sig figs in the answer. Consider doing the following operations with a calculator.

$$\frac{0.04 \times 2.546 \times (3.12 \times 10^6)}{7.0012} = 4.5384 \times 10^4 \quad \text{Calculator answer}$$

$$= 5 \times 10^4 \quad \text{Correct answer}$$

Because one of the multipliers (0.04) has only 1 sig fig, the correct answer to report must have only 1 sig fig. In this case, the answer would be 5×10^4.

••• problem 9

Assuming that the following numbers are measurements, perform the calculation and report the answer to the correct number of sig figs.

$$\frac{(3.8 \times 48.20)}{7.81} =$$

When rounding off larger numbers to a small number of sig figs, it is best to express the answer in scientific notation. For example, 1,903 rounded to 2 sig figs is 1.9×10^2. If you were to express the answer as 1,900, it would be ambiguous. The number 1,900 may indicate 2, 3, or 4 sig figs. When written with a decimal point, e.g., 1,900., it indicates 4 sig figs are present. As you can see, scientific notation offers the best expression.

The idea of sig figs applies only to *measured* numbers. Some numbers are *defined* or *exact* numbers and, when used in a calculation, do not affect the number of sig figs allowed in the answer. For example, one dozen is defined as 12 eggs. To determine the number of eggs in 11 dozen, you multiply

$$11\ \cancel{\text{dozen}} \times \left(\frac{12\ \text{eggs}}{1\ \cancel{\text{dozen}}}\right) = 132\ \text{eggs}$$

You do not round off here because you are dealing with exact numbers, not measured numbers.

All equalities given in Tables 2.2 and 2.3 are exact or defined. For example, 1 m is defined as exactly 100 cm, just as one dozen is defined as 12 eggs. As a result, when you use them in calculations, they will not affect the number of significant figures allowed in your answer. Some equalities are inexact because they are measured numbers with a limited number of sig figs. For example, for 1 lb = 454 g, the 1 is exact or defined and will not affect a calculation, but 454 is measured and will limit a calculated answer to 3 sig figs.

To summarize, *only measured numbers and inexact equalities limit the number of sig figs in a calculated answer.*

Addition and subtraction. In the course of other calculations, you will be given numbers that must be added and subtracted. Determining the number of significant figures in the sum or difference requires a different rule from that used for multiplication and division. In adding and subtracting measurements, the final answer must not contain any more digits to the right of the decimal point than does the measured number with the *least number of digits to the right of the decimal point.*

For example, in adding 937.3 g + 15.224 g + 71.04 g, the number 937.3 governs the sig figs in the answer; in other words, there may be only *one digit* to the right of the decimal point.

$$\begin{array}{r} 937.3\ \ \ \ \text{g} \\ 15.224\ \text{g} \\ 71.04\ \ \text{g} \\ \hline 1{,}023.564\ \text{g} \end{array} = 1{,}023.6\ \text{g}$$

Because one mass is known only to the nearest tenth of a gram, the answer must be rounded off to tenths.

---

sample exercise 18

Report the result of the following subtraction to the correct number of sig figs.

$$2.572\ \text{m} - 0.41\ \text{m} =$$

solution

$$\begin{array}{r} 2.572\ \text{m} \\ -0.41\ \ \ \text{m} \\ \hline 2.162\ \text{m} \end{array}$$

← This is the limiting number because there are fewer digits to the right of the decimal point.

The answer must be rounded off to 2.16 m.

---

problem 10

Add the following weights and report the sum to the proper number of sig figs.

$$2.2332\ \text{g} + 1004.12\ \text{g} + 26.557\ \text{g} =$$

The solutions to all sample exercises in this book (from Chapter 2 on) and the solutions to the problems given in Appendix E are reported to the correct number of sig figs. You should always round off your final answers to the proper number of sig figs.

chapter accomplishments

After completing this chapter, you should be able to define all key terms and do the following.

2.1 What is measurement?

- ❑ State the role of a reference standard in measurement.

2.2 Scientific notation

- ❑ Indicate the base and exponent of an exponential number.
- ❑ Write any decimal number in scientific notation.
- ❑ Convert a number in scientific notation to an ordinary decimal number.
- ❑ Add, subtract, multiply, and divide two numbers in scientific notation.

2.3 Using scientific notation on a hand calculator

- ❑ Use a scientific hand calculator for arithmetic with numbers in scientific notation.

2.4 Metric system

- ❑ State the metric base units used for length, mass, and volume measurements.
- ❑ State the meanings of the common metric prefixes (Table 2.1).
- ❑ State the equalities among common metric units (Table 2.3).
- ❑ Distinguish between mass and weight.
- ❑ State the advantage of measuring mass rather than weight.

2.5 Temperature

- ❑ Given a temperature in degrees Fahrenheit or degrees Celsius, convert from the given scale to the other scale.

2.6 Graphing

- ❑ Construct a graph from a given data table.
- ❑ Given a graph and a particular *x* value, determine the corresponding *y* value.
- ❑ Given a graph and a particular *y* value, determine the corresponding *x* value.

2.7 What are significant figures?

- ❑ Given a measuring device, state the number of sig figs that can be reported in a measurement.
- ❑ Given a measured number, state the number of sig figs it has.

2.8 Rounding off

- ❑ Round off numbers to a given number of sig figs.

2.9 Sig figs in calculations

- ❑ Express the result of the multiplication or division of measured numbers to the correct number of sig figs.
- ❑ Distinguish between measured numbers and defined or exact numbers.
- ❑ Express the result of the addition or subtraction of measured numbers to the correct number of sig figs.

key terms

scientific notation
metric system
SI (system of units)
length
kilogram
gram
mass
weight
volume
temperature
degrees Fahrenheit
degrees Celsius
graph
significant figures
round off

problems

2.1 What is measurement?

11. Describe in your own words what is meant by the measurement process.
12. List five measuring devices used in your everyday activities.

2.2 Scientific notation

13. Write the following numbers in scientific notation.
    a. 58.7
    b. 0.082
    c. 631,000,000
    d. 3,012
    e. 73.98
    f. 0.000000000718
14. Calculate the following and express your answers in scientific notation.
    a. $(3.0 \times 10^{8})\ (2.0 \times 10^{3})$
    b. $(8.39 \times 10^{-5})\ (3.21 \times 10^{4})$
    c. $(2.14 \times 10^{-6})\ (1.39 \times 10^{-9})$
    d. $(7.15 \times 10^{-9})\ (3.81 \times 10^{2})$
    e. $(9.6 \times 10^{16})\ (4.5 \times 10^{-1})$
15. Calculate the following and express your answers in scientific notation.
    a. $\frac{6.4 \times 10^{6}}{3.2 \times 10^{2}}$
    b. $\frac{3.2 \times 10^{-2}}{6.4 \times 10^{-6}}$
    c. $\frac{1.8 \times 10^{5}}{9.3 \times 10^{4}}$
    d. $\frac{1.9 \times 10^{4}}{4.2 \times 10^{5}}$
    e. $\frac{2.3 \times 10^{-12}}{0.7 \times 10^{-9}}$
    f. $\frac{6.0 \times 10^{-23}}{3.0 \times 10^{9}}$
16. Add or subtract the following and express your answers in scientific notation.
    a. $(3.8 \times 10^{4}) + (1.9 \times 10^{4}) =$
    b. $(8.3 \times 10^{4}) - (2.4 \times 10^{3}) =$
    c. $(7.4 \times 10^{4}) + (8.9 \times 10^{3}) =$
    d. $(4.7 \times 10^{7}) - (5.2 \times 10^{6}) =$

2.4 Metric system

17. Give five common prefixes of the metric system and their mathematical meanings.

18. State the equalities between the following metric units.
    a. Grams and kilograms
    b. Liters and milliliters
    c. Meters and centimeters
    d. Micrometers and meters
    e. Milliliters and cubic centimeters
19. Give the names of the metric system units that have the following abbreviations.
    a. mm
    b. mL
    c. g
    d. mg
    e. km
20. Arrange each of the following units in sequence from smallest to largest.
    a. Meter, centimeter, kilometer
    b. Milligram, microgram, gram
    c. Liter, milliliter, cubic centimeter
21. Express the following metric prefixes as a power of 10.
    a. Centi-
    b. Kilo-
    c. Milli-

2.5 Temperature

22. Do the following temperature conversions.
    a. 75°F to degrees Celsius
    b. –5.0°F to degrees Celsius
    c. 18°C to degrees Fahrenheit
    d. –80°C to degrees Fahrenheit
    e. 15.45°F to degrees Celsius
23. Which is the higher temperature, –25°C or –15°F?
24. A typical antifreeze mixture raises the boiling point of the water in your radiator to 240°F. What is this in degrees Celsius?
25. The highest recorded weather temperature on the earth is 136.4°F. What is this in degrees Celsius?
26. Which is warmer, 265°C or 425°F?

2.6 Graphing

27. a. Make a plot of the Fahrenheit versus Celsius temperature scales from the following measured data.

Temperature (°F)	Temperature (°C)
–32	–35
22	–5
32	0
54	12
70	21
78	26

    b. According to your graph, what is the Fahrenheit equivalent of 18°C?
    c. According to your graph, what is the Celsius equivalent of 68°F?

28. a. Make a plot of the following measured mass and volume data.

Mass (g)	Volume (cm^3)
35	12
54	18
64	21
80	27
92	31
98	33

    b. What is the volume corresponding to a mass of 75 g?
    c. What is the mass corresponding to a volume of 15 cm^3?
    d. What is the mass corresponding to a volume of 0 cm^3?
29. The circumference and diameter of a variety of circles were measured and recorded below.

Circumference (cm)	Diameter (cm)
1.6	0.5
3.1	1.0
4.7	1.5
6.3	2.0
7.9	2.5
9.4	3.0

    a. What is the circumference of a circle corresponding to a diameter of 1.8 cm?
    b. What is the diameter of a circle that has a circumference of 8.0 cm?
30. Make a plot of the following measured volume and Celsius temperature data.

Volume (mL)	Temperature (°C)
10.5	10.0
10.7	20.0
11.1	30.0
11.5	40.0
11.9	50.0
12.2	60.0

    a. What happens to the volume as the temperature increases?
    b. Does the volume double when the temperature doubles?
    c. What is the volume when the temperature is 25°C?

2.7 What are significant figures?

31. How many sig figs are in each of the following measured numbers?
    a. 1.09
    b. 0.1
    c. 0.3040
    d. 9.2×10^{-3}
    e. 0.000001
    f. 11×10^{7}
    g. 0.0230
    h. 100.

2.8 Rounding off

32. Round off each of the following to 4 sig figs.
    a. 235.674
    b. 10.528 mL
    c. 4.0534×10^{-4} L
    d. 7.2457 g/mL
    e. 0.000328730 kg

2.9 Sig figs in calculations

33. Do the following multiplications or divisions and express your answers to the proper number of sig figs.
    a. 100.1×0.094
    b. 3×2.95
    c. $\frac{0.04}{18.1}$
    d. $\frac{2.00 \times 10^{1}}{3.9 \times 10^{3}}$
    e. 0.00903×21.18
34. Do the following additions or subtractions and express your answers to the proper number of sig figs.
    a. 12.0140 + 11.91
    b. 84.938 + 7.452 + 0.9
    c. 28.75 – 1.4
    d. 0.9523 – 6.540
35. Convert 3,182 g to pounds in each case and report each answer to the proper number of sig figs.
    a. Given 1 lb = 453.6 g
    b. Given 1 lb = 454 g
36. What is the total volume accumulated in a beaker by successive additions of liquid in the amounts 50.40 mL, 1.5 mL, 8.9 mL, and 0.123 L? Report the answer to the proper number of sig figs.

chapter 3

applying measurements to chemical calculations

Routine density determination of a liquid is performed with a top-loading analytical balance and picnometer (flask with a fixed volume).

3.1 introduction

Chemists often describe laws, theories, and experimental results in quantitative (numerical) terms and therefore find it necessary to do calculations. This chapter will review those math skills that you will need to perform chemical calculations and apply them to specific examples. It will provide the basis for your success with calculations throughout the rest of the book. If you look carefully, you will find that there are only a few basic mathematical concepts and skills that you need for chemical calculations, and some of those you probably already have.

Most chemical calculations involve the unit conversion method, which will be introduced in Section 3.3. The unit conversion method shows you how to use conversion factors to solve chemical problems. Once you understand the unit conversion method, you will realize that, as soon as you are given the necessary conversion factors, you are ready to perform many of the calculations required in chemistry. The purpose of this chapter is to make you comfortable with chemical calculations, including the unit conversion method and simple algebraic equations. As you progress through the rest of this textbook, you will learn what conversion factors are needed, what definitions can be written as algebraic equations, and how chemists arrived at both.

To use the unit conversion method, you must understand the meaning of a fraction and be able to multiply fractions successfully. Section 3.2 will provide a review of fractions so that you may "brush up" if you feel "rusty."

3.2 fractions

A fraction relates two numbers in terms of their proportion or ratio and is calculated by taking the number of parts of some quantity and dividing it by the total number of parts. A large pizza is usually cut into 12 equal pieces. One piece is 1 part out of 12 parts, or 1/12 of the pizza.

Most objects and ideas can be divided into parts, and the relationship of the part to the whole is a fraction; for example, this chapter is one of 17 chapters in this book and may be thought of as 1/17 of the text.

A fraction a/b, where a is called the numerator and b is called the denominator, is an arithmetic expression meaning $a \div b$. Every fraction can be expressed as a decimal number by simply carrying out the required division.

$$\frac{3}{8} = 3 \div 8 = 0.375$$

Every number, such as 16 or 4.2, can be written as a fraction, with 1 as the denominator.

$$16 = \frac{16}{1} \qquad 4.2 = \frac{4.2}{1} \qquad 0.78 = \frac{0.78}{1}$$

A symbol, such as *d,* can also be written as *d*/1.

Equality of two fractions. If two fractions *a*/*b* and *c*/*d* are equal, the following is true.

$$a \times d = c \times b$$

In other words, the products of the numerator of one fraction times the denominator of the other fraction are equal. Look at this again, more carefully.

$$\frac{a}{b} = \frac{c}{d}$$

means

a	$\times$	d	$=$	c	$\times$	b
Numerator (left)		Denominator (right)		Numerator (right)		Denominator (left)

Later in this chapter, you will use the relationship for density, $d = m/V$, which may also be written as $d/1 = m/V$. Following the equality of two fractions, $d \times V = m \times 1$ or $m = d \times V$. You remember from Section 2.9 that density, a physical property of a substance, may be calculated by dividing the mass of the sample, *m,* by its volume, *V.*

Multiplying two fractions. To multiply two fractions, first cancel any common factors in the numerator and the denominator, including any common units. Second, multiply the numerators together, including any uncanceled units, and then multiply the denominators together, including uncanceled units. Place the product of the numerators over the product of the denominators, as shown:

$$\left(\frac{a}{b}\right)\left(\frac{c}{d}\right) = \frac{a \times c}{b \times d}$$

Then convert fractional answers to decimal numbers by dividing the numerator of the answer by the denominator of the answer.

More than two fractions may be multiplied the same way, by multiplying all numerators and then by dividing by all denominators, as shown:

$$\left(\frac{a}{b}\right)\left(\frac{c}{d}\right)\left(\frac{e}{f}\right)\left(\frac{h}{i}\right) = \frac{a \times c \times e \times h}{b \times d \times f \times i}$$

Using a scientific hand calculator to evaluate this product is particularly convenient because you can complete all of the multiplication and division in one calculation. You will apply this type of multiplication to the unit conversion method discussed in the following section and used in solving chemical problems throughout this book.

sample exercise 1

Carry out the following.

$$\frac{13}{6} \times \frac{3}{4} =$$

solution

1. The denominator of the first fraction may be written as a product of its factors, 2×3; therefore, 13/6 may be written as $13/(2 \times 3)$, permitting cancellation of the common factor, 3, from the numerator and denominator.

$$\frac{13}{2 \times \cancel{3}} \times \frac{\cancel{3}}{4} = \frac{13}{2} \times \frac{1}{4}$$

There are no other common factors.

2. Multiply the numerators together, then multiply the denominators.

$$\frac{13}{2} \times \frac{1}{4} = \frac{13 \times 1}{2 \times 4} = \frac{13}{8}$$

3. Convert the fractional answer, 13/8, to a decimal number by dividing 13 by 8.

$$\frac{13}{8} = 1.625$$

---

sample exercise 2

Carry out the following.

$$\frac{2.54\ \text{cm}}{1.00\ \text{in.}} \times \frac{12.0\ \text{in.}}{1.00\ \text{ft}} =$$

solution

1. The only obvious common factor is the unit inches, which you cancel:

$$\frac{2.54\ \text{cm}}{1.00\ \cancel{\text{in.}}} \times \frac{12.0\ \cancel{\text{in.}}}{1.00\ \text{ft}}$$

2. Multiply the numerators, then the denominators. Remember to include any remaining units in these products and carry these units into your answer.

$$\frac{2.54\ \text{cm}}{1.00} \times \frac{12.0}{1.00\ \text{ft}} = \frac{2.54\ \text{cm} \times 12.0}{1.00 \times 1.00\ \text{ft}} = \frac{30.5\ \text{cm}}{1.00\ \text{ft}} = 30.5\ \frac{\text{cm}}{\text{ft}}$$

Dividing two fractions. When confronted by a division problem such as

$$\frac{a/b}{c/d} = \frac{a}{b} \div \frac{c}{d} =$$

Dividend Divisor

you solve the problem by converting the division operation into a multiplication. To divide a fraction (a/b) by a fraction (c/d),

1. Invert the denominator or divisor (c/d).

$$\frac{c}{d} \text{ inverted becomes } \frac{d}{c}$$

2. Multiply the numerator or dividend (a/b) by the inverted denominator (d/c), as described earlier in this section under "multiplying two fractions."

$$\frac{a/b}{c/d} = \underset{\text{Dividend}}{\frac{a}{b}} \div \underset{\text{Divisor}}{\frac{c}{d}} = \frac{a}{b} \times \underset{\text{Inverted divisor}}{\frac{d}{c}} = \frac{a \times d}{b \times c}$$

sample exercise 3

Carry out the following.

$$\frac{3}{8} \div \frac{1}{4} =$$

solution

$$\frac{3}{8} \div \frac{1}{4} = \frac{3}{8} \times \frac{4}{1} =$$

$$\frac{3}{_2\cancel{8}} \times \frac{^1\cancel{4}}{1} = \frac{3 \times 1}{2 \times 1} \times \frac{3}{2} = 1.5$$

sample exercise 4

Carry out the following.

$$3.2\text{ g} \div \frac{0.80\text{ g}}{1.0\text{ mL}} =$$

solution

You may always write 3.2 g as the fraction 3.2 g/1. The fraction 0.80 g/1.0 mL must be inverted and multiplied by 3.2 g/1. Then,

$$\frac{3.2\cancel{\text{g}}}{1} \times \frac{1.0\text{ mL}}{0.80\cancel{\text{g}}} = \frac{3.2\text{ mL}}{0.80} = 4.0\text{ mL}$$

3.3 unit conversion

If you measure the length of the line marked off below, the measurement you report might be 12.7 cm, 127 mm, or 5.00 in., depending on the ruler used and the scale used. Notice that the number reported is meaningless without the proper unit.

Measure this distance.

Usually you measure a quantity in units that are most convenient at the time. For example, if you had a ruler calibrated in inches, you would report 5 in. as your result for the length of the line above. However, you do not need to remeasure the same line if you want to know its length in centimeters or any other unit. What is necessary is that you are able to convert your measurement to different units. The most direct method to perform these conversion calculations is the *unit conversion method.* It is well worth your time to master this method because later you can apply it to a variety of chemical problems.

table 3.1 equalities between the metric and english systems*

Length	Mass	Volume
1 in. = 2.54 cm	2.20 lb = 1 kg	1.06 qt = 1 L
39.4 in. = 1 m	1 lb = 454 g	
1 mi = 1.61 km		

*Except for 1 in. = 2.54 cm, these equalities are inexact and will affect the number of significant figures allowed in a calculated answer (see Section 2.9).

mathematical background

The unit conversion method depends on two mathematical facts: (1) any equality can be used to write a fraction equal to 1, and (2) like quantities (common factors) in the numerators and denominators of fractions can be canceled out as discussed in Section 3.2.

Like quantities in the numerator and denominator can be canceled out because any quantity divided by itself is equal to 1. For example, clearly 8 ft / 8 ft = 1. The equality 1 m = 100 cm tells us that 1 m and 100 cm represent exactly the same distance. Therefore, dividing 1 m by 100 cm is the same as dividing 1 m by itself, and therefore the fraction

$$\frac{1\ \text{m}}{100\ \text{cm}} = 1$$

Similarly,

$$\frac{100\ \text{cm}}{1\ \text{m}} = 1$$

Any equality can be made into a fraction equal to 1; we call that fraction a conversion factor. The following fraction, which has a meterstick in both the numerator and denominator, demonstrates this visually. Both metersticks are the same length, even though the numerator is calibrated with 100 cm and the denominator is calibrated as 1 m, so their fraction must be 1 (Figure 3.1).

Useful equalities relating the metric and British systems are given in Table 3.1. Each of these equalities can be made into a conversion factor that you will be able to use to change a measurement in one set of units into the other.

sample exercise 5

Write conversion factors, that is, fractions equal to 1, from the equalities 1 kg = 1,000 g and 1,000 cm^3 = 1,000 mL.

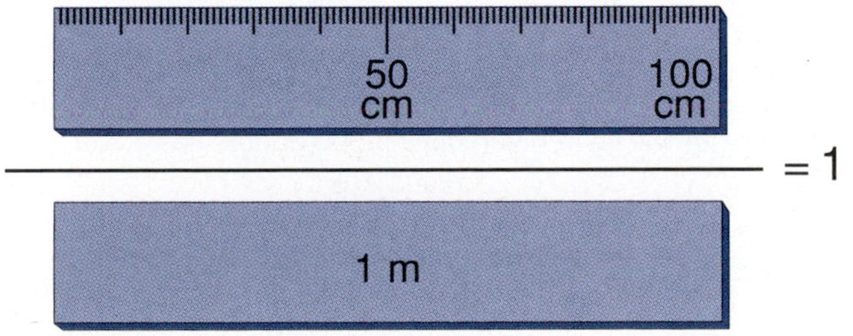

Figure 3.1
This fraction of metersticks has a value of 1 because the numerator and denominator have identical lengths of 1 m.

solution

Both $\frac{1\ \text{kg}}{1{,}000\ \text{g}} = 1$ and $\frac{1{,}000\ \text{g}}{1\ \text{kg}} = 1$ because 1 kg and 1,000 g represent an identical mass.

Both $\frac{1{,}000\ \text{cm}^3}{1{,}000\ \text{mL}} = 1$ and $\frac{1{,}000\ \text{mL}}{1{,}000\ \text{cm}^3} = 1$ because 1,000 cm^3 and 1,000 mL represent an identical volume. Notice also that 1,000 can be canceled out, so that

$$\frac{\cancel{1{,}000}\ \text{cm}^3}{\cancel{1{,}000}\ \text{mL}} = \frac{1\ \text{cm}^3}{1\ \text{mL}}$$

problem 1

Write conversion factors for the following equalities.

a. 1 kg = 2.2 lb

b. 1 L = 1,000 mL

multiplication by 1

Now that you know how to write conversion factors, multiplying with them is the same as multiplying by 1, since each conversion factor equals 1. As you know, multiplication by 1 does not change the quantity that is being multiplied, for example, 8 ft × 1 = 8 ft. However, units can be changed when multiplying by conversion factors. For example, if you multiply 8 kg by a factor equal to 1, that is, 1,000 g/1 kg, you get 8,000 g:

$$8\ \cancel{\text{kg}} \times \frac{1{,}000\ \text{g}}{1\ \cancel{\text{kg}}} = 8{,}000\ \text{g}$$

You can be certain that 8,000 g is the same quantity as 8 kg because the multiplication is by 1; you have done a *unit conversion,* i.e., changed the units from kilograms to grams.

For this simple example, of course, we could have used the following reasoning: if 1 kg is 1,000 g, then 8 kg must be 8,000 g. Frequently, however, unit conversions are more complex, and so we recommend the general method of problem solving that follows.

The unit conversion method

The steps to be followed in reading a problem and setting up a calculation by the unit conversion method are given here and applied in Sample Exercise 6.

1. Identify the *given quantity* and *unit* and write them down.
2. Identify the *new quantity* to be determined and write down the *new units* it is to have.
3. Determine the conversion factor(s) that will change the given into the new quantity and unit. This factor will have given units in the denominator and new units in the numerator.
4. Set up the calculation according to the following format.

Given quantity and unit × Conversion factor(s) = New quantity and unit

$$\_\_\_\_\ \cancel{\text{given unit}} \times \frac{(\_\_\_\_\ \text{new unit})}{(\_\_\_\_\ \cancel{\text{given unit}})} = \_\_\_\_\ \text{new unit}$$

sample exercise 6

How many meters are there in 76 cm?

solution

Follow the steps just outlined.

Step 1. The given quantity is 76 and the unit is cm; 76 cm.

Step 2. The new quantity to be determined is numbers of meters. Thus, the new unit is the meter.

Step 3. The equality relating meters and centimeters can be found in Table 2.3 as 1 m = 100 cm. Thus, the conversion factor to be used must be either 1 m/100 cm or 100 cm/1 m. Choose the one that has the given units in the denominator, new units/given units, which, in this case, is 1 m/100 cm. This conversion factor is chosen so that the given units will cancel out.

Step 4. Set up the proper format.

Given quantity and unit × Conversion factor = New quantity and unit

$$76\ \cancel{\text{cm}} \times \frac{1\ \text{m}}{100\ \cancel{\text{cm}}} = \underline{\qquad}\ \text{m}$$

Notice that the given units cancel to give the desired new unit, meters. The setup of the fractions tells you to divide 76 by 100, that is,

$$76\ \cancel{\text{cm}} \times \frac{1\ \text{m}}{100\ \cancel{\text{cm}}} = \frac{76\ \text{m}}{100} = 0.76\ \text{m}$$

Note that your answer is limited to 2 sig figs by the original quantity of 76 cm. The conversion factor, 1 m = 100 cm, contains exact numbers and does not limit the number of sig figs in your answer.

multistep conversions

Many times in converting units, you do not have a single equality that leads from one set of units to the other. However, you may have several equalities that, when combined, will enable you to make the desired conversion. You then make a conversion factor from each equality and multiply all of the conversion factors together so that the units all cancel except for those of interest.

This method may always be used for more complex conversions. The idea in such a case is to examine tables of equalities and assemble all the equalities you need to write the necessary conversion factors. When you multiply the conversion factors times the given quantity and unit, all units will cancel except for the ones to which you are converting.

sample exercise 7

How many milligrams are there in 0.53 kg?

solution

Follow the preceding steps.

Step 1. The given quantity and unit is 0.53 kg.

Step 2. The new unit to be determined is milligrams.

Step 3. Table 2.3 does not offer a direct equality between kilograms and milligrams. However, it does relate kilograms to grams and grams to milligrams. Therefore, if we use the conversion factors corresponding to these equalities, we should be able to convert kilograms to grams to milligrams. The factors are always in the form new unit/given unit, so, in this case, we have 1,000 g/1 kg and 1,000 mg/1 g.

Step 4. The setup is then

Given quantity and unit	×	Conversion factors	=	New quantity and unit
$0.53\ \not{kg}$	×	$\frac{1{,}000\ \not{g}}{1\ \not{kg}} \times \frac{1{,}000\ \text{mg}}{1\ \not{g}}$	=	$530{,}000\ \text{mg} = 5.3 \times 10^5\ \text{mg}$

The units cancel to give milligrams, and the setup tells you to multiply $0.53 \times 1{,}000 \times 1{,}000$. Note that the answer has been written in scientific notation to show the correct number of sig figs, which is limited by the original 0.53 kg. Both conversion factors contain exact numbers and do not limit the number of sig figs in the answer.

From the review of multiplication of fractions in Section 3.2, remember that you may calculate the final result on your scientific hand calculator by multiplying all numerators and then dividing by all denominators–*all in one process*–without having to write down any intermediate results.

••• problem 2

How many milliliters are in 0.439 L?

With the help of Table 3.1, you can also do conversions between the metric and British systems. For example, you can now prove that the three measurements reported at the very beginning of Section 3.3 are all correct. The three values given for the distance between the two lines were 12.7 cm, 127 mm, and 5.00 in.

$$12.7\ \not{cm} \times \frac{1\ \not{m}}{100\ \not{cm}} \times \frac{1{,}000\ \text{mm}}{1\ \not{m}} = \frac{12.7 \times 1{,}000\ \text{mm}}{100} = 127\ \text{mm}$$

$$12.7\ \not{cm} \times \frac{1\ \text{in.}}{2.54\ \not{cm}} \times \frac{12.7 \times 1\ \text{in.}}{2.54} = 5.00\ \text{in.}$$

If you have confidence in the unit conversion method, you should be able to apply it to conversions with which you are totally unfamiliar, such as those in later chapters of this book. For example, if you are given the following equality, you can construct a conversion factor and use it in Sample Exercise 8.

Equality: 1 mole of Cu = 64 g of Cu

---

sample exercise 8

How many moles of Cu are there in 10.0 g of Cu?

solution

Following the guidelines on page 66.

Step 1. The given quantity and unit is 10.0 g of Cu.

Step 2. The new unit to be determined is moles of Cu.

Step 3. The equality relating grams of Cu and moles of Cu is given before this sample exercise. Selecting the conversion factor of the form new units/given units yields 1 mole of Cu/64 g of Cu.

Step 4. The setup is then

Given quantity and unit	×	Conversion factor	=	New quantity and unit
$10.0\ \not{g\ of\ Cu}$	×	$\frac{1\ \text{mole of Cu}}{64\ \not{g\ of\ Cu}}$	=	0.16 moles of Cu

Please remember that the equality used in this exercise will be explained in a later chapter. If you understand this application of the unit conversion method now, however, you have mastered your approach to even the most challenging problems ahead. If you find this exercise difficult, you are at least aware of the challenges toward which to set your goals. Also note that the answer above is limited to 2 sig figs because of the 2 sig figs in the conversion factor (64 g of Cu).

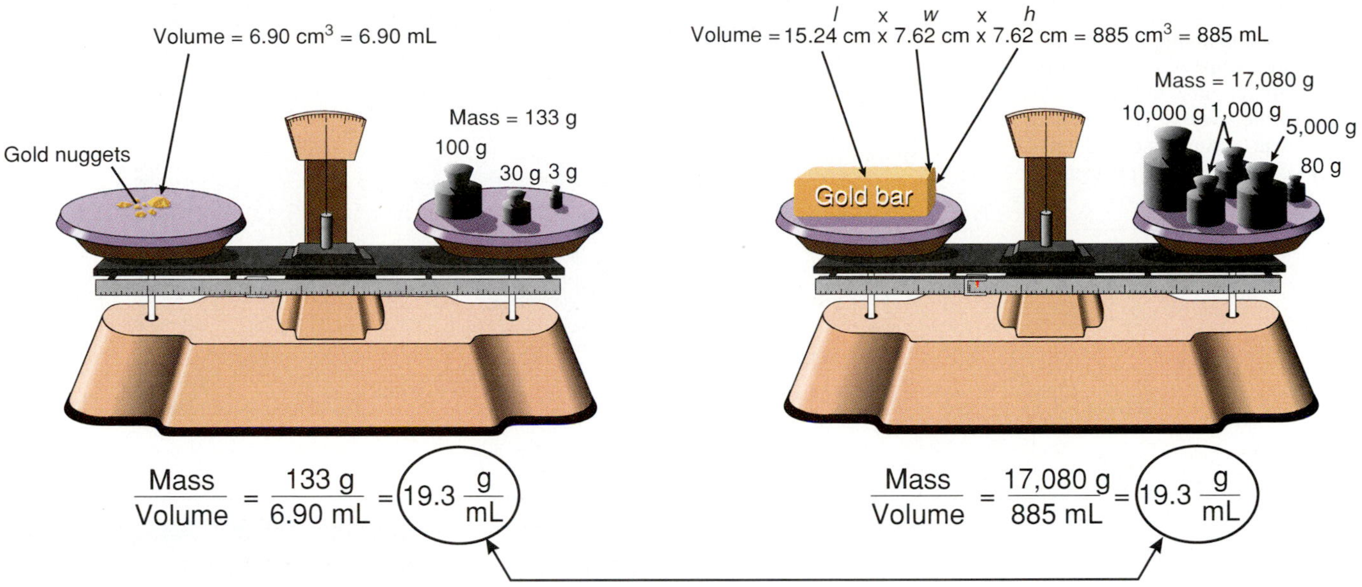

Figure 3.2

The masses and volumes of the two gold samples are very different, but the ratio of mass to volume is a constant characteristic property. This property is called density.

• • • problem 3

How many grams of Cu are in 1.5 moles of Cu?

The unit conversion method is a powerful tool that is the basis of most of the chemical calculations in this book. Remember,

1. Whenever you have an equality, you can construct a fraction equal to 1.
2. Fractions equal to 1 can be used as conversion factors to convert units.

3.4 density

From Section 1.4, you recall that neither the mass nor the volume of a substance is a characteristic property because both mass and volume vary with the size of the sample. For example, consider two samples of gold, one with small gold nuggets and the other a gold bar from Fort Knox (Figure 3.2). Clearly the gold bar has a larger volume (occupies more space) and weighs more (has a larger mass), but if you calculate the ratio of the mass to the volume, that is, mass / volume, the answer is the same for both samples of gold and will be the same constant value for any sample of gold that you measure.

Figure 3.3 is the graph you drew for Problem 7 in Chapter 2. You plotted the mass versus volume for several samples of the same substance. A straight line on this graph is a visual representation of a *constant* proportion or a *constant* ratio when mass is divided by volume. This constant ratio, mass / volume, is called **density,** a characteristic property that can be used to identify gold or any specific material.

Each material will have its own unique density. Solid lead, iron, and gold all have their own individual densities, and those densities are listed in chemical handbooks. Liquids, too, can be identified by their individual densities. Water has a density of 1 g/mL at 20°C, which is different from the density of corn oil or liquid mercury. A copper penny will float on liquid mercury because the density of copper, Cu, is less than that of mercury, Hg. Likewise, fat will float on the surface of soup (Figure 3.4a) because its density is less than the density of the soup. A mixture of oil and vinegar, a common salad dressing, will separate into two liquid layers (a physical process)

Figure 3.3

The straight line on the graph is a visual representation of a constant value when the mass of the sample is divided by the volume. This constant ratio of mass / volume is called the density—a characteristic physical property of the material.

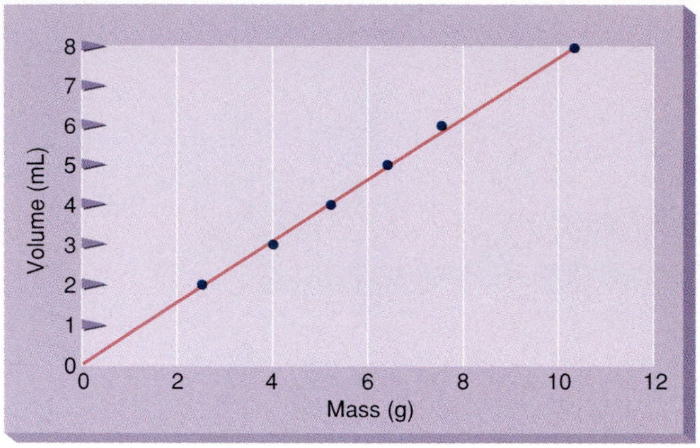

Figure 3.4

When one material floats on another, it has a lower density. (a) Fat floats on the surface of soup because it is less dense than water. Fat may be separated from the soup by skimming it off the surface to produce leaner soup; separation by skimming is a physical process. (b) You can conclude a cork is less dense than liquid water, which is less dense than an iron nut, which in turn, is less dense than liquid mercury.

(a) (b)

because the liquids are not soluble in each other; the oil will float on top of the water because oil has a lower density than water. As densities of liquids are compared with the density of water, densities of gases are compared with the density of air. Balloons filled with helium will rise in air because helium has a lower density. A balloon filled with pure oxygen will sink.

Density is determined in the laboratory by measuring both the mass and the volume of a sample and then dividing the mass by the volume.

$$\text{Density} = \frac{\text{Mass}}{\text{Volume}}$$

Because the mass is usually measured in grams and the volume in milliliters, density has the unit grams per milliliter, g / mL or

$$\frac{\text{g}}{\text{mL}}.$$

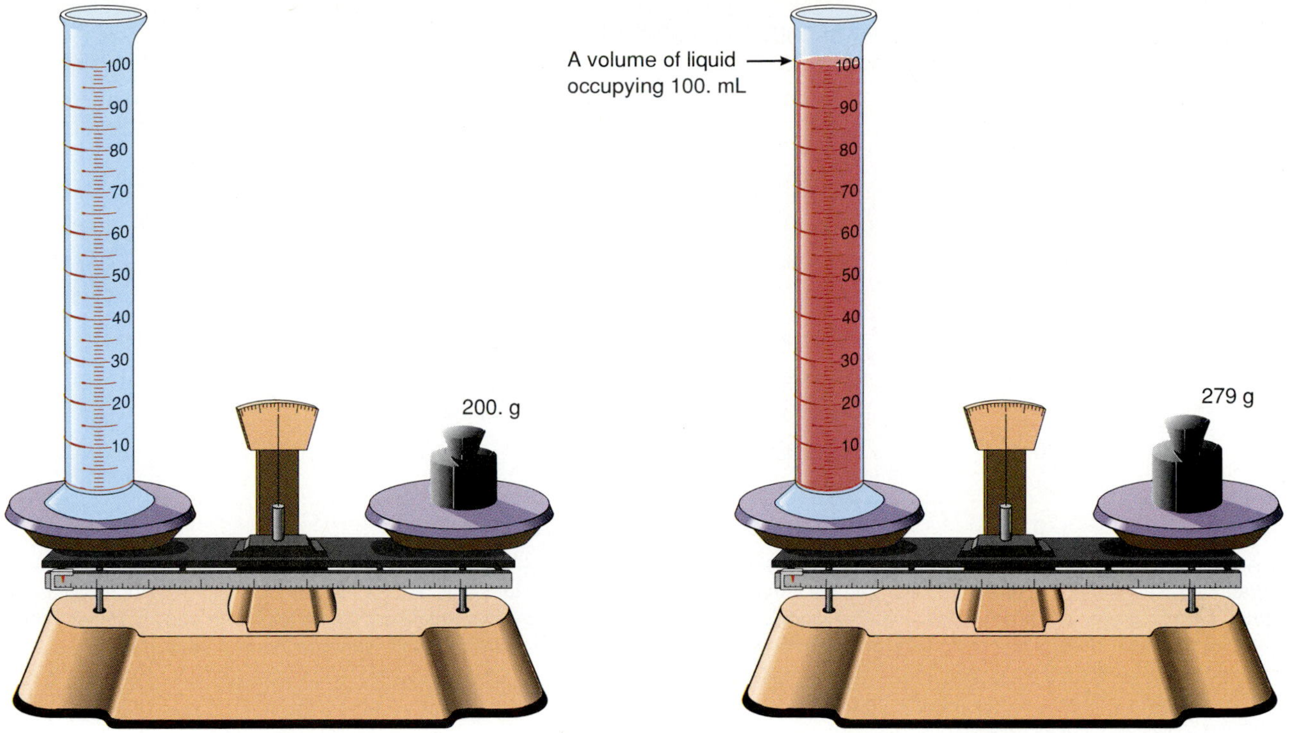

Empty graduated cylinder
Mass = 200. g

Cylinder filled to the 100. mL mark
Mass = 279 g

Mass of liquid = 279 g − 200. g = 79 g
Volume of liquid = 100. mL

$$\text{Density} = \frac{\text{Mass}}{\text{Volume}} = \frac{79\ \text{g}}{100.\ \text{mL}} = 0.79\ \frac{\text{g}}{\text{mL}}$$

Figure 3.5
All matter has both mass and volume. Mass is measured by the double-pan balance. For a liquid, one way of measuring volume is to use a graduated cylinder calibrated in milliliters.

If you were to measure the masses and volumes of the two gold samples shown in Figure 3.2, you would find very different values, as you would expect.

	Gold Nuggets	**Gold Bar**
Mass	133 g	17,080 g
Volume	6.90 cm^3 = 6.90 mL	885 cm^3 = 885 mL

But consider the mass/volume ratios for the two samples:

$$\text{Density (gold nuggets)} = \frac{\text{Mass}}{\text{Volume}} = \frac{133\ \text{g}}{6.90\ \text{mL}} = 19.3\ \frac{\text{g}}{\text{mL}}$$

$$\text{Density (gold bar)} = \frac{\text{Mass}}{\text{Volume}} = \frac{17{,}080\ \text{g}}{885\ \text{mL}} = 19.3\ \frac{\text{g}}{\text{mL}}$$

You see that the density is constant. A density of 19.3 g / mL is a characteristic property of any sample of gold of any size.

table 3.2 densities of common materials

Solid	Density (g/mL, 20°C)	Liquid	Density (g/mL, 20°C)	Gas	Density (g/L, 0°C*)
Gold	19.3	Water	1.00	Air	1.29
Lead	11.3	Gasoline	0.67	Oxygen	1.43
Copper	8.92	Milk	1.03	Hydrogen	0.090
Iron	7.86	Sea water	1.03	Helium	0.178
Aluminum	2.70	Blood	1.06	Carbon dioxide	1.96
Salt	2.16	Mercury	13.6		
Paper	0.70	Olive oil	0.92		
Balsa	0.20	Alcohol	0.79		
Redwood	0.44	Vinegar	1.01		
Rubber	1.1	Ether	0.70		
Ice	0.92	Carbon tetrachloride	1.59		

*Notice that, for gases, the unit is grams per liter and the temperature is 0°C, whereas, for solids and liquids, the unit is grams per milliliter and the temperature is 20°C. Gases are much less dense than liquids or solids and need a more convenient unit to express their density values. The temperature at which the density has been determined must be specified because the value of density varies with temperature.

sample exercise 9

A student wants to know the density of ethyl alcohol, the intoxicating component of fermented liquors. She carefully measures out 100. mL of the liquid in a graduated cylinder. Then she carefully weighs the 100. mL and finds the mass to be 79 g (Figure 3.5). What is the density of the ethyl alcohol?

solution

To find density, divide the mass of a sample by its volume. In this case,

$$\text{Density} = \frac{79\ \text{g}}{100.\ \text{mL}} = 0.79\ \frac{\text{g}}{\text{mL}}$$

••• problem 4

What is the density of olive oil if 21 mL weighs 19 g?

In stating that density is constant, we have assumed that the temperature of our sample has not changed. Density does vary slightly with temperature. Therefore, when densities are reported, the temperature at which the measurement was done is usually noted. Table 3.2 lists the densities of some common materials; solids and liquids are at 20°C and gases are at 0°C.

When working in the laboratory, you will find that it is particularly useful to know the densities of liquids. When you wish to know the mass of a liquid, it is generally easier to measure the volume and use the known density to calculate the mass. Density relates mass and volume and allows the calculation of one from the other either by algebraic manipulation of the equation density = mass/volume or by regarding density as a conversion factor relating mass and volume units.

Density is a conversion factor for converting units of mass into units of volume, or vice versa. Notice in Sample Exercise 9 that 100. mL of ethyl alcohol and 79 g of ethyl alcohol represent the identical sample; therefore, the fraction 79 g/100. mL, or the reduced form 0.79 g/mL, is equivalent to 1 and can be used as a conversion factor. This is true of all densities.

sample exercise 10

What is the mass of 3.0 mL of ether?

solution

Approach this by the unit conversion method.

Step 1. The given quantity and unit is 3.0 mL.

Step 2. The new quantity to be determined is the mass, that is, the number of grams of sample.

Step 3. Density relates grams and milliliters. The density of ether from Table 3.2 is 0.70 g/mL.

Step 4. Set up the calculation in the proper format.

Given quantity and unit	×	Conversion factor	=	New quantity and unit
$3.0\ \cancel{mL}$	×	$\frac{0.70\ g}{1.0\ \cancel{mL}}$	=	$\frac{3.0 \times 0.70\ g}{1.0} = 2.1\ g$

sample exercise 11

What is the volume of 27.2 g of mercury, a liquid metal employed in thermometers to measure the temperature?

solution

Step 1. The given quantity and unit is 27.2 g.

Step 2. The new quantity is the volume, that is, the number of milliliters.

Step 3. Density relates grams and milliliters. The density of mercury is 13.6 g/mL, which can also be written as 1.00 mL/13.6 g.

Step 4. The setup is as follows.

Given quantity and unit	×	Conversion factor	=	New quantity and unit
$27.2\ \cancel{g}$	×	$\frac{1.00\ mL}{13.6\ \cancel{g}}$	=	2.00 mL

••• problem 5

What volume (in milliliters) will 84 g of carbon tetrachloride occupy?

Notice that, for gases, the unit is grams per liter and the temperature is 0°C. This is because gases are much less dense than liquids or solids and their volumes change more dramatically with temperature. The density of gases increases as temperature decreases.

CHEMLAB

Laboratory exercise 3: density

Determination of the density of a solid material is one of the first experiments in which you develop your quantitative skills with laboratory equipment. To calculate density, a characteristic physical property, you need to measure the mass and the volume of your sample and then divide these values. Remember, neither the mass nor the volume of a substance is a characteristic property because both vary with the size of the sample.

Before using any balance to measure mass, you must always make sure that the balance is zeroed before placing your sample on the pan. The mass of solid aluminum wire, which was measured on a top-loading balance, was found to be 11.034 g (Figure 3.6).

Figure 3.6
Aluminum wire is measured on a top-loading balance. Its mass is read directly from the balance.

Experimentally, you may determine the volume of the Al wire in terms of the volume of water it displaces. Fill a 10-mL graduated cylinder approximately half full of water and read the volume it contains. In this case, you find that the graduated cylinder contains 4.90 mL. After placing the Al wire in the graduated cylinder, you find that the water level has risen to 9.03 mL. The difference in the level of water (9.03 mL – 4.90 mL = 4.13 mL) represents the volume of water that the Al wire has displaced (Figure 3.7).

Questions:

1. What is the density of the wire sample as determined from the above experiment?
2. How does your experimental value of the density compare with the density of Al in Table 3.2?
3. If you did not know that your wire was made of Al, how could you identify its composition from the above experiment?
4. If you had used a smaller piece of wire with a mass of 6.50 g, what volume would your sample have had?

Measuring the density of a liquid sample is more direct than measuring that of a solid. If the liquid is poured into an empty graduated cylinder, you may read the volume directly. Suppose your liquid sample has a volume of 8.5 mL. The mass of the liquid can be determined by measuring the mass of the graduated cylinder plus the liquid (35.9 g) and then subtracting the mass of the empty graduated cylinder (28.1 g).

5. What is the density of the liquid sample measured above?
6. Use Table 3.2 to identify the liquid sample.
7. Would your unknown liquid float on or sink below water?

Density measurement of a liquid routinely uses equipment that has less uncertainty in its values than your laboratory glassware and is shown in the photo at the beginning of this chapter. Instead of using a graduated cylinder to measure volume, a picnometer is used. A picnometer is a stoppered flask that has a carefully calibrated volume so that any liquid completely filling it will have the same known volume. The mass of the liquid is still obtained by the difference in the weight of the picnometer plus liquid and the empty picnometer. The temperature of the liquid sample is kept constant using a constant-temperature water bath.

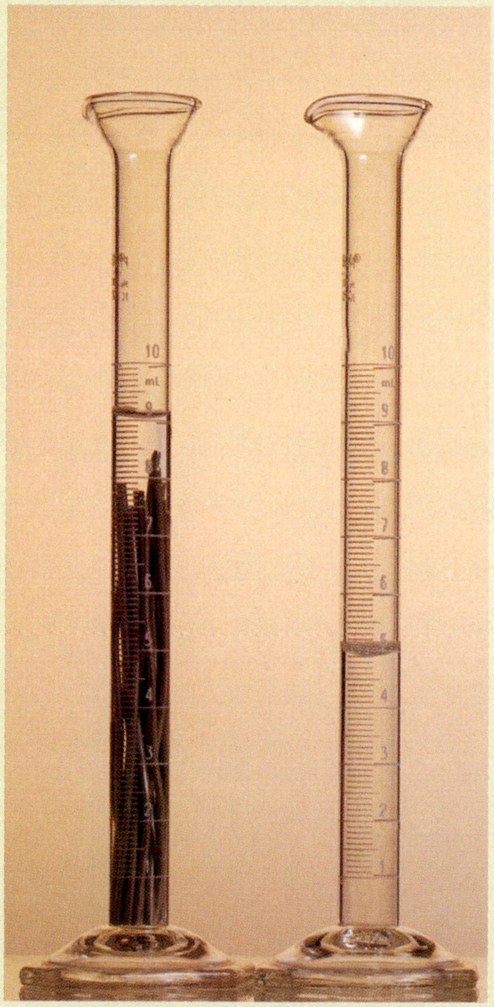

Figure 3.7
The volume of the solid sample is equal to the volume of water it displaces. The graduated cylinder originally contained 4.90 mL of water. After adding the solid sample, the water level rose to 9.03 mL. The difference of 4.13 mL between the readings represents the volume of water displaced by the sample.

3.5 algebraic manipulations

As discussed previously, problems in this text will usually be solved by the unit conversion method (Section 3.3). However, you will find that there are some problems that can be solved more easily by the use of an unknown x and the basic techniques of algebra. Only the most basic techniques of algebra will be employed. Once you have reviewed the basic techniques of algebra, your math skills should be fully prepared to master any of the chemical problems in this book.

The key step in solving an algebraic equation for an unknown is to isolate the unknown to one side of the equation. You manipulate the equation by adding, subtracting, multiplying, and dividing, keeping in mind the primary rule that *whatever is done to one side of the equation must be done to the other.*

Let us look at some examples of simple algebraic manipulations. If you are given the equation, $x + 3 = 9$, and asked to solve for x, you concentrate on getting x isolated on one side of the equation. If you remove (subtract) the 3 from the left side of the equation, you will have x alone and the equation is solved. According to the primary rule, you can subtract 3 from the left side so long as you also subtract it from the right side.

$$x + 3 = 9$$

$$x + \underbrace{3 - 3}_{} = \underbrace{9 - 3}_{}$$

$$x = 6$$

To solve another equation, $6x + 5 = 29$, for x, once again you begin by subtracting from each side. Note that $6x$ means that x is multiplied by 6.

$$6x + \underbrace{5 - 5}_{} = \underbrace{29 - 5}_{}$$

$$6x = 24$$

To complete the solution, recognize that $6x$ must be divided by 6 to isolate x, so we divide each side by 6.

$$\frac{\cancel{6}x}{\cancel{6}} = \frac{24}{6}$$

$$x = 4$$

---

sample exercise 12

Solve for t in $5t - 12 = 4 - 3t$.

solution

This problem has the unknown, t, on both sides of the equation. Apply the primary rule to the equation in order to isolate t to the left side.

Add $3t$ to both sides of the equation and sum together terms with the unknown t.

$$\underbrace{5t - 12 + 3t}_{} = \underbrace{4 - 3t + 3t}_{}$$

$$8t - 12 = 4$$

Add 12 to both sides.

$$8t - \underbrace{12 + 12}_{} = \underbrace{4 + 12}_{}$$

$$8t = 16$$

Divide both sides by 8.

$$\frac{\cancel{8}t}{\cancel{8}} = \frac{16}{8}$$

$$t = 2$$

In Chapter 2, you learned the relationship between the Celsius and Fahrenheit temperature scales as given in Equations 2.1 and 2.2, which are repeated in the following text. Now that you are gaining confidence using your algebra skills, you can take either of these two equations and rearrange to the other. Equation 2.1 prescribes how to calculate degrees Fahrenheit if you know the temperature in degrees Celsius. To obtain the form that is more convenient for calculating degrees Celsius from degrees Fahrenheit, you may apply the primary algebraic rule to isolate °C. You begin by subtracting 32 from each side of Equation 2.1.

$$°F = 1.8\,(°C) + 32 \tag{2.1}$$

$$\underbrace{°F - 32} = 1.8\,(°C) + \underbrace{32 - 32} = 1.8\,(°C)$$

Divide each side by 1.8.

$$\frac{(°F - 32)}{1.8} = \frac{\cancel{1.8}\,(°C)}{\cancel{1.8}} = °C$$

Rewrite the equation, putting degrees Celsius on the left.

$$°C = \frac{(°F - 32)}{1.8} \tag{2.2}$$

Note that 32 and 1.8 are exact numbers and do not limit the number of significant figures in a calculation.

---

sample exercise 13

Body temperature is 98.6°F. What is body temperature in degrees Celsius?

solution

You are given the number of degrees Fahrenheit; therefore, substitute the value into Equation 2.2.

$$°C = \frac{(°F - 32)}{1.8} \tag{2.2}$$

Substitution yields

$$°C = \frac{(98.6 - 32)}{1.8}$$

To complete this calculation, you must subtract before dividing because the minus sign is inside the parentheses.

$$°C = \frac{66.6}{1.8} = 37.0$$

Answer: 98.6°F = 37.0°C

alternate solution

You really need to know only one equation to solve for temperature. You can also use Equation 2.1 to solve this problem.

$$°F = 1.8\,(°C) + 32 \tag{2.1}$$

Substitute 98.6 for degrees Fahrenheit because that is the given.

$$98.6 = 1.8\,(°C) + 32$$

Subtract 32 from each side.

$$\underbrace{98.6 - 32}_{} = 1.8\,(°C) + \underbrace{32 - 32}_{}$$
$$66.6 = 1.8\,(°C)$$

Divide each side by 1.8.

$$\frac{66.6}{1.8} = \frac{\cancel{1.8}\,(°C)}{\cancel{1.8}}$$
$$37.0 = °C$$

Your choice of solution will depend on how comfortable you feel with algebraic manipulations.

The general method of problem solving for temperature conversions is outlined as follows.

1. Identify the kind of degrees given.
2. Select a correct equation relating degrees Celsius and degrees Fahrenheit, and substitute for the degrees given.
3. Complete the calculation, being careful to multiply or add, subtract or divide, in the proper order. Notice that temperature conversion cannot be done by the unit conversion method because the relationship between degrees Celsius and degrees Fahrenheit is not strictly multiplicative. One must always add or subtract 32 at some point in the calculation.

A type of algebraic problem other than temperature conversion will appear in Chapter 11 with regard to the behavior of ideal gases. The ideal gas law relates the pressure of a gas, P, to its temperature, T, volume, V, and the number of moles of gas, n, according to the equation, $PV = nRT$. This equation can be rearranged to solve for any of the properties, P, V, or T. In Sample Exercise 14, the equation is rearranged to solve for temperature.

---

sample exercise 14

Solve for T in $PV = nRT$.

solution

You may divide both sides of the equation by n and then divide both sides by R, or, in one step, you may divide both sides by the product nR. Let's divide both sides by nR.

$$PV = nRT$$
$$\frac{PV}{nR} = \frac{\cancel{nR}T}{\cancel{nR}}$$
$$\frac{PV}{nR} = T$$

Note that it does not matter whether the unknown is isolated to the left or to the right; convenience and simplicity dictate the choice.

In Chapter 2, you learned that density is frequently used as a conversion factor to solve chemical problems, but that algebraic rearrangement of the density equation is sometimes convenient. It will be to your advantage to be familiar with both methods. Algebraic rearrangement of the density equation is illustrated in the following sample exercise.

sample exercise 15

Solve for the volume, V, in the density equation, $d = m/V$.

solution

When the unknown is in a denominator, it is usually best to begin by clearing the fraction, which is done by cross multiplying the fractions, as shown in our review of fractions in Section 3.2. Remember the density equation can be written as

$$\frac{d}{1} = \frac{m}{V}$$

So,

$$d \times V = m$$

Now divide each side by d to isolate V.

$$\frac{\cancel{d} \times V}{\cancel{d}} = \frac{m}{d}$$

Canceling the common factor of d,

$$V = \frac{m}{d}$$

••• problem 6

Solve for the unknown x in each of the following.

a. $6 - 4x = 10$

b. $6 + 16x = 33 + 7x$

c. $13x = A$

d. $109 + 3x = 43$

3.6 percentage calculations

Occasionally, it will be necessary to do calculations that involve percent (%). In Chapter 1, you learned the chemical formula for several compounds, such as H_2O for water. Chemists often analyze the quantity of each element in a compound and report the fraction of each element as a percentage by weight.

When measuring the amount of product actually formed in a chemical reaction, chemists report it as a percent of the maximum amount of product that could form theoretically.

"Percent" literally means "parts per 100 parts." A percentage of a quantity may be found using a special fraction in which the total number of parts is always 100 (refer to Section 3.2). Thirty percent of a quantity means taking "30 parts per 100"; seven percent means taking "7 parts per 100." You can evaluate any percentage using a fraction with 100 in the denominator.

$$30\% \text{ means } \frac{30}{100}$$

$$7\% \text{ means } \frac{7}{100}$$

$$0.45\% \text{ means } \frac{0.45}{100}$$

From these fractions, you may easily express the percentage in decimal form by dividing through by 100.

$$30\% \text{ means } \frac{30}{100} = 0.30$$

$$7\% \text{ means } \frac{7}{100} = 0.07$$

$$0.45\% \text{ means } \frac{0.45}{100} = 0.0045$$

Notice that, in each case, you get from the numerical value of percent to the decimal form by moving the (understood) decimal two places to the left.

$$30.\% = 0.30$$

$$07.\% = 0.07$$

$$00.45\% = 0.0045$$

---

problem 7

Convert the following percentages to decimal form.

a. 4.5%
b. 33%
c. 0.092%

If you were asked to calculate the number of male students in a class of 120 students, given that the class is 30% male, you probably know from previous experience that to do this you "take 30% of 120." Since "of" means multiply, you replace 30% by its decimal form, 0.30, and multiply by 120 to find the answer.

30% of 120 students = Number of male students

0.30×120 students = 36 male students

"Taking a percent" always involves this reasoning: You must multiply

Percent in decimal form × Given total

---

sample exercise 16

If water is 11% hydrogen by weight, calculate the number of pounds of hydrogen in 40. lb of water.

solution

In decimal form, 11% is 11., which is 0.11. The *given total* is 40. lb of water. Therefore,

Pounds of hydrogen = $0.11 \times 40.$ lb = 4.4 lb

Suppose you are given data about the makeup of something that is composed of parts. For example, you are told that, in a class of 75 students, there are 36 female students. How do you determine the percentage of female students? The method is to divide the parts by the total parts and multiply by 100. In this example,

$$\frac{36 \text{ female students}}{75 \text{ students}} \times 100 = 48\%$$

Determination of a percent, given numbers of parts and total parts, involves the following operation.

$$\textbf{Percent} = \frac{\textbf{Parts}}{\textbf{Total parts}} \times \textbf{100}$$

sample exercise 17

Calculate the percentage of copper in a 79.0-lb sample of copper ore that contains 3.75 lb of copper.

solution

$$\text{Percent} = \frac{\text{Parts}}{\text{Total parts}} \times 100 = \frac{3.75\ \cancel{\text{lb}}}{79.0\ \cancel{\text{lb}}} \times 100 = 4.75\%$$

••• problem 8

A sugar mixture is found to contain 3.50% water. How many pounds of water are in 178 lb of the sugar mixture?

••• problem 9

What is the percentage of gold in a sample in which there is 5.0 oz of gold in every 9.0 oz of sample?

3.7 energy

As you learned in Section 1.1, matter is defined as anything that occupies space and has mass. In addition to mass, all matter has energy. **Energy** is the ability to do work, or we might say the ability to cause some change. The change may be in such things as the position of an object or its temperature.

There are different types or forms of energy: heat energy, electric energy, solar energy, and atomic energy, to name a few. All forms of energy can be classified as either kinetic or potential energy. **Kinetic energy** is the energy of matter in motion. The kinetic energy of a moving car is the energy of the car in motion. As a result of the car's kinetic energy, people may be moved or a trailer may be pulled. Notice that this energy effects a change in position. In Figure 3.8, the kinetic energy of a large meteor compressed and displaced an enormous volume of soil, forming a crater 2/3 of a mile across and 600 feet deep.

Figure 3.8

The kinetic energy of a large meteor did work on the ground by compressing and displacing a large volume of soil. The crater measures approximately 1,100 m across (2/3 of a mile) and is 180 m (600 feet) deep.

In chemistry, we talk more about **potential energy,** which is the energy matter possesses because of its position or composition (Figure 3.9). A rock at the top of a hill has potential energy because of its position. If it begins to roll down the hill, the potential energy is converted into kinetic energy. The farther it rolls, the more kinetic energy it acquires because it moves faster. As the rock moves down the hill, its potential energy is decreasing since it has less potential for converting its position into kinetic energy. Notice that as the kinetic energy increases, the potential energy decreases. The total energy, which is the sum of the kinetic and potential energies, remains the same. At the top of the hill, the total energy is all potential and no kinetic. At the bottom of the hill, the total energy is all kinetic and no potential. Along the hill, the total energy is a mixture of both types of energy, but the total amount is always the same; that is, total energy is constant. Another way of saying the same thing is that total energy is conserved. Scientists summarize this finding in terms of the law of conservation of energy.

You contain a great deal of potential energy. Move your arm. You just changed some of your potential energy into kinetic energy. Your potential energy comes from the composition of the matter in your body. All matter contains **chemical energy,** the potential energy that comes from the way the matter is put together. Gasoline contains chemical energy, which is released when it burns. Note that chemical energy and potential energy are *stored energy*. Stored energy is used by changing it to a form of kinetic energy.

The energy changes that occur within an automobile cylinder offer good examples of what is meant by a change in the form of energy. The gasoline-air mixture, rich in chemical energy, is introduced into the cylinder. Electric energy produces the spark (light energy) that ignites the gasoline, converting the chemical energy of gasoline into heat energy and the mechanical energy of the moving piston. Thus, the potential energy of gasoline is converted into the kinetic energy of the moving piston and moving car.

Although energy can change form, the total amount of energy in the universe is constant. This is one way of stating the law of conservation of energy. Another way to state this law is to say energy is neither created nor destroyed in any change in matter. Both statements tell you that you can never make new energy or destroy what energy you have.

What is meant, then, when you hear on the news that you must save energy? The answer is that saving energy means keeping it in a useful, potential (stored) form. The message is that you must save fuels, rich sources of chemical energy. If we burn all our fuel, the universe will still contain the same amount of energy, but it will not be in a form that is useful to us. Understanding graphs such as the projected reduction in oil consumption in Figure 2.12 helps you to make informed decisions about our national energy policy.

Figure 3.9
This diver has only potential energy the moment this picture was taken. Her potential energy will be converted into kinetic energy as she falls toward the water below.

3.8 units of energy

For the chemist, the most useful and easiest measurement of energy is the heat change in a chemical or physical process. Heat is a form of energy. Temperature measures the hotness or coldness of a substance and indicates the tendency of heat energy to be transferred. Heat energy flows from objects of higher temperature to objects of lower temperature.

The unit of energy traditionally used by chemists is the **calorie** (abbreviated cal). When nutritionists speak of Calories (with a capital C, abbreviated Cal) as applied to foods, they mean kilocalories (abbreviated kcal). Note that 1 Cal = 1,000 cal = 1 kcal. However, chemists have now converted to using the **joule** (pronounced "jewel" and abbreviated J), which is the unit of energy in the SI system. The joule is the energy unit you will encounter in your future scientific studies, and you will use the joule throughout the remainder of this text. Today, you may talk about your diet in terms of the number of Calories contained in different foods. Soon, everyone will be talking about the number of joules or kilojoules (kJ) in their diet.

The prefix *kilo-* is used with units of energy in the usual manner, that is, 1 kilojoule (kJ) = 1,000 joules (J). Because energy can change form, we can use joules and kilojoules as the units for any form of energy.

One calorie is the amount of heat necessary to raise the temperature of 1 g of water by 1°C. The convenience of this unit is obvious by the ease with which you could measure the temperature change for a given number of grams of water. By definition, 1 cal = 4.184 J and 1 kcal = 4.184 kJ, so you can construct conversion factors from these equalities to change calories to joules or joules to calories. A few conversion factors are given below.

$$\frac{1\ \text{cal}}{4.184\ \text{J}} \quad \text{or} \quad \frac{4.184\ \text{J}}{1\ \text{cal}} \quad \text{or} \quad \frac{1\ \text{kcal}}{4.184\ \text{kJ}} \quad \text{or} \quad \frac{4.184\ \text{kJ}}{1\ \text{kcal}}$$

sample exercise 18

A hamburger contains 409 kcal (409 Cal). How many kilojoules does this represent?

solution

The given quantity and unit is 409 kcal. The desired unit is kilojoules. The conversion factor is 4.184 kJ/1 kcal.

The setup is

Given quantity and unit × **Conversion factor** = **New quantity and unit**

So,

$$409\ \cancel{\text{kcal}} \times \frac{4.184\ \text{kJ}}{1\ \cancel{\text{kcal}}} = 1.71 \times 10^3\ \text{kJ}$$

Note that there are always about 4 times more kilojoules than kilocalories.

••• problem 10

If a jogger burns 2,200 kJ during her 5-mile run, how many kilocalories did she use up? How many hamburgers will it take to replace those kilojoules she burned? (Refer to Table 3.3.)

Notice that temperature does *not* measure heat energy. The amount of heat energy contained in a body depends on

1. The nature of the substance
2. The mass of the substance
3. The temperature of the substance

These three items are all reflected in the definition of a joule.

For every pure substance, you may measure a physical property called the **specific heat** of that substance. Specific heat tells you the amount of heat (energy) necessary to raise the temperature of 1 g of material by 1°C. Just as density can be used as a conversion factor in chemical problems, so can specific heat. Examples are given in the exercises below.

Table 3.4 lists specific heats for some common materials. Notice that the unit of specific heat is J/g°C. The specific heat of water is 4.184 J/g°C; that is, it takes 4.184 J to raise 1 g of water by 1°C. For iron, the specific heat is 0.46 J/g°C; that is, it takes only 0.46 J to raise 1 g of iron by 1°C.

table 3.3 energy values of some common foods

Food	Typical serving size	Energy content (kcal)	Energy content (kJ)
Apples (raw)	1 apple (175 g)	105	440
Beer	12 oz	150	630
Bread (white, enriched)	1 slice	68	285
Cheese (cheddar)	1 oz	133	556
Chocolate	1 oz	156	653
Eggs	1 egg	80	330
Hamburger	1/4 lb	409	1,710
Milk	1 glass	160	670
Peanuts	1 oz	159	665
Sugar	2 tsp	25	105

table 3.4 specific heats of some common substances

Substance	Specific heat (J/g°C)
Aluminum	0.92
Copper	0.39
Ethyl alcohol	2.13
Iron	0.46
Lead	0.13
Olive oil	2.09
Silver	0.23
Table salt	0.88
Water	*4.184*

You will note from Table 3.4 that water has an unusually high specific heat. A high specific heat means that a large amount of heat must be absorbed to raise the temperature of water in comparison with raising the temperature of other substances. Conversely, when water cools a few degrees, a large amount of heat must be released into the surroundings. The high specific heat of water is responsible for the tempering of the environment near an ocean or large lake. In spring, the water warms slowly, but in the fall, it cools slowly as well.

Solar water heaters installed in private homes use sunlight to warm the water used to shower and to wash clothes and dishes (Figure 3.10). Since heating water requires a large amount of energy, the use of sunlight saves burning a nonrenewable fuel, such as oil or natural gas, to obtain the energy needed to heat the water. Solar thermal power plants, which produce electricity, are being developed as alternate energy sources. A 360-megawatt facility is now in operation in California.

Figure 3.10
Sunlight provides an alternative energy source for heating water in private homes without consuming a nonrenewable fuel such as oil or natural gas.

sample exercise 19

How much heat energy must be applied to raise the temperature of 10.0 g of water by 3.00°C?

solution

It requires 4.184 J to raise the temperature of 1 g of water by 1°C. Therefore, 41.84 J must be applied to raise the temperature of 10.0 g of water by 1.00°C. But, since you want the temperature to go up 3.00°C, you need 3 times the 41.84 J or 126 J.

alternate solution

You may also do this problem by the unit conversion method. Follow the usual steps.

Step 1. The given quantities and units are 10.0 g of water and 3.00° C change in temperature.

Step 2. The new unit to be determined is joules, the unit of heat energy.

Step 3. The conversion factor relating joules, grams, and degrees Celsius is the specific heat of the given substance, in this case water.

Step 4. The setup is therefore

Given quantities and units	×	Conversion factor	=	New quantity and unit
$10.0\,\cancel{g}$	×	$4.184\,\frac{J}{\cancel{g}\,^\circ\cancel{C}} \times 3.00\,^\circ\cancel{C}$	=	126 J

In general, the heat energy change for a given mass, *m*, of substance undergoing a temperature change from t_1 to t_2 is given by the following equation. Specific heat is abbreviated SH.

$$m \times \text{SH} \times (t_2 - t_1) = \text{Heat energy change}$$

$$\cancel{\text{Grams}} \times \frac{\text{Joules}}{\cancel{\text{Gram}} \times \cancel{^\circ C}} \times \cancel{^\circ C} = \text{Joules}$$

••• problem 11

How much heat energy must be applied to raise the temperature of 36 g of copper 18°C?

To determine the specific heat of a substance experimentally, you measure the temperature change that occurs when a known amount of heat is applied to or lost from a known mass of material. The experimental data are then used in the equation

$$\text{SH} = \frac{\text{Heat energy in joules}}{\text{Mass in g} \times \text{Temperature change in °C}}$$

---

sample exercise 20

When 30.0 g of gold absorbs 20.0 J, the temperature of the gold changes from 21.3°C to 26.4°C. What is the specific heat of gold?

solution

Use the preceding equation for specific heat (SH). Remember to use the change in temperature, 26.4 − 21.3 = 5.1°C.

$$\text{SH of gold} = \frac{20.0 \text{ J}}{30.0 \text{ g} \times 5.1\text{°C}}$$

Doing the indicated multiplication and division, you obtain

$$\text{SH of gold} = 0.13 \frac{\text{J}}{\text{g°C}}$$

The energy content (Calorie content) in food is measured by completely burning the food and measuring the heat given off by the burning food. This amount of heat energy is exactly equivalent to the amount of energy the food can potentially provide to the body for moving muscles, conducting electrical impulses along nerves, maintaining body temperature, etc. Table 3.3 shows the energy content of some common foods in both kilocalories and kilojoules.

chapter accomplishments

After completing this chapter, you should be able to define all key terms and do the following.

3.2 Fractions

- ❑ State the meaning of a fraction in terms of the division operation.
- ❑ Derive a multiplication expression from the equality of two fractions.
- ❑ Multiply a set of fractions.
- ❑ Divide one fraction by another.

3.3 Unit conversion

- ❑ Construct conversion factors from equalities.
- ❑ Beginning with a given quantity and unit, use conversion factors to calculate some new quantity and unit.

3.4 Density

- ❑ State the usual unit in which density is expressed.
- ❑ Given the mass and volume of a sample, calculate the density of the sample.
- ❑ Recognize density as a conversion factor.
- ❑ Given the density, convert the volume of a sample to the mass of the sample.
- ❑ Given the density, convert the mass of a sample to the volume of the sample.

3.5 Algebraic manipulations

- ❑ Manipulate an algebraic equation to isolate an unknown to one side.

3.6 Percentage calculations

- ❑ Convert a given percentage to a fraction and/or a decimal form.
- ❑ Given the percent of one component in a given amount of the total, calculate the amount of that component.
- ❑ Given the parts of one component in a given amount of the total, calculate the percent of that component.

3.7 Energy

- ❑ Distinguish between potential and kinetic energy.
- ❑ Given a description of an energy transformation, identify the energy forms as kinetic or potential.
- ❑ State the law of conservation of energy.

3.8 Units of energy

- ❑ State the SI unit of energy commonly used by chemists.
- ❑ Given a table of specific heats, calculate the amount of heat energy necessary to raise the temperature of a given mass a given number of degrees.
- ❑ Given appropriate experimental data, calculate the specific heat of a substance.

key terms

density	kinetic energy	chemical energy	joule (J)
energy	potential energy	calorie (cal)	specific heat

problems

3.2 Fractions

12. Calculate the quotient of the following divisions.
    a. $(-16) \div (4) =$
    b. $(-4) \div (-16) =$
    c. $-8.2 \div 1.9 =$
    d. $-81.9 \div -5.4 \div -3.9 =$
    e. $-81.9 \div 5.4 \div -3.9 =$
13. Calculate the following.
    a. $(4)(9) \div 3 =$
    b. $(-13.4)(2.1) \div 7.5 =$
    c. $(75.4 \div 4)(3) =$
    d. $(-45.7 \div (1.1))(-2.3) =$
14. Show that the two fractions $\frac{2}{17}$ and $\frac{6}{51}$ are equal.
15. Show that the two fractions $\frac{3}{19}$ and $\frac{8}{57}$ are *not* equal.
16. A pizza pie is cut into eight pieces. If Jack ate three slices, what part of the pie, expressed as a decimal, did Jack eat?
17. Calculate the following.
    a. $\left(\frac{7}{2}\right)\left(\frac{2}{14}\right) =$
    b. $\left(\frac{5}{8}\right)\left(\frac{-3}{4}\right)\left(\frac{2}{9}\right) =$
    c. $\left(\frac{3\text{ ft}}{1\text{ m}}\right)\left(\frac{1\text{ m}}{100\text{ cm}}\right) =$
    d. $\left(\frac{24\text{ hours}}{1\text{ day}}\right)\left(\frac{365\text{ days}}{1\text{ year}}\right) =$
    e. $\left(\frac{-12}{9}\right)\left(\frac{-2}{3}\right)\left(\frac{-8}{3}\right) =$
18. Calculate the following and express your answers in decimal form.
    a. $\frac{2}{3} \div 3 =$
    b. $\frac{\frac{2}{3}}{\frac{1}{3}} =$
    c. $\frac{\frac{2}{3}}{-\frac{3}{4}} =$
    d. $18 \div \frac{2}{3} =$
    e. $\frac{\frac{1}{4}}{\frac{3}{8}} =$
    f. $-11 \div \frac{9}{3} =$

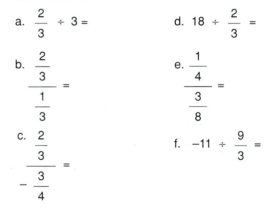

3.3 Unit conversion

19. Using the unit conversion method, carry out the following one-step conversions within the metric system.
    a. 18.9 m to kilometers
    b. 37.5 cm to meters
    c. 0.145 L to milliliters
    d. 452 mL to liters
    e. 645 cm^3 to milliliters
    f. 0.089 kg to grams
    g. 1.9 L to milliliters
    h. 38.4 g to milligrams

20. Using Table 3.1, carry out these one-step conversions between the English and metric systems.
    a. 26.0 mi to kilometers
    b. 3.6 kg to pounds
    c. 5.00 in. to centimeters
    d. 4.00 L to quarts
    e. 1.92 lb to kilograms
    f. 125 g to pounds
    g. 1.00 qt to liters
21. Carry out the following two-step conversions within the metric system.
    a. 0.053 km to centimeters
    b. 38.5 cm to kilometers
    c. 3190. mg to kilograms
    d. 82 mm to centimeters
22. Using Table 3.1, carry out the following two-step conversions between the English and metric systems.
    a. 3.00 ft to meters
    b. 1.42 qt to milliliters
    c. 78 mm to inches
23. Carry out the following multistep conversions.
    a. 0.52 mi to centimeters
    b. 400. mL to gallons
    c. 1.50 oz to grams (16 oz = 1 lb)
24. Do the following conversions.
    a. 1/4 lb of salami to kilograms
    b. 9.3×10^7 mi to kilometers
    c. 4.50 yd to millimeters
    d. 42.0 in. to centimeters
    e. 6.80 mg to kilograms
    f. 750. mL to quarts
25. A run of 100. m corresponds to how many yards?
26. Which represents more gold, 1.9 oz or 39 g?
27. A highway sign tells you that you are 341 km away from home. How many miles away are you?
28. How long is a standard football field (100. yd) in meters?
29. Men's shirt sizes are based on neck circumference measured in inches. Size 16 means the neck band measures 16 in. What is the metric size (in centimeters) corresponding to size 16?
30. Trouser sizes are based on waist and inseam (leg length) measurements in inches. What is the metric size (in centimeters) corresponding to a size of 38 waist, 32 length?
31. The distance from home plate to the pitcher's mound on a standard baseball field is 60 ft 6 in. How far is it in meters?
32. A basketball must weigh at least 20. oz. What is the minimum weight in grams (16 oz = 1 lb)?
33. A swimmer swims 50. lengths of a pool each day. The pool is 30. m long. What distance does she swim each day (a) in meters and (b) in miles?
34. In France, gasoline is sold by the liter. If you wanted 12 gallons (gal) of gasoline, how many liters should you ask for (1 gal = 4 qt)?
35. If gasoline costs $1.14 per gallon, what is the price in cents per liter?
36. A speeder is found to be moving at 82 mi/h. What is his speed in kilometers per hour?
37. If your car gets 26 mi/gal, how many kilometers per liter does it get?
38. On a long trip, you travel 565 mi in 14 h and use 19 gal of gasoline.
    a. What is your average speed in miles per hour?
    b. What is your gas mileage in miles per gallon?
39. On another trip you travel 308 km in 3.5 h and use 32 L of gasoline.
    a. What is your average speed in kilometers per hour?
    b. What is your gas mileage in kilometers per liter?
40. Reconsider the trip described in Problem 38.
    a. What is the average speed in kilometers per hour?
    b. What is the gas mileage in kilometers per liter?
41. Reconsider the trip described in Problem 39.
    a. What is the average speed in miles per hour?
    b. What is the gas mileage in miles per gallon?
42. A person's body measurements are weight = 152 lb, height = 5 ft 8 in., chest = 39 in., and waist = 34 in. Convert the weight measurement to kilograms and the length measurements to centimeters.
43. The moon is 238,860. mi from the earth.
    a. How far away is this in meters? (Use scientific notation.)
    b. Given that a radio signal travels 3.00×10^8 m/s, how many seconds does it take for the radio signal to travel from the moon to the earth?
44. A fish tank is 30. in. long, 20. in. deep, and 12 in. high.
    a. How many liters of water are required to fill it?
    b. How many gallons are needed?

3.4 Density

45. Calculate the density of a liquid, 31.90 mL of which weighs 22.38 g.
46. A piece of metal weighs 59.24 g and occupies a volume of 6.64 mL.
    a. What is the metal's density?
    b. Consult Table 3.2 and identify the metal.
47. Use Table 3.2 to calculate the mass in grams of each of the following samples.
    a. 35.0 mL of gasoline
    b. 989.0 mL of vinegar
    c. 18.0 mL of olive oil

48. Use Table 3.2 to calculate the volume of each of the following samples.
    a. 65 g of alcohol
    b. 98.2 g of gold
    c. 454 g of mercury
49. How much does 1 pt of milk weigh in pounds (see Table 3.2; 1 qt = 2 pt)?
50. Which liquid sample will occupy the greater volume, 100. g of water or 100. g of carbon tetrachloride? Show the reasoning you use to find your answer.
51. Consult Table 3.2 and calculate the mass in pounds of 1.0 qt of the following.
    a. Blood
    b. Gasoline
52. You are given three liquid samples, *A, B,* and *C,* and told that one is water, one is alcohol, and one is ether. Each sample is 10.00 mL. The masses of the samples are *A,* 10 g; *B,* 7 g; and *C,* 8 g. Identify *A, B,* and *C.*
53. A cube of lead measures 2.50 cm on each edge and has a mass of 176 g. Calculate the density of lead.
54. A flask has a mass of 54.12 g when empty and 180.27 g when completely filled with water.
    a. Using the density of water at 20°C (1.00 g/mL), calculate the volume of the flask.
    b. When the same flask is filled with an unknown liquid, the total mass is 142.48 g. What is the density of the unknown liquid?
    c. Identify the unknown liquid by consulting Table 3.2.
55. What is the mass of air in a room measuring 10. m × 5.0 m × 3.0 m?
56. A shiny, yellow stone of volume 7.0 mL has a mass of 56.0 g. Could this stone be gold? Explain.

3.5 Algebraic manipulations

57. Solve for C in $A = \dfrac{B}{C}$.
58. Solve for x in $3x + 4 = 16$.
59. Solve for C in $A = \dfrac{B}{C} + 9$.
60. Solve for °F in $°\text{C} = \dfrac{(°\text{F} - 32)}{1.8}$.
61. Solve for x in $\dfrac{2x - 10}{8} = 15$.
62. Solve for n in $PV = nRT$.
63. Solve for x in $0.40x + 3 = 11$.
64. Solve for x in $(6.5)(2x) = 39$.
65. Find the value of H, including its unit, in $H = (32\text{ g})(28°\text{C})\dfrac{(5.0\text{ J})}{\text{g°C}}$.
66. Solve for Δt, including its unit, in $(50.\text{g})(1.0\dfrac{\text{J}}{\text{g}}) = \dfrac{4.18\text{ J}}{\text{g°C}}(5.0\text{ g})(\Delta t)$.

3.6 Percentage calculations

67. Calculate the percentage of nitrogen in a sample of air in which 100. oz of air contains 76 oz of nitrogen.
68. A sample of ore was found to contain 1.8% Au. How many pounds of gold are present in 573 lb of ore?
69. What is the percentage of copper in 137 lb of copper ore containing 113 lb of copper?
70. What is the percentage of alcohol in a beer in which 12 oz of beer contains 0.39 oz of alcohol?
71. Calculate the percentage of sugar in a liquid mixture containing 6.0 g of sugar and 30.0 g of water.
72. Convert 34.9% to a decimal number.
73. Which number is larger, $\dfrac{3}{8}$ or 28.9%?

3.7 Energy

74. What name do we give to the ability of matter to do work?
75. What are the two major classifications of energy?
76. For each of the following, state whether the energy described is potential or kinetic.
    a. A book is poised at the edge of a table.
    b. The book is falling.
    c. You are walking.
    d. Food.
    e. A stretched rubber band.
77. a. Give an example of potential energy as a result of position.
    b. Give an example of potential energy as a result of composition.
    c. Can a substance have both types of potential energy, position and composition? Explain.
78. Give two examples of potential energy changing into kinetic energy.
79. State the law of conservation of energy.
80. What is the source of energy for humans?
81. What happens if humans take in more energy than they use up?

3.8 Units of energy

82. What does the amount of heat energy within a sample depend on?
83. How many joules are there in 43.5 kJ?
84. What is 1,032 J expressed in kilojoules?
85. How much heat must be applied to raise the temperature of 84 g of olive oil 75°C (see Table 3.4)?
86. How much heat must be removed to cool 76 g of iron from 95°C to 16°C (see Table 3.4)?
87. 6.28 kJ is applied to a sample of water and the temperature rises from 22.3°C to 28.9°C. What is the mass of the water sample?
88. When a piece of graphite weighing 3.00 g absorbs 62.8 J, the temperature changes from 25.0°C to 54.0°C. What is the specific heat of graphite?

89. When a 75.4-g sample of Ni loses 1.43 kJ, its temperature falls from 88.5°C to 45.8°C. What is the specific heat of Ni?

90. Consult Table 3.4 and decide which sample below is the hottest (highest temperature). Both materials were originally at room temperature.
    a. 10. g of water absorbs 314 J.
    b. 10. g of iron absorbs 314 J.

91. Consult Table 3.4 and decide which sample is the coldest. Both materials were originally at room temperature.
    a. 2 g of olive oil loses 84 J.
    b. 5 g of aluminum loses 126 J.

92. A cup of yogurt contains 150. Calories.
    a. How many calories is this?
    b. If this energy were used to heat 100. lb of water originally at 98.6° F, what would be the final temperature of the water?

Additional problems

93. Carry out the following two-step conversions.
    a. 0.16 km to centimeters
    b. 57 mm to centimeters
    c. 3,340. mg to kilograms
    d. 3.00 ft to meters
    e. 764 mg to pounds
    f. 0.34 qt to milliliters

94. Carry out the following multistep conversions.
    a. 3.00 mi to centimeters
    b. 570. mL to gallons
    c. 2.1 oz to grams (16 oz = 1 lb)

95. Diamonds are measured in carats and 1 carat = 0.200 g. The density of diamond is 3.51 g/cm^3. What is the volume of a 4.0-carat diamond?

96. Describe how you would measure the density of an irregularly shaped object using a balance and a graduated cylinder.

97. A hospital patient has an oral temperature of 39.5°C and weighs 190. lb. He is to receive a drug, the dosage of which is 50. mg/kg of body weight. The drug is dissolved in water (25 mg/mL).
    a. What is his temperature in degrees Fahrenheit?
    b. What is his metric weight?
    c. What mass of pure drug should he receive?
    d. How many milliliters of the drug solution should he drink?

98. The total cylinder volume of cars used to be expressed in cubic inches. Compute the total volume in liters of a 475-$in.^3$ engine.

99. A light year is the distance light travels in 1 year. The speed of light is 3.00×10^{10} cm/s. Compute the distance in kilometers and in miles that light travels (a) in 1 year and (b) in 1 s.

100. How many atoms are in 0.25 moles of Cu if 1 mole of Cu = 6.02×10^{23} atoms of Cu?

chapter 4

elements and their invisible structures

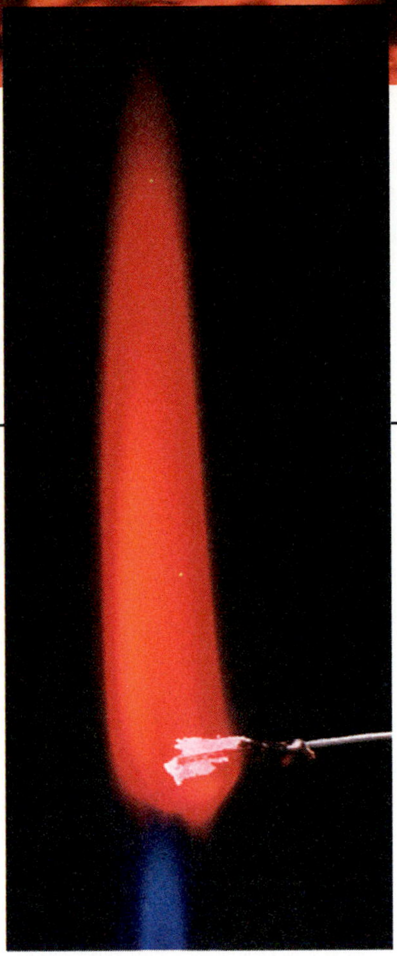

A clue to the invisible structure of elements that will unfold in this chapter can be observed in the characteristics spectra that elements emit when heated in a flame. The red color of this flame identifies strontium (Sr).

4.1 introduction

You have learned of the physical and chemical properties of elements, compounds, and mixtures in Chapter 1. You have also learned how to measure some of those properties and how to apply them to chemical problems in Chapters 2 and 3. In Chapter 4, you will learn that individual atoms are composed of smaller particles—electrons, protons, and neutrons. Further, you will learn that the number of protons in an atom identifies it as an element and that the combined number of protons and neutrons in an atom approximates the mass of the atom. In Chapter 5, you will learn how the detailed electron structure in each element determines the chemical properties of that element. By understanding the detailed structure of each element, you will acquire the ability to predict trends in the chemical and physical properties of elements and how those trends are grouped on the periodic table.

Thus far, the discussion of matter has been in *macroscopic* terms. That is, you have classified and identified matter by examining large (*macro-* means "large") "chunks" of matter that you can see and hold. For a clearer and more predictive understanding of matter, it must be explored in *microscopic* (*micro-* means "small") terms. We should really say *sub*microscopic terms, because the smallest particles of matter cannot be seen even by the most sensitive electron microscopes. In Figure 4.1, you see that modern tunneling electron microscopes enable you to distinguish the elongated spherical shape of each of 28 carbon monoxide molecules that have been arranged in the outline of a person.

As soon as you learn the basic makeup of *atoms,* the tiniest particles of matter, you will be able to understand such environmental, health, and safety topics as radioactivity, nuclear medicine, and nuclear power plants.

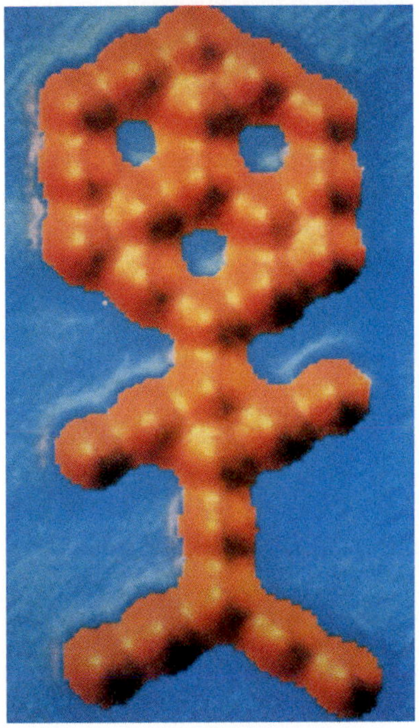

Figure 4.1

This molecular person is made of 28 individual carbon monoxide molecules that have been moved into position on a single crystal platinum surface using a scanning tunneling electron microscope tip. This electron micrograph illustrates the resolution and control available in state-of-the-art electron microscopes.

4.2 atomic theory

In Section 1.9, you learned that all matter is made up of tiny particles called atoms and that atoms may combine in fixed proportions to form molecules. Atoms are the building blocks of elements, and molecules are the building blocks of compounds.

Philosophically, there are only two ways matter could be constructed: (1) It might be *continuous,* that is, divisible into ever smaller parts so that no smallest particle exists, or (2) it might be *discrete,* i.e., made up of characteristic small particles that, if further divided, no longer have the properties of the matter under examination. Experiments indicate that matter is discrete and made up of atoms.

The concept of the atom is an old one. The word was coined by the Greek Democritus, who first suggested the atomic idea in 400 B.C. Scientists and philosophers generally did not believe in atoms until the early nineteenth century. At that time, an English school teacher and experimental chemist, John Dalton, stated in a clear manner the atomic theory of matter, which explained many known scientific laws and experiments and also accounted for the classification of matter into elements, compounds, and mixtures. Some of the key statements of Dalton's theory appear below.

1. All matter is made up of atoms, which are indestructible by ordinary means.
2. All atoms of a given element are identical in their chemical properties.
3. Atoms of different elements have different chemical properties.
4. Atoms of different elements can chemically combine in simple, whole-number ratios to form compounds.

You have used all of these statements in classifying and identifying matter throughout Chapter 1, so let us review and expand our definitions of elements, compounds, and mixtures in terms of their atomic makeup.

An **element** is matter that is composed of one kind of atom.

A **compound** is matter that is composed of different kinds of atoms chemically combined in simple, whole-number ratios.

A **mixture** is matter that is a physical combination of particles of elements or compounds and has a variable composition.

sample exercise 1

How does the atomic definition of an element explain the former definition given in Section 1.9 (that an element is "a pure substance that cannot be broken down into simpler substances by ordinary chemical changes")?

solution

Elements are pure substances, which means that they have a constant composition and unique set of properties because they are collections of just one kind of atom. Elements cannot be broken down because atoms are essentially indestructible.

• • • problem 1

Explain how the atomic definition of a compound relates to the former definition given in Section 1.9.

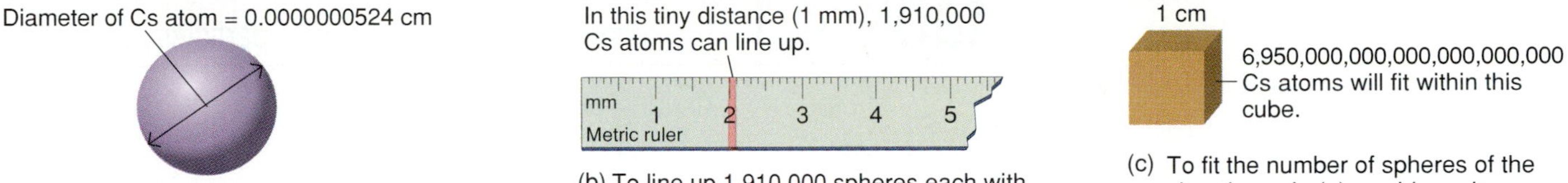

(a) This drawing has been scaled up almost *40 million* times the actual size of the Cs atom.

(b) To line up 1,910,000 spheres each with a diameter of 2 cm (approximately the size shown in [a]) would require 38 km.

(c) To fit the number of spheres of the size shown in (a) would require a cube 380 km on its edge. The volume of the cube (5.5×10^7 km^3) then would be a volume corresponding to the water content of 15 Mediterranean Seas!

Figure 4.2

Although cesium atoms have the largest diameter of any element, they are still very, very small.

4.3 picturing atoms and elements

In the early nineteenth century, John Dalton pictured atoms as tiny spherical particles just as you have seen in Figure 1.22, and for some purposes, this is still a convenient representation for understanding the composition of matter. In later chapters, you will find that this is not a completely accurate picture, but the modifications you will learn later do not affect the properties or classification of macroscopic matter.

It is difficult to picture how very, very small an atom is and thus to realize the extremely large number of atoms present in any sample of an element. Although it is not the heaviest atom, cesium, Cs, has the largest diameter of any of the 109 known elements. One cesium atom has a diameter of only 0.0000000524 cm. This means that between adjacent millimeter-marking lines on a meterstick, we could line up 1,910,000 Cs atoms. A 1-cm cube of cesium metal (smaller than a sugar cube) would contain approximately 6,950,000,000,000,000,000,000 Cs atoms. Once again, you can see why we need to be able to express numbers in scientific notation. Discussing sizes and numbers of atoms involves very small and very large numbers (Figure 4.2).

••• problem 2

Refer to the preceding paragraph. Express in scientific notation the diameter of a cesium atom, the number of cesium atoms along a 1-mm line, and the number of cesium atoms in a 1-cm cube.

Of course, because atoms are so very small in size, the mass of one atom is also very small. One Cs atom has a mass of only 2.21×10^{-22} g, so it would take an enormous number of atoms to have the mass equivalent to a person. If you, for example, weigh 130 lb, then 2.67×10^{26} Cs atoms correspond to your mass.

4.4 inside the atom

Atoms are not really hard little balls as shown in Figure 1.22. Actually, atoms are mostly empty space because of the arrangement of the *subatomic particles* of which they are made. That is, by the end of the nineteenth century, it had become clear that atoms were not the smallest particles. Rather, atoms can be subdivided into smaller subatomic particles. Today, atomic physicists have identified hundreds of indescribably small and often short-lived particles. Luckily, to understand chemistry, you need to be aware of only the three major types of subatomic particles: protons, electrons, and neutrons. Atoms of different elements differ in the *numbers* of these particles they contain. The **protons** and **neutrons** are located in a tiny region at the center of the atom called the **nucleus,** while the less massive **electrons** are distributed in the space around the nucleus as shown in Figure 4.3. The electrons are in constant motion, moving in paths around the nucleus. Since the nucleus is extremely small at the center, the entire atom is seen to be mostly empty space.

table 4.1 characteristics of subatomic particles

Particle	Symbol	Relative charge	Relative mass
Proton	p^+	1^+	1
Electron	e^-	1^-	$0 \left(\frac{1}{1,840}\right)$
Neutron	${}^1_0n$	0	1

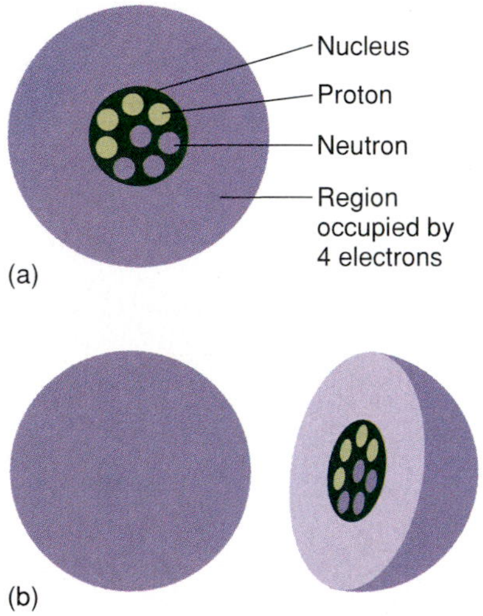

Figure 4.3
(a) A cross section of an atom shows the protons and neutrons clustered in the nucleus which is proportionally much smaller than shown here. The electrons are small and are in constant motion about the nucleus in the region shown. (b) A three-dimensional representation of the atom in which the region occupied by the electrons can be thought of as a cloud because the electrons are moving so fast. Cutting the cloud in half reveals the nucleus in the center.

Characteristics of subatomic particles. The important characteristics of protons, electrons, and neutrons are their relative masses and electrical charges. Protons and neutrons have approximately the same mass. Electrons are much lighter; the electron mass is only 1/1,840 the mass of a proton. To express these mass relationships conveniently, we assign the proton a mass of 1, then the neutron mass must be 1, and the electron mass must be 1/1,840. Because the electron mass is so much less than the mass of a proton or neutron, we may generally ignore it. Electrons contribute practically no mass to an atom. Therefore, you assign the electron a relative mass of zero. In Section 4.9, you will assign units to these relative masses. A useful summary of the masses of protons, electrons, and neutrons is given in Table 4.1.

Arrangement of subatomic particles. The arrangement of protons, electrons, and neutrons is shown in Figure 4.3. Notice that the mass of an atom is concentrated in the center because the protons and neutrons are clustered there in the nucleus. The nucleus is described as being dense or compact because most of the mass of the atom is contained in an extremely small volume. The nucleus is much smaller than depicted in Figure 4.3; if drawn to scale in this figure, the nucleus would be too small to show the individual protons and neutrons. In addition, the nucleus bears a positive charge; because each proton carries a single positive charge and each neutron carries no charge, the sum of the positive charges on the protons equals the charge on the nucleus.

Moving around the dense, positive nucleus are "weightless" electrons, each of which carries a single negative charge. In Chapter 5, you will learn more details about the arrangement of electrons in each element and how they influence the chemical properties of elements. For now, it is sufficient to know the following.

1. Electrons are outside the nucleus.
2. Electrons are constantly moving.
3. Because of the space between electrons and between the electrons and the nucleus, atoms are mostly empty space.
4. Because electrons bear negative charges and are constantly moving, they form a negatively charged cloud with fuzzy borders.

When you learn more about the number of protons and electrons in each element (Sections 4.5 and 4.6), you will understand why atoms are electrically neutral, that is, they have no net charge. This is so because, for every electron with a 1− charge, there is a proton with a 1+ charge. Just as adding signed numbers of equal magnitude and opposite sign results in zero (Appendix C), equal numbers of positive protons and negative electrons result in zero charge (electric neutrality) (Figure 4.4). A summary of the charge characteristics of protons, electrons, and neutrons is also given in Table 4.1.

As a means of visualizing an atom to scale, look out your window at the clouds. Pick out a nice round cloud. Imagine that the cloud is negatively charged and that buried in the center of the cloud is a tiny positively charged pebble. The size of the

Figure 4.4

Each atom has no net charge (electrically neutral) because the sum of the positive charges on the protons and the negative charges on the electrons add up to zero.

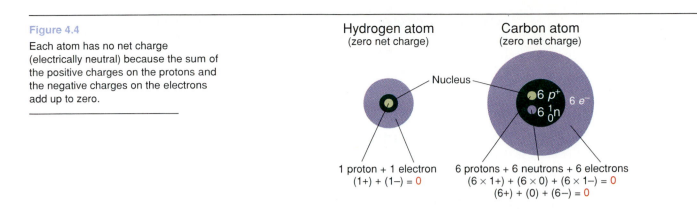

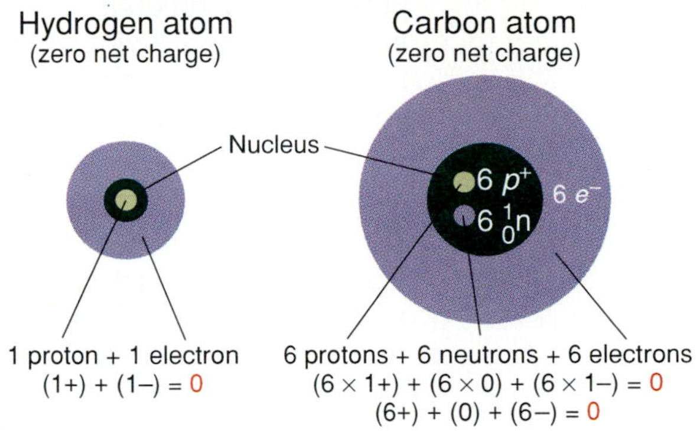

CHEMLAB

Laboratory exercise 4: Rutherford experiment

Rutherford used alpha particles (helium nuclei, which consist of two protons and two neutrons) to bombard a thin piece of gold (Au) foil. Even though the foil looked "solid," Rutherford found that most of the alpha particles passed straight through the foil. From this observation, he concluded that the atoms composing the gold foil were mostly empty space, since there was little mass in the way of the alpha particles to deflect them. By measuring the fraction of particles that were deflected and the angles into which they were deflected, however, he was able to calculate that the diameter of the gold nucleus was about 10^{-12} cm and that it had a positive charge.

To help you understand Rutherford's experiment and the procedure used to determine the size of the Au atom, you may perform an experiment based on the same principles in which you determine the diameter of a marble. In this experiment, you will not measure the marble directly, but indirectly in a scattering experiment similar in concept to that Rutherford used to measure the size of a Au atom.

The apparatus you use to measure the size of a marble is shown in Figure 4.5. It consists of a simple ramp with nine identical marbles placed within a fixed target region below. The width of the ramp and the width of the target region are the same, but they can be any convenient distance, *d*, that you make.

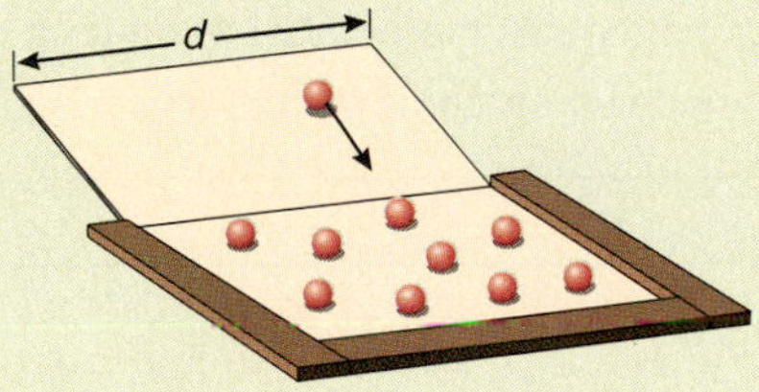

Figure 4.5

This simple apparatus is used to measure the diameter of a marble in a scattering experiment similar in concept to the one Rutherford used to measure the size of a gold atom.

To determine the diameter of one target marble, randomly position another identical marble on the ramp and let it roll down, watching to see if it collides with one of the nine target marbles. Repeat this procedure several hundred times. Record the total number of collisions or hits, *H*, the total number of tries, *T*, and the width, *d*, you measured for the ramp and target region.

The diameter, *D*, of the marble can be calculated according to the formula below, where *N* is the number of target molecules you used. (You may find this formula derived in the lab manual accompanying this text.)

$$D = \frac{Hd}{2NT}$$

Notice that *D*, the diameter of the marble, has the same units as *d*, the width of the target region, since *H*, *N*, and *T* are unitless. So, if you measure *d* in centimeters, the diameter of the marble will also be given in centimeters.

Questions:

1. If you had nine target marbles with a target width measuring 30.5 cm and observed 295 hits out of 500 tries, what is the diameter of each of your marbles based on the results of your scattering experiment?
2. Because the marbles are macroscopic, they may be measured directly. Place all ten of your marbles (nine target marbles and one bombarding marble) in the groove of an open book or along a ruler edge. Have the marbles touching with no space between them. Using a ruler, the length of the ten marbles (that is, the length of ten diameters) is measured to be 10.1 cm. What is the average diameter of each marble when measured directly?
3. How does the diameter of each marble as measured in your scattering experiment compare with the diameter measured as the average value of ten marbles?
4. What conclusion might you draw about the ability of Rutherford to indirectly measure the size of a gold nucleus with a scattering experiment?

pebble is proportional to the size of the nucleus in an atom, and the cloud represents the region in which electrons can be moving around the nucleus. The cloud helps you to develop a mental model of an atom in which you can understand the extent of empty space within the atom.

4.5 atomic number

The subatomic structure of an atom determines a new property that can be used to identify the element that it forms. That property is called the **atomic number** and is equal to the number of protons in the atom.

Look at the periodic table inside the front cover. The number above each atomic symbol is the atomic number of that element, that is, the number of protons in atoms of that element. Any atom with six protons must be carbon. An atom with just one proton must be hydrogen (Figure 4.4). For an element to be gold, there must be 79 protons in the nucleus of each atom.

problem 3

a. What are the atomic numbers of Mg, S, and Ag?
b. How many protons are there in Mg, S, and Ag?

problem 4

Identify the elements for which the following information is given.

a. Atomic number 17
b. 20 protons

As stated earlier, atoms are electrically neutral because, for every proton (with a charge of 1+), there is an electron (with a charge of 1–). Therefore, the atomic number also tells you the number of electrons in an atom. Carbon atoms with atomic number 6 have six protons and six electrons. Hydrogen atoms with atomic number 1 have just one proton and one electron.

sample exercise 2

A neutral atom is known to have 14 electrons.

a. How many protons does it have?
b. What is its atomic number?
c. What is the name of the element with this atomic number?

solution

a. It has 14 protons because the number of protons equals the number of electrons.
b. This is also 14 because atomic number equals the number of protons.
c. Silicon; only the element silicon is made up of atoms with atomic number 14.

problem 5

How many electrons are there in atoms of Al, K, and Cu?

problem 6

a. What are the atomic numbers of elements with 3, 18, and 53 electrons in their neutral atoms?
b. What are the names of these atoms?

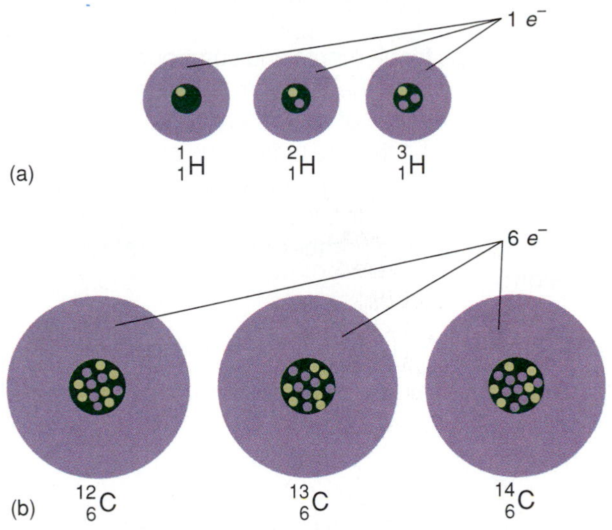

Figure 4.6
Isotopes of an element differ in the number of neutrons they possess. (a) Three isotopes of hydrogen with zero, one, and two neutrons (lavender dot(s)), respectively. In each case, the atom has one proton (light green dot); this makes it hydrogen. (b) Three isotopes of carbon with six, seven, and eight neutrons (lavender dots), respectively. In each case, the atom has six protons (light green dots); this makes it carbon.

4.6 isotopes

Whereas the number of protons (atomic number) identifies atoms of a particular element—that is, the number of protons is always the same in the same element—the number of neutrons may be different in the same element. For example, *all* carbon atoms have six protons and most carbon atoms have six neutrons. But about 1% of carbon atoms contain seven neutrons, and there are some carbon atoms with eight neutrons. The six protons in the atom identify that atom as the element carbon; the six, seven, or eight neutrons identify the isotope of carbon. Atoms with the same number of protons but different numbers of neutrons are called **isotopes**

There are three isotopes of hydrogen (Figure 4.6). All isotopes of hydrogen have one proton in the nucleus. Each nucleus of the most abundant isotope (protium) has no neutrons. Each nucleus of the other two isotopes, deuterium and tritium, have one and two neutrons, respectively. Tritium is radioactive, that is, its nucleus is unstable and gives off a particle spontaneously.

The word *isotope* may frequently be used for the word *atom.* Notice: "All *atoms* of nitrogen contain seven protons. All *isotopes* of nitrogen contain seven protons." Keep this idea in mind as further discussions of isotopes arise.

Since all isotopes of the same element have the same number of protons, their atoms also have the same number of electrons. In Section 4.1, we mentioned that, in Chapter 5, you will learn that the number of electrons determines the chemical properties of atoms. Since all isotopes have the same number of electrons, they will also have the same chemical properties.

The property of isotopes that will vary is their mass number, which will be discussed in the next section. Remember that essentially all of the mass of an atom is found in the nucleus. Since each proton and each neutron has the same mass, the different number of neutrons in the nucleus of each isotope will result in different total masses for the atoms of each isotope.

••• problem 7

How many protons are in the nucleus of all isotopes of magnesium?

4.7 mass number

To establish mass relationships for elements and compounds in the following sections, you need to predict the total mass for the atoms of any element. The **mass number** for any atom is the sum of the number of protons and the number of neutrons in the nucleus. Remember from Table 4.1 that the mass of an atom is almost totally accounted for by its protons and neutrons because the mass of an electron is effectively zero.

If you know the atomic number and the mass number for any isotope, you can use the definitions to figure out the number of protons, electrons, and neutrons in that atom. For any atom,

Atomic number = Number of protons = Number of electrons

Mass number = Number of protons + Number of neutrons

You can substitute atomic number for number of protons. Then,

Mass number = Atomic number + Number of neutrons

Now subtract atomic number from each side of the equation, and you get

Mass number − Atomic number = Number of neutrons

sample exercise 3

How many protons, electrons, and neutrons are there in isotopes of sodium that have atomic number 11 and mass number 23?

solution

Atomic number = Number of protons = Number of electrons

11 = Number of protons = Number of electrons

Mass number − Atomic number = Number of neutrons

23 − 11 = Number of neutrons

12 = Number of neutrons

••• problem 8

How many protons, neutrons, and electrons are there in atoms which have the following?

a. Atomic number 8, mass number 17
b. Atomic number 15, mass number 31
c. Atomic number 27, mass number 60

Atomic number and mass number data are conveniently written together with the element symbol. For example, the symbol that represents atoms of carbon with six protons and six neutrons is ${}^{12}_{6}C$. The atomic number is written as a *sub*script to the left of the symbol for carbon. The mass number is written as a *super*script to the left of the symbol. This method can be represented in general by the following illustration, in which Sy stands for the symbol of any element.

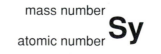

$${}^{\text{mass number}}_{\text{atomic number}}\mathbf{Sy}$$

The symbol ${}^{12}_{6}C$ can also be read "carbon-12." "Carbon-13" is ${}^{13}_{6}C$. Remember, all atoms of carbon have the same atomic number. But different isotopes of carbon will have different mass numbers because the number of neutrons varies. You can also define *isotopes* as atoms with the same atomic number but different mass numbers.

Often a particular isotope of an element is indicated only by its name or symbol and mass number, for example, carbon-12, ${}^{12}C$, or uranium-238, ${}^{238}U$. It is not necessary to include the atomic number since the name or symbol is sufficient to identify the element with a unique atomic number.

The basic arrangement of the subatomic particles within the atom has been illustrated with diagrams in Figure 4.4 showing the *numbers of protons and neutrons in a small circle* representing the nucleus, surrounded by a larger circle within which reside a specific and characteristic number of electrons. Now you are able to construct such a diagram for any chemical symbol. For example, for ${}^{65}_{30}Zn$, the diagram would be that shown in Figure 4.7. You recall, however, that these diagrams are representational, that the nucleus is not drawn to scale, and that the entire atom is mostly empty space.

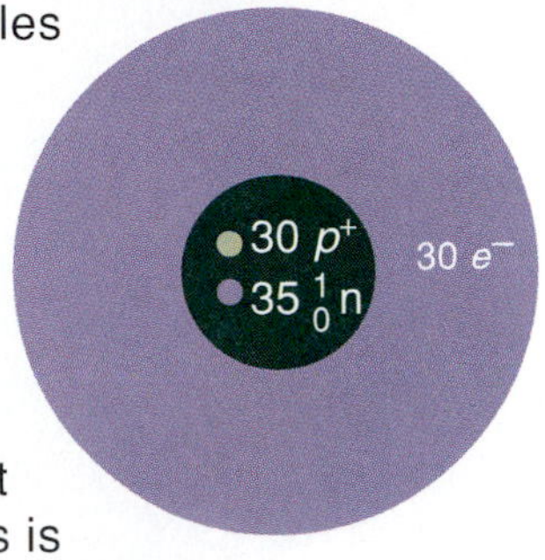

Figure 4.7
Representation of the subatomic structure of the zinc atom showing the number of protons and neutrons in the nucleus and the specific and characteristic number of electrons moving around the nucleus.

sample exercise 4

Determine the number of protons (p^+), electrons (e^-), and neutrons (${}^{1}_{0}n$) in the atoms indicated. Also, identify the element represented by each symbol, X.

a. ${}^{65}_{30}X$ b. ${}^{27}_{13}X$ c. ${}^{28}_{13}X$ d. ${}^{28}_{14}X$ e. ${}^{64}_{29}X$

solution

${}^{\text{mass number}}_{\text{atomic number}}Sy$	Atomic number = $p^+ = e^-$	Mass number − Atomic number = ${}^{1}_{0}n$
a. ${}^{65}_{30}X$	$30 = p^+ = e^-$	$65 - 30 = 35\ {}^{1}_{0}n$
b. ${}^{27}_{13}X$	$13 = p^+ = e^-$	$27 - 13 = 14\ {}^{1}_{0}n$
c. ${}^{28}_{13}X$	$13 = p^+ = e^-$	$28 - 13 = 15\ {}^{1}_{0}n$
d. ${}^{28}_{14}X$	$14 = p^+ = e^-$	$28 - 14 = 14\ {}^{1}_{0}n$
e. ${}^{64}_{29}X$	$29 = p^+ = e^-$	$64 - 29 = 35\ {}^{1}_{0}n$

The atomic number identifies the atoms. Consult the periodic table to see that the atoms are Zn, Al, Al, Si, and Cu in the order they are given.

sample exercise 5

Refer to Sample Exercise 4. Which of the atoms are isotopes of the same element?

solution

${}^{27}_{13}X$ and ${}^{28}_{13}X$ are isotopes of Al. They are atoms with the same number of protons and different numbers of neutrons. They are atoms with the same atomic number and different mass numbers.

sample exercise 6

Show a pictorial diagram of the two isotopes of aluminum ^{27}Al and ^{28}Al.

solution

Note that the only difference between these two isotopes is the number of neutrons in the nucleus. The symbols ^{27}Al and ^{28}Al are often read as aluminum-27 (Al-27) and aluminum-28 (Al-28), which are just alternate ways of reading the atomic symbols.

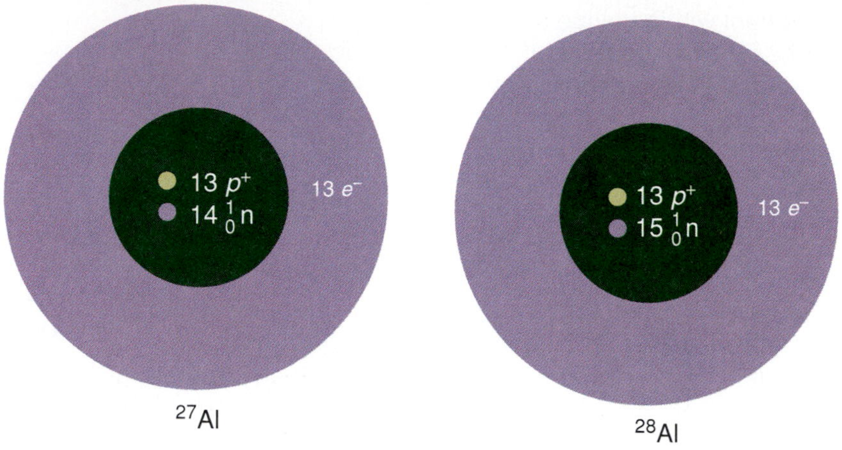

problem 9

Consider the atoms $^{32}_{16}X$, $^{32}_{15}X$, $^{127}_{53}X$, $^{31}_{15}X$, and $^{130}_{53}X$.

a. Indicate the numbers of protons, neutrons, and electrons in each.
b. Identify each.
c. Show a pictorial diagram of each.
d. Pick out any sets of isotopes.

4.8 ions: atoms that have lost or gained electrons

As we discussed in Sections 4.1 and 4.6 and will examine in greater detail in Chapter 5, the chemical properties of elements depend on the number and arrangement of electrons around the nucleus. In the formation of new compounds, atoms of each element often gain or lose electrons. When this happens, charged particles called **ions** result because the number of protons and the number of electrons are no longer equal. It is important to realize that *atoms* and *ions* of an element are very different in physical, chemical, and biological properties. For example, sodium atoms compose the metal sodium, which is so reactive that it must be protected from moist air as shown in Figure 1.14. Sodium ions, on the other hand, are the form sodium takes in sodium chloride or common table salt, which is also shown in Figure 1.14. In Chapter 1, you saw how different the properties of sodium chloride, a white, crystalline solid, were from metallic sodium.

Understanding the subatomic change from sodium atoms to sodium ions will help you understand why the properties of sodium ions differ from those of sodium atoms. In chemical reactions, sodium atoms (atomic number 11) tend to lose one electron and become sodium ions, Na^{+}. Sodium ions have a 1+ charge because the algebraic sum of 11 protons (11+) and 10 electrons (10–) results in an overall 1+. Ions with a *positive* charge are called **cations**. Cations are formed when a

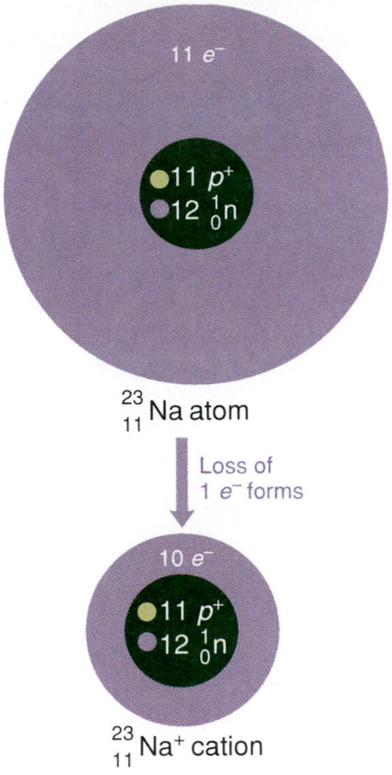

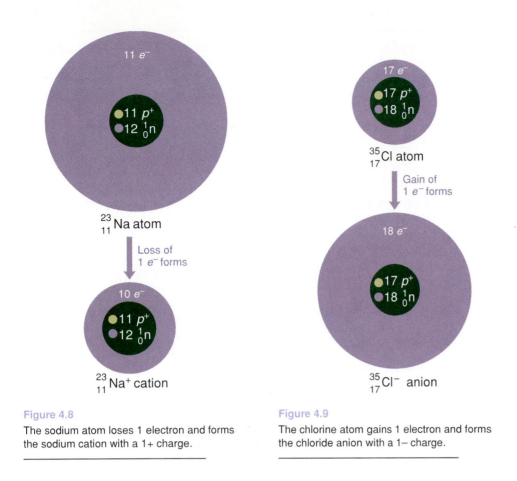

Figure 4.8
The sodium atom loses 1 electron and forms the sodium cation with a 1+ charge.

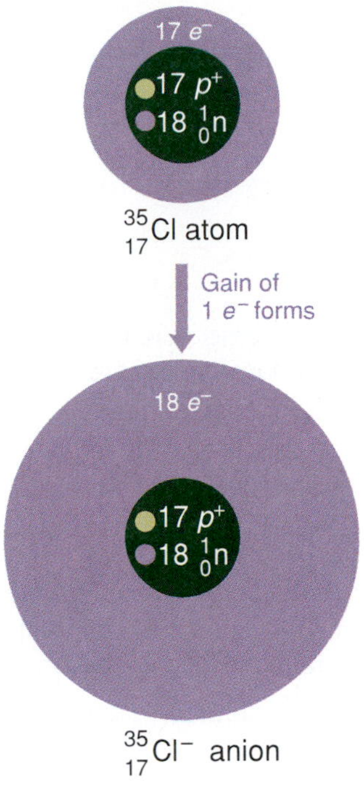

Figure 4.9
The chlorine atom gains 1 electron and forms the chloride anion with a 1− charge.

neutral atom *loses* electrons. Remember, losing electrons means losing negative charge; therefore, the ion is positive because the number of protons in the ion is greater than the number of electrons. Pictorially, sodium atoms become cations as shown in Figure 4.8.

Metals like sodium tend to lose one or more electrons in many chemical processes. Nonmetals like chlorine, however, tend to gain one or more electrons. Chlorine atoms (atomic number 17), in particular, often gain one electron and become chloride ions, Cl^-. Pictorially, chloride ions are represented in Figure 4.9.

Chlorine atoms are electrically neutral because the 17+ charge from 17 protons is canceled by the 17− charge from 17 electrons. Chloride ions have a net charge of 1− because of the one extra electron in the chloride ion ((17 +) + (18 −) = 1−). *The properties of chlorine atoms and chloride ions are very different.* Once again, if you refer to Figure 1.14, you will recall that chlorine atoms compose the diatomic gas, Cl_2, which is a yellow-green, reactive gas. Chloride ions are negatively charged anions chlorine forms when it chemically combines with sodium to form sodium chloride (table salt). Consistent with the differences between sodium ions and sodium atoms, chloride ions and chlorine atoms have very different properties because of their characteristic number of electrons.

Ions with a *negative* charge are called **anions.** Anions are formed when a neutral atom *gains* electrons. Remember, gaining electrons means gaining negative charge.

The charge on an ion is written as a superscript to the right of the symbol, with the magnitude first and the sign following, for example, Cl^{1-}. When the magnitude is 1, however, the 1 is usually omitted, so that a chloride ion is written as simply Cl^-.

sample exercise 7

Calculate the charges on the following ions and indicate whether the ions are cations or anions.

Ion	*Protons*	*Electrons*
X	35	36
Y	3	2
Z	20	18

solution

To find the charge, take the algebraic sum of the number of protons (assigned a plus sign) and the number of electrons (assigned a minus sign).

X:	Charge = (35+) + (36–) = 1–	Anion (negative charge)
Y:	Charge = (3+) + (2–) = 1+	Cation (positive charge)
Z:	Charge = (20+) + (18–) = 2+	Cation (positive charge)

••• problem 10

Calculate the charges on ions with the following numbers of protons and electrons and indicate whether the ions are cations or anions.

Ion	*Protons*	*Electrons*
Q	21	18
R	12	10
T	9	10

Let us review the atomic symbolism we have seen so far and expand it to include the charge of an ion.

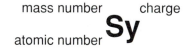

$$^{\text{mass number}}_{\text{atomic number}}\mathbf{Sy}^{\text{charge}}$$

Atomic number = Number of protons

Mass number = Number of protons + Number of neutrons

Charge = Algebraic sum of the protons, assigned a plus sign, and the electrons, assigned a minus sign

For neutral atoms in which the number of p^+ = the number of e^- so that the charge = 0, the 0 is usually omitted. All atomic symbols in which no charge is explicitly given are assumed to represent neutral atoms.

sample exercise 8

Calculate the number of protons, neutrons, and electrons in the ions $^{24}_{12}Mg^{2+}$ and $^{14}_{7}N^{3-}$.

solution

Refer to the immediately preceding section, which reviews this symbolism.

	$^{24}_{12}Mg^{2+}$	$^{14}_{7}N^{3-}$
Number of p^+ = Atomic number	12	7
Number of $^{1}_{0}n$ = Mass number – Atomic number	12	7

Saying that the charge is the algebraic sum of the number of protons, assigned a + (because they are positively charged), and the number of electrons, assigned a – (because they are negatively charged), can be most conveniently expressed as

$$\text{Charge} = \text{Number of protons} - \text{Number of electrons}$$

Subtracting the number of protons from each side:

$$\text{Charge} - \text{Number of protons} = -\text{Number of electrons}$$

Therefore, for Mg^{2+},

$$(2+) - 12 = -10$$
$$10 \text{ electrons in } Mg^{2+}$$

and for N^{3-},

$$(3-) - 7 = -10$$
$$10 \text{ electrons in } N^{3-}$$

4.9 relative atomic mass

The most conveniently measured property of matter is its mass. Every chemical laboratory has a balance with which the mass of a sample may be determined. In this section, we will examine the relative masses of atoms of one element to those of another element.

A single atom, by itself, is much too tiny to be weighed. Fortunately, scientists realized a long time ago that it is not necessary to weigh individual atoms. What you need to know are the masses of atoms in one element *relative* to the masses of atoms in other elements. Knowing which atoms are more massive than others and by what factor enables you to establish all needed mass relationships for elements and compounds.

In Section 2.1, we discussed the idea that all measurements are made relative to some standard. (Refer also to Section 2.3, which discussed the various standards in the metric system.) The standard chosen for the atomic mass scale (often called the atomic weight scale) is the most abundant isotope of carbon, ${}^{12}_{6}C$ (carbon-12). The most recent revision of the atomic mass scale occurred in 1961 when carbon-12 was assigned a mass of 12 **atomic mass units** (amu). In comparison with that standard, experiments determined how much more massive atoms of other elements were. From these experiments, the atomic mass scale for all 109 known elements was established.

For example, it was found that helium atoms are 1/3 as heavy as carbon-12 atoms, and titanium atoms are 4 times as heavy. This means that the mass of a helium atom is

$$\frac{1}{3} \times 12 \text{ amu} = 4 \text{ amu}$$

and the mass of a titanium atom is

$$4 \times 12 \text{ amu} = 48 \text{ amu}$$

These multiplying factors can be found for all atoms, and thus relative atomic masses can be assigned to all atoms.

Using the assigned mass of carbon-12 as the standard, the protons and neutrons have a mass of approximately 1 amu. Because this is true, the mass number of an atom is approximately equal to its atomic mass in atomic mass units. The approximation is usually good to at least three significant figures.

sample exercise 9

Atoms of elements X and Y are, respectively, 1/2 and 6.25 times as heavy as carbon-12 atoms. What are the atomic masses of these atoms?

solution

Multiply the relative mass factor times 12 amu, the mass of ${}^{12}_{6}C$.
For X atoms, $1/2 \times 12$ amu = 6 amu. For Y atoms, 6.25×12 amu = 75 amu.

problem 11

Atoms of element Q are $1\frac{1}{3}$ times as heavy as ^{12}C atoms. What is the atomic mass of these atoms?

sample exercise 10

What are the approximate atomic masses of the isotopes ${}^{1}_{1}H$, ${}^{19}_{9}F$, and ${}^{11}_{5}B$?

solution

$$\text{Mass number} = \text{Protons} + \text{Neutrons}$$

Because each proton and neutron weighs approximately 1 amu, the mass number also gives the approximate mass of the atom in atomic mass units. Therefore,

${}^{1}_{1}H$	1 amu
${}^{19}_{9}F$	19 amu
${}^{11}_{5}B$	11 amu

4.10 average atomic mass (weight)

In Section 4.6, you learned that an element may be a mixture of two or more isotopes. Since each isotope has a different mass number, the average mass of any sample will have a value between the values of the individual isotopes.

Look at the square devoted to the element carbon in the periodic table shown in Figure 4.10. Above the symbol, the number 6 is the atomic number. Below the symbol is the **average atomic mass,** 12.01. The average atomic mass (often referred to as the *average atomic weight*)[1] reflects the fact that carbon is made up of atoms with different masses, i.e., isotopes with mass 12, 13, and 14 atomic mass units. In the case of carbon, most carbon atoms are carbon-12, mass = 12 amu. But about 1% of carbon atoms are carbon-13, mass = 13 amu, and there are a very small number of carbon-14 atoms, mass = 14 amu. The weighted average turns out to be 12.01 amu. It takes into account the amount of each isotope as well as the mass of each isotope.

Notice the other atomic masses (*weights*) in the periodic table; they are shown as the number below each element symbol. They are not whole numbers because they are weighted averages of the masses of all isotopes of that element.

How do you determine an average atomic mass? To do this you need to know (1) which isotopes exist for a given element and (2) the percentage abundance of each isotope.

For example, the element magnesium is made up of three isotopes: 78.7% ^{24}Mg, 10.1% ^{25}Mg, and 11.2% ^{26}Mg. The mass contributed by each isotope is its fractional

1. The use of the term *atomic weight* is gradually being replaced by the correct term, *atomic mass*. Usage of the term *atomic weight* has been so widespread that you should recognize it in everyday language.

Representative Elements (*s* Series)

Representative Elements (*p* Series)

Key

1	Atomic Number
Hydrogen	Name
H	Symbol
1.0079	Atomic Weight

Transition Metals (*d* Series of Transition Elements)

Period	IA	IIA	IIIB	IVB	VB	VIB	VIIB	VIIIB	VIIIB	VIIIB	IB	IIB	IIIA	IVA	VA	VIA	VIIA	VIIIA
1	1 Hydrogen H 1.0079																	2 Helium He 4.0026
2	3 Lithium Li 6.941	4 Beryllium Be 9.0122											5 Boron B 10.811	6 Carbon C 12.0112	7 Nitrogen N 14.0067	8 Oxygen O 15.9994	9 Fluorine F 18.9984	10 Neon Ne 20.179
3	11 Sodium Na 22.989	12 Magnesium Mg 24.305											13 Aluminum Al 26.9815	14 Silicon Si 28.086	15 Phosphorous P 30.9738	16 Sulfur S 32.064	17 Chlorine Cl 35.453	18 Argon Ar 39.948
4	19 Potassium K 39.098	20 Calcium Ca 40.08	21 Scandium Sc 44.956	22 Titanium Ti 47.90	23 Vanadium V 50.942	24 Chromium Cr 51.996	25 Manganese Mn 54.938	26 Iron Fe 55.847	27 Cobalt Co 58.933	28 Nickel Ni 58.71	29 Copper Cu 63.546	30 Zinc Zn 65.38	31 Gallium Ga 69.723	32 Germanium Ge 72.59	33 Arsenic As 74.922	34 Selenium Se 78.96	35 Bromine Br 79.904	36 Krypton Kr 83.80
5	37 Rubidium Rb 85.468	38 Strontium Sr 87.62	39 Yttrium Y 88.905	40 Zirconium Zr 91.22	41 Niobium Nb 92.906	42 Molybdenum Mo 95.94	43 Technetium Tc (99)	44 Ruthenium Ru 101.07	45 Rhodium Rh 102.905	46 Palladium Pd 106.4	47 Silver Ag 107.868	48 Cadmium Cd 112.40	49 Indium In 114.82	50 Tin Sn 118.69	51 Antimony Sb 121.75	52 Tellurium Te 127.60	53 Iodine I 126.904	54 Xenon Xe 131.30
6	55 Cesium Cs 132.905	56 Barium Ba 137.34	*57 Lanthanum La 138.91	72 Hafnium Hf 178.49	73 Tantalum Ta 180.948	74 Tungsten W 183.85	75 Rhenium Re 186.2	76 Osmium Os 190.2	77 Iridium Ir 192.2	78 Platinum Pt 195.09	79 Gold Au 196.967	80 Mercury Hg 200.59	81 Thalium Tl 204.37	82 Lead Pb 207.19	83 Bismuth Bi 208.980	84 Polonium Po (209)	85 Astatine At (210)	86 Radon Rn (222)
7	87 Francium Fr (223)	88 Radium Ra (226)	**89 Actinium Ac (227)	104 Rutherfordium Rf (261)	105 Hahnium Ha (262)	106 Seaborgium Sg (263)	107 Neilsbohrium Ns (261)	108 Hassium Hs (265)	109 Meitnerium Mt (266)									

Inner Transition Elements (*f* Series)

*Lanthanides	4f	58 Cerium Ce 140.12	59 Praseodymium Pr 140.907	60 Neodymium Nd 144.24	61 Promethium Pm 144.913	62 Samarium Sm 150.35	63 Europium Eu 151.96	64 Gadolinium Gd 157.25	65 Terbium Tb 158.925	66 Dysprosium Dy 162.50	67 Holmium Ho 164.930	68 Erbium Er 167.26	69 Thulium Tm 168.934	70 Ytterbium Yb 173.04	71 Lutetium Lu 174.97
**Actinides	5f	90 Thorium Th 232.038	91 Protactinium Pa (231)	92 Uranium U 238.03	93 Neptunium Np (237)	94 Plutonium Pu 244.064	95 Americium Am (243)	96 Curium Cm (247)	97 Berkelium Bk (247)	98 Californium Cf 242.058	99 Einsteinium Es (254)	100 Fermium Fm 257.095	101 Mendelevium Md 258.10	102 Nobelium No 259.101	103 Lawrencium Lr 260.105

Figure 4.10

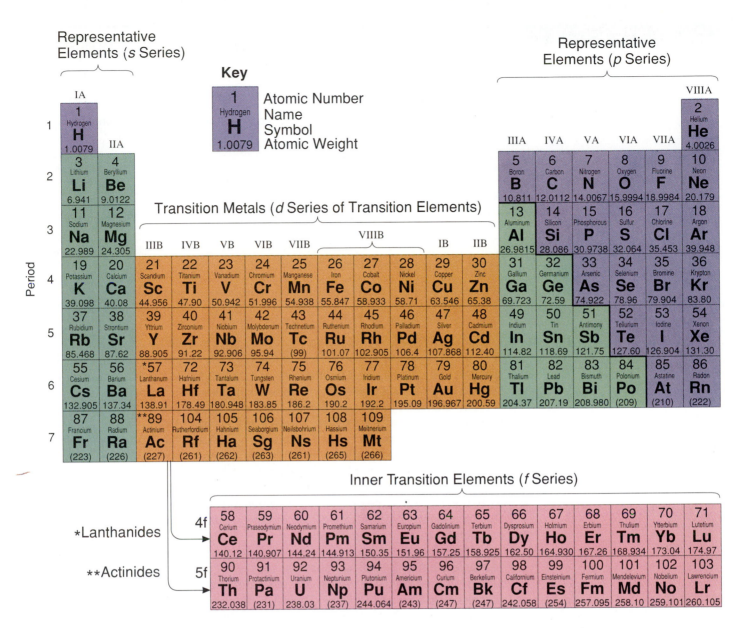

A periodic table arranges elements in rows in increasing atomic number. Below the symbol for each element is its atomic mass, which is the weighted average of the masses of all isotopes of that element.

Names and symbols for elements 104–109 follow the 1994 nomenclature recommendations of the American Chemical Society and may be subject to change.

abundance (i.e., the percentage in fractional or decimal form; see Section 2.6) multiplied by its isotopic mass. Therefore, the mass contributions for magnesium are

	Fractional abundance	×	Isotopic mass	=	Mass contribution
^{24}Mg contribution	$\frac{78.7}{100}$	×	24.0	=	18.9
^{25}Mg contribution	$\frac{10.1}{100}$	×	25.0	=	2.53
^{26}Mg contribution	$\frac{11.2}{100}$	×	26.0	=	2.91
Average atomic mass					24.34

The average atomic mass (weight) is the sum of the contributions. In this example, the average atomic mass (weight) of Mg is 24.3 amu. (Note that 3 sig figs are all that we are allowed in the final sum.)

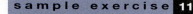

sample exercise 11

Calculate the average atomic mass (weight) of potassium, given that potassium consists of 93.1% ^{39}K and 6.9% ^{40}K isotopes.

solution

Fractional abundance is the percentage divided by 100. Therefore, the fractional abundances are 0.931 for ^{39}K and 0.069 for ^{40}K.

^{39}K contribution	$0.931 \times 39.0 =$	36.3
^{40}K contribution	$0.069 \times 40.0 =$	2.8
Average atomic mass	=	39.1

••• problem 12

Calculate the average atomic mass (weight) of lithium. The element lithium is 7.42% ^{6}Li and 92.58% ^{7}Li.

4.11 periodic table

The periodic table contains a wealth of information that will be applied in later chapters. In Section 1.9, you saw how to identify metals and nonmetals, and in this chapter, you have learned how atomic number and atomic mass (atomic weight) are related to the subatomic structure of each element.

The elements in the periodic table are arranged in order of increasing atomic number, with the number placed above each symbol. The average atomic mass (*atomic weight*) appears below the symbol. Arranging the elements in this order produces the surprising and useful result that elements appearing in the same vertical column have similar physical and chemical properties. Since you have learned that properties of elements are dependent on their electrons, you might hypothesize that there is some relationship between the electrons in elements of the same vertical column, also called a **group** or family. Each horizontal row of elements in the periodic table is called a **period.**

Notice that the groups (columns) are designated by Roman numerals and a capital letter A or B. You will consider mostly A group elements, which are more regular in their behavior. There are special names for some of the groups. The names of the A groups appear in Table 4.2. While scientists have proposed alternative designations and numbering of groups in the periodic table, the designations shown in the periodic table in this book are most convenient for beginning students.

Periods are horizontal rows numbered from 1 to 7 beginning at the top of the periodic table. The first period contains only hydrogen and helium. The second period begins with lithium in Group IA and ends with neon in Group VIIIA. The third-period sequence begins again in IA and terminates with argon in VIIIA.

Because the members of a group show similar properties, learning about one member of a group means that you have learned something about the other members. For example, when you learn how sodium reacts, you will then know that lithium and potassium will react similarly. This is how the periodic table can help you classify the properties of many elements more efficiently.

4.12 periodic groups

Since the periodic table helps you classify the properties of the elements, it will be informative to learn the characteristic properties of each of the A group elements, since those elements will be used more frequently throughout this text. You recall from Section 1.9 that groups to the left on the periodic table are metals and groups

table 4.2 names of groups in the periodic table

Group	Name
IA	Alkali metals
IIA	Alkaline earth metals
IIIA	Boron family
IVA	Carbon family
VA	Nitrogen family
VIA	Oxygen family (chalcogens)
VIIA	Halogens
VIIIA	Noble (inert) gases

to the right, nonmetals. Metals usually have the characteristic properties of shiny, malleable, conducting solids. Nonmetals, by contrast, have opposite properties; they may be gases, not solids, at room temperature, they are brittle if they are solids, they generally lack luster, and they do not conduct electricity. As you proceed from left to right across the periodic table, you will find that along the stepped, diagonal line dividing metals from nonmetals are a few elements, Si, Ge, As, Sb, Te, and At, that have properties between metals and nonmetals and are called semimetals or **metalloids**

Metallic characteristics of elements decrease when going across a period from left to right; they also decrease when going up any vertical column or group in the periodic table.

alkali metals—group IA

With the exception of hydrogen, all elements in Group IA (**alkali metals**) are soft metals that react violently with water, producing hydrogen gas and another compound, called a metal hydroxide, which will be discussed extensively in Chapter 15. Because of this vigorous reaction, the alkali metals are stored under oil or kerosene to protect them from moist air (see Figure 1.14).

Because of their high reactivity, the alkali metals are never found free in nature, but always as an ion combined in a compound. Sodium and potassium are the most abundant alkali metals and the most important in the human body, where they occur as ions (Na^+ and K^+) that play an important role in nerve transmission and maintenance of proper fluid balance between the cells and tissues of the body. Sodium ion in your diet, from sources such as table salt and food preservatives, will be restricted if you have high blood pressure.

Hydrogen, although in Group IA, is not a metal, but it does react and form compounds of a composition similar to that of the alkali metals. Hydrogen, a diatomic gas at room temperature, is the most abundant element in the universe. The energy radiated from the sun is due to a nuclear reaction fusing together hydrogen nuclei to form helium nuclei.

alkaline earth metals—group IIA

All elements in Group IIA (**alkaline earth metals**) are less chemically reactive than the alkali metals. They do react with oxygen in the air, but the compound formed (an oxide) acts as a coating and protects the metal from further attack. Thus, elemental magnesium can be used in a mixture with other metals as a low-density, structural material. Calcium is the most abundant metal in the body, occurring in the ionic

(a)

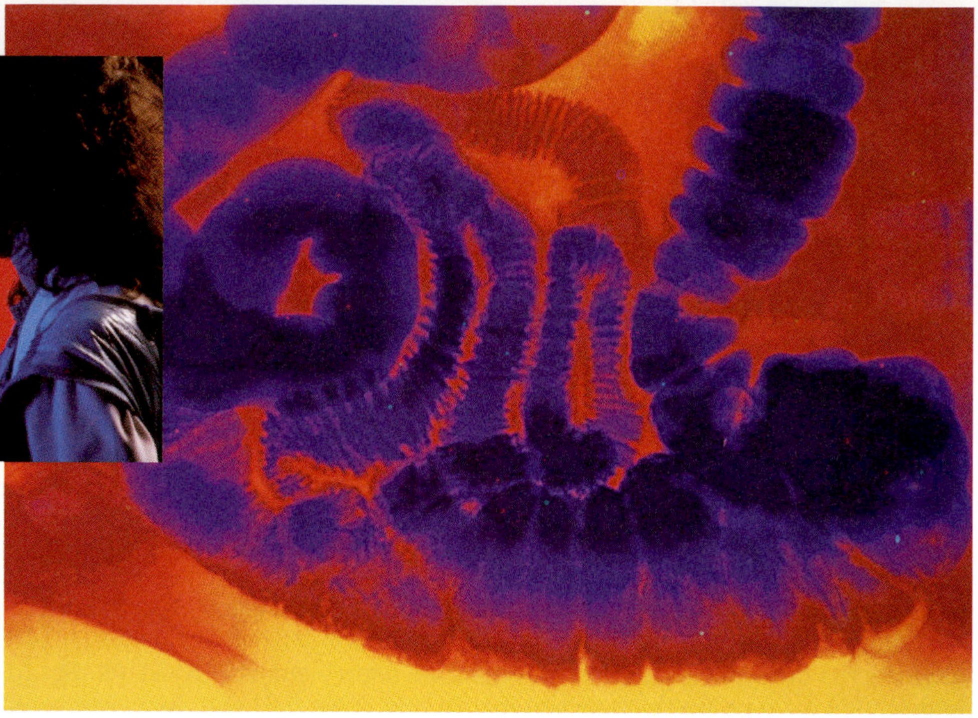

(b)

Figure 4.11

(a) A patient drinks a barium cocktail which then (b) coats the gastrointestinal tract of this patient (shown horizontally) in order to make it opaque to X rays. As a result, doctors may observe abnormalities that may be present.

Figure 4.12

The nonmetallic properties of boron (B), can be seen in its grayish-black crystalline form. In contrast, aluminum (Al), gallium (Ga), and indium (In), have a lustrous, metallic appearance.

form Ca^{2+}, principally in bones and teeth. Barium may be familiar to you because of its presence in the "barium cocktail," a milky suspension that patients drink to make the gastrointestinal tract opaque to X rays (Figure 4.11).

boron family—group IIIA

At the top of this periodic group is the nonmetal boron; proceeding down the group, the other four members, aluminum (Al), gallium (Ga), indium (In), and thallium (Tl), are metals. This reflects a general feature of all groups, namely, that metallic character increases as you proceed down a group (Figure 4.12).

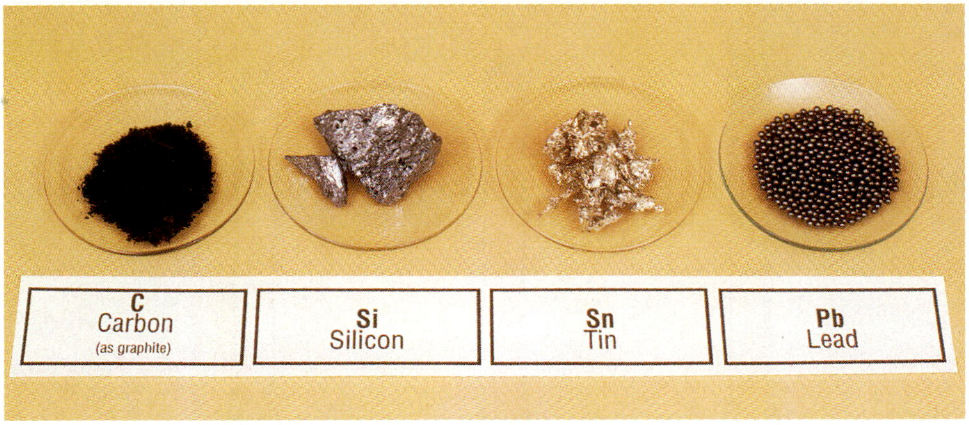

Figure 4.13

The trend of increasing metallic character can be seen as you look down the carbon group elements from nonmetallic carbon (C) to semimetallic (metalloid) silicon (Si) to metallic tin (Sn) and lead (Pb).

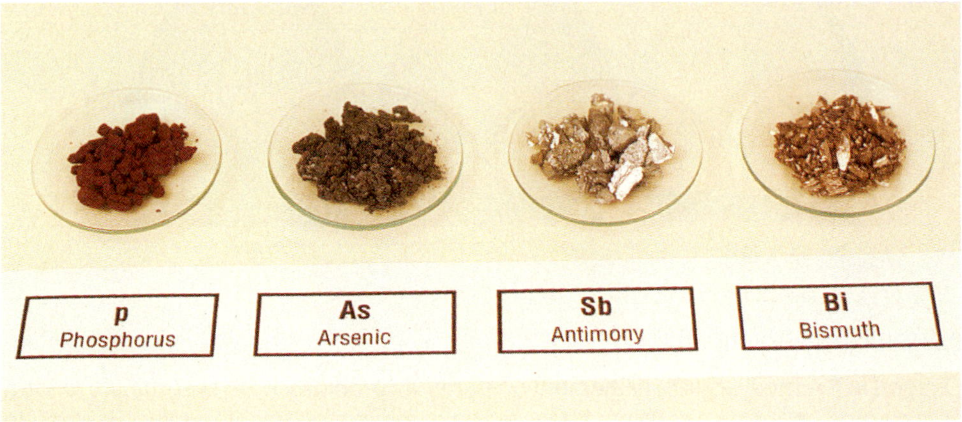

Figure 4.14

The nitrogen group has only one true metal, bismuth (Bi). Two metalloids, arsenic (As) and antimony (Sb), are also present, phosphorus is a nonmetal.

carbon family—group IVA

Carbon, a nonmetal, is found in its elemental form in graphite, the "lead" of pencils, and in diamond. The ten million or so compounds of carbon make up the field of organic chemistry, which includes all compounds found in living systems. Silicon (Si) is a dark-gray, hard solid. The metalloid character of silicon confers special electrical, semiconducting properties, making it a basic component of transistors and integrated circuits. Again, moving down the group, metallic character tends to increase with tin (Sn) and lead (Pb), which display all of the common properties of metals (Figure 4.13).

nitrogen family—group VA

Nitrogen, a diatomic gas (N_2) (see Sections 4.13 and 6.11) at room temperature, is a nonmetal that makes up about 79% of air. Nitrogen and phosphorus are key components of many physiological compounds. Arsenic (As) is a metalloid having a steel-gray, crystalline form, but it is brittle, not malleable like a true metal. Antimony (Sb), too, is a metalloid, appearing as a lustrous, silvery, blue-white solid, but it is extremely brittle and exhibits a flaky, crystalline texture. Antimony is also a poor conductor of electricity compared with most metals. Bismuth (Bi) has metallic character and is used extensively as an additive to steel, cast iron, and aluminum (Figure 4.14).

Figure 4.15

(a) Chlorine (Cl) is a greenish-yellow gas at room temperature. (b) Bromine (Br) is a red liquid with a reddish-orange vapor as shown here. (c) Iodine (I) is a grayish-black solid that sublimes to a dense violet vapor. A periodic trend can be seen in that, proceeding from the top to the bottom of the halogen group, the color of the vapor darkens.

oxygen family—group VIA

Oxygen is the most abundant element on earth. It combines with most metals to form compounds called *oxides*. Elemental oxygen is a diatomic gas, O_2, that is inhaled in that form by the human body and eventually combines with hydrogen ions to form water and supply the body with energy. Sulfur (S), a nonmetallic solid, has many industrial applications, including vulcanization, the addition of sulfur to rubber to make rubber less brittle. Tellurium (Te), a metalloid, and polonium (Po), a metal, are relatively scarce and not found in common chemical processes or products.

halogens—group VIIA

All elements in Group VIIA (**halogens**) are diatomic. Fluorine, F_2, and chlorine, Cl_2, are gases; bromine, Br_2, is a liquid; and iodine, I_2, is a solid. At this end of the periodic table, all elements within the group are nonmetals. Chlorine is present as an anion (Cl^-) with the sodium cation (Na^+) in the familiar substance, table salt or sodium chloride (NaCl). Iodine, I_2, sublimes to yield a characteristic purplish vapor; a mixture of iodine dissolved in alcohol has been used as a disinfectant (Figure 4.15).

noble (inert) gases—group VIIIA

All elements in Group VIIIA (**noble (inert) gases**) are simple, monatomic gases, and all display a lack of chemical reactivity. Helium gas has a density less than air and has been used as a lifting gas for airships and balloons in meteorological investigations. Its nonreactive behavior makes it safer than using hydrogen gas, which is potentially explosive. Because of its nonreactivity, argon gas is commercially used in incandescent lamps and fluorescent lamps to minimize vaporization of the filaments. Neon and xenon gases are used in lamps that emit a specific color of light; for example, neon emits the familiar orange light of neon signs and xenon emits a beautiful blue glow (Figure 4.16). Radon (Ra) is the only radioactive element that is a gas. Because radon gas may escape from concrete, its accumulation in closed buildings is carefully monitored to reduce the risk of lung cancer.

4.13 elemental makeup

You have learned the subatomic structure for individual atoms of each element and how that structure accounts for the position of elements in the periodic table. The elements are not always composed of separate atoms, however. While the noble-gas elements (Group VIIIA) are made up of uncombined atoms, very few other elements actually are. The atoms of pure elements may be combined in groups of two for many gases or in extremely large arrays for many solids.

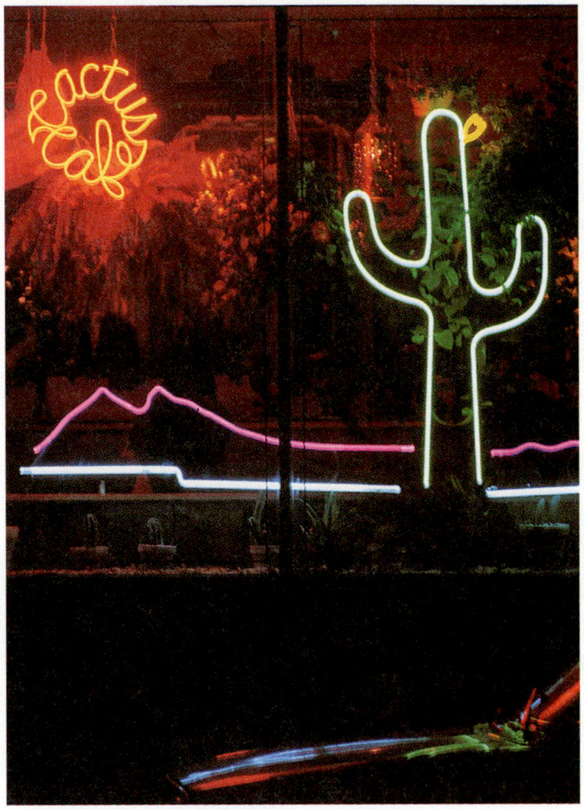

Figure 4.16
Neon (Ne) and xenon (Xe) are used in lamps that emit a specific light color. Neon emits orange light and xenon emits blue light.

Many common elements exist as diatomic molecules, which you learned about in Section 1.11. You remember that the gases H_2, N_2, O_2, F_2, and Cl_2, the liquid Br_2, and the solid I_2 all exist as diatomic molecules.

problem 13

The above diatomic molecules are nonmetals. What properties would you expect them to display?

The element carbon has long been recognized in two pure forms—graphite and diamond. Both forms involve a continuous network of carbon atoms bonded in different geometric arrangements. Since 1990, large quantities of another form of pure carbon, called fullerene,[2] have been prepared and its properties investigated. As you can see from the atomic models in Figure 4.17a and b, graphite and diamond have their own structural arrangements of atoms, and their different structures account directly for their different physical properties shown in Figure 4.17d and e. The three-dimensional structure of C atoms in diamond creates an extremely hard, crystalline solid that you know for its durability in gemstones and cutting tools. The two-dimensional structure of C atoms in graphite creates a solid composed of sheet-like layers that is soft and can easily be used as a pencil tip to write on paper. The third form of pure carbon, fullerene, is also shown in Figure 4.17c and f. Fullerene is composed of discrete units shaped like soccer balls, each made up of 60 carbon atoms. This novel shape and the fitting together of these soccer-ball molecules give rise to unique and useful properties such as superconductivity. The three structures of pure, solid carbon arise because the C electrons may form bonds in various ways. The detailed arrangement of electrons about atoms and how the electrons participate in the formation of bonds will be the topics of the next two chapters.

2. Fullerene was named after the renowned designer and architect, Buckminster Fuller, who is famous for developing the geodesic dome architecture used in wind-resistant buildings, large auditoriums, and playground equipment.

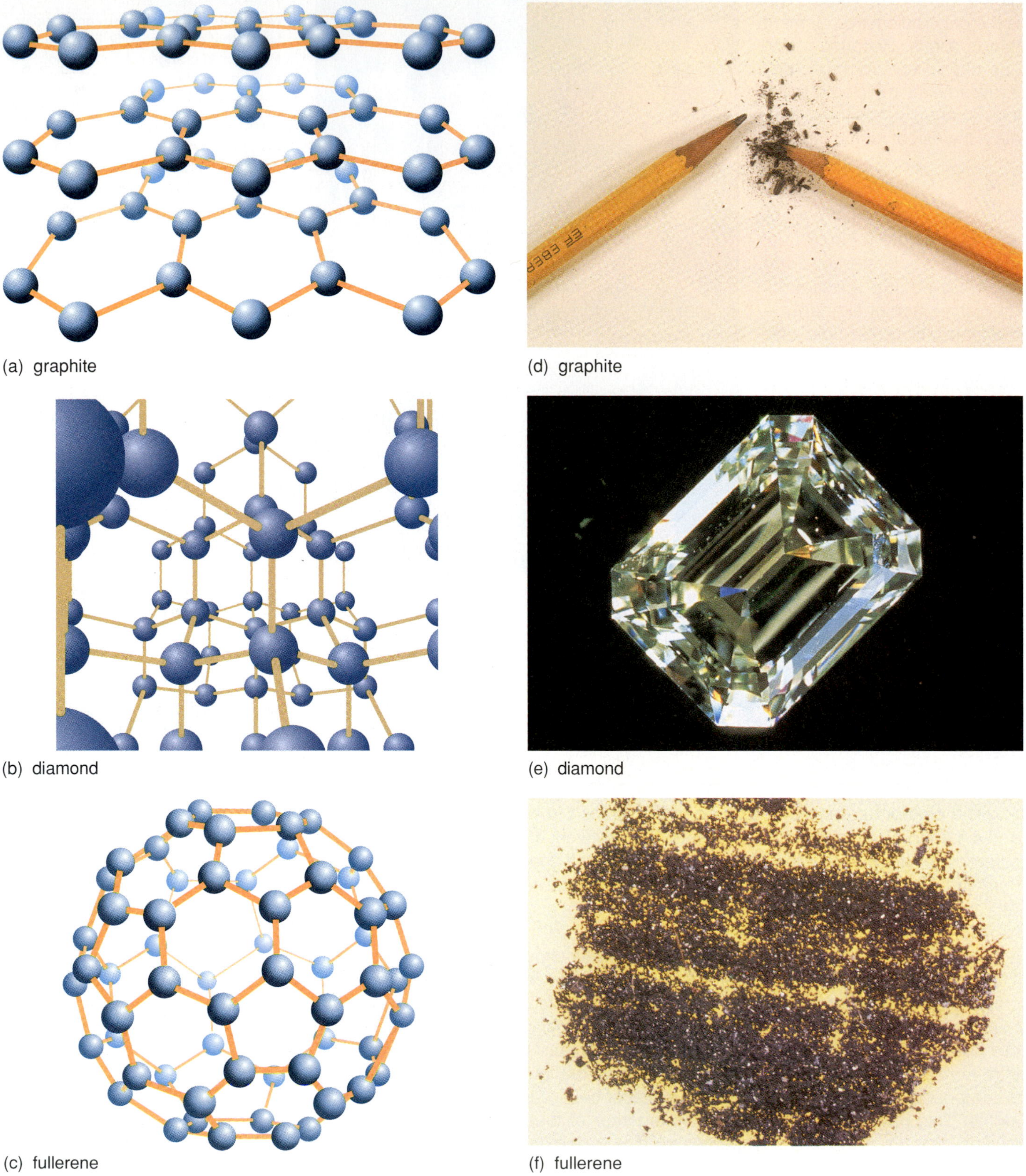

(a) graphite
(b) diamond
(c) fullerene
(d) graphite
(e) diamond
(f) fullerene

Figure 4.17

In graphite, carbon atoms form two-dimensional layers of a chicken-wire pattern. In diamond, carbon atoms are arranged in a three-dimensional crystal array. Each fullerene molecule, C_{60}, is a soccer-ball shape. Arrangements of C atoms in three different forms of the element carbon—graphite, diamond, and fullerene. Left: (a-c) Atomic models illustrate the different arrangement of atoms in graphite, diamond, and fullerene. Right (d-f): Actual samples of graphite, diamond, and fullerene show the differences in appearance and structure for the three forms of carbon.

chapter accomplishments

After completing this chapter, you should be able to define all key terms and do the following.

4.2 Atomic theory

- ❑ Distinguish elements, compounds, and mixtures in terms of their atomic makeup.

4.3 Picturing atoms and elements

- ❑ Form a mental picture of atoms and elements.

4.4 Inside the atom

- ❑ State the names and symbols of the three subatomic particles.
- ❑ State the relative charges and masses of the subatomic particles.
- ❑ Form a mental picture of atoms, showing the arrangement of the subatomic particles.

4.5 Atomic number

- ❑ Given the atomic number, determine the number of protons and electrons in an atom.
- ❑ Given the number of protons or electrons in a neutral atom, write the atomic number.

4.6 Isotopes

- ❑ Given the number of protons and neutrons in two atoms, indicate whether the atoms are isotopes.

4.7 Mass number

- ❑ Given the atomic and mass numbers, determine the number of subatomic particles in an atom.
- ❑ Given a periodic table and the number of protons and neutrons in an atom, write a correct symbol showing the atomic and mass numbers.
- ❑ Given the atomic and mass numbers of a set of atoms, indicate which atoms are isotopes.

4.8 Ions: atoms that have lost or gained electrons

- ❑ Indicate how anions are formed.
- ❑ Indicate how cations are formed.
- ❑ Given the number of protons and electrons in an ion, calculate the charge on the ion.
- ❑ Given the symbol for an ion, including the mass number, charge, and atomic number, calculate the number of protons, neutrons, and electrons in the ion.
- ❑ Given the number of protons, neutrons, and electrons in an ion, write a correct symbol for the ion, including the mass number, atomic number, and charge.

4.9 Relative atomic mass

- ❑ Given the relative mass factor between an atom X and a ^{12}C atom, calculate the atomic mass of that atom.

4.10 Average atomic mass (weight)

- ❑ Calculate the average atomic mass (weight) of an element given the percentage abundances for the isotopes of that element.

4.11 Periodic table

- ❑ Use the periodic table to determine atomic numbers and atomic masses (weights).
- ❑ Explain the significance of elements appearing in the same group in the periodic table.
- ❑ Given a periodic table, indicate the name of the family for an A group element.

4.12 Periodic groups

- ❑ Describe the trend of nonmetallic or metallic character among elements within a group.

4.13 Elemental makeup

- ❑ List the elements that exist as diatomic molecules.

key terms

element	electrons	anions	metalloids
compound	atomic number	atomic mass units	alkali metals
mixture	isotopes	average atomic mass	alkaline earth metals
protons	mass number	group	halogens
neutrons	ions	period	noble (inert) gases
nucleus	cations		

problems

4.2 Atomic theory

14. In what way do atoms of different elements differ?
15. In what way are atoms of different elements similar?
16. How do elements and compounds differ in terms of atomic theory?

4.3 Picturing atoms and elements

17. Draw a picture of an atom the way you visualize it.

4.4 Inside the atom

18. Name the three subatomic particles.
19. Which subatomic particle has zero mass?
20. Which subatomic particle has zero charge?
21. Compare the relative masses of the three subatomic particles.
22. Compare the relative charges of the three subatomic particles.
23. Which subatomic particles account for the charge on the nucleus?
24. How can an atom that contains charged subatomic particles be neutral?
25. Describe the location of each of the three subatomic particles.
26. What is between the nucleus and the electrons of an atom?

4.5 Atomic number

27. What is the difference between the atomic number and the number of electrons in an atom?
28. An atom has 12 protons and 12 neutrons.
    a. What is its atomic number?
    b. What is the number of electrons?
    c. What is the name of this atom?
29. What are the names of the elements with the following atomic numbers?
    a. 7
    b. 11
    c. 18
    d. 2
30. What are the names of the elements with the following numbers of electrons in their atoms?
    a. 14
    b. 13
    c. 20
    d. 35

4.6 Isotopes

31. Two isotopes of the same element are alike in what way? In what way do they differ?
32. Give a possible set of protons, electrons, and neutrons for an atom that is an isotope of the following.
    a. A carbon atom with 6 neutrons
    b. An atom with 11 protons, 11 electrons, and 12 neutrons
    c. An atom with 19 electrons and 20 neutrons
33. What is the restriction on the number of protons for two isotopes of the same element?

4.7 Mass number

34. Complete the following table for the neutral atoms indicated.

	Element Name	*Atomic Number*	*Atomic Mass*	*Number of Protons*	*Number of Neutrons*	*Number of Electrons*
a.		35	80			
b.		4			5	
c.				40	51	
d.			59			28

35. Tell the numbers of protons and electrons in atoms that have the following mass numbers and numbers of neutrons.

	Mass Number	*Number of Neutrons*	*Number of Protons*	*Number of Electrons*
a.	51	28		
b.	28	14		
c.	19	10		

36. With the help of the periodic table, determine the number of protons, neutrons, and electrons in the following isotopes.
    a. Carbon-13
    b. Cobalt-59
    c. Bromine-79
    d. Strontium-87
37. Complete the following table.

	Symbol	*Number of Protons*	*Number of Neutrons*	*Number of Electrons*
a.	$^{9}_{4}Be$			
b.	$^{127}_{53}I$			
c.	$^{31}_{15}P$			
d.	$^{40}_{18}Ar$			

38. Write symbols for atoms with the following numbers of protons and neutrons.

	Symbol	Number of Protons	Number of Neutrons
a.		11	12
b.		2	2
c.		17	18

39. Which of the following atoms are isotopes of each other?

a. $^{28}_{14}X$ d. $^{15}_{7}X$
b. $^{14}_{7}X$ e. $^{31}_{15}X$
c. $^{45}_{21}X$ f. $^{30}_{14}X$

40. Consult the periodic table and identify each of the elements represented in Problem 39.

41. An atom contains 19 protons, 20 neutrons, and 19 electrons.
 a. Write the symbol for this atom.
 b. Write the symbol for a possible isotope of this atom.

42. An isotope of neon has 11 neutrons in its nucleus. Write the symbol of the isotope.

43. The mass number of an isotope of nitrogen is 15. Write the symbol for the isotope.

44. Write the symbol for an isotope of $^{16}_{8}O$ that contains the following.
 a. One more neutron
 b. Two more neutrons

45. Is there any difference between the representation $^{13}_{6}C$ and carbon-12? Between $^{13}_{6}C$ and carbon-13? Explain.

4.8 Ions: atoms that have lost or gained electrons

46. Calculate the charges on the following ions and indicate whether the ion is a cation or anion.

	Ion	Number of Protons	Number of Electrons	Charge	Cation or Anion
a.	F	9	10		
b.	Al	13	10		
c.	P	15	18		
d.	Ag	47	46		

47. Calculate the numbers of protons, neutrons, and electrons in the following ions.

	Ion	Number of Protons	Number of Neutrons	Number of Electrons
a.	$^{88}_{38}Sr^{2+}$			
b.	$^{23}_{11}Na^{+}$			
c.	$^{14}_{7}N^{3-}$			
d.	$^{127}_{53}I^{-}$			

48. Consult the periodic table and name the ions that have the following.
 a. 35 protons and 36 electrons
 b. 30 protons and 28 electrons

49. Write a symbol, including mass number, atomic number, and charge, for ions containing the following numbers of protons, neutrons, and electrons.

	Number of Protons	Number of Neutrons	Number of Electrons	Symbol
a.	15	16	18	
b.	19	20	18	
c.	1	0	0	
d.	9	10	10	
e.	37	48	36	

50. Explain how cations are formed from atoms.

51. Explain how anions are formed from atoms.

52. Complete the following table.

	Isotope	Symbol	Atomic Number	Mass Number	Number of Protons	Number of Neutrons	Number of Electrons
a.	Lithium-7						3
b.		$^{59}_{27}Co$					
c.			26			30	
d.				40	18		
e.		$^{16}_{8}O^{2-}$					
f.		$^{138}_{56}Ba^{2+}$					

53. Give the symbol and charge of an ion having 10 electrons and the following numbers of protons.
 a. 9 protons
 b. 12 protons
 c. 11 protons
 d. 8 protons
 e. 7 protons

54. What is the difference between the representations Mg and Mg^{2+}?

4.9 Relative atomic mass

55. A certain isotope of Co is 5 times as heavy as ^{12}C. What is the atomic mass of this isotope?

56. What are the approximate atomic masses of the atoms symbolized in Problem 39?

57. What is the mass ratio of a magnesium-24 atom to a carbon-12 atom?

58. If a carbon atom had a mass of 6 instead of 12 amu, what would be the mass of a helium atom? Titanium?

4.10 Average atomic mass (weight)

59. The element sulfur is made up of 95.0% sulfur-32, 0.76% sulfur-33, and 4.22% sulfur-34. What are the fractional abundances of the three isotopes?

60. Refer to Problem 59 and calculate the average atomic mass (weight) of sulfur.

61. Neon is 90.9% ^{20}Ne, 0.257% ^{21}Ne, and 8.82% ^{22}Ne. Calculate the atomic mass (weight) of neon to 3 sig figs.

62. The element iron is made up of the isotopes ^{54}Fe (5.82%), ^{56}Fe (91.7%), ^{57}Fe (2.19%), and ^{58}Fe (0.330%). Calculate the average atomic mass (weight) of iron.

4.11 and 4.12 Periodic table and groups

63. Consult the periodic table and arrange the following elements in groups: oxygen, sodium, sulfur, phosphorus, potassium, arsenic, and lithium.

64. Give the number and name of the group containing iodine.

65. Give the symbol of the element that is found in the following location in the periodic table.
   a. Period 2, Group IIA
   b. Period 5, Group VIA
   c. Period 2, Group VIIA
   d. Period 4, Group VIIIA

66. Give the symbol of an element that would have properties similar to the following.
   a. Al
   b. N
   c. Si
   d. Ar

67. Which of the following elements would you predict to have similar properties?
   a. Atomic number 13
   b. Atomic number 37
   c. Atomic number 11
   d. Atomic number 55

68. Elements in Groups IA, IIA, and IIIA form compounds with oxygen that have the following general formulas.

IA	*IIA*	*IIIA*
M_2O	MO	M_2O_3

   a. Write a formula for the compound of calcium and oxygen.
   b. Write a formula for a compound of aluminum and oxygen.

69. Which element in Group IIIA do you predict to be most nonmetallic? Which is the most metallic?

4.13 Elemental makeup

70. Why are the noble (inert) gas elements more likely than other elements to be made up of uncombined atoms?

71. Hydrogen is a gas at room temperature. Is it one of the noble gases? Explain.

Additional problems

72. Two isotopes of iodine are used in the diagnosis of thyroid malfunction. One is $^{125}_{53}I$ and the other is $^{131}_{53}I$. How many electrons, protons, and neutrons are there in each of these atoms?

73. Fill in the blanks in the following table.

Symbol	*Atomic Number*	*Number of Neutrons*	*Mass Number*
$^{39}_{19}K$			
$^{\_\_}_{\_\_}Ca$			48
	11	12	
	36		84

74. Three isotopes of hydrogen occur naturally: protium (mass number 1), deuterium (mass number 2), and tritium (mass number 3). Tritium is radioactive. Write symbols for these three isotopes that distinguish among them.

75. Are graphite and diamond made up of different types of carbon atoms? Explain.

76. How can an atom become charged?

77. What is the maximum positive charge that can occur on the following?
   a. Hydrogen ion
   b. Beryllium ion

78. In principle, is there a limit to the negative charge that can occur on an ion?

chapter 5

electronic structure of the atom

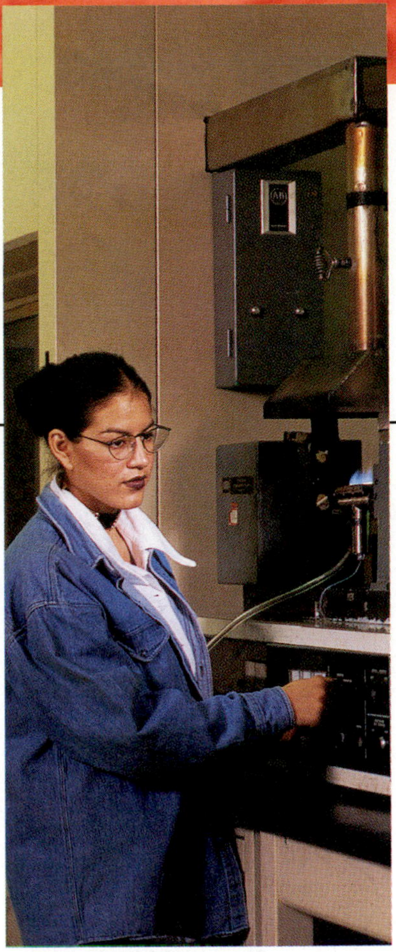

The concentration of ions in foods and physiological fluids can be measured by a procedure known as atomic absorption spectroscopy. Electrons in the sample being studied are excited when the sample is injected into the flame. Light absorbed by the atoms is used to identify the kind and amount of ions present in the sample.

5.1 introduction

You learned in Sections 4.11 and 4.12 that elements are arranged into groups in the periodic table and that all elements in a group have similar properties. In this chapter, you will see how electrons determine the chemical properties and some of the physical properties of each element. To account for the similarity in properties, scientists have formulated the hypothesis that elements in the same group must have common features in how their electrons are arranged about the nucleus. This arrangement of electrons is called **electron structure.** Learning the similarities of electron structure among elements in a group will enable you to use the power of the atomic theory you have been studying. From your understanding of how electrons are configured about the nucleus of each element, you will be able to predict periodic trends in the properties of the elements as observed experimentally and summarized in the periodic table.

Now that you can identify each element from its electron, proton, and neutron composition, you will look into the detailed structure of the electron arrangement that characterizes the chemical properties of each element. Since the electrons compose the outermost structure of each atom, they easily interact with other atoms. The ease with which elements gain, lose, or share electrons determines the chemical nature of each element and its likelihood to form compounds. Compound formation will be discussed in Chapter 6, and the discussions there will rely heavily on the principles you learn in this chapter. The chemical reactions that you encounter in daily life involve changes only in the electron structure of atoms; the nucleus is unaltered. In Chapter 17, reactions in which the nucleus undergoes change, called **nuclear reactions,** will be considered, but until then the chemistry we discuss is a consequence of electronic changes only.

5.2 energy revisited

The arrangement of electrons about the nucleus of an atom governs the energy of the atom. Different amounts of energy are associated with different electron arrangements. As the electron structure changes, you can observe changes in the energy content of matter.

In Section 3.7 *energy* was defined as the ability to do work, and we discussed the idea that energy can be witnessed only when work is done and some *energy change* occurs. It was also pointed out that all matter has energy stored within it. The amount of stored energy depends on how the sample of matter is constructed. One important aspect of the construction is its electron structure.

Figure 5.1
The ground and first two excited states for the electrons of a typical atom. An arrow pointing upward means that the electrons, starting from the lower-energy state at the base of the arrow, are gaining energy and being excited to the higher-energy state at the tip of the arrow. An arrow pointing downward means that the electrons, starting from the higher-energy state at the base of the arrow are losing energy and falling to the lower-energy state at the tip of the arrow.

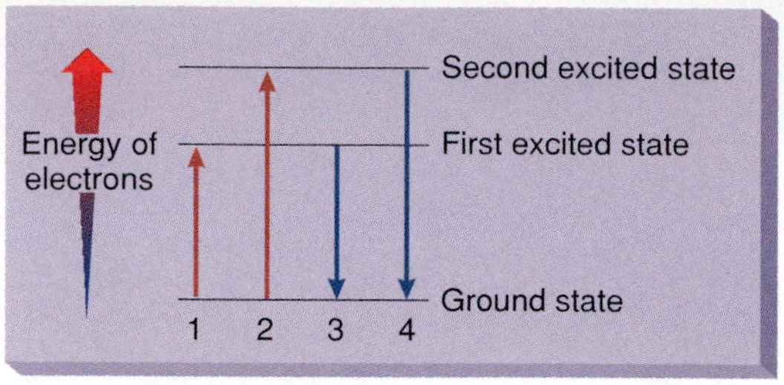

5.3 concept of minimum energy

Scientists talk about **energy states** of atoms; these states describe the total energy of each atom in terms of the sum of the energies of the electrons in it. Atoms in a high-energy state have a great deal of energy. If atoms of an element lose energy, then their electrons will be in a lower-energy state. If atoms of an element gain energy, then their electrons will be in a higher-energy state. The change in energy state of the atoms tells you how much energy the electrons have lost or gained and is represented pictorially by each arrow in Figure 5.1.

The lowest energy state of the electrons in an atom is called the **ground state** of the electrons and all other states are called excited states; all excited states have an energy greater than the ground state.

In Figure 5.1, you can see how much energy must be gained by each electron of an atom when it goes from its ground state to its first excited state (arrow 1) or from its ground state to its second excited state (arrow 2). Conversely, you can also see how much energy must be lost or given off by an electron of an atom when it goes from the first excited state (arrow 3) to the ground state or from its second excited state to the ground state (arrow 4). From Figure 5.1, you can also see clearly from the length of the arrows that the energy change is the same between two states, whether the electron is being excited by gaining energy or is being de-excited by losing energy. Arrows 1 and 3 have the same length, as do arrows 2 and 4.

It is a principle of nature that all things try to reach a minimum-energy state. For example, lay your pencil at the edge of your desk. In terms of energy states, the pencil on the desk is in a high-potential-energy state relative to the floor. If the pencil is disturbed, even slightly, it will roll off the desk and fall onto the floor. When the pencil hits the floor, it has reached a minimum-energy state for the room and, in so doing, has lost the higher potential energy it possessed when perched at the edge of the desk. In going from a higher- to a lower-energy state, the pencil lost energy, which it gave off as work when it struck the floor.

For the pencil to return to the higher-energy state that it occupied while it lingered at the edge of the desk, energy must be put in. If you were to pick the pencil up and replace it on the desk, you would be transferring energy from yourself to the pencil. When the pencil is back on the desk, it has the same likelihood, if disturbed, to fall again to the lower state.

This natural tendency toward minimum energy is also true for the chemical energy of matter. Atoms, making up matter, are generally found in their ground state. Although atoms can be excited into a higher-energy state temporarily by exciting their electrons with outside energy, they exhibit a natural tendency to return to their lowest energy or ground state. Because atoms will stay in their ground state if left alone, their condition in this state is labeled a **stable** condition. Something that is stable or has *stability* is in a low-potential-energy state. Higher-energy states are less stable. This idea that *high energy corresponds to low stability and low energy corresponds to high stability* is very important but can be confusing. You will find it well worth the effort to clarify these terms for yourself now to save confusion later.

Think about the pencil at the edge of the desk. The pencil is **unstable** (exhibits low stability) because it has high potential energy. It will fall to the ground state, or lowest-energy state, and become more stable.

sample exercise 1

Consider a woman who is standing on a diving board.

a. Does she possess more potential energy before or after diving?
b. Is she in a more stable condition when standing on the diving board or after hitting the water?
c. Sketch an energy diagram similar to Figure 5.1 to illustrate the potential-energy state of the woman before and after diving.

solution

a. Her potential energy is higher when she is poised on the board. When she dives, some potential energy is converted into kinetic energy, so she has less potential energy when she hits the water.
b. While she is in the water, her condition is more stable since it is a lower-energy state.
c.

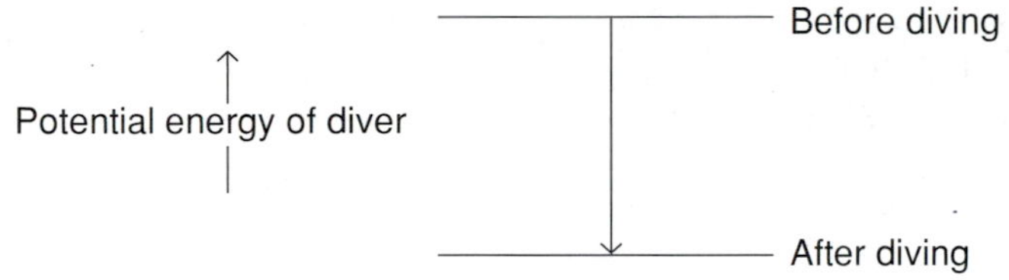

sample exercise 2

When gasoline burns, the products of the combustion are carbon dioxide (CO_2) and water (H_2O), both of which are more stable than gasoline.

a. Describe the relative energy contents of gasoline, CO_2, and H_2O.
b. Sketch an energy diagram similar to Figure 5.1 to illustrate the chemical energy states of the reactants and products in the combustion of gasoline.

solution

a. You are told that gasoline is *less* stable than CO_2 and H_2O; therefore, it must possess *more* energy than CO_2 and H_2O.
b.

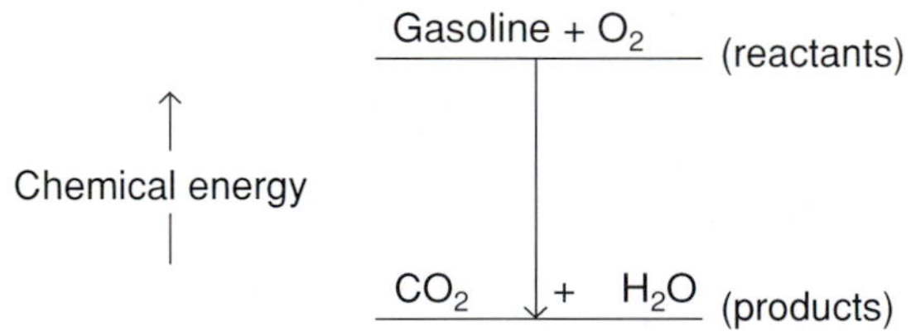

Note that when gasoline burns, it loses or gives off energy in the form of heat energy. The amount of heat given off is found by subtracting the lower-energy state of the products from the higher-energy state of the reactants. In your automobile engine, the heat energy released is converted into the work used to move your car.

••• problem 1

Food contains more energy than the products of metabolism into which it is converted in the body. Is food more or less stable than its metabolic products? Sketch an energy diagram showing the higher- and lower-energy states and the change in energy from metabolism.

5.4 minimum energy in the atom

In terms of energy, electrons in atoms obey the same rules as all other matter in the universe; they prefer to be arranged in a condition (state) of minimum (lowest) energy and hence maximum (highest) stability.

electrostatics

Let us divert our attention from atoms for a moment to discuss some aspects of electrostatics that will be useful in discovering the minimum-energy electron structure (arrangement) of the atom. **Electrostatics** deals with interactions between charged particles. The concept of "opposites attract and likes repel" is well established. In electrostatics, *objects with opposite electrical charge* (*one* + *and one* –) *attract each other; objects with identical charge* (*both* + *or both* –) *repel each other.* This attraction or repulsion increases as the objects come closer together.

In terms of energy states, oppositely charged bodies are in a low-energy, stable state when they remain close to each other; work must be performed and energy thereby increased in order for them to move farther apart. The low-energy, stable condition for identically charged bodies occurs when they are far apart. Work must be performed and energy increased in order to push them closer together. The principle of minimum energy is fundamental to the structure of ions in solids such as table salt (NaCl), where the cations and anions are arranged for maximum attraction and minimum repulsion. The arrangement of ions in solids will be discussed further in Chapter 12.

As a result of electrostatics, *the low-energy arrangement of electrons in an atom must be based on the ideas that*

1. The close approach of a negatively charged electron to the positive nucleus is a low-energy condition.
2. The far separation of two negatively charged electrons is a low-energy condition.

To obtain the minimum-energy condition for the atom as a whole, these two conditions must be met simultaneously. That is, electrons prefer to be close to the nucleus, but they must avoid other electrons. These two simple ideas lay the entire foundation for the following discussion of the electronic arrangements in atoms.

---

sample exercise 3

Two positively (+) charged tennis balls are held at opposite ends of a room 3 m long, 3 m wide, and 3 m high (volume = 3 m × 3 m × 3 m = 27 m^3). Is their energy state in this room higher or lower than when they were in a cubic container that measured 30 cm along an edge? Sketch an energy-state diagram to illustrate the difference in energy of the tennis balls under the two conditions.

solution

The energy state in the room is lower because the two like-charged objects are farther apart there than when they were in the container with the smaller volume.

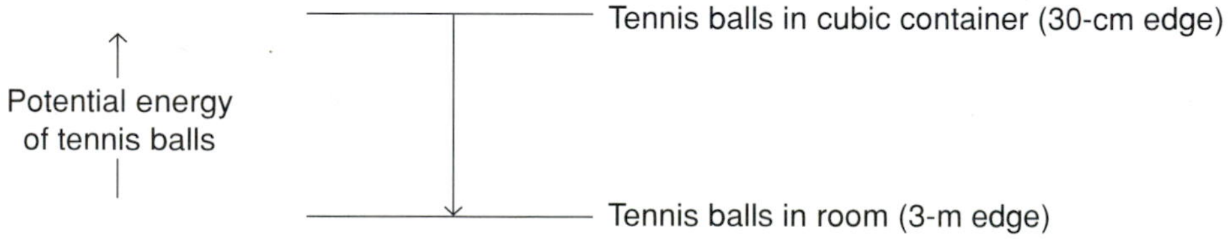

---

sample exercise 4

If the tennis balls in Sample Exercise 3 were oppositely charged (one + and one −), then would their energy be lower in the room or in the container? Sketch an energy-state diagram to illustrate their respective energies under the new condition.

solution

The energy state in the container would be lower because the positively charged ball and the negatively charged ball can get closer together.

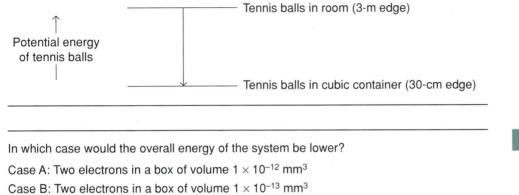

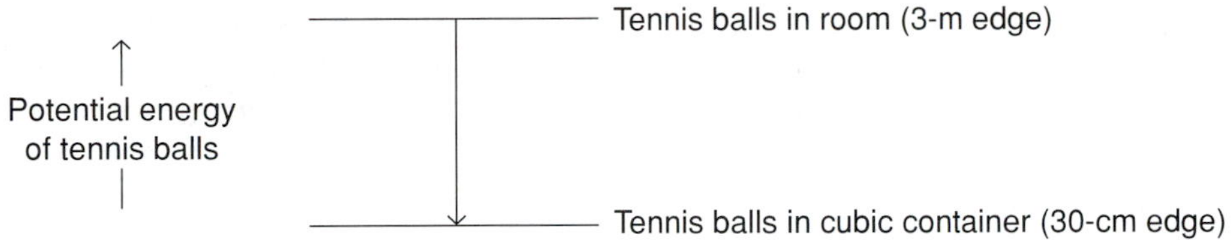

problem 2

In which case would the overall energy of the system be lower?

Case A: Two electrons in a box of volume 1×10^{-12} mm^3

Case B: Two electrons in a box of volume 1×10^{-13} mm^3

hydrogen, the smallest atom

Consider the H atom with its one positive proton in the nucleus and its one negative electron outside the nucleus. From the foregoing discussions, you might immediately conclude that the electron must "fall into" the nucleus, i.e., draw as close as possible to the proton. But you must accept a puzzling but firmly established fact: Although the electron may get very close to the nucleus, it never "falls in" or combines with the proton to form a neutron. This contradiction of classical electrostatics really had scientists baffled early in this century. The person who offered the first explanation to the problem was Niels Bohr, the Danish physicist who eventually won a Nobel prize for this and other brilliant work.

Bohr's model of the atom required that the electron could only have certain definite energies and therefore could only occupy certain definite orbits that are characterized by a specific energy and located at a specific distance from the nucleus—just like the orbits that planets have around the sun. The concept of an electron in an orbit of specific energy was a breakthrough in the model of the atom and had important consequences for the amount of energy required by an electron to change its energy state. The Bohr atomic model with its possible orbits for the electron is shown in Figure 5.2a. The energy states of the electron were actually calculated by Bohr and are shown in the energy-state diagram in Figure 5.3. As a result of his calculated energy states, Bohr could also calculate the change in energy needed for an electron to be excited to a higher-energy state or for an excited electron to be dropped to a lower-energy state. This change in energy of the electron was like climbing or descending a ladder in which each rung was an energy state. The electron could go up or down any number of rungs, but it could only rest on one of the energy rungs and not any place between.

The ground state (lowest-energy state) for the hydrogen atom would occur when its electron was in orbit 1, the one closest to the nucleus. If the electron were in an orbit farther from the nucleus, the atom would exhibit higher energy and be in an excited state. The electron can never be in the regions between orbits.

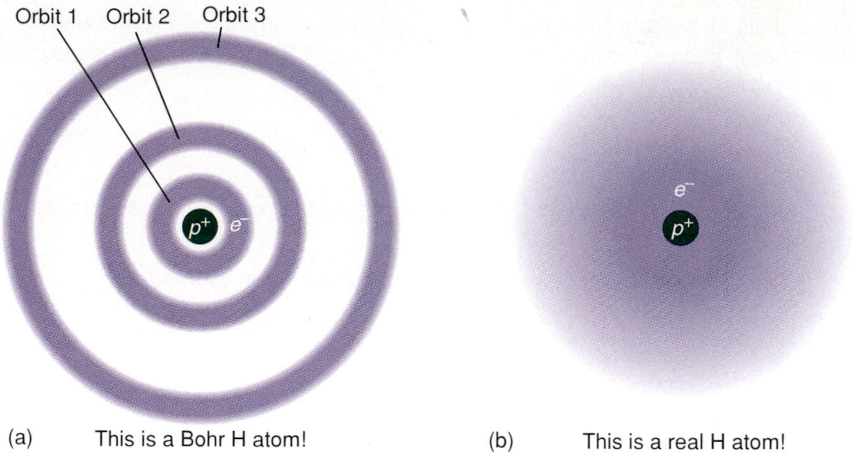

Figure 5.2

(a) Bohr's model of a H atom. The electron is restricted to definite orbits at specific distances from the nucleus. The figure shows hydrogen's electron in the ground-state orbit closest to the nucleus. (b) A real H atom. In the real atom, the electron moves about the nucleus within the spherical volume of a negatively charged cloud.

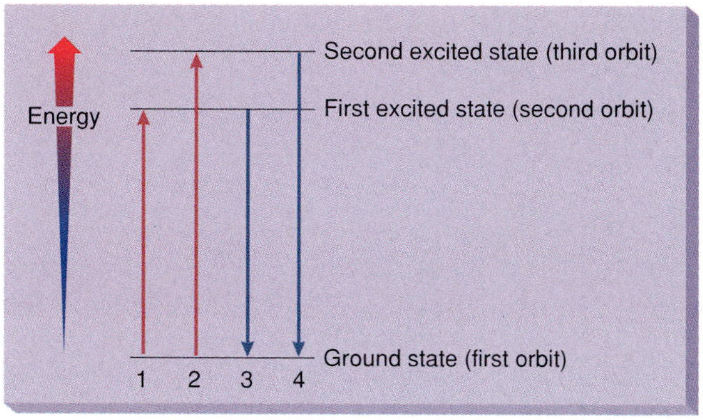

Figure 5.3

Energy states of the electron in the hydrogen atom have been calculated from the Bohr model and compared closely with the energy states measured experimentally.

Bohr's visualization of the electron distribution around the nucleus is not entirely correct. The movement of the hydrogen electron is dispersed around the nucleus within a more fuzzy cloud, as shown in Figure 5.2b. Also, the Bohr model did not agree with experimental observations for atoms containing more than one electron and hence required further modification. However, this idea that electrons possess only certain definite energies associated with certain regions in space is correct and is very important. This is the idea of **quantization.** Electron energies in atoms are quantized; that is, there are definite allowed energy states that electrons may occupy. Electrons never have intermediate energies. Bohr's introduction of quantization eventually led to the development of **quantum mechanics,** which fully describes electronic energies and arrangements mathematically. It turns out from this mathematical description that it is not possible to predict the precise positions of electrons; rather, one can only predict the likelihood of electrons being within some volume of space.

Even though the mathematical description is the only precise one, most people find it easier and more useful to develop a mental picture of the atom. Happily, it is possible to do so by sticking to the ideas already discussed. To recap: (1) the close approach of a negative electron to the positive nucleus is a low-energy condition, and (2) the far separation of two negative electrons is a low-energy condition. The only exception to these two principles that we need to remember is that electrons do not enter the nucleus.[1] If we picture electrons as continuously moving very rapidly, then a snapshot at slow shutter speed would capture them as a fuzzy cloud. The cloud is negatively charged because electrons each bear a negative charge. Figure 5.2b shows the picture of a hydrogen atom implied by these ideas. The one negative electron has a definite lowest-energy state that confines it to movement within a region close to the positive nucleus.

The fuzzy cloud, or more scientifically, the volume, in which the electron is likely to be found is called an **orbital.** An orbital in general is defined as a volume in space in which an electron or a pair of electrons is found. As you continue reading, you will find that no more than two electrons can occupy an orbital and that there is a specific energy associated with every orbital.

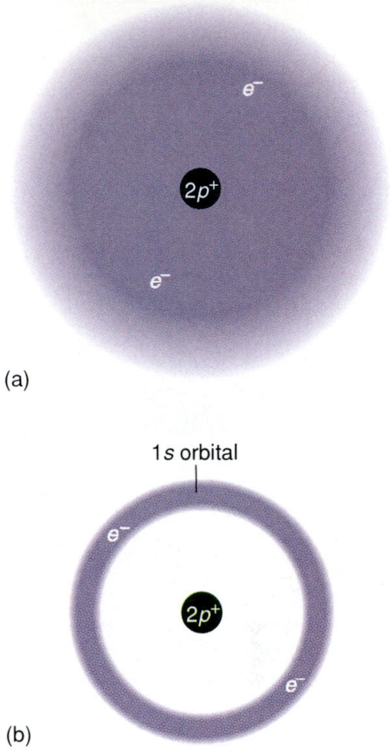

Figure 5.5
The He 1*s* orbital: (a) the spherical volume occupied by the two electrons; (b) a cross section of the orbital.

helium atoms

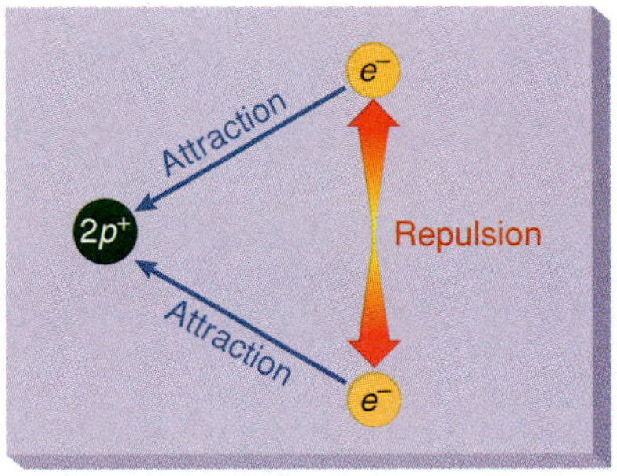

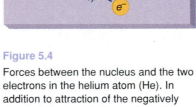

Figure 5.4
Forces between the nucleus and the two electrons in the helium atom (He). In addition to attraction of the negatively charged electrons to the nucleus, the electrons themselves repel each other.

Let us go on to the next largest atom, helium, which has two protons (2+) in its nucleus and two electrons (2–), as seen in Figure 5.4. There are attractive forces between the nucleus and each of the electrons, and these attractive forces will attempt to keep each electron close to the nucleus. However, there is an additional force in this system of charges: the repulsive force operating between the two like-charged (–) electrons. To answer the question of how the electrons are arranged in the helium atom, you must ask: What is the arrangement of minimum energy of this three-component system that includes the two electrons (each negatively charged) and the positive nucleus?

It turns out that the ground-state orbital described earlier for the hydrogen atom can accommodate another electron. Placing the two electrons in a spherical volume, or orbital, surrounding the nucleus allows the electrons to remain well separated, yet satisfies the two principles that lead to a minimum-energy condition: namely, proximity to the nucleus and the ability to avoid one another (Figure 5.5). This spherical volume within which the two electrons of lowest energy in any atom may reside is called the *1s orbital.*

lithium atoms

What is the minimum-energy picture for the lithium atom, Li, which contains three electrons? If all three electrons are put in the 1*s* orbital, calculations show that the repulsion between electrons in that small volume is greater than the attractive energy from the pull of the nucleus on the electrons. Minimum energy can be obtained only when the *third* electron is located in a spherical volume around the nucleus that is larger than the volume of the 1*s* orbital. This new spherical volume is called the 2*s* orbital. You can envision the relationship between the 1*s* and 2*s* orbitals as a softball (1*s*) within a basketball (2*s*). The 2*s* orbital, like the 1*s*, can contain a maximum of two electrons (Figure 5.6).

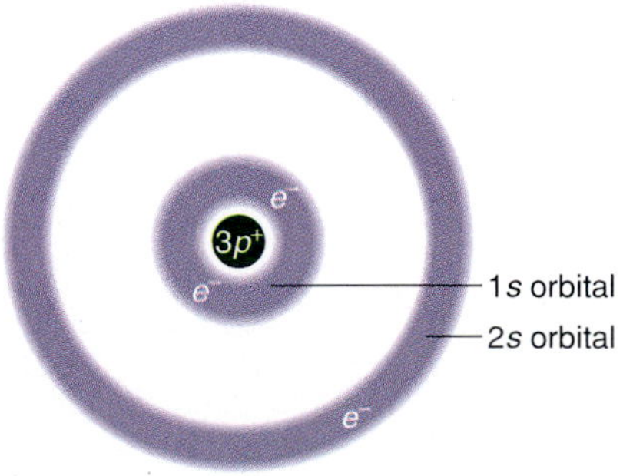

Figure 5.6
Cross section of the lithium (Li) atom. The third electron occupies the 2*s* orbital to minimize electron repulsion.

1. You may be wondering how so many positive particles (the protons) can cluster in the nucleus. This question is one of the major puzzles that physicists are trying to unravel. Fortunately for us, we need to know only that protons do cluster, and we can regard the nucleus as one clump of positive charge.

Figure 5.7
The three $2p$ orbitals are directed along the x, y, and z axes—three perpendicular axes that define three-dimensional space.

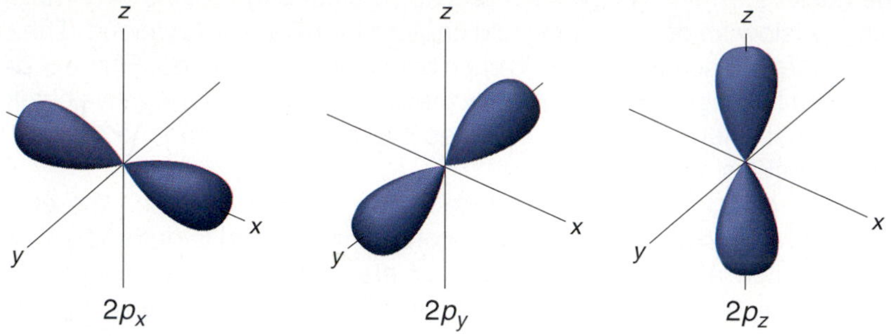

sample exercise 5

How are the four electrons of Be arranged?

solution

Two electrons are in the $1s$ orbital and two are in the $2s$ orbital. This arrangement maintains the electrons close to the nucleus, but does not exceed the limit of two electrons per orbital.

orbitals have various shapes

What is the minimum-energy picture for the boron atom, B, containing five electrons? The fifth electron cannot be in the $2s$ orbital because of repulsive forces with the two electrons already in that orbital. However, to minimize energy, it is always desirable to keep electrons close to the nucleus rather than move them significantly farther away than the $2s$ orbital. An orbital of a different shape accomplishes the juggling act of keeping the fifth electron close to the nucleus, but at the same time away from other electrons. However, unlike the $1s$ and $2s$ orbitals, which occur by themselves, the new orbital belongs to a set of three identical orbitals, each of which is called a $2p$ orbital and is shaped like a long balloon pinched in the middle (Figure 5.7). You can think of electrons in these $2p$ orbitals as being only slightly farther from the nucleus, on the average, than they would be if they were in the $2s$ orbital. However, repulsion between the $2s$ electrons and $2p$ electrons is minimized by the fact that they are located in different three-dimensional volumes. Since each of the three $2p$ orbitals is the same distance from the nucleus, each represents the same energy.

The three dimensions of space are represented mathematically by three mutually perpendicular axes, x, y, and z (Figure 5.7). There are three $2p$ orbitals: $2p_x$, $2p_y$, and $2p_z$, where the subscript tells you along which axis in space the orbital is oriented. Each $2p$ orbital can hold a maximum of two electrons, just as the $1s$ and $2s$ orbitals each hold a maximum of two electrons. Altogether, the three $2p$ orbitals can hold a maximum of six electrons, and electrons in all three of the $2p$ orbitals have the same energy.

orbitals and energy levels

By now you have begun to appreciate the arrangement of electrons in atoms of different elements. As you consider elements of increasing atomic number, there are more electrons to be accommodated around the nucleus. As you have seen for helium and lithium, the additional electrons will be located in orbitals with the lowest

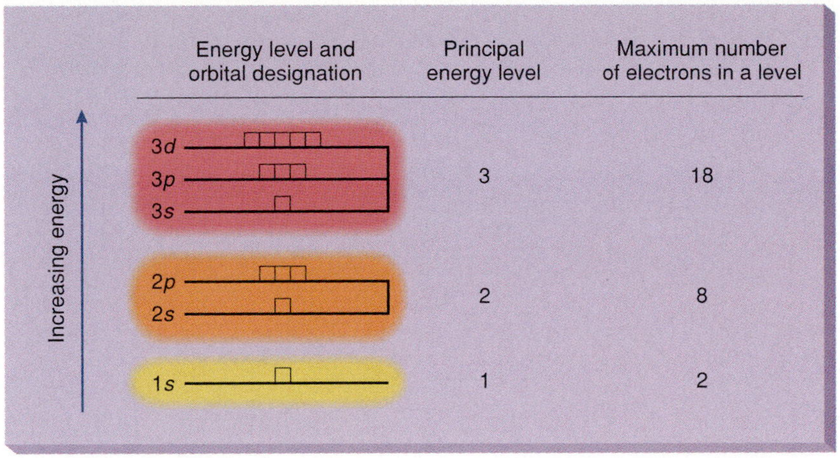

Figure 5.8
Electrons in an atom fill orbitals (boxes) of increasing energy. Orbitals in the same sublevel have the same energy; that is, the three 2*p* orbitals have the same energy, as do the three 3*p* orbitals and the five 3*d* orbitals. The energy spacings between the levels vary for different elements.

possible energy. The lowest energy orbital is the 1*s*. Since each orbital can contain no more than two electrons, additional electrons will occupy the orbital with the next lowest energy, which is the 2*s* orbital. Once that orbital is filled with two electrons, additional electrons will occupy one or more of the three 2*p* orbitals until all those orbitals are filled.

Since there are 109 elements in the periodic table, chemists must account for the possible arrangement of up to 109 electrons around the nucleus of the element whose atomic number is 109. Your task now is to learn what orbitals are available and how those orbitals are ordered with respect to energy so that you can understand the arrangement of electrons of any element.

The trends of the periodic table will soon become evident if the orbitals containing the electrons are grouped into levels based on energy (**energy levels**). With the information given so far about 1*s,* 2*s,* and 2*p* orbitals, it is possible to predict the proper location for the ten electrons of lowest energy in any atom. The lowest-energy level, which is closest to the nucleus, is called the *first principal energy level* and contains just a 1*s* orbital and a maximum of two electrons. The next energy level, which is more distant from the nucleus, is called the *second principal energy level* and contains both 2*s* and 2*p* orbitals. Since the 2*s* orbital may contain up to two electrons and the three 2*p* orbitals may each contain up to two electrons, the second energy level may contain a maximum of eight electrons. The first and second principal energy levels may thus contain a total of ten electrons—two electrons in the first level and eight in the second.

Because the second principal energy level contains two types of orbitals of slightly different energy, 2*s* and 2*p,* it is said to consist of sublevels, with the lower-energy 2*s* sublevel consisting of one 2*s* orbital and the slightly higher-energy 2*p* sublevel consisting of the three perpendicular 2*p* orbitals. You will see shortly that a third principal energy level contains three sublevels, the 3*s,* 3*p,* and 3*d* orbitals (Figure 5.8).

the third energy level

Electrons 11 and 12 in any atom are found to occupy a 3*s* orbital, another spherical orbital similar in shape to the 1*s* and 2*s,* but farther away from the nucleus. The next six electrons, electrons 13 to 18, are placed in a 3*p* sublevel (consisting of $3p_x$, $3p_y$, and $3p_z$ orbitals). The 3*p* orbitals have the same shape as the 2*p,* but on the

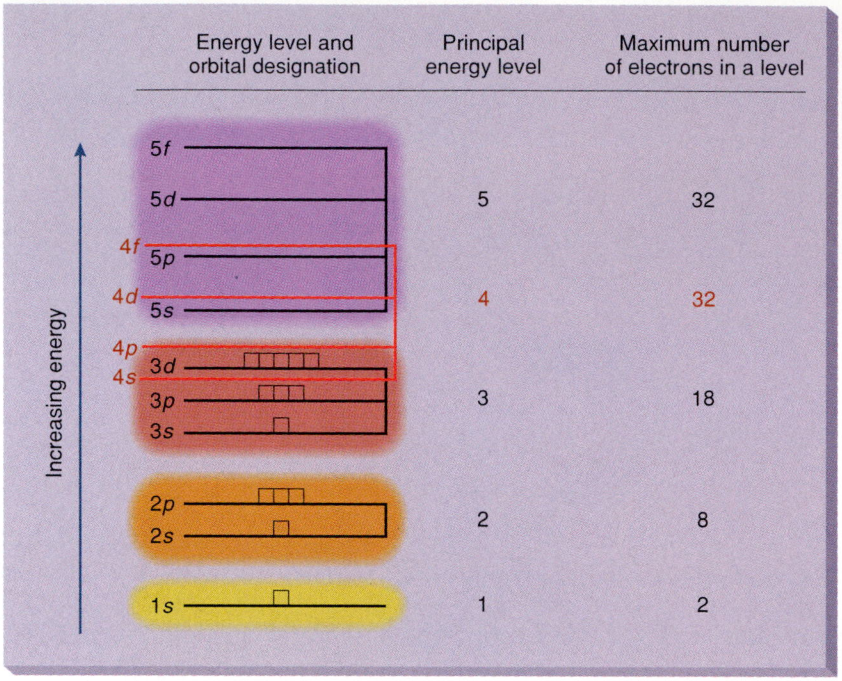

Figure 5.9

For higher-energy electrons, the principal energy levels overlap. The 4*s* orbital, the lowest sublevel in the fourth principal energy level, is lower in energy than the 3*d* orbital, the highest sublevel in the third principal energy level. Similarly, the 5*s* and 5*p* orbitals in the fifth principal level have lower energy than upper sublevels in the fourth principal energy level.

average they are farther away from the nucleus. Each orbital may still contain a maximum of two electrons. Therefore, the 3*p* sublevel may contain a total of six electrons among the three orbitals. Within the third energy level is another sublevel called the 3*d* sublevel, which consists of a set of five orbitals with unique shapes. Since the 3*d* sublevel may contain up to ten electrons (five orbitals with two electrons in each), the 3*p* sublevel may contain up to six electrons (three orbitals with two electrons in each), and the 3*s* orbital may contain up to two electrons (one orbital with up to two electrons), the third principal energy level may contain up to a total of 18 electrons.

It turns out, however, that the 3*d* orbitals are not the next orbital higher in energy than the 3*p* orbital. Rather, the lowest orbital in the next principal level, the 4*s* orbital, penetrates closer to the nucleus than the 3*d*, making the overall energy of an electron in a 4*s* orbital lower than one in a 3*d* orbital. Consequently, the total energy of an atomic system is found to be at a minimum when electrons 19 and 20 are contained in a 4*s* orbital.

As you can see in Figure 5.9, the third and fourth principal energy levels overlap; that is, the energy of the 4*s* is lower than the energy of the 3*d* orbital, so electrons 19 and 20 will fill the 4*s* orbital rather than the 3*d*.

Once the 4*s* is filled, the next ten electrons, electrons 21 to 30, occupy the five 3*d* orbitals. As you can see in Figure 5.10, the shapes of these orbitals are quite complex. These shapes need not concern you. The important idea is that there are five *d* orbitals in the *d* sublevel and that each orbital can hold a maximum of two electrons. As you go to higher principal energy levels, such as the fourth, fifth, and sixth principal energy levels, there are also *f* sublevels (Figure 5.9), each containing a set of seven orbitals or a maximum of 14 electrons. Knowing all of the sublevels enables you to specify the electron configuration for atoms of any element. Throughout this book, however, we will concentrate on the electron configurations of elements with low atomic numbers.

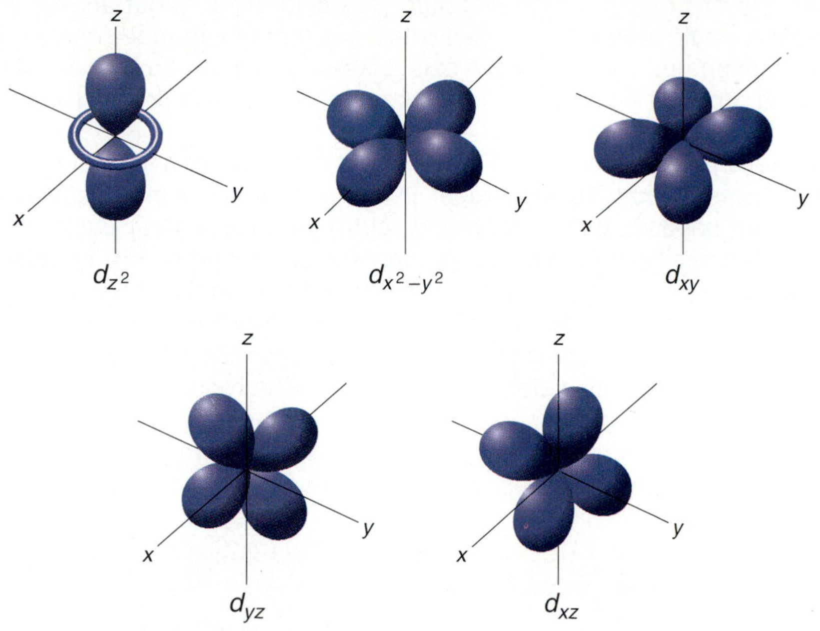

Figure 5.10

The shapes of the five *d* orbitals are complex, but they are the same whether they are *d* orbitals in the third, fourth, fifth, or sixth principal energy level. However, the size of the orbitals increases; that is, the average distance of the electron from the nucleus increases when the orbital belongs to a higher principal energy level. Each of the five *d* orbitals may contain up to two electrons for a maximum of ten electrons in a *d* sublevel.

sample exercise 6

For each of the following atoms, describe how the electrons are arranged.

a. Oxygen (8 electrons)
b. Aluminum (13 electrons)
c. Calcium (20 electrons)

solution

a. The 1*s* and 2*s* sublevels can each hold two electrons for a total of four. That leaves four more electrons, which can occupy the 2*p* sublevel, since the 2*p* can hold up to six electrons.
b. Two electrons in the 1*s,* two in the 2*s,* six in the 2*p,* and two in the 3*s* account for 12 electrons (2 + 2 + 6 + 2 = 12). Electron 13 must go into the 3*p.*
c. Two in the 1*s,* two in the 2*s,* six in the 2*p,* two in the 3*s,* and six in the 3*p* account for 18 electrons (2 + 2 + 6 + 2 + 6 = 18). Electrons 19 and 20 must go into the 4*s.*

problem 3

For each of the following atoms, describe how the electrons are arranged.
a. Sulfur (16 electrons)
b. Phosphorus (15 electrons)
c. Magnesium (12 electrons)

spin is a property of electrons

Scientists have found that electrons produce very small magnetic fields and act almost as if they were tiny bar magnets. This property leads us to believe that electrons are spinning, either clockwise or counterclockwise, because it has been known for a long time that any charged particle that spins produces a magnetic field. An **electron spin** in one direction, say clockwise, about an axis through the electron would produce a magnetic field like that of a magnet whose positive pole is pointing upwards along the axis. Likewise, a counterclockwise electron spin would produce a magnetic field whose positive pole is pointing in the opposite direction.

Magnetic poles behave just like electrical charges behave in electrostatics. Opposite poles attract; like poles repel. In energy terms, two magnets have lower energy when opposite poles (north and south) are placed near each other than they do when similar poles (north and north or south and south) are placed together. *When two electrons reside in one orbital, energy will be lower when the two electrons have opposite spin rather than when their spins are the same.* This requirement of opposite spins to ensure low energy also accounts for the existence of a maximum of two electrons per orbital. The requirement that electrons minimize their energy in the same orbital by having different spins is called the **Pauli exclusion principle.** If more than two electrons are placed in an orbital, then two electrons would necessarily have the same spin and the total energy would increase.

Summary of electron arrangement

Let us summarize briefly the picture of electrons in atoms you should have developed by now.

For the negatively charged electrons, low energy is linked with being close to a positively charged nucleus. At the same time, electrons repel one another and must remain as far apart as possible. To minimize their total energy, electrons are located within different orbitals (three-dimensional shapes) as close to the nucleus as allowed by their need to avoid other electrons. Each orbital can contain two electrons with opposite spins.

5.5 electron configuration notation

Chemists employ a system of shorthand notation to show the **electron configuration** of an atom, i.e., how the electrons within a given atom are arranged in orbitals within energy sublevels. For example, the fact that the one electron of hydrogen occupies the lowest-energy 1*s* orbital is abbreviated $1s^1$. The superscript 1 refers to the one electron in the 1*s* orbital. Helium has two electrons in the 1*s,* and the superscript 2 is used:

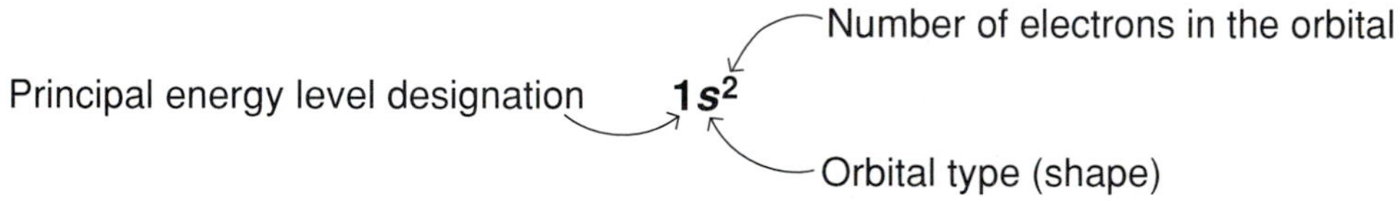

Now consider fluorine, which has nine electrons to be arranged. The lowest-energy arrangement calls for two electrons in the 1*s* orbital, $1s^2$, two electrons in the 2*s* orbital, $2s^2$, and the last five electrons distributed among the $2p_x$, $2p_y$, and $2p_z$ orbitals (abbreviated $2p^5$). The full fluorine configuration is abbreviated $1s^2$ $2s^2$ $2p^5$.

To write an electron configuration for any atom, begin by determining the atom's atomic number. Since, in a neutral atom, the number of protons always equals the number of electrons, the atomic number tells you the number of electrons in a neutral atom as well as the number of protons. Place the electrons in orbitals of increasing energy (Figure 5.9) until the total number of electrons is accounted for. In this notation system, *p* designates the set of three *p* orbitals (p_x, p_y, and p_z), or more rigorously, the sublevel of energy that distinguishes *p* orbitals from *s* orbitals. Consequently, the superscript of *p* may be as large as 6 if all three orbitals are occupied. Similarly, the superscript for *d* may be as high

table 5.1 electron configuration of first 18 elements in the periodic table

Element	Atomic number	Notation
Hydrogen	1	$1s^1$
Helium	2	$1s^2$
Lithium	3	$1s^2\,2s^1$
Beryllium	4	$1s^2\,2s^2$
Boron	5	$1s^2\,2s^2\,2p^1$
Carbon	6	$1s^2\,2s^2\,2p^2$
Nitrogen	7	$1s^2\,2s^2\,2p^3$
Oxygen	8	$1s^2\,2s^2\,2p^4$
Fluorine	9	$1s^2\,2s^2\,2p^5$
Neon	10	$1s^2\,2s^2\,2p^6$
Sodium	11	$1s^2\,2s^2\,2p^6\,3s^1$
Magnesium	12	$1s^2\,2s^2\,2p^6\,3s^2$
Aluminum	13	$1s^2\,2s^2\,2p^6\,3s^2\,3p^1$
Silicon	14	$1s^2\,2s^2\,2p^6\,3s^2\,3p^2$
Phosphorus	15	$1s^2\,2s^2\,2p^6\,3s^2\,3p^3$
Sulfur	16	$1s^2\,2s^2\,2p^6\,3s^2\,3p^4$
Chlorine	17	$1s^2\,2s^2\,2p^6\,3s^2\,3p^5$
Argon	18	$1s^2\,2s^2\,2p^6\,3s^2\,3p^6$

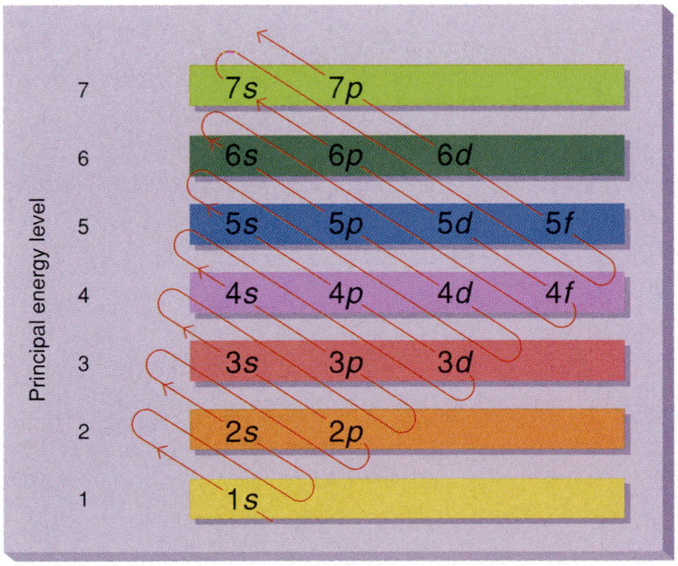

Figure 5.11
This ordering diagram is a shortcut to learning the order in which electrons fill orbitals in an atom. All the orbitals needed for each principal energy level are given. Follow the arrows from tail to head to find the orbital with the next higher energy. You will predict the correct electron configuration for most elements of the periodic table using this shortcut scheme.

as 10 because *d* designates the set of five *d* orbitals. Table 5.1 shows the lowest-energy arrangement of electrons for the first 18 elements.

Remembering the orbitals or sublevels in order of increasing energy is important for the first three principal energy levels. Fortunately, there is a device shown in Figure 5.11 for easily determining the energy order for levels above the third principal level. You may refer to Figure 5.11 in the future should you need to determine electron configurations of elements of higher atomic number.

sample exercise 7

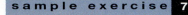

Write out the electron configuration of sodium.

solution

The atomic number of sodium is 11, which indicates that there are 11 electrons in the sodium atom. Place these electrons in orbitals in order of increasing energy, in accord with the maximum number that can fit into each sublevel: two electrons in *s*, six electrons in *p*, and ten electrons in *d* when necessary.

$$\text{Na} \quad 1s^2\,2s^2\,2p^6\,3s^1$$

The sum of the superscripts should equal the number of electrons in the atom, in this case 11.

problem 4

Write out the electron configuration for silicon (without looking at Table 5.1).

aufbau principle

The orderly placement of electrons, first in lower-energy orbitals and then in higher-energy orbitals, is known as the **Aufbau principle.** *Aufbau* means "buildup" in German. Another way of showing the electron configuration is by using boxes for orbitals and arrows for electrons. The boxes for the orbitals are arranged in order of increasing energy with the lowest-energy orbitals on the left and the highest on the right. Electrons in each orbital (box) are shown with arrows. When two electrons reside in the same orbital, the arrows pointing in opposite directions show that the two electrons have opposite spins. Consider Na, atomic number 11, as a first example.

Na	$1s^2$	$2s^2$	$2p^6$	$3s^1$
	[↑↓]	[↑↓]	[↑↓][↑↓][↑↓]	[↑]

Increasing potential energy as electron distance from the nucleus increases →

The separation of boxes indicates differences in energy. The drawing of the three 2*p* boxes together indicates that the three orbitals all have the same energy, i.e., the 2*p* sublevel contains three orbitals all of the same energy. Consider Cl, atomic number 17, as another example.

Cl	$1s^2$	$2s^2$	$2p^6$	$3s^2$	$3p^5$
	[↑↓]	[↑↓]	[↑↓][↑↓][↑↓]	[↑↓]	[↑↓][↑↓][↑]

N, atomic number 7, presents us with another consideration.

N	$1s^2$	$2s^2$	$2p^3$
	[↑↓]	[↑↓]	[↑][↑][↑]

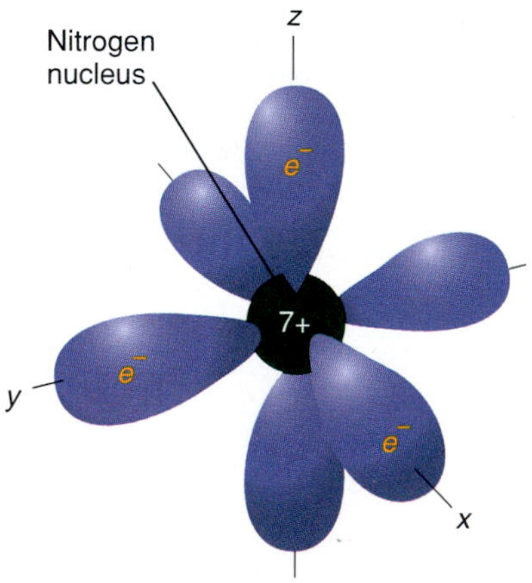

Figure 5.12
The three *p* orbitals are all the same distance from the nucleus, and the one electron in each will feel the same attractive force to the nucleus. However, by residing in separate orbitals, each electron is as far from another electron as it can possibly be.

Notice that the electrons enter the 2*p* orbitals singly, with the same spin, before any pairing takes place. This is completely consistent with our idea that low energy is associated with maximum separation of electrons, but at the minimum distance from the nucleus. The principle that electrons enter orbitals of equal energy singly before they become paired is called **Hund's rule.** The lowest-energy arrangement for the three *p* electrons in nitrogen is shown in Figure 5.12, where one electron is in each of the p_x, p_y, and p_z orbitals.

At this point, you should be able to make a diagram, complete with boxes and arrows, for the electronic configuration of any A group element. Remember:

1. The atomic number of the element tells you the total number of electrons in the atom.
2. Minimization of energy tells you that lower-energy sublevels are completely filled with electrons before electrons enter higher-energy sublevels.
3. Hund's rule tells you that one electron goes into each orbital of equal energy, such as in *p* orbitals and *d* orbitals, before pairing of electrons in the orbitals takes place.

sample exercise 8

How many single (unpaired) electrons are there in oxygen?

solution

The distribution of electrons in oxygen is described in Sample Exercise 6a and shown in Table 5.1 to be $1s^2\,2s^2\,2p^4$. Showing the boxes-and-arrows diagram and following Hund's rule, you see

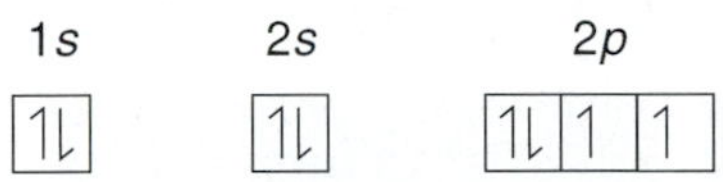

Thus, you predict *two* single (unpaired) electrons in the 2*p* sublevel. The presence of two single (unpaired) electrons in the oxygen atom may also be observed experimentally. When unpaired electrons are present in the atoms of an element, the element exhibits paramagnetic behavior; that is, the material is attracted to a magnetic field. Oxygen atoms are paramagnetic and you are able to predict this behavior directly from their electron configuration.

••• problem 5

How many single (unpaired) electrons are there in phosphorus?

5.6 quantum numbers

Everyone is familiar with being identified by a set of numbers. For example, no doubt you have a Social Security number. Electrons are identified by a set of four *quantum numbers* that come from the quantum mechanics we mentioned in Section 5.4 as the basic theory of electron arrangement. Every electron in an atom has a set of four quantum numbers, each of which reveals specific information about a different aspect of the electron.

You are already familiar with the four criteria for identifying each electron in an atom: the principal energy level, the energy sublevel, and the orbital within which the electron resides account for three criteria, and the spin of the electron accounts for the fourth criterion. Quantum mechanics simply assigns a number to each of these criteria for convenience in handling a large number of electrons. If you continue with more advanced chemistry courses, you will learn the definition of the quantum numbers so that you can handle more complicated elements with ease. But for now, you have sufficient knowledge to predict the chemical properties and trends within the periodic table based on the electron configurations of the elements.

5.7 evidence and uses of electron energy levels

You have now seen that the arrangement of electrons in an atom is based on the buildup of electrons in discrete energy levels, with low-energy levels close to the nucleus and high-energy levels farther from the nucleus. In the undisturbed ground state, the electrons occupy the lowest-energy levels consistent with the ideas of energy minimization presented in Section 5.4.

Figure 5.13
Excited electrons in the six elements (a) lithium (Li), (b) sodium (Na), (c) potassium (K), (d) calcium (Ca), (e) strontium (Sr), and (f) barium (Ba) each give off a characteristic visible light when they lose energy and fall back to their ground state.

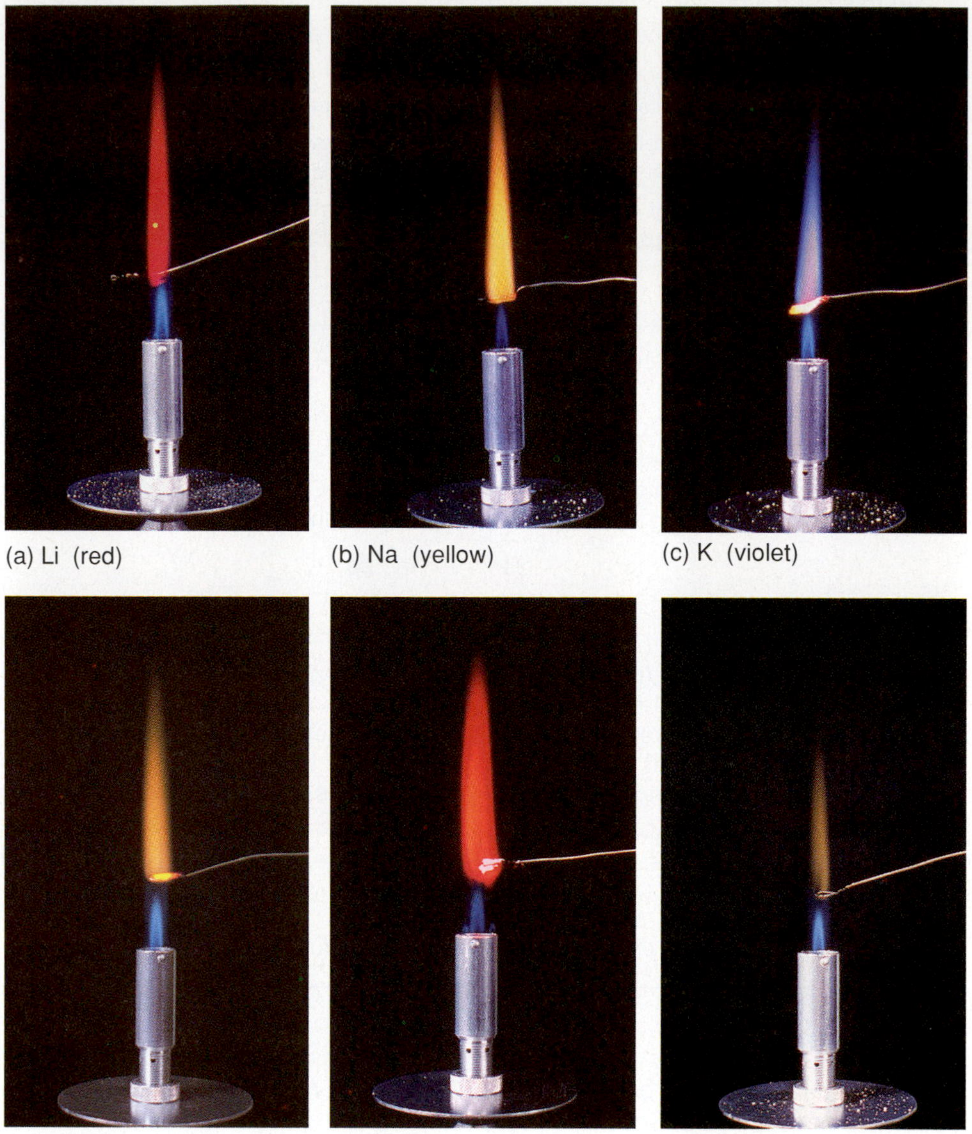

(a) Li (red) (b) Na (yellow) (c) K (violet)

(d) Ca (yellow) (e) Sr (red) (f) Ba (yellow)

However, energy can be put into a ground-state atom and the electrons will move to an excited energy state, which is identified with an electron configuration of higher energy. For example, if we heat atoms in a flame, we are adding heat energy. As a result of this added energy, one or more electrons may be raised to a higher-energy level and the atom is then said to be in an excited state. Remember that any electron configuration that does not have the lowest energy is called an excited state and that each atom has various possible excited states. Because an excited state is unstable, the electrons in an excited state will fall back to the lower levels of the ground state within a very short period of time. When the electrons lose energy, the lost energy will be given off (or emitted) as light, which you can detect. The colors of light that each element emits are a property of that element and can be used to identify that element.

In Figure 5.13, six different elements have been placed on the end of a nichrome wire and heated in a Bunsen burner flame. As a result of being heated, the ground-state electrons of each element are raised to a higher-energy, excited state. When the excited electrons fall back to their ground state, they emit a specific color light

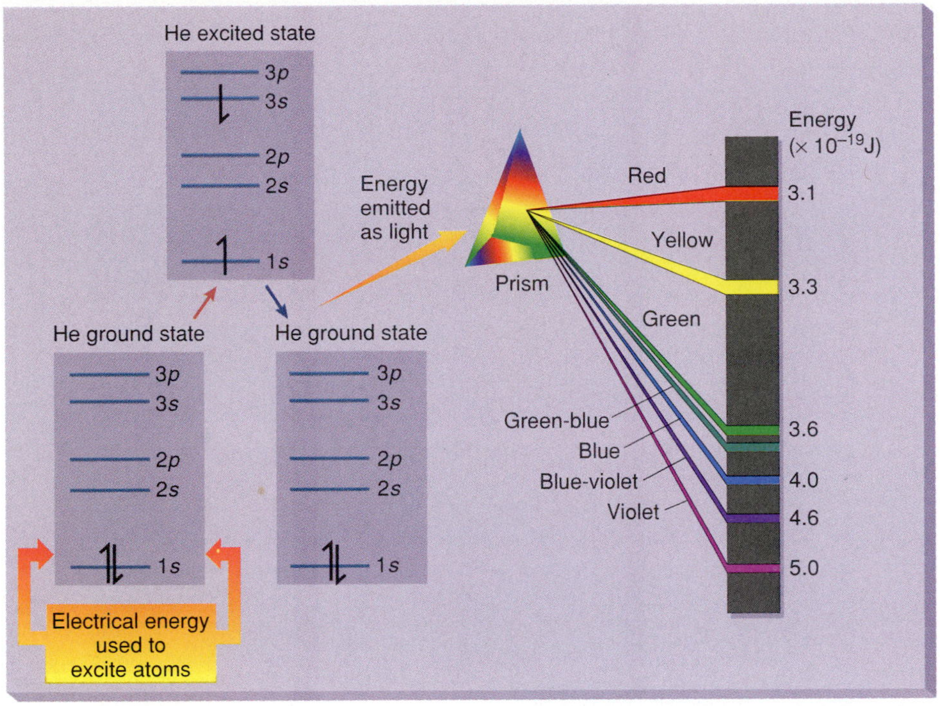

Figure 5.14

An electrical discharge excites helium atoms, promoting electrons to higher-energy levels. In this example, a 1*s* electron is promoted to the 3*s* orbital by the input of electrical energy. The light emitted as electrons from various excited states fall back to their ground state has been separated by a prism.

that is characteristic of that element. Lithium and strontium emit red light, sodium, calcium, and barium emit yellow light, and potassium emits violet. Even though two elements emit red light and three elements emit yellow light, there is considerable variation in these red and yellow lights, so that under more careful color (spectral) analysis each element can be identified uniquely.

Light is a form of radiant energy or radiation. You know it is energy because it is capable of doing work or effecting a change. For example, light provides you with a suntan or sunburn, or it can expose photographic film. Visible light from the sun or a light bulb is usually white light. A prism can divide white light into a rainbow. Scientists call the array of colors in the rainbow a *spectrum.* (Raindrops in the atmosphere sometimes act as prisms in the sky and produce a rainbow.) What the prism and rainbow (spectrum) tell us is that white light is a mixture of light rays of different colors. Each color of visible light has associated with it a distinct energy; violet light corresponds to higher energies and red light corresponds to lower energies. The definition of a **spectrum** is that it is an array of energies.

Excited atoms of an element do not emit the total spectrum of energies (white light). Rather, they emit a spectrum of discrete colors, each with its own energy, that is unique to that element, as you saw in Figure 5.13. The unique spectrum of each element results because the energy levels of each element (lithium, sodium, potassium, etc.) are uniquely spaced and electrons can occupy only these states. When the excited electrons of a particular element fall back to their ground state, the difference in energy is unique to the spacing of energy levels in that element. Therefore, a particular excited element will emit its own unique spectrum as electrons fall back to the ground state. This, in turn, means that there will be a unique spectrum of discrete colors in the visible light that is emitted. The emitted radiation can be precisely analyzed with an instrument known as a **spectroscope,** which separates the components of the visible light into distinct bands of color.

The use of a prism as a spectroscope is shown in Figure 5.14, which presents a detailed description of the light emitted by excited helium atoms as they fall back to their ground state. In this figure, you see that the light emitted by helium atoms is a

Figure 5.15

White light, shown at the bottom of this figure, is a continuous spectrum of energies. The spectrum of discrete color for Na, H, Ca, Hg, and Ne is unique for each of these elements. Consequently, discrete spectra are used to identify the kind and amount of these and other elements on distant stars, as well as in samples of food, water, and body fluids.

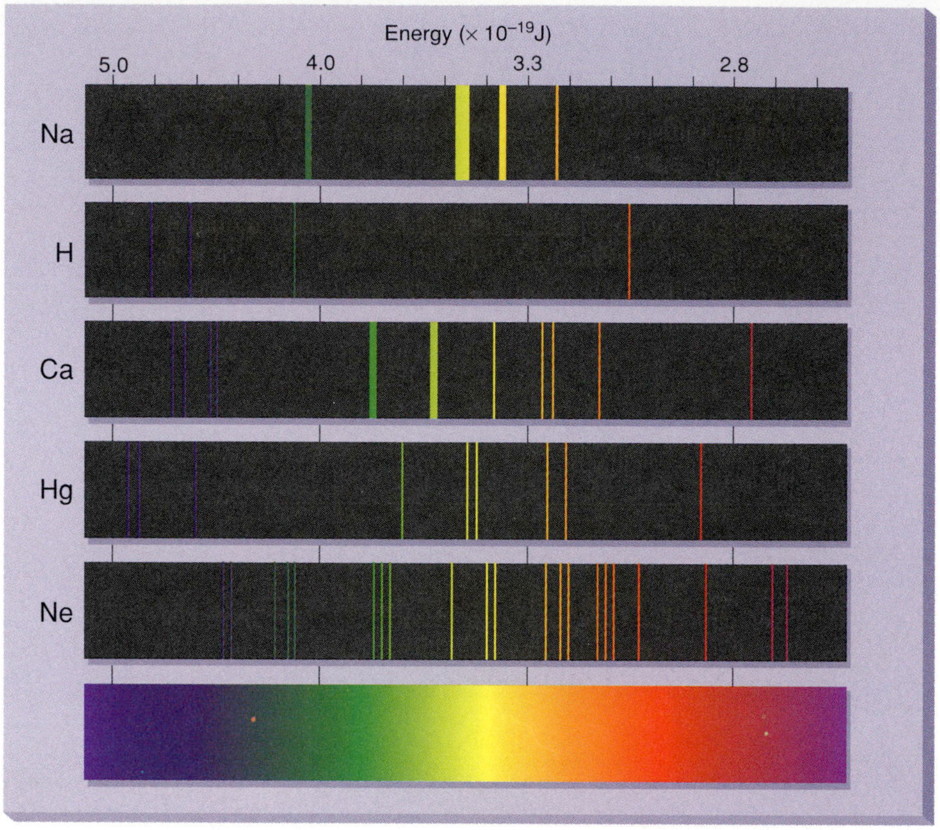

mixture of seven distinct colors, each with its own energy. The combination of these colors defines a visible spectrum unique to helium, and whenever that same combination occurs in visible light, chemists can be sure that helium atoms are present. By analyzing the spectrum of light coming from the sun or other stars, chemists can identify the presence or absence of helium by determining whether the helium spectrum is contained in the spectrum of light they are observing. Similarly, the presence of other elements can be confirmed if the known spectrum of each particular element can be found within a sample of light being studied (Figure 5.15).

Also, from Figure 5.15, you can understand why the light emitted by sodium and calcium appears as different shades of yellow to your eye in Figure 5.13. The sodium spectrum contains seven distinct bands of color, while the calcium spectrum contains 16 distinct bands of color. The spectra of both elements appear predominantly yellow to your eye, but result from a very different mixture of colors.

Routine medical and toxicological tests use emission spectra to identify the presence of elements. The concentration of sodium ion in blood plasma and in various foodstuffs (after being made water-soluble) can be readily determined by flame emission spectroscopy. A solution containing the sodium ion is sprayed into a high-intensity flame, which excites the sodium. The ion then emits its light spectrum. The emitted light passes through a filter that is selective for the yellow light in the sodium spectrum. The intensity of the light passing through the filter is proportional to the concentration of the sodium ion in the original solution.

The concentrations of many other ions in physiological fluids can be measured by a procedure known as atomic absorption spectroscopy. Here, a selected amount of energy, which can be varied for different elements, is pumped into a sample; electrons are promoted to higher levels by this energy. The amount of energy absorbed by the sample can be quantitatively measured and related to the concentration of the ion in question.

CHEMLAB

Laboratory exercise 5: emission spectra of elements

As you have learned in Section 5.7, emission spectra of elements may be used to identify the presence of elements in samples. For this laboratory exercise, you will use gas discharge tubes, each containing a different element. The electrons of each gas will be excited by a high-voltage power supply.

Record the color of the light emitted by each gas, as seen by your eye, as the excited electrons of the gas fall back to their electronic ground state. Then, use a spectroscope to observe the distinct bands of color composing the spectrum of that gas (Figure 5.16). Record the colors of the spectrum for each gas. Identify each of your samples by comparing your recorded spectrum to spectra of known elements such as those in Figure 5.15.

Questions:

1. Use Figure 5.15, a chart showing selected spectra of elements, to identify the following samples.
   a. Four distinct bands of color were observed: two in the blue-violet range, one green, and one reddish-orange.
   b. Seven distinct bands of color were observed: two green and five in the yellow-orange range.
2. When observed with the naked eye, both calcium and sodium emit yellow light. How would you experimentally distinguish a sample of calcium from a sample of sodium using their emission spectra?
3. Neon lights are used in many illuminated signs and billboards. How could you experimentally confirm that neon is responsible for the orange color of the light?

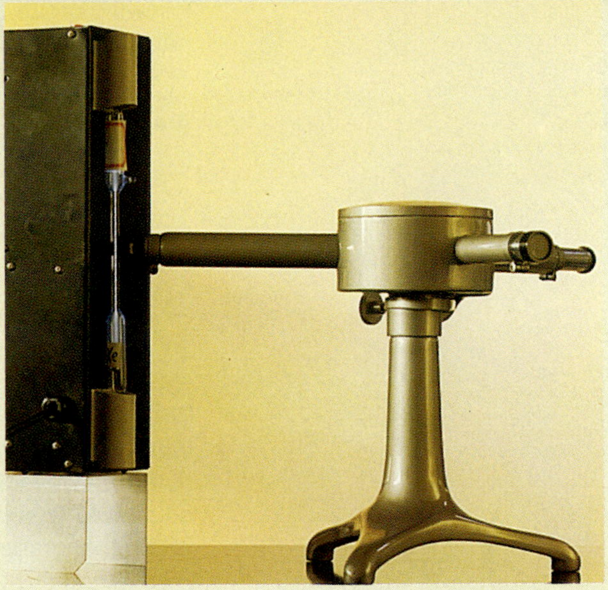

Figure 5.16
The blue glow from xenon gas in the gas discharge tube on the left may be observed in greater detail with the spectroscope shown on the right.

Using these two analytical methods, medical and food-testing laboratories can quickly and precisely obtain concentrations of almost every metal ion in physiological fluids, foodstuffs, and other samples of biological matter. Forensic specialists and ecologists also employ these techniques extensively.

5.8 electron configurations in the periodic table

Can the knowledge of electron configuration provide a deeper understanding of the periodic table and periodic trends? To find out, let's write the electron configurations of lithium, sodium, potassium, and rubidium, elements from Group IA of the periodic table.

Atomic number	Element	Electron configuration
3	Li	$1s^2\,2s^1$
11	Na	$1s^2\,2s^2\,2p^6\,3s^1$
19	K	$1s^2\,2s^2\,2p^6\,3s^2\,3p^6\,4s^1$
37	Rb	$1s^2\,2s^2\,2p^6\,3s^2\,3p^6\,4s^2\,3d^{10}\,4p^6\,5s^1$

These four elements exhibit very similar chemical and physical properties (Section 4.11). Can you detect the similarity in their electron configurations?

In each of these elements, there is one electron in the outermost energy level. The outermost levels for Li, Na, K, and Rb are $n = 2$, 3, 4, and 5, respectively. Could it be that the properties of an element are determined by the number of electrons in its outermost energy level? After all, the outermost electrons are farthest from the nucleus and most likely to interact with other atoms. Let us examine the electron configurations of some elements in Group VIIA of the periodic table. Notice that in the following group, the sublevels are not arranged in the order in which they are filled; rather, the sublevels within each major energy level are arranged together in order of increasing distance from the nucleus.

Atomic number	**Element**	**Electron configuration**
9	F	$1s^2\,2s^2\,2p^5$
17	Cl	$1s^2\,2s^2\,2p^6\,3s^2\,3p^5$
35	Br	$1s^2\,2s^2\,2p^6\,3s^2\,3p^6\,3d^{10}\,4s^2\,4p^5$
53	I	$1s^2\,2s^2\,2p^6\,3s^2\,3p^6\,3d^{10}\,4s^2\,4p^6\,4d^{10}\,5s^2\,5p^5$

If chemical tests are performed on these elements as discussed in Section 4.12, it is found that fluorine, chlorine, bromine, and iodine have very similar chemical properties. Each of these elements has seven electrons in the outermost level. The outermost levels for F, Cl, Br, and I are $n = 2, 3, 4$, and 5, respectively. The idea that the chemistry of an element is determined by the number of electrons in its outermost level seems to be correct, and moreover, the group number in the periodic table is simply the number of electrons in the outermost level for elements within that group. Electrons in the outermost level are called **valence electrons.**

Stating this principle somewhat differently, lithium, sodium, potassium, and rubidium are in the same group in the periodic table because they have the same number of valence electrons (one). *In general, for the A group elements, all elements within a group have the same number of valence electrons. The number of valence electrons is the group number.*

sample exercise 9

How many electrons are there in the outermost level of Group VA elements?

solution

Five, the group number tells you. Check any member of the group, for example, N, $1s^2\,2s^2\,2p^3$. There are five electrons in the outermost second level.

Note also for this group that all the elements will have three unpaired electrons in their ground-state configuration. Each of the three $2p$ electrons will occupy one of the three $2p$ orbitals according to Hund's rule.

••• problem 6

How many electrons are there in the outermost level of Group VIIIA elements?

You may check this generalization further by looking at a modified periodic table (Figure 5.17). You find that all elements within a particular A group have the same outer-level electron configuration. In Groups IA and IIA (*s*-block elements), the *s* sublevel is being filled, whereas in Groups IIIA through VIIIA (*p*-block elements), the *p* sublevel is being filled. In Chapter 4, the discussion of the periodic table pointed out that the properties of elements repeat themselves at certain fixed intervals. Here in Chapter 5, it appears that the underlying basis for the periodic repetition of chemical properties is that the valence-electron configuration is repeated at certain fixed intervals.

By noting the period number and the group number of a particular A group element, you can immediately write down the outer-level electron configuration. The period number is the level number. The group number is the total number of

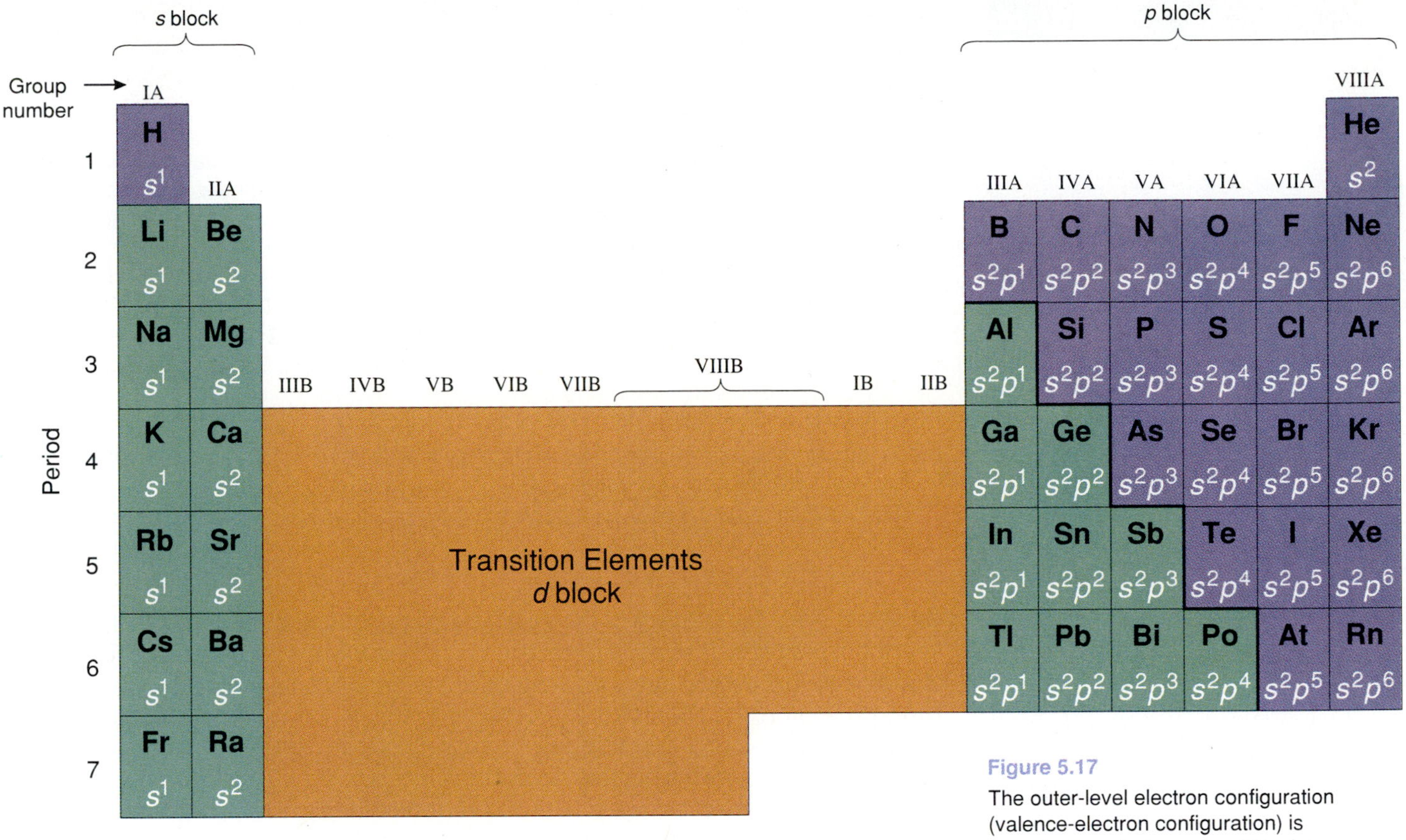

Figure 5.17
The outer-level electron configuration (valence-electron configuration) is repeated for all elements within the same group of the periodic table.

electrons that must be distributed between the *s* and *p* sublevels, with two going into the *s* and the rest into the *p* sublevel. For example, for nitrogen, the period number is 2 and the group number is VA:

$$\text{N:}\quad \underbrace{2s^2\ 2p^3}_{\text{Total of 5 valence electrons}}$$

Total of 5 valence electrons

N: $2s^2$ $2p^3$

↑ Period number

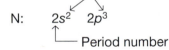

In Chapter 6, the key role that the valence electrons play in the bonding properties of an atom will be seen. You should be able to write the outer-level electron configuration for any A group element by simply noting its position in the periodic table.

Notice that these rules are restricted to the groups in the periodic table labeled A. The elements in these groups are known as the *representative* (or main-group) *elements.* For these A group elements, all sublevels in the levels below the outermost level are completely filled.

For the *nonrepresentative* (*or transition*) B group elements, the inner levels are not always completely filled. For example, the electron configuration of vanadium (atomic number 23) is

$$\text{V:}\quad 1s^2\,2s^2\,2p^6\,3s^2\,3p^6\,3d^3\,4s^2$$

You can see that the inner 3*d* sublevel is only partially filled with three electrons. Since the *d* sublevel has five orbitals, it could contain up to ten electrons. We will discuss the consequences of an unfilled inner sublevel later in Section 5.10.

sample exercise 10

Write the electron configuration for the valence electrons of sulfur.

solution

From Figure 5.17, note that sulfur is in the third period, which means that the valence electrons must be in the $n = 3$ level. Sulfur is in Group VIA, which means that there are six valence electrons. Filling the six electrons into the third level produces the arrangement $3s^2$ $3p^4$ for the valence electrons of sulfur. Once again, note that sulfur atoms, like other elements in this same group, have two unpaired electrons. Three of the four *p* electrons each enter one of the 3*p* orbitals; once one electron is in each, the fourth electron pairs up with an electron in one of these orbitals, leaving two *p* orbitals with a single electron in each.

problem 7

Using the periodic table in the inside cover of your text, write out the electron configuration for the valence electrons of oxygen.

5.9 Lewis electron dot structures

Because of the importance of the valence electrons, chemical symbols that represent them are often used and are especially important when you consider the bonding of atoms to form compounds in Chapter 6. These symbols are called **Lewis electron dot structures** after the theoretical chemist G. N. Lewis, who developed them.

A Lewis dot structure for an element consists of the element's symbol surrounded with dots to represent the number of valence electrons. To write a Lewis dot structure, write down the symbol of the element and surround the symbol with a number of dots corresponding to the number of valence electrons. Remember that, for an A group element, the group number is equal to the number of valence electrons (the number of electrons in the outermost level). Dots representing valence electrons are placed around the symbol, one on each side, until all four sides are occupied if necessary. For elements with more than four valence electrons, a second dot is placed on each side until all valence electrons have been distributed. No more than two dots are allowed at any given edge of the symbol. Note that single dots may remain on any side of the symbol. For example, the two single dots in oxygen may appear on any two sides of the O symbol, not just on the sides shown below.

H·　　:Ö·　　K·　　:F̤̈:　　·Ȧl·

problem 8

Write Lewis dot structures for the following elements: Na, N, Br, C, Ne, and Mg.

5.10 periodic trends

The arrangement of electrons with which you are now familiar not only explains the existence of chemical groups, it also accounts for other observable trends in properties. Look at the special periodic table in Figure 5.18. This periodic table tells the atomic size and ionization energy for every representative group element. The top and side of this periodic table are marked with the trends in these properties among the elements.

Ionization energy is defined as the amount of energy needed to remove one electron from each of the atoms in a gaseous sample. For example, the ionization energy of sodium is 494 kJ for a 23-g sample. Notice in Figure 5.18 that the ionization energy increases as you proceed across a period from left to right; that is, it becomes more

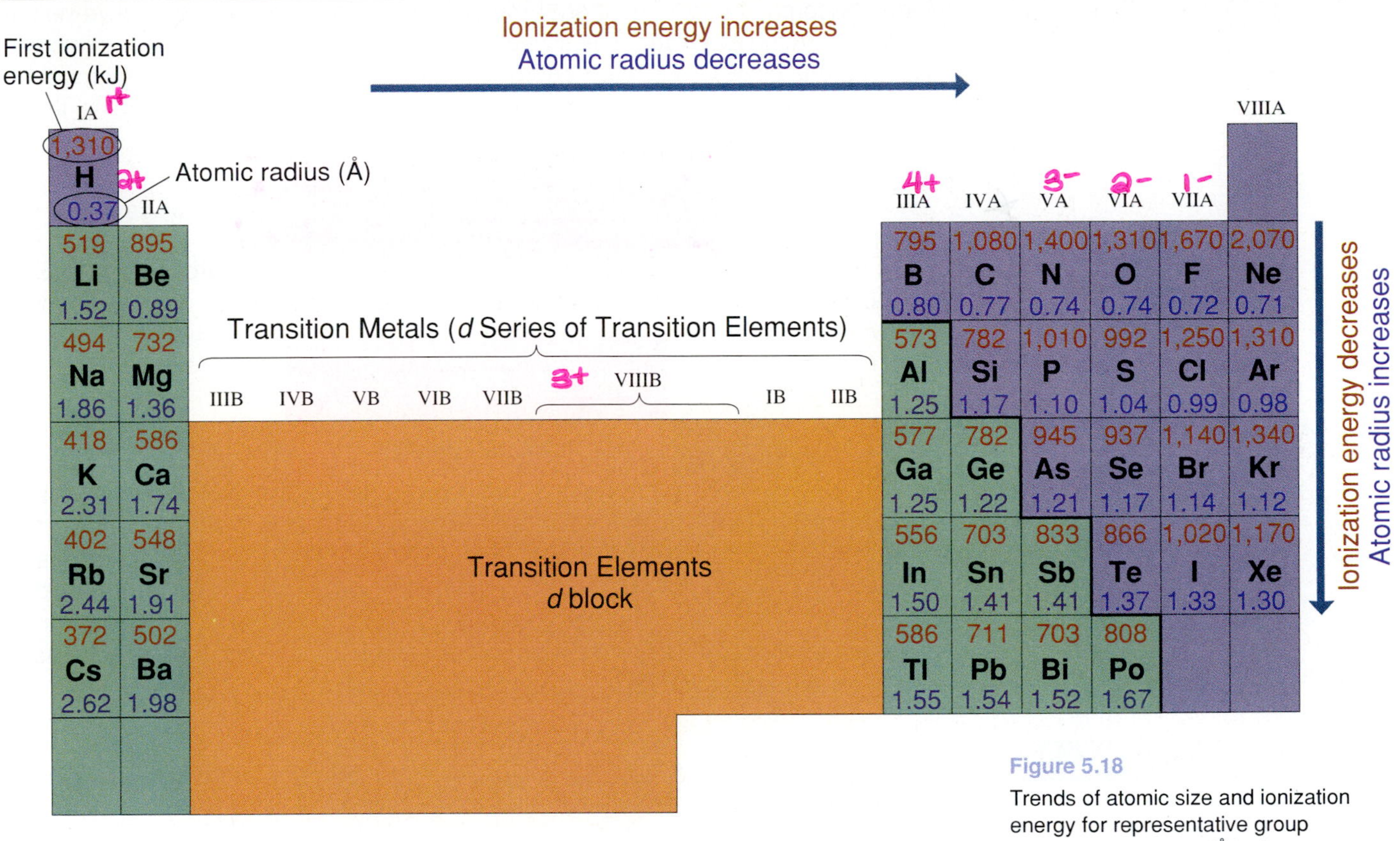

First ionization energy (kJ)

Ionization energy increases
Atomic radius decreases

Atomic radius (Å)

IA													IIIA	IVA	VA	VIA	VIIA	VIIIA
1,310																		
H	IIA																	
0.37																		
519	895												795	1,080	1,400	1,310	1,670	2,070
Li	**Be**												**B**	**C**	**N**	**O**	**F**	**Ne**
1.52	0.89												0.80	0.77	0.74	0.74	0.72	0.71
494	732												573	782	1,010	992	1,250	1,310
Na	**Mg**	IIIB	IVB	VB	VIB	VIIB		VIIIB		IB	IIB		**Al**	**Si**	**P**	**S**	**Cl**	**Ar**
1.86	1.36												1.25	1.17	1.10	1.04	0.99	0.98
418	586												577	782	945	937	1,140	1,340
K	**Ca**												**Ga**	**Ge**	**As**	**Se**	**Br**	**Kr**
2.31	1.74												1.25	1.22	1.21	1.17	1.14	1.12
402	548												556	703	833	866	1,020	1,170
Rb	**Sr**												**In**	**Sn**	**Sb**	**Te**	**I**	**Xe**
2.44	1.91												1.50	1.41	1.41	1.37	1.33	1.30
372	502												586	711	703	808		
Cs	**Ba**												**Tl**	**Pb**	**Bi**	**Po**		
2.62	1.98												1.55	1.54	1.52	1.67		

Transition Metals (*d* Series of Transition Elements)

Transition Elements
d block

Ionization energy decreases
Atomic radius increases

Figure 5.18

Trends of atomic size and ionization energy for representative group (Group A) elements. (Å is the abbreviation for angstrom, a unit of length equivalent to 1×10^{-8} cm.)

difficult to remove an electron as you move across the period from metals to nonmetals. In addition, the atomic size of elements shrinks from left to right across a period.

Both of these effects are based on the same feature of electron structure: the nuclear charge (atomic number) increases from left to right, but the added electrons are being placed in the same outermost or valence level. For example, the buildup of electrons in period 3 can be represented as follows.

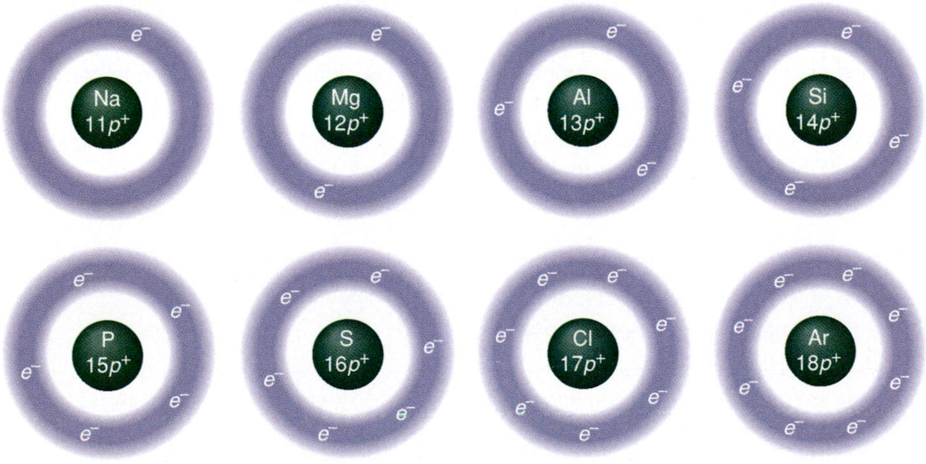

Because the valence electrons are all in the same principal energy level, the distance between the positively charged nucleus and the valence electrons is not significantly increased going across a period; however, the positive charge on the nucleus becomes larger. The bigger the charge, the stronger the attractive pull of the nucleus on the electrons. The size of the atoms decreases because the nucleus pulls electrons closer. Ionization energy increases because more energy must be applied to remove electrons from the stronger nuclear pull. Summarizing these horizontal trends, you find proceeding left to right across a period:

Decreasing atomic size →
Increasing ionization energy

Let us now examine vertical trends, that is, trends within a group. Figure 5.18 also shows you that atomic size increases and ionization energy decreases moving down a group:

Increasing atomic size	↓	Decreasing ionization energy

These trends arise because, proceeding down a group, the valence electrons of each element are found in a new principal energy level that is farther away from the nucleus. For example, in Group IA, the outer level for Na is 3; for K, 4; for Rb, 5; and for Cs, 6. The atomic size increases because electrons are found at larger distances from the nucleus. It is easier to remove electrons from an atom when they are farther away from the nucleus and feel the nuclear pull less strongly. As a result, ionization energy decreases as atoms become larger.

sample exercise 11

Using only the periodic table on the inside cover of this text, predict which atom in each of the following sets is larger and which has the higher ionization energy.

a. As and N
b. Ca and Br
c. Cl and Rb

solution

Refer to the preceding summarization of horizontal and vertical trends.

a. These elements are in the same group. Arsenic is larger because it is below nitrogen in the periodic table. The ionization energy of arsenic is smaller for the same reason.
b. These elements are in the same period. Bromine is smaller than calcium because it is farther to the right. Bromine has a larger ionization energy for the same reason.
c. Rubidium is larger because it is both below and to the left of chlorine. The ionization energy of rubidium is smaller for the same reason.

••• problem 9

a. Which of the A group elements have the lowest ionization energy?
b. Are these metals or nonmetals?

••• problem 10

In general, across a period, which are the larger atoms, metals or nonmetals?

Consulting *only* the periodic table on the inside cover, predict the following.

a. Which atom has the highest ionization energy?

b. Which has the lowest ionization energy?

problem 11

You saw in Section 5.8 that nonrepresentative (transition) B group elements may contain an unfilled inner energy level. In general, for the nonrepresentative elements, a *d* sublevel in a principal energy level below the outermost is unfilled even though there are two electrons in the *s* sublevel of the outermost energy level. The consequence of this is that these elements do not exhibit the dramatic periodic trends seen for A group elements. For example, there is little variation in ionization energy among the elements with atomic numbers 21 to 30 (Sc to Zn). The inner 3*d* sublevel is filled as you go from $_{21}Sc$ to $_{30}Zn$, but there is no change in outer electron configuration and essentially no change in ionization energy. You see a similar pattern as the 4*d* orbitals fill in $_{39}Y$ through $_{48}Cd$ (Figure 5.19).

Many of the B group elements (nonrepresentative or transition elements), which are all metals, have important industrial applications and are required in body chemistry. For example, iron is processed into steel and also occurs in hemoglobin, a component of red blood cells essential to the oxygen-carrying capacity of blood.

As you progress in your study of chemistry, you will find numerous other uses for this unique compilation known as the periodic table. It should be as helpful to you in studying chemical properties as your scientific hand calculator is in performing numerical calculations.

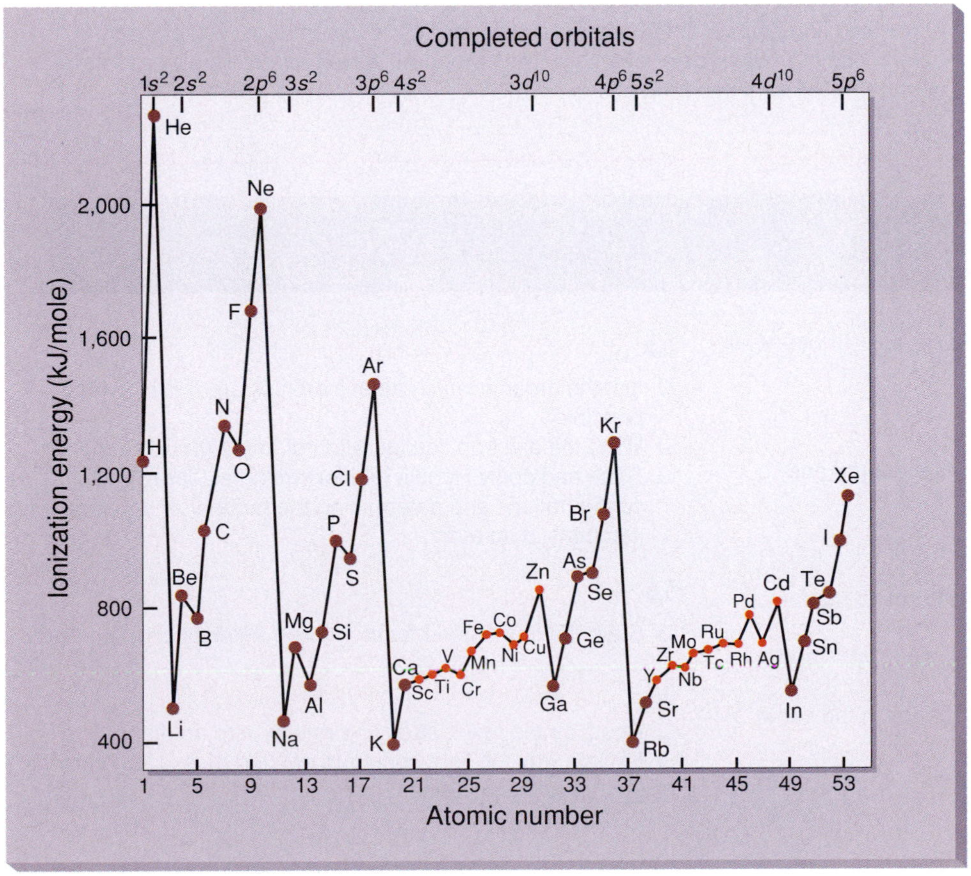

Figure 5.19

The ionization energy increases strongly with increasing atomic number for elements within the same period. For example, see the increase in ionization energy for period 2 (Li through Ne) and period 3 (Na through Ar). The ionization energy of transition metal elements (nonrepresentative elements) does not increase dramatically with increasing atomic number because of the partially filled *d* sublevel. For example, see period 3 (Sc through Zn) and period 4 (Y through Cd).

CHEMQUEST

Exciting electrons to a higher-energy state is a common occurrence in everyday life. For example, heat produced from combustion is used to excite electrons of the elements contained in fireworks. When the excited electrons of each type of element in the fireworks lose energy, they emit a distinctive color of visible light, characteristic of that element, resulting in a captivating visual display.

Both elements and compounds may also have their electrons excited by means other than combustion. The energy from visible light, shining on an element or compound, may be absorbed by the electrons to excite them to higher energy levels. In this case, visible light is being *absorbed* not *emitted.* As with elements, each compound will absorb one or more specific colors of light, each color corresponding to a specific energy state to which its electrons may be excited.

The principle of compounds absorbing specific colors of light has been fundamental to the commercial dye and paint industries. When you observe blue jeans under white light (sunlight), the jeans appear blue because only blue light is being reflected back to your eye from the jeans. The other colors in white light have been absorbed by the compounds, called dyes, attached to the surface of the jeans. In general, dyes are compounds, or mixtures of compounds, which absorb specific colors of visible light. They are prepared by chemists who either synthesize these compounds or extract them from natural sources such as lichens, insects, seashells, and plants.

Some compounds absorb ultraviolet radiation, a component of sunlight that has a higher energy than visible light. The exposure of your skin to ultraviolet radiation over a long period may be related to the development of skin cancer. The exposure of rubber, vinyl, and leather in your automobile to ultraviolet radiation hastens the deterioration of these components. As a preventative, you may apply sunscreen to your skin and a protective coating to your car parts.

Report on the types of compounds used in sunscreens and vinyl protectors. Explain how these compounds act as a preventative with as much detail as you can. What determines the rating factor of a sunscreen product? What proven effectiveness has been reported for these products?

chapter accomplishments

After completing this chapter, you should be able to define all key terms and do the following.

5.3 Concept of minimum energy

- ❑ Explain the relationship between minimum energy and maximum stability.

5.4 Minimum energy in the atom

- ❑ Describe the electrostatic factors that determine how electrons are arranged in atoms.
- ❑ State the relationships between orbitals, sublevels, and main energy levels.
- ❑ State the spin relationship between electrons in the same orbital.

5.5 Electron configuration notation

- ❑ List the order in which atomic orbitals are filled (Aufbau principle).
- ❑ Write the electron configuration of any A group element.
- ❑ State and apply Hund's rule for writing electron configurations and determining the number of unpaired electrons in an atom.

5.6 Quantum numbers

- ❑ State the four criteria for identifying an electron in an atom.

5.7 Evidence and uses of electron energy levels

- ❑ Describe the result of putting energy into an atom.
- ❑ Explain why the light spectrum emitted by excited atoms is unique for each element.

5.8 Electron configurations in the periodic table

- ❑ Explain the relationship between electronic arrangement and the periodic table.
- ❑ Use the periodic table to obtain the number of valence electrons for any A group element.
- ❑ Use the periodic table to write the valence-electron configuration of any A group element.

5.9 Lewis electron dot structures

- ❑ Write the Lewis dot structure for any element, given the number of valence electrons.

5.10 Periodic trends

- ❑ Explain the trends in ionization energy and atomic size across a period and within a group of the periodic table.
- ❑ Given a periodic table, recognize the transition metals.
- ❑ State the distinguishing difference in electron arrangement between the A and B group elements.

key terms

electron structure
nuclear reactions
energy states
ground state
stable
unstable
electrostatics
Bohr's model
quantization
quantum mechanics
orbital
energy levels
electron spin
Pauli exclusion principle
electron configuration
Aufbau principle
Hund's rule
spectrum
spectroscope
valence electrons
Lewis electron dot structures
ionization energy

problems

5.2 Energy revisited

12. a. Give an example of a chemical change that releases energy.
    b. Where did this released energy come from?

5.3 Concept of minimum energy

13. Explain the following in terms of an energy change.
    a. Apples fall from a tree to the ground.
    b. Water runs down a hill, not up a hill.
    c. Electrons try to remain close to the nucleus.
14. Compare the stabilities of two states, one of which has a high-energy content and the other a low-energy content.

5.4 Minimum energy in the atom

15. a. What forces must be overcome to remove an electron from an atom?
    b. What forces would be relieved by removing an electron from an atom?
    c. Is energy required or released in removing an electron from an atom?
    d. Is the type of force in part a greater or less than that in part b? Explain how you arrived at your answer.
16. What forces must be balanced in an electrostatically neutral atom?
17. a. Describe the attractive force in a hydrogen atom.
    b. Describe the attractive forces that could exist within and between two hydrogen atoms.
    c. Describe the repulsive forces that could exist between two hydrogen atoms.
    d. What happens to these attractive and repulsive forces as we change the distance between the two hydrogen atoms?
    e. Could a minimum-energy structure exist for two hydrogen atoms? What would you call this structure?
18. Why is there only one *s*-type orbital, but three *p*-type orbitals at any given level (above $n = 1$)?
19. Describe the geometry of an *s* orbital and a *p* orbital.
20. What is the angular relationship between any two of the three 2*p* orbitals?
21. What requirement is there on the spin relationship between any two electrons in the same orbital?
22. a. Which generally contains a greater number of electrons: an orbital, a sublevel, or an energy level?
    b. Is there any main energy level that can only contain two electrons?
23. Within a particular energy level, which sublevel is of lowest energy?

5.5 Electron configuration notation

24. Using only the periodic table on the inside front cover, write the electron configurations of the following atoms.
    a. Li d. O
    b. F e. P
    c. B
25. Using only the periodic table on the inside front cover, write the electron configurations of the following atoms.
    a. Calcium d. Bromine
    b. Arsenic e. Indium
    c. Potassium
26. Correct any errors in the following ground-state electron configurations.
    a. $1s^2\,1p^6\,2s^2$
    b. $1s^2\,2s^2\,2p^6\,3s^1\,3p^2$
    c. $1s^2\,2s^2\,2p^6\,3s^2\,3d^7$
    d. $1s^2\,2s^2\,2p^6\,3s^2\,3p^5\,4s^1$
27. Give the symbols of all sublevels within the third main energy level.
28. Use the boxes-and-arrows notation to show the electron configurations of the following atoms.
    a. Sodium d. Carbon
    b. Silicon e. Fluorine
    c. Beryllium f. Cobalt
29. How many unpaired electrons are there in each of the atoms in Problem 28?
30. Why does nitrogen have three unpaired electrons in its *p* orbitals?

5.6 Quantum numbers

31. a. State the four criteria for identifying an electron in an atom.
    b. Of the criteria you identified in part a, which relate to the energy of an electron in an atom?

5.7 Evidence and uses of electron energy levels

32. Fill in the blanks with the word *emitted* or *absorbed.* For an electron to be promoted to an excited state, energy must be ______________ . When an electron falls back to its ground state, energy is ______________ .
33. Explain in your own words the difference between absorption and emission spectra.
34. When you see a yellow flame, what are you seeing on an atomic level?
35. After being heated by a Bunsen burner, a crucible may glow cherry-red. To what do you attribute the red color?

5.8 Electron configurations in the periodic table

36. What similarity exists among the electron configurations of elements in Group VIA? Group VIIA?
37. The atomic numbers of the elements within a certain set are 7, 15, 33, 51, and 83. What similarity exists among the electron configurations of these elements?
38. Write the electron configurations of the following.
    a. Strontium
    b. Iodine
    c. Bismuth
    d. A synthetic element with atomic number 104
39. Give the symbols of the elements with the following electron configurations.
    a. $1s^2\,2s^2\,2p^1$
    b. $1s^2\,2s^2\,2p^6\,3s^2\,3p^6\,4s^1$
    c. $1s^2\,2s^2\,2p^6\,3s^2\,3p^6\,4s^2\,3d^7$
    d. $1s^2\,2s^2\,2p^6\,3s^2\,3p^6\,4s^2\,3d^{10}\,4p^1$
    e. $1s^2\,2s^2\,2p^6\,3s^2\,3p^6\,4s^2\,3d^{10}\,4p^6$
40. Consulting only the periodic table on the inside cover, predict the orbital type for the last electron added to the following elements.
    a. Al d. C
    b. Cl e. S
    c. Ca
41. Consulting only the periodic table on the inside cover, write the electron configurations for the valence electrons of the following.
    a. Sulfur
    b. Phosphorus
    c. Iodine
    d. Krypton
42. Give the symbols of the elements with the following valence-electron configurations.
    a. $2s^2$ d. $1s^1$
    b. $4s^1$ e. $3s^2\,3p^4$
    c. $2s^2\,2p^5$ f. $5s^2$
43. Consulting only the periodic table on the inside cover, give the symbol of an element that has chemical properties similar to an element with each of the following valence-electron configurations.
    a. $3s^2\,3p^4$ d. $3s^1$
    b. $5s^2$ e. $1s^2$
    c. $2s^2\,2p^5$
44. a. Which groups in the periodic table are the representative elements?
    b. Give the symbol of a nonrepresentative element.

5.9 Lewis electron dot structures

45. Write Lewis dot structures for the following elements.
    a. Li d. Ar
    b. Si e. P
    c. I f. Ba

5.10 Periodic trends

46. a. Why does atomic size decrease proceeding across a period from left to right?
    b. Why does ionization energy increase as atomic size decreases?
47. a. Where in the periodic table are elements with low ionization energies?
    b. Where in the periodic table are elements with high ionization energies? What type of elements are these?
    c. Which group in the periodic table contains elements with the highest ionization energies? What does this say about the stability of the outermost electron arrangement for this group?
48. Arrange each set of elements in order of increasing ionization energy.
    a. Na, Li, Rb, Cs, K
    b. Br, Ge, As, Ga, K, Ca
    c. Se, Br, K
49. Give the correct choice in each set.
    a. Highest ionization energy: Na, K, C, Si
    b. Smallest size: C, Li, Ge, F
    c. Largest size: K, Cs, Se
    d. Smallest ionization energy: K, Cs, Se
50. What is the main difference in electron configuration between a representative and a transition element?
51. Define the term *transition metal.*
52. Which elements tend to have higher ionization energies, metals or nonmetals?
53. For atoms of any given element, state at least five characteristics that you can glean from a periodic table.

Additional problems

54. a. What is the maximum number of electrons in the first principal energy level?
    b. In the second principal energy level?
    c. In the third principal energy level?
55. How many valence electrons are present in the following?
    a. C
    b. Sn
    c. F
    d. P
56. Classify each of the following as a representative (main-group) element or a transition element. Further classify each main-group element by identifying its group and whether it is a metal or nonmetal.
    a. Lithium
    b. Carbon
    c. Argon
    d. Iron
    e. Copper
    f. Zinc
57. Write the electron configuration for the valence electrons of each of the following.
    a. Si
    b. Cl
    c. K
    d. Ca
58. Group the following electron configurations into pairs that would represent similar chemical properties.
    a. $1s^2\,2s^2\,2p^6\,3s^2\,3p^5$
    b. $1s^2\,2s^2\,2p^6\,3s^2$
    c. $1s^2\,2s^2\,2p^3$
    d. $1s^2\,2s^2\,2p^6\,3s^2\,3p^3$
    e. $1s^2\,2s^2\,2p^6\,3s^2\,3p^6\,4s^2\,3d^{10}\,4p^6$
    f. $1s^2\,2s^2$
    g. $1s^2\,2s^2\,2p^6$
    h. $1s^2\,2s^2\,2p^5$
59. The first ionization energies of Ge, As, and Se are 782, 945, and 937 kJ, respectively. Explain the variation in these values in terms of electron configuration.
60. a. What is meant by the term *ground state?*
    b. What is meant by the term *excited state?*
61. Explain in your own words why elements produce a characteristic pattern of colors when there is an emission of energy from an excited state to a ground state?
62. a. What are the similarities between a 1*s* orbital and a 2*s* orbital?
    b. What are the differences?
63. What is the maximum number of electrons that can be accommodated in a 4*p* sublevel? 2*s?* 3*d?*
64. a. What are the similarities between a $2p_x$ and a $2p_y$ orbital?
    b. What is the difference between them?
65. Specify the group in the periodic table in which you would expect to find an element with the following valence shell configuration.
    a. $3s^1$
    b. $3s^2\,3p^2$
    c. $3s^2\,3p^5$
    d. $4s^1$
66. Which of the following electron configurations correspond to an excited state? For each excited-state configuration write the ground-state configuration.
    a. $1s^2\,2s^2\,2p^6\,3p^1$
    b. $1s^1\,3s^2$
    c. $1s^2\,2s^2\,2p^6\,3s^1$
    d. $1s^2\,2s^2\,3s^1\,3p^5$

chapter 6

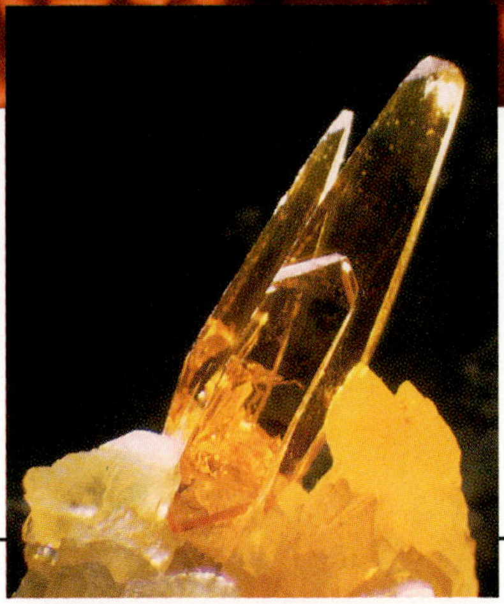

chemical bonding

6.1 introduction

Elements are rarely found in uncombined form. Rather, they are combined with other elements in fixed proportions as new compounds. Knowledge of the electron structure of elements places you at the heart of modern chemistry, for it gives you the ability to predict the formation of new compounds and their properties. With such knowledge, chemists are able to build or synthesize new materials.

You are familiar with the use of new materials as replacements for other, more costly materials in automobiles, appliances, and innumerable everyday items. Reinforced plastics are now being used to replace metal in automobile bodies, doors and other panels, bumper covers, wheel covers, and most interior trim. Even metal roller-skate frames have been replaced with frames made of nylon resin, a new synthetic material (Figure 6.1).

The use of new materials is possible because of extensive research and development into the properties of new compounds. These properties, such as hardness, elasticity (ability to stretch and return to original shape), and melting point

The regular arrangement of ions in a solid is called a lattice structure. Because the positive and negative ions alternate so exactingly, layers of ions can be seen in ionic models and those layers are responsible for the geometry of the bulk crystal samples. The ionic lattice structure gives the material a definite shape with angles characteristic of the material.

Figure 6.1
Nylon resin replaces steel in the in-line roller-skate frames worn by this skater.

table 6.1 electron configurations of the noble (inert) gases

Element	Symbol	Electron configuration (with valence electrons highlighted)
Helium	He	$1s^2$
Neon	Ne	$1s^2 2s^2 2p^6$
Argon	Ar	$1s^2 2s^2 2p^6 3s^2 3p^6$
Krypton	Kr	$1s^2 2s^2 2p^6 3s^2 3p^6 3d^{10} 4s^2 4p^6$
Xenon	Xe	$1s^2 2s^2 2p^6 3s^2 3p^6 3d^{10} 4s^2 4p^6 4d^{10} 5s^2 5p^6$
Radon	Rn	$1s^2 2s^2 2p^6 3s^2 3p^6 3d^{10} 4s^2 4p^6 4d^{10} 4f^{14} 5s^2 5p^6 5d^{10} 6s^2 6p^6$

can be varied by controlling how electrons interact to form a chemical **bond,** that is, the force that holds elements together in compounds. While a full discussion of bond formation is beyond the scope of this text, you may begin to appreciate how chemists can apply their understanding of the electron structure of elements as you study bonding in the simplest compounds.

Compounds may be classified into two major types: (1) *ionic compounds,* in which the ionic bond holds elements together, and (2) *molecular compounds,* in which the covalent bond is the holding force. In this chapter, you will explore the nature of the ionic bond and the formulas of ionic compounds. You will learn why there are two types of compounds and then explore the nature of the covalent bond and molecular formulas.

In Chapter 5, the most stable state of an *element* was determined by the lowest-energy electron configuration of each constituent atom. Similarly, in this chapter, the most stable state of a *compound* will be determined by the lowest-energy electron configuration of its ions or molecules.

6.2 how can atoms achieve lower-energy states?

In Chapter 5, you learned that there is a natural tendency for matter to attain a lower-energy state. An atom left to itself does the best it can, given the number of electrons that it has, to arrange those electrons in such a way as to produce the lowest possible energy state for that atom. Since the atoms of most elements combine with atoms of other elements to form compounds, their new atomic arrangement as compounds enables them to form electron arrangements of lower energy than they could achieve as individual atoms. The stability of this interaction leads to compound formation. Thus, compound formation can be explained by an extension of the same principles of energy minimization that you have seen for elements.

There is a group of elements in the periodic table, Group VIIIA, the **noble (inert) gases,** that is particularly stable and has essentially no tendency to interact with other elements to form compounds.[1] Because the noble gases are a periodic *group,* you know that their electron configurations are similar. As you can see in Table 6.1, all noble (inert) gases have a consistent lowest-energy electron configuration in which the outermost energy level is completely filled with paired electrons. For helium, this corresponds to two valence electrons in the $1s$ orbital. For all other noble gases, having completely filled outermost s and p orbitals requires eight valence electrons. Eight valence electrons in the same principal level are often referred to as an **octet.**

The electron configurations of the noble gases are apparently the most stable (possessing the lowest energy) of all elements, since these gases have so little tendency to undergo change in their electron configuration. They are stable just as they are.

1. Since 1962, several compounds of the noble gases xenon, radon, and krypton have been prepared. However, these compounds are exceptions to the rule. In this book, we will regard the noble gases as inert.

Atoms of other elements do not have completely filled outer-level electron configurations, however. You recall from Section 5.8 that sodium atoms have an electron configuration of $1s^2\ 2s^2\ 2p^6\ 3s^1$. If a sodium atom lost one electron to form a sodium ion, Na^+, its electron configuration would become $1s^2\ 2s^2\ 2p^6$ because the outermost (highest-energy) 3*s* electron would be lost. The sodium ion then has an electron configuration that is exactly the same as neon, a noble (inert) gas (Table 6.1). Chemically, the sodium ion might be predicted to behave like neon, being so stable that it does not react readily with other materials. Your prior knowledge confirms just such a prediction. In Sections 4.11 and 4.12, you learned that neutral atoms of sodium, Na, react vigorously with water, whereas sodium ions, Na^+, which are a part of the compound sodium chloride (table salt), are stable and unreactive in water.

Throughout this chapter, you will learn not only about how atoms lose or gain electrons, but also how they share electrons as they bond in new compounds. However, from the change in the chemical properties of sodium as it becomes an ion, you should not be surprised by the following generalization. *Most elements react to form compounds through a process whereby their atoms gain, lose, or share valence electrons in order to achieve the highly stable (octet) electron configuration of a noble-gas element.* Atoms can gain, lose, or share electrons and form compounds because, by so doing, they acquire lower-energy electron arrangements.

By gaining, losing, or sharing electrons, an element can acquire the electron configuration of the noble gas to which it is closest in the periodic table. For H, Li, and Be, this means that they strive to attain the He valence-electron configuration of $1s^2$, two valance electrons that, in this case, form a stable **duet.** Other elements strive to achieve eight valence electrons. This is often called the rule of eight or the **octet rule.**[2] In acquiring inert electron configurations, atoms that lose electrons become positively charged and atoms that gain electrons become negatively charged.

6.3 metals lose electrons to form positively charged ions (cations)

The trends in ionization energy described in Section 5.10 reveal that electrons are lost from metals more readily than from nonmetals. In this section, you will see that the number of electrons that a metal readily loses is the number that will give the metal the stable electron configuration of a noble gas. Electron loss always produces a positively charged metal ion. This is so because, upon electron loss from an atom, the positively charged protons in the nucleus outnumber the negatively charged electrons. A positive ion is called a **cation** (Section 4.8).

Ionization energy is the amount of energy needed to completely remove an electron from a gaseous atom or ion. In Section 5.10, we discussed only the first ionization energy, the energy needed to remove the outermost electron from a neutral atom. This first electron removed is the one that is the least tightly held by the nucleus because it is the farthest away. Considering aluminum as an example, 577 kJ is required to remove one electron from each atom in a 27 g sample[3] of gaseous aluminum, forming Al^+ ions.

If you put enough energy in, you can remove an electron from the Al^+ ion. This is called the *second ionization energy* of aluminum, which turns out to be 1,812 kJ for the same 27 g sample originally described. Note that removing the second electron requires more than 3 times the energy to remove the first electron.

You can continue the process and measure the third and fourth ionization energies, and so on. The second ionization energy for a given element is always greater

2. Elements do not always form a noble-gas electron configuration, but that is far and away the general rule. An exception of minor importance is boron, which strives to attain six valence electrons.

3. When comparing ionization energies, a sample with the same number of atoms is always used. A sample of Al whose mass is the same as its atomic weight has the same number of atoms as any other element whose mass is the same as its atomic weight. The significance of this sample size will be defined in Chapter 8.

Figure 6.2
More and more energy is required as successive electrons are removed from an atom.

than the first, since, in the case of the second, an electron is being removed from an already positively charged species. Remember that separating positively and negatively charged ions requires energy (Section 5.4). For similar reasons, the third ionization energy is greater than the second (Figure 6.2).

Table 6.2 shows the change in electron configuration when atoms of three A group elements lose electrons. The electron configuration of each metal atom is given beside the electron configuration of the nearest noble gas. Comparison of these electron configurations will help you to see how many valence electrons must be lost by each element to acquire the noble-gas electron configuration of its most stable ion. Once Li has lost one electron (the same number as its group number) to form the Li^+ ion, it has acquired the electron configuration of the noble gas, helium. Once Mg has lost two electrons (the same number as its group number) to form the Mg^{2+} ion, it has acquired the electron configuration of the noble gas, neon. Once Al has lost three electrons (the same number as its group number) to form the Al^{3+} ion, it has also acquired the electron configuration of the noble gas, neon.

That the ions with a noble-gas electron configuration are particularly stable can be seen from the experimental data graphed in Figure 6.3. For any element, there is a gradual increase in ionization energy as you remove more electrons, just as the student portrayed in Figure 6.2 works harder to remove succeeding electrons. However, once the atom acquires a noble-gas electron configuration, there is a very large increase in energy needed to remove an additional electron. The exceptionally large

table 6.2 noble-gas configurations attained by electron loss in sample metals

Metal	Group number	Metal configuration		Noble-gas configuration achieved	Number of electrons lost*
Li	IA	$1s^22s^1$	⟶	He: $1s^2$	1
Mg	IIA	$1s^22s^22p^63s^2$	⟶	Ne: $1s^22s^22p^6$	2
Al	IIIA	$1s^22s^22p^63s^23p^1$	⟶	Ne: $1s^22s^22p^6$	3

*Notice how the number of electrons lost always equals the group number.

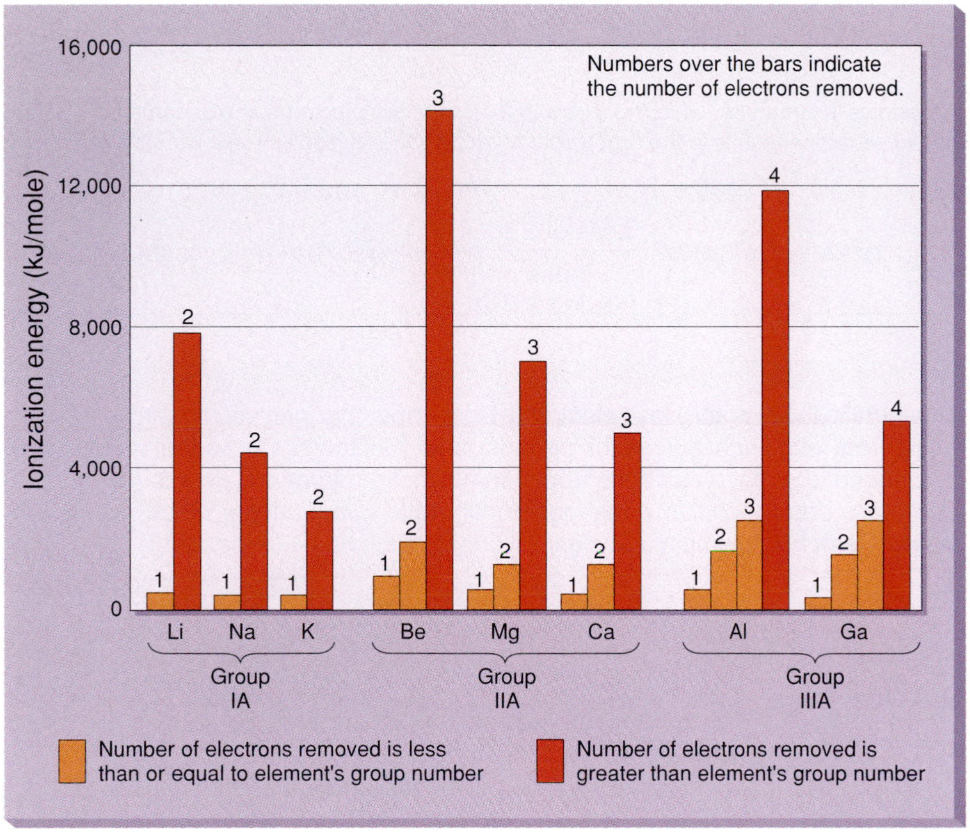

Figure 6.3
Ionization energy increases dramatically when an electron one beyond the element's group number is removed. This confirms the stability of the ion with a noble-gas electron configuration formed from each A group element. For Group IA elements, the most stable ion occurs when one electron is lost; for Group IIA elements, when two electrons are lost; for Group IIIA elements, when three electrons are lost.

ionization energy for the electron beyond the group number of the element confirms that the ion with the noble-gas electron configuration is unusually stable.

For any A group metal, losing a number of electrons that is equal to the group number produces a noble-gas configuration. The A group metals generally form cations with positive charges equal to the group number.

When two particles (atoms or ions) have the same electron configuration, they are **isoelectronic.** In the examples in Table 6.2, Li^+ is isoelectronic with He, and Mg^{2+}, Al^{3+}, and Ne are isoelectronic because they all have the same electron configuration. However, isoelectronic species are not identical. They differ from one another in the numbers of protons and neutrons in their nuclei. For example, compare Mg^{2+} with Ne in Figure 6.4. Because each particle has ten electrons, they are isoelectronic, but their nuclei contain different numbers of protons and neutrons. Because the number of protons characterizes an element, the Mg^{2+} ion has different properties from the Ne atom. However, because it is isoelectronic with the noble gas neon, the Mg^{2+} ion is stable and does not tend to further change its electron arrangement.

Figure 6.4

The cation, Mg^{2+}, and the atom, Ne, are isoelectronic because they have the same electron configuration.

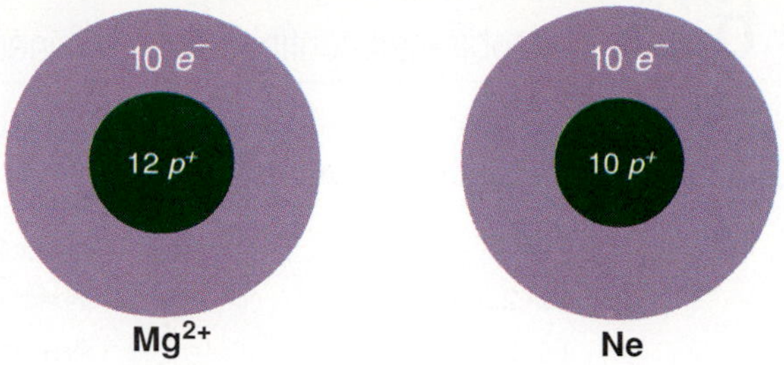

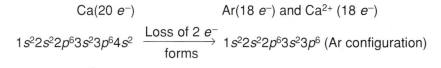

sample exercise 1

How many electrons must Ca lose to achieve a noble-gas configuration? With which noble gas will the Ca cations be isoelectronic?

solution

Calcium is in Group IIA, which tells us that it will lose two electrons and become Ca^{2+}. A loss of two electrons results in the argon configuration. Ca^{2+} is isoelectronic with argon.

$$\text{Ca}(20\ e^-) \qquad\qquad \text{Ar}(18\ e^-) \text{ and } \text{Ca}^{2+}\ (18\ e^-)$$

$$1s^2 2s^2 2p^6 3s^2 3p^6 4s^2 \xrightarrow[\text{forms}]{\text{Loss of 2 } e^-} 1s^2 2s^2 2p^6 3s^2 3p^6 \text{ (Ar configuration)}$$

6.4 nonmetals gain electrons to form negatively charged ions (anions)

Nonmetals do not readily lose electrons. Remember that ionization energy increases going from left to right across the periodic table (Section 5.10), so that more energy is required to remove electrons from nonmetals than from metals. Also, nonmetals can more easily achieve noble-gas configurations by gaining electrons than by losing them. For example, consider chlorine, which has 17 electrons:

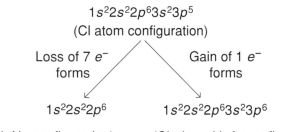

$$1s^2 2s^2 2p^6 3s^2 3p^5$$
(Cl atom configuration)

Loss of 7 e^- forms ↙ $1s^2 2s^2 2p^6$ (Cl^{7+} ion with Ne configuration)

Gain of 1 e^- forms ↘ $1s^2 2s^2 2p^6 3s^2 3p^6$ (Cl^- ion with Ar configuration)

It is much easier to gain one electron than to lose seven. **Electron affinity** is the name given to the energy change that occurs when an electron is added to a gaseous atom or ion.

The halogens (Group VIIA), which need only one electron to reach a stable, noble-gas electron configuration, show the highest electron affinity or release of energy when an electron is added to a gaseous atom. For example, when 35 g of gaseous chlorine atoms gain electrons to become Cl^- ions, 351 kJ is liberated.

Nonmetals gain electrons to form ions with a noble-gas configuration, that is, to gain an octet (eight electrons). For an A group nonmetal, the number of electrons gained is equal to the difference between eight and the group number (8 – group number). Chlorine is in Group VIIA, so it gains 8 – 7 = 1 electron. Fluorine, also in Group VIIA, gains one electron as well. Phosphorus, in Group VA, gains 8 – 5 = 3 electrons (Table 6.3).

Electron gain always produces negatively charged nonmetal ions. A negative ion is called an **anion** (Section 4.8). The charge is negative because the gain causes

table 6.3 noble-gas configurations attained by electron gain in sample nonmetals

Nonmetal	Group number	8 – Group number	Nonmetal configuration	Noble-gas configuration achieved	Number of electrons gained*
F	VIIA	8 – 7 = 1	$1s^22s^22p^5$	Ne: $1s^22s^22p^6$	1
O	VIA	8 – 6 = 2	$1s^22s^22p^4$	Ne: $1s^22s^22p^6$	2
P	VA	8 – 5 = 3	$1s^22s^22p^63s^23p^3$	Ar: $1s^22s^22p^63s^23p^6$	3

*Notice how the number of electrons gained always equals 8 – the group number.

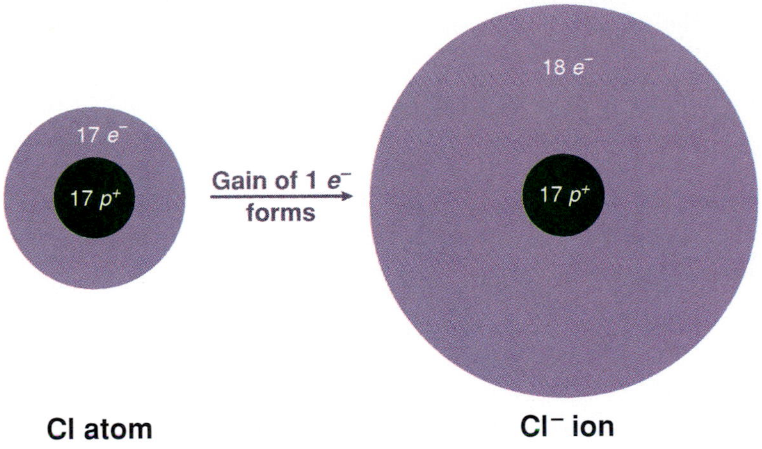

Figure 6.5
A chlorine atom, Cl, gains one electron to form a chloride anion, Cl^-.

the electrons to outnumber the protons in the nucleus. The difference between eight and the nonmetal anion's group number also tells you the magnitude of the negative charge on a nonmetal anion. A gain of one electron produces an ion with a 1– charge, as exemplified by chlorine in Figure 6.5.

sample exercise 2

How many electrons must N gain to achieve a noble-gas configuration? With which noble gas will the N anions be isoelectronic?

solution

Nitrogen is in Group VA, which tells us that it will gain 8 – 5 = 3 electrons and become N^{3-}. A gain of three electrons results in the Ne configuration. N^{3-} is isoelectronic with neon.

$$\text{N } (7\ e^-) \xrightarrow[\text{forms}]{\text{Gain of 3 } e^-} \text{N}^{3-}\ (10\ e^-)$$

$$1s^22s^22p^3 \qquad\qquad 1s^22s^22p^6 \text{ (Ne configuration, 10 } e^-)$$

••• problem 1

a. How many electrons must aluminum lose to achieve a noble-gas configuration? With which noble gas will the aluminum ions be isoelectronic?

b. How many electrons must sulfur gain to achieve a noble-gas configuration? With which noble gas will the sulfur ions be isoelectronic?

c. Identify the ions in parts a and b as cations or anions and give the magnitude of each ionic charge.

table 6.4 common ions of the main group elements*

IA	IIA		IIIA	IVA	VA	VIA	VIIA
H^+							
Li^+	Be^{2+}	Transition			N^{3-}	O^{2-}	F^-
Na^+	Mg^{2+}	Elements	Al^{3+}		P^{3-}	S^{2-}	Cl^-
K^+	Ca^{2+}		Ga^{3+}				Br^-
Rb^+	Sr^{2+}		In^{3+}	(Sn^{4+})†			I^-
Cs^+	Ba^{2+}		Tl^{3+}	(Pb^{4+})†			

*Atoms such as carbon, which do not commonly form ions, are omitted from the table.
†Tin and lead are shown in parentheses because they more commonly form 2+ ions (Table 6.5).

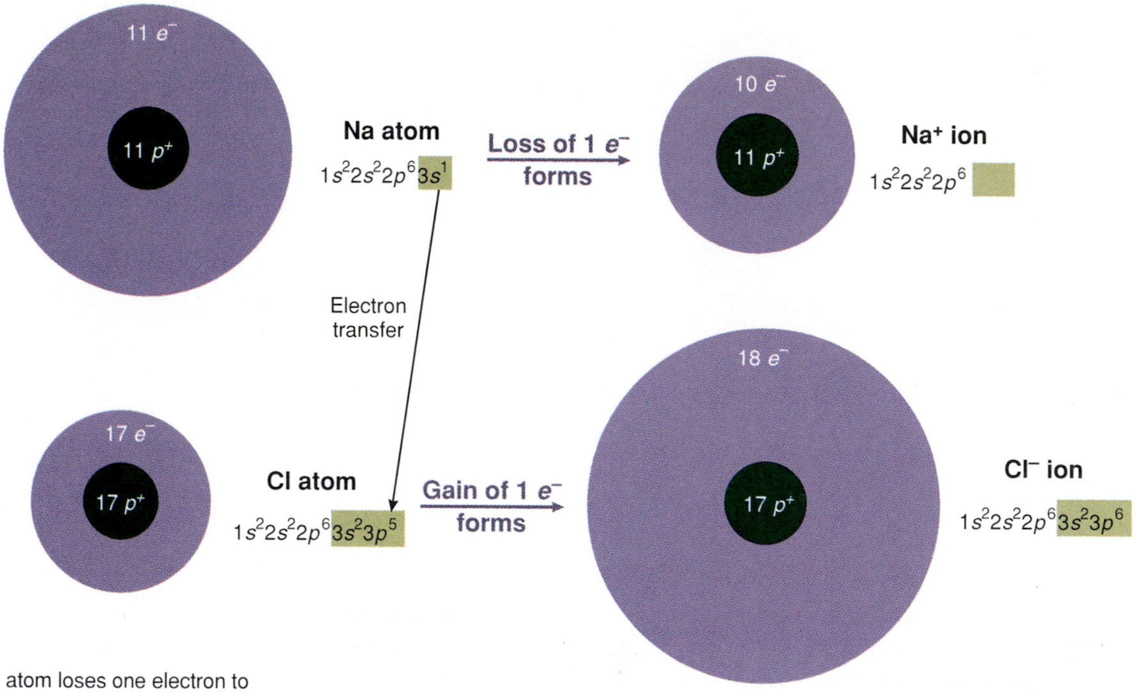

Figure 6.6

Each sodium atom loses one electron to become a cation, while each chlorine atom gains one electron to become an anion. As a result, the oppositely charged ions form in a fixed proportion of one Na^+ ion to one Cl^- ion. The resultant compound, sodium chloride, shows this proportion in its formula, NaCl.

You will find it necessary to know the ionic charges that common metals and nonmetals acquire. Table 6.4 shows that with a periodic table in hand there is no need to memorize the charges of the A group ions. Simply apply the relationships among group number, number of electrons gained or lost, and ionic charge acquired.

6.5 electron transfer results in the formation of ionic compounds

Because metals tend to achieve a noble-gas configuration by losing electrons and nonmetals tend to do this by gaining electrons, there is a perfect setup for them to *transfer electrons* to each other in a chemical reaction. For example, sodium atoms (Group IA) could each lose an electron to form sodium cations. Chlorine atoms (Group VIIA) could, in turn, each gain an electron to form chloride anions. Each sodium atom loses one electron and each chlorine atom gains one electron, so the number of ions that may form is in a fixed proportion: one chloride anion forms for each sodium cation (Figure 6.6).

Once the oppositely charged cations and anions are formed with noble-gas electron configurations, they are stable. These stable ions are attracted to each

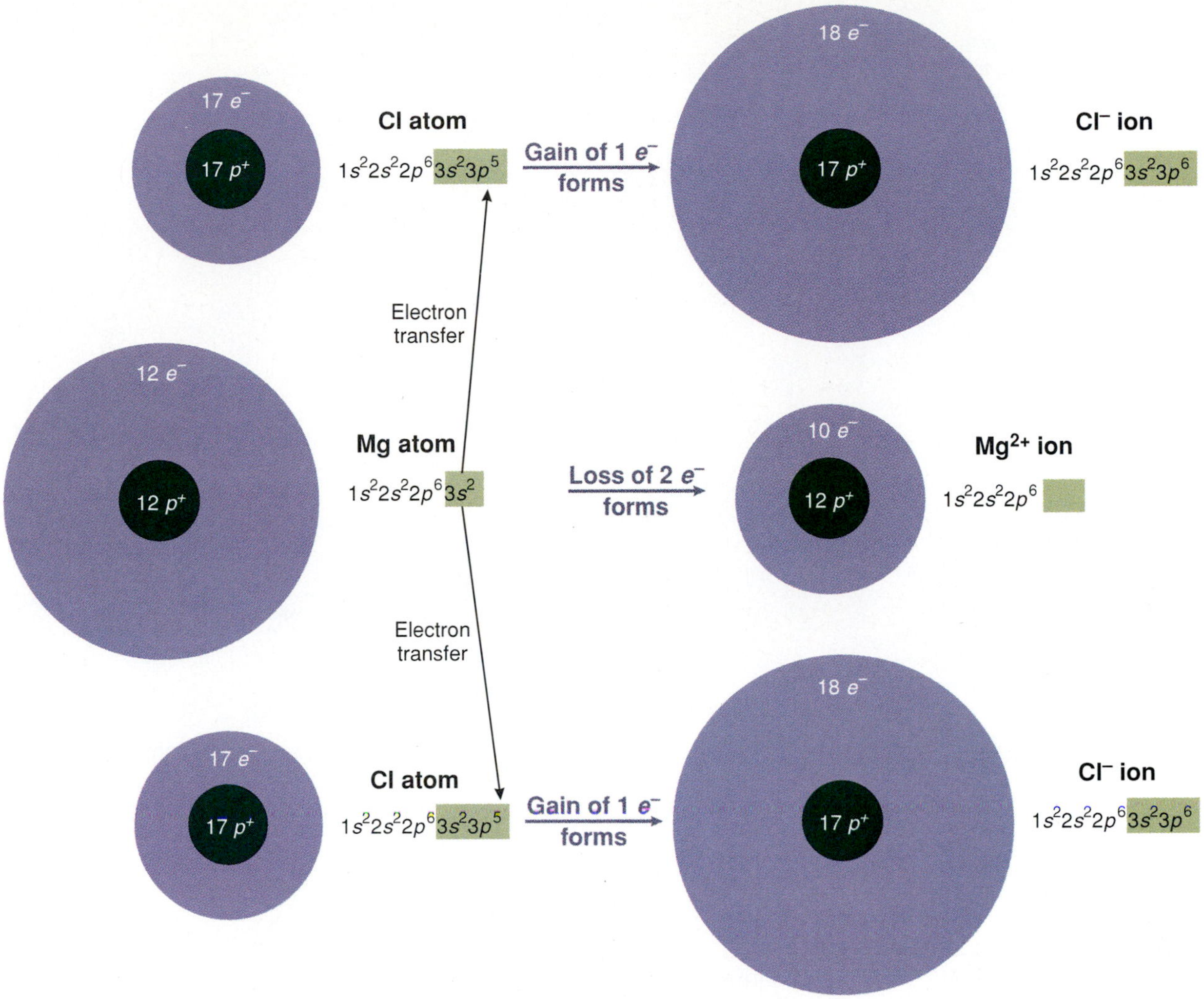

Figure 6.7
Each magnesium atom (Group IIA), loses two electrons to become a cation, while two chlorine atoms each gain one electron. The oppositely charged ions form in a fixed proportion of one Mg^{2+} ion to two Cl^- ions. The electrostatic attraction between cations and anions holds the compound magnesium chloride, $MgCl_2$, together with ionic bonds.

other by electrostatic forces to form the new compound sodium chloride, NaCl, also known as table salt. **Ionic compounds,** such as sodium chloride, are those formed by the electrostatic attraction of oppositely charged ions. The bond formed between oppositely charged ions is called an **ionic bond.**

Just like the example with sodium, a metal, and chlorine, a nonmetal, the combination of any metal and nonmetal employs the ionic bond to form an ionic compound through electron transfer. Metals give electrons to nonmetals, thus enabling each to form ions of opposite charge. The total number of electrons lost by the metal *must* equal the total number gained by the nonmetal so that electrical neutrality is maintained.

Because of electrical neutrality, magnesium and chlorine will form ions in a different fixed proportion than sodium and chlorine. Each magnesium atom, a Group IIA metal, loses two electrons to form one magnesium cation, Mg^{2+}, but each chlorine atom can gain only one electron to form a chloride anion, Cl^-. Therefore, two chlorine atoms, each of which gains one electron, are needed to receive the two electrons lost by the magnesium atom. This transfer is seen pictorially in Figure 6.7. Notice that after the electron transfer, each ion has a noble-gas configuration; Mg^{2+} has the same electron configuration as Ne, and Cl^- has the Ar electron configuration. It is the electrostatic attraction between the cations and anions that is the ionic bond that holds ionic compounds together in the fixed proportion of one magnesium cation to two chloride anions. The compound magnesium chloride has the formula $MgCl_2$ (Figure 6.8).

Figure 6.8

Shiny magnesium metal reacts with chlorine, a greenish-yellow gas, to form magnesium chloride, a white crystalline solid.

General rules for ionic compound formation

1. *Metals lose* electrons to attain a stable, noble-gas configuration. For A group metals, the number of electrons lost equals the group number.
2. *Nonmetals gain* electrons to attain a stable, noble-gas configuration. For A group nonmetals, the number of electrons gained equals the difference between eight and the group number.
3. The total number of *electrons lost always equals* the total number of *electrons gained.*

In Figures 6.6 and 6.7, the atoms and their corresponding ions have been drawn to scale. Careful observation of these figures will reveal that each cation is drawn smaller and each anion is drawn larger than their respective atoms. The change in size is consistent with the electron structure of each atom, as you have learned in Chapter 5. For example, when an atom of the metal sodium, Na, loses its 3*s* electron to become a sodium cation, Na^+ (Figure 6.6), it shrinks in size because it no longer has a valence electron in its third principal energy level or third shell. (A principal energy level is often referred to as a shell.) The outermost electrons of the sodium cation are in its filled second shell and are isoelectronic with the noble gas neon (Section 6.3). Similarly, when an atom of the metal magnesium, Mg, loses its two 3*s* electrons to become a magnesium cation, Mg^{2+} (Figure 6.7), it shrinks in size because it, too, no longer has valence electrons in its third shell and is also isoelectronic with both Na^+ and the noble gas neon. On the contrary, when an atom of the nonmetal chlorine, Cl, gains an electron to become a chloride anion, Cl^- (Figures 6.6 and 6.7), the filling of its second shell makes it isoelectronic with the noble gas argon and results in a considerable size increase. The relative sizes of atoms in two groups of the periodic table and their corresponding ions are shown in Figure 6.9. In each example, the cation is smaller and the anion is larger than the corresponding atom.

In Figure 6.10, the relative sizes of ions are compared throughout the periodic table. For A group elements, both cations and anions shrink in size as you go from left to right across a period of elements in the table because of increasing nuclear charge. For example, period 3 cations decrease in size from Na^+ to Mg^{2+} to Al^{3+}. All three of these cations are isoelectronic; that is, they have the same noble-gas electron configuration. However, the higher nuclear charge of Al^{3+} from 13 protons in its nucleus, compared to 12 protons in the nucleus of Mg^{2+} and 11 protons in the

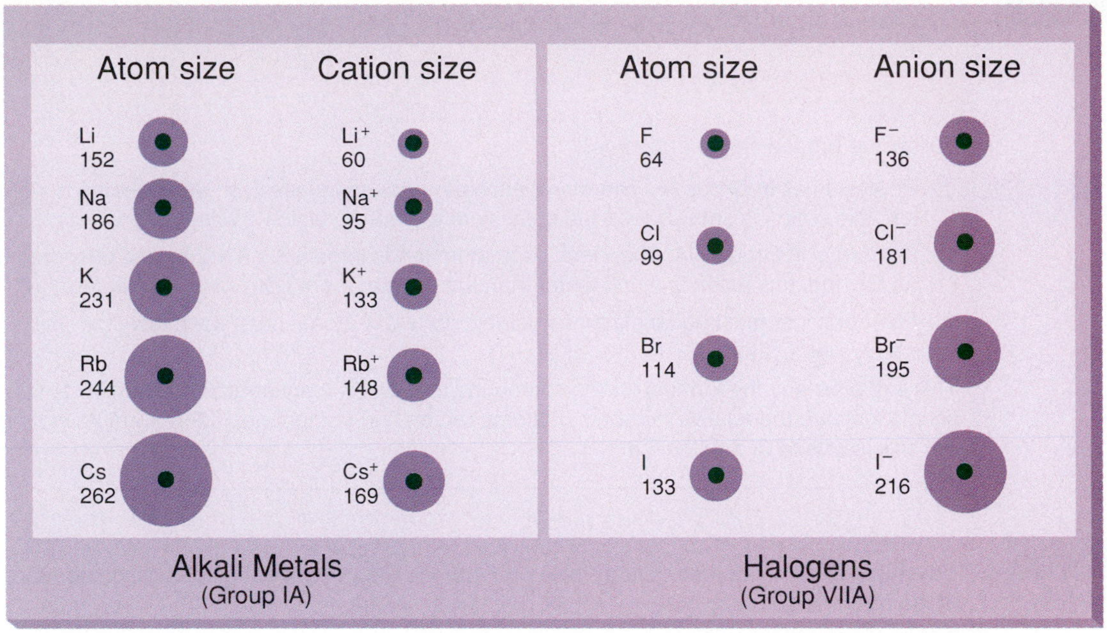

Figure 6.9

The relative sizes of atoms in two groups of the periodic table and their corresponding ions. In each example the cation is smaller and the anion is larger than the corresponding atom. Sizes of ions are given in meters $\times 10^{12}$.

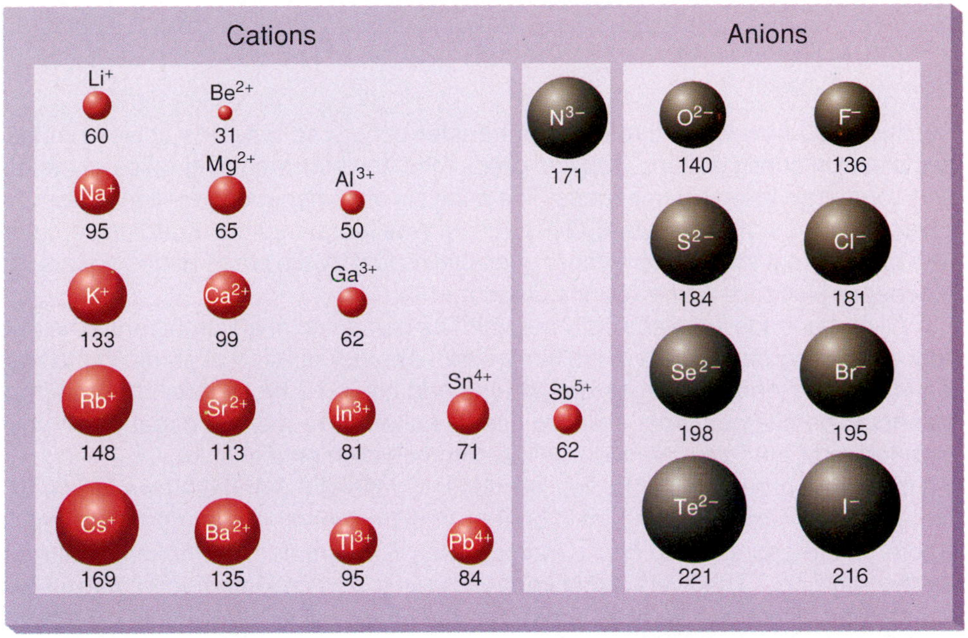

Figure 6.10

Sizes of ions are compared throughout the periodic table. For A group elements, both cations and anions shrink in size as you go from left to right across a period of elements in the table because of increasing nuclear charge. Sizes of ions are given in meters $\times 10^{12}$.

nucleus of Na^+, attracts the electrons more strongly and pulls them closer to the nucleus. As a result, Al^{3+} is smaller than Mg^{2+}, which is still smaller than Na^+. A similar trend can be seen for period 2 anions, which decrease in size from N^{3-} to O^{2-} to F^-. All three of these anions are isoelectronic. However, the higher nuclear charge of F^- from nine protons in its nucleus, compared to eight protons in the nucleus of O^{2-} and seven protons in the nucleus of N^{3-}, attracts the electrons more strongly and pulls them closer to the nucleus.

The size of ions in ionic compounds determines how closely they may fit together in the solid state. The smaller the ions, the more that may fit into a given volume and, therefore, the higher the density of the solid. In addition, the relative size of cations and anions determines the geometric arrangement of the ions about each other and, consequently, the shape of the crystalline solid (Figures 6.9 and 6.10).

sample exercise 3

Describe how the ionic compound composed of potassium and oxygen forms.

solution

Follow the preceding general rules.

1. Potassium is in Group IA. Therefore, each atom tends to lose 1 e^- and become a K^+ ion; this cation is smaller than the potassium atom from which it was formed.
2. Oxygen is in Group VIA. Therefore, each atom tends to gain $8 - 6 = 2$ e^- and becomes an O^{2-} ion; this anion is much larger than the oxygen atom from which it was formed.
3. Electrons lost must equal electrons gained, so two K atoms must each lose 1 e^- to satisfy one O atom.

This explains why the formula for the compound is K_2O. As mentioned in Section 1.11, subscripts indicate the relative numbers of atoms (or ions) in compounds. More formula writing will be discussed in Section 6.8.

••• problem 2

Develop a pictorial representation similar to Figure 6.7 for the formation of ionic compounds made from the following.

a. Calcium and oxygen
b. Aluminum and chlorine

6.6 ionic charges

Because you now know that electron transfer occurs so that ions achieve noble-gas electron configurations, you can predict the correct combining ratios of metals and nonmetals in ionic compounds. The analysis of compound formation in Sample Exercise 3 led to the formula K_2O because it was found that two K atoms combine with one O atom so that the number of electrons lost by the two K atoms equals the number of electrons gained by the O atom.

All compounds are electrically neutral. The balancing of electron transfer is one way of deriving the formula of the compound. Another method of predicting formulas is to focus on the charges on the ions and combine those ions in a ratio that balances the charges. Your ability to use this alternative method depends on your knowledge of the ionic charges that the common elements acquire.

For the A group elements, you learned in Sections 6.3 and 6.4 how to use the group number to predict the ionic charge acquired when an element acquires a stable, noble-gas electron configuration. Table 6.4 summarized the ionic charges for the A groups. There are exceptions, however, and cations of lead and tin are often found with a charge of 2+ as well as the charge of 4+ that you would predict from their group number. Other common exceptions are listed in Table 6.5.

••• problem 3

What is the charge on the ion formed from phosphorus?

For B group metals (transition metals), it is not possible to predict the magnitude of the positive charges on their cations from the periodic table. Fortunately, there are only a few cations of the B group metals you need to know, and these are also listed in Table 6.5. *Notice in this table that many so-called transition-metal elements (B group) and some A group elements form ions with more than one charge.* The charges on these ions can be explained with a deeper understanding

table 6.5 common transition-metal ions and A group ions with more than one charge*

1+ Charge	2+ Charge	3+ Charge	4+ Charge
Copper(I) Cu^{+} (cuprous)	Copper(II) Cu^{2+} (cupric)	Chromium Cr^{3+}	Tin(IV) Sn^{4+} (stannic)
Silver Ag^{+}	Iron(II) Fe^{2+} (ferrous)	Iron(III) Fe^{3+} (ferric)	Lead(IV) Pb^{4+} (plumbic)
	Lead(II) Pb^{2+} (plumbous)		
	Mercury(II) Hg^{2+} (mercuric)		
	Tin(II) Sn^{2+} (stannous)		
	Zinc Zn^{2+}		

*Roman numerals in parentheses are used to distinguish between different ions from the same metal. The Roman numeral is the same as the charge on the ion. This is discussed in Section 7.3. The words in parentheses are archaic names used to distinguish metal ions.

of how electrons partially fill electronic sublevels. However, rather than diverting to explain a few exceptions to the rules you know, it will suffice for you to refer to Table 6.5 when necessary.

6.7 polyatomic ions

The ions you have learned so far are *simple* ions, formed from one atom of one element, such as Li^{+}, Mg^{2+}, Al^{3+}, Cl^{-}, and S^{2-}. Simple ions are also called **monatomic ions** because the prefix *mono* means "one" (the terminal *o* in *mono*- is dropped before a vowel).

Not all ions are composed from only one atom, however. There are numerous polyatomic ions, so called because the prefix *poly*- means "many." Thus, a **polyatomic ion** is a charged species made from many atoms. Table 6.6 lists the names and formulas of the more common polyatomic ions.

Let us look at a specific example, the formula for sulfate, SO_4^{2-}, and examine its atomic structure. This formula tells us that four oxygen atoms and one sulfur atom are combined in such a way that the group of atoms exists as a unit and has a 2– charge. Just as you learned in Section 1.11 for neutral molecules, subscripts in any formula always refer to the number of atoms in a unit; when no subscript appears, a 1 is understood. The whole sulfate group is the ion; the component sulfur and oxygen atoms are inseparable. Within the ion are covalent bonds (Section 6.13) that hold the sulfur and oxygens together. However, the internal bonding within the ion is less important right now than the fact that the entire group of atoms stays together as a charged unit and may combine as a unit with an oppositely charged ion to form a compound. Polyatomic ions are groups of covalently bonded atoms that have a charge (Table 6.6).

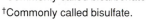

table 6.6 polyatomic ions

1+ Charge	1– Charge	2– Charge	3– Charge
Ammonium, NH_4^+	Acetate, $C_2H_3O_2^-$	Carbonate, CO_3^{2-}	Phosphate, PO_4^{3-}
	Chlorate, ClO_3^-	Chromate, CrO_4^{2-}	
	Cyanide, CN^-	Dichromate, $Cr_2O_7^{2-}$	
	Dihydrogen phosphate, $H_2PO_4^-$	Hydrogen phosphate, HPO_4^{2-}	
	Hydrogen carbonate,* HCO_3^-	Sulfate, SO_4^{2-}	
	Hydrogen sulfate,† HSO_4^-	Sulfite, SO_3^{2-}	
	Hydroxide, OH^-		
	Nitrate, NO_3^-		
	Nitrite, NO_2^-		
	Perchlorate, ClO_4^-		
	Permanganate, MnO_4^-		

*Commonly called bicarbonate.
†Commonly called bisulfate.

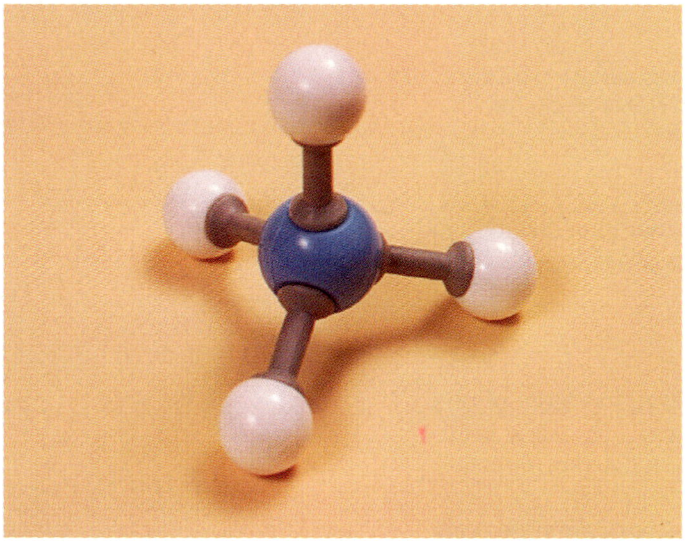

(a)

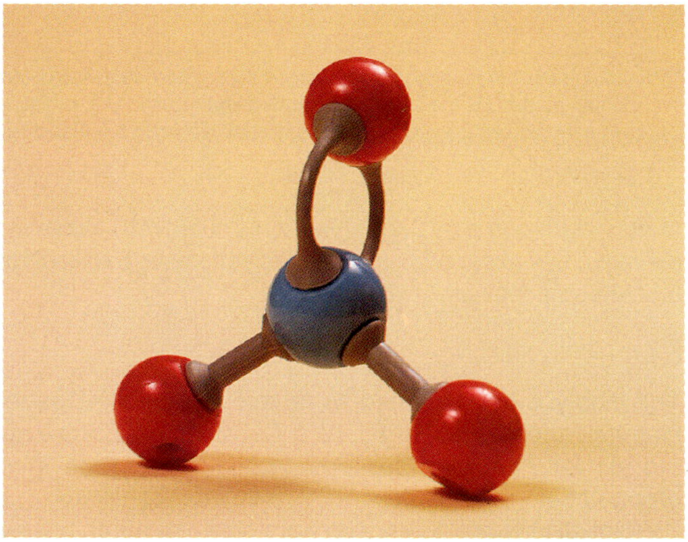

(b)

Figure 6.11
Molecular models illustrate the structural arrangements of atoms in the ammonium and nitrate ions. (a) A nitrogen atom and four hydrogen atoms exist as a unit with a charge of 1+ in the ammonium ion. (b) In the nitrate ion, a nitrogen atom and three oxygen atoms exist as a unit with a charge of 1–. Each of these ions can form ionic bonds with other ions of opposite charge.

Just as a simple ion attracts another ion of opposite charge to form an ionic bond in new compounds, polyatomic ions may also attract another ion (either simple or polyatomic) to form new compounds. For example, the ammonium ion, NH_4^+, is a polyatomic ion of charge 1+ that may form an ionic bond with a chloride ion, Cl^-, to form ammonium chloride, NH_4Cl, in the fixed proportion of one ammonium ion to one chloride ion (Figure 6.11).

Many beginning chemistry students have difficulty recognizing polyatomic ions. Please *memorize* Table 6.6 and become acquainted with the idea that a total group of atoms has a charge associated with it. For example, NO_3^- represents the nitrate anion in which one nitrogen and three oxygens have the charge 1–; HPO_4^{2-} represents the hydrogen phosphate anion in which one hydrogen, one phosphorus, and four oxygens have a 2– charge. Many common compounds such as baking soda, $NaHCO_3$, and vinegar, $HC_2H_3O_2$ (dissolved in water), contain polyatomic ions; baking soda contains the hydrogen carbonate ion (commonly called bicarbonate), HCO_3^-, and vinegar contains the acetate ion, $C_2H_3O_2^-$.

Figure 6.12
Calcium chloride is a white solid used as a drying agent because it absorbs moisture from the gas passing through the drying tube.

6.8 formulas for ionic compounds

Since you know the ionic charges of simple ions and have memorized the charges of polyatomic ions, you can readily write correct formulas for ionic compounds. Using the principle of electrical neutrality (Section 6.6), cations (+) and anions (–) combine in such a way that the magnitude of total positive charge equals the magnitude of total negative charge. For example, the proper formula for the combination of the elements potassium and bromine in the sedative potassium bromide is KBr. The fixed proportion of one K^+ cation and one Br^- anion results in electrical neutrality (Table 6.4).

In writing a formula, you always write the positive, metallic ion first and then the negative, nonmetallic ion. Charges on the individual ions are not shown in the formula. Even though the compound KBr is made up of K^+ and Br^- ions, the formula does not show the charges explicitly. You indicate with a subscript after each ion the number of those ions required in the formula. For example, in Sample Exercise 3, two potassium ions were indicated in the formula K_2O. If no subscript is written, then it is understood to be 1, as for both ions in KBr or for the oxygen ion in K_2O.

sample exercise 4

What is the formula for the compound, calcium chloride, formed from Ca^{2+} and Cl^-? Calcium chloride is used as a drying agent because it absorbs moisture from the air or gas surrounding it (Figure 6.12).

solution

The combination of one Ca^{2+} ion and one Cl^- ion would leave the compound with a 1+ charge: (2+) + (1–) = 1+. Instead, one Ca^{2+} cation and two Cl^- anions must combine: (2+) + 2(1–) = 0. The formula thus is $CaCl_2$. Note the use of the subscript 2 to indicate two chloride ions. $CaCl_2$ is made up of one Ca^{2+} and two Cl^- ions.

When polyatomic ions are involved, the rules work very much the same way. What is the formula for the compound formed from Mg^{2+} and OH^-? To balance the charges, you will need one magnesium ion combining with two hydroxide ions. But where can you write the subscript to indicate two hydroxide ions clearly? The formula is written $Mg(OH)_2$. Whenever you have more than one of a polyatomic ion, surround the formula for the polyatomic ion with parentheses and indicate with a subscript after the parentheses the number of polyatomic ions present. The compound $Mg(OH)_2$ is made up of one Mg^{2+} and two OH^- ions.

Figure 6.13
Sodium hydroxide is a caustic, white solid, available as pellets. It is commonly used as a laboratory reagent in aqueous solution and in the manufacture of soap and rayon. In aqueous solution, it is known as lye. Sodium hydroxide pellets absorb moisture and become very wet looking if left exposed to moist air. The pellets on the left watch glass have just been poured from the bottle. The pellets on the right watch glass have been sitting in air and have absorbed a great deal of moisture.

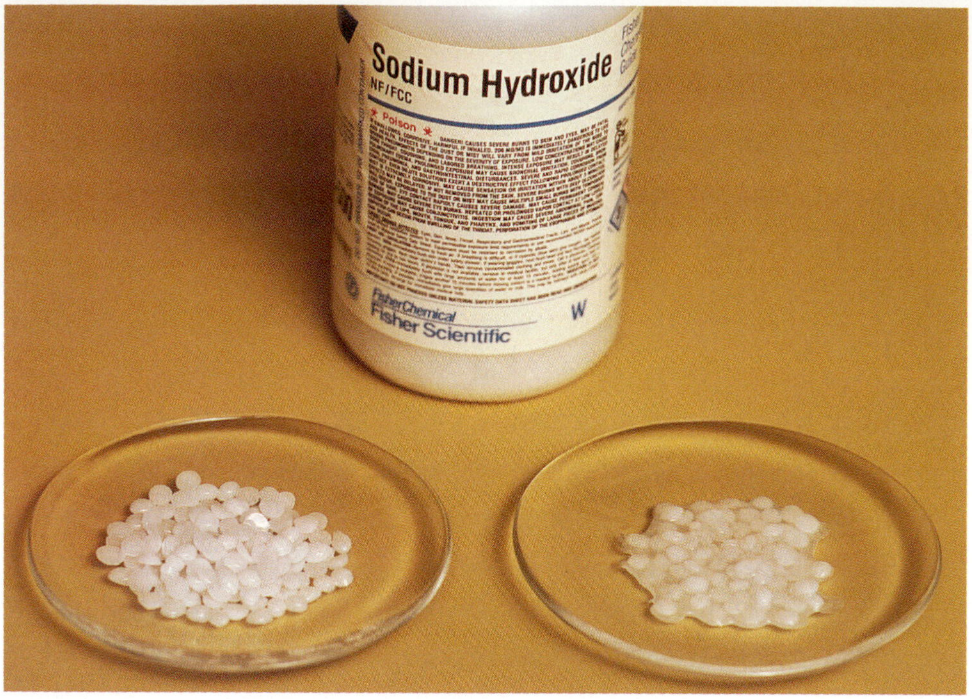

sample exercise 5

What is the formula for the compound, sodium hydroxide (Figure 6.13), formed from Na^+ and OH^-?

solution

To balance the charges, we simply combine one Na^+ and one OH^-. The formula will then be NaOH. We do *not* have to enclose the hydroxide ion within parentheses because there is only one of this polyatomic ion present in NaOH. However, you must be able to recognize the unit OH^- in the absence of parentheses.

sample exercise 6

What is the formula for the compound, aluminum sulfate, formed from Al^{3+} and $SO_4{}^{2-}$? Aluminum sulfate is a component of baking powder.

solution

How are we to balance the charges 3+ and 2–? Begin by remembering that the common denominator of 3 and 2 is 6. If we take two 3+ and three 2– ions, the total positive charge is 6+ and the total negative charge is 6–; the total positive and negative charges are now balanced. Therefore, the correct formula is $Al_2(SO_4)_3$. This formula tells us that the ratio of ions in this ionic compound is two aluminum ions to three sulfate ions.

••• problem 4

Write formulas for the ionic compounds formed from the following sets of ions.

a. Na^+ and I^-
b. $NH_4{}^+$ (ammonium) and S^{2-}
c. Al^{3+} and $NO_3{}^-$ (nitrate)

d. Al^{3+} and CO_3^{2-}(carbonate)
e. Al^{3+} and PO_4^{3-}(phosphate)
f. NH_4^+ (ammonium) and PO_4^{3-}(phosphate)

---

problem 5

What is the ratio of cations to anions present in each of the following ionic compounds? Also name any polyatomic ion present in the compounds below.

a. CaO, also called lime, a white solid used in plaster and cement
b. $Mg(OH)_2$, also called milk of magnesia, a slightly soluble, white solid used in water as an antacid and laxative
c. Na_3PO_4, sold in hardware and paint stores as the detergent TSP (trisodium phosphate)
d. $KC_2H_3O_2$, a white, crystalline solid used in the analysis of other chemicals
e. $Ca_3(PO_4)_2$, used both as a fertilizer and food additive as well as in baking powder and toothpaste
f. $KClO_3$, a white, crystalline solid used in matches, fireworks, bleaches, and disinfectants

---

Suppose you were asked to write a formula for the combination of calcium and sulfur. This time we have not indicated the charges as in previous examples, so you must first ask yourself what ions do these elements form and then proceed as usual. Calcium is in Group IIA; therefore, it forms Ca^{2+}. Sulfur is in Group VIA; therefore, it is S^{2-}. The correct formula is CaS.

---

sample exercise 7

Write the correct formula for the compound, zinc chloride, formed by combining zinc and chlorine. Zinc chloride is a white, crystalline solid used as a wood preservative, disinfectant, and antiseptic.

solution

The compound formed from zinc and chlorine must be an ionic compound since zinc is a metal and chlorine a nonmetal. Remember to look up the group number (IIB) of the zinc ion, which tells you that the charge is Zn^{2+}. Chlorine is in Group VIIA; therefore, its charge is 1–. To form a neutral compound, the correct combination is one cation and two anions. Thus, we get $ZnCl_2$.

If an element can form ions with more than one charge (Table 6.5), you must be given additional information as to which ion is actually forming in the particular problem at hand.

---

sample exercise 8

Write a correct formula for the compound, iron(III) chloride, formed from iron and chlorine.

solution

You are told that in this problem iron forms the Fe^{3+} ion, and you already know that in all ionic compounds chlorine forms the Cl^- ion. The correct combination to form a neutral compound is one cation and three anions. Thus, the formula is $FeCl_3$. Iron(III) chloride is used in engraving, in deodorizing sewage, and in medicine as an astringent and styptic.

General rules for writing formulas for ionic compounds

1. Determine the ionic charges of the elements that are combining.
2. Choose a combination of the ions such that the sum of the ionic charges equals zero.
3. Write the symbol for the positive ion first, followed by the negative ion's symbol. Do not include their charges.
4. Use a subscript after an ion symbol to indicate two or more of that kind of ion present in the compound. Polyatomic ions require parentheses when there is more than one. The subscript 1 is understood and not written.

••• problem 6

Write a correct formula for the compounds formed from the following.

a. Magnesium and sulfur
b. Chromium and chlorine
c. Calcium and nitrogen
d. Lead (assume it forms a 4+ ion) and oxygen

There is a shortcut method for writing ionic formulas that you should feel free to use. However, it is a good idea to always remember that the basis for ion combination is the formation of a neutral (uncharged) compound. See how this method works for the compound formed from Al^{3+} and S^{2-}.

Shortcut for writing ionic formulas

1. Write the two ions next to each other with their charges as superscripts.

$$Al^{3+} \quad S^{2-}$$

2. Then bring the 2 of the sulfur down as the subscript for Al and the 3 from the aluminum down as the subscript for S; that is, "crisscross" the charges as subscripts without signs. Always remember to omit the signs in the subscripts.

$$Al^{(3+)} \times S^{(2-)} \quad \text{gives} \quad Al_2S_3$$

3. Now check to see that the subscripts are the smallest possible whole-number ratio. If the subscript ratio can be converted into a smaller whole-number ratio, this must be done. In this example, the subscript ratio 2:3 cannot be converted into a smaller whole-number ratio.

If the first two rules are applied to the combination of Ca^{2+} and O^{2-},

$$Ca^{(2+)} \times O^{(2-)} \quad \text{gives} \quad Ca_2O_2$$

But the subscript ratio 2:2 can be converted into the smaller whole-number ratio 1:1 by dividing both subscripts by 2. This gives the correct formula, CaO. Similarly, for Pb^{4+} combining with O^{2-},

$$Pb^{(4+)} \times O^{(2-)} \quad \text{gives} \quad Pb_2O_4$$

which should be written PbO_2.

Do not forget the parentheses around polyatomic ions. For the combination of Ba^{2+} and NO_3^-,

$$Ba^{(2+)} \times NO_3^{(-)} \quad \text{gives} \quad Ba(NO_3)_2$$

The subscript ratio of 1:2 cannot be converted into a smaller whole-number ratio.

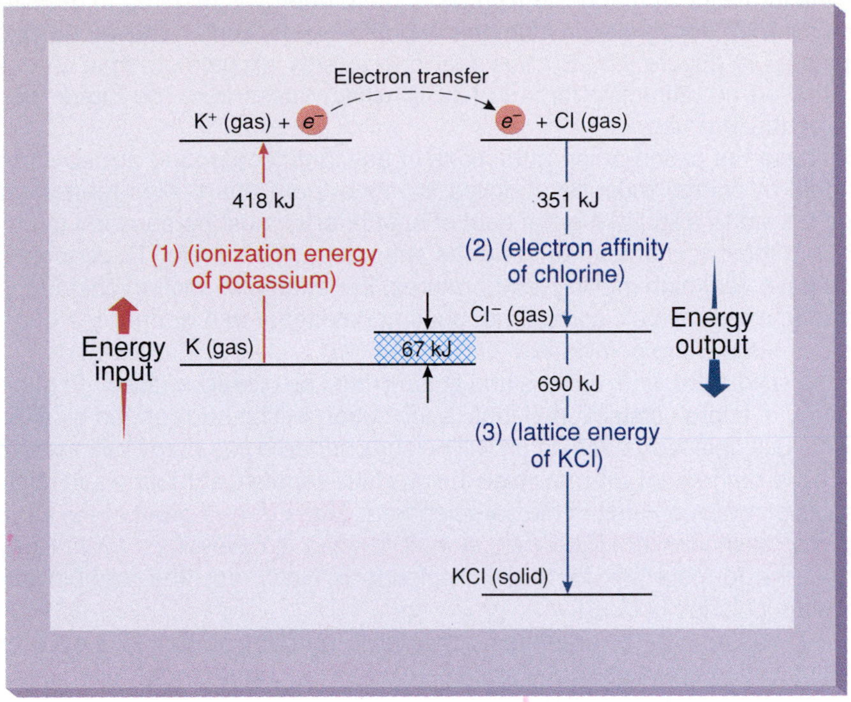

Figure 6.14

This energy-level diagram illustrates the ionic bonding in KCl. The total energy released from the sum of (2), the electron affinity of chlorine atoms, and (3), the lattice energy of K^+ and Cl^- ions, is greater than (1), the ionization energy of K atoms. Therefore, KCl, an ionic solid, will form.

• • • problem 7

Redo Problem 4 using the shortcut method for formula writing.

6.9 the nature of the ionic bond

What is called the **ionic bond** is the *electrostatic attraction between cations* (+) *and anions* (–) in an ionic compound. You know from Chapter 5 that bringing together plus and minus charges results in a lower-energy state. It is the fact that the cation-anion attraction results in a lowering of energy of the separated ions that enables the ionic bond to form.

Remember that cation formation requires an input of energy. For example, the ionization of $K \rightarrow K^+ + e^-$ requires 418 kJ per 39 g of potassium (step 1 in Figure 6.14). Also remember that anion formation releases energy, called *electron affinity,* and the formation of Cl^- ions from 35 g of Cl atoms[4] releases 351 kJ of energy (step 2 in Figure 6.14). For the ionization of K to occur, more energy must be released than is needed for the ionization. Since only 351 kJ are released by the formation of Cl^-, you are 67 kJ short (hatched area in Figure 6.14) because a minimum of 418 kJ are needed to form K^+. However, the attractive force between the formed K^+ and Cl^- is strong and releases 690 kJ (step 3 in Figure 6.14) as the ions bond to each other in the solid state. The energy released from the attraction between a gaseous cation and a gaseous anion to form an ionic compound in the ionic state is called the **lattice energy.** The additional 690 kJ released is more than enough energy to enable the ionic bonding process to occur.

To form an ionic compound from atoms in the gas phase, there must be a balance among the (1) ionization energy of the metal atom, (2) the electron affinity of the nonmetal atom, and (3) the energy released from the net electrostatic force between ions in the solid state (lattice energy). In other words, *you may expect ionic compounds to form when the ionization energy is relatively low (energy input), the electron affinity is relatively high (energy release), and the lattice energy from the net attractive force is also relatively high (energy release).*

4. You will see in Chapter 10 that these masses are the required amounts to have no leftover K or Cl.

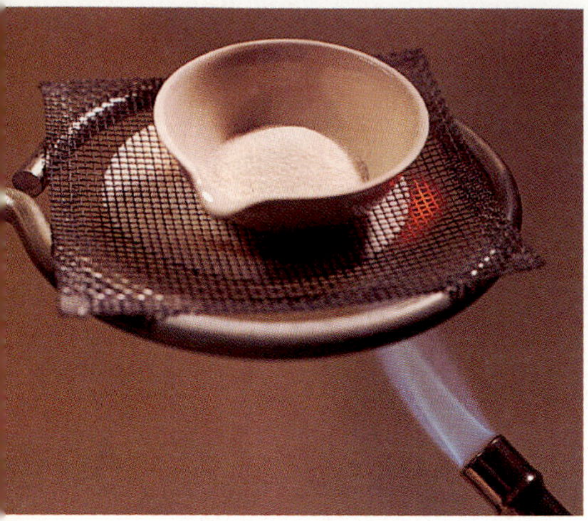

Figure 6.15

Sodium chloride, an ionic compound, does not melt when heated by a Bunsen burner. The strong ionic bond is responsible for its high melting point (801°C).

In the formation of magnesium chloride, more energy is required to remove two electrons from the magnesium atom than would be required to remove one electron from potassium (Figure 6.3). But this additional energy will be more than returned in the increased net attractive force in $MgCl_2$, which arises from the higher positive charge on the Mg^{2+} ion.

Because the cation-anion attractions in any ionic compound are so strong, a great deal of energy must be supplied to move them apart. This means that to change a solid to a liquid, a great deal of heat energy must be supplied to pull the ions out of their rigid shape and into the shapeless liquid state. Thus, ionic compounds have very high melting temperatures. For example, sodium chloride, NaCl (table salt), a typical ionic compound, does not melt in a pan or near the burner of your stove. Its melting point is 801°C (Figure 6.15).

In the solid state, both cations and anions must be closely spaced. To minimize the energy of a large collection of ions, each cation will be surrounded by as many anions as possible, and each anion will be surrounded by as many cations as possible. This arrangement will maximize the positive-negative attraction and minimize the repulsion between ions of the same charge. The smallest grouping of ions corresponding to the correct fixed ratio in their formula is called a **formula unit**. For example, the formula unit for sodium chloride is NaCl, and the formula unit for magnesium chloride is $MgCl_2$.

The lowest-energy arrangement for a large number of ions is a systematic, alternating cation-anion pattern (Figure 6.16). The regular arrangement of ions in a solid is called a **lattice structure**. Because the ions alternate so exactingly, layers of ions can be seen in an ionic model (Figure 6.16) and those layers are responsible for the geometry of the bulk crystal. Thus, the ionic lattice structure gives the material a definite shape with angles characteristic of that structure.

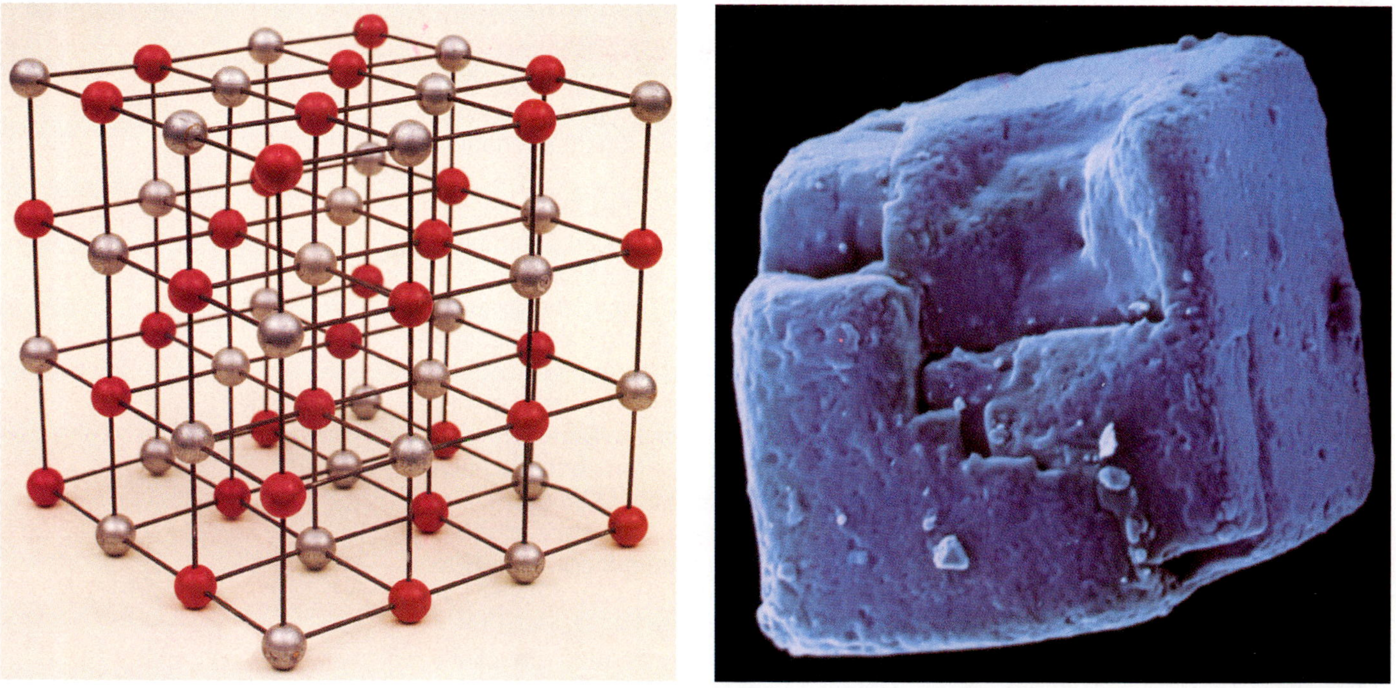

(a) (b)

Figure 6.16

(a) The ionic model of sodium chloride, NaCl, shows the regular arrangement of Na^+ and Cl^- ions and how the layers of ions build upon each other in a fixed geometry. (b) The crystalline shape of NaCl is cubic, with all angles being 90° no matter how large or how small the crystal. This photograph of a small NaCl crystal was taken with a scanning electron microscope.

The strong ionic bond, i.e., the attraction between cations and anions, gives ionic compounds their characteristic properties. As previously mentioned, ionic compounds typically have high melting points and crystalline structures with definite shape and characteristic angles. By contrast, the molecular compounds discussed in the next section have much weaker bonds. As a result of their weaker bonds, molecular compounds have much lower melting points and form solids that are irregular in shape (Figure 6.17).

Figure 6.17
Sulfur is a nonionic element. Its melting point at 113°C is much lower than ionic NaCl. As a solid, it lacks the clearly defined shape of cubic NaCl.

6.10 covalent compounds

So far in this chapter, you have explored the nature of the ionic bond and the formulas of ionic compounds. The transfer of electron(s) between metals and nonmetals leads to the formation of ions, which permits the formation of ionic bonds in ionic compounds. The transfer of electrons between two nonmetals is not possible, however, because both nonmetals strive to gain electrons and neither is willing to lose one. As a result, in nonmetals, electrons may not be transferred and ions cannot form. Nonmetals *do* bond to one another by minimizing the energy of their valence electrons, but instead of transferring electrons to form ions, they *share* electrons between themselves. Nonmetallic atoms become linked together by forces known as covalent bonds in units called **molecules.** Since the compounds formed are made of molecules, they are called **molecular compounds.**

Even though there are fewer nonmetals in the periodic table than metals, the number of molecular compounds they form is no less than the number of ionic compounds formed with metals. There are millions of molecular compounds. Carbon compounds, which form the basis of study for organic chemistry and are the building blocks of life, are primarily molecular compounds. In addition, such abundant compounds as water, ammonia, natural gas (methane), petroleum, and plastics are all molecular compounds.

The atoms in molecules are held together by shared pairs of electrons; that is, a pair may be considered part of the valence-electron configuration of two atoms. A shared pair of electrons is called a **covalent bond,** and the compounds formed with covalent bonds are called **covalent compounds.** As in the case of the formation of ionic bonds, covalent bonds form so that each atom in the bond can achieve a noble-gas configuration.

Thus, compound formation in general occurs so that atoms can achieve lower-energy, noble-gas electron configurations.

Both ionic and covalent compounds form because there are two ways to achieve the noble-gas configuration:

1. *Transfer of electrons* from metals to nonmetals leads to the formation of ions, ionic bonds, and ionic compounds.
2. *Sharing of electrons* between nonmetals leads to molecules that are held together by covalent bonds. Molecular compounds are also called covalent compounds.

In the next five sections, you will focus on the nature and characteristics of covalent bonds. As mentioned above, these bonds are enormously important in a great variety of molecules, especially those that make up living systems.

6.11 electron sharing in diatomic molecules

You have known since Sections 1.11 and 4.13 that certain elements (H_2, O_2, N_2, and the halogens F_2, Cl_2, Br_2, and I_2) exist as diatomic molecules. These elements offer an excellent example of atoms sharing electrons in order for each atom to achieve a noble-gas electron configuration. Since both atoms are identical nonmetals, they each strive to gain electrons equally.

For example, consider two individual chlorine atoms (Group VIIA), for which you write the Lewis electron dot structures (Section 5.9):

Seven dots represent the seven valence electrons in the third main level of $_{17}Cl$, $1s^22s^22p^63s^23p^5$. Each atom can acquire eight valence electrons (and consequently greater stability and lower energy) if the two come together to share a pair of electrons. As the two atoms come close together, each atom contributes one unpaired electron to form a shared pair between the atoms. The shared pair belongs equally to both atoms, so each now has a complete octet of valence electrons, that is, a noble-gas configuration. The use of shared pairs of the electrons to form octets was developed by G. N. Lewis and is referred to as Lewis theory.

Shared pair of electrons, a covalent bond

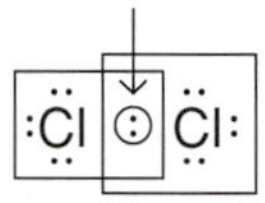

Each Cl has 8 e^- around it, forming a stable, noble-gas electron configuration.

The shared pair (covalent bond) can be indicated more simply by a dash (—) placed between the two symbols representing the atoms. The shared pair is called a **bonding pair of electrons.**

A covalent bond consisting of a shared or bonding pair of electrons

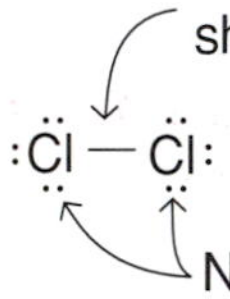

Nonbonding electron pairs

The valence electrons not involved in bonding are called **nonbonding electrons**. They are also called **lone pairs**.

All other halogens (Group VIIA) form covalent bonds in exactly the same way:

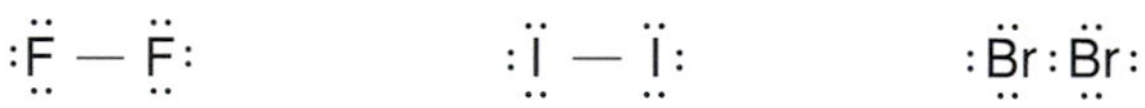

Remember, the dash and the shared pair of dots between symbols mean exactly the same thing.

H_2 molecules are also held together by a shared pair. The two H atoms come together to form H:H (H—H). Each H atom now has the helium duet of two electrons. This is a lower-energy, more stable electron configuration corresponding to that of the noble-gas element, helium.

A single covalent bond cannot provide a stable electron configuration for all diatomic molecules, and double and triple covalent bonds are found. For example, when we try to account for the bonding in O_2 with a single pair of shared electrons, the oxygen atoms do not obtain a complete octet. Since oxygen is in Group VIA, each atom has six valence electrons:

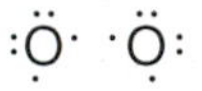 Two individual O atoms

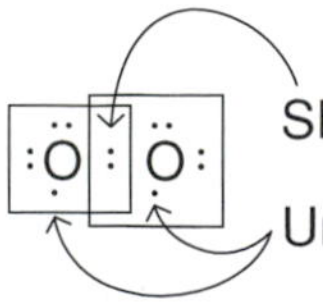

Sharing one pair

Unpaired electrons

Not stable

table 6.7 bond-dissociation energies for diatomic molecules

X_2 (gas)	Bond in X_2	Bond-dissociation energy (kJ/mol) ($X_2(g) \rightarrow 2\,X(g)$)
N_2	N≡N	946
O_2	O═O	493
H_2	H—H	435
Cl_2	Cl—Cl	243
Br_2	Br—Br	192
F_2	F—F	159
I_2	I—I	151

Each O has only seven electrons (count within the boxes around the atoms), but if the oxygens share their remaining unpaired electrons, octets will be attained.[5]

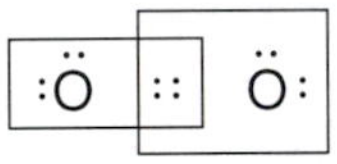
or :Ö═Ö:

Stable   Sharing two pairs

Now each O has eight electrons. Two shared pairs of electrons are called a double covalent bond, or simply a **double bond**.

For each nitrogen in N_2 to have a noble-gas configuration, three shared pairs of electrons are required. This is known as a **triple bond**.

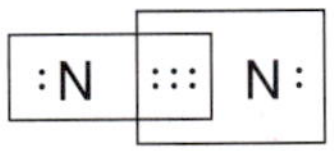
or :N≡N: Sharing three pairs

The more pairs of electrons that are shared, the stronger and shorter the covalent bond. In general, triple bonds are stronger and shorter than double bonds, and double bonds are stronger and shorter than single bonds. It takes more energy to break the double bond in O_2 (493 kJ) than to break the single bond in F_2 (159 kJ), and it takes even more energy to break the triple bond in N_2 (946 kJ) (Table 6.7). The strength of bonds becomes a determining factor for an atom's participation in chemical reactions, as discussed later in the text. From the relative bond strengths in Table 6.7, you may already predict the property that fluorine gas is more chemically reactive than oxygen, and oxygen is more reactive than nitrogen.

6.12 the nature of the covalent bond

You have seen that it is the electrostatic attraction between cations and anions that lowers energy in the formation of ionic bonds. The formation of covalent bonds leads to a lowering of energy because of the attraction between the electrons (−) of one atom and the nucleus (+) of another atom. Let us examine how energy is lowered in the H_2 molecule (relative to two separate H atoms) through covalent bond formation.

In two separate H atoms, the lone valence electron of each isolated atom is attracted to its respective nucleus (Figure 6.18). As the separated atoms come closer together to form the H_2 molecule, there are two new attractive forces (between

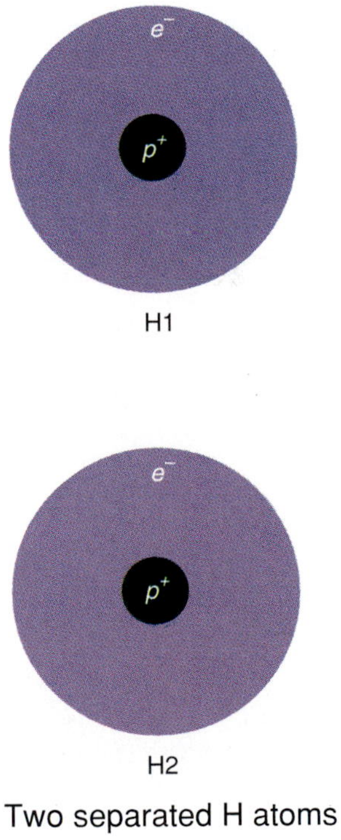

Figure 6.18
In two separate H atoms, the lone valence electron of each isolated atom is attracted to its respective nucleus. As the separated atoms come closer together, each electron is attracted to both H nuclei.

5. This electronic structure for oxygen does not account for some of its observed properties and, consequently, represents a failure of Lewis theory to predict structure. However, it offers a simple example of the concept of a double bond.

Figure 6.19

Graphical illustration of how energy changes for two H atoms at different distances of separation. The separation for minimum energy is called the bond length of the H_2 molecule and is located where the yellow energy arrow changes from pointing downward to pointing upward.

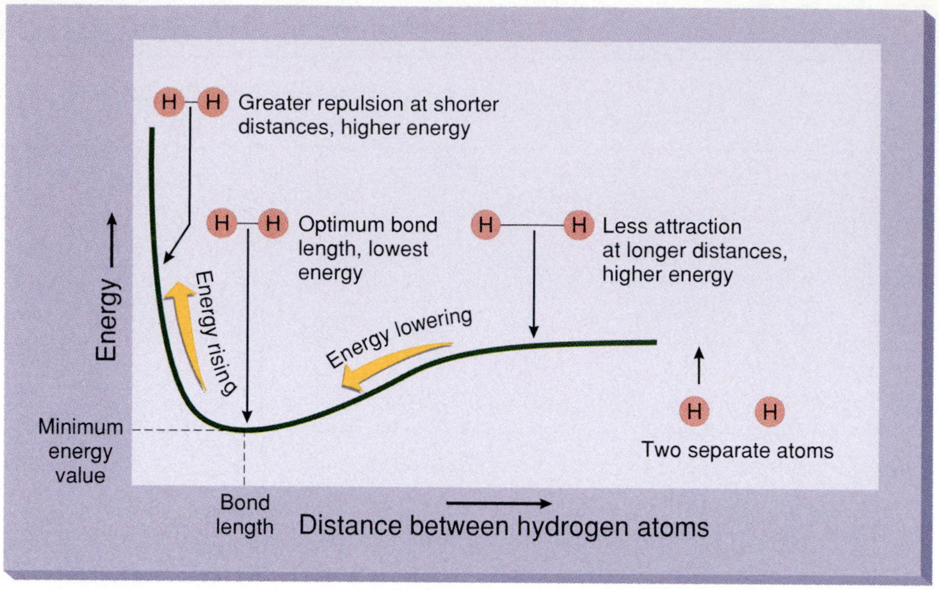

electron$_1$ to nucleus$_2$, electron$_2$ to nucleus$_1$) and two new repulsive forces (between electron$_1$ and electron$_2$, nucleus$_1$ and nucleus$_2$). As the atoms continue to approach each other, the energy of the H_2 molecule continues to decrease until it reaches a minimum at a certain characteristic distance called the **bond length**. Figure 6.19 illustrates graphically the lowering of energy for the H_2 molecule compared with two separate H atoms. The yellow arrow points downward as the atoms approach each other until it reaches the optimum bond length of lowest energy. If the atoms come closer than the bond length, the repulsive forces become larger than the attractive forces, so the total energy increases, as illustrated by the yellow arrow pointing upward. At the bond length, the two H atoms have reached maximum stability. Whether they move closer together or farther apart, their energy is raised and their stability is decreased.

Experimentally, the stability of the covalent bond in the hydrogen molecule is demonstrated by the fact that it takes 435 kJ to convert 2 g of hydrogen molecules into hydrogen atoms.

$$H_2\text{ (gas)} + 435\text{ kJ} \rightarrow 2\text{ H (gas)}$$

This energy is known as the **bond-dissociation energy**. The larger the bond-dissociation energy, the stronger the bond. Table 6.7 lists bond-dissociation energies for several diatomic molecules. The numbers are measured in kilojoules per a mass amount equal to the gram atomic mass in grams.[6] For each element, the bond-dissociation energy represents breaking the same number of bonds. As mentioned in Section 6.11, covalent bonds are not all of equal strength. Experimentally measured bond-dissociation energies confirm that, in general, triple bonds (N_2) are stronger than double bonds (O_2), which in turn are stronger than single bonds (F_2).

• • • problem 8

a. Which of the halogens has the strongest bond?

b. Which of the halogens has the weakest bond?

c. What kinds of substances are formed when the bond in a diatomic molecule is broken by the input of bond-dissociation energy?

6. This mass amount called a mole will be discussed fully in Chapter 8.

6.13 Lewis electron dot structures for molecules

You have seen how atoms may share valence electrons to achieve a stable, noble-gas electron configuration and form covalent bonds in simple diatomic molecules. If you understand the principle of electron sharing as outlined by G. N. Lewis (Figure 6.20), you may predict the most stable electron configuration of many other molecules whether or not they are diatomic. Knowing the electron configuration of any covalent molecule adds further to your predictive powers, because the distribution of electrons among the atoms will enable you to predict trends in physical properties such as melting point, boiling point, solubility (the ability to dissolve in a liquid), and in some cases, the chemical property of reactivity. For example, simply by knowing the Lewis electron dot structures for a water molecule and for an oil molecule, such as gasoline, you will be able to explain in the following sections why water has a higher boiling point than gasoline, why water is less volatile than gasoline, why gasoline and water do not dissolve in each other, and why ionic compounds may dissolve in water but not in gasoline (Figure 6.21).

Figure 6.20

Gilbert Newton Lewis (1875–1946) was an American chemist best known for his theory of electron sharing in covalent bonds.

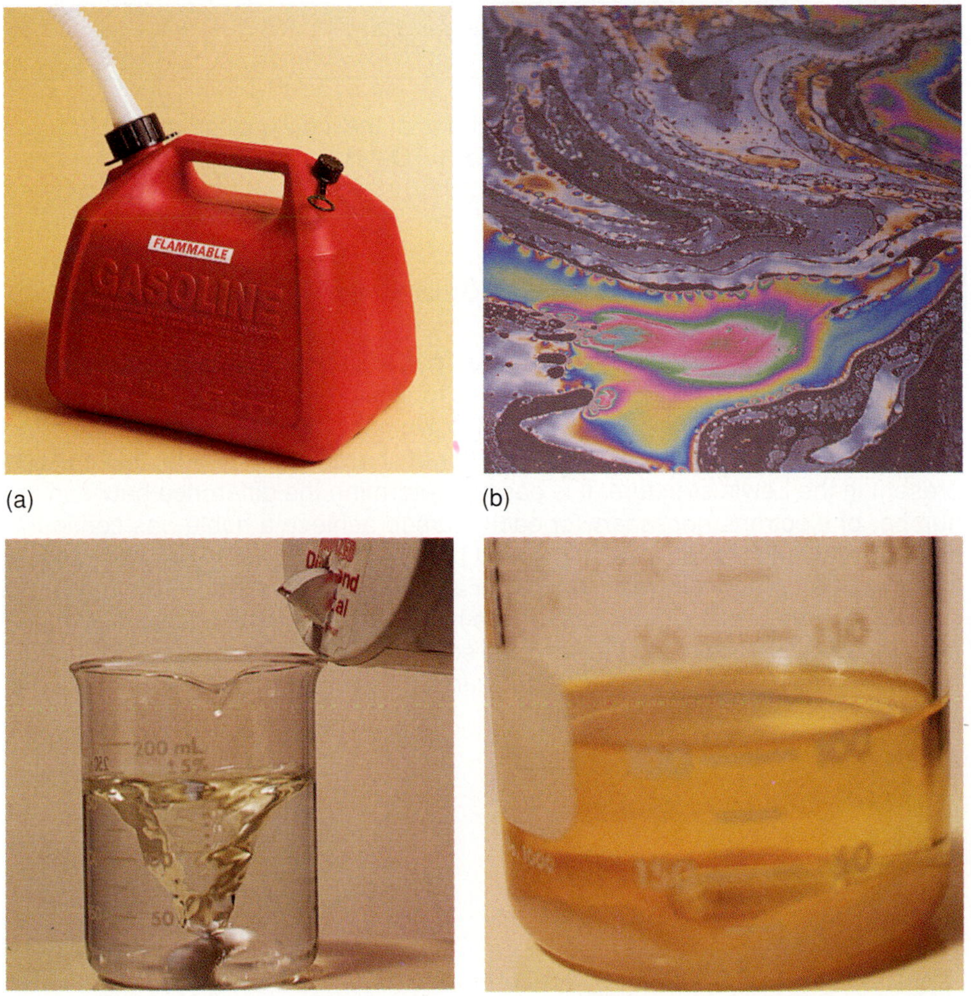

(a) (b) (c) (d)

Figure 6.21

(a) Gasoline must be stored in a labeled container because its vapor is highly volatile and flammable. (b) Gasoline floats on the surface of water. It is less dense than water and is not soluble in water. (c) Table salt, an ionic compound, dissolves readily in water. (d) Table salt is not soluble in gasoline. It settles to the bottom of this beaker because it is more dense than water.

Another simple example of how to write a Lewis electron dot structure can be seen by considering the bonding in HCl. In this diatomic molecule, the two atoms are not the same, and in Section 6.15, you will see that the electrons are not shared equally. For now, you must be able to expand your skills in predicting the electron configurations for diverse molecules. First write the Lewis dot structures for the atoms in HCl:

Notice how the unpaired electrons can be shared so that both H and Cl attain a noble-gas configuration. This is achieved easily for this small molecule.

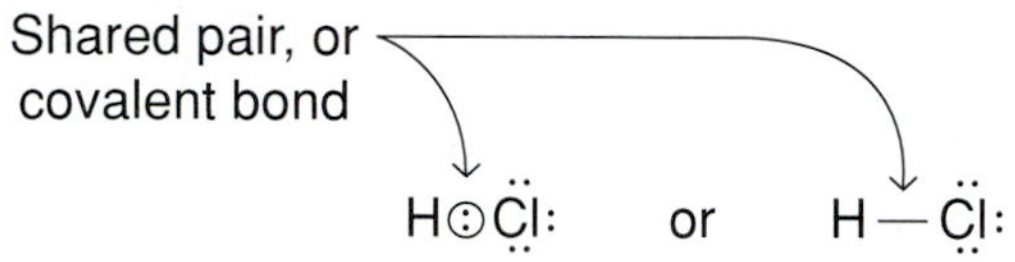

Whether you write the formula horizontally, vertically, or at an angle makes no difference.

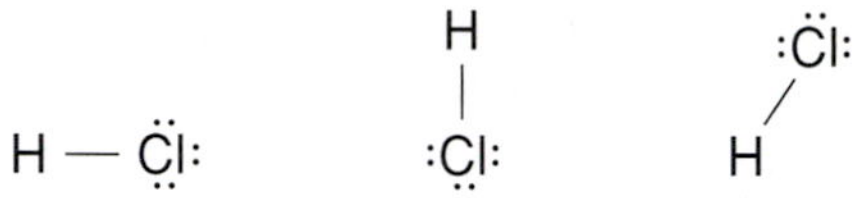

H has a duet, and Cl has an octet in all of these arrangements.

Lewis electron dot structures are also called **Lewis structures** or simply structural formulas because they tell you how the molecule is *constructed* or built.

---

problem 9

Write the Lewis electron dot structures for HBr, HF, and HI.

By sharing the unpaired valence electrons of the individual atoms of H_2O, you would figure out that the Lewis structure is H—O—H.

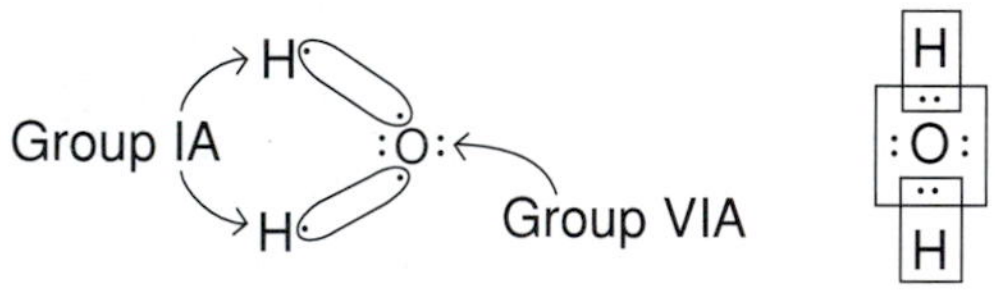

For more complicated molecules, however, particularly those with many atoms, it would be helpful to have a set of guidelines for writing Lewis structures instead of having to just guess. The guidelines set down here allow you to determine the central atom in the molecule and the number of shared pairs (bonds) that must be present in the Lewis structure. It is easy to determine the difference between the number of electrons *necessary* for each atom to achieve a noble-gas configuration (called *t* in the guidelines on the next page) and the number of valence electrons *actually present* in the molecule (called *v*). Half this difference (called *b*) reveals the number of shared pairs in the molecule. After making this determination, you will still need to juggle electrons somewhat, but the job should be greatly simplified. Read through the guidelines carefully and see how they are applied in the sample exercises that follow them.

Guidelines for writing Lewis structures

The following terms are used in the guidelines.

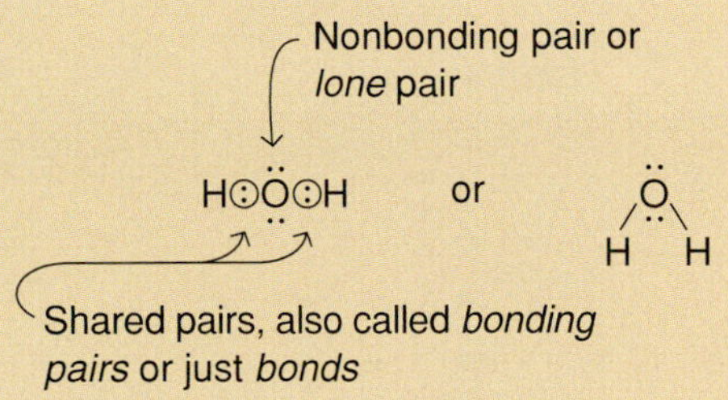

1. *Determine the number of bonding pairs (b) based on the number of electrons (t) and the number of valence electrons (v) in the molecule or polyatomic ion.* Remember, $b = (t - v)/2$.

   a. Determine the number of electrons (t) that the molecule or ion would need for each atom to achieve a noble-gas configuration, that is, for each hydrogen, H (if present), to have two electrons and for each other atom to have eight electrons.[7]

   $$t = 2\text{ (number of H atoms)} + 8\text{ (number of all other atoms)}$$

   For example, for H_2O,

   $$t = 2(2) + 8(1) = 12$$

   b. Find the number of valence electrons (v) for all atoms in the molecule or ion. For example, for H_2O,

2 H (1 valence e^- each)	$= 2\ e^-$
1 O (6 valence e^-)	$= 6\ e^-$
	$v = 8\ e^-$

   For polyatomic ions, a negative charge indicates extra valence electrons and a positive charge indicates fewer valence electrons. For example, in the negatively charged hydroxide ion, OH^-,

1 H (1 valence e^-)	$= 1\ e^-$
1 O (6 valence e^-)	$= 6\ e^-$
1– charge means 1 extra e^-	$= 1\ e^-$
	$v = 8\ e^-$

   In the positively charged ammonium ion, NH_4^+,

4 H (1 valence e^-)	$= 4\ e^-$
1 N (5 valence e^-)	$= 5\ e^-$
1+ charge means 1 less e^-	$= -1\ e^-$
	$v = 8\ e^-$

   c. Calculate the number of bonding pairs using the formula $b = (t - v)/2$. For example, for H_2O,

   From step 1a   From step 1b

   $$b = \frac{12 - 8}{2} = \frac{4}{2} = 2$$

   Two bonds are predicted.

7. Boron is an exception: each boron atom, if present, should achieve only six electrons.

2. *Determine the central atom in the molecule or polyatomic ion.*

   a. For covalent compounds containing only two elements and for polyatomic ions, the central atom is the one that appears only once in the formula (H is *never* a central atom).

Formula	**Central atom**
CH_4	C
SO_3	S
PCl_3	P
SF_4	S
NH_3	N
PO_4^{3-}	P
SO_4^{2-}	S
NH_4^+	N
CO_3^{2-}	C
H_2O	O
HSO_4^-	S

   b. For covalent compounds containing H, O, and a third element, the atom other than H or O is the central atom.

   c. Carbon is the central atom in any carbon-containing compound.

Formula	**Central atom**
H_2SO_4	S
HNO_3	N
$HClO_3$	Cl
HCN	C
$COCl_2$	C

3. *Write down the central atom and arrange the other atoms around the central one using the number of bonded pairs you calculated in step 1.* Not all atoms will necessarily be connected to the central atom. In compounds containing H, O, and a third element, distribute the oxygens around the central atom (i.e., avoid O—O bonds) and for most cases connect hydrogens to the oxygens.

   Continuing with H_2O as our example, O is central and we must have two bonds, thus H—O—H.

4. *Fill in the rest of the structure with nonbonding pairs until each hydrogen atom (if present) has two electrons and every other atom has eight electrons.* For H_2O, the O requires two nonbonding pairs to complete its octet,

$$\mathrm{H:\ddot{\underset{..}{O}}:H.}$$

5. *Recheck to be sure that the total number of electrons shown in your structure equals v.* The structure we have written for H_2O shows eight electrons and $v = 8$.

sample exercise 9

Write a Lewis electron dot structure for ammonia, NH_3, a chemical used in household cleaning.

solution

Step 1. *Determine the number of bonding pairs (b) based on the number of electrons (t) and the number of valence electrons (v) in the molecule.*

   a. Determine *t*, the number of electrons (two for H and eight for each other atom).

3 H	$3 \times 2 = 6\ e^-$
1 N	$1 \times 8 = 8\ e^-$
	$t = 14\ e^-$

b. Determine *v,* the number of valence electrons.

3 H	$3 \times 1 = 3\ e^-$
1 N	$1 \times 5 = 5\ e^-$
	$v = 8\ e^-$

c. Determine *b,* the number of bonding pairs.

$$b = \frac{t - v}{2}$$

$$b = \frac{14\ e^- - 8\ e^-}{2} = 3 \text{ bonding pairs}$$

Step 2. *Determine the central atom in the molecule.*
In NH_3, the central atom is N (guideline 2a: H is never a central atom).

Step 3. *Write down the central atom and arrange the other atoms around the central one using the number of bonded pairs you calculated in step 1.*

```
H : N : H
    ··
    H
```

Step 4. *Fill in the rest of the structure with nonbonding pairs until each hydrogen atom (if present) has two electrons and every other atom has eight electrons.*

```
    ··
H : N : H
    ··
    H
```

Step 5. *Recheck to be sure that the total number of electrons shown in your structure equals v, which in this case is 8.*

---

sample exercise 10

Write a Lewis dot structure for hydrogen cyanide, HCN, a highly toxic liquid that boils at room temperature and is used as a fumigant in agriculture.

solution

Step 1. *Determine the number of bonding pairs (b) based on the number of electrons (t) and the number of valence electrons (v) in the molecule.*

a. Determine *t,* the number of electrons (two for H and eight for each other atom).

1 H	$1 \times 2 = 2\ e^-$
1 C	$1 \times 8 = 8\ e^-$
1 N	$1 \times 8 = 8\ e^-$
	$t = 18\ e^-$

b. Determine *v,* the number of valence electrons.

1 H	$1 \times 1 = 1\ e^-$
1 C	$1 \times 4 = 4\ e^-$
1 N	$1 \times 5 = 5\ e^-$
	$v = 10\ e^-$

c. Determine *b,* the number of bonding pairs.

$$b = \frac{(t - v)}{2}$$

$$b = \frac{18\ e^- - 10\ e^-}{2} = 4 \text{ bonding pairs}$$

Step 2. *Determine the central atom in the molecule.*

The central atom in HCN is C (guideline 2c: C is the central atom in any carbon-containing compound).

Step 3. *Write down the central atom and arrange the other atoms around the central one using the number of bonded pairs you calculated in step 1.*

Write down H C N. You must use four bonding pairs. You should recognize that hydrogen can participate in only one bonding pair because one pair gives H its noble-gas configuration of two electrons. Therefore, the other three bonding pairs must be bonding from C to N.

Step 4. *Fill in the rest of the structure with nonbonding pairs until each hydrogen atom (if present) has two electrons and every other atom has eight electrons.*

N needs a nonbonding pair to complete its octet.

H : C ::: N:

Step 5. *Recheck to be sure the total number of electrons shown in your structure equals v.*

Ten electrons are showing, which equals *v*.

H : C ::: N:

---

sample exercise 11

Write a Lewis electron dot structure for sulfuric acid, H_2SO_4, a very corrosive acid that dissolves most metals and is used extensively in the manufacture of chemicals, fertilizers, and explosives. Sulfuric acid is the most widely manufactured chemical in the world.

solution

Step 1. *Determine the number of bonding pairs (b) based on the number of electrons (t) and the number of valence electrons (v) in the molecule.*

a. Determine *t*, the number of electrons (two for H and eight for each other atom).

2 H	$2 \times 2 = 4\ e^-$
1 S	$1 \times 8 = 8\ e^-$
4 O	$4 \times 8 = 32\ e^-$
	$t = 44\ e^-$

b. Determine *v*, the number of valence electrons.

2 H	$2 \times 1 = 2\ e^-$
1 S	$1 \times 6 = 6\ e^-$
4 O	$4 \times 6 = 24\ e^-$
	$v = 32\ e^-$

c. Determine *b*, the number of bonding pairs.

$$b = \frac{(t - v)}{2}$$

$$b = \frac{44\ e^- - 32\ e^-}{2} = 6 \text{ bonding pairs}$$

Step 2. *Determine the central atom in the molecule.*

The central atom is S (guideline 2b: the atom other than O or H is the central atom).

Step 3. *Write down the central atom and arrange the other atoms around the central one using the number of bonded pairs you calculated in step 2.*

Write down S, arrange the O's around it, and attach H's to O's.

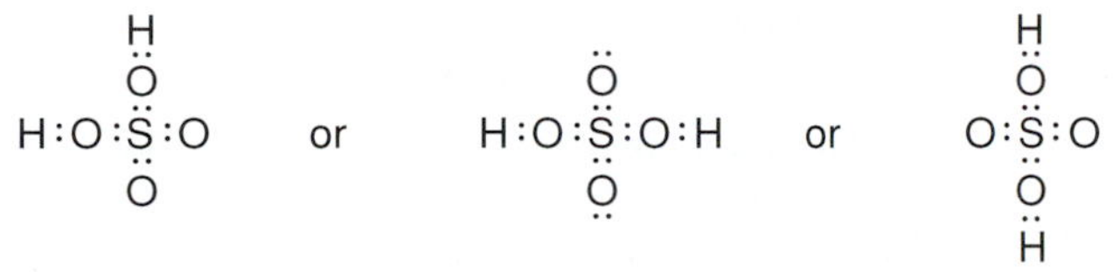

Note that these three forms are the same.

Step 4. *Fill in the rest of the structure with nonbonding pairs until each hydrogen atom (if present) has two electrons and every other atom has eight electrons.*

The oxygens require nonbonding pairs to complete their octets.

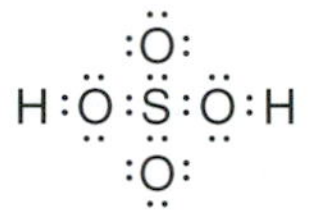

This is more easily and more clearly written as

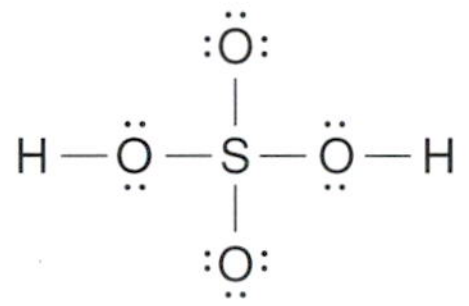

Step 5. *Recheck to be sure that the total number of electrons shown in your structure equals v.*

Thirty-two electrons are showing, which equals *v*.

---

sample exercise 12

Write a Lewis dot structure for the carbonate anion, CO_3^{2-}.

solution

Step 1. *Determine the number of bonding pairs (b) based on the number of electrons (t) and the number of valence electrons (v) in the ion.*

a. Determine *t*, the number of electrons (two for H and eight for each other atom).

1 C	$1 \times 8 = 8\ e^-$
3 O	$3 \times 8 = 24\ e^-$
	$t = 32\ e^-$

b. Determine *v*, the number of valence electrons.

1 C	$1 \times 4 = 4\ e^-$
3 O	$3 \times 6 = 18\ e^-$
2⁻ charge means 2 extra e^- =	$2\ e^-$
	$v = 24\ e^-$

c. Determine *b*, the number of bonding pairs.

$$b = \frac{(t - v)}{2}$$

$$b = \frac{32\ e^- - 24\ e^-}{2} = 4 \text{ bonding pairs}$$

Step 2. *Determine the central atom in the ion.*

The central atom is C (guideline 2c: C is the central atom in any carbon-containing compound).

Step 3. *Write down the central atom and arrange the other atoms around the central one using the number of bonded pairs you calculated in step 2.*

Write down the central C, arrange the O's around it, and fill in four bonding pairs. There must be two bonding pairs between C and one of the O's; i.e., there must be a double bond.

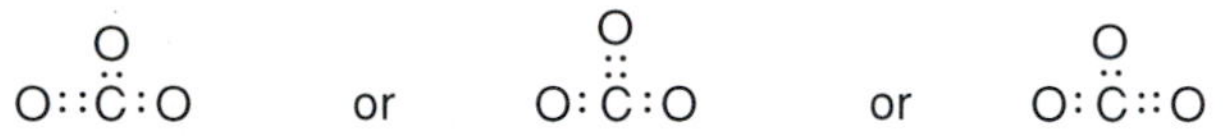

All three forms are equivalent.

Step 4. *Fill in the rest of the structure with nonbonding pairs until each hydrogen atom (if present) has two electrons and every other atom has eight electrons.*

The C is now surrounded by eight electrons. Fill in nonbonding pairs so that each oxygen will have an octet. Because the substance in this example is an ion, write its Lewis structure within brackets with the charge as a superscript.

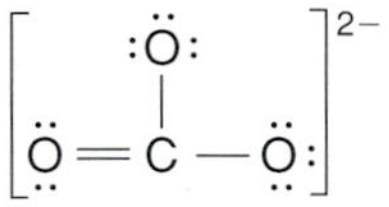

Step 5. *Recheck to be sure that the total number of electrons shown in your structure equals v.*

Twenty-four electrons are showing, which equals *v.* The negative carbonate ion can bond to cations by an ionic bond. However, the bonding within the ion between the oxygens and the carbon is covalent. The bonding *within* all polyatomic ions is *covalent.*

••• problem 10

Write a Lewis electron dot structure for each of the following.

a. CH_4, methane or natural gas, used in gas appliances and furnaces
b. H_3PO_4, phosphoric acid, used in soft drinks, flavored syrups, and pharmaceuticals
c. SO_4^{2-}, the sulfate anion
d. CH_2O, formaldehyde, an embalming fluid and disinfectant

6.14 coordinate covalent bonds

Before examining a special way some covalent bonds may form, let's review briefly. When you first wrote Lewis structures for diatomic molecules, you did so by pairing up unpaired electrons. For example,

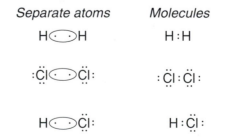

Separate atoms *Molecules*

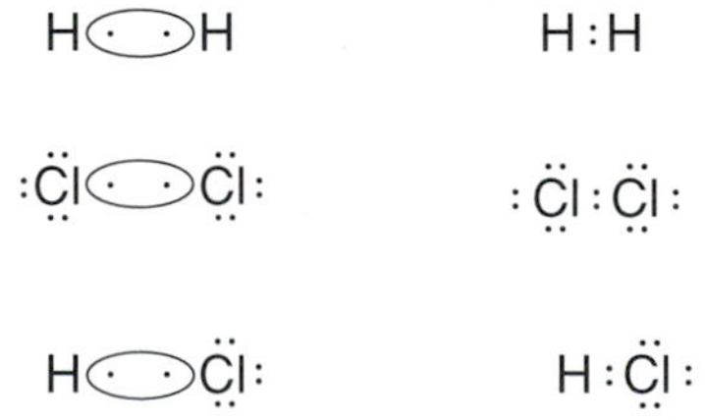

When using the general guidelines described in Section 6.13, you do not look for unpaired electrons in writing Lewis structures, but, in fact, *most covalent bonds are formed by each bonded atom contributing one electron to the pair.* This can be seen in the preceding examples concerning H_2, Cl_2, and HCl. Looking back at NH_3, for which the Lewis structure was determined in Sample Exercise 9, you see that the N—H bonds in NH_3 can easily be envisioned as arising from the contribution of one electron from N and one from H.

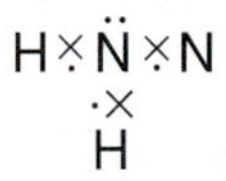

N has five valence electrons represented here as dots (·).
Each H has one valence electron represented here as an x.

It is also possible to form a covalent bond in which one of the bonded atoms furnishes both shared electrons. Such a bond is called a **coordinate covalent bond.** For example, the ammonium ion, NH_4^+, can be viewed as being formed from a coordinate covalent bond in which the two nonbonding electrons on NH_3 form a bond with an H^+ ion that has no electrons to contribute.

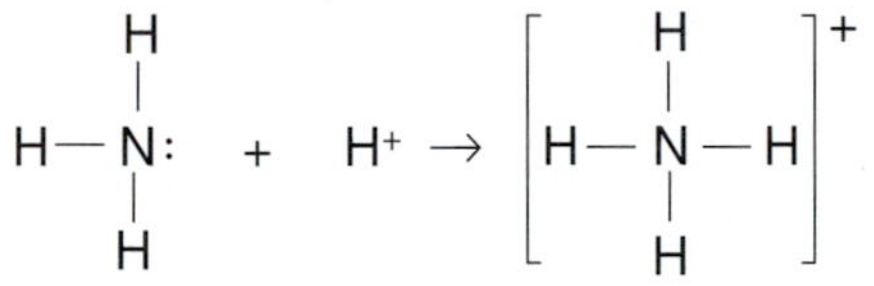

The four N—H bonds in NH_4^+ are identical. Once formed, a coordinate covalent bond has the same properties as any other covalent bond.

The concept of a coordinate covalent bond is particularly useful in envisioning the formation of hydronium ion, H_3O^+, an ion always present in any aqueous solution. H^+ ions from acidic compounds (Section 7.5) may combine with water molecules to form H_3O^+ ions, and the number of these ions in solution is used to determine how acidic the solution is (Chapter 15).

---

sample exercise 13

Write a Lewis structure for the hydronium ion, H_3O^+.

solution

Step 1. *Determine the number of bonding pairs (b) based on the number of electrons (t) and the number of valence electrons (v) in the ion.*

a. Determine *t*, the number of electrons (two for H and eight for each other atom).

3 H	$3 \times 2 = 6\ e^-$
1 O	$1 \times 8 = 8\ e^-$
	$t = 14\ e^-$

b. Determine *v*, the number of valence electrons.

3 H	$3 \times 1 = 3\ e^-$
1 O	$1 \times 6 = 6\ e^-$
1+ charge means 1 missing e^-	$= -1\ e^-$
	$v = 8\ e^-$

c. Determine *b*, the number of bonding pairs.

$$b = \frac{(t - v)}{2} = \frac{(14 - 8)}{2} = 3 \text{ bonds}$$

Step 2. *Determine the central atom in the ion.*

The central atom is O (guideline 2a: H is never the central atom).

Step 3. *Write down the central atom and arrange the other atoms around the central one using the number of bonded pairs you calculated in step 2.*

Write down the central O, surrounded by three bonds to three H's.

$$\begin{array}{c} H:O:H \\ \cdot\cdot \\ H \end{array}$$

Step 4. *Fill in the rest of the structure with nonbonding pairs until each hydrogen atom (if present) has two electrons and every other atom has eight electrons.*

$$\left[\begin{array}{c} H:\ddot{O}:H \\ \cdot\cdot \\ H \end{array}\right]^{+}$$

Step 5. *Recheck to be sure that the total number of electrons shown in your structure equals v.*

There are eight electrons shown and $v = 8$.

problem 11

Show how the ion H_3O^+ can be formed by coordinate covalent bond formation between H_2O and H^+.

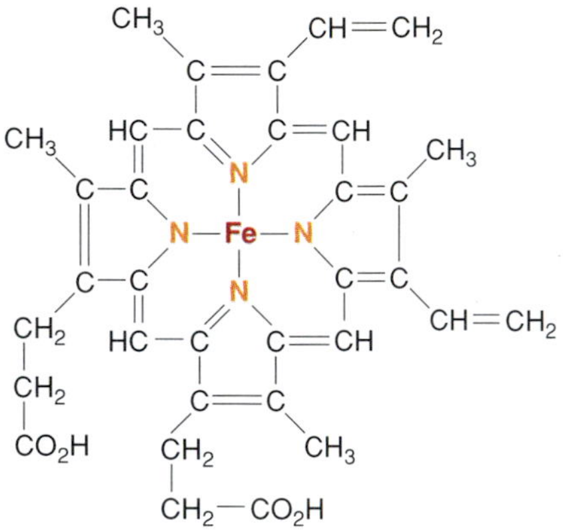

Figure 6.22

Heme is an important part of the hemoglobin molecule. The iron ion (Fe^{2+}) is bonded to four nitrogen atoms in a complex molecule known as the protoporphyrin IX. Each five-membered, nitrogen-containing ring is called a porphyrin ring. The Fe^{2+} ion is capable of accepting a pair of electrons from an oxygen molecule to form a coordinate covalent bond.

The hydronium ion, shown in Sample Exercise 13, will be a particularly important chemical species when you study acids in Chapter 15. All acids produce H_3O^+ ions in water solution, and the amount of these ions in water determines the acidity of the solution.

Another notable example of a coordinate covalent bond is that between molecular oxygen, O_2, and the iron in the molecule hemoglobin formed in red blood cells. It is this bond that allows the hemoglobin in your blood to carry oxygen from the air to the cells of your body. While the molecular structure of hemoglobin is quite complex, both bonding electrons for joining oxygen to iron are supplied by oxygen; hence, a coordinate covalent bond is formed (Figure 6.22). Iron is also capable of forming coordinate covalent bonds with other electron pairs. In the case of blood hemoglobin, if another compound binds to the iron instead of O_2, there would be a serious problem with oxygen transport to the cells and asphyxiation would result. Carbon monoxide ($:C{\equiv}O:$), a toxic gas found in automobile exhaust, can bond to iron in hemoglobin through its lone pairs of electrons. Because this bond is stronger than the iron-oxygen bond, oxygen is prevented from linking up with iron. As a result, oxygen is not carried to your body's cells; this is what makes CO toxic. When working near automobiles with the engine running, it is very important to make sure that you have good ventilation. Carbon monoxide is odorless and can cause loss of consciousness and death without your realizing any symptoms (Figure 6.23).

6.15 electronegativity of atoms and polarity of bonds

So far, you have pictured the electron pair in a covalent bond as if it were shared equally between the two atoms. This is definitely the case when the two bonded atoms are identical, as with H_2 or Cl_2. However, in most bonds, one bonded atom attracts the bonding electrons more strongly. For example, in H—Cl, the chlorine has a stronger attraction for electrons than does the hydrogen.

Figure 6.23

Servicing an automobile with its engine running requires proper venting of the exhaust gases, which contain carbon monoxide, CO. Also, a well-tuned engine produces less CO so less is released into the atmosphere. Carbon monoxide forms a coordinate covalent bond with the iron in the hemoglobin in your blood and blocks the possibility of oxygen in the air bonding with the same iron. As a result, oxygen from the air cannot be transported throughout your body.

Figure 6.24

Linus Carl Pauling (1901–94) was an American chemist who introduced concepts that helped explain the bonding forces of molecules. He was awarded the Nobel prize in chemistry in 1954 and the Nobel Peace Prize in 1962.

The attractive force that an atom exerts on shared electrons in a chemical bond is known as its **electronegativity.** A scale of relative electronegativities has been devised by the Nobel Laureate Linus Pauling (1954 Nobel prize in chemistry, 1962, Peace Prize) (Figure 6.24). Electronegativity values for the representative elements are given in Table 6.8. When you examine this table, you will see that electronegativity is a periodic property. In general, it increases from left to right across a period; it also increases as you go up a group from bottom to top. This means that nonmetals have high electronegativities with the highest value (4.0) assigned to fluorine, F, in the upper right corner. Metals have low electronegativities because they prefer to lose electrons than to gain them. Because metals have low electronegativities, they are said to be *electropositive.*

The difference in electronegativities between two bonded atoms offers an indication of the **polarity** in the bond. Polarity is a measure of the inequality in the distribution of bonding electrons. A **nonpolar bond** is one in which there is *equal*

table 6.8 electronegativities of some representative elements

IA	IIA	IIIA	IVA	VA	VIA	VIIA
H 2.1	Increasing electronegativity →					
Li 1.0	Be 1.5	B 2.0	C 2.5	N 3.0	O 3.5	F 4.0
Na 0.9	Mg 1.2	Al 1.5	Si 1.8	P 2.1	S 2.5	Cl 3.0
K 0.8	Ca 1.0	Ga 1.6	Ge 1.8	As 2.0	Se 2.4	Br 2.8
Rb 0.8	Sr 0.9	In 1.7	Sn 1.8	Sb 1.9	Te 2.1	I 2.5
Cs 0.7	Ba 0.9	Tl 1.8	Pb 1.8	Bi 1.9	Po 2.0	At 2.2
Fr 0.7	Ra 0.9					

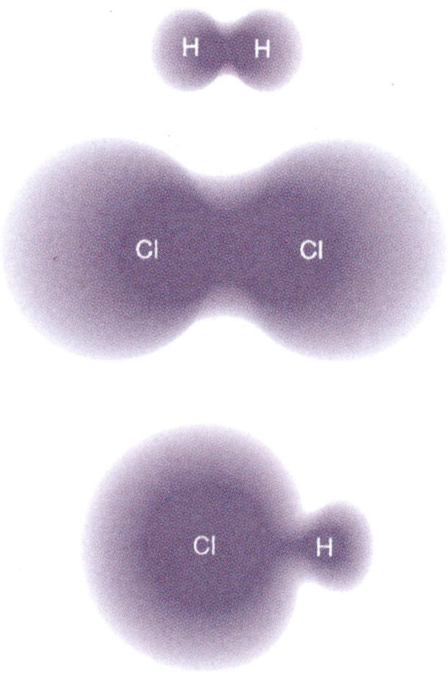

Figure 6.25

The fuzzy cloud around the atoms is the shared pair of electrons. In H_2 and Cl_2, the cloud is evenly distributed around both nuclei. In HCl, the cloud is drawn toward the more electronegative Cl atom, which comes to bear a slightly negative charge. This distortion of the electron distribution exposes the positively charged hydrogen nucleus, giving this end of the molecule a slightly positive charge.

sharing of bonding electrons. In a **polar bond,** there is *unequal sharing;* the more electronegative atom pulls the electrons closer. Figure 6.25 shows polar and nonpolar bonds pictorially.

To obtain a rough measure of the polarity of a bond, take the mathematical difference between the electronegativities of the bonded atoms, subtracting the smaller electronegativity from the larger one. The larger the difference, the more polar the bond. For example, in the case of H_2 (H—H), the electronegativity difference (2.1 − 2.1) is zero; thus, the bond is a *nonpolar covalent bond,* or simply a *covalent bond.* The two electrons in a nonpolar covalent bond are equally distributed between the atoms. In H—Cl, the electronegativity difference between H (2.1) and Cl (3.0) is 0.9. Because of this electronegativity difference, the H—Cl bond is called a *polar covalent bond.*

If you look more closely at the consequences of unequal sharing of electrons, you will find how this word *polar* originates. In H—Cl, because the bonding electrons are closer to Cl, Cl acquires a partial negative charge (remember that electrons are negative). H acquires a partial positive charge because the electrons neutralizing the nuclear charge are being drawn away. This is written as

$$\overset{\delta^-}{\text{Cl}} : \overset{\delta^+}{\text{H}}$$

where the symbol δ means "partial." The bond has developed a positive pole and a negative pole, just as we speak of the positive and negative poles of a battery. That is why the bond is said to be polar.

Because there are *two* poles (one + and one −), you can also say that a **dipole** exists in the bond. The prefix *di-* always means "two." A crossed arrow ($\leftarrow\!\!\!+\!\!\!\rightarrow$) pointing toward the more electronegative element is used to indicate the presence of a dipole.

$$\overset{+\!\!\longrightarrow}{\text{H—Cl}}$$

The greater the electronegativity difference, the more polar the bond and the larger the dipole.

When molecules have more than one polar bond, the entire molecule may or may not be polar depending on its Lewis structure (Section 6.13) and the particular three-dimensional shape that characterizes the molecule (Sections 12.2 and 12.3). The molecular polarity of the water molecule, as shown below,

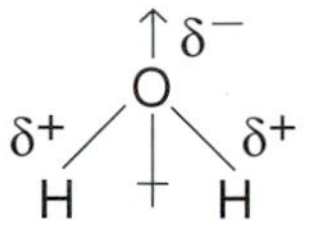

is a result of the two O—H bond polarities and the bent shape of the molecule.

Because the water molecule is polar, one end of the molecule carries a partial negative charge and the other end, a partial positive charge. As a result, the *positive* end of the molecule may attract or be attracted to other *negative* charges, such as the negative end of other polar molecules or anions from ionic compounds. Similarly, the *negative* end of the water molecule may attract or be attracted to other *positive* charges, such as the positive end of other polar molecules or cations from ionic compounds.

The attraction between molecules of polar compounds means that they would be soluble in each other. By the same reasoning, since there is little attraction between polar and nonpolar molecules, you would predict that polar compounds are not soluble in nonpolar liquids. This is consistent with your observation that gasoline or oil (nonpolar compounds) and water (a polar compound) do not mix.

The attraction between molecules of polar compounds also suggests that more energy would be needed to separate the polar molecules from each other when they melt or vaporize. As a result, polar molecules will have higher melting points and boiling points than nonpolar molecules, which do not carry partial charges. Hence, nonpolar gasoline molecules vaporize much more readily (are more volatile) than polar water molecules; furthermore, both the melting point and the boiling point of gasoline are much lower than those for water.

The charged ends of polar water molecules are attracted to charged ions in ionic compounds. You remember from Figures 6.12 and 6.13 that gaseous water molecules were removed from air by ionic compounds such as $CaCl_2$ and NaOH. This drying effect was possible because the polar water molecules were attracted to the ions of the solids. Furthermore, when excess water molecules are available, cations and anions may be separated from ionic compounds to dissolve in water solution. By contrast, uncharged, nonpolar molecules are not attracted to ionic solids so that nonpolar liquids such as gasoline will not dissolve ionic substances. Hence, the ionic compound sodium chloride, NaCl, dissolves readily in water, but remains undissolved when added to gasoline (Figure 6.21d).

---

sample exercise 14

For the bonds indicated below, (1) decide whether they are nonpolar or polar covalent, (2) mark any dipoles with the crossed-arrow symbol, and (3) decide which of the three is the most polar.

a. H—I

b. The N—H bond in NH_3

c. The O—H bond in H_2O

solution

1. Calculate the difference in electronegativities between bonded atoms (Table 6.8). Subtract the smaller electronegativity from the larger one.

   a. 2.5 − 2.1 = 0.4
      (I) (H)

   Although the difference is small, 0.4, you classify this bond as polar based on the discussion so far.

   b. 3.0 − 2.1 = 0.9
      (N) (H)

Classify this bond as polar, because of the electronegativity difference.

c. $3.5 - 2.1 = 1.4$
   (O) (H)

Again, you classify the bond as polar.

2. The dipole is marked by a crossed arrow pointing in the direction of the more electronegative element.

   a. $\overset{+\!\!\longrightarrow}{\text{H—I}}$

   b. $\overset{\longleftarrow\!\!+}{\text{N—H}}$

   c. $\overset{\longleftarrow\!\!+}{\text{O—H}}$

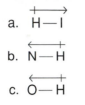

3. The O—H bond is the most polar because the electronegativity difference (1.4) is largest.

6.16 recognizing ionic versus molecular compounds

A simple generalization enables us to readily distinguish between ionic and molecular (covalent) compounds. *Most ionic compounds are formed from metals and nonmetals,* whereas *most molecular compounds are formed from combinations of nonmetals.* This rule is very convenient for purposes of formula writing and, as you will see (Section 7.2), for naming compounds, because metal-nonmetal combinations are named according to the rules for ionic compounds, and compounds containing exclusively nonmetals usually are not.

However, in considering the properties of a compound, its ionic or covalent (molecular) nature depends on the electronegativity differences between the bonded atoms. See Table 6.9 for a comparison of the properties of ionic versus covalent compounds. There is a gradual transition from covalent to polar covalent to ionic bonding as electronegativity differences increase. The properties of a compound depend on this bonding character. The following ranges of electronegativity differences may be used to predict bonding character.

Electronegativity difference	Bond type
0 to 0.5	Covalent or slightly polar covalent
0.5 to 1.7	Polar covalent
Over 1.7	Ionic

Because the transition in properties is gradual, the cutoff points of 0.5 and 1.7 are somewhat arbitrary, and bonds with electronegativity differences within 0.2 of these values are expected to be intermediate in character.

The rule that a metal-nonmetal combination yields an ionic compound works very well for most examples you encounter. For example, from Table 6.8 you can obtain the electronegativities for Na and Cl as 0.9 and 3.0, respectively. The difference of 2.1 between these two numbers tells us that the bond between Na and Cl is ionic, and of course, we know from experience that sodium chloride (table salt) exhibits the properties of an ionic compound, as indicated in Table 6.9.

In contrast, gasoline exhibits the properties of a covalent compound. Its formula, C_8H_{18}, which shows only nonmetals in the compound, and the small electronegativity difference (0.4) between C and H both predict that this should be so.

You remember from the nature of the ionic bond in Section 6.9 that ionic compounds have high melting points (in the range of 500°C to 1,000°C). Because of

table 6.9 comparison of properties of ionic and covalent compounds

Property	Ionic compounds	Covalent compounds
Physical state at room temperature	Solids	Gases, liquids, or low-melting-point solids
Electrical conductivity of the molten state	Conductors	Nonconductors
Water solubility	Many are soluble	Most are not soluble
Electrical conductivity of a water solution	Conductors	Most are nonconductors (unless they react with water)
Melting point	Very high	Much lower than ionic compounds
Shape of a solid sample	Crystalline with characteristic angles	May be crystalline, but lacking characteristic angles

the strength of attraction between ions in the solid state, much heat energy must be supplied to disrupt the lattice structure and convert it to a liquid. Molecular solids do not have ions present and, as a result, have much weaker forces between their molecules. Their melting points are characteristically low, less than 300°C. Simply looking at the melting point of an unknown may give you a quick indication of the classification of the sample as ionic or molecular.

In the solid state, the rigid structure of the ions in the crystalline lattice prevents the ions from moving; as a result, solid, ionic compounds do not conduct electricity. In the molten state (liquid), however, the ions become sufficiently mobile that they may conduct electricity. Conductivity of the molten solid is an experimental test that distinguishes between ionic and covalent solids. Since molecular (covalent) solids do not have ions present, they do not conduct electricity either as a solid or in the molten state.

Another experimental test to distinguish ionic and covalent compounds is their solubility (ability to dissolve) in water. Ionic solids are frequently, but not always, soluble in water. When soluble, their water solutions conduct electricity very well because of the presence of ions in the solution. Ionic solids are rarely soluble in liquids other than water. Molecular solids, on the other hand, are less likely to be soluble in water and more likely to be soluble in organic (carbon-containing) solvents such as dichloromethane (CH_2Cl_2) or ethanol (C_2H_5OH). If molecular solids do dissolve in water, their solutions sometimes conduct electricity, but usually much more weakly than ionic solids.

The property of flammability, introduced in Section 1.7, may also give an indication of the type of bonding in an unknown compound. Not all molecular compounds burn, but a flammable material is much more likely to be molecular than ionic. Remember, many of our common fuels are organic (molecular) compounds.

Based on the nature of ionically and covalently bonded materials, you now understand the structural basis for several physical and chemical properties. By measuring or observing these properties in the laboratory, you will be able to classify the type of bonding in most compounds.

6.17 law of definite composition revisited

You first encountered the law of definite composition in Section 1.9 as part of the first discussion of the properties of compounds. This law and the idea of fixed, combining weight ratios are hard to appreciate before you know about the existence of atoms and how they combine. Now that you have explored bonding, you

CHEMLAB

Laboratory exercise 6: effects of bond type on properties

Laboratory experiments to classify solid samples as ionic or molecular are relatively straightforward and have been discussed in the previous section. An apparatus that measures melting points of solid samples is shown and discussed in the introductory photo of Chapter 2. Determining the flammability of a sample by placing it near a flame or observing the solubility of a sample by placing a small amount into a few milliliters of water, ethanol, or dichloromethane are self evident. The only new procedures introduced in this chapter are determining the conductivity of a water solution and a molten solid. A simple apparatus is shown in Figure 6.26. Note that the bare electrodes illustrated in this figure are dangerous and are intended for use only as a demonstration by your instructor.

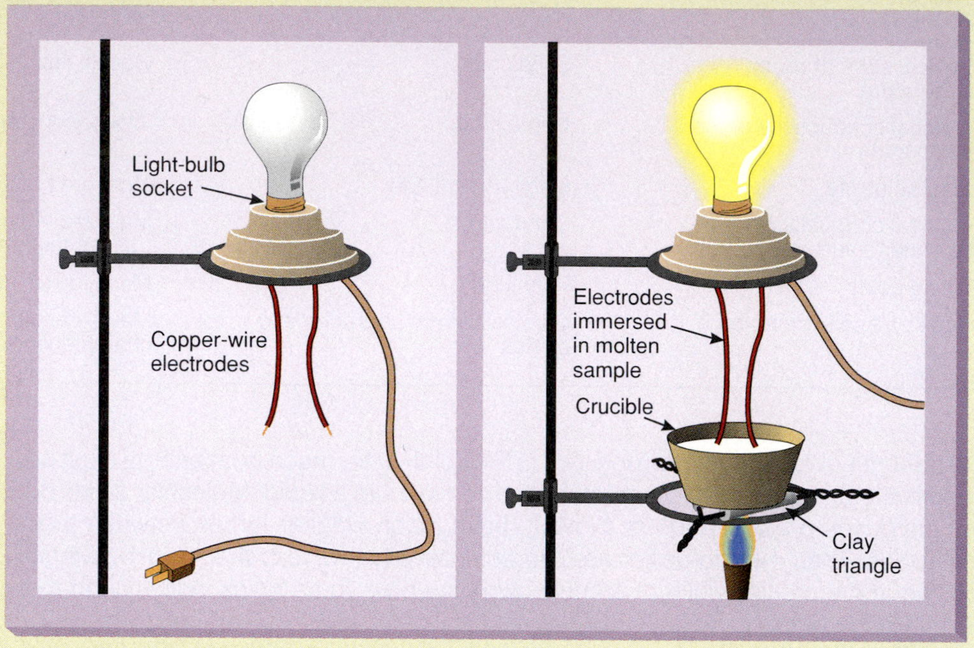

Figure 6.26

Apparatus to determine electrical conductivity of an unknown sample. Two separated electrodes are immersed into the sample, which must be molten or in a water solution, and the apparatus is plugged in. If the sample conducts, the circuit will be complete and the light will glow. If the sample does not conduct, the bulb will not light.

Question:

1. Use the following data to classify the following materials as ionic or molecular.

Material	Flammability	MP (°C)	Solubility	Molten conductivity	Classification
X	No	151	Dissolves in CH_2Cl_2	No	
Y	Yes	74	Dissolves in C_2H_5OH	No	
Z	No	745	Dissolves in H_2O	Yes	

know that elements always show some definite combining ratio of atoms or ions, and this inevitably leads to some definite combining weight ratio. For water (H_2O), the combining atomic ratio is two hydrogens to one oxygen because this ratio allows all atoms to have octets (O) or duets (H) of electrons. For sodium chloride, the combining ionic ratio is Na:Cl = 1:1.

Because atoms (and ions) have definite atomic masses, a definite combining ratio of atoms (or ions) means a definite combining ratio by mass or weight. Thus, the weight ratio or fixed proportion of elements in H_2O is always 2.0:16 = 1.0:8.0 because two hydrogens have a mass of 2.0 amu and one oxygen has a mass of 16 amu. Similarly the weight ratio for NaCl is always 23.0:35.5. Notice that the mass of an atomic ion is the same as the mass of an atom. This is so because ions arise by gains or losses of electrons, which have negligible mass.

sample exercise 15

Bunsen burner gas, methane, has the combining ratio of one carbon atom to four hydrogen atoms. What is the weight ratio for methane?

solution

Carbon has a relative mass of 12. Hydrogen has a relative mass of 1. Therefore, one atom weighing 12 combined with four atoms each weighing 1 leads to a weight ratio of 12:4, or 3:1.

••• problem 12

Acetylene gas, a compound burned in welding torches, has the combining ratio of two carbon atoms to two hydrogen atoms. What is the weight ratio for acetylene?

••• problem 13

Hydrogen sulfide, a noxious gas that smells like rotten eggs, has the combining ratio of two hydrogen atoms to one sulfur atom. What is the weight ratio for hydrogen sulfide?

sample exercise 16

In fluoridated "hard" water, calcium ions (Ca^{2+}) combine with fluoride ions (F^-) in the ratio 1:2. What is the weight ratio in this ionic compound?

solution

Calcium has a relative mass of 40. Fluorine has a relative mass of 19. Therefore, one ion weighing 40 combined with two ions each weighing 19 ($2 \times 19 = 38$) leads to a weight ratio 40:38, or 20:19.

••• problem 14

Strontium sulfide, a compound used in depilatories and luminous paints, has the combining ratio of one strontium ion to one sulfide ion. What is the weight ratio in this ionic compound?

chapter accomplishments

After completing this chapter, you should be able to define all key terms and do the following.

6.1 Introduction

- ❑ Give the general names for the two major types of compounds.

6.2 How can atoms achieve lower-energy states?

- ❑ State why most elements interact to form compounds.
- ❑ State the common pattern in the electron configuration of noble gases that is responsible for their stability.
- ❑ State the type of electron configuration atoms strive to obtain when elements combine to form compounds.

6.3 Metals lose electrons to form positively charged ions (cations)

- ❑ Determine the number of electrons an A group metal will lose to obtain a noble-gas configuration.

6.4 Nonmetals gain electrons to form negatively charged ions (anions)

- ❑ Determine the number of electrons an element in Group VA, VIA, or VIIA will gain to obtain a noble-gas configuration.
- ❑ Compare the electron configuration of an A group ion with that of a noble gas.
- ❑ Using a periodic table, predict the charge on an ion.

6.5 Electron transfer results in the formation of ionic compounds

- ❑ Develop a pictorial representation for the formation of ions from atoms.
- ❑ State the relationship between the number of electrons lost by a metal and the number of electrons gained by a nonmetal in the formation of an ionic compound.

6.6 Ionic charges

- ❑ Write the symbol and charge of each A group ion, given a periodic table.
- ❑ Given the name or the symbol, state the charge(s) of each ion listed in Table 6.5.

6.7 Polyatomic ions

- ❑ Distinguish between the terms *monatomic ion* and *polyatomic ion.*
- ❑ Given the name, write the formula and charge of a polyatomic ion, or given the formula, name each polyatomic ion in Table 6.6.

6.8 Formulas for ionic compounds

- ❑ Write a correct formula for an ionic compound, given the combining elements and a periodic table.
- ❑ Using a periodic table, write the formula for an ionic compound, given two combining polyatomic ions or a polyatomic ion and a combining element.

6.9 The nature of the ionic bond

- ❑ Explain why an ionic bond lowers the energy of a compound relative to the individual ions or atoms.

6.10 Covalent compounds

- ❑ Explain why some elements combine to form ionic bonds and others combine to form covalent bonds.

6.11 Electron sharing in diatomic molecules

- ❑ Distinguish between single, double, and triple covalent bonds.

6.12 The nature of the covalent bond

- ❑ State why there is a lowering of energy relative to the separate atoms when a covalent bond forms.

6.13 Lewis electron dot structures for molecules

- ❑ Write Lewis electron dot structures for molecular compounds.
- ❑ Write Lewis electron dot structures for polyatomic ions.

6.15 Electronegativity of atoms and polarity of bonds

- ❑ Define electronegativity.
- ❑ Describe the trends in electronegativity across a row and down a column of the periodic table.
- ❑ State the difference between polar and nonpolar bonds.
- ❑ Given a table of electronegativities, indicate whether a bond formed between two atoms will be nonpolar or polar.
- ❑ Use the crossed arrow ($\leftarrow\!\!\!+\!\!\rightarrow$) to label the dipole in a polar covalent bond.

6.16 Recognizing ionic versus molecular compounds

- ❑ Distinguish between ionic and molecular compounds for purposes of nomenclature and formula writing.
- ❑ Given a table of electronegativities, distinguish between covalent, polar covalent, and ionic bonds.
- ❑ Compare the properties of ionic and molecular compounds.

6.17 Law of definite composition revisited

- ❑ Predict the weight ratio of elements in a compound from the combining ratio of atoms or ions in the compound.

key terms

bond
noble (inert) gases
octet
duet
octet rule
cation
ionization energy
isoelectronic
electron affinity
anion
ionic compounds
ionic bond
monatomic ions
polyatomic ion
lattice energy
formula unit
lattice structure
molecules
molecular compounds
covalent bond
covalent compounds
bonding pair of electrons
nonbonding electrons
lone pairs
double bond
triple bond
bond length
bond-dissociation energy
Lewis structures
coordinate covalent bond
electronegativity
polarity
nonpolar bond
polar bond
dipole

problems

6.1 Introduction

15. What is a chemical bond?
16. List two types of chemical bonds.

6.2 How can atoms achieve lower-energy states?

17. Describe the characteristic pattern(s) in a noble-gas electron configuration.
18. What feature of the structure of an atom is altered when atoms combine to form compounds?
19. Describe the two ways by which atoms can obtain a noble-gas electron configuration.

6.3 Metals lose electrons to form positively charged ions (cations)

20. Explain why Group IA metals form a cation with a charge of 1+.
21. Explain why Ca does not form a Ca^{3+} ion.
22. a. Write the electron configuration of the K^{+} ion.
    b. With which noble gas is this ion isoelectronic?
23. a. How many electrons must a barium atom lose to form an ion with a noble-gas configuration?
    b. With which noble gas will the barium ion be isoelectronic?
24. Give the symbols of two cations that will be isoelectronic with xenon.
25. State the noble gas that is isoelectronic with each ion below.
    a. Sr^{2+}
    b. Li^{+}
    c. Rb^{+}

6.4 Nonmetals gain electrons to form negatively charged ions (anions)

26. Explain why nonmetals gain rather than lose electrons in forming ionic bonds.
27. Explain why Group VIIA nonmetals form an anion with a charge of 1–.
28. a. How many electrons must N gain to form a noble-gas configuration?
    b. With which noble gas will the N ion be isoelectronic?
29. a. Write the electron configuration of the S^{2-} ion.
    b. With which noble gas is this ion isoelectronic?
30. Give the symbols of three anions that would be isoelectronic with krypton.
31. State the noble gas that is isoelectronic with each ion below.
    a. Se^{2-}
    b. P^{3-}
    c. I^{-}

6.5 Electron transfer results in the formation of ionic compounds

32. Explain why two sodium atoms are needed to react with one sulfur atom in the formation of the ionic compound sodium sulfide.
33. a. Develop a pictorial representation similar to Figure 6.7 for the formation of an ionic compound from sodium and fluorine.
    b. Explain how your diagram accounts for the formula NaF.
34. Develop a pictorial representation similar to Figure 6.7 for the formation of the ionic compound K_3N.
35. a. Develop a pictorial representation similar to Figure 6.7 for the formation of an ionic compound from aluminum and oxygen.
    b. Explain how your diagram accounts for the formula Al_2O_3.

6.6 Ionic charges

36. Write the symbol and charge of the ion that could form from each of the following elements.
    a. Bromine
    b. Strontium
    c. Cesium
    d. Phosphorus
    e. Aluminum
    f. Lithium

6.7 Polyatomic ions

37. Give the formula and charge of the following polyatomic ions.
    a. Carbonate
    b. Hydrogen carbonate
    c. Phosphate
    d. Ammonium
    e. Sulfite

6.8 Formulas for ionic compounds

38. Write formulas for the compounds formed from the following ions.
    a. Cs^+ and F^-
    b. Ba^{2+} and I^-
    c. Ba^{2+} and O^{2-}
    d. NH_4^+ and CO_3^{2-}
    e. Al^{3+} and SO_4^{2-}
    f. Na^+ and MnO_4^-
    g. Li^+ and PO_4^{3-}
    h. K^+ and $C_2H_3O_2^-$
    i. Mg^{2+} and N^{3-}
    j. Zn^{2+} and HPO_4^{2-}
    k. Pb^{4+} and SO_4^{2-}
    l. Fe^{3+} and O^{2-}
    m. Ca^{2+} and HCO_3^-
39. Write a correct formula for the compounds formed by combining the following.
    a. Lithium and oxygen
    b. Barium and chlorine
    c. Sodium and sulfur
    d. Zinc and nitrogen
    e. Silver and bromine
    f. Iron (the 2+ ion) and oxygen
    g. Iron (the 3+ ion) and oxygen
    h. Tin (the 2+ ion) and chlorine
    i. Tin (the 4+ ion) and chlorine
    j. Gallium and sulfur
    k. Calcium and iodine
40. Write the formula of the ionic compound that could form from elements X and Y if X has two valence electrons and Y has five valence electrons.
41. Write the formula of the ionic compound that could form from elements X and Y if X has one valence electron and Y has six valence electrons.
42. What is the ratio of cations to anions in each of the following ionic compounds?
    a. CaO
    b. MgF_2
    c. Na_3PO_4
    d. $KC_2H_3O_2$
    e. $Ca_3(PO_4)_2$

6.9 The nature of the ionic bond

43. Which step in ionic bond formation releases the greatest amount of energy?
44. Which compound, CaO or KF, will have the strongest electrostatic attractions between the ions? Why?
45. Determine the total number of ions in one formula unit of each of the following.
    a. $CaCl_2$
    b. $Ca_3(PO_4)_2$
    c. $Al(HCO_3)_3$
    d. $(NH_4)_2SO_4$
46. Determine the total number of atoms in one formula unit of each of the compounds in Problem 45.

6.10 Covalent compounds

47. Distinguish between ionic and covalent bonds based on the way that they form.

6.11 Electron sharing in diatomic molecules

48. Write Lewis electron dot structures for the following diatomic molecules.
    a. ICl
    b. BrCl
    c. ClF
    d. HF
49. Explain why helium exists as a monatomic gas but hydrogen exists as diatomic molecules.
50. Explain why chlorine molecules have a single covalent bond whereas nitrogen molecules have a triple covalent bond.
51. a. How many nonbonding electrons surround each chlorine in a chlorine molecule?
    b. How many nonbonding electrons surround each nitrogen in a nitrogen molecule?

6.12 The nature of the covalent bond

52. a. Write the Lewis symbols for a hydrogen and a fluorine atom.
    b. Indicate the new attractive and repulsive forces that arise as the two atoms are brought together to form a covalent bond.
53. According to Figure 6.19, what is true about the energy content of a covalent bond at the following distances?
    a. Separation distances greater than the bond length
    b. Separation distances less than the bond length
    c. A separation distance equal to the bond length
54. The bond-dissociation energies in kilojoules for the hydrogen halides are HF = 565, HCl = 431, HBr = 368, and HI = 297.
    a. Which compound has the strongest bond?
    b. Which compound is the least stable?
55. The bond-dissociation energy of a chlorine molecule is 243 kJ per 71 g of chlorine. Is a chlorine molecule more or less stable than two separated chlorine atoms? Explain your answer.

6.13 Lewis electron dot structures for molecules

56. Determine the number of bonding pairs of electrons in each of the following molecules or ions.
    a. H_2S
    b. $CHCl_3$
    c. H_2SO_3
    d. NO_2^-
57. Determine the central atom for each of the molecules or ions in Problem 56.
58. Write the Lewis structures showing all bonding and nonbonding valence electrons for each of the molecules or ions in Problem 56.

59. Write Lewis structures for the following molecules.

a. CO_2
b. $HClO_3$
c. CO
d. C_2H_6
e. OF_2
f. H_2O_2
g. CH_4S
h. H_2CO_3
i. SO_3

60. Write Lewis structures for the following polyatomic ions.

a. OH^-
b. HS^-
c. SO_3^{2-}
d. PO_4^{3-}
e. ClO^-
f. NH_4^+
g. HCO_3^-
h. $C_2H_3O_2^-$

61. a. Write Lewis structures for SO_2 and O_3 (ozone).
b. Indicate the similarities in their structure.
c. Experiments show that all the bonds in SO_2 are exactly the same. A similar result is found for O_3. Do your structures agree with these results?

6.14 Coordinate covalent bonds

62. a. Write Lewis structures for H^+, H, and H^-.
b. Which two could form a coordinate covalent bond?

63. a. Write a Lewis structure for PH_3.
b. Indicate a substance with which PH_3 could form a coordinate covalent bond.
c. Write a Lewis dot structure for the complex formed when PH_3 is joined by a coordinate covalent bond to the substance you indicated in part b.

64. In general, what kinds of substances could act as the electron-pair donor in the formation of a coordinate covalent bond?

65. Which type of electrons, bonding or nonbonding, can participate in forming a coordinate covalent bond?

6.15 Electronegativity of atoms and polarity of bonds

66. Define electronegativity.

67. Describe the trends in electronegativity found in the periodic table.

68. a. Where in the periodic table do we find elements with the highest electronegativity?
b. Where do we find elements with the lowest electronegativity?

69. Using Table 6.8 and Section 6.15, classify the following bonds as nonpolar or polar.

a. P—Br
b. H—O
c. I—Br
d. N—H

70. Using Table 6.8, arrange the bonds given in Problem 69 in order of increasing polarity.

71. For each polar bond that you found in Problem 69, use the crossed-arrow symbol ($\mapsto$) to indicate the dipole in the bond.

72. a. Write a Lewis structure for CF_4.
b. Use the symbol $\mapsto$ to indicate the polarity in each of the bonds in CF_4.
c. Where is the center of the δ^- polarity in CF_4?
d. Where is the center of the δ^+ polarity in CF_4?

73. Use the symbol $\mapsto$ to indicate any bond polarities in the Lewis structures written for Problem 58.

6.16 Recognizing ionic versus molecular compounds

74. Using Table 6.8 and the guidelines in Section 6.16, indicate whether the following bonds are covalent, polar covalent, or ionic.

a. H—Br
b. H—F
c. Li—O
d. Mg—Cl
e. Al—Cl
f. C—Br

75. Indicate whether the following compounds are ionic or molecular, and state whether the smallest basic grouping is a molecule or a formula unit.

a. H_2O
b. NaBr
c. Al_2O_3
d. CH_4O
e. $PbClO_4$
f. HBr
g. K_2CO_3
h. CS_2
i. $CaCl_2$

76. Compound A is a high-melting-point, water-soluble solid. Compound B is a liquid that does not conduct electricity. Indicate the likely ionic or covalent nature of compounds A and B.

77. Using the guidelines in Section 6.16, predict the type of bond (ionic, polar covalent, or covalent) that would form between atoms of the following elements.

a. C and H
b. K and Br
c. N and H
d. O and S
e. Ca and Cl
f. Al and F

78. Predict the order of increasing polarity for the bonds in the following molecules: H_2S, H_2O, H_2Se, and H_2Te.

6.17 Law of definite composition revisited

79. Propane, a fuel used in mobile homes, has a combining ratio of three carbon atoms to eight hydrogen atoms. What is the weight ratio for propane?

80. Describe how Dalton's atomic model provides an explanation for the law of definite composition.

81. When 64.0 g of sulfur dioxide, SO_2, is decomposed, 32.0 g of sulfur and 32.0 g of oxygen are formed. When 15.0 g of sulfur dioxide is decomposed, 7.50 g of oxygen is formed. How much sulfur will be formed in this decomposition?

82. Could a compound have a combining ratio of 1.5 atoms to 2 atoms? Explain your answer.

83. When 25.0 g of ammonia, NH_3, is decomposed, 20.6 g of nitrogen and 4.41 g of hydrogen are formed. What is the weight ratio of nitrogen to hydrogen in ammonia?

Additional problems

84. Write the formula for the ionic compound that could form from elements with atomic numbers 20 and 17.
85. Explain why ionic solids tend to have high melting points.
86. What would likely be the charge on an ion formed from an atom with the following electron configuration: $1s^22s^22p^63s^23p^63d^{10}4s^24p^5$?
87. Why is Na^- not a stable ion?
88. How does a polyatomic ion differ from a monatomic ion?
89. Write structural formulas for the following.
    a. C_2H_4
    b. $CHBr_3$
    c. C_2H_3Cl
    d. CH_4O
    e. C_2HCl
90. A—B is a diatomic molecule where A is less electronegative than B. Is the A—B molecule polar? If so, indicate the dipole in the molecule.
91. Connect the carbons in a "straight chain" and draw Lewis structures for the following.
    a. Propane, C_3H_8
    b. Butane, C_4H_{10}
92. Draw *all* possible Lewis structures for $C_2H_4Cl_2$.
93. Formic acid, the irritating ingredient in the bite of ants, has a chemical formula HCOOH. Draw its Lewis structure.
94. Four elements, A, B, C, D, have electronegativities A = 3.8, B = 3.3, C = 2.8, and D = 1.3. These elements form the compounds AB, AD, BD, and AC. Arrange these compounds in order of increasing ionic bond character.
95. If the ionization energy of A is larger than the electron affinity of B, how can the ionic compound formation between A and B be an energy-lowering process?
96. Which substance would you predict would conduct electricity in the molten state, ICl or KCl?
97. a. Write Lewis structures for the following four species CO, NO^+, CN^-, and N_2.
    b. What is common to these structures?

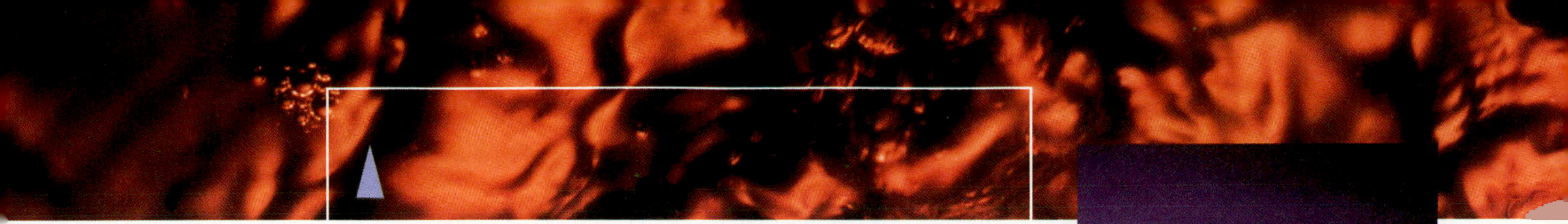

chapter 7

nomenclature, weight, and percentage composition

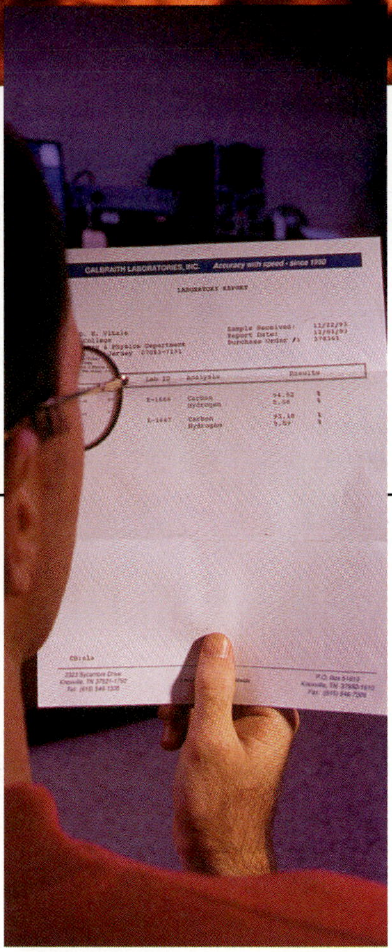

Analytical laboratories routinely measure the percentage composition of elements in compounds. Knowledge of the percentage composition enables chemists to identify the formula for new materials or unknown samples.

7.1 introduction

You have learned the basic principles of compound formation and mastered the skills of formula writing for both ionic and molecular compounds. These skills will enable you to name compounds in a systematic manner. The rules for naming many compounds follow from their classification as ionic or molecular compounds. Since you can now recognize compounds (Section 6.16) in both these categories from their formulas, you are prepared to apply the rules of nomenclature. Also, your new familiarity with formulas will allow you to extend your knowledge of atomic weights for elements (Section 4.10) to formula weights and molecular weights for compounds. The *formula weight* of an ionic compound and the *molecular weight* of a molecular compound represent two new conversion factors (Section 3.3) that you will use to predict the quantity of products formed in a chemical reaction (Chapters 8 and 9). Before studying chemical reactions, however, it is extremely useful in identifying substances to calculate the percentage composition of elements within a compound. Simply by knowing or hypothesizing the formula of a compound, you may predict the percentage composition by weight (mass) of each of its elements. Without knowing the chemical formula, the percentage composition of the elements may be determined with experiments in the laboratory. Agreement between the experimentally measured percentages and those you have predicted from its formula helps you to verify the identity of a sample.

7.2 nomenclature

There are more than 10 million known compounds today. If each one had a separate arbitrarily assigned name, called a **common name,** we would have a totally impossible situation. To alleviate this, sets of rules for systematically naming compounds as ionic, molecular, or as acids have been developed. These naming rules are known as **nomenclature.** Despite the advantages of using names systematically, some compounds have been in use so long and are so widely used that they are often called by their common names even by chemists. Table 7.1 lists some common-name compounds along with their systematic names and common uses, and Figure 7.1 shows these common household chemicals, most of which you'll recognize.

Two of the compounds listed in Table 7.1, washing soda and borax, have "$10H_2O$" at the end of their formulas and have the word *decahydrate* (*deca-* means "ten" and *hydrate* means "water") at the end of their systematic names. Ionic compounds formed with a specific number of water molecules attached to each formula unit are called

table 7.1 some household chemicals and their common and systematic names

Common name	Formula	Systematic name	Use
Lye	$NaOH$	Sodium hydroxide	Unclogs drains
Baking soda	$NaHCO_3$	Sodium hydrogen carbonate (sodium bicarbonate)	Makes cakes and breads rise by giving off CO_2 when heated
Washing soda	$Na_2CO_3 \cdot 10H_2O$	Sodium carbonate decahydrate	General cleanser
Milk of magnesia	$Mg(OH)_2$	Magnesium hydroxide	Antacid and laxative
Borax	$Na_2B_4O_7 \cdot 10H_2O$	Sodium tetraborate decahydrate	Cleanser
Cream of tartar	$KHC_4H_4O_6$	Potassium hydrogen tartrate	Mixed with baking soda in cake making
Lime	CaO	Calcium oxide	Making plaster and mortar
Slaked lime	$Ca(OH)_2$	Calcium hydroxide	Making plaster and mortar
Table salt	$NaCl$	Sodium chloride	Salty taste in foods

Figure 7.1
Common household chemicals are familiar to you by their common names. As a chemist, you are becoming familiar with their systematic names and chemical formulas. Systematic naming of chemical compounds enables you to learn the names of thousands of compounds by memorizing only a few simple rules.

hydrates. For both washing soda and borax, ten water molecules are included with each formula unit; for other compounds, the number of water molecules may vary.

To a great extent, success in learning nomenclature depends on recognizing ionic versus molecular compounds. Although there is no sharp dividing line between ionic and covalent bonds or between ionic and molecular compounds (Section 6.16), for purposes of nomenclature we use the simple generalization that metal-nonmetal combinations are ionic, and combinations of nonmetals are molecular. Compounds containing polyatomic ions (Table 6.6) are always ionic and follow the rules of ionic-compound nomenclature regardless of whether a metal is present. For example, ammonium carbonate is an ionic compound containing the ammonium ion, NH_4^+, and the carbonate ion, CO_3^{2-}, even though both ions contain no metals.

table 7.2 naming simple anions

Element	Root	Anion name (root + *-ide*)
Oxygen	Ox-	Oxide
Chlorine	Chlor-	Chloride
Bromine	Brom-	Bromide
Iodine	Iod-	Iodide
Fluorine	Fluor-	Fluoride
Phosphorus	Phosph-	Phosphide
Nitrogen	Nitr-	Nitride
Sulfur	Sulf-	Sulfide

7.3 naming ionic compounds

In naming any ionic compound, name the cation first, followed by a separate word for the anion:

Cation	Anion
First name	Last name

Cations are named exactly the same as the metals from which they come. Simple monatomic anions are named by using the root names given in Table 7.2 and adding the suffix *-ide*. So, to name NaI, you start with the cation name *sodium* and follow it with the word *iodide* (root *iod* + *ide*). Thus, NaI is sodium iodide. To name $MgCl_2$, you start with the cation name *magnesium* and follow it with the word *chloride*. Note that the name of an ionic compound does not mention the number of cations or anions indicated by subscripts in the formula, even when the subscripts for the cations and anions differ as they do in $MgCl_2$. Since you know the charge on each ion, it is assumed that you know the number of cations and anions that must be present to maintain electrical neutrality of the compound.

sample exercise 1

Name the following ionic compounds.

a. MgO
b. K_2S
c. $AlCl_3$

solution

a. MgO is magnesium oxide, a white powder used in making heat-resistant materials, cosmetics, pharmaceuticals, and insulation. It is also a component of milk of magnesia, an antacid and laxative.
b. K_2S is potassium sulfide, a red, crystalline solid used in medicine and depilatories. Note that no mention is made in the name of the subscript 2 after potassium. This is always true for *ionic* compounds. *The name contains no mention of subscript numbers.*
c. $AlCl_3$ is aluminum chloride, a yellow-white, crystalline solid. It fumes in air and reacts explosively with water. It is used in cracking (breaking down) petroleum and in manufacturing rubber and lubricants.

••• problem 1

Name the following compounds.

a. NaCl (table salt)

b. CaF_2 (used in etching glass)

c. BaO (dehydrating agent used to remove water from materials)

In the case of cations with variable charge, such as iron, copper, tin, and lead (Table 6.5), common names are sometimes employed to distinguish the lower and higher charges, with the suffix *-ous* used for the lower charge and *-ic* used for the higher charge. To form the systematic name for an ionic compound with a cation of variable charge, you should follow the name of the cation by a Roman numeral placed inside parentheses to indicate the magnitude of the charge. For example, Fe^{2+} combines with Cl^- to give the compound $FeCl_2$. This is named iron(II) chloride to distinguish it from the compound that results from the combination of Fe^{3+} and Cl^-, $FeCl_3$ or iron(III) chloride. The common names for cations of variable charge are given in Table 6.5 and may be substituted directly for the cation name. Since the common names for the Fe^{2+} and Fe^{3+} ions are ferrous and ferric, the compounds $FeCl_2$ and $FeCl_3$ are also called ferrous chloride and ferric chloride, respectively.

sample exercise 2

Name the following ionic compounds.

a. CuBr

b. $CuBr_2$

c. $SnCl_4$

solution

a. CuBr is copper(I) bromide. Note here the need to indicate the magnitude of the charge on copper. The charge on Cu is 1+, since Br as an anion is always 1− and there is only one copper ion to balance the 1− charge. From Table 6.5, you find that the common name for the Cu^+ ion is cuprous; hence, the common name for this compound is cuprous bromide. Copper(I) bromide is a white, crystalline solid used as a catalyst (which increases reaction speed) in carbon-containing reactions.

b. $CuBr_2$ is copper(II) bromide. There are two Br^- ions, and therefore the one copper ion must have a 2+ charge. From Table 6.5, you find that the common name for the Cu^{2+} ion is cupric; hence, the common name for this compound is cupric bromide. Copper(II) bromide is a black, crystalline solid used in photography as an indicator of humidity, and as a wood preservative.

c. $SnCl_4$ is tin(IV) chloride. There are four Cl^- ions, and therefore the one tin ion must have a 4+ charge. From Table 6.5, you find that the common name for the Sn^{4+} ion is stannic; hence, the common name for this compound is stannic chloride. Tin(IV) chloride is a fuming, caustic liquid used as a stabilizer for the colors and perfumes in soaps; it is also used in dyeing fabrics and tinning containers.

••• problem 2

Name the following compounds.

a. FeO (a jet-black powder used in the manufacture of steel and heat-absorbing glass)

b. Fe_2O_3 (also called rust, a dark-red solid used chiefly as a pigment and as a coating for magnetic recording tape)

c. CuO (a brownish-black, crystalline powder used as a pigment and in the manufacture of rayon)

So far in this section, all of the compounds named have been binary ionic compounds. A **binary compound** is one that contains only two kinds of elements. It may contain more than two atoms, but it must contain only two elements. Look back and convince yourself that this is so. For example, aluminum chloride, $AlCl_3$, has four atoms in its formula unit but only two kinds of elements, aluminum and chlorine.

When you see a formula for an ionic compound that is *not* binary, that is, it contains more than two kinds of elements, you should recognize the presence of a polyatomic ion (review Section 6.7 and Table 6.6). For example, Na_2SO_4 is not binary because it contains three elements, Na, S, and O. To name this compound, you must recognize that the cation is sodium and the anion is the polyatomic sulfate ion. Thus, the compound is named sodium sulfate. Similarly, NH_4Cl is ammonium chloride and NH_4OH is ammonium hydroxide.

---

sample exercise 3

Name the following ionic compounds.

a. Na_3PO_4 c. $(NH_4)_2SO_4$

b. K_2SO_3 d. $CuCO_3$

solution

a. Na_3PO_4 is sodium phosphate. Again, there is no mention of subscript numbers. Sodium phosphate is a heavy-duty cleaning agent sold under the commercial name of TSP® (trisodium phosphate).

b. K_2SO_3 is potassium sulfite, a white, crystalline solid used in photographic developers.

c. $(NH_4)_2SO_4$ is ammonium sulfate, a white, crystalline solid used chiefly as a fertilizer.

d. $CuCO_3$ is copper(II) carbonate. Carbonate has a 2– charge. The positive ion must therefore have a total positive charge of 2+. Copper(II) carbonate is a toxic, green powder used in pigments, as a fungicide, and as a feed additive.

••• problem 3

Name the following compounds.

a. K_2CO_3 (a white powder used in brewing, ceramics, explosives, and fertilizers)

b. NH_4NO_3 (a colorless, crystalline solid that is a stable explosive and a material used in fertilizers)

c. $Pb_3(PO_4)_2$ (a poisonous, white powder used as a stabilizer in plastics)

The number of actual polyatomic ions is considerably larger than that given in Table 6.6; however, in learning formulas and names, you do not have to rely completely on memory. There are many examples of the same two elements combining in different ratios to form different polyatomic anions. Examples of such series are

SO_4^{2-}	PO_4^{3-}	ClO_4^-
SO_3^{2-}	PO_3^{3-}	ClO_3^-
		ClO_2^-
		ClO^-

In each series of the same two elements, the number of oxygen atoms varies, but the number of the other nonmetal is fixed at one. In addition, the charge of the polyatomic anion remains the same throughout the series. If there are only two ions in the series, then the name of the one with the greater number of oxygen atoms ends in *-ate,* while the name of the one with less oxygen atoms ends in *-ite.* For example, PO_4^{3-} is phosphate and PO_3^{3-} is phosphite.

sample exercise 4

Name SO_4^{2-} and SO_3^{2-}.

solution

Since SO_4^{2-} is sulfate (Table 6.6), from the preceding discussion it follows that SO_3^{2-} is sulfite.

In the series of four anions, the ion with the greatest number of oxygen atoms is given the prefix *per-* as well as the suffix *-ate.* The ion with the least number of oxygens is given the prefix *hypo-* as well as the suffix *-ite.*

sample exercise 5

Name the ions in the series ClO_4^-, ClO_3^-, ClO_2^-, and ClO^-.

solution

ClO_4^- is *per*chlor*ate* (Table 6.6).

ClO_3^- is chlor*ate.*

ClO_2^- is chlor*ite.*

ClO^- is *hypo*chlor*ite.*

---

writing correct formulas from names

Writing correct formulas from the names of ionic compounds requires the skills described in Section 6.8. Review especially Sample Exercises 7 and 8 in Chapter 6. From this review, you should recognize that the compound calcium chloride, for example, is made up of calcium ions and chloride ions. Write the symbols for the ions with their charges, and achieve a neutral combination by the criss-cross method:

$$Ca^{2+} \; Cl^{-} \quad \text{gives} \quad CaCl_2$$

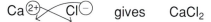

Given the name copper(I) bromide, notice that the Roman numeral tells you that it is Cu^+ that is combined in this compound. Bromide is always Br^-. Thus, the correct formula is determined:

$$Cu^{+} \; Br^{-} \quad \text{gives} \quad CuBr$$

sample exercise 6

Write the formula for the compound ammonium sulfate.

solution

The name ammonium sulfate is associated with a compound made up of the ammonium ion, NH_4^+, and the sulfate ion, SO_4^{2-}. Write the two ions:

$$NH_4^+ \qquad SO_4^{2-}$$

and balance the charges by taking two ammonium ions and one sulfate ion, giving

$$(NH_4)_2SO_4$$

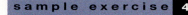

Since the ammonium ion is a polyatomic ion, you must enclose it within parentheses when the subscript is greater than 1. Once again, you see that the name of the compound does

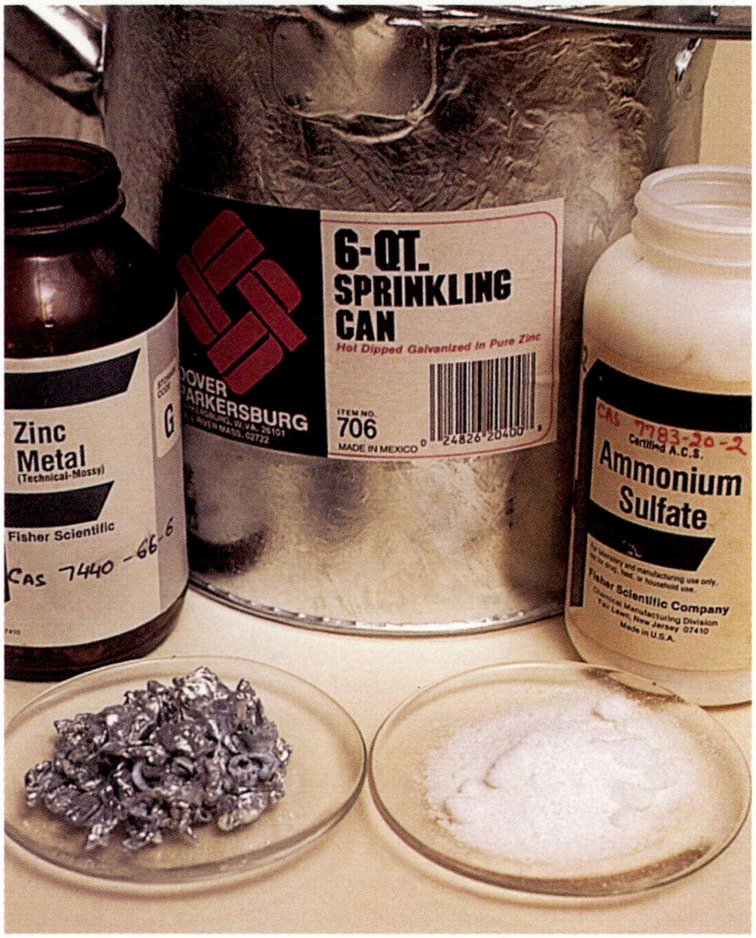

Figure 7.2
Ammonium sulfate is used in galvanizing iron (plating it with zinc) to produce a rustproof coating.

not directly tell you the number of each ion involved. It does tell you what ions are involved, and if you have memorized the charges on these ions, then you can immediately write a correct formula with the proper subscripts.

Ammonium sulfate is a white, crystalline material used in flameproofing fabrics and paper and in plating iron with zinc to keep it from rusting. You are familiar with galvanized screws and watering cans, which are rustproofed by plating iron with zinc (Figure 7.2).

7.4 naming binary molecular compounds

Binary molecular compounds, like their ionic counterparts, contain only two elements. Unlike binary ionic compounds, which contain one metal and one nonmetal, binary molecular compounds contain two nonmetals, such as in natural gas, CH_4, or laughing gas, N_2O.

••• problem 4

Identify the binary compounds in the following list.

a. H_2O
b. P_4O_{10}
c. NH_3
d. $C_6H_6O_6$
e. CH_2O
f. $CHCl_3$
g. CCl_4

table 7.3 numerical prefixes

Number of atoms	Numerical prefix
1	Mono-
2	Di-
3	Tri-
4	Tetra-
5	Penta-
6	Hexa-
7	Hepta-
8	Octa-
9	Nona-
10	Deca-

writing correct names from formulas

Naming binary molecular compounds differs from naming ionic compounds in that the number of atoms of each element combining to form the compound is indicated by a prefix such as *mono-* (one), *di-* (two), or *tri-* (three), listed in Table 7.3.

The complete name for a binary molecular compound is composed of the name of the first element (on the left) preceded by the prefix indicating the number of atoms of that element and followed by the name of the second element, which is constructed of a prefix for the number of atoms plus the element's root name (Table 7.2) plus the suffix *-ide.* For example, P_4O_{10} is tetraphosphorus decoxide.

numerical prefix + **element name**	*numerical prefix* + **root name** + *-ide*
First name	Last name

Where there is only one atom of the *first* element, it is common to omit the prefix *mono-* for that element. For example, NO_2 is nitrogen dioxide, not mononitrogen dioxide. When there is only one atom of the *second* element, the prefix *mono-* for that element must be included whenever the same two elements form more than one compound. For example, CO is carbon monoxide, not carbon oxide, to distinctly differentiate it from CO_2, carbon dioxide.

Binary molecular compounds containing *hydrogen as the first element symbolized in the formula* are an exception to the prefix rule. Like ionic compounds, they are named without the use of prefixes for either element. For example, H_2S is called hydrogen sulfide. The reason for the exception is that, whereas most pairs of nonmetals, for example, P and Cl, can form more than one compound (PCl_3 and PCl_5), H and another nonmetal generally yield only one compound. PCl_3 and PCl_5 must be distinguished as phosphorus trichloride and phosphorus pentachloride to indicate their combining ratios; no such distinction is usually necessary for hydrogen compounds. In the case of the combination of hydrogen with oxygen, two compounds, H_2O and H_2O_2, can be formed, and they are distinguished by the common names *water* and *hydrogen peroxide.*

sample exercise 7

Write names for the following binary molecular compounds.

a. N_2O_5 d. NCl_3
b. SO_2 e. HCl
c. CO f. H_2S

solution

The names are constructed of a first name and a last name as just described.

a. N_2O_5 is [di|nitrogen] [pent|ox|ide]

- di → Prefix for 2; nitrogen → Element name
- pent → Prefix for 5; ox → Root name; ide → Suffix

To avoid awkwardness, the *a* of the prefix *penta-* is dropped when the element root begins with a vowel.

b. SO_2 is sulfur dioxide. Note the absence of the prefix *mono-* before sulfur, the first element in the compound. Sulfur dioxide is a colorless gas used in preserving fruits and vegetables (Figure 7.3); it is also used as a disinfectant and bleach.

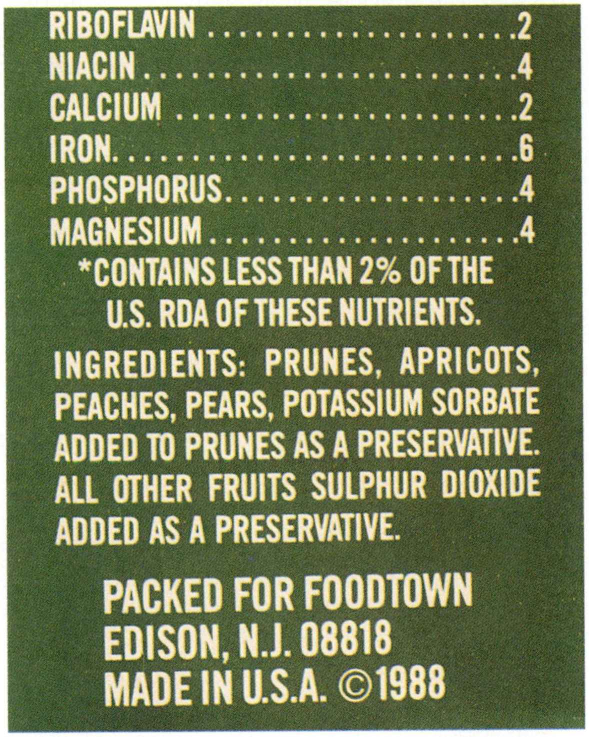
RIBOFLAVIN 2
NIACIN 4
CALCIUM 2
IRON.......................6
PHOSPHORUS.......................4
MAGNESIUM 4
*CONTAINS LESS THAN 2% OF THE
U.S. RDA OF THESE NUTRIENTS.
INGREDIENTS: PRUNES, APRICOTS, PEACHES, PEARS, POTASSIUM SORBATE ADDED TO PRUNES AS A PRESERVATIVE. ALL OTHER FRUITS SULPHUR DIOXIDE ADDED AS A PRESERVATIVE.
PACKED FOR FOODTOWN
EDISON, N.J. 08818
MADE IN U.S.A. ©1988

Figure 7.3
Sulfur dioxide is widely used to preserve fruits and vegetables.

c. CO is carbon monoxide. One atom of the second element (oxygen) in this binary molecular compound must be indicated by *mono-*. You remember from Section 6.14 that carbon monoxide is a toxic component in automobile exhaust. It is also used productively in the Mond process to produce high-purity nickel.

d. NCl_3 is nitrogen trichloride, a yellow, thick, oily liquid that evaporates rapidly in air and is used to bleach flour.

e. HCl is hydrogen chloride. A binary molecular compound containing hydrogen as the first element is named without prefixes. It is an odorless, corrosive gas used in the manufacture of pharmaceuticals and industrial chemicals.

f. H_2S is hydrogen sulfide. It is another binary molecular compound containing hydrogen as the first element and thus is named without prefixes. A flammable, poisonous gas with the distinctive odor of rotten eggs, H_2S is a natural contaminant of crude oil. It is used in the manufacture of other chemicals.

••• problem 5

Write names for the following binary molecular compounds.

a. CO_2 (a colorless, odorless gas that, along with water, is the end product of human metabolism and the starting material of plant photosynthesis; it is used in the carbonation of beverages, in fire extinguishers, and as dry ice in refrigeration)

b. CCl_4 (a colorless, toxic liquid formerly used to clean clothes)

c. PCl_3 (a colorless, fuming liquid used in the manufacture of saccharin)

d. N_2O_4 (a colorless gas usually present in combination with reddish-brown NO_2 and used in the production of nitric and sulfuric acids)

e. OF_2 (an unstable, colorless gas)

f. HF (a colorless, corrosive, poisonous gas used in the manufacture of fluorine-containing refrigerants, in the aluminum industry, and in etching glass)

writing correct formulas from names

Because the prefixes in the names of molecular compounds tell the numbers of atoms, it is particularly easy to write formulas from names of binary molecular compounds. For example, carbon dioxide must have the formula CO_2 (one carbon atom because no prefix indicates that *mono-* has been omitted, and two oxygen atoms because of the prefix *di-*).

••• problem 6

Write formulas for the following.

a. Sulfur trioxide (a toxic, irritating liquid used to produce sulfur-containing chemicals)

b. Hydrogen bromide (a colorless, corrosive gas used to produce bromine-containing chemicals)

c. Tetraphosphorus decoxide (a flammable, soft, white powder used as a dehydrating agent and in medicine and sugar refining)

d. Sulfur hexafluoride (a colorless, odorless, very dense gas used in making electronic components)

A **ternary compound** contains three elements. There may be more than three atoms, but the number of elements is limited to three. The only ternary molecular compounds discussed in this text are the ternary oxoacids discussed in the next section.

7.5 naming acids

Certain molecular compounds made of hydrogen and nonmetals are classified as **acids** because of the properties of their water solutions, which contain an abundance of H^+, hydrogen ions. Acids that are components of foods are familiar to you by their sour taste. For example, citrus fruits owe their sour taste to citric acid and vinegar owes its sour taste to acetic acid. The oxoacids discussed in this section are not components of foods and should never be tasted, but do have a sour taste also. Acids will be discussed in detail in Chapter 15. At this point, you should become familiar with the naming system used for acids so that a given name will suggest a formula and vice versa. It is easy to recognize acids because their formulas usually begin with H. However, although it has the formula H_2O, water is not usually thought of as an acid. For purposes of naming, it is convenient to classify acids into two categories based on whether the compound contains oxygen.

table 7.4 nonoxoacids

Formula	Name of the pure gas	Name of the solution formed by dissolving the gas in water
H_2S	Hydrogen sulfide	Hydrosulfuric acid
HCN	Hydrogen cyanide	Hydrocyanic acid
HF	Hydrogen fluoride	Hydrofluoric acid
HCl	Hydrogen chloride	Hydrochloric acid
HBr	Hydrogen bromide	Hydrobromic acid
HI	Hydrogen iodide	Hydriodic acid

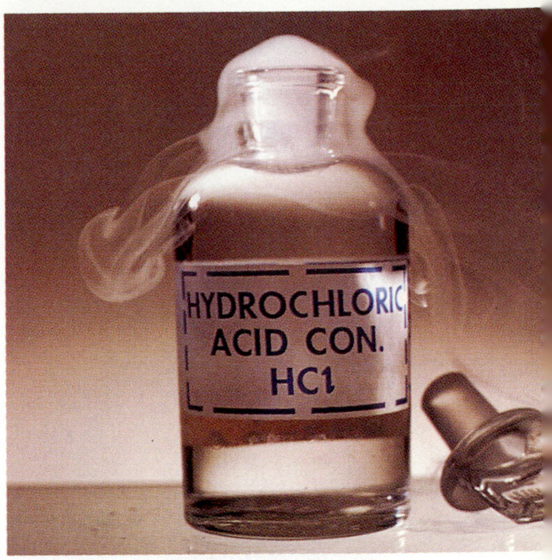

Figure 7.4

Concentrated HCl fumes in moist air.

nonoxoacids

Nonoxoacids are water solutions of molecular compounds composed of hydrogen and some nonmetal other than oxygen (or carbon). A listing of some nonoxoacids and their names is given in Table 7.4.

Notice that all of the binary compounds listed in Table 7.4 are gases. These gases themselves are named as binary molecular compounds without prefixes. For example, the formula HF is read as hydrogen fluoride. The naming of the acid, i.e, the solution formed when the gas is dissolved in water, is quite different. To name a nonoxoacid, begin with the prefix *hydro-* (for hydrogen), then use the root name for the other element (Table 7.2), attach the suffix-*ic,* and follow this word with *acid.* In summary:

hydro- + **root name** + *-ic* + acid

This prescription works well for all of the acids listed with the exception of HI and H_2S. For HI, the root name of *iod-* begins with a vowel, so the *o* from *hydro-* is dropped in combining it with the root in naming the acid. Thus, HI in water solution becomes hydriodic acid, not hydroiodic acid, which is commonly misused. For H_2S, the root used in acid nomenclature is *sulfur-,* not *sulf-* as shown in Table 7.2. As a result, H_2S in water solution becomes hydrosulfuric acid, not hydrosulfic acid.

The compounds listed in Table 7.4 are usually named as acids (i.e., it is assumed they are dissolved in water), *not* by their pure-gas names.

sample exercise 8

Name the acid HCl.

solution

You begin with *hydro-,* attach the root *chlor-,* add the suffix *-ic,* then add the word *acid,* and the complete name is hydrochloric acid. It is widely used in chemical laboratories, in acidifying oil wells, in obtaining metals from their ores, and in cleaning metals (Figure 7.4).

oxoacids

An **oxoacid** is the water solution of a molecular compound made up of hydrogen, some nonmetal, and oxygen. Although oxoacids are truly molecular and therefore not composed of ions, when they are dissolved in water, they break apart into ions. The anions are recognizable as the polyatomic anions you have previously encountered.

sample exercise 9

What polyatomic anion can be found in each of the following acid formulas.

a. H_2SO_4
b. HNO_3
c. H_3PO_4
d. $HClO_4$

Figure 7.5
Sulfuric and phosphoric acids are oxoacids that you depend on daily. Sulfuric acid is needed to produce the current in your car battery; phosphoric acid is likely a component of your favorite soft drink.

solution

a. H_2SO_4 yields the sulfate anion (SO_4^{2-}). Sulfuric acid is a colorless, oily liquid used in the manufacture of chemicals, fertilizers, explosives, and in petroleum refining. It is the acid in your car battery that you test to determine the charge (Figure 7.5).
b. HNO_3 yields the nitrate anion (NO_3^-). Nitric acid is a colorless liquid with a choking odor used to manufacture nitrates and *nitro-* compounds for fertilizers, dyes, explosives, and many organic (carbon-containing) chemicals.
c. H_3PO_4 yields the phosphate anion (PO_4^{3-}). Phosphoric acid is a crystalline solid used in fertilizers, in carbonated soft drinks and flavor syrups to give them tang (Figure 7.5), in detergents, and in toothpastes.
d. $HClO_4$ yields the perchlorate anion (ClO_4^-). Perchloric acid is a corrosive, colorless liquid used in medicine, electropolishing, explosives, and chemical analysis.

writing correct names of oxoacids from formulas

Oxoacids are named according to the polyatomic ion they yield in solution. The nonmetal other than oxygen in the polyatomic ion determines the oxoacid's root name. For example, HNO_3 yields NO_3^- in solution. Therefore, its name is based on the root *nitr-* for N. For Cl, Br, and I, the roots are *chlor-, brom-,* and *iod-* as given in Table 7.2. Other roots are *sulfur-* for S, *phosphor-* for P, *carbon-* for C, and *acet-* for the ion $C_2H_3O_2^-$.

The polyatomic ion also determines the oxoacid's suffix. If the anion ends in *-ate,* the ending of the acid is *-ic* followed by the word *acid.* If the anion ends in *-ite,* the acid ends in *-ous* followed by the word *acid.*

-ate ions → *-ic* acids
-ite ions → *-ous* acids

For example, the carbon*ate* anion, CO_3^{2-}, is produced by carbon*ic* acid, H_2CO_3. The nitr*ite* anion, NO_2^- is produced by nitr*ous* acid, HNO_2.

If the anion contains the prefix *per-* or *hypo-,* then these prefixes are also used in the names of the acid. The prefix *hydro-* is *never* found in an oxoacid.

In summary, oxoacids are named as follows.

polyatomic anion root + *-ic* or *-ous* + acid

sample exercise 10

Name the following acids.

a. H_2SO_4
b. HNO_3
c. H_3PO_4
d. $HClO_2$
e. $HClO$

solution

a. H_2SO_4 is *sulfuric* acid. The sulfate ion determines the root *sulfur-* and the suffix *-ic* because of the anion ending *-ate.*
b. HNO_3 is *nitric* acid, from the root *nitr-* and suffix *-ic.*
c. H_3PO_4 is *phosphoric* acid, from the root *phosphor-* and the suffix *-ic.*
d. $HClO_2$ is *chlorous* acid, from the root *chlor-* and the suffix *-ous.* It is *-ous* here because the anion ClO_2^-, chlorite, ends in *-ite.* Chlorous acid is an acid existing only in solution and is used to produce chlorite compounds.

Figure 7.6

Hypochlorous acid, HClO, is formed when chlorine is used to purify water or disinfect swimming pools. The hypochlorite anions, ClO^-, formed in water destroy bacteria.

e. HClO is *hypochlorous* acid. The hypochlorite anion gives us the root *hypochlor-,* to which we add *-ous* because of the ending *-ite*. Hypochlorous acid is an unstable acid existing only in solution and is used as a disinfectant; it is formed when chlorine is used to purify water and disinfect swimming pools (Figure 7.6).

problem 7

Name the following oxoacids.

a. H_2CO_3
b. $HC_2H_3O_2$
c. H_2SO_3
d. $HClO_3$
e. $HClO_4$

writing correct formulas from acid names

To write an acid's formula from its name, begin by deciding whether the name is that of an oxoacid or a nonoxoacid. The prefix *hydro-* tells you an acid is a nonoxoacid, i.e., a binary compound of hydrogen and another nonmetal. The root name indicates the nonmetal combined with hydrogen. Thus, hydriodic acid is a combination of hydrogen and iodine. In the correct formula, the number of hydrogens (H^+) depends on the number needed to neutralize the negative charge on the anion. For hydriodic acid, I^- requires one H^+. Therefore, the formula is HI. Hydrogen is always written to the left.

Oxoacids are recognized by the word *acid* and the absence of the prefix *hydro-*. The root name tells you the nonmetal that is combined with oxygen in the polyatomic ion. If the acid suffix is *-ic,* the polyatomic ion must end in *-ate;* whereas if the acid suffix is *-ous,* a polyatomic ion ending in *-ite* must be present. An appropriate number of H^+ ions is combined with the polyatomic ion to yield a balanced formula.

sample exercise 11

Give formulas for the following acids.

a. Sulfurous acid
b. Hydrobromic acid
c. Chloric acid

solution

a. Sulfurous acid is H_2SO_3. The absence of *hydro-* indicates an oxoacid. The root *sulfur-* indicates that the polyatomic anion must be a combination of S and O. The *-ous* suffix then tells you that the polyatomic anion is SO_3^{2-}, the polyatomic ion of sulfur with three, not four, oxygen atoms. Combining H^+ and SO_3^{2-} leads to the formula H_2SO_3. Sulfurous acid has never been isolated; it exists only in extremely small amounts in water solution.
b. Hydrobromic acid is HBr. The prefix *hydro-* tells you the acid is a binary nonoxoacid. The root *brom-* indicates the anion Br^-. Combining H^+ and Br^- yields HBr. Hydrobromic acid is a faintly yellow, corrosive liquid.
c. Chloric acid is $HClO_3$, an oxoacid formed from the polyatomic ion ClO_3^-. Chloric acid is another acid that is only known to occur in small amounts in water solution.

problem 8

Give formulas for the following acids.

a. Chlorous acid
b. Hydrofluoric acid
c. Nitric acid

7.6 nomenclature summary

As you have seen in this chapter, there are a variety of rules for naming ionic and molecular compounds, including acids. It is useful to follow a classification scheme to decide on the appropriate rule to use for naming a given compound. One such scheme that follows the organization of sections within this chapter is summarized with rules for nomenclature in Figure 7.7.

General rules for nomenclature

1. Classify the compound as ionic (metal + nonmetal) or molecular (nonmetals only).
2. If the compound is ionic, determine whether it is a binary compound or contains a polyatomic ion and follow the nomenclature rules in Section 7.3 as summarized in Figure 7.7.
3. If the compound is molecular, classify it as an acid (H *begins* the formula) or nonacid (no H).
4. If it is an acid, classify it as a nonoxoacid or oxoacid and follow the nomenclature rules in Section 7.5 as summarized in Figure 7.7.
5. If it is a nonacid, follow the nomenclature rules in Section 7.4 as summarized in Figure 7.7.

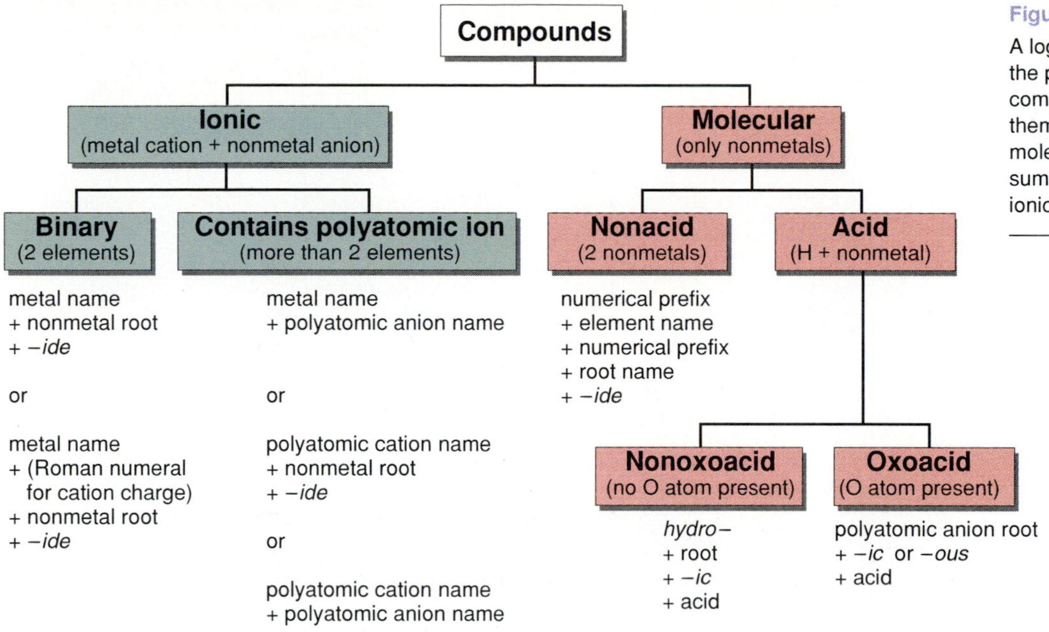

Figure 7.7

A logical classification of compounds for the purpose of naming them. Naming compounds depends on first classifying them into two major groups: ionic and molecular. Rules of nomenclature are summarized beneath each subgroup of ionic and molecular compounds.

sample exercise 12

Name the following compounds.

a. HI
b. Na_2CO_3
c. HNO_3
d. KBr
e. $Ca(OH)_2$
f. N_2O_4
g. $CaSO_4$

solution

a. HI is molecular because it contains nonmetals only; it is an acid because of the H and a nonoxoacid because there's no oxygen. Therefore, its name is *hydriodic acid* according to the rules in Section 7.5 (or hydrogen iodide according to Section 7.4). Hydrogen iodide is a colorless, acrid gas that dissolves in water to form hydriodic acid. Hydriodic acid is used in the manufacture of pharmaceuticals and disinfectants.

b. Na_2CO_3 is ionic because it contains a metal and nonmetals. Therefore, its name is *sodium carbonate* according to Section 7.3. A white powder also known as washing soda, it is used in cleansers and for washing textiles and bleaching linen and cotton (Table 7.1).

c. HNO_3 is a molecular compound and an oxoacid. Its name is *nitric acid* according to Section 7.5.

d. KBr is an ionic compound called *potassium bromide* according to Section 7.3. It is a white, crystalline solid used in medicine, soaps, and photography.

e. $Ca(OH)_2$ is an ionic compound called *calcium hydroxide* according to Section 7.3. It is a white, crystalline solid used in cements, mortar, and plaster (Table 7.1).

f. N_2O_4 is a molecular compound and a nonacid. It is named *dinitrogen tetroxide* according to Section 7.4.

g. $CaSO_4$ is an ionic compound called *calcium sulfate* according to Section 7.3. Calcium sulfate is a desiccant widely used to keep other materials dry by absorbing moisture from the air in a closed environment (Figure 7.8).

Figure 7.8
In the laboratory, chemicals are routinely stored in a desiccator, a sealed glass jar containing calcium sulfate, to keep them dry. Calcium sulfate is a desiccant because it absorbs the moisture from the air within the desiccator.

••• problem 9

Name the following compounds.

a. $MgCl_2$
b. HBr
c. H_2SO_4
d. H_2SO_3
e. $Ca_3(PO_4)_2$
f. OF_2

7.7 molecular weight and formula weight

Knowing chemical formulas for compounds and being able to recognize them from chemical names are powerful skills that enable you to find the mass of any compound. The nomenclature skills you have developed in this chapter enable you to write a formula from a given name. Since each compound is composed of elements in fixed proportions, its formula tells you the number of each type of atom in each molecule or formula unit. Just as the total mass (weight)[1] of any group of objects is the sum of the masses (weights) of all the objects, the mass assigned to any compound is the sum of the masses of the atoms or ions that compose the compound.

The formula for the molecular compound water, H_2O, indicates two hydrogen atoms and one oxygen atom held together by covalent bonds. The formula for the ionic compound sodium sulfate, Na_2SO_4, indicates the ratio of two sodium ions (Na^+) to one sulfate ion (SO_4^{2-}). The sulfate ion itself is made up of one sulfur atom and four oxygen atoms held together by covalent bonds. Chapter 4 discussed the atomic masses of atoms. Section 6.17 pointed out that because ions are formed from atoms by the loss or gain of electrons and because electrons have essentially no mass, the masses of ions are the same as those of the atoms from which they form. Calculating the mass of an ionic compound is no different from calculating the mass of a molecular compound. Both are computed by adding the masses of their atoms or ions.

1. As noted in Section 4.10, the terms *mass* and *weight* are often used interchangeably by chemists. Even though chemists determine mass by balancing compared to a standard, the process historically has been called weighing.

molecular weight

Since a water molecule is made up of two hydrogen atoms and one oxygen atom and you know the mass of these atoms in atomic mass units, you can assign a mass to one H_2O molecule by simply adding the masses of the atoms involved. The mass of one H_2O molecule is called a **molecular weight**.[2] For H_2O, the molecular weight is 2×1.01 amu (for two hydrogens) plus 16.0 amu (for one oxygen), or 18.0 amu. If you could weigh one H_2O molecule, it would be 18.0/12.0 = 1.50 times as heavy as one carbon-12 atom. Molecular weights are the sums of atomic masses. Each atomic mass is a mass compared with the mass of one carbon-12 atom (Section 4.9).

sample exercise 13

Determine the molecular weight of HNO_3, nitric acid.

solution

HNO_3 contains	Mass of each atom	×	Number of that atom present	=	Total mass of each element
1 H atom	1.01 amu	×	1	=	1.01 amu
1 N atom	14.0 amu	×	1	=	14.0 amu
3 O atoms	16.0 amu	×	3	=	48.0 amu
			Total mass of HNO_3	=	63.0 amu

One HNO_3 molecule has a mass of 63.0 amu. Its molecular weight reported to 3 sig figs is 63.0 amu. One HNO_3 molecule is 63.0/12.0 = 5.25 times as heavy as one carbon-12 atom.

formula weight

Ionic compounds do not contain molecules (Section 6.9); rather, the formula unit tells the smallest grouping of ions corresponding to the correct fixed ratio of ions in that compound. For this reason, the mass of a formula unit of an ionic compound is called a **formula weight** (for traditional reasons). Because ions have the same masses as the atoms from which they are made, the calculation of formula weight is mechanically identical to the calculation of molecular weight. That is, you sum the masses of all ions (atoms) present.

sample exercise 14

What is the formula weight of Na_2SO_4, sodium sulfate, a crystalline compound used in making soaps, detergents, glass, and in dyeing and printing textiles?

solution

Sodium ions are formed from sodium atoms with no mass change. Sulfate ions are formed from sulfur atoms and oxygen atoms with no mass change.

Na_2SO_4 contains	Mass of each atom (ion)	×	Number of that atom (ion) present	=	Total mass of each element
2 Na^+ ions	23.0 amu	×	2	=	46.0 amu
1 SO_4^{2-} ion: 1 S atom	32.1 amu	×	1	=	32.1 amu
4 O atoms	16.0 amu	×	4	=	64.0 amu
			Total mass of Na_2SO_4	=	142.1 amu

The formula weight of Na_2SO_4 reported to 4 sig figs is 142.1 amu. If you could weigh one formula unit of Na_2SO_4, it would be 142.1/12.0 = 11.8 times as heavy as one carbon-12 atom.

2. Technically, this term should be molecular mass, but it is traditionally referred to as a weight.

sample exercise 15

What is the formula weight of $Mg(OH)_2$, magnesium hydroxide, a nearly insoluble, white powder used in water as an antacid and laxative (also called milk of magnesia)?

solution

The method is the same as in Sample Exercise 14. In the formula for this ionic compound, notice the multiplication of *both* atoms O and H within the parentheses.

$Mg(OH)_2$ contains	Mass of each atom (ion)	×	Number of that atom (ion) present	=	Total mass of each element
1 Mg^{2+} ion	24.3 amu	×	1	=	24.3 amu
2 OH^- ions: 2 O atoms	16.0 amu	×	2	=	32.0 amu
2 H atoms	1.01 amu	×	2	=	2.02 amu
			Total mass in $Mg(OH)_2$	=	58.3 amu

The formula weight of $Mg(OH)_2$ is 58.3 amu. Note that when an ion is enclosed in parentheses in a formula, the subscript after the parentheses indicates the number of ions in each formula unit. You must remember that the number outside the parentheses multiplies *all* atoms inside the parentheses.

••• problem 10

Determine the molecular weights of SO_2, sulfur dioxide, N_2O_4, dinitrogen tetroxide, and H_2SO_4, sulfuric acid.

••• problem 11

Determine the formula weights of NaOH, sodium hydroxide, $CaCl_2$, calcium chloride, and $Al_2(CO_3)_3$, aluminum carbonate.

7.8 percentage composition

When something has more than one component and you divide the size of each part by the size of the whole and multiply by 100, you calculate the percentage that each part contributes to the whole. (For a review of percentage calculations, see Section 3.6.) The chemical formula for a compound enables you to find the mass of each element present in the compound. The molecular or formula weight tells you the total mass in the compound, whether it be ionic or molecular. By dividing the mass of each element by the molecular or formula weight and then multiplying by 100, you have calculated the *percentage by mass* of each element in the compound, called the **percentage composition** or **mass composition.** Traditionally, the percentage composition is also referred to as the *percentage by weight.*

In general, the percentage composition of any element may be calculated in the following manner.

$$\text{Percentage of an element in a compound} = \frac{\text{Total mass of the element in one molecule (}\textit{or}\text{ formula unit)}}{\text{Molecular weight (}\textit{or}\text{ formula weight)}} \times 100$$

For example, consider the molecular compound carbon dioxide, CO_2.

CO_2 contains	Mass of each atom	×	Number of that atom present	=	Total mass of each element
1 C atom	12.0 amu	×	1	=	12.0 amu
2 O atoms	16.0 amu	×	2	=	32.0 amu
			Molecular weight	=	44.0 amu

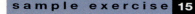

$$\text{percentage of C in } CO_2 = \frac{\text{Mass of C}}{\text{Molecular weight of } CO_2} \times 100 = \frac{12.0\ \cancel{amu}}{44.0\ \cancel{amu}} \times 100 = 27.3\%$$

$$\text{percentage of O in } CO_2 = \frac{\text{Mass of O}}{\text{Molecular weight of } CO_2} \times 100 = \frac{32.0\ \cancel{amu}}{44.0\ \cancel{amu}} \times 100 = 72.7\%$$

Note that the sum of the percentage compositions of all the elements in any compound should equal 100%. In the example of CO_2, the percentage of C, 27.3%, and the percentage composition of O, 72.7%, total to exactly 100.0%. In actual practice, the total may deviate slightly from 100.0% because of uncertainty in measurements and rounding of the data.

Calculating percentage composition of the elements from a known or hypothesized formula is an extremely helpful tool to chemists in identifying a substance. Since the percentage composition of the elements in a pure sample of a compound may be measured experimentally without any knowledge of what the compound is, the chemist may compare the experimentally determined percentage compositions with those predicted from a known or hypothesized formula. An example of experimentally measuring the percentage composition of one element in a compound is described in Laboratory Exercise 7.

---

sample exercise 16

Calculate the percentage of sodium, sulfur, and oxygen in sodium sulfate, Na_2SO_4.

solution

The formula weight of Na_2SO_4 is 142 amu (Sample Exercise 14).

$$\text{Percentage of Na} = \frac{\text{Total mass of Na in one formula unit}}{\text{Formula weight}} \times 100$$

$$= \frac{46.0\ \cancel{amu}}{142\ \cancel{amu}} \times 100 = 32.4\%$$

(Note that the unit amu in the numerator cancels with the unit amu in the denominator; percent does not have a unit associated with it.)

$$\text{Percentage of S} = \frac{\text{Total mass of S in one formula unit}}{\text{Formula weight}} \times 100$$

$$= \frac{32.1\ \cancel{amu}}{142\ \cancel{amu}} \times 100 = 22.6\%$$

$$\text{Percentage of O} = \frac{\text{Total mass of O in one formula unit}}{\text{Formula weight}}$$

$$= \frac{64.0\ \cancel{amu}}{142\ \cancel{amu}} \times 100 = 45.1\%$$

To check, you can add the percentages of all the parts. The sum should be 100.0% to the uncertainty in the last digit.

Percentage of Na	=	32.4%
Percentage of S	=	22.6%
Percentage of O	=	45.1%
		100.1%

---

problem 12

Calculate the percentage of each element in water, H_2O.

CHEMLAB

Laboratory exercise 7: percentage of oxygen in potassium chlorate

You have learned in this chapter that the percentage by mass of each element in a compound is called the percentage composition. If the formula of a compound is known, then the percentage by mass of each element in the compound can be calculated in a straightforward manner.

In the laboratory, the percentage composition of a compound can be determined experimentally without a knowledge of the compound's formula. Research chemists routinely send samples to an analytical laboratory for determination of their percentage composition. The percentage composition of the elements can then be used to help establish the correct formula for a new or unknown compound.

In this experiment, you will measure the mass of oxygen gas formed from the decomposition of a known mass of solid potassium chlorate. Upon heating potassium chlorate, gaseous oxygen escapes and the residue remaining after heating is potassium chloride. Measuring the mass of the residue and subtracting this mass from the mass of the original sample of potassium chlorate gives the mass of oxygen lost in the decomposition.

Potassium chlorate	→	Potassium chloride	+	Oxygen
$KClO_3$ (solid)		KCl (solid)		O_2 (gas)

From these experimental measurements, the percentage of oxygen may be determined by using the following equation.

$$\text{Percentage of O in } KClO_3 = \frac{\text{Mass of sample} - \text{Mass of residue}}{\text{Mass of sample}} \times 100$$

$$= \frac{\text{Mass of O formed in decomposition}}{\text{Mass of sample}} \times 100$$

Place a clean, dry, porcelain crucible with its cover ajar on a clay triangle that is resting on an iron ring as shown in Figure 7.9. Heat the crucible for 3 to 4 minutes with a Bunsen burner flame that has been adjusted so that the hottest spot (the tip of the inner blue cone) of the flame touches the bottom of the crucible. Using tongs, remove the crucible to a heat-resistant surface and allow it to cool (Figure 7.10). Weigh the cooled crucible and cover as accurately as possible and record this mass. Add approximately 2.3 g of potassium chlorate ($KClO_3$) to the crucible (Figure 7.11). Weigh the crucible, cover, and $KClO_3$

Figure 7.9

Heating the empty crucible dries it thoroughly so that no moisture is included with its weight. The glowing red color of the white, porcelain crucible indicates that it has reached a high temperature.

chapter accomplishments

After completing this chapter, you should be able to define all key terms and do the following.

7.2 Nomenclature

- ❑ Recognize household chemicals by their common names.

7.3 Naming ionic compounds

- ❑ Recognize a binary compound.
- ❑ Name a binary ionic compound, given a chemical formula.
- ❑ Name an ionic compound containing a polyatomic ion, given a chemical formula.
- ❑ Write a correct formula for an ionic compound, given the name of that compound.

7.4 Naming binary molecular compounds

- ❑ Classify a molecular compound as binary or ternary.
- ❑ Name binary molecular compounds, given a chemical formula.
- ❑ Write the formula of a binary molecular compound, given the name.

as accurately as possible and record this mass. Obtain the mass of the $KClO_3$ by difference of these two masses.

Place the covered crucible containing $KClO_3$ back on the triangle and heat gently for about 5 minutes, gradually increasing the heat by adjusting the position of the crucible in the flame. Heat strongly for about 15 minutes so that the bottom of the crucible takes on a dull red color. Use tongs to remove the crucible to a heat-resistant surface, allow it to cool, and weigh again. Subtract the mass of the crucible and cover from the mass of the crucible, cover, and residue to obtain the mass of the residue.

Questions:

1. If your dry crucible and cover weighs 25.459 g and your crucible, cover, and $KClO_3$ sample weighs 27.773 g, what is the mass of your $KClO_3$ sample?
2. If your crucible, cover, and sample after heating weighs 26.867 g, what is the mass of your residue?
3. What is the mass of the oxygen gas that was lost in the decomposition of $KClO_3$?
4. Calculate the percentage of oxygen in the original sample of $KClO_3$.

Figure 7.10
Use tongs to gently move the crucible to a heat-resistant surface so it may cool before weighing.

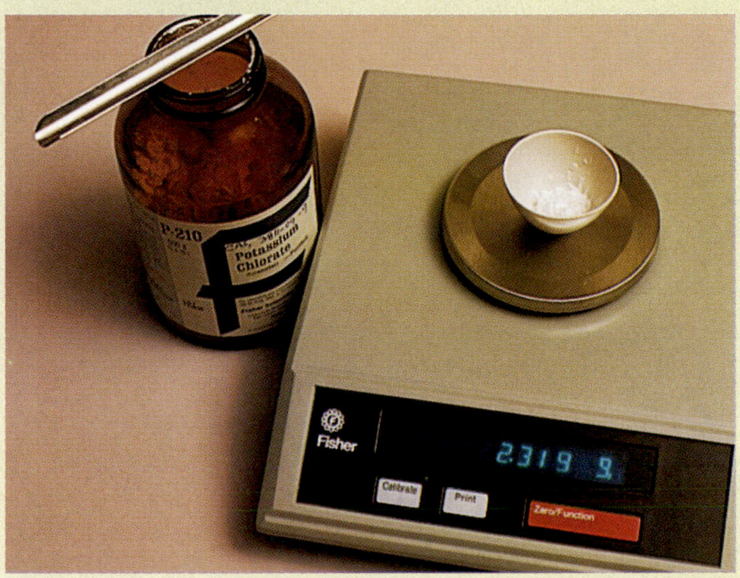

Figure 7.11
This balance was tared (adjusted to zero) after the empty crucible was placed on it. The reading of 2.319 g indicates the weight (mass) of the $KClO_3$ solid that has been added to the crucible.

7.5 Naming acids

- ❑ Distinguish acids from other compounds.
- ❑ Recognize nonoxoacids.
- ❑ Name nonoxoacids, given the formulas.
- ❑ Write the formulas of nonoxoacids, given the names.
- ❑ Recognize oxoacids.
- ❑ Name oxoacids, given the formulas.
- ❑ Write the formulas of oxoacids, given the names.

7.6 Nomenclature summary

- ❑ Name any ionic compound, binary molecular compound, or acid, given a chemical formula.

7.7 Molecular weight and formula weight

- ❑ Given a periodic table and a correct molecular formula, calculate the molecular weight of a molecular compound.
- ❑ Given a periodic table and a corrrect ionic formula, calculate the formula weight of an ionic compound.

7.8 Percentage composition

- ❑ Calculate the percentage composition of any component in a molecular or ionic compound, given the formula of that compound and a periodic table.
- ❑ Calculate the percentage composition of a component in a molecular or ionic compound, given experimental decomposition data.

key terms

common name
nomenclature
hydrates
binary compound
ternary compound
acids
nonoxoacids
oxoacid
molecular weight
formula weight
percentage composition (mass composition)

problems

7.2 Nomenclature

13. Give the systematic names for the following common chemicals.
    a. Table salt
    b. Lye
    c. Baking soda

7.3 Naming ionic compounds

14. a. Indicate which of the following metals can form cations with different charges: Ba, Ca, Sn, and Zn.
    b. State the charges that can form on the cation of each variable-charge metal in part a.
15. Indicate which of the following compounds are binary.
    a. $AlCl_3$
    b. $AlPO_4$
    c. NH_4Cl
    d. Fe_2O_3
    e. NaOH
    f. Li_3N
16. Name the following binary ionic compounds.
    a. NaBr
    b. BaI_2
    c. CaO
    d. $PbCl_2$
    e. Mg_3N_2
    f. Fe_2O_3
    g. PbO_2
    h. CsF
17. Name the binary ionic compound formed by combining the following.
    a. Potassium and oxygen
    b. Aluminum and chlorine
    c. Calcium and bromine
    d. Silver and oxygen
    e. Copper (2+ ion) and iodine
    f. Copper (1+ ion) and chlorine
    g. Tin (2+ ion) and bromine
    h. Tin (4+ ion) and bromine
    i. Gallium and sulfur
    j. Barium and iodine
    k. Nickel and nitrogen
18. Name the following ionic compounds.
    a. $CaCO_3$
    b. K_2SO_4
    c. NH_4Cl
    d. $(NH_4)_2CO_3$
    e. $Mg(HCO_3)_2$
    f. $NaC_2H_3O_2$
    g. $ZnHPO_4$
    h. $Zn_3(PO_4)_2$
    i. $KMnO_4$
    j. $NaClO_4$
19. Name the ionic compounds formed by combining the following ions.
    a. Ca^{2+} and HCO_3^-
    b. Li^+ and PO_4^{3-}
    c. Sn^{2+} and Cl^-
    d. Fe^{3+} and OH^-
    e. Mg^{2+} and ClO_3^-
20. Write formulas for the following compounds.
    a. Silver oxide
    b. Potassium chloride
    c. Magnesium hydrogen carbonate
    d. Aluminum dihydrogen phosphate
    e. Sodium phosphate
    f. Barium sulfate
    g. Copper(II) iodide
    h. Lead(IV) sulfide
    i. Copper(I) oxide
    j. Tin(II) nitride
21. The polyatomic anion IO_3^- is called iodate. Name the polyatomic anions in the series: IO_4^-, IO_3^-, IO_2^-, and IO^-.
22. What is the meaning of the Roman numeral found in parentheses in the names of some ionic compounds?
23. Name the compound formed by combining each cation in the left column with each anion in the top row.

	OH^-	SO_4^{2-}	PO_4^{3-}
NH_4^+			
Fe^{2+}			
Al^{3+}			

7.4 Naming binary molecular compounds

24. Classify the following compounds as binary or ternary.

a. H_2O	e. H_2SO_4
b. $HClO$	f. H_2S
c. NH_3	g. CO_2
d. CBr_4	h. CCl_2Br_2

25. Name the following binary molecular compounds.
    a. PCl_5
    b. SO_2
    c. NO
    d. HF (as pure gas)
    e. CBr_4
    f. Cl_2O_7
26. What kinds of elements make up molecular compounds?
27. Name each compound in the series: N_2O, NO, N_2O_3, NO_2, N_2O_4, and N_2O_5.
28. Write the formula of the following molecular compounds.
    a. Nitrogen monoxide
    b. Carbon tetriodide
    c. Phosphorus trichloride
    d. Hydrogen chloride
    e. Chlorine monobromide
    f. Dinitrogen oxide
    g. Sulfur dioxide
    h. Tetraphosphorus decoxide
29. Complete the following table of molecular compounds.

Name	Formula
a. Sulfur tetrafluoride	
b.	NCl_3
c. Boron trichloride	
d. Dinitrogen trioxide	
e.	CBr_4

7.5 Naming acids

30. Classify the following as acids or nonacids.

a. H_2SO_4	c. $NaOH$
b. CH_4	d. $HC_2H_3O_2$

31. Classify the following as oxo- or nonoxoacids.

a. H_3PO_4	d. H_2S
b. HF	e. HBr
c. H_2SO_3	f. HNO_3

32. Name the following acids.
    a. HBr
    b. HF
    c. H_2S
33. Name the following acids.

a. H_2SO_4	e. $HClO_3$
b. H_2SO_3	f. $HClO_4$
c. HNO_3	g. H_2CO_3
d. H_3PO_4	h. $HC_2H_3O_2$

34. Write the formulas of the following acids.
    a. Sulfuric acid
    b. Hydrosulfuric acid
    c. Acetic acid
    d. Hydrobromic acid
    e. Chlorous acid
    f. Hypochlorous acid

7.6 Nomenclature summary

35. Name each of the following substances.

a. K_2CrO_4	f. CO
b. NaF	g. HCl
c. FeO	h. $HgCl_2$
d. Cu_2CO_3	i. $HClO_2$
e. $LiOH$	j. Cl_2O

36. Write a chemical formula for each of the following substances.
    a. Ammonium nitrate
    b. Silver chloride
    c. Sulfur trioxide
    d. Copper(I) sulfate
    e. Hydriodic acid
    f. Nitric acid
    g. Phosphoric acid
    h. Carbon tetrachloride
    i. Magnesium hydrogen carbonate
    j. Potassium permanganate

7.7 Molecular weight and formula weight

37. Calculate the molecular or formula weights of the following compounds.

a. I_2	f. $(NH_4)_3PO_4$
b. NH_3	g. H_2SO_3
c. $(NH_4)_2CO_3$	h. $HClO_4$
d. K_2SO_4	i. $KMnO_4$
e. Cl_2O_7	j. $Ca(HCO_3)_2$

7.8 Percentage composition

38. Calculate the percentage composition of each element in the following.
    a. H_3PO_4 3 + 19 + 64
    b. $Ca(HCO_3)_2$ 86
    c. Potassium chromate
    d. Barium nitrate 40 + 2 + 24 + 96
    e. Na_2CO_3 162
39. What is the percentage composition of silver in silver nitrate?
40. When 100. g of tin(IV) oxide is decomposed, 78.8 g of tin and 21.2 g of oxygen are formed. Calculate the percentages of tin and oxygen in tin(IV) oxide.
41. When 100. g of tin(II) oxide is decomposed, 88.1 g of tin and 11.9 g of oxygen are formed. Calculate the percentages of tin and oxygen in tin(II) oxide.
42. When 5.000 g of vitamin C is decomposed, 2.045 g of carbon, 0.2290 g of hydrogen, and 2.725 g of oxygen are formed. Calculate the percentages of carbon, hydrogen, and oxygen in vitamin C.
43. When 25.00 g of an unknown compound is decomposed, 19.97 g of copper and 5.030 g of oxygen are formed. Calculate the percentages of copper and oxygen in this unknown compound.

Additional problems

44. Name the following compounds.
    a. $Sn_3(PO_4)_2$
    b. PF_3
    c. HBr (gas)
    d. KClO
    e. K_2CO_3
    f. HCl (in water)
45. Write formulas for the following compounds.
    a. Ammonium sulfate
    b. Potassium sulfide
    c. Bromic acid
    d. Perbromic acid
    e. Iron(III) chromate
    f. Mercuric oxide
46. Name the following compounds.
    a. Li_3P
    b. NO_2
    c. N_2O_4
    d. Al_2O_3
    e. CS
    f. CS_2
47. Calculate the molecular or formula weight of each of the following compounds.
    a. H_2O_2
    b. $C_6H_8O_6$
    c. $K_2Cr_2O_7$
    d. Al_2Cl_6
    e. C_8H_{18}
    f. $C_{56}H_{88}O_2$
48. Ethyl alcohol, the alcohol in alcoholic beverages, is a compound of carbon, hydrogen, and oxygen. A 3.00-g sample of ethyl alcohol contains 1.56 g of C and 0.40 g of H. Calculate the mass percentages of C, H, and O in ethyl alcohol.
49. Ordinary table sugar is sucrose and has the molecular formula $C_{12}H_{22}O_{11}$. Calculate the percentages of C, H, and O in sucrose.
50. All of the substances below are fertilizers that contribute nitrogen to the soil. Which of these has the richest source of nitrogen on a mass percentage basis?
    a. Urea, $(NH_4)_2CO$
    b. Guanidine, CN_3H_5
    c. Ammonium nitrate, NH_4NO_3
    d. Ammonia, NH_3
51. A compound containing only C and H is burned in air to form 2.20 g of CO_2 and 0.450 g of H_2O. What are the mass percentages of C and H in this compound?

chapter 8

the mole concept

The mole is the link or bridge between the microscopic world of atoms, ions, molecules, and formula units and the macroscopic world of elements and compounds. Mole quantities of the elements Cu, Fe, C (graphite), and Hg and of the compounds salt, water, and aspirin are shown above.

8.1 mole as a very large "package"

Throughout the previous chapters, you have developed a mental picture of individual atoms, ions, molecules, and formula units. A neutral atom consists of a positively charged nucleus that contains essentially all the mass of the element and that has negatively charged electrons surrounding it to balance the total charge (Figure 8.1a). An **ion** is an atom that has gained electrons to become a negatively charged anion or has lost electrons to become a positively charged cation (Figure 8.1b). A **formula unit** is the smallest unit of an ionic compound with at least one anion and one cation chemically combined in a fixed proportion (Figure 8.1c). A **molecule** is the smallest unit of a molecular compound with two or more atoms chemically combined in a fixed proportion (Figure 8.1d).

You are also aware that atoms, ions, or combinations of atoms or ions, are extremely tiny; you cannot see, let alone hold, any one of them. Samples of matter that you handle are collections of enormous numbers of these basic units. Aluminum foil is a huge collection of Al atoms, a glass of water contains a gigantic number of H_2O molecules, and table salt is a huge collection of equal numbers of Na^+ ions and Cl^- ions.

Keeping track of such enormous numbers of basic chemical units becomes essential when you work with quantities of elements and compounds large enough to be measured in the laboratory. Historically, the relationship between numbers of particles in a sample and its mass was worked out in the early nineteenth century, long before there was any detailed mental picture of the particles of matter. Now that you have a mental picture to relate to each chemical unit, large quantities of such units become easier to comprehend and keep track of in chemical reactions.

Chemists relate number of particles to mass by defining a unit of matter called a *mole* (abbreviated mol), which contains a definite number of particles. Think of a mole as a "package" containing a definite number of particles (atoms, ions, molecules, or formula units), just as a dozen is a package containing 12 items, whether the items are eggs, doughnuts, or apples. The difficulty in understanding the mole is that the number of items is so large (6.022×10^{23} or 602 sextillion) that it is nearly impossible to visualize just how many atoms, ions, molecules, or formula units it contains. But the concept is the same as that of the dozen. Moles are packages of atoms, ions, molecules, or formula units; dozens are packages of various items, such as eggs. You may speak of a mole of copper atoms, a mole of water molecules, or a mole of sodium chloride formula units just as you may speak of a dozen eggs or a dozen table-tennis balls.

Unlike an atom or molecule, you can see a table-tennis ball and you can imagine having a dozen of them on your desk. But a mole of balls, that is, 6.022×10^{23}

Figure 8.1

(a) A neutral atom consists of a positively charged nucleus that contains essentially all of the mass of the element and has negatively charged electrons surrounding it to balance the total charge. (b) An ion is an atom that has gained electrons to become a negatively charged anion or lost electrons to become a positively charged cation. (c) A formula unit is the smallest unit of an ionic compound with at least one anion and one cation chemically combined in a fixed proportion. (d) A molecule is the smallest chemical unit of a molecular compound with two or more atoms chemically combined in a fixed proportion.

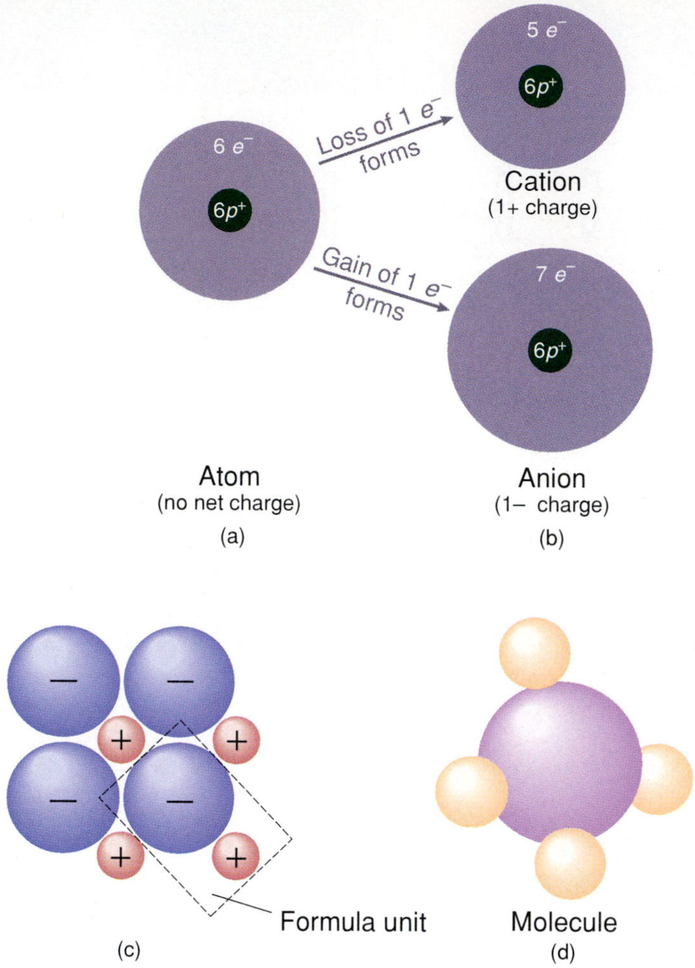

table-tennis balls, would not fit on your desk or even in the Grand Canyon. One mole of table-tennis balls would cover the entire surface of the earth, including the surface of the oceans, to a depth of 28.5 miles!

Perhaps the illustration in Figure 8.2 will help you to understand just how enormous a mole really is. And, conversely, perhaps you can begin to understand just how small an atom or molecule must be when a few grams of an element contains 1 mol of atoms. For example, 64 g of copper, which is the amount of copper found in about 24 pennies, contains 1 mol of copper atoms (Figure 8.3). To imagine the size of a copper atom, you must shrink the volume occupied by 1 mol of table-tennis balls down to the size of about 24 pennies. Then, each ball would be small enough to represent one Cu atom.

The mole concept bridges the gap between the *microscopic* world of atoms, ions, molecules, and formula units, which you cannot see and handle, and the *macroscopic* world of elements and compounds, which you can see and hold. In this chapter, you will see how you can keep track of the numbers of particles in a sample, and furthermore, how numbers of particles are related to the mass of a sample. Knowledge of this relationship is absolutely essential if you use chemical reactions in the laboratory, as discussed in detail in Chapter 9.

Before seeing how the mole concept can be related to the mass of a sample, it might be useful for you to review the concepts of atomic, molecular, and formula weights discussed in Sections 4.9, 4.10, and 7.7.

Figure 8.2

A mole of table-tennis balls would cover the entire surface of the earth, including the surface of the oceans, to a depth of 28.5 miles. The circle on the satellite photo indicates the height above the surface of the earth that 1 mol of table-tennis balls would reach if the balls were evenly distributed.

Figure 8.3

Twenty-four pennies contain 1 mol of copper or 602 sextillion Cu atoms.

8.2 relative masses of atoms

In Section 4.9, you learned that the atomic mass (weight) of the atoms of each element in the periodic table is known relative to the atomic mass of carbon-12 and that the values of atomic mass are given in atomic mass units. The concept of relative masses leads directly to the relationship between numbers of particles, packages of particles, and mass of a sample. As an example, let us choose the elements oxygen and carbon to demonstrate the relationship between numbers of atoms and mass of a sample.

From the periodic table, we see that the average atomic masses of oxygen and carbon are 16.00 amu and 12.01 amu, respectively. As a matter of practicality,[1] we can accept these numbers as the relative masses of individual atoms. The ratio of atomic weights is

$$\frac{1 \text{ atom O}}{1 \text{ atom C}} = \frac{16.00 \cancel{\text{amu}}}{12.01 \cancel{\text{amu}}} = 1.332$$

In Table 8.1, the atomic masses of an O atom and a C atom have each been multiplied by 1, 10, 10^6, and 6.022×10^{23} to obtain the total mass of that many O atoms and that many C atoms. As Table 8.1 shows, the mass ratio for any

1. Section 4.10 carefully explains that average atomic mass is a weighted average of the masses of all isotopes of an element. The weighting accounts for the natural abundance of each isotope. No single atom actually has the average atomic mass. Because you must deal with atoms "on the average," you routinely use the periodic table of average atomic masses for the masses of single atoms.

table 8.1 mass ratios for varying numbers of oxygen and carbon atoms

Atomic ratio		Mass ratio				
$\frac{1 \text{ atom O}}{1 \text{ atom C}}$	$\frac{1 \times 16.00 \text{ amu}}{1 \times 12.01 \text{ amu}}$	=	$\frac{16.00}{12.01}$	=	1.332	= Ratio of atomic masses
$\frac{10 \text{ atoms O}}{10 \text{ atoms C}}$	$\frac{10 \times 16.00 \text{ amu}}{10 \times 12.01 \text{ amu}}$	=	$\frac{160.0}{120.1}$	=	1.332	
$\frac{10^6 \text{ atoms O}}{10^6 \text{ atoms C}}$	$\frac{10^6 \times 16.00 \text{ amu}}{10^6 \times 12.01 \text{ amu}}$	=	$\frac{1.600 \times 10^7}{1.201 \times 10^7}$	=	1.332	
$\frac{6.022 \times 10^{23} \text{ atoms O}}{6.022 \times 10^{23} \text{ atoms C}}$	$\frac{6.022 \times 10^{23} \times 16.00 \text{ amu}}{6.022 \times 10^{23} \times 12.01 \text{ amu}}$	=	$\frac{9.635 \times 10^{24}}{7.232 \times 10^{24}}$	=	1.332	

number of O and C atoms remains the same so long as the number of O atoms equals the number of C atoms. Notice in this table that the units of mass cancel out because they are identical and therefore equal to 1. The *mass ratio* has no units associated with it.

What Table 8.1 establishes is that for equal numbers of atoms, the mass ratio is always the same as the ratio of the atomic masses given in the periodic table. The reverse of this statement is also true and is of great importance: *If the mass ratio of two samples corresponds to the ratio of atomic masses, then the two samples must contain equal numbers of atoms.* This is true of all the examples in Table 8.1 and will always be true, regardless of the units of mass, because identical mass units cancel out.

sample exercise 1

Do 98.07 g of nitrogen atoms and 7.056 g of hydrogen atoms contain equal numbers of atoms?

solution

The ratio of the given masses is

$$\frac{\text{Nitrogen}}{\text{Hydrogen}} = \frac{98.07 \text{ g}}{7.056 \text{ g}} = 13.90$$

From the periodic table, the ratio of the average atomic masses is

$$\frac{14.01 \text{ amu}}{1.008 \text{ amu}} = 13.90$$

Because the ratio of the sample masses equals the ratio of the average atomic masses, the numbers of nitrogen and hydrogen atoms will be equal. You have no information as to how many nitrogen or hydrogen atoms are present, but you can be sure that however many nitrogen atoms are present, there must be an equal number of hydrogen atoms.

8.3 moles of atoms

You are now thoroughly familiar with the concept that the atomic mass of any element represents a "package" of atoms of that element. The number of atoms in that package is the same regardless of the element you choose. So the atomic mass gives you a very convenient way of comparing samples of different elements with the same number of atoms.

Since the unit of mass in the metric system is the gram, it is particularly convenient to consider packages of an element corresponding to the atomic mass of the material in grams, or the **gram atomic weight** (GAW).[2] The GAW for the element carbon is 12.01 g. For the element oxygen, the GAW is 16.00 g. You know from Table 8.1 that 16.00 g/12.01 g = 1.332; therefore, the number of atoms in 12.01 g of carbon must be equal to the number of atoms in 16.00 g of oxygen. Or, in general, *the number of atoms in the GAW of one element is always equal to the number of atoms in the GAW of any other element.*

It turns out that the number of atoms in a GAW of any element is 6.022×10^{23}. Unfortunately, the experiments by which this actual number is determined are too complicated to be described in this text. We ask you to accept this number without proof. However, as a chemist, it is extremely important for you to be able to weigh a sample of an element and to know how many atoms are in it, just as we were able to weigh copper pennies in Figure 8.3 and to know that they contain 1 mol of Cu atoms. The concept that enables you to do this is

1 GAW of any element = 6.022×10^{23} atoms of that element, the same number of atoms as for any other element

This number, 6.022×10^{23}, is called Avogadro's number to honor the Italian physicist Amadeo Avogadro, who in the early nineteenth century recognized the importance of being able to relate numbers of particles to measurable quantities of matter. Avogadro never knew the value of his number, but he did know the principle. You have the advantage of knowing both the principle and the value of his number.

Now you know what the unit or package called the mole contains. Because of the reasoning just detailed, a **mole** is defined as Avogadro's number of constituent units, and this corresponds to easily assignable gram atomic weights. In the following sections, the constituent units of a mole will be extended from just atoms to include molecules and formula units as well. For *atoms,*

$$\textbf{1 mol atoms} = \mathbf{6.022 \times 10^{23}}\ \textbf{atoms} = \textbf{1 GAW}$$

Very similar relationships for moles of molecules and formula units will be discussed in Section 8.5. But for now, it is important that you recognize and apply the powerful conversion factors that the above equations give you for relating moles of atoms to grams of an element and the reverse procedure of converting grams of an element to moles of atoms.

You recall from Section 3.3 that any equality can be made into a fraction equaling 1, called a conversion factor. From the equalities above, you should quickly be able to write the following conversion factors, which you have used in Sample Exercise 8 and Problem 3 in Chapter 3:

$$\frac{\text{1 GAW}}{\text{1 mol atoms}} = \frac{\text{1 mol atoms}}{\text{1 GAW}} = 1$$

Converts from moles to grams	Converts from grams to moles

The following exercise will illustrate the conversion of moles of an element to grams. As a result of the conversion, you should be able to weigh out on a balance any number of moles of an element.

2. Technically, this term should be called gram atomic mass, but for historical reasons it is referred to as gram atomic weight.

Figure 8.4

24.1 g of magnesium contains 1 mol of magnesium, that is 6.022×10^{23} atoms of Mg; 8.02 g of sulfur contains 0.250 mol of sulfur, that is $0.250 \times 6.022 \times 10^{23}$ atoms of S equals 1.51×10^{23} atoms of S. Note that 1 mol of most solid elements corresponds to a volume that will fit comfortably in the palm of your hand.

sample exercise 2

a. How many grams of Mg are there in exactly 1 mol of Mg? Magnesium is a low-density, silver-white metal that burns with a brilliant white light and is used in flares, fireworks, and flash bulbs.

b. How many grams of S are there in exactly 2 mol of S? Sulfur is a yellow, nonmetallic solid used in gunpowder, matches, and medicine.

c. How many grams of S are there in 0.250 mol of S?

solution

a. Consult the periodic table and find that the atomic mass of magnesium is 24.31 amu. Therefore, its GAW is 24.31 g (Figure 8.4).

$$1 \text{ mol Mg} = \text{GAW} = 24.31 \text{ g Mg}$$

b. The GAW of S is 32.06 g. Therefore, by proportional reasoning,

$$1 \text{ mol S} = 32.06 \text{ g S}$$

$$\text{so,} \quad 2 \text{ mol S} = 64.12 \text{ g S}$$

Using the unit conversion method,

Given quantity × Conversion factor = New quantity

$$2 \, \cancel{\text{mol S}} \times \frac{32.06 \text{ g S}}{1 \, \cancel{\text{mol S}}} = 64.12 \text{ g S}$$

Note that the answer, 64.12 g of S, has four significant figures, the same number as given in the GAW of S. If you need exactly 2 mol of S in the laboratory, you would weigh out 64.12 g of S on a balance.

c. Using the unit conversion method and the GAW of S given in part b,

Given quantity × Conversion factor = New quantity

$$0.250 \, \cancel{\text{mol S}} \times \frac{32.06 \text{ g S}}{1 \, \cancel{\text{mol S}}} = 8.02 \text{ g S}$$

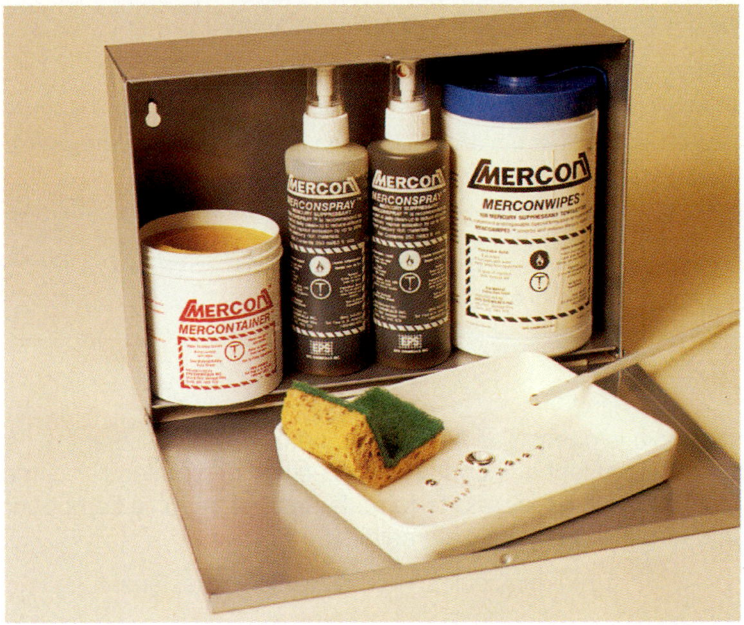

Figure 8.5

Mercury is a liquid metal at room temperature. It is highly toxic and must be cleaned up immediately if spilled from a bottle or lost from a broken thermometer. Kits, like the one shown here, contain sprays, powdered sulfur, and sponges for cleaning up spills of mercury. The disposal of mercury is a toxic environmental hazard, and mercury batteries must be treated as hazardous waste. Some dry cells no longer use mercury to produce electricity.

Note that the answer, 8.02 g of S, has three significant figures because it is limited by the three significant figures in the value of 0.250 mol of S. If you need 0.250 mol of S in the laboratory, you would weigh out 8.02 g of S on a balance (Figure 8.4).

The last exercise in this section will illustrate the reverse process of converting grams of a sample to moles. As a result of the conversion, you know the number of moles of an element from its mass.

sample exercise 3

a. How many moles of Hg are there in 200.6 g of Hg? Mercury is a high-density, liquid metal that is used to measure the pressure of the atmosphere in a barometer and the temperature in a thermometer.
b. How many moles of Hg are there in 454 g of Hg?

solution

a. Consult the periodic table and find that the atomic mass of mercury is 200.6 amu. Therefore, its GAW is 200.6 g.

$$200.6 \text{ g Hg} = 1 \text{ GAW} = 1 \text{ mol Hg}$$

b. The GAW of Hg is 200.6 g. Therefore, by proportional reasoning, using the unit conversion method,

$$\textbf{Given quantity} \times \textbf{Conversion factor} = \textbf{New quantity}$$

$$454 \cancel{\text{ g Hg}} \times \frac{1 \text{ mol Hg}}{200.6 \cancel{\text{ g Hg}}} = 2.26 \text{ mol Hg}$$

Note that the answer, 2.26 mol of Hg, has three significant figures because it is limited by the three significant figures in the mass. If you have a 1-lb (454 g) bottle of mercury, you have 2.26 mol of Hg. Safety precautions must be observed when handling Hg (Figure 8.5).

problem 1

What are the gram atomic weights (GAWs) of the following?

a. Al, a metal commonly used as a foil for cooking and wrapping food
b. P, a solid existing in three different forms (allotropes) and used in forming smoke screens
c. Si, a metalloid solid that is the second most abundant element in the earth's crust, is used in the production of steel, transistors, and glass

problem 2

How many grams are in 3 mol of Al?

8.4 moles of compounds

Because the mole is defined as Avogadro's number of constituent units, you may easily extend the concept of the mole to molecules of a molecular compound or a diatomic element such as N_2 and to formula units of an ionic compound. For molecular compounds and diatomic elements,

$$1 \text{ mol molecules} = 6.022 \times 10^{23} \text{ molecules}$$

and for ionic compounds,

$$1 \text{ mol formula units} = 6.022 \times 10^{23} \text{ formula units}$$

Just as 1 mol of atoms = Avogadro's number of atoms = 1 GAW for elements, it can be seen that for molecular compounds,

$$1 \text{ mol molecules} = 6.022 \times 10^{23} \text{ molecules} = 1 \text{ GMW}$$

and for ionic compounds,

$$1 \text{ mol formula units} = 6.022 \times 10^{23} \text{ formula units} = 1 \text{ GFW}$$

where GMW, the **gram molecular weight**, is the molecular weight expressed in grams, and the GFW, the **gram formula weight**, is the formula weight expressed in grams.

In summary, you may relate moles and grams for any material. For elements, you must look up the gram atomic weight; for molecular compounds and diatomic elements, you must calculate the gram molecular weight; for ionic compounds, you must calculate the gram formula weight (Section 7.7). The number of grams in 1 GAW, 1 GMW, or 1 GFW is equal to 1 mol.

Elements:	1 mol atoms = 1 GAW
Molecular compounds and diatomic elements:	1 mol molecules = 1 GMW
Ionic compounds:	1 mol formula units = 1 GFW

Before using the above relationships in converting moles of molecules or formula units to grams of a compound, you must recall what you learned about chemical formulas. Chemical formulas identify matter and indicate the number of atoms or ions of each element contained in the compound. Throughout Chapters 6 and 7, and especially in Sections 6.8, 6.13, and 7.7, you used chemical formulas and subscripts within the formula to specify the number of each type of atom (or ion) in a molecular compound (or ionic compound).

For example, the subscripts in its formula tell you that the molecule HNO_3 is made up of one H atom, one N atom, and three O atoms. In an ionic compound such as Al_2S_3, the subscripts of the formula tell you that there are two Al^{3+} ions and three S^{2-} ions in one formula unit. In $Mg(OH)_2$, the subscripts indicate that one formula unit contains one Mg^{2+} ion and two OH^- ions. You also know that the 2 outside the parentheses in $Mg(OH)_2$ is a multiplier for both O and H atoms inside the parentheses.

Figure 8.6

NO_2, nitrogen dioxide, is a reddish-brown gas that gives photochemical smog its characteristic color.

sample exercise 4

How many grams are in 1 mol of each of the following?

a. NO_2, nitrogen dioxide, the reddish-brown gas that gives smog its characteristic color (Figure 8.6)
b. Na_2CO_3, sodium carbonate, commonly called washing soda, a general cleanser
c. O, atomic oxygen, a highly reactive species that cannot be isolated; it may react with molecular oxygen, O_2, to form ozone, O_3, another pure form of elemental oxygen
d. O_2, molecular oxygen, the stable, gaseous form of the element oxygen at 25°C and 1 atm of pressure

solution

a. To find the molecular weight (Section 7.7), sum the atomic masses of all atoms present. For NO_2, this is

$$\begin{aligned} 1 \times \text{N} &= 1 \times 14.01 \text{ amu} = 14.01 \text{ amu} \\ 2 \times \text{O} &= 2 \times 16.00 \text{ amu} = \underline{32.00 \text{ amu}} \\ & \qquad\qquad\qquad\qquad 46.01 \text{ amu} \end{aligned}$$

The molecular weight of NO_2 is 46.01 amu. The molecular weight in grams, or the gram molecular weight (GMW), is 46.01 g.

$$1 \text{ mol } NO_2 = 46.01 \text{ g } NO_2$$

To obtain 1 mol of NO_2, you would fill a container with 46.01 g of NO_2 gas.

b. To find the formula weight (Section 7.7), sum the masses of all ions present in one formula unit. For Na_2CO_3, this is

$$\begin{aligned} 2 \times \text{Na}^+ &= 2 \times 22.99 \text{ amu} = 45.98 \text{ amu} \\ 1 \times \text{CO}_3^{2-} &= 1 \times 60.01 \text{ amu} = \underline{60.01 \text{ amu}} \\ & \qquad\qquad\qquad\qquad 105.99 \text{ amu} \end{aligned}$$

The formula weight of Na_2CO_3 is 105.99 amu. The formula weight in grams, or the gram formula weight (GFW), is 105.99 g.

$$1 \text{ mol } Na_2CO_3 = 105.99 \text{ g } Na_2CO_3$$

To obtain 1 mol of Na_2CO_3, you must weigh out 105.99 g of solid Na_2CO_3 on a balance.

c. The atomic mass of oxygen is 16.00 amu. The atomic mass in grams, or the gram atomic weight (GAW), is 16.00 g.

$$1 \text{ mol O} = 16.00 \text{ g O}$$

O atoms are not stable, but may be formed in gas discharge tubes. If you have 1 mol of O atoms, they will weigh 16.00 g.

d. The molecular weight of oxygen is

$$2 \times \text{O} = 2 \times 16.00 \text{ amu} = 32.00 \text{ amu}$$

The molecular weight in grams, or the gram molecular weight (GMW), is 32.00 g.

$$1 \text{ mol } O_2 = 32.00 \text{ g}$$

To obtain 1 mol of O_2, you would fill a container with 32.00 g of O_2 gas.

• • • problem 3

How many grams are there in 1 mol of $CaCl_2$ (a drying agent or desiccant)?

8.5 gram-to-mole conversions for compounds

The conversion factors that you will use most often in chemistry will be those relating moles of atoms, molecules, or formula units to grams of an element or compound through the GAW, GMW, or GFW. These conversion factors will be particularly useful in Chapter 10 when you learn how to determine how many moles or grams of products will be produced in a chemical reaction.

Because the GAW, GMW, and GFW all refer to the mass of 1 mol of material, we can use the more general term **molar mass** in referring to any of the three. The molar mass of a monatomic element is the element's GAW; the molar mass of a molecular compound is its GMW; and the molar mass of an ionic compound is its GFW.

$$\text{Molar mass} \begin{array}{l} = \text{GAW} \\ = \text{GMW} \\ = \text{GFW} \end{array}$$

Frequently we will use the term *molar mass* instead of GAW, GMW, or GFW. The molar mass is the mass of 1 mol of anything. Figures 8.3 and 8.4 illustrate the molar mass of several common chemical substances.

In Section 8.3, you constructed conversion factors that enabled you to convert grams of *atoms* to moles of *atoms* and to convert moles of *atoms* to the corresponding mass in grams. In this section, you will extend that concept to construct conversion factors that enable you to convert grams of a compound to moles of *molecules or formula units* and the reverse conversion of moles of *molecules or formula units* to grams of the compound.

$$\frac{1 \text{ GMW}}{1 \text{ mol molecules}} \quad = \quad \frac{1 \text{ mol molecules}}{1 \text{ GMW}} \quad = 1$$

$$\frac{1 \text{ GFW}}{1 \text{ mol formula units}} \quad = \quad \frac{1 \text{ mol formula units}}{1 \text{ GFW}} \quad = 1$$

Converts from moles to grams   Converts from grams to moles

The following exercise will illustrate the conversion of moles of molecules or formula units to grams of compound. As a result of these conversions, you should be able to weigh out on a balance any number of moles of a compound.

sample exercise 5

How many grams are in each of the following samples?

a. 0.390 mol of O_3, ozone, a highly reactive form (allotrope) of gaseous oxygen with a pungent odor; it is used to purify drinking water and deodorize air and sewage gases. Ozone in the upper atmosphere protects life on earth by absorbing harmful ultraviolet rays from the sun before they reach the surface of the earth.

b. 2.50 mol of $Ca_3(PO_4)_2$, a solid occurring naturally in some rocks and animal bones; it is the primary source of elemental phosphorus.

solution

In both parts of this exercise, you will use the unit conversion method and one or more of the conversion factors resulting from the equality

$$1 \text{ mol} = 1 \text{ molar mass}$$

a. Step 1. The given quantity and unit is 0.390 mol of O_3.
Step 2. The new unit is grams of O_3.
Step 3. The equality relating the given and new units is

$$1 \text{ mol} = 1 \text{ molar mass}$$

$$1 \text{ mol } O_3 = 48.00 \text{ g } O_3$$

Step 4. Given quantity and unit × Conversion factor = New quantity and unit

$$\frac{(\text{new units})}{(\text{given units})}$$

Therefore,

$$0.390 \cancel{\text{mol } O_3} \times \frac{48.00 \text{ g } O_3}{1 \cancel{\text{mol } O_3}} = 18.7 \text{ g } O_3$$

b. Step 1. The given quantity and unit is 2.50 mol of $Ca_3(PO_4)_2$.
Step 2. The new unit is grams of $Ca_3(PO_4)_2$.
Step 3. The equality relating the given and new units is

$$1 \text{ mol} = 1 \text{ molar mass}$$

$$1 \text{ mol } Ca_3(PO_4)_2 = 310.18 \text{ g } Ca_3(PO_4)_2$$

Step 4. Given quantity and unit × Conversion factor = New quantity and unit

$$\frac{(\text{new units})}{(\text{given units})}$$

Therefore,

$$2.50 \cancel{\text{mol } Ca_3(PO_4)_2} \times \frac{310.18 \text{ g } Ca_3(PO_4)_2}{1 \cancel{\text{mol } Ca_3(PO_4)_2}} = 775.45 \text{ g } Ca_3(PO_4)_2 = 775 \text{ g } Ca_3(PO_4)_2$$

(3 sig figs)

To obtain 2.50 mol of $Ca_3(PO_4)_2$, you must weigh out 775 g of solid $Ca_3(PO_4)_2$ on a balance.

The following exercise will illustrate the conversion of grams of a compound to moles of molecules or formula units. As a result of these conversions, you should be able to determine how many moles of a compound are in any given sample.

sample exercise 6

a. How many moles correspond to 50.5 g of H_2S, hydrogen sulfide (a poisonous gas that has the odor of rotten eggs and is a natural component of raw petroleum; Figure 8.7)?

b. 14.9 g of $Mg(OH)_2$ is equivalent to how many moles of $Mg(OH)_2$ (milk of magnesia)?

Figure 8.7
The distinct odor of rotten eggs is occasionally detected when raw petroleum products are being transferred. The odor is H_2S, a naturally occurring component of raw petroleum.

solution

In both parts of this exercise, we will be using the unit conversion method and one or more of the conversion factors resulting from the equality

$$1 \text{ mol} = \text{molar mass}$$

a. Step 1. The given quantity and unit is 50.5 g of H_2S.
   Step 2. The new unit is moles of H_2S.
   Step 3. The equality relating the given and new units is

$$1 \text{ mol} = 1 \text{ molar mass}$$

$$1 \text{ mol } H_2S = 34.1 \text{ g } H_2S$$

   Step 4. Given quantity and unit × Conversion factor = New quantity and unit

$$\frac{(\text{new units})}{(\text{given units})}$$

   Therefore,

$$50.5 \cancel{\text{g } H_2S} \times \frac{1 \text{ mol } H_2S}{34.1 \cancel{\text{g } H_2S}} = 1.48 \text{ mol } H_2S$$

   A 50.5 g sample of H_2S will contain 1.48 mol of H_2S.

b. Step 1. The given quantity and unit is 14.9 g of $Mg(OH)_2$.
   Step 2. The new unit is moles of $Mg(OH)_2$.
   Step 3. The equality relating the given and new units is

$$1 \text{ mol} = 1 \text{ molar mass}$$

$$1 \text{ mol } Mg(OH)_2 = 58.33 \text{ g } Mg(OH)_2$$

Step 4. Given quantity and unit × Conversion factor = New quantity and unit

$$\frac{\text{(new units)}}{\text{(given units)}}$$

Therefore,

$$14.9\ \cancel{\text{g Mg(OH)}_2} \times \frac{1\ \text{mol Mg(OH)}_2}{58.33\ \cancel{\text{g Mg(OH)}_2}} = 0.255\ \text{mol Mg(OH)}_2$$

A 14.9 g sample of $Mg(OH)_2$ will contain 0.255 mol of $Mg(OH)_2$.

---

problem 4

How many moles of lithium sulfide correspond to 75.8 g of lithium sulfide?

8.6 moles of atoms and ions within moles of compounds

In Section 8.4, you reviewed the meaning of chemical formulas and how they indicate the number of atoms or ions of each element contained in the compound. For example, one molecule of HNO_3 contains one atom of hydrogen, one atom of nitrogen, and three atoms of oxygen. Similarly, one formula unit of $Mg(OH)_2$ contains one Mg^{2+} ion and two OH^- ions.

To extend the concept of moles to atoms within molecules or ions within formula units, it is first necessary to construct conversion factors relating the numbers of atoms or ions within a molecule or formula unit to the whole molecule or formula unit. You can use your knowledge of subscripts in chemical formulas to write the conversion factors for HNO_3.

$$\frac{1\ \text{H atom}}{1\ \text{HNO}_3\ \text{molecule}} \qquad \frac{1\ \text{N atom}}{1\ \text{HNO}_3\ \text{molecule}} \qquad \frac{3\ \text{O atoms}}{1\ \text{HNO}_3\ \text{molecule}}$$

Similarly, for the formula unit $Mg(OH)_2$, you can write

$$\frac{1\ \text{Mg}^{2+}\ \text{ion}}{1\ \text{Mg(OH)}_2\ \text{formula unit}} \qquad \frac{2\ \text{OH}^-\ \text{ions}}{1\ \text{Mg(OH)}_2\ \text{formula unit}}$$

---

sample exercise 7

Write the conversion factors relating numbers of atoms within a molecule or numbers of ions within a formula unit for the following.

a. N_2O_5

b. $CaCl_2$

c. Na_2SO_4

d. $Mg_3(PO_4)_2$

solution

a. This is a molecular compound because N and O are both nonmetals; therefore,

$$\frac{2\ \text{N atoms}}{1\ \text{N}_2\text{O}_5\ \text{molecule}} \qquad \frac{5\ \text{O atoms}}{1\ \text{N}_2\text{O}_5\ \text{molecule}}$$

The compounds in parts b, c, and d are all ionic because they are combinations of metals (Ca, Na, and Mg) and nonmetals.

b. $$\frac{1\ \text{Ca}^{2+}\ \text{ion}}{1\ \text{CaCl}_2\ \text{formula unit}} \qquad \frac{2\ \text{Cl}^-\ \text{ions}}{1\ \text{CaCl}_2\ \text{formula unit}}$$

c. $$\frac{2\ \text{Na}^+\ \text{ions}}{1\ \text{Na}_2\text{SO}_4\ \text{formula unit}} \qquad \frac{1\ \text{SO}_4^{2-}\ \text{ion}}{1\ \text{Na}_2\text{SO}_4\ \text{formula unit}}$$

d. $$\frac{3\ \text{Mg}^{2+}\ \text{ions}}{1\ \text{Mg}_3(\text{PO}_4)_2\ \text{formula unit}} \qquad \frac{2\ \text{PO}_4^{3-}\ \text{ions}}{1\ \text{Mg}_3(\text{PO}_4)_2\ \text{formula unit}}$$

When a coefficient is placed before a chemical formula, it multiplies the entire formula. For example, 2 N_2O_5 means two molecules of N_2O_5; therefore, four N atoms and ten O atoms must be present:

$$2\ \cancel{N_2O_5\ \text{molecules}} \times \frac{2\ \text{N atoms}}{1\ \cancel{N_2O_5\ \text{molecule}}} = 4\ \text{N atoms}$$

$$2\ \cancel{N_2O_5\ \text{molecules}} \times \frac{5\ \text{O atoms}}{1\ \cancel{N_2O_5\ \text{molecule}}} = 10\ \text{O atoms}$$

Similarly, 3 Na_2SO_4 means three formula units of Na_2SO_4, and therefore a total of six Na^+ ions and three SO_4^{2-} ions are present:

$$3\ \cancel{Na_2SO_4\ \text{formula units}} \times \frac{2\ Na^+\ \text{ions}}{1\ \cancel{Na_2SO_4\ \text{formula unit}}} = 6\ Na^+\ \text{ions}$$

$$3\ \cancel{Na_2SO_4\ \text{formula units}} \times \frac{1\ SO_4^{2-}\ \text{ion}}{1\ \cancel{Na_2SO_4\ \text{formula unit}}} = 3\ SO_4^{2-}\ \text{ions}$$

The ideas introduced in this section, namely, the meaning of coefficients and the construction of conversion factors relating the number of constituent atoms or ions to their molecules or formula units will be used repeatedly.

••• problem 5

How many PO_4^{3-} ions are represented by 4 $Mg_3(PO_4)_2$ (a compound used in dental cement)?

When you write 1 mol of NaCl, this means 6.022×10^{23} formula units of NaCl, and hence 6.022×10^{23} Na^+ ions and 6.022×10^{23} Cl^- ions.

$$6.022 \times 10^{23}\ \text{formula units of NaCl contains} \begin{cases} 6.022 \times 10^{23}\ Na^+\ \text{ions} \\ \text{and} \\ 6.022 \times 10^{23}\ Cl^-\ \text{ions} \end{cases}$$

So, you see that 1 mol of NaCl contains 1 mol of Na^+ ions (6.022×10^{23} ions) and 1 mol of Cl^- ions (6.022×10^{23} ions).

$$1\ \text{mol NaCl contains} \quad \begin{array}{l} 1\ \text{mol}\ Na^+\ \text{ions} \\ 1\ \text{mol}\ Cl^-\ \text{ions} \\ \hline 2\ \text{mol ions} \end{array}$$

Some students have difficulty visualizing 2 mol of particles coming from 1 mol. Remembering that the mole concept is like the dozen concept might be helpful in this case. In one dozen eggs there are one dozen yolks and one dozen whites. So we see two dozen subunits (one dozen yolks and one dozen whites) within one dozen of the whole eggs. As another example, in 1 mol of married couples, there are 1 mol of males and 1 mol of females, that is, 2 mol of human beings within 1 mol of couples.

Just as the subscripts in a chemical formula tell you the numbers of atoms or ions within a molecule or formula unit, the *subscripts also tell you the numbers of moles of atoms or ions contained within 1 mol of the compound.* For NaCl, the unexpressed subscripts are both 1, so 1 mol of NaCl holds within it 1 mol of Na^+ ions and 1 mol of Cl^- ions. In 1 mol of Na_2SO_4, there are 2 mol of Na^+ ions (because the subscript is 2) and 1 mol of SO_4^{2-} ions. In 1 mol of N_2O_5, there are 2 mol of N atoms and 5 mol of O atoms.

Conversion factors can be written that relate the moles of the components of a compound to the moles of the compound itself. For example, for $Mg(OH)_2$,

$$\frac{1 \text{ mol } Mg^{2+}}{1 \text{ mol } Mg(OH)_2} \quad \text{and} \quad \frac{2 \text{ mol } OH^-}{1 \text{ mol } Mg(OH)_2}$$

are two possible conversion factors. Using these conversion factors, you can calculate the number of moles of ions in any number of moles of formula units.

When an ionic compound dissolves completely in water, the solution formed consists entirely of ions. For example, when 1 mol of solid sodium hydroxide, NaOH, dissolves in water, it produces 1 mol of Na^+ ions and 1 mol of OH^- ions. In Chapter 13, you will use the conversion factors you have just constructed to determine the number of moles of ions in aqueous solutions.

---

sample exercise 8

How many moles of Li^+ ions and CrO_4^{2-} ions are there in 3.75 mol of Li_2CrO_4 (a corrosion inhibitor used in water-cooled nuclear reactors)?

solution

Given quantity and unit × Conversion factor = Number of moles of each ion

$$3.75 \cancel{\text{ mol } Li_2CrO_4} \times \frac{2 \text{ mol } Li^+}{1 \cancel{\text{ mol } Li_2CrO_4}} = 7.50 \text{ mol } Li^+$$

$$3.75 \cancel{\text{ mol } Li_2CrO_4} \times \frac{1 \text{ mol } CrO_4^{2-}}{1 \cancel{\text{ mol } Li_2CrO_4}} = 3.75 \text{ mol } CrO_4^{2-}$$

---

sample exercise 9

How many moles of nitrogen atoms and oxygen atoms are contained in 3.00 mol of N_2O_4 (a compound that has been used as a rocket propellant in the U.S. space program)?

solution

Given quantity and unit × Conversion factor = Number of moles of each atom

$$3.00 \cancel{\text{ mol } N_2O_4} \times \frac{2 \text{ mol N atoms}}{1 \cancel{\text{ mol } N_2O_4}} = 6.00 \text{ mol N atoms}$$

$$3.00 \cancel{\text{ mol } N_2O_4} \times \frac{4 \text{ mol O atoms}}{1 \cancel{\text{ mol } N_2O_4}} = 12.0 \text{ mol O atoms}$$

Of course, as you already saw in Chapter 6 (especially Section 6.13), there are no *free* N or O atoms in N_2O_4. In this example, all 6.00 mol of N atoms and 12.0 mol of O atoms are bonded together to form 3.00 mol of N_2O_4 molecules.

••• problem 6

How many moles of K^+ ions and PO_4^{3-} ions are there in 4.25 mol of K_3PO_4 (a compound used as a farm fertilizer)?

8.7 empirical formulas represent the simplest formulas for compounds

There are other kinds of chemical formulas besides the molecular or ionic formulas that we have been discussing. The simplest formula of all, and the one that is obtained directly from experimental data, is the **empirical formula.** In science, *empirical* is just another word for *experimental.*

The empirical formula of a compound is one in which the subscripts are in the form of the simplest whole-number ratio. For example, for the compound dinitrogen tetroxide, for which the molecular formula is N_2O_4, the empirical formula is NO_2, because NO_2 represents the *simplest* whole-number ratio of atoms. In N_2O_4, the ratio 2:4 can be reduced to the simpler ratio 1:2. Dinitrogen tetroxide only contains molecules in which two nitrogen and four oxygen atoms are bonded together. However, the simplest ratio of the bonding atoms, called the empirical formula, is 1:2. Similarly, the compound benzene has the molecular formula C_6H_6, but the empirical formula would be CH because the ratio 6:6 can be reduced to 1:1. Molecular and empirical formulas are often identical. For example, for the compound carbon dioxide, for which the molecular formula is CO_2, the empirical formula is also CO_2, since the ratio 1:2 cannot be reduced further. For ionic compounds, the ionic formula is almost always identical to the empirical formula. Consider, for example, Na_2SO_4, Li_2S, and Al_2O_3.

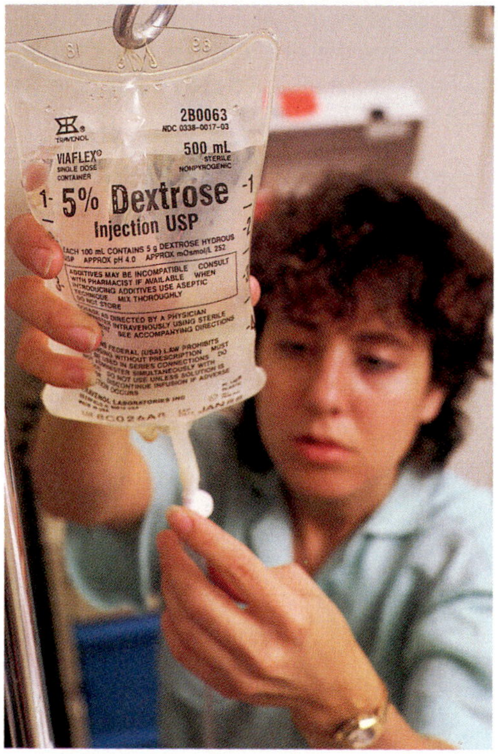

Figure 8.8
Glucose solution (also called dextrose) is used medicinally for intravenous feeding.

sample exercise 10

Give the empirical formulas for each of the following compounds, which are represented by their molecular formulas.

a. Glucose, $C_6H_{12}O_6$, a sugar found in honey, the juices of many fruits, and animal blood; it is used commercially as a sweetening agent and medicinally for intravenous feeding (Figure 8.8)
b. Chloroform, $CHCl_3$, a liquid formerly used as an anesthetic
c. Ethylene glycol, $C_2H_6O_2$, the most widely used automobile antifreeze

solution

a. 6:12:6 can be reduced to 1:2:1 by dividing through by 6. Therefore, the empirical formula is CH_2O.
b. $CHCl_3$ cannot be further simplified. $CHCl_3$ is both the molecular and empirical formula.
c. 2:6:2 can be reduced to 1:3:1 by dividing through by 2. Therefore, the empirical formula is CH_3O.

Remember that the subscripts in a formula tell you not only the ratio of atoms in a molecule (or ions in a formula unit), but also the ratio of the moles of the constituent components of a compound. For example, the ratio 1:2 in CO_2 represents the ratio of one carbon atom to two oxygen atoms in one CO_2 molecule and also, the ratio of 1 mol of carbon to 2 mol of oxygen in 1 mol of CO_2.

sample exercise 11

Give the molar ratio of the elements in the following compounds.

a. NH_3, ammonia, a gas used in water solution as a cleaning agent
b. CaF_2, calcium fluoride, a white solid used as a decay preventive in toothpaste
c. AlN, aluminum nitride, a bluish-white solid used in semiconductor electronics and in the manufacturing of steel

solution

a. The subscripts are in the ratio 1 : 3; therefore, the molar ratio is 1 mol of N to 3 mol of H.
b. 1 mol of Ca to 2 mol of F
c. 1 mol of Al to 1 mol of N

8.8 calculation of empirical formulas

In Section 7.8, you calculated the percentage composition of elements in a compound. Now that you also understand moles of atoms and the ratio with which they combine in empirical formulas, you can understand how chemists use experimental data obtained in the laboratory to determine correct chemical formulas for compounds.

The number of grams of each element that combine into a compound can be measured in an experiment, and that information may be reported directly or converted to percentage composition. Once you know the number of grams of each element, the molar mass provides you with the conversion factor to calculate the corresponding number of moles. The simplest whole-number ratio of moles of the elements in the compound defines the *empirical formula,* that is, the formula in which each element has the smallest possible value for its subscript. In this section, a detailed, stepwise procedure will help you determine empirical formulas.

Calculation of an empirical formula from a given percentage composition by mass

1. Calculate the number of grams of each element present in some number of grams (usually choose exactly 100 g) of the compound.

$$\begin{matrix}\text{Percentage of element} \\ \text{(in decimal form)}\end{matrix} \times \text{Grams of compound} = \text{Grams of element}$$

   (If 100 g of compound is chosen, then the percentage of an element equals the grams of the element.)

2. Calculate the number of moles of each element in the compound.

$$\text{Grams of element} \times \frac{1 \text{ mol element}}{\text{Molar mass}} = \text{Moles of element}$$

   This will establish a molar ratio of the elements in the compound.

3. Simplify the molar ratio by dividing through by the smallest number.
4. Round off the simplified ratio to 2 sig figs.
5. If the ratio turns out to be all whole numbers to 2 sig figs, you have the answer. If not, proceed to step 6.
6. If the ratio is not integral to 2 sig figs, find a number that, when multiplied by the simplified experimental ratio, converts that ratio to a ratio of whole numbers to 2 sig figs.

Carefully follow through Sample Exercises 12 and 13 which apply this stepwise procedure.

sample exercise 12

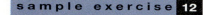

A compound used in soaps, detergents, and glass is found to contain 32.38% Na, 22.58% S, and 45.07% O. What is the empirical formula of the compound?

solution

Follow the procedure just described for the calculation of empirical formulas.

Step 1. Choose 100 g sample of the compound to simplify the calculation:

	Percentage of element (In decimal form)	×	Grams of compound	=	Grams of element
Na:	0.3238	×	100	=	32.38 g Na
S:	0.2258	×	100	=	22.58 g S
O:	0.4507	×	100	=	45.07 g O

Notice that the gram amounts equal the percentages given. This is because we chose 100 g of compound.

Step 2. Convert grams of each element to moles of the element.

$$\text{Na:}\quad 32.38\,\cancel{\text{g Na}} \times \frac{1\text{ mol Na}}{22.99\,\cancel{\text{g Na}}} = 1.408\text{ mol Na}$$

$$\text{S:}\quad 22.58\,\cancel{\text{g S}} \times \frac{1\text{ mol S}}{32.06\,\cancel{\text{g S}}} = 0.7043\text{ mol S}$$

$$\text{O:}\quad 45.07\,\cancel{\text{g O}} \times \frac{1\text{ mol O}}{16.00\,\cancel{\text{g O}}} = 2.817\text{ mol O}$$

Step 3. The molar ratio is

$$\text{Na} : \text{S} : \text{O} = 1.408 : 0.7043 : 2.817$$

Simplify by dividing through by the smallest number, which in this case is 0.7043.

$$\frac{1.408}{0.7043} = 1.999 \qquad \frac{0.7043}{0.7043} = 1.000 \qquad \frac{2.817}{0.7043} = 4.000$$

Now

$$\text{Na} : \text{S} : \text{O} = 1.999 : 1.000 : 4.000$$

Step 4. Rounding off the ratio to 2 sig figs, we get

$$\text{Na} : \text{S} : \text{O} = 2.0 : 1.0 : 4.0$$

Step 5. The numbers of the ratio are all whole numbers, so we have the subscripts for the empirical formula: Na_2SO_4.

sample exercise 13

A compound that coats magnetic recording tape is found to contain 69.94% Fe and 30.06% O. Determine its empirical formula.

solution

Step 1. In 100 g of compound, there are

$$0.6994 \times 100 = 69.94\text{ g Fe}$$
$$0.3006 \times 100 = 30.06\text{ g O}$$

Step 2. The numbers of moles of elements in 100 g of the compound are

$$69.94\ \cancel{\text{g Fe}} \times \frac{1\ \text{mol Fe}}{55.85\ \cancel{\text{g Fe}}} = 1.252\ \text{mol Fe}$$

$$30.06\ \cancel{\text{g O}} \times \frac{1\ \text{mol O}}{16.00\ \cancel{\text{g O}}} = 1.879\ \text{mol O}$$

Step 3. The molar ratio is Fe : O = 1.252 : 1.879, which is simplified by division by 1.252 to give Fe:O = 1.000 : 1.501.

Step 4. Rounding off to 2 sig figs, we get Fe : O = 1.0 : 1.5.

Step 5. The numbers of the ratio are not whole numbers, so we must proceed to step 6.

Step 6. If we multiply the simplified experimental ratio by 2, then we will have achieved a whole-number ratio to 2 sig figs:

$$2(1.000 : 1.501) = 2.000 : 3.002 = 2.0 : 3.0$$

Therefore, the subscripts are 2 and 3, and the empirical formula is Fe_2O_3.

Very often the molar ratio is integral and we can stop at step 5 (e.g., Sample Exercise 12). When step 6 is necessary (e.g., Sample Exercise 13), the multiplier is usually 2, 3, or 4. Just keep multiplying until whole numbers are achieved.

---

sample exercise 14

The molar ratio of elements in a compound is found to be 1.00 : 1.33. Convert this to a whole-number ratio.

solution

Apply steps 4, 5, and 6.

Step 4. Round off the ratio to 2 sig figs: 1.0 : 1.3.

Step 5. Because the ratio is not composed of whole numbers, you must proceed to step 6.

Step 6. Try 2 as a multiplier.

$$2(1.00 : 1.33) = 2.000 : 2.666 = 2.0 : 2.7$$

Obviously, multiplication by 2 does not achieve a whole-number ratio to 2 sig figs. Try 3 as a multiplier.

$$3(1.00 : 1.33) = 3.00 : 3.99 = 3.0 : 4.0$$

This time we have achieved a whole-number ratio to 2 sig figs.

••• problem 7

Calculate the empirical formulas for compounds with the following percentage compositions.

a. 52.4% K and 47.6% Cl
b. 37.2% C, 7.82% H, and 55.0% Cl

Sometimes experimental data about the elemental makeup of a compound are given in a form other than percentage composition. For example, you might be told that a compound is made up of exactly 17.9 g of C and 4.50 g of H. These types of data simplify the calculation of the empirical formula. In this case, you are given the grams of each element in the compound and can proceed directly to the calculation of the molar ratio.

$$17.9 \cancel{\text{g C}} \times \frac{1 \text{ mol C}}{12.0 \cancel{\text{g C}}} = 1.49 \text{ mol C}$$

$$4.50 \cancel{\text{g H}} \times \frac{1 \text{ mol H}}{1.01 \cancel{\text{g H}}} = 4.46 \text{ mol H}$$

$$\text{C} : \text{H} = 1.49 : 4.46 = 1.00 : 2.99$$

To 2 sig figs, this is 1.0 : 3.0, and the empirical formula is CH_3.

problem 8

In the laboratory, 56.00 g of oxygen is found to combine with 3.528 g of hydrogen to form a compound. The compound is found to have properties that are distinct from those of water. What is the empirical formula of this unknown compound?

8.9 molecular formulas

You have seen in Section 8.7 that the difference between a molecular formula and an empirical formula is that the molecular formula represents the actual number of atoms in one molecule whereas the empirical formula gives only the smallest whole-number ratio of the atoms in one molecule. Sample Exercise 10 demonstrated that the relationship between the subscripts of a molecular formula and the subscripts of an empirical formula was given by some whole-number (integral) multiplier.

Molecular formula = Whole number × Empirical formula

Experimental percentage-composition data enable us to calculate an empirical formula. To determine the appropriate whole-number multiplier and hence the molecular formula from the empirical formula, we also need to know the experimental molecular weight of the compound.

Guidelines for calculating molecular formula from empirical formula

1. Calculate the **empirical weight,** the mass of the empirical formula in grams.
2. Divide the molecular weight by the empirical weight to find the integral multiple relating them.
3. Multiply the subscripts of the empirical formula by the multiplier determined in step 2 (Table 8.2).

sample exercise 15

A compound of hydrogen and oxygen that is used as an antiseptic and bleaching agent has an empirical formula HO and a molecular weight of 34. What is the molecular formula?

solution

Step 1. Since the empirical formula is HO, the empirical weight is 17 (1.0 for H and 16 for O).

table 8.2 empirical and molecular formulas

Name of substance	Empirical formula	Empirical weight	Molecular weight	$\frac{\text{Molecular weight}}{\text{Empirical weight}}$ = Integral multiple	Molecular formula
Benzene	CH	13	78	6	C_6H_6
Hydrogen peroxide	HO	17	34	2	H_2O_2
Dinitrogen tetroxide	NO_2	46	92	2	N_2O_4
Propylene	CH_2	14	42	3	C_3H_6
Sulfur dioxide	SO_2	64	64	1	SO_2

Step 2. $\frac{\text{Molecular weight}}{\text{Empirical weight}} = \frac{34}{17} = 2$

Step 3. Multiply each subscript in the empirical formula by the integral multiple to obtain the molecular formula.

$$\text{Molecular formula} = H_{(1\times2)}O_{(1\times2)} = H_2O_2$$

If you are asked to find both the empirical and molecular formulas, remember always to find the empirical formula first. This is illustrated in the following sample exercise.

---

sample exercise 16

Acetylene, a compound used in welding, has a percentage composition of 92.24% C and 7.742% H and a molecular weight of 26.04. What is the molecular formula of acetylene?

solution

First determine the empirical formula:

Step 1. In 100.0 g of acetylene, there are 92.24 g of C and 7.742 g of H.

Step 2. Determine the number of moles of C and H:

$$92.24\ \cancel{\text{g C}} \times \frac{1\text{ mol C}}{12.01\ \cancel{\text{g C}}} = 7.680\text{ mol C}$$

$$7.742\ \cancel{\text{g H}} \times \frac{1\text{ mol H}}{1.008\ \cancel{\text{g H}}} = 7.681\text{ mol H}$$

Step 3. The molar ratio is C : H = 7.680 : 7681, which is simplified to C : H = 1.0 : 1.0. Thus, the empirical formula is CH.

Now use the empirical formula to find the molecular formula:

Step 1. Calculate the empirical weight for CH:

$$12.01 + 1.008 = 13.02$$

Step 2. $\frac{\text{Molecular weight}}{\text{Empirical weight}} = \frac{26.04}{13.02} = 2$

Step 3. Thus, the molecular formula is $C_{(1\times2)}H_{(1\times2)}$ or C_2H_2.

CHEMLAB

Laboratory exercise 8: empirical formula of an oxide of tin

In this experiment, you will form an oxide of tin from the chemical reaction of tin with nitric acid. You will determine the gram amounts of tin and oxygen that combine to form a measured amount of tin oxide. These masses of tin and oxygen are each converted to moles, and the simplest mole ratio of tin:oxygen is determined. This ratio represents the empirical formula of tin oxide.

Place a crucible with its lid ajar in a clay triangle mounted on an iron ring that is connected to a ring stand as shown in Figure 8.9. Strongly heat the crucible using the hottest point of the flame for 10 minutes until the crucible glows a cherry-red color. Heating removes any moisture present in the crucible and renders any contaminants on the surface of the crucible inert. Remove the crucible to a heat-resistant surface and allow it to cool to room temperature.

Weigh the cooled, empty crucible and lid as accurately as your balance allows. Add about 1.5 g of Sn to the crucible and weigh the crucible, lid, and tin sample. By difference, you may then calculate the mass of your initial sample of tin.

Be sure you are *wearing your safety goggles* and using a hood; add 4 mL of 10 M HNO_3 (10 molar HNO_3)[3] dropwise to the sample of Sn in the crucible. Place the Bunsen burner, ring stand, and clay triangle in a hood to capture any fumes. Wait 2 minutes and begin to heat the crucible and nitric acid gently, with the crucible lid slightly ajar to avoid any splattering of the nitric acid. Continue to heat gently for about 5

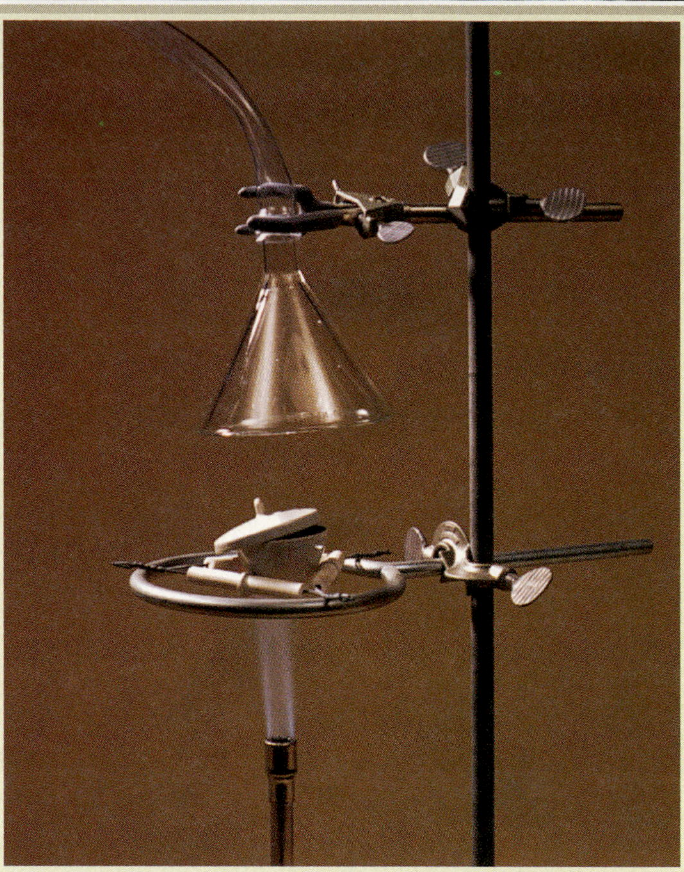

Figure 8.9
Heat a crucible gently at first to avoid cracking it. Gradually increase the temperature until the crucible glows a cherry-red color.

3. The term *10 M* (*10 molar*) is a measure of the strength of the nitric acid solution and will be discussed further in Chapter 12. It means that there are 10 mol of HNO_3 dissolved in 1 L of the solution.

chapter accomplishments

After completing this chapter, you should be able to define all key terms and do the following.

8.1 Mole as a very large "package"

- ❑ Compare the use of the mole as a package unit to the use of the dozen as a package unit.

8.2 Relative masses of atoms

- ❑ Recognize that equal numbers of atoms are present in any two samples of elements if the mass ratio of the samples corresponds to the atomic-mass ratio of the elements.

8.3 Moles of atoms

- ❑ Recognize that the acronym GAW stands for gram atomic weight.
- ❑ State the relationship between the number of atoms in a GAW of one element and the number of atoms in a GAW of any other element.
- ❑ State Avogadro's number.
- ❑ Relate 1 mol of atoms to the number of atoms in a mole.
- ❑ Relate 1 mol of atoms of an element to the mass of the element.

minutes and begin to raise the heat gradually until the tip of the inner blue cone is touching the bottom of the crucible. When brown fumes are no longer visible, heat strongly for 25 minutes.

Remove the crucible with lid to a heat-resistant surface and allow it to cool to room temperature. Weigh the crucible, lid, and tin oxide product (Figure 8.10). By subtracting the mass of the crucible, lid, and original Sn sample from the mass of the crucible, lid, and tin oxide product, you have determined the mass of the oxygen in the tin oxide product.

The number of moles of Sn and O combined to form tin oxide can be calculated as follows.

Step 1. The number of grams of each element present in the tin oxide has been determined experimentally above.

Step 2. The number of moles of Sn and O in the product can be calculated as follows.

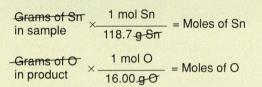

$$\begin{array}{l}\cancel{\text{Grams of Sn}}\\ \text{in sample}\end{array} \times \frac{1\ \text{mol Sn}}{118.7\ \cancel{\text{g Sn}}} = \text{Moles of Sn}$$

$$\begin{array}{l}\cancel{\text{Grams of O}}\\ \text{in product}\end{array} \times \frac{1\ \text{mol O}}{16.00\ \cancel{\text{g O}}} = \text{Moles of O}$$

Step 3. Simplify the molar ratio by dividing through by the smallest number.

Step 4. Round off the simplified ratio to 2 sig figs.

Step 5. If the ratio contains all whole numbers to 2 sig figs, you have the ratio of Sn and O in the tin oxide.

Step 6. If the ratio is not integral to 2 sig figs, find a number that, when multiplied by the simplified experimental ratio, converts that ratio to a ratio of whole numbers to 2 sig figs.

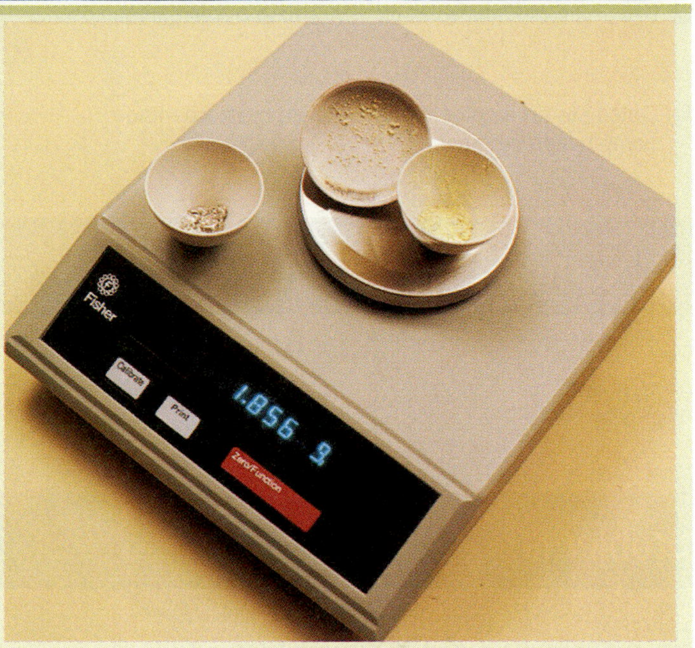

Figure 8.10
The crucible containing the tin oxide must be cool to obtain an accurate weighing.

Question:

1. The empty crucible and lid weighed 21.222 g. The crucible, lid, and original Sn sample weighed 22.724 g. The crucible, lid, and tin oxide product weighed 22.769 g. Find the empirical formula for the tin oxide.

8.4 Moles of compounds

- ❑ Relate 1 mol of a molecular compound to a number of molecules.
- ❑ Relate 1 mol of an ionic compound to a number of formula units.
- ❑ Recognize that the acronym GMW stands for gram molecular weight.
- ❑ Recognize that the acronym GFW stands for gram formula weight.
- ❑ Relate 1 mol of a compound to the mass of the compound.

8.5 Gram-to-mole conversions for compounds

- ❑ Given the formula of a substance, convert a given number of moles of that substance to grams of that substance.
- ❑ Given the formula of a substance, convert a given number of grams of that substance to moles of that substance.

8.6 Moles of atoms and ions within moles of compounds

- ❑ Construct conversion factors that relate the numbers of atoms in a molecule to the molecule that contains them.
- ❑ Construct conversion factors that relate the numbers of ions in a formula unit to the formula unit that contains them.
- ❑ Correctly interpret coefficients that are placed before chemical formulas.
- ❑ Interpret subscripts of chemical formulas in terms of moles of constituent particles.
- ❑ Construct conversion factors that relate moles of atoms to moles of molecular compounds.
- ❑ Construct conversion factors that relate moles of ions to moles of ionic compounds.

8.7 Empirical formulas represent the simplest formulas for compounds

- ❑ Given a molecular formula, determine the empirical formula.

8.8 Calculation of empirical formulas

- ❑ Calculate an empirical formula from the percentage composition by mass.
- ❑ Calculate an empirical formula from information about combining masses.

8.9 Molecular formulas

- ❑ Calculate a molecular formula from the empirical formula and the molecular weight.

key terms

ion	gram atomic weight	gram formula weight	empirical formula
formula unit	mole	molar mass	empirical weight
molecule	gram molecular weight		

problems

8.1 Mole as a very large "package"

9. What kinds of packages, other than a dozen, are you familiar with?
10. a. Give an example of an object found in the microscopic world.
    b. Give an example of an object found in the macroscopic world.
    c. Describe one difference between the microscopic and macroscopic worlds.

8.2 Relative masses of atoms

11. Calculate the mass ratio of a neon atom to a hydrogen atom.
12. Refer to Problem 11. Which atom has greater mass, Ne or H? How many times more massive?
13. a. Will the number of sodium atoms in 22.99 g of sodium be equal to the number of sulfur atoms in 22.99 g of sulfur?
    b. Which mass, the sodium or the sulfur, will contain the greater number of atoms?
14. Do 64.00 tons of molecular oxygen and 48.04 tons of carbon contain equal numbers of atoms (to 4 significant figures)?

8.3 Moles of atoms

15. Give the gram atomic weight to 4 sig figs for the following elements.

a. He	d. Au
b. Se	e. Li
c. F	f. Ge

16. a. Give the gram atomic weight of helium to 4 sig figs.
    b. How many helium atoms are present in the mass of helium described in part a?
    c. Calculate the mass in grams of one helium atom.
17. What would be the mass of Avogadro's number of magnesium atoms?
18. How many grams are there in 2.00 mol of potassium?
19. How many grams are there in 3.95 mol of sulfur?
20. How many grams are there in 5.99 mol of copper?
21. How many moles are there in 5.19 g of magnesium?
22. How many moles are there in 87.9 g of aluminum?
23. How many moles are there in 2.19 g of iron?
24. How many moles are there in 211 mg of lithium?
25. How many moles of chromium do we need in order to have as many chromium atoms as there are iron atoms in 1.75 mol of iron?
26. You have 7.5 mol of silicon. How many moles of carbon should be measured out to ensure that you have an equal number of Si atoms and C atoms?
27. You already possess 0.250 mol of krypton. How many grams of argon would ensure a number of argon atoms that equals the number of atoms in the krypton sample?
28. a. A student has weighed out 14.5 g of zinc. How many grams of nickel should the student weigh out to ensure equal numbers of zinc and nickel atoms?
    b. Suppose the student needs twice as many nickel atoms as zinc atoms. How many grams of nickel should be weighed out?

29. Complete the following table.

Element	Number of moles	Mass (g)
Lithium	1.50	
Calcium	1.00	
Phosphorus		12.9
Silicon		88.8
Zinc	3.90	

30. Arrange the following in order from highest *mass* to lowest *mass*.
   a. 1.00 mol of magnesium
   b. 1.00 mol of iron
   c. 1.00 mol of neon

8.4 Moles of compounds

31. Compute the molar mass of each of the following.
   a. KCl
   b. SO_3
   c. H_2SO_4
   d. $(NH_4)_2S$
   e. $NaHCO_3$
   f. $Mg_3(PO_4)_2$
32. How many grams are there in 1.00 mol of each of the following?
   a. N_2
   b. I_2
   c. $C_6H_6O_6$
   d. Na^+
   e. CO_3^{2-}
   f. $Al(HCO_3)_3$
33. a. Suppose you wish to combine equal numbers of H_2 molecules and I_2 molecules. If you have 1.50 mol of H_2, how many moles of I_2 should you measure out?
   b. If you had 4.04 g of H_2, how many moles of I_2 would be required to ensure equal numbers of H_2 and I_2 molecules? How many grams of I_2 is this?
34. What is the mass of Avogadro's number of MgO formula units?
35. What is the mass of Avogadro's number of P_4O_{10} molecules?

8.5 Gram-to-mole conversions for compounds

36. Calculate the number of moles in the following.
   a. 64.0 g of O_2
   b. 58.9 g of Cl
   c. 0.205 g of NH_3
   d. 1.91×10^2 g of $Ba_3(PO_4)_2$
   e. 90.5 g of $(NH_4)_2SO_4$
   f. 24.3 g of Mg^{2+}
37. Calculate the number of grams contained in each of the following.
   a. 0.500 mol of F_2
   b. 9.75 mol of H_2
   c. 2.50×10^{-2} mol of NH_3
   d. 11.3 mol of $BaSO_4$
   e. 3.011×10^{23} formula units of LiF
   f. 3.90 mol of S^{2-}
38. Arrange the following in order from largest *mass* to smallest *mass*.
   a. 20.0 g of H_2O
   b. 1.55 mol of H_2O
   c. 9.05×10^{23} molecules of H_2O
39. The density of carbon tetrachloride is 1.595 g/mL. What is the volume of 1.00 mol of CCl_4?

8.6 Moles of atoms and ions within moles of compounds

40. Construct conversion factors that relate the numbers of hydrogen, nitrogen, and oxygen atoms to one molecule of HNO_3.
41. Construct conversion factors that relate the number of calcium ions and the number of phosphate ions to one formula unit of $Ca_3(PO_4)_2$.
42. Construct conversion factors that relate the numbers of atoms of each element in C_3H_7NO to one molecule of that compound.
43. What is the total number of ions that can be found in one formula unit of $(NH_4)_2CO_3$? What are the ions?
44. How many aluminum ions are present in 6 $AlCl_3$? How many chloride ions?
45. Construct a conversion factor that relates the number of oxygen atoms to one molecule of oxygen.
46. State the meaning of each expression in terms of either moles of molecules, moles of formula units, or moles of atoms that are present.
   a. 5 H_2
   b. 4 Mg
   c. 16 P_4O_{10}
   d. 8 H_2O
   e. 4 O
   f. 3 MgO
47. State the number of moles of each ion contained in each of the following.
   a. 5.25 mol of KI
   b. 4.00 mol of $AlCl_3$
   c. 1.10×10^{-3} mol of $Ca(NO_3)_2$
   d. 0.250 mol of $Ba_3(PO_4)_2$
48. Calculate the number of moles of *each atom* contained in the given amount of each of the following molecules.
   a. 1.50 mol of Br_2
   b. 3.50 mol of H_2
   c. 3.00 mol of H_2O
   d. 1.25×10^{-1} mol of $C_6H_6O_6$
   e. 0.450 mol of H_2SO_4
49. a. What is the molar mass of SO_2 (to 4 sig figs)?
   b. How many SO_2 molecules are present in the mass of SO_2 from part a?
   c. Calculate the mass in grams of one SO_2 molecule.
50. a. What is the molar mass of Na_2SO_4 (to 4 sig figs)?
   b. How many Na^+ ions are present in the mass of Na_2SO_4 from part a?

8.7 Empirical formulas represent the simplest formulas for compounds

51. Distinguish between empirical and molecular formulas.
52. Each of the following compounds is represented by its molecular formula; state its empirical formula.
    a. Ethanol, C_2H_6O
    b. Methylene chloride, CH_2Cl_2
    c. Lindane, $C_6H_6Cl_6$
    d. Sucrose, $C_{12}H_{22}O_{11}$

8.8 Calculation of empirical formulas

53. Calculate the empirical formula of each compound from the mass-composition data given below.
    a. 46.7% N and 53.3% O
    b. 92.3% C and 7.70% H
    c. 75.0% C and 25.0% H
    d. 11.6% N and 88.4% Cl
    e. 60.0% C, 13.4% H, and 26.6% O
    f. 68.3% Pb, 10.6% S, and 21.1% O
    g. 45.9% K, 16.5% N, and 37.6% O
    h. 42.1% Na, 18.9% P, and 39.0% O
54. When 9.16 g of copper and 2.31 g of oxygen combine, they form a copper oxide. What is the empirical formula of this compound?
55. When 7.56 g of iron and 4.34 g of sulfur combine, they form an iron sulfide. What is the empirical formula of this compound?
56. Carbon and oxygen can combine to form more than one compound.

	Mass of carbon	Mass of oxygen
Compound 1	0.168 g	0.448 g
Compound 2	0.515 g	0.686 g

Calculate the empirical formula of each compound.

8.9 Molecular formulas

57. Analysis of a compound indicates that its mass composition is 80.0% C and 20.0% H. Its molecular weight is found to be 30.0. What is its molecular formula?
58. Acetone, a liquid often used as a nail polish remover, is found to obtain 62.0% carbon, 10.4% hydrogen, and 27.5% oxygen. Its molecular weight is found to be 58.1. What is the molecular formula of acetone?
59. A liquid of molecular weight 60.0 was found to contain 40.0% C, 6.7% H, and 53.3% O by mass. What is the molecular formula of the compound?
60. Vitamin C is a compound that upon analysis is found to contain 40.92% C, 4.58% H, and 54.51% O. The molecular weight of vitamin C is 176.1. What is the molecular formula of vitamin C?
61. Upon analysis, histidine, an amino acid found in proteins, yields the following percentage composition: 46.38% C, 5.90% H, 27.01% N, and 20.71% O. The molecular weight of histidine is found to be 155.1. What is its molecular formula?
62. When 125.0 g of a compound is decomposed, it is found to yield 50.00 g of C, 8.25 g of H, and 66.75 g of O. The molecular weight is found to be 90.0. What is the molecular formula of this compound?
63. A compound forms by combining 22.65 g of carbon, 1.90 g of hydrogen, and 8.73 g of nitrogen. The molecular weight of that compound is found to be 106. What is its molecular formula?

Additional problems

64. a. There are approximately 5 billion (5×10^9) people on the earth. The budget of the U.S. government is approximately 1.5 trillion dollars ($\$1.5 \times 10^{12}$). How many dollars would be distributed if every person on earth were to receive an amount of money equal to the entire budget of the U.S. government?
    b. How many dollars would be distributed if a mole of people each received one penny?
    c. In which case, part a or b, would the greater number of dollars be distributed?
65. a. An atom of "starwarsane" (a new element) has a mass of 4.367×10^{-22} g. If a robot requires Avogadro's number of starwarsane atoms for smooth running, how many grams does the robot require?
    b. What is the GAW of starwarsane?
66. a. You just inherited Avogadro's number of dollars. Assuming that you can spend or give money away at a rate of 1 million dollars per second, how many years would it take to spend your inheritance?
    b. The estimated age of the earth is 4.6 billion years. Is your spending time longer or shorter than the earth's age?
67. The average mass of an adult male is 70. kg. What is the mass of Avogadro's number of average adult males? Compare this mass to the mass of the earth (6.0×10^{24} kg).
68. Isopropyl alcohol, also known as rubbing alcohol, has a percentage composition of 60.0% carbon, 13.4% hydrogen, and 26.6% oxygen. Calculate the empirical formula of isopropyl alcohol.
69. Copper metal combines with chlorine gas to form a copper chloride. When 2.46 g of copper is reacted with chlorine, it is completely converted to 5.22 g of copper chloride. Determine the empirical formula of this copper chloride.
70. Fill in the blanks. A 100.0 g sample of $AlPO_4$ contains ________ mol of $AlPO_4$, ________ mol of ions, and ________ mol of atoms.
71. Platinum can form two different compounds with chlorine. One compound contains 26.7% Cl and the other contains 42.1% Cl. Determine the empirical formula of each of these compounds.

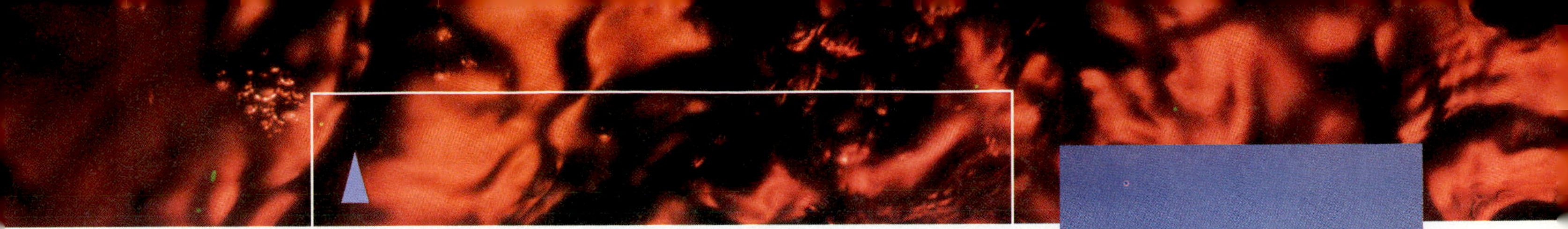

chapter 9

chemical reactions

Many chemical reactions can be classified into four major types: combination, decomposition, single replacement, and double replacement. The figure above illustrates a titanium airplane shell. Titanium metal is produced in a single-replacement reaction in which magnesium replaces the titanium in titanium chloride.

9.1 recognizing chemical reactions

Your excitement for chemistry builds when you combine materials and watch chemical reactions occur. Within your body, chemical reactions occur constantly. The digestion of food, the repair of body tissues, and the storage of chemical energy in your muscles all result from chemical reactions that are continually taking place (Figure 9.1). The engine in your car hums as a consequence of the chemical reaction between gasoline and air, and smog forms when sunlight reacts with the products of gasoline combustion in your automobile exhaust. Batteries provide electrical energy to light your flashlight because of chemical reactions (Figure 9.2). Nails rust and silver tarnishes as a result of chemical reactions. In all of these cases, **reactants** (starting materials) are converted into **products** (new materials). As you understand chemical reactions, you appreciate how chemists use them to create new commodities, such as longer lasting batteries or less polluting fuels, and to synthesize new materials, such as lightweight insulating fibers (Figure 9.3) or ceramics for high-temperature applications.

Figure 9.1
The digestion of food and the storage of chemical energy in muscles are examples of chemical reactions that are continually taking place in your body.

Figure 9.2
Long-life batteries are an example of how chemists optimize chemical reactions to produce electrical energy.

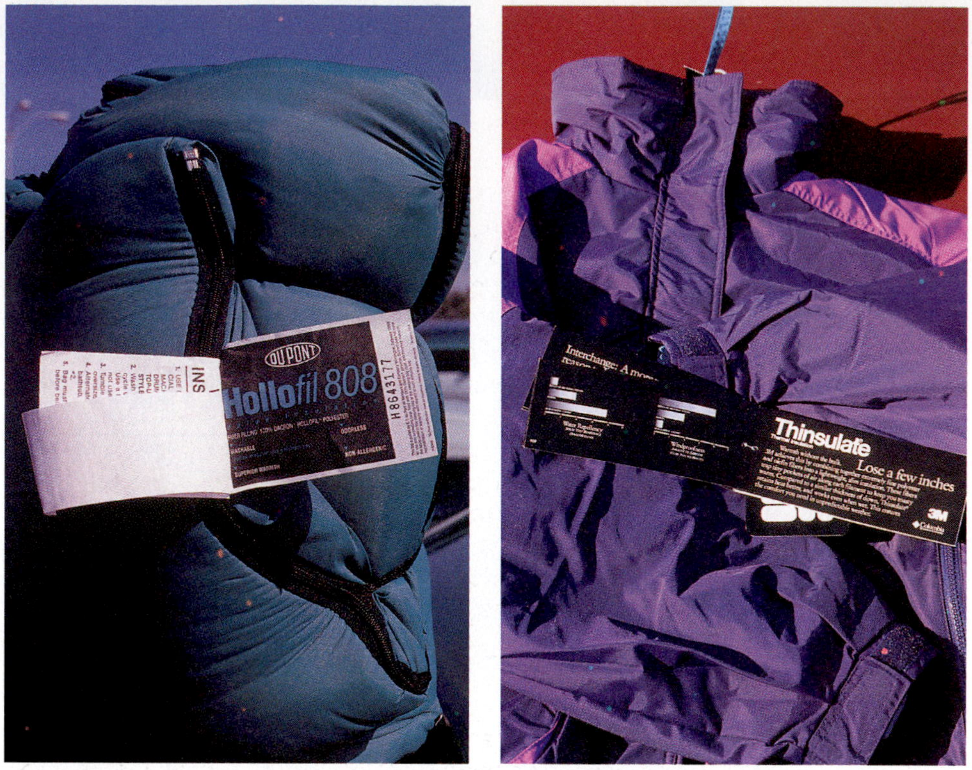

(a) (b)

Figure 9.3

Hollofil and Thinsulate are examples of insulating fibers used to produce thin, lightweight clothing and sleeping bags. The fibers are synthesized in controlled chemical reactions.

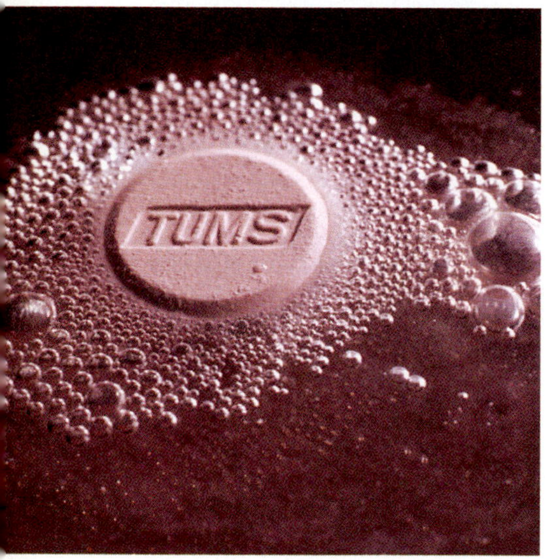

Figure 9.4

The appearance of bubbles indicates the formation of CO_2 gas in the chemical reaction of TUMS with stomach acid. According to the chemical equation, one molecule of CO_2 gas is formed for every two molecules of HCl used up and for every one formula unit of $CaCO_3$ consumed.

In Chapter 1, you learned about chemical reactions—how elements and compounds rearrange their atoms, molecules, or formula units into new materials with distinctly different physical and chemical properties. You recall from Figure 1.14 that sodium, a shiny, silver metal that burns violently in moist air, reacts with chlorine, a yellow-green, corrosive gas, to produce sodium chloride (table salt). Sodium chloride is a white, crystalline solid that has a high melting point, dissolves readily in water, and is used to enhance the flavor of food. Sodium metal and chlorine gas are the reactants, which exhibit physical and chemical properties distinctly different from the product, sodium chloride.

You recognize chemical reactions by identifying the appearance of new materials with physical and chemical properties different from the properties of the materials with which you started. In this chapter, you will learn to predict the products of a few simple reactions. But first, you must learn to represent chemical reactions through accurate chemical equations.

9.2 chemical equations

In previous chapters, you have dealt with chemicals one compound or element at a time. In doing so, you have become thoroughly familiar with how to represent elements with their *atomic symbols* and compounds with their *chemical formulas.* Knowing the atomic symbols and chemical formulas of reactants and products in a chemical reaction enables you to express the chemical change as a *chemical equation.* The atomic symbols or chemical formulas of reactants are written on the left and connected by an arrow leading to the atomic symbols or chemical formulas of products written on the right.

For example, if you swallow an antacid tablet like TUMS for indigestion, the reaction that occurs would be described in English as follows: Calcium carbonate (TUMS) reacts with hydrochloric acid (stomach acid) to yield calcium chloride, carbon dioxide gas, and water (Figure 9.4). Expressing this reaction as a chemical equation is clearly shorter than its English description:

$$CaCO_3 + 2\ HCl \longrightarrow CaCl_2 + CO_2 + H_2O$$

In addition to being shorter, the chemical equation contains more useful information than the English sentence does. It tells you how many molecules or formula units of each reactant are needed for the reaction to occur; it also tells you how many molecules or formula units of each product are formed.

In the next chapter, you will learn to extend the number of molecules (or formula units) to moles of reactants and products. From moles of material, you can readily convert to the number of grams through the molar mass (Section 8.5). You will have ample opportunity to use these conversions later, but for now, they may help you to appreciate how useful a chemical equation can be. The chemical equation enables you to determine just how many grams of each of the reactants you need to weigh out for the reaction to occur. In addition, you can determine how many grams of each of the products will form. All synthetic chemistry, whether it is creating a new material in the research laboratory or producing a pharmaceutical drug industrially, depends on chemical equations to predict the amount of material that can be produced in the chemical reaction.

From Section 1.11 you know that chemical equations always have the form:

$$\begin{array}{ccc} \text{Reactants} & \longrightarrow & \text{Products} \\ \text{Starting materials} & & \text{Final materials} \end{array}$$

The arrow means "yields" or "gives"; it can also be regarded as an equal sign (=).

You may translate any English sentence that describes a chemical reaction into a chemical equation by following the guidelines summarized below.

Guidelines for writing a chemical equation

1. Write correct formulas that correspond to the names of the chemical substances mentioned.
2. Decide which formulas represent reactants and which represent products.
3. Follow the preceding format, writing reactant formulas to the left of the arrow and product formulas to the right of the arrow.

You are ready to apply these guidelines to any chemical reaction for which you know the formulas of the reactants and products. The most important step in writing chemical equations is to write *correct chemical formulas* for the reactants and products.

---

sample exercise 1

Translate the following English sentences into chemical equations.

a. Charcoal (carbon) reacts with oxygen in your barbecue grill to yield carbon dioxide (Figure 9.5).
b. Sodium chloride reacts with silver nitrate to yield silver chloride and sodium nitrate (Figure 9.6).

solution

a.

Step 1. Establish correct atomic symbols or formulas. C for carbon, O_2 for oxygen (the element oxygen occurs in air as a diatomic molecule), and CO_2 for carbon dioxide.

Step 2. "Carbon reacts with oxygen" tells you that C and O_2 are the reactants; "to yield carbon dioxide" tells you that CO_2 is the product.

Figure 9.5
The burning of carbon into carbon dioxide yields only carbon dioxide gas, which is colorless and not observed. The best indication that the reaction is occurring is the intense heat released by the glowing charcoal. Since charcoal is not pure carbon, the charcoal does not completely disappear in the reaction; impurities other than pure carbon will remain as the ash residue.

Figure 9.6
Solutions of sodium chloride and silver nitrate are clear and colorless. When mixed, the formation of silver chloride as a product can be seen because it precipitates from the solution as a white solid.

Step 3. Thus, the equation is $C + O_2 \rightarrow CO_2$.

b.

Step 1. Establish correct formulas. Note that all of the materials follow nomenclature for ionic compounds.

Sodium chloride, $NaCl$ (Na^+, Cl^-)
Silver nitrate, $AgNO_3$ (Ag^+, NO_3^-)
Silver chloride, $AgCl$ (Ag^+, Cl^-)
Sodium nitrate, $NaNO_3$ (Na^+, NO_3^-)

Step 2. $NaCl$ and $AgNO_3$ are reactants; $AgCl$ and $NaNO_3$ are products.

Step 3. Thus, the equation is $NaCl + AgNO_3 \rightarrow AgCl + NaNO_3$.

••• problem 1

Translate the following English sentence into a chemical equation: Lead reacts with oxygen to yield lead(IV) oxide.

9.3 the meaning of balanced chemical equations

The few chemical equations that you have seen up to now have been carefully selected so that they are balanced. **Balanced chemical equations** have the same number of atoms (or ions) of each element on both sides of the arrow.

One of the fundamental principles of chemistry is that atoms may not be created or destroyed in a chemical reaction; rather, a chemical reaction conserves atoms as they rearrange themselves to form new materials. This principle of conserving atoms in a chemical equation is called the **law of conservation of matter**.

Examine the equations in Sample Exercise 1 to see what is meant by "balanced":

C	$+$	O_2	$\longrightarrow$	CO_2
1 carbon atom		2 oxygen atoms bonded together in a diatomic molecule		1 carbon and 2 oxygen atoms bonded together in 1 carbon dioxide molecule

This equation is balanced because there are one carbon and two oxygen atoms *on each side of the arrow*. What has changed is how the three atoms are bonded to each other. As reactants, the carbon atom is separate and the two oxygen atoms are bonded together in a diatomic molecule. As products, however, one carbon and two oxygen atoms are bonded together in one carbon dioxide molecule (Figure 9.7). This reaction is an example in which the two reactive elements combine or join together to form one product; other examples of this type of reaction will be discussed in Section 9.8.

The second example in Sample Exercise 1 is also balanced. In this case, the balanced equation can be seen by counting the same number of atoms (or ions, including polyatomic ions) of each type on both sides of the equation. When the same ion appears on both sides of a chemical equation, it too is conserved according to the law of conservation of matter.

$NaCl$	$+$	$AgNO_3$	$\longrightarrow$	$AgCl$	$+$	$NaNO_3$
1 Na^+ ion		1 Ag^+ ion		1 Ag^+ ion		1 Na^+ ion
1 Cl^- ion		1 NO_3^- ion		1 Cl^- ion		1 NO_3^- ion

Each side of this equation contains one of each ion, but the cations and anions have switched partners in moving from reactants to products. This reaction is an example in which the anions and cations of the reactants replace each other to form new products; other examples of this type of reaction will also be discussed in Section 9.8.

In the above examples, all of the elements and compounds in the balanced chemical equations have a coefficient of 1. Notice in the Tums equation, introduced in Section 9.2 and shown again below, there is a whole-number coefficient of 2 before the HCl molecule, indicating that two molecules rather than one are participating in the reaction. This coefficient is essential to have a balanced equation, i.e., to have equal numbers of each kind of atom on both sides of the equation.

Balanced: $CaCO_3 + 2\,HCl \longrightarrow CaCl_2 + CO_2 + H_2O$

1 Ca 1 C 3 O | 2 H 2 Cl | 1 Ca 2 Cl | 1 C | 2 H 3 O

Without the coefficient of 2 before the reactant HCl molecule, there would be one H atom on the left side of the equation and two H atoms on the right side. Similarly, there would be only one Cl atom on the left side and two Cl atoms on the right side.

Unbalanced: $CaCO_3 + HCl \longrightarrow CaCl_2 + CO_2 + H_2O$

1 H 1 Cl | 2 Cl | 2 H

The translation of an English sentence into a chemical equation by the three guidelines outlined in Section 9.2 does not lead automatically to a balanced equation. When it does not, balancing is accomplished by inserting proper, whole-number coefficients before the reactant or product formulas contributing additional atoms or ions. Practice in balancing equations is given throughout the next section.

9.4 balancing equations

Writing chemically correct equations means always checking to be sure that each equation is *balanced.* As long as you keep in mind how to count atoms or ions in molecules and formula units (Section 8.4), you should have no difficulty balancing chemical equations.

As an example of balancing an equation, consider that hydrogen gas reacts with oxygen gas to yield liquid water. Translating this English description into a chemical equation does not lead automatically to a *balanced* chemical equation for this reaction:

Step 1. The correct formulas are H_2 and O_2 for the elements hydrogen and oxygen and H_2O for the compound water.

Step 2. The reactants are H_2 and O_2. The product is H_2O.

Step 3. Thus, $H_2 + O_2 \rightarrow H_2O$.

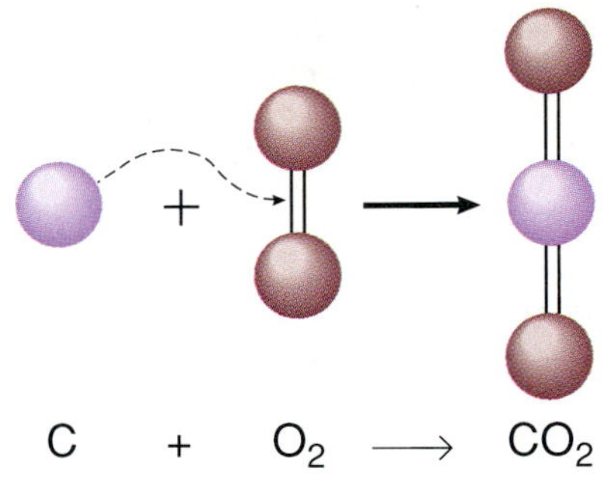

Figure 9.7
Chemical reactions are rearrangements of atoms in which atoms are neither created nor lost. In this example, the bonding between atoms changes. A C atom inserts itself between the two atoms of the O_2 molecule to yield a new molecule, CO_2.

The next step is to determine whether the equation is balanced. This is done by counting the numbers of each kind of atom on each side of the equation and comparing them. For each molecule or formula unit, you determine the number of atoms of a kind by multiplying

$$\textbf{Coefficient} \times \textbf{Subscript} = \textbf{Number of atoms}$$

Remember that when there is no coefficient, the coefficient is understood to be 1.

Unbalanced:

$$H_2 \quad + \quad O_2 \longrightarrow H_2O$$

H_2	O_2	H_2O	H_2O
$1 \times 2 = 2$ H	$1 \times 2 = 2$ O	$1 \times 2 = 2$ H	$1 \times 1 = 1$ O
↑ Coefficient × ↑ Subscript	↑ Coefficient × ↑ Subscript	↑ Coefficient × ↑ Subscript	↑ Coefficient × ↑ Subscript

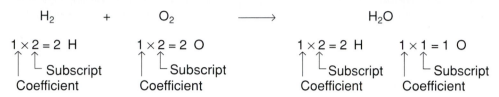

You see that this equation is *not* balanced because, whereas there are two oxygen atoms on the left, there is only one oxygen atom on the right. To balance the equation, you must insert *coefficients* that simply change the number of atoms, molecules, or formula units of a reactant or product. *Never alter subscripts* because that would change the chemical formulas. Different chemical formulas represent different chemical materials with different physical and chemical properties. If you introduce new formulas, you have inadvertently changed the chemical reaction!

In the preceding unbalanced equation, there are two O atoms on the left side, but only one O atom on the right. You begin the balancing procedure by placing the coefficient 2 before H_2O to increase the number of O atoms on the right to two. The O atoms are then balanced.

Unbalanced: $H_2 + O_2 \longrightarrow 2\,H_2O$

H_2	O_2	$2\,H_2O$	$2\,H_2O$
2 H	2 O	$2 \times 2 = 4$ H	$2 \times 1 = 2$ O
		↑ Coefficient × ↑ Subscript	↑ Coefficient × ↑ Subscript

Even though there are now two O atoms on each side of the equation, the hydrogen balance is upset. There are two H atoms on the left side and four H atoms on the right side. To correct this, we put a 2 before H_2:

Balanced: $2\,H_2 + O_2 \longrightarrow 2\,H_2O$

$2\,H_2$	O_2	$2\,H_2O$	$2\,H_2O$
$2 \times 2 = 4$ H	$1 \times 2 = 2$ O	$2 \times 2 = 4$ H	$2 \times 1 = 2$ O
↑ Coefficient × ↑ Subscript	↑ Coefficient × ↑ Subscript	↑ Coefficient × ↑ Subscript	↑ Coefficient × ↑ Subscript

Now the equation is balanced; there are two O atoms and four H atoms on each side of the equation.

To review, balance an unbalanced equation by inserting coefficients before the molecules or formula units. *Never alter subscripts* in chemical formulas. You might think that the preceding equation, $H_2 + O_2 \rightarrow H_2O$, could be "balanced" by writing $H_2 + O_2 \rightarrow H_2O_2$. But this would be wrong because H_2O_2 is not water; H_2O_2 is hydrogen peroxide, a bleach and antiseptic with chemical and physical properties very different from water. If you alter subscripts, the formula has changed and the equation no longer represents the same chemical reaction.

Figure 9.8

About 120 million tons of ammonia are produced each year in plants using the Haber process. Ammonia is an important raw material used in making fertilizers and explosives. It is also used in dilute water solution for housekeeping tasks.

••• problem 2

Decide whether the following chemical equations are balanced.

a. $KOH + H_2SO_4 \rightarrow K_2SO_4 + H_2O$

b. $2\ Na + H_2O \rightarrow 2\ NaOH + H_2$

c. $H_2 + Cl_2 \rightarrow 2\ HCl$

Guidelines for equation writing and balancing

1. Write the equation after identifying reactants and products with the correct chemical formulas. Use the stepwise procedure described in Section 9.2.
2. Determine whether the equation is balanced by multiplying the coefficient by the subscript for each atom in each formula on either side and comparing.
3. If the equation is *not* balanced, balance it by inserting the correct coefficients. When an equation contains unbalanced atoms other than H and O, it is most readily balanced by first inserting coefficients to balance these other atoms and then by balancing H and O.
4. When you think the equation is balanced, check it by again multiplying the coefficient by the subscript for each atom in each formula on either side.

sample exercise 2

Write a balanced equation for the reaction between nitrogen and hydrogen, which yields ammonia (NH_3), a substance that is the primary source of nitrogen fertilizer (Figure 9.8). This reaction, known as the Haber process, produces significant amounts of ammonia when conducted at high temperature with a solid, iron catalyst (a material that increases the speed of a chemical reaction).

solution

Follow steps 1 through 4 as outlined in the guidelines immediately preceding this exercise.

Step 1. $N_2 + H_2 \rightarrow NH_3$

Step 2.

Reactants	Products
Coefficient × Subscript	**Coefficient × Subscript**
1 × 2 = 2 N	1 × 1 = 1 N
1 × 2 = 2 H	1 × 3 = 3 H

Both N and H are not balanced since there are different numbers of N and H atoms on each side of the equation.

Step 3. Begin by balancing N, which can be accomplished by placing a coefficient of 2 before NH_3:

$$N_2 + H_2 \rightarrow 2\ NH_3$$

This coefficient multiplies the entire NH_3 formula, so that you now have 2 × 3 = 6 H atoms on the right. You can obtain 6 H atoms on the left side by using the coefficient 3 for H_2:

$$N_2 + 3\ H_2 \rightarrow 2\ NH_3$$

Step 4.

Reactants	Products
Coefficient × Subscript	**Coefficient × Subscript**
1 × 2 = 2 N	2 × 1 = 2 N
3 × 2 = 6 H	2 × 3 = 6 H

The equation is balanced because you have exactly 2 N atoms and 6 H atoms on either side.

• • • problem 3

Write a balanced equation for the reaction between hydrogen and bromine, which yields hydrogen bromide.

In Section 9.3, you learned that when the same ion appears on both sides of a chemical equation, it must be conserved according to the law of conservation of matter. When a polyatomic ion appears on *both* sides of an equation, it can be counted and balanced as a whole unit. It is usually much easier for you to do this than to balance the atoms making up the polyatomic ion. In the following example, you can balance the sulfate anion, SO_4^{2-}, rather than balancing S and O atoms individually.

Al	+	H_2SO_4	→	$Al_2(SO_4)_3$	+	H_2
Count:						
1 × 1 = 1 Al		1 × 2 = 2 H 1 × 1 = 1 SO_4		1 × 2 = 2 Al 1 × 3 = 3 SO_4		1 × 2 = 2 H
Balance:						
2 Al	+	3 H_2SO_4	→	$Al_2(SO_4)_3$	+	3 H_2
Count:						
2 × 1 = 2 Al		3 × 2 = 6 H 3 × 1 = 3 SO_4		1 × 2 = 2 Al 1 × 3 = 3 SO_4		3 × 2 = 6 H

sample exercise 3

Balance the equation $MgCl_2 + K_3PO_4 \rightarrow Mg_3(PO_4)_2 + KCl$. Magnesium phosphate, $Mg_3(PO_4)_2$, is a component of dental cements used to bind materials to teeth.

solution

Begin with step 2 of the balancing procedure, since the equation is given.

Step 2.

Reactants	Products
Coefficient × Subscript	**Coefficient × Subscript**
$1 \times 1 = 1$ Mg	$1 \times 3 = 3$ Mg
$1 \times 2 = 2$ Cl	$1 \times 1 = 1$ Cl
$1 \times 3 = 3$ K	$1 \times 1 = 1$ K
$1 \times 1 = 1$ PO_4	$1 \times 2 = 2$ PO_4

All atoms are unbalanced.

Step 3. Proceed to balance in an orderly fashion one atom or ion at a time going from left to right.

To balance Mg, use the coefficient 3 before $MgCl_2$:

$$3\ MgCl_2 + K_3PO_4 \rightarrow Mg_3(PO_4)_2 + KCl$$

There are now $3 \times 2 = 6$ Cl atoms on the left, so the coefficient 6 is required before KCl on the right:

$$3\ MgCl_2 + K_3PO_4 \rightarrow Mg_3(PO_4)_2 + 6\ KCl$$

There are now three K atoms on the left and six on the right, so use the coefficient 2 in front of K_3PO_4:

$$3\ MgCl_2 + 2\ K_3PO_4 \rightarrow Mg_3(PO_4)_2 + 6\ KCl$$

There are now two PO_4^{3-} ions on each side of the equation.

Step 4. Check:

Reactants	Products
Coefficient × Subscript	**Coefficient × Subscript**
$3 \times 1 = 3$ Mg	$1 \times 3 = 3$ Mg
$3 \times 2 = 6$ Cl	$6 \times 1 = 6$ Cl
$2 \times 3 = 6$ K	$6 \times 1 = 6$ K
$2 \times 1 = 2$ PO_4	$1 \times 2 = 2$ PO_4

Since there are the same number of each type of atom or ion on each side of the equation, the equation is balanced.

---

sample exercise 4

Balance the equation for the decomposition of potassium chlorate. In Laboratory Exercise 7, this reaction was used to find the percentage of the element oxygen in the compound potassium chlorate. The reactants are heated in the presence of a solid catalyst, MnO_2.

$$KClO_3 \rightarrow KCl + O_2$$

solution

Begin with step 2.

Step 2.

Reactants	Products
Coefficient × Subscript	**Coefficient × Subscript**
$1 \times 1 = 1$ K	$1 \times 1 = 1$ K
$1 \times 1 = 1$ Cl	$1 \times 1 = 1$ Cl
$1 \times 3 = 3$ O	$1 \times 2 = 2$ O

Clearly, oxygen requires balancing.

Step 3. When you see three of a kind on one side and two on the other, you must use the idea of the lowest common denominator (6 in this case) in order to balance. Thus,

$$2\ KClO_3 \longrightarrow KCl + 3\ O_2$$
$$2 \times 3\ O = 6\ O \qquad 3 \times 2\ O = 6\ O$$

The 2 in front of $KClO_3$ requires a 2 before KCl. Thus,

$$2\ KClO_3 \longrightarrow 2\ KCl + 3\ O_2$$

Step 4. Check:

Reactants	Products
Coefficient × Subscript	**Coefficient × Subscript**
2 × 1 = 2 K	2 × 1 = 2 K
2 × 1 = 2 Cl	2 × 1 = 2 Cl
2 × 3 = 6 O	3 × 2 = 6 O

Since there are the same number of each type of atom on each side of the equation, the equation is balanced.

The only way to learn to balance equations comfortably is through practice. Proceed slowly, atom by atom, and keep track by writing down the numbers of atoms on either side of the equation at each step. Always make a final check. If you proceed systematically, you will always obtain a balanced equation, and you can always check your result.

••• problem 4

Balance the following equations.

a. $Zn + Pb(NO_3)_2 \rightarrow Zn(NO_3)_2 + Pb$
b. $Al + Cl_2 \rightarrow AlCl_3$
c. $Ag_2O \rightarrow O_2 + Ag$

9.5 using correct formulas

You will find that most of the chemical equations you encounter can be balanced quite easily. If you experience great difficulty in balancing an equation, it is probably because you have not used the correct formulas. Always remember that *correct equations require correct formulas.* For example, suppose you were asked to write an equation to show that aluminum chloride reacts with potassium carbonate to yield aluminum carbonate and potassium chloride. Begin by writing correct formulas according to the rules in Sections 6.8 and 6.13.

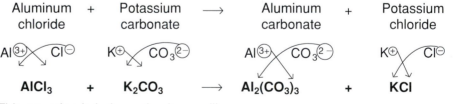

Aluminum chloride + Potassium carbonate ⟶ Aluminum carbonate + Potassium chloride

$$\mathbf{AlCl_3 + K_2CO_3 \longrightarrow Al_2(CO_3)_3 + KCl}$$

This equation is balanced quite readily:

1. 2 $AlCl_3$ balances the Al.
2. 6 KCl then balances the Cl.
3. 3 K_2CO_3 then balances the K and CO_3.

$$2\ AlCl_3 + 3\ K_2CO_3 \rightarrow Al_2(CO_3)_3 + 6\ KCl$$

But suppose that an incorrect formula had been used accidentally. For example,

$$AlCl_3 + K_2CO_3 \rightarrow \mathit{AlCO_3} + KCl$$
Wrong

It is *impossible* to balance this. A few attempts should convince you. Try

$$2\ AlCl_3 + 3\ K_2CO_3 \rightarrow 2\ AlCO_3 + 6\ KCl$$

Al, Cl, and K are balanced, but not CO_3. Try

$$2\ AlCl_3 + 3\ K_2CO_3 \rightarrow 3\ AlCO_3 + 6\ KCl$$

Cl, K, and CO_3 are balanced, but not Al. Write to us if you find a way to balance this with coefficients only. Remember, to balance, you use coefficients only. The subscripts are supposed to be those of a correct formula to begin with.

problem 5

Write a balanced equation for the reaction between calcium bromide and silver nitrate, which yields calcium nitrate and silver bromide, a compound used in photographic film.

9.6 helpful hints for balancing equations

As you practice balancing, you will develop confidence that a systematic approach always works. You may even find some shortcuts of your own. In this section, you will learn some simplifying devices that others have noticed.

Hint 1. Balance polyatomic ions as single entities if they appear on both sides of an equation. You have seen this idea used in Section 9.4.

Hint 2. If hydroxide (OH^-) appears on *only one* side of an equation and water on the other, it is convenient to write water (H_2O) as HOH and regard it as being composed of H^+ and OH^-. For example,

$$Mg(OH)_2 + HCl \rightarrow MgCl_2 + H_2O$$

can be balanced more easily when it is written

$$Mg(OH)_2 + HCl \rightarrow MgCl_2 + HOH$$

Count:

1 Mg	1 Mg
2 OH	1 OH
1 H	1 H
1 Cl	2 Cl

Balance:

OH	by	2 HOH on right
Cl	by	2 HCl on left

$$Mg(OH)_2 + 2\ HCl \rightarrow MgCl_2 + 2\ HOH$$

Count:

1 Mg	1 Mg
2 OH	2 OH
2 H	2 H
2 Cl	2 Cl

Hint 3. Begin balancing an equation with an unbalanced atom that has an ionic charge greater than 1. You may see the difficulties of not following this hint; for example, if you begin to balance the equation

$$K_2S + AlCl_3 \rightarrow KCl + Al_2S_3$$

by starting with K (K^+), then you would put a 2 in front of KCl.

$$K_2S + AlCl_3 \rightarrow 2\ KCl + Al_2S_3$$

Then, when you consider S and find that 3 K_2S is required, you would have to change 2 KCl to 6 KCl. If you had begun with S (S^{2-}) or Al (Al^{3+}), no extra change of coefficients would have been required. Beginning with S, you would proceed:

$$3\ K_2S + AlCl_3 \rightarrow KCl + Al_2S_3$$

Then fix K:

$$3\ K_2S + AlCl_3 \rightarrow 6\ KCl + Al_2S_3$$

Then Al:

$$3\ K_2S + 2\ AlCl_3 \rightarrow 6\ KCl + Al_2S_3$$

The job is done!

Hint 4. When there is an *even* number of one atom on one side of an equation and an *odd* number of that same atom on the other side, begin balancing by multiplying the formula with the odd number by 2. For example, if you are asked to balance

$$KClO_3 \rightarrow KCl + O_2$$

notice the three O atoms (odd) on the left and the two O atoms (even) on the right. Begin by multiplying the formula with the odd number by 2.

$$2\ KClO_3 \rightarrow KCl + O_2$$

Now count:

2 K	1 K
2 Cl	1 Cl
6 O	2 O

Balance:

$$2\ KClO_3 \rightarrow 2\ KCl + 3\ O_2$$

This is an alternative to the least common multiple idea discussed previously. Another example of the application of Hint 4 is seen in balancing

Count:

$$\begin{array}{ccccc} CO & + & O_2 & \rightarrow & CO_2 \\ \uparrow \nwarrow & & \uparrow & & \uparrow \nwarrow \\ 1\ C & & 1 + 2 = 3\ O & & 1\ C \quad 2\ O \end{array}$$

Noticing that there is an odd number of O on the left, you begin by "evening up" by multiplying CO by 2.

$$2\ CO + O_2 \longrightarrow CO_2$$

Now balance C:

$$2\ CO + O_2 \longrightarrow 2\ CO_2$$

Count:

2 C	2 C
2 O + 2 O = 4 O	4 O

• • • problem 6

Balance the following equations by using the appropriate helpful hints.

a. $K + H_2O \rightarrow KOH + H_2$

b. $NO + O_2 \rightarrow NO_2$

c. $Na_2CO_3 + Fe(NO_3)_2 \rightarrow NaNO_3 + FeCO_3$

9.7 designating physical states and special conditions in reactions

The physical state of each reactant and product may be indicated by certain symbols chemists use in a chemical equation. For example, the following equation indicates the physical state of each material as it is observed when the reaction occurs in the laboratory at a standard temperature of 25°C.

$$\underset{\text{Solid}}{Ca(s)} \quad + \quad \underset{\text{Liquid}}{2\,H_2O(l)} \longrightarrow \underset{\text{Gas}}{H_2(g)} \quad + \quad \underset{\text{Aqueous}}{Ca(OH)_2(aq)}$$

Reactants and products may be solids (s), liquids (l), or gases (g), and each phase is labeled accordingly. More often, reactions are conducted in water (aqueous) solution so that some of the reactants or products are dissolved rather than being in some pure physical state. A reactant or product that is in water solution is designated by (aq), short for **aqueous,** a word derived from *aqua,* the Latin word for water.

The formation of a product that is a gas or solid may also be indicated in other ways. The previous example, which illustrates the formation of a gaseous product, could also have been written as

$$Ca(s) + 2\,H_2O(l) \longrightarrow H_2\uparrow + Ca(OH)_2(aq)$$

The arrow pointing upward ($\uparrow$) indicates that a gaseous product is escaping from the reaction mixture (Figure 9.9). The reaction of silver nitrate solution with sodium chloride, which illustrates the formation of a solid product, could also be written in two ways.

$$AgNO_3(aq) + NaCl(aq) \longrightarrow AgCl(s) + NaNO_3(aq)$$

or

$$AgNO_3(aq) + NaCl(aq) \longrightarrow AgCl\downarrow + NaNO_3(aq)$$

The arrow pointing downward ($\downarrow$) indicates the formation of a solid product that is *not* soluble in water and, therefore, "falls out" of solution. An insoluble solid that falls out of solution is called a **precipitate** (Figure 9.6). Remember that the symbols $\uparrow$ for a gas and $\downarrow$ for a solid are used only for *products,* i.e., only on the right side of the equation.

Additional information about the conditions under which a reaction occurs is given by symbols written above or below the yield arrow ($\rightarrow$). For example, the symbol Δ represents heat. When written above the yield arrow as seen in the following equation, it means that heat must be applied for this reaction to occur; it will not take place at the standard temperature of 25°C.

$$CaCO_3(s) \xrightarrow{\Delta} CaO(s) + CO_2\uparrow$$

Similarly, a chemical formula written above the yield arrow indicates a catalyst is present. A **catalyst** is a substance that increases the speed of a chemical reaction without itself being changed in the course of the reaction. Since it does not change, it is neither a reactant nor a product, and so it appears above the arrow rather than on either side (Table 9.1). In the following reaction, MnO_2 is a catalyst (Figure 9.10).

$$2\,KClO_3(s) \xrightarrow[\Delta]{MnO_2} 2\,KCl(s) + 3\,O_2\uparrow$$

Many catalysts, like manganese dioxide, are solids and remain as solids throughout the reaction. Even when a solution is present, some catalysts are insoluble and do not dissolve in the reaction mixture. They increase the speed (rate) of the reaction because reactants first adsorb on (stick to) their surface and, in so doing, weaken bonds that participate in the reaction. When bonds of reactant molecules have been weakened, they break more rapidly and, hence, increase the rate of the reaction.

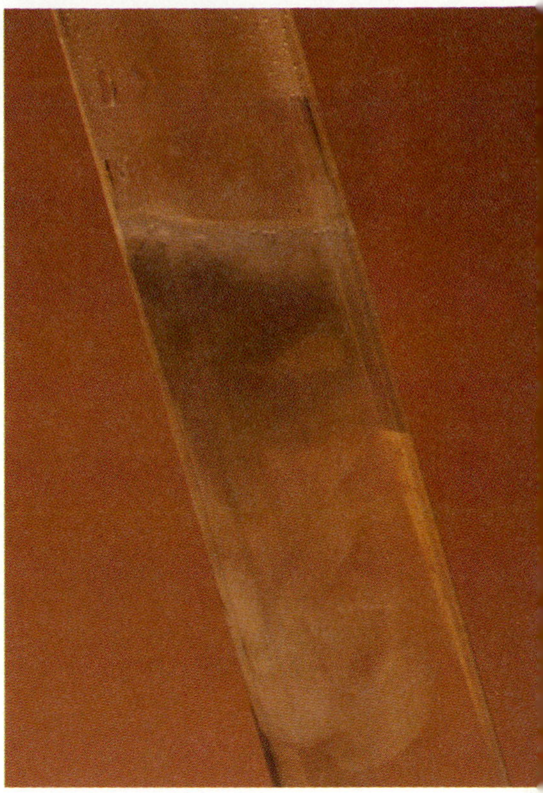

Figure 9.9
The appearance of bubbles escaping from the solution indicates the formation of H_2 gas from the reaction of calcium metal with water.

Figure 9.10

Potassium chlorate decomposes upon heating with the catalyst MnO_2 to form KCl and O_2 gas. The oxygen gas produced in the reaction causes a glowing wood splint to burn vigorously.

table 9.1 summary of special symbols used in chemical equations

Symbol	Meaning
(s)	Solid reactant or product
(l)	Liquid reactant or product
(g)	Gaseous reactant or product
(aq)	Reactant or product in water (aqueous) solution
$\uparrow$	Gaseous product
$\downarrow$	Solid product (precipitate)
$\xrightarrow{\Delta}$	Heat applied to force reaction
$\xrightarrow{\text{Formula}}$	Catalyst used to speed reaction

9.8 types of reactions

In all of the previous examples, you have been shown the reactants and the products of a chemical reaction because you did not know how to predict the products that form. Once you recognize that many chemical reactions can be classified into only four major types, you are well on your way to predicting possible products. This ability to predict products takes a great deal of experience. Realistically, at the end of this chapter, your ability will be very limited. However, you will be prepared to develop this skill as you proceed in the chapters that follow.

Four major categories that classify many chemical reactions

1. Combination reactions
2. Decomposition reactions
3. Single-replacement reactions
4. Double-replacement reactions

The name of each category indicates the process that helps you to predict reaction products. In the following discussion, a model chemical equation will illustrate each reaction type. Understanding the model equation for each type of reaction will enable you to classify a given reaction into one of these types. You can then predict the kinds of products to be expected and often the specific compounds that are formed.

combination reactions

As the name implies, in **combination reactions,** two or more reactants *combine* or join together to form one product.

Combination model equation

$A + B$	$\rightarrow$	AB
Two reactants		One product

All reactions in which a compound is formed from its elements are combination reactions. Notice how the following examples fit the model equation.

$$C(s) + O_2(g) \rightarrow CO_2(g)$$

$$Mg(s) + Cl_2(g) \rightarrow MgCl_2(s)$$

$$2\ Na(s) + Br_2(l) \rightarrow 2\ NaBr(s)$$

In each example, *two elements* react to give *one product* (a compound). These are the combination reactions you will encounter most frequently. Now, whenever you see two elements reacting, you can predict that the product will be the compound that they form. For the combination of metallic and nonmetallic elements, you can complete the equation by writing a correct formula for the ionic compound formed according to the rules of Section 6.8. For example, given the uncompleted equation,

$$Al + Cl_2 \rightarrow ?$$

you should recognize the reactants as the metallic element aluminum (a solid) and the nonmetallic element chlorine (a gas) and realize that the product will be the ionic compound that they form, namely, aluminum chloride (a solid). The product, $AlCl_3$, is a white solid with properties different from the shiny, metallic reactant aluminum. As a result, its formation should be observable in the laboratory. Complete the equation for this reaction by writing the correct formula for aluminum chloride and balancing the equation.

Complete:

$$Al(s) + Cl_2(g) \rightarrow AlCl_3(s) \qquad (Al^{3+} \times Cl^{1-})$$

Balance:

$$2\ Al(s) + 3\ Cl_2(g) \rightarrow 2\ AlCl_3(s)$$

If the metallic element can form more than one cation (Section 6.6), then you need to be told which cation is forming in the given reaction. In the above reaction, aluminum only forms one cation, so the product may be determined directly. Since aluminum chloride is an ionic solid, it has a high melting point (Section 6.9); as a result, it must be a solid at the standard temperature of 25°C.

---

sample exercise 5

Complete the equation: $Ca + O_2 \rightarrow ?$

solution

1. Recognize that the reactants are both elements, which signals that this is a combination reaction.
2. Recognize Ca as a metal (which is a solid) and oxygen as a nonmetal (which is a gas). The combination of a metal and nonmetal tells you that they will form an ionic compound (a solid).
3. Write a correct formula for calcium oxide and complete the equation.

$$Ca^{2+} \times O^{2-} \qquad Ca_{\cancel{2}}O_{\cancel{2}} \qquad CaO$$

$$Ca(s) \quad + \quad O_2(g) \quad \longrightarrow \quad CaO(s)$$

4. Balance the equation according to the procedure outlined earlier.

$$2\ Ca(s) + O_2(g) \rightarrow 2\ CaO(s)$$

For the reaction of *non*metallic elements, you can predict that the product will be a molecular compound, but you cannot complete the equation unless you are told the formula for the molecular compound. In many cases, the same nonmetallic elements can combine to form more than one compound. Similarly, you cannot indicate the physical state of the molecular compound without being told whether it is a solid, liquid, or gas. Unlike ionic compounds, molecular compounds cannot be assumed to exist in the solid state at 25°C. To verify the reaction in the laboratory, you would look for the appearance of a new material with physical properties different from those of the reactants.

$$2\ P(s) + 3\ Cl_2(g) \rightarrow 2\ PCl_3(l)$$

$$2\ P(s) + 5\ Cl_2(g) \rightarrow 2\ PCl_5(s)$$

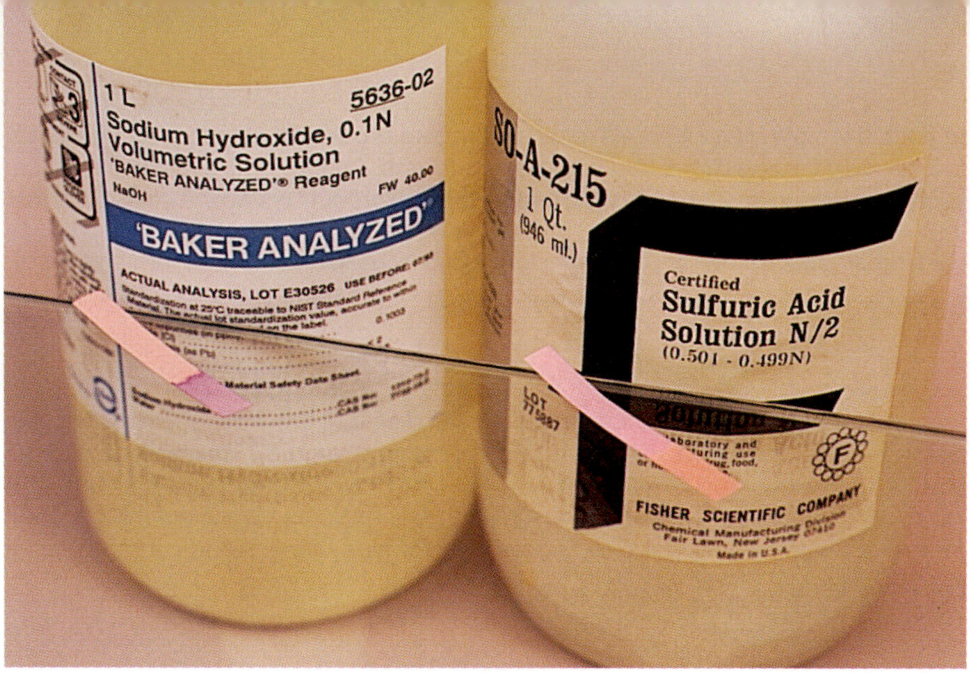

Figure 9.11
The formation of a metal hydroxide can be tested with red litmus paper, which will be turned blue by the hydroxide base. The formation of an oxoacid can be tested in the laboratory with blue litmus paper, which will be turned red by the acid.

Another example of a combination reaction is the reaction between water and an oxide compound. Notice how the following examples also fit the combination model equation, even though compounds rather than elements are combining.

$H_2O(l)$	+	$SO_3(l)$	$\longrightarrow$	$H_2SO_4(l)$
Water		Nonmetallic oxide		Oxoacid
$H_2O(l)$	+	$MgO(s)$	$\longrightarrow$	$Mg(OH)_2(s)$
Water		Metallic oxide		Metal hydroxide, a base

In each example, two compounds react to give one product (a new compound). If the reacting oxide is *nonmetallic,* the product is always an *oxoacid,* a type of molecular compound you reviewed in Section 7.5. If the reacting oxide is *metallic,* the product is always a metal hydroxide. Metal hydroxides are also called **bases**. The formation of these products can be tested in the laboratory with litmus paper. Litmus paper comes as pink or blue paper that changes color in the presence of an acid or base. For the above reactions, an oxoacid would turn blue litmus paper red; a metal hydroxide would turn red litmus paper blue (Figure 9.11).

decomposition reactions

As the name implies, in **decomposition reactions**, one reactant *decomposes,* or breaks apart, into two or more products.

Decomposition model equation

XY	$\longrightarrow$	$X + Y$
One reactant		Two (or more) products

Examples of decomposition are given below for metal carbonates, metal oxides, and chlorates. For these specific sets of compounds, the gaseous product can be predicted.

$$CaCO_3(s) \xrightarrow{\Delta} CaO(s) + CO_2\uparrow$$
Metal carbonates decompose when heated to yield $CO_2(g)$.

$$2\ HgO(s) \xrightarrow{\Delta} 2\ Hg(l) + O_2\uparrow$$
Some metal oxides decompose when heated to yield $O_2(g)$.

$$2\ KClO_3(s) \xrightarrow{\Delta} 2\ KCl(s) + 3\ O_2\uparrow$$
Chlorates decompose when heated to yield $O_2(g)$.

Smelling salts, $(NH_4)_2CO_3$, a carbonate compound, work because of their decomposition to the products shown:

$$(NH_4)_2CO_3(s) \rightarrow 2\ NH_3\uparrow + CO_2\uparrow + H_2O(l)$$

The pungency of the liberated NH_3 gas (ammonia) revives a semiconscious person.

If oxygen is the gas produced in a decomposition reaction, it can be easily verified by placing a glowing splint near the escaping gas. The presence of O_2 gas will cause the splint to burn vigorously (see Figure 9.10).

Chemical equations representing decomposition reactions are very easy to recognize because they are the only type of equation in which there is *only one* reactant. But it is difficult to predict the products of decomposition because there are no general rules. There are only specific rules for the products of decomposition of metal carbonates, metal oxides, and chlorates, which are listed in the preceding examples. You should be able to recognize decomposition reactions in general; however, if the reactions do not involve the specific sets of compounds listed above, you will be given the products of the decomposition.

single-replacement reactions

In **single-replacement reactions**, an element reacts with a compound in such a way that the element replaces one of the existing elements in the compound. An analogy can be made to a single person cutting in on dance partners, producing a new couple and a different single person.

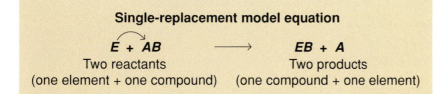

Single-replacement model equation

$E + AB$	$\longrightarrow$	$EB + A$
Two reactants		Two products
(one element + one compound)		(one compound + one element)

Some metallic elements replace other metal cations in ionic compounds. You recall that ionic compounds are made up of ions even though their chemical formulas do not show the ionic charges.

$Zn(s)$	+	$CuSO_4(aq)$	$\longrightarrow$	$ZnSO_4(aq)$	+	$Cu(s)$
Metallic element		Ionic compound		Ionic compound		Metallic element
		$(Cu^{2+}SO_4^{2-})$		$(Zn^{2+}SO_4^{2-})$		

Notice how this example fits the model equation with Zn replacing Cu in the sulfate compound. When examined in greater detail, you notice also that zinc atoms (Zn) are becoming zinc ions (Zn^{2+}); they are replacing the copper cations (Cu^{2+}), which are becoming copper atoms (Cu). In the laboratory, you may readily observe these changes: When Zn atoms become aqueous Zn^{2+} ions, the solid piece of Zn metal partially disappears. When aqueous Cu^{2+} ions become Cu atoms, the Cu^{2+} ions precipitate out of solution and appear as solid, metallic Cu.

Other examples show these same features, namely, one metallic element is changed from atoms to ions, while the other metallic element undergoes the change from ions to atoms. Reactions such as these in which an element changes charge are also called **oxidation-reduction reactions** and will be discussed further in Chapter 16. Particularly note the last equation given below, which is the equation used to produce the metal titanium (see introductory photo for this chapter).

$$2\ K(s) + \underset{(Pb^{2+})}{Pb(NO_3)_2(aq)} \longrightarrow \underset{(K^+)}{2\ KNO_3(aq)} + Pb(s)$$

$$Sn(s) + \underset{(Ag^+)}{2\ AgNO_3(aq)} \longrightarrow \underset{(Sn^{2+})}{Sn(NO_3)_2(aq)} + 2\ Ag(s)$$

$$2\ Mg(l) + TiCl_4(g) \xrightarrow{\Delta} Ti(s) + 2\ MgCl_2(l)$$

Figure 9.12

The test tube on the left contains a strip of copper metal in a clear solution of zinc sulfate. The shiny, untarnished appearance of the metal and the clarity of the solution indicate that there is no reaction between the materials. The test tube on the right contains a strip of zinc metal in copper(II) sulfate solution. The portion of the metal that is submerged has darkened as it reacts with the solution. The particles at the bottom of the test tube are Cu particles that have been produced in the reaction and precipitated out of solution.

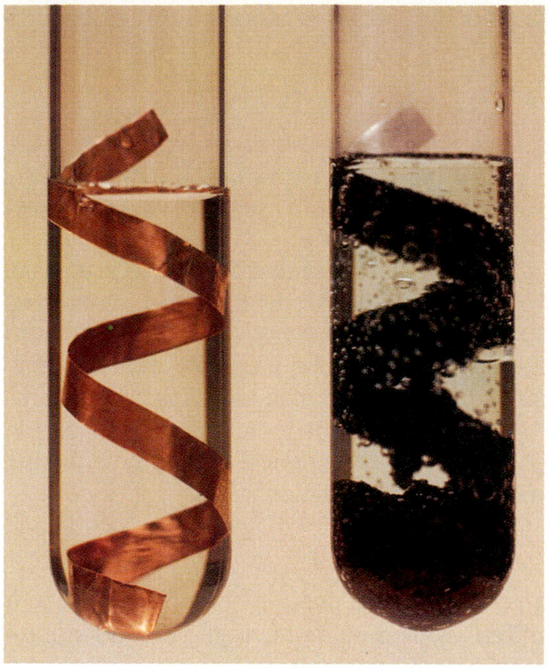

Whereas zinc will replace copper in a compound, as shown earlier, copper will not replace zinc; that is, mixing together Cu(*s*) and $ZnSO_4$(*aq*) results in no reaction (Figure 9.12). We say that Zn(*s*) is more reactive than Cu(*s*). *More reactive metals replace less reactive metals in compounds.* Table 9.2, the activity series of metals, tells you which are the more reactive metals. Look back at the other preceding examples and notice that K(*s*) can replace Pb^{2+}(*aq*) because K(*s*) is higher than Pb(*s*) in the activity series. Similarly, Sn(*s*) is above Ag(*s*) in the activities series, so Sn(*s*) can replace Ag^+ from $AgNO_3$; Mg(*s*) is above Ti(*s*), so Mg(*s*) can replace Ti^{4+} from $TiCl_4$.

sample exercise 6

Predict whether the following replacement reactions will occur.

a. Na(*s*) + $AgNO_3$(*aq*) $\rightarrow$?
b. Al(*s*) + KCl(*aq*) $\rightarrow$?
c. Pb(*s*) + $ZnSO_4$(*aq*) $\rightarrow$?

solution

Consult the activity series (Table 9.2). Replacement occurs if the metallic element is higher in the series than the metallic cation in the compound.

a. Na is above Ag in the activity series. The single-replacement reaction will occur.
b. No reaction. Al is below K in the activity series.
c. No reaction. Pb is below Zn in the activity series.

Hydrogen appears in the activity series shown in Table 9.2 even though it is not a metal. It turns out that active metals, those above hydrogen in the activity series, such as Zn and Mg, can replace the hydrogen in acids, whereas those metals below hydrogen in the activity series, such as Cu, Hg, Ag, and Au, will not replace hydrogen in acids.

Metallic element		Acid		Ionic compound		Element (hydrogen gas)
Zn(*s*)	+	H_2SO_4(*aq*)	$\longrightarrow$	$ZnSO_4$(*aq*)	+	$H_2 \uparrow$
Mg(*s*)	+	2 HCl(*aq*)	$\longrightarrow$	$MgCl_2$(*aq*)	+	$H_2 \uparrow$

Notice how these examples fit the model equation. Metals above H in the series replace it; metals below do not.

The evolution of H_2 gas in these reactions is readily observed in the laboratory. Gas bubbles appear steadily in the reaction mixture. A common test for the presence of hydrogen gas is to place a burning wooden splint into the mouth of a test tube containing the reaction mixture. For safety, the test tube should be in a hood behind a safety shield. If you hear an audible pop or see a flash, then H_2 gas is present.

table 9.2 activity series of metals

Symbol	Element	
Li	Lithium	**Most** reactive
K	Potassium	↑
Ba	Barium	
Ca	Calcium	
Na	Sodium	
Mg	Magnesium	
Al	Aluminum	
Ti	Titanium	
Zn	Zinc	Reactivity increases as you go up the series
Fe	Iron	
Cd	Cadmium	
Ni	Nickel	
Sn	Tin	
Pb	Lead	
(H)	Hydrogen	
Cu	Copper	
Hg	Mercury	
Ag	Silver	
Au	Gold	**Least** reactive

••• problem 7

Predict whether the following replacement reactions will occur.

a. $Au(s) + HCl(aq) \rightarrow ?$

b. $Al(s) + HNO_3(aq) \rightarrow ?$

The order of metals in the activity series depends on the ease with which the metal atoms lose electrons to become metal cations. The more easily they lose electrons to form cations, the more reactive they are. An important factor that indicates the ease with which a metal loses electrons is the ionization energy of the metal (Section 6.3). The lower the ionization energy of the metal, the easier it loses electrons. You will study this further in Section 16.8.

Whenever you see a metal reacting with either an ionic compound or an acid, you should now be able to predict the products with the help of Table 9.2.

sample exercise 7

Complete the following equation: $Ca(s) + AgNO_3(aq) \rightarrow ?$

solution

1. Recognize that there is a metal, $Ca(s)$, reacting with an ionic compound. This signals a single-replacement reaction.
2. Check the activity series (Table 9.2). In this case, Ca is above Ag, so a reaction will occur. (If Ca were below Ag, you would write "no reaction" and proceed no further.)

3. Write correct formulas for the products, which will be Ag(*s*) metal and the ionic compound that forms from Ca^{2+} and NO_3^-, namely, $Ca(NO_3)_2$.

$$Ca(s) + AgNO_3(aq) \rightarrow Ca(NO_3)_2(aq) + Ag(s)$$

4. Balance the equation according to the procedure in Section 9.4.

$$Ca(s) + 2\ AgNO_3(aq) \rightarrow Ca(NO_3)_2(aq) + 2\ Ag(s)$$

---

sample exercise 8

Complete the following equation: Al(*s*) + HCl(*aq*) → ?

solution

1. Recognize that this is a metal, Al(*s*), reacting with an aqueous acid. This signals a single-replacement reaction.
2. Check the activity series. In this case, Al is above H, so a reaction will occur. (If Al were below H, you would write "no reaction" and proceed no further.)
3. Write correct formulas for the products. One product of the reaction of an active metal and an acid is always H_2. (Remember, the element hydrogen exists as diatomic molecules.) The other product is the ionic compound that forms from Al^{3+} and Cl^-, namely, $AlCl_3$.

$$Al(s) + HCl(aq) \rightarrow AlCl_3(aq) + H_2 \uparrow$$

4. Balance the equation.

$$2\ Al(s) + 6\ HCl(aq) \rightarrow 2\ AlCl_3(aq) + 3\ H_2 \uparrow$$

You will study these reactions again in Chapter 13, where you will learn to write these equations in terms of the ions that exist in aqueous solution. You will see that ionic compounds that are soluble in water dissociate (separate) into ions in aqueous solution.

double-replacement reactions

In **double-replacement reactions,** two compounds react with each other to form two different compounds.

Double-replacement model equation

$$AB + XY \rightarrow AY + XB$$

Two reactants (two compounds)	Two products (two compounds)

There is a double replacement (two replacements) in the sense that *A* replaces *X* in *XY* and *X* replaces *A* in *AB*. It is perhaps easier to view the reaction as a "switching of partners." The *AB* and *XY* partnerships are replaced with *AY* and *XB* partnerships.

Ionic compounds reacting in aqueous solution are excellent examples of double-replacement reactions.

$$\begin{matrix} NaCl(aq) & + & AgNO_3(aq) & \rightarrow & NaNO_3(aq) & + & AgCl \downarrow \\ (Na^+, Cl^-) & & (Ag^+, NO_3^-) & & (Na^+, NO_3^-) & & (Ag^+, Cl^-) \end{matrix}$$

Notice how this example fits the model equation. The cations and anions switch partners to form new ionic compounds. Look at some more examples:

$$\begin{array}{ccccccc} K_2S(aq) & + & MgSO_4(aq) & \rightarrow & K_2SO_4(aq) & + & MgS\downarrow \\ (K^+, S^{2-}) & & (Mg^{2+}, SO_4{}^{2-}) & & (K^+, SO_4{}^{2-}) & & (Mg^{2+}, S^{2-}) \end{array}$$

$$\begin{array}{ccccccc} Na_2CO_3(aq) & + & CaCl_2(aq) & \rightarrow & 2\ NaCl(aq) & + & CaCO_3\downarrow \\ (Na^+, CO_3{}^{2-}) & & (Ca^{2+}, Cl^-) & & (Na^+, Cl^-) & & (Ca^{2+}, CO_3{}^{2-}) \end{array}$$

Not all ionic compounds actually react to yield the products shown exclusively, neither do they necessarily yield products in an easily isolated form. One driving force that ensures that the products form as written is the occurrence of one of the products as a precipitate (↓). Please notice that one product in each of the above double-replacement examples is a precipitate. Precipitates form when the ions combine into a product that is insoluble in water. The formation of a precipitate is readily observed in the laboratory. Usually, the precipitate will settle to the bottom of the test tube or beaker holding the reaction mixture. If the precipitate is fine and remains suspended in a cloudy solution, place the solution in a centrifuge for about 30 seconds. The precipitate, if present, will quickly settle out of the solution (Figure 9.13). In Section 13.10, you will learn solubility rules to help you predict which compounds will precipitate in double-replacement reactions. In this chapter, you will be given examples of ionic compounds that do react in double-replacement reactions and asked to complete the equations correctly for these reactions by switching cation-anion partners. To do this, you must be able to write correct formulas for ionic compounds according to the principles in Section 6.8.

Figure 9.13

If a precipitate is fine, it may not settle out of the reaction mixture, but remain suspended and appear as a cloudy solution. To verify that a precipitate is present, place the solution in a centrifuge for about 30 seconds. The precipitate, if present, will settle to the bottom as a distinct solid layer.

sample exercise 9

Complete the following equation: $MgCl_2(aq) + K_2CO_3(aq) \rightarrow ?$

solution

1. Recognize that the reactants are two ionic compounds. This signals a double-replacement reaction.
2. Identify the cation-anion pairs in the reactants. $MgCl_2$ is made up of Mg^{2+} cations and Cl^- anions, and K_2CO_3 is made up of K^+ cations and $CO_3{}^{2-}$ anions.
3. Switch the cation-anion partners and write correct formulas for the new cation-anion pairs. Mg^{2+} teams up with $CO_3{}^{2-}$ to give $MgCO_3$, a precipitate, and K^+ teams up with Cl^- to give KCl.

$$MgCl_2(aq) + K_2CO_3(aq) \rightarrow MgCO_3\downarrow + KCl(aq)$$

4. Balance the equation.

$$MgCl_2(aq) + K_2CO_3(aq) \rightarrow MgCO_3\downarrow + 2\ KCl(aq)$$

Magnesium carbonate is often used as an additive to table salt to keep it free flowing.

sample exercise 10

Complete the following equation: $Na_3PO_4(aq) + Fe_2(SO_4)_3(aq) \rightarrow ?$

solution

1. Recognize the signal for a double-replacement reaction: the reactants are two ionic compounds.
2. Identify the cation-anion pairs in the reactants. Na_3PO_4 is made up of Na^+ cations and PO_4^{3-} anions, and $Fe_2(SO_4)_3$ is made up of Fe^{3+} cations and SO_4^{2-} anions. Note: You can ascertain the ionic charges by using a reverse crisscross procedure:

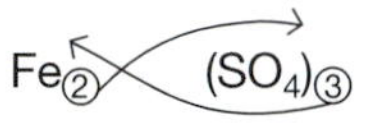

3. Switch the cation-anion partners and write correct formulas for the new cation-anion pairs. Na^+ teams up with SO_4^{2-} to give Na_2SO_4, and Fe^{3+} teams up with PO_4^{3-} to give $FePO_4$, a precipitate.

$$Na_3PO_4(aq) + Fe_2(SO_4)_3(aq) \rightarrow Na_2SO_4(aq) + FePO_4\downarrow$$

4. Balance the equation.

$$2\ Na_3PO_4(aq) + Fe_2(SO_4)_3(aq) \rightarrow 3\ Na_2SO_4(aq) + 2\ FePO_4\downarrow$$

Sodium sulfate has been used as a filler in household laundry detergents because it added bulk to the active agents. Manufacturers found that consumers preferred a cupful of detergent rather than a smaller amount of concentrate. However, widespread problems of waste disposal have led detergent manufacturers to remove fillers and to produce concentrated detergent in the smallest possible package.

Another commonly encountered double-replacement reaction is the reaction between an aqueous acid and a base (metal hydroxide) to produce water and an ionic compound. For example,

$HCl(aq)$	+	$NaOH(aq)$	$\longrightarrow$	$NaCl(aq)$	+	$HOH(l)$
Acid		Base		Ionic compound		Water

HCl and NaOH behave like ionic compounds because they separate into ions in aqueous solution. HCl is an acid because it forms H^+ ions and NaOH is a base because it forms OH^- ions in water. The H^+ and OH^- ions combine to form water and, consequently, neutralize each other. The aqueous sodium chloride solution formed is neither acidic nor basic. In the laboratory, you may test the acidic, aqueous HCl solution with blue litmus paper and see that the paper turns red. You may test the basic, aqueous NaOH solution with red litmus paper and see that the paper turns blue. As you slowly add NaOH to a test tube containing HCl, the reaction mixture will gradually become less acidic and finally neutral, at which point neither the red nor the blue litmus paper changes color (Figure 9.14).

In Chapter 15, you will explore in more detail which acids and bases separate into ions in aqueous solution. But, in general, any aqueous acid and base react to neutralize each other by forming an ionic compound and water. You will recognize neutralization reactions as a special category of double-replacement reactions that form an ionic compound and water. For now, Table 9.3 summarizes the features of the four types of reactions with which you should be familiar.

sample exercise 11

Identify the type of reaction represented by each equation below.

a. $2\ Li(s) + Ni(NO_3)_2(aq) \rightarrow 2\ LiNO_3(aq) + Ni(s)$
b. $2\ HgO(s) \rightarrow 2\ Hg(l) + O_2\uparrow$
c. $N_2(g) + 3\ H_2(g) \rightarrow 2\ NH_3(g)$
d. $2\ KCl(aq) + Pb(NO_3)_2(aq) \rightarrow 2\ KNO_3(aq) + PbCl_2\downarrow$

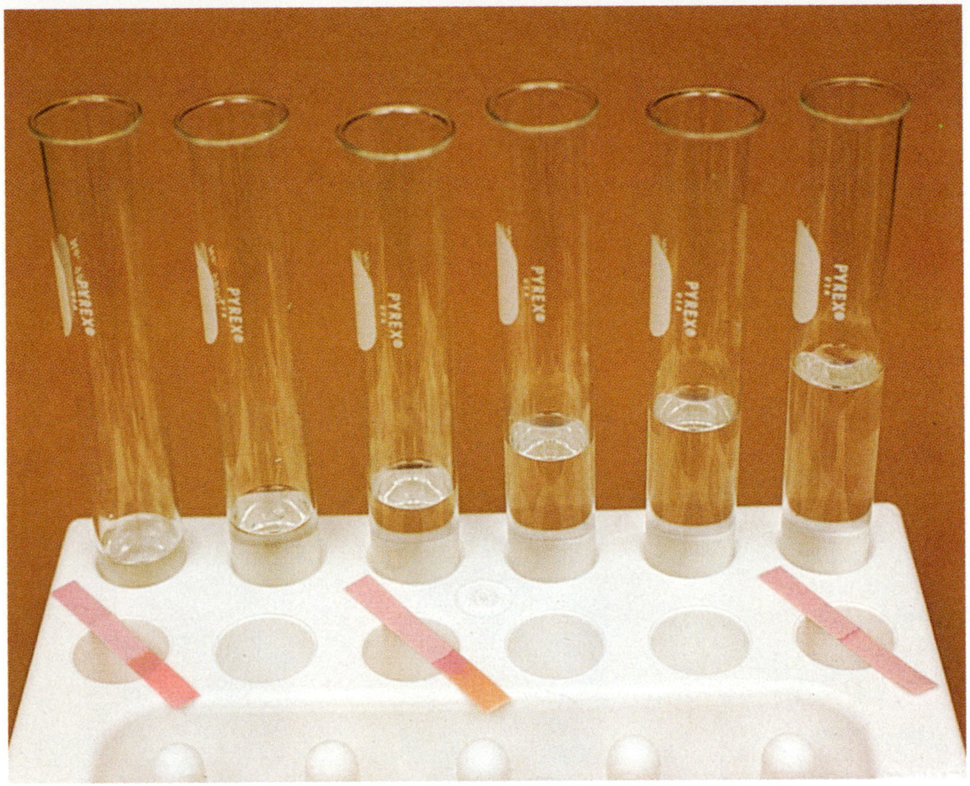

Figure 9.14

Originally, all of these test tubes contained the same volume of HCl. An increasing amount of NaOH was added to each test tube from left to right. The total volume in each test tube indicates the added volume of NaOH. You can see that blue litmus paper becomes less red as more base is added to neutralize the acid.

table 9.3 summary of the four reaction types

Reaction type	Model equation	Number of reactants	Number of products	Important examples*
Combination	$A + B \rightarrow AB$	2	1	Metal + Nonmetal → Ionic compound Nonmetal + Nonmetal → Covalent compound
Decomposition	$XY \rightarrow X + Y$	1	2 (or more)	See page 258
Single-replacement	$E + AB \rightarrow EB + A$	2	2	Metal + Ionic compound → Ionic compound + Metal Metal + Acid → Ionic compound + H_2
Double-replacement	$AB + XY \rightarrow AY + XB$	2	2	2 Ionic compounds → 2 Ionic compounds Acid + Base → Ionic compound + Water

*You must be able to recognize and distinguish among all four reaction types. In addition, you should be able to complete equations for the reactions listed under "Important Examples."

solution

Compare each equation to the model equations to see which one fits. Start by counting the number of reactants and products.

a. Two reactants → two products; therefore, this must be a replacement reaction. It is a *single-replacement reaction:* Li replaces Ni.

b. One reactant → two products; therefore, it is a *decomposition reaction.*

c. Two reactants → one product; therefore, it is a *combination reaction.*

d. Two reactants → two products; therefore, this must be a replacement reaction. It is a *double-replacement reaction* in which there is a switch of cation-anion pairs.

CHEMLAB

Laboratory exercise 9: types of chemical reactions

Questions:

For each of the four experiments described in this exercise, determine the following.

1. Classify the reaction type.
2. Complete and balance your predicted chemical equation.
3. Predict the experimental observations.
4. Describe the experimental test(s) you would use to confirm your predictions.
5. Compare your experimental results with your predictions.

Experiment 1:

Half-fill a 6-in. test tube with copper(II) sulfate solution. Place a small piece of Al wire or foil in the $CuSO_4$ solution and allow the solution to sit for 3 to 5 minutes.

Since aluminum is an element and copper(II) sulfate is an ionic compound, their reaction would follow the single-replacement model since Al is above Cu in the activity series (Table 9.2).

$$Al(s) + CuSO_4(aq) \rightarrow Cu(s) + Al_2(SO_4)_3(aq)$$

Balancing this equation gives

$$2\ Al(s) + 3\ CuSO_4(aq) \rightarrow 3\ Cu(s) + Al_2(SO_4)_3(aq)$$

You should observe aluminum disappearing while solid, metallic copper is deposited from the reaction mixture. The appearance of Cu and the disappearance of Al confirm the reaction as predicted (Figure 9.15).

Figure 9.15

The single-replacement reaction between aluminum and copper(II) sulfate results in the deposition of Cu on the tube and the disappearance of metallic Al, which has dissolved into the solution.

Experiment 2:

Half-fill a test tube with dilute HNO_3. Add a small piece of Zn to the acid solution. Since zinc is an element and HNO_3 is an acid, their reaction would follow the single-replacement model in which the metal replaces hydrogen from the acid. This will occur since Zn is above H in the activity series (Table 9.2).

$$Zn(s) + 2\ HNO_3(aq) \rightarrow Zn(NO_3)_2(aq) + H_2 \uparrow$$

You should observe Zn disappearing while hydrogen gas bubbles from the reaction mixture. You may test for the presence of H_2 gas by placing a burning splint in the mouth of the test tube containing the reaction mixture. If H_2 is present, you will hear an audible pop or see a flash (Figure 9.16).

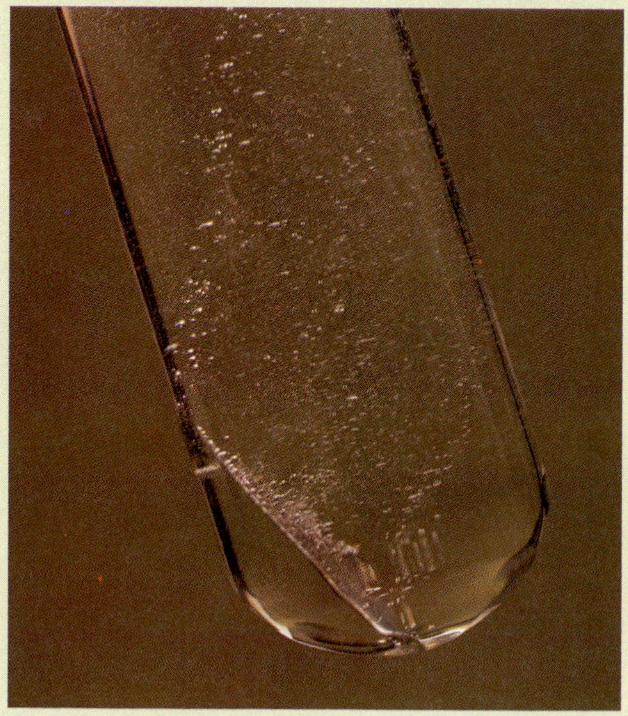

Figure 9.16

The formation of a gas in the single-replacement reaction of Zn with HNO_3 can be seen by the formation of bubbles in the reaction mixture. If a burning splint gives an audible pop or visible flash when placed in the mouth of the test tube, H_2 gas is present.

Experiment 3:

Place 2 mL of an aqueous solution of $Pb(NO_3)_2$ in a test tube and add 5 to 10 drops of NaI solution.

Since both reactants are ionic compounds, their reaction should follow the double-replacement model with cations and anions exchanging partners. You must be told for now that one of the possible products, PbI_2, is not soluble in water, so the chemical equation will be

$$Pb(NO_3)_2(aq) + NaI(aq) \rightarrow Na(NO_3)(aq) + PbI_2 \downarrow$$

Balancing this equation gives

$$Pb(NO_3)_2(aq) + 2\ NaI(aq) \rightarrow 2\ Na(NO_3)(aq) + PbI_2 \downarrow$$

Since PbI_2 is not soluble in water, it will precipitate from the reaction mixture. The formation of a precipitate confirms the double-replacement reaction predicted (Figure 9.17).

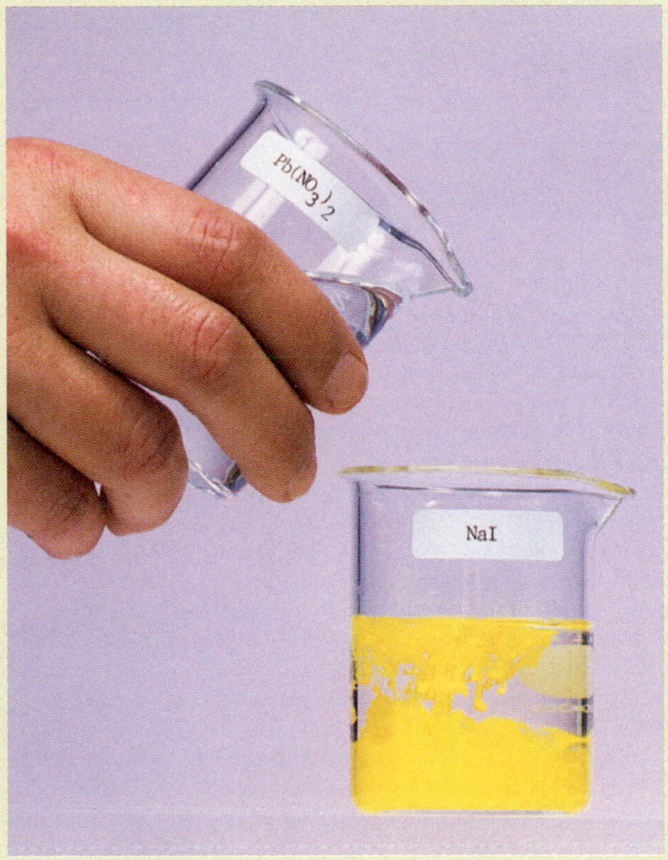

Figure 9.17

A yellow precipitate (PbI_2) forms from two clear, colorless solutions when $Pb(NO_3)_2$ and NaI are mixed. The reaction follows the double-replacement model.

Experiment 4:

Place a small amount of solid HgO in a test tube and heat gently. This experiment should be performed only *as a demonstration* by your instructor in a well-vented hood so that you are not exposed to toxic mercury vapor. If the tube is heated gently so that the upper portion of the tube remains cool, the mercury vapor should condense within the test tube and not escape.

Since HgO is a compound and it is the only one present, it must decompose into its elements, Hg and O_2. The chemical equation for the decomposition would be

$$HgO \rightarrow Hg + O_2 \uparrow$$

Balancing this equation gives

$$2\ HgO \rightarrow 2\ Hg + O_2 \uparrow$$

Droplets or a film of liquid mercury should appear on the cool part of the test tube containing the HgO after it has been heated. The evolution of O_2 gas can be tested by placing a glowing splint in the mouth of the test tube. If O_2 gas is present, the splint will burn vigorously (Figure 9.18).

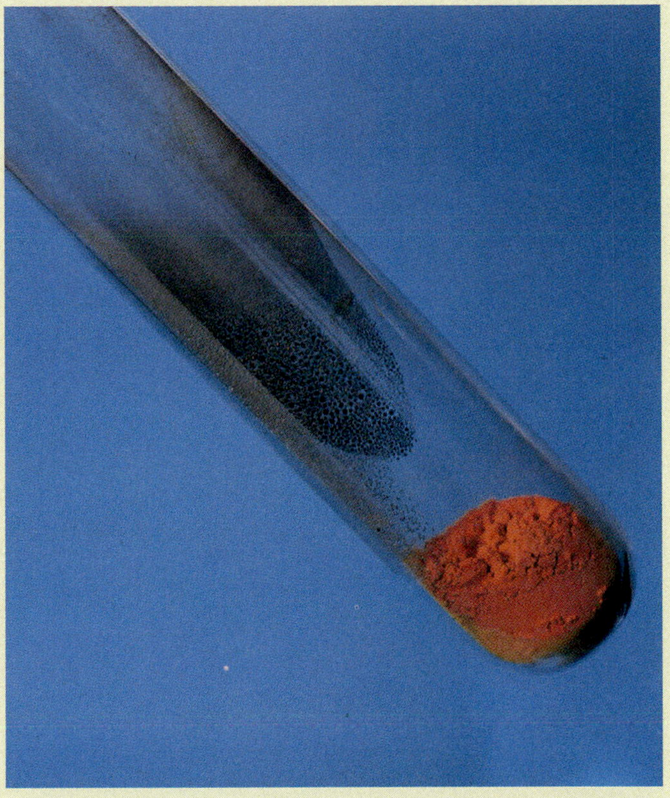

Figure 9.18

After heating mercuric oxide, HgO, a red solid and a silver-colored film of Hg appear. The Hg is deposited on the cooler portion of the test tube, above where HgO was heated. By placing a glowing splint in the mouth of the test tube, the presence of O_2 gas could also be confirmed.

sample exercise 12

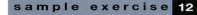

Complete and balance the following equations.

a. $Zn(s) + HNO_3(aq) \rightarrow$?

b. $Zn(s) + O_2(g) \rightarrow$?

c. $ZnCl_2(aq) + K_2CO_3(aq) \rightarrow$? (Note that zinc carbonate is an insoluble compound.)

solution

Decide on the reaction type and then proceed as previously described for that type.

a. This fits the model

E	+	*AB*	$\rightarrow$
Element		Compound	

Therefore, it is a *single-replacement reaction.* This is the reaction of an active metal plus an acid. See Sample Exercise 8.

$$Zn(s) + 2\ HNO_3(aq) \rightarrow Zn(NO_3)_2(aq) + H_2 \uparrow$$

b. This is the *combination* of two elements to give a compound. See Sample Exercise 5.

$$2\ Zn(s) + O_2(g) \rightarrow 2\ ZnO(s)$$

c. Two ionic compounds as reactants signal a *double-replacement reaction.* See Sample Exercises 9 and 10. Note that $ZnCO_3$ is insoluble and forms a precipitate in the reaction.

$$ZnCl_2(aq) + K_2CO_3(aq) \rightarrow ZnCO_3 \downarrow + 2\ KCl(aq)$$

CHEMQUEST

Burning fuels such as coal or petroleum that contain impurities of sulfur produce sulfur dioxide gas, $SO_s(g)$, according to the combustion reaction, $S(s) + O_2(g) \rightarrow SO_2(g)$. Sulfur dioxide then reacts with oxygen in the air to form sulfur trioxide, $SO_3(g)$, in the reaction $2\ SO_2(g) + O_2(g) \rightarrow 2\ SO_3(g)$. The sulfur trioxide subsequently combines with moisture in the air to form sulfuric acid, $H_2SO_4(aq)$, and acid rain by $H_2O(l) + SO_3(g) \rightarrow H_2SO_4(aq)$. Note that all three of the previous reactions are combination reactions. A similar sequence of environmentally pertinent reactions lead to the formation of nitric acid, $HNO_3(aq)$, also as a component of acid rain.

The harmful effects of acid rain on buildings, flora, and fauna has become well documented. The acidity of some lakes has increased such that fish may no longer survive. Acres of forest have been defoliated with conifers particularly susceptible. Buildings and statues made from marble and limestone, whose primary constituent is calcium carbonate, $CaCO_3(s)$, slowly dissolve according to the double-replacement reactions with acid, $2\ HNO_3(aq) + CaCO_3(s) \rightarrow Ca(NO_3)_2(aq) + CO_2(g) + H_2O(l)$. Note that the products of this reaction include carbon dioxide gas and water, which form from unstable carbonic acid, $H_2CO_3(aq)$ and soluble calcium nitrate, which washes away from the surface.

Acid rain also dissolves toxic, heavy metal compounds in the soil that would remain insoluble without the presence of acid. Normally insoluble lead carbonate, $PbCO_3(s)$, will dissolve in the presence of nitric acid to produce soluble lead nitrate, $Pb(NO_3)_2(aq)$, according to the double replacement reaction, $2\ HNO_3(aq) + PbCO_3(s) \rightarrow Pb(NO_3)_2(aq) + CO_2(g) + H_2O(l)$. Consequently, the toxic, heavy metal ions of lead may be leached from the soil into your drinking water.

Many simple reactions may be used to analyze the quality of your environment. For example, every hardware store now has a simple kit for testing your drinking water for the presence of lead. Research the chemicals provided in such a test kit and how they determine whether lead is present. How small an amount of lead can they detect?

chapter accomplishments

After completing this chapter, you should be able to define all key terms and do the following.

9.2 Chemical equations

- ❑ Translate English sentences that describe chemical reactions into chemical equations.
- ❑ Recognize the importance of correct chemical formulas in chemical equations.

9.3 The meaning of balanced chemical equations

- ❑ State the law of conservation of matter.
- ❑ State the meaning of the word *balanced* as it refers to a chemical equation.

9.4 Balancing equations

- ❑ Recognize whether an equation is unbalanced or balanced.
- ❑ Balance a chemical equation.

9.5 Using correct formulas

- ❑ Recognize chemical equations with an improper chemical formula.

9.6 Helpful hints for balancing equations

- ❑ Recognize and balance common polyatomic ions on both sides of chemical equation.

9.7 Designating physical states and special conditions in reactions

- ❑ State the meaning of the special symbols in Table 9.1.

9.8 Types of reactions

- ❑ Name the four major types of reactions.
- ❑ Recognize a combination reaction.
- ❑ Predict the nature of the product that will be formed from the combination of any two elements.
- ❑ Given metal and nonmetal reactants, write a balanced chemical equation for their chemical reaction.
- ❑ Recognize a decomposition reaction.
- ❑ Recognize a single-replacement reaction.
- ❑ Given an activity series, predict whether a given single-replacement reaction will occur.
- ❑ Given an activity series and metal and ionic compound reactants, write a balanced chemical equation for their chemical reaction.
- ❑ Given an activity series and metal and acid reactants, write a balanced equation for their reaction.
- ❑ Recognize a double-replacement reaction.
- ❑ Given two ionic compound reactants, write a balanced equation for their reaction.

key terms

reactants
products
balanced chemical equations
law of conservation of matter
aqueous
precipitate
catalyst
combination reactions
bases
decomposition reactions
single-replacement reactions
oxidation-reduction reactions
double-replacement reactions

problems

9.1 Recognizing chemical reactions

8. Define or explain the following terms.
   a. Chemical reaction
   b. Chemical equation
   c. Reactant
   d. Product

9.2 Chemical equations

9. What symbol do chemists use to represent the word *yields?*
10. Associate the terms *reactant* and *product* with the words *right* and *left.*
11. $Zn + 2\ HCl \rightarrow ZnCl_2 + H_2$
    a. Write the names of the reactants in the above equation.
    b. Write the names of the products for the above reaction.
12. Translate the following English sentences into chemical equations.
    a. Iron reacts with sulfur to yield iron(II) sulfide.
    b. Hydrochloric acid reacts with sodium hydroxide to yield sodium chloride and water.
    c. Tin reacts with oxygen to form tin(IV) oxide.
    d. Nitric acid reacts with lithium carbonate to yield lithium nitrate, water, and carbon dioxide.

13. Translate these equations into English sentences.
   a. $BaCl_2 + Na_2SO_4 \rightarrow 2\ NaCl + BaSO_4$
   b. $Si + O_2 \rightarrow SiO_2$
   c. $2\ K + I_2 \rightarrow 2\ KI$
   d. $H_2SO_4 + Ca(OH)_2 \rightarrow CaSO_4 + 2\ H_2O$

9.3 The meaning of balanced chemical equations

14. What is the relationship between the law of conservation of matter and a balanced chemical equation?
15. Identify each of the following equations as balanced or unbalanced.
   a. $2\ KI + Br_2 \rightarrow 2\ KBr + I_2$
   b. $2\ P + 5\ O_2 \rightarrow 2\ P_2O_5$
   c. $Al + 3\ HBr \rightarrow AlBr_3 + 3\ H_2$
   d. $2\ NH_3 + H_2SO_4 \rightarrow (NH_4)_2SO_4$
   e. $H_2O_2 \rightarrow H_2O + O_2$
   f. $H_2 + Cl_2 \rightarrow HCl$
16. A chemist carries out a reaction in which $A + B \rightarrow C$. She finds that 18 g of product form when she uses 8 g of reactant *A*. What is the minimum amount of reactant *B* that the chemist must have combined with reactant *A?*

9.4 Balancing equations

17. Fill in the blanks: To balance chemical equations, one should use __________. In balancing equations, __________ in chemical formulas should never be altered.
18. Balance the equations that you identified as unbalanced in Problem 15.
19. Balance the following equations.
   a. $KI + Cl_2 \rightarrow KCl + I_2$
   b. $Cu + O_2 \rightarrow CuO$
   c. $Li_2O + H_2O \rightarrow LiOH$
   d. $SO_2 + O_2 \rightarrow SO_3$
   e. $H_2 + N_2 \rightarrow NH_3$
   f. $Na + ZnSO_4 \rightarrow Na_2SO_4 + Zn$
20. Balance the following equations.
   a. $N_2 + O_2 \rightarrow NO$
   b. $NaBr + Cl_2 \rightarrow NaCl + Br_2$
   c. $P + Cl_2 \rightarrow PCl_5$
   d. $BaCl_2 + (NH_4)_2SO_4 \rightarrow BaSO_4 + NH_4Cl$
   e. $K_2O + H_2O \rightarrow KOH$
   f. $Fe + O_2 \rightarrow Fe_2O_3$
   g. $CaC_2 + H_2O \rightarrow C_2H_2 + Ca(OH)_2$
   h. $Zn + HNO_3 \rightarrow Zn(NO_3)_2 + H_2$
   i. $NH_4NO_2 \rightarrow N_2 + H_2O$
   j. $PbO + O_2 \rightarrow PbO_2$
21. Balance the following equations.
   a. $CH_4 + O_2 \rightarrow CO_2 + H_2O$
   b. $FeS + O_2 \rightarrow FeO + SO_2$
   c. $P_4O_{10} + H_2O \rightarrow H_3PO_4$
   d. $Cl_2 + H_2O \rightarrow HCl + HClO$
   e. $(NH_4)_2Cr_2O_7 \rightarrow Cr_2O_3 + N_2 + H_2O$
   f. $Al + CuSO_4 \rightarrow Al_2(SO_4)_3 + Cu$
22. Balance the following equations.
   a. $C_2H_6 + O_2 \rightarrow CO_2 + H_2O$
   b. $C_4H_{10} + O_2 \rightarrow CO_2 + H_2O$
   c. $IBr + NH_3 \rightarrow NH_4Br + NI_3$
   d. $Fe(OH)_3 + H_2SO_4 \rightarrow Fe_2(SO_4)_3 + H_2O$
   e. $Al + Sn(NO_3)_2 \rightarrow Al(NO_3)_3 + Sn$
   f. $NaOH + H_3PO_4 \rightarrow Na_3PO_4 + H_2O$
23. Translate the following English sentences into balanced chemical equations.
   a. Carbon reacts with chlorine to form carbon tetrachloride.
   b. Potassium reacts with nitrogen to form potassium nitride.
   c. Barium nitrate reacts with sulfuric acid to form insoluble barium sulfate and nitric acid.
   d. Calcium hydroxide decomposes to calcium oxide and water.
   e. Phosphorus and oxygen combine to yield diphosphorus pentoxide.
24. Translate the following English sentences into balanced chemical equations.
   a. Sodium chloride reacts with silver nitrate to form sodium nitrate and insoluble silver chloride.
   b. Iron reacts with hydrochloric acid to yield iron(III) chloride and hydrogen.
   c. Sodium bicarbonate reacts with acetic acid to form sodium acetate, carbon dioxide, and water.
   d. Copper reacts with sulfur to form copper(II) sulfide.

9.5 Using correct formulas

25. Find and correct the translation error(s) in each of the following.
   a. Sentence: Sodium reacts with iodine to form sodium iodide.

      Equation: $Na + I_2 \rightarrow NaI_2$
   b. Sentence: Silver oxide decomposes to silver and oxygen.

      Equation: $Ag_2O \rightarrow 2\ Ag + O$
   c. Sentence: Lead(II) nitrate reacts with potassium chloride to form insoluble lead(II) chloride and potassium nitrate.

      Equation: $Pb(NO_3)_2 + KCl \rightarrow PbCl + K(NO_3)_2$
   d. Sentence: Silicon reacts with oxygen to form silicon dioxide.

      Equation: $S + O_2 \rightarrow SO_2$

26. The following equations are incorrect, although they are balanced. Identify what is wrong with each one and write a correct equation.

a. $K_2CO_3 + CaCl_2 \rightarrow CaCO_3 + K_2Cl_2$

b. $N + 3\ H \rightarrow NH_3$

c. $C + Cl_4 \rightarrow CCl_4$

d. $K_2O + H_2O \rightarrow K_2(OH)_2$

27. The following equations are *incorrect.* Identify what is wrong with each one and write a correct equation.

a. $Al + H_2SO_4 \rightarrow AlSO_4 + H_2$

b. $N_2 + H_2 \rightarrow NH_4$

c. $BaCl_2 + Na_2SO_4 \rightarrow Na_2Cl_2 + BaSO_4$

d. $Mg + O_2 \rightarrow MgO_2$

9.6 Helpful hints for balancing equations

28. Balance the following equations.

a. $Fe(OH)_3 + H_2SO_4 \rightarrow Fe_2(SO_4)_3 + H_2O$

b. $C_2H_6 + O_2 \rightarrow CO_2 + H_2O$

c. $NaOH + Al(OH)_3 \rightarrow NaAlO_2 + H_2O$

d. $P_4O_{10} + H_2O \rightarrow H_3PO_4$

e. $K_2O + P_4O_{10} \rightarrow K_3PO_4$

f. $MgI_2 + H_2SO_4 \rightarrow HI + MgSO_4$

g. $PCl_5 + H_2O \rightarrow HCl + H_3PO_4$

h. $Al + Sn(NO_3)_2 \rightarrow Al(NO_3)_3 + Sn$

29. Balance the following equations.

a. $Cl_2O_7(l) + H_2O(l) \rightarrow HClO_4(aq)$

b. $NH_4NO_3(s) \rightarrow N_2O(g) + H_2O(l)$

c. $BaCl_2(aq) + (NH_4)_2CO_3(aq) \rightarrow BaCO_3 \downarrow + NH_4Cl(aq)$

d. $Ca(s) + HCl(g) \rightarrow CaCl_2(s) + H_2 \uparrow$

30. Translate into balanced equations.

a. Solid lithium reacts with liquid water to yield aqueous lithium hydroxide and hydrogen gas.

b. Heating solid tin(II) carbonate produces solid tin(II) oxide and carbon dioxide gas.

c. Aqueous potassium chloride reacts with aqueous silver nitrate, producing a precipitate of silver chloride and aqueous potassium nitrate.

31. Translate into balanced equations.

a. Iodine crystals react with chlorine gas to form solid iodine trichloride.

b. Solid zinc reacts with hydrochloric acid to yield aqueous zinc chloride and hydrogen gas.

c. Heating solid silver oxide produces metallic silver and oxygen gas.

d. Aluminum metal reacts with aqueous copper(II) sulfate to form aqueous aluminum sulfate and copper metal.

9.7 Designating physical states and special conditions in reactions

32. What are the meanings of the symbols (*s*), (*l*), (*g*), (*aq*), Δ, $\uparrow$, and $\downarrow$?

9.8 Types of reactions

33. What are the four types of chemical reactions discussed in this chapter?

34. Write a model equation using *A*s, *B*s, *X*s, and *Y*s for each of the four reaction types.

35. Look through the equations in Problems 11 through 31 and find one example of each reaction type.

36. Identify each of the following reactions as a combination, decomposition, single-replacement, or double-replacement.

a. $3\ Na(s) + Al(NO_3)_3(aq) \rightarrow 3\ NaNO_3(aq) + Al(s)$

b. $Na_3PO_4(aq) + Al(NO_3)_3(aq) \rightarrow AlPO_4 \downarrow + 3\ NaNO_3(aq)$

c. $2\ H_2O_2(l) \rightarrow 2\ H_2O(l) + O_2 \uparrow$

d. $BaO(s) + SO_3(g) \rightarrow BaSO_4 \downarrow$

e. $2\ NaClO(aq) \rightarrow 2\ NaCl(aq) + O_2 \uparrow$

f. $Cl_2(g) + 2\ KI(s) \rightarrow 2\ KCl(s) + I_2(s)$

g. $Ba(OH)_2(aq) + 2\ HNO_3(aq) \rightarrow Ba(NO_3)_2(aq) + 2\ H_2O(l)$

37. Fill in the blank: Elements react in combination reactions to form _______________.

38. Complete and balance the following.

a. $Mg + I_2 \rightarrow ?$

b. $Li + O_2 \rightarrow ?$

c. $Al + S \rightarrow ?$

d. $K + P \rightarrow ?$

e. $Ca + N_2 \rightarrow ?$

39. Write and balance a full chemical equation for each of the following.

a. Calcium reacts with bromine to yield _________.

b. Sulfur dioxide combines with water to form __________.

c. Magnesium reacts with nitrogen to yield __________.

d. Calcium oxide combines with water to form __________.

40. Give the formula of a reactant that could decompose to yield the following products.

a. $CO_2(g) + H_2O(l)$

b. $MgO(s) + CO_2(g)$

c. $Ag(s) + O_2(g)$

d. $NaCl(aq) + O_2(g)$

41. Consult the activity series in Table 9.2 and predict whether the following single-replacement reactions will occur.
   a. $Mg(s) + CuSO_4(aq) \rightarrow ?$
   b. $Mg(s) + CaSO_4(aq) \rightarrow ?$
   c. $Ba(s) + HCl(aq) \rightarrow ?$
   d. $Ag(s) + HNO_3(aq) \rightarrow ?$
   e. $Al(s) + Ni(NO_3)_2(aq) \rightarrow ?$
   f. $Ca(s) + H_3PO_4(aq) \rightarrow ?$
   g. $Al(s) + SnCl_2(aq) \rightarrow ?$
   h. $Au(s) + H_2SO_4(aq) \rightarrow ?$
42. Complete and balance those equations in Problem 41 for which reactions do occur.
43. a. Write an example of a balanced chemical equation illustrating the replacement of a metal by a more active metal.
   b. Write an example of a balanced chemical equation illustrating a metal replacing H_2 from an acid.
44. Complete and balance the following equations (write "NR" to indicate no reaction).
   a. $Cd(s) + AgNO_3(aq) \rightarrow ?$
   b. $Hg(l) + HCl(aq) \rightarrow ?$
   c. $Li(s) + H_3PO_4(aq) \rightarrow ?$
45. Complete and balance the following equations representing double-replacement reactions.
   a. $K_2SO_4(aq) + Ba(NO_3)_2(aq) \rightarrow ?$
   b. $(NH_4)_2CO_3(aq) + MgCl_2(aq) \rightarrow ?$
   c. $(NH_4)_3PO_4(aq) + Ca(NO_3)_2(aq) \rightarrow ?$
   d. $FeCl_2(aq) + K_3PO_4(aq) \rightarrow ?$
   e. $Na_2S(aq) + Ni(NO_3)_2(aq) \rightarrow ?$
46. Complete and balance the following double-replacement reactions.
   a. $Ba(OH)_2(aq) + HNO_3(aq) \rightarrow ?$
   b. $Fe(NO_3)_3(aq) + NaOH(aq) \rightarrow ?$
   c. $(NH_4)_2S(aq) + BaI_2(aq) \rightarrow ?$
   d. $H_2SO_3(aq) + Al(OH)_3(aq) \rightarrow ?$
   e. $(NH_4)_3PO_4(aq) + Ni(NO_3)_2(aq) \rightarrow ?$
47. Classify each of the following reactions as one of the four basic types, then complete and balance.
   a. $Li(s) + AuCl_3(aq) \rightarrow ?$
   b. $Al(NO_3)_3(aq) + K_2CO_3(aq) \rightarrow ?$
   c. $Ba(s) + F_2(g) \rightarrow ?$
   d. $Ba(s) + SnF_2(aq) \rightarrow ?$
   e. $Mg(s) + P(s) \rightarrow ?$
   f. $SnCl_2(aq) + Na_3PO_4(aq) \rightarrow ?$
48. Classify each of the following reactions as one of the four basic types, then complete and balance.
   a. $BaCO_3(s) \rightarrow ?$
   b. $Pb(NO_3)_2(aq) + Na(s) \rightarrow ?$
   c. $HClO_3(aq) + KOH(aq) \rightarrow ?$
   d. $Li(s) + O_2(g) \rightarrow ?$
   e. $Al(s) + Br_2(g) \rightarrow ?$

Additional problems

49. Baking soda (sodium bicarbonate) is used in cakes and breads to make them rise. Its action results from the fact that solid sodium bicarbonate decomposes when heated to form gaseous carbon dioxide, gaseous water, and solid sodium carbonate. Write a balanced equation for this reaction.
50. Rusting involves the chemical reaction of iron, water, and oxygen to form iron(III) hydroxide. Write a balanced equation for this description of the rusting process.
51. Zinc metal can be obtained by the reaction of zinc oxide with hydrogen gas. The other product is a familiar substance. Complete and balance the equation for the reaction of zinc oxide and hydrogen.
52. The chemical reaction that supplies current in an automobile battery is the reaction of lead plus lead(IV) oxide plus sulfuric acid to give lead(II) sulfate and water. Write a balanced equation for this reaction.
53. The combustion of gasoline is the reaction between octane (C_8H_{18}) and oxygen to form carbon dioxide and water. Write a balanced equation for the combustion.
54. Photosynthesis is the combination of carbon dioxide and water in the presence of light to form glucose ($C_6H_{12}O_6$) and oxygen. Write a balanced equation for photosynthesis.
55. Sulfur dioxide, a gaseous pollutant of air, readily combines with oxygen to form sulfur trioxide, which in turn dissolves in rainwater to form sulfuric acid. The sulfuric acid reacts with calcium carbonate, the chemical composition of marble structures, to form calcium sulfate, carbon dioxide, and water. Write balanced chemical equations for each of these three reactions.
56. Balance the following equations.
   a. $P_2 + O_2 \rightarrow P_2O_5$
   b. $NO + Br_2 \rightarrow NOBr$
   c. $Li_2O + H_2O \rightarrow LiOH$
   d. $N_2H_4 \rightarrow NH_3 + N_2$
   e. $H_3PO_4 + CaCO_3 \rightarrow Ca_3(PO_4)_2 + CO_2 + H_2O$
   f. $MgO + SiO_2 \rightarrow MgSiO_3$
   g. $ZnS + O_2 \rightarrow ZnO + SO_2$
   h. $Ba + HNO_3 \rightarrow Ba(NO_3)_2 + H_2$
   i. $Ca_3P_2 + H_2O \rightarrow Ca(OH)_2 + PH_3$
   j. $NH_3 + O_2 \rightarrow NO + H_2O$
57. Complete and balance the following equations.
   a. $Ca(s) + O_2(g) \rightarrow ?$
   b. $Li_2O(s) + H_2O(l) \rightarrow ?$
   c. $NaClO_3(s) \xrightarrow{\Delta} ?$
   d. $SO_3(g) + H_2O(l) \rightarrow ?$
   e. $MgO(s) + CO_2(g) \rightarrow ?$
58. Complete and balance the following equations (write "NR" for no reaction where appropriate).
   a. $Na(l) + AlCl_3(s) \rightarrow ?$
   b. $Cu(s) + Fe_2O_3(s) \rightarrow ?$
   c. $Pb(s) + H_2O(l) \rightarrow ?$
   d. $Ca(s) + HBr(aq) \rightarrow ?$

59. Complete and balance the following double-replacement reactions.
   a. $Pb(NO_3)_2 + KOH \rightarrow ?$
   b. $H_2SO_4 + Ca(OH)_2 \rightarrow ?$
   c. $HCl + Al(OH)_3 \rightarrow ?$
   d. $LiBr + Pb(ClO_4)_2 \rightarrow ?$
60. Limestone is made up primarily of calcium carbonate. Write a balanced equation for the reaction of limestone with sulfuric acid.
61. Common table sugar is sucrose, which has the molecular formula $C_{12}H_{22}O_{11}$. Write a balanced equation for the metabolism of sucrose in the body; that is, write its reaction with oxygen to yield carbon dioxide and water.
62. Write a balanced equation for the reaction of potassium metal with water.
63. When 36 g of water are decomposed, 4 g of hydrogen are formed. What is the maximum amount of oxygen that can form?
64. Write a balanced equation for the following reaction: 2 mol of iron metal react with 2 mol of water and 1 mol of oxygen (dissolved in water) to yield 2 mol of solid iron(II) hydroxide.
65. When octane, C_8H_{18}, the main component of gasoline, burns in a limited amount of oxygen, the products are water and the toxic gas, carbon monoxide. Write a balanced equation for this reaction.
66. Explain why copper, silver, and gold are commonly used as coinage materials.
67. When bromine is bubbled into an aqueous solution of sodium iodide, sodium bromide and iodine are produced. This reaction is sometimes used for the commercial production of iodine from sources concentrated in sodium iodide.
   a. Write a balanced chemical equation for this reaction.
   b. Which model equation is this reaction an example of?
   c. When bromine is bubbled into aqueous sodium chloride, no reaction occurs. What can you say about the displacement activity of chlorine, bromine, and iodine? How is this displacement activity related to their positions in Group VIIA? Compare the trend in displacement activity for a nonmetal group to the trend for a metallic group (that is, Group IA).

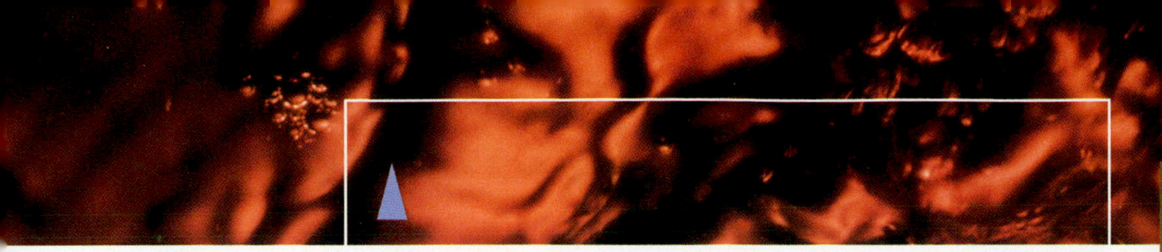

chapter 10

stoichiometry

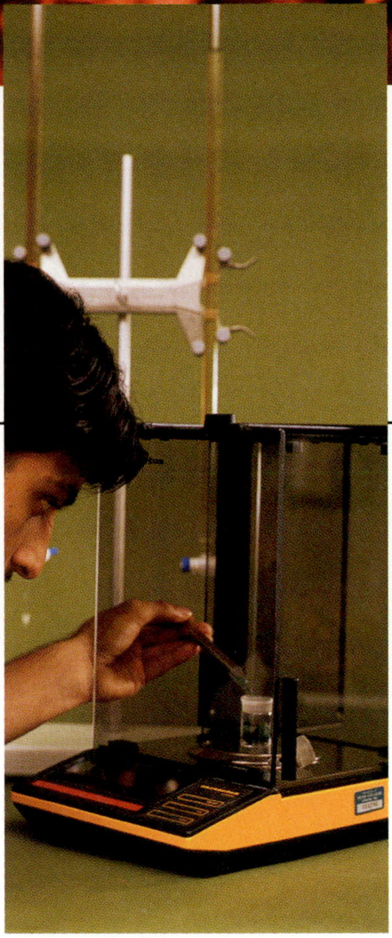

Stoichiometry relates the amount of reactants and products in chemical reactions. It enables you to weigh out the correct number of grams of each reactant and to determine the maximum number of grams of each product that may be formed.

10.1 what is stoichiometry?

In Chapter 9, you learned how to classify chemical reactions into four basic types. Knowledge of the four types of reactions often enables you to predict what reactants will be consumed and what new materials may be produced in a chemical reaction. For example, you know that rust will be produced when oxygen and iron (the major component of steel) react in a combination reaction. Conversely, if you observe the appearance of rust, a new material, you may deduce that the reactants, oxygen and iron, were present for the reaction to have occurred.

In Chapter 9, you also learned how to balance a chemical equation to make sure that the mass of each type of atom or ion is conserved in every chemical reaction. For the reaction of iron, Fe, with oxygen, O_2, to form iron(III) oxide, Fe_2O_3,[1] the balanced chemical equation

$$4\ Fe(s) + 3\ O_2(g) \rightarrow 2\ Fe_2O_3(s)$$

tells you that four atoms of iron react with three molecules of oxygen to form two formula units of iron(III) oxide. With this information, you are but a short step from knowing how to predict the masses of products that result from any given mass of chemical reactant; similarly, you are also a short step from knowing how to predict the mass of reactants needed to produce a given amount of chemical product (Figure 10.1).

1. Iron(III) oxide is a major constituent of rust. Rust is predominantly a mixture of iron(III) oxide, Fe_2O_3, and ferric hydroxide, which in some sources is now being written as FeO(OH). For simplicity in this discussion, we will assume that our sample of rust is mostly iron(III) oxide.

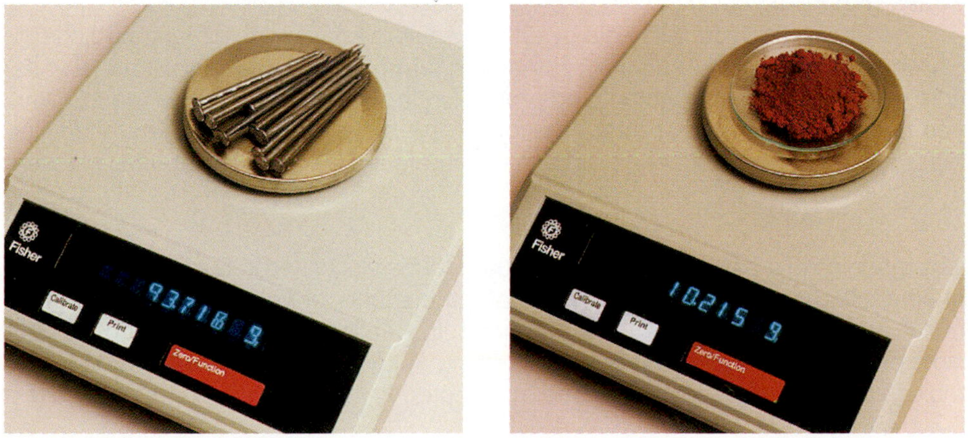

(a) (b)

Figure 10.1
From your study of stoichiometry, you will be able to answer questions such as the following: (a) How many grams of rust can be formed from 93.716 g of these iron nails? (b) How many grams of nails are needed to form 10.215 g of rust?

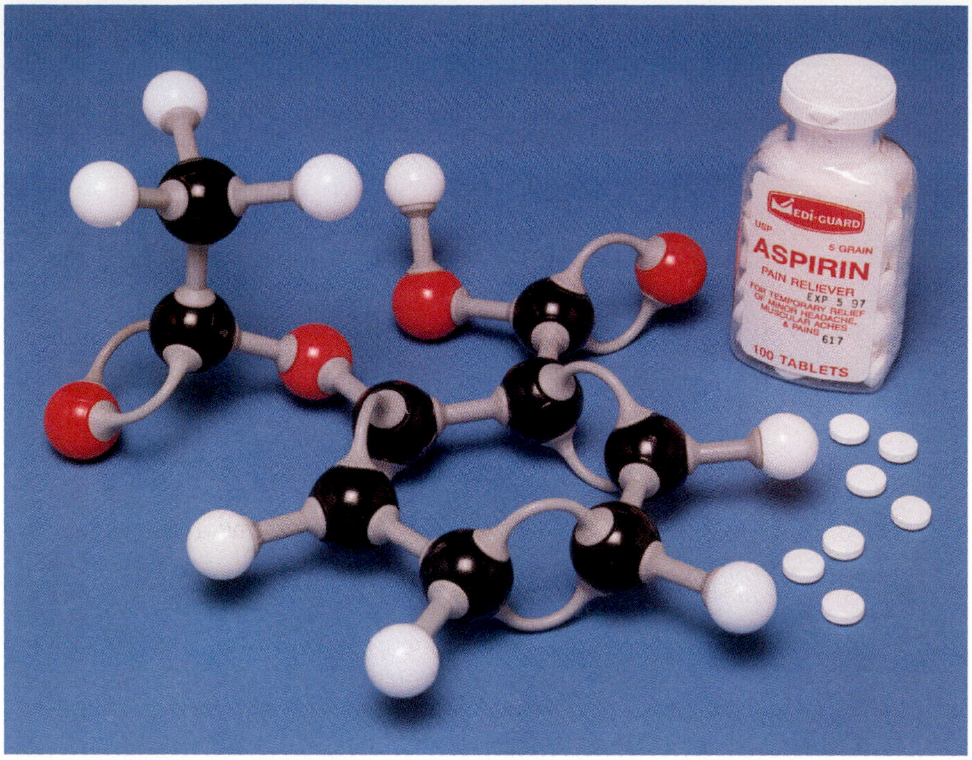

Figure 10.2
Approximately 15,000 tons of aspirin are produced annually in the United States. Special controls must be made in manufacturing pharmaceuticals to ensure high purity.

Stoichiometry, pronounced "stoy´-key-ahm´-eh-tree," is defined as calculations relating the amounts of reactants and products in chemical reactions. Stoichiometry applies the mole concept (Chapter 8) to the balanced chemical equation (Chapter 9). With your previous mastery of Chapters 8 and 9, you should find this chapter fairly easy and exciting.

When you finish this chapter, you will be fully prepared to go into a laboratory and measure out the proper amounts of reactants required for a particular reaction. You will be able to predict how much product should be produced from your weighed reactants; conversely, if you need a specific amount of product, you will know how much of each reactant you should weigh out to produce it. If you work in an industrial chemical-producing plant, you will be able to predict the number of tons of polymer fiber or pharmaceutical that you may produce from the chemical reactants on hand. Or, if you are in the purchasing department of a company, you will be able to calculate how many kilograms (or tons) of reactants you must order if you are to synthesize the approximately 2 million tons of polyester fiber that are woven into textiles or the 15,000 tons of aspirin that are marketed each year in the United States (Figure 10.2).

In this chapter, you will begin by extending your knowledge of balanced chemical equations from the number of particles (atoms, molecules, or formula units) corresponding to each reactant and product to the number of moles of each. Once you understand the molar interpretation of the balanced chemical equation, the mass ratios for reactants and products may be found by using their molar masses.

10.2 molar interpretation of the balanced equation

The meaning of a balanced chemical equation was discussed in Section 9.3 in terms of interacting particles, that is, in terms of the numbers of atoms, molecules, or formula units that combine and are produced. In Section 10.1, the numerical coefficients in the balanced equation for the rusting of iron help to review the atomic and molecular interpretation you learned in Chapter 9. In that example, the numerical coefficients (also called **stoichiometric coefficients**) told you that *four* atoms of the element iron, Fe, combine with *three* molecules of the element oxygen, O_2, to yield *two* formula units of the ionic compound iron(III) oxide, Fe_2O_3.

$$\mathbf{4}\ Fe(s) + \mathbf{3}\ O_2(g) \rightarrow \mathbf{2}\ Fe_2O_3(s)$$

This balanced chemical equation is not limited to specific numbers of Fe atoms or O_2 molecules, however. It may be applied to the rusting of iron no matter how many Fe atoms or O_2 molecules are present as long as they are present in the proportions specified by the coefficients in the balanced chemical equation. The stoichiometric coefficients are interpreted to mean that every four Fe atoms require exactly three O_2 molecules to react according to this equation. And, for every four Fe atoms or every three O_2 molecules that react, exactly two formula units of Fe_2O_3 are produced. Since the relationships given by the coefficients in the balanced chemical equation are fundamental for stoichiometric calculations in later sections, they are summarized below.

4 Fe atoms	require	3 O_2 molecules
4 Fe atoms	yield	2 Fe_2O_3 formula units
3 O_2 molecules	require	4 Fe atoms
3 O_2 molecules	yield	2 Fe_2O_3 formula units

To apply the balanced chemical equation to other quantities of atoms or molecules, you may multiply the coefficients by any convenient number as long as you multiply all the coefficients by that same number. In balancing equations, you have always found the smallest set of whole-number coefficients. However, any exact multiple of the simplest set of coefficients still gives a balanced equation. For example, the coefficients of the equation above can be multiplied by 2 to give

$$\underset{8\ Fe\quad 12\ O}{8\ Fe + 6\ O_2} \longrightarrow \underset{8\ Fe\ 12\ O}{4\ Fe_2O_3}$$

This is still a balanced equation because there are 8 Fe atoms and 12 O atoms on each side, and the total mass of each type of atom is conserved. Likewise, the equation representing the Haber process for the production of ammonia

$$\underset{6\ H\quad 2\ N}{3\ H_2 + N_2} \longrightarrow \underset{2\ N\ 6\ H}{2\ NH_3}$$

may be multiplied by 3 to give a balanced equation with 18 H and 6 N atoms on each side.

$$\underset{18\ H\quad 6\ N}{9\ H_2 + 3\ N_2} \longrightarrow \underset{6\ N\ 18\ H}{6\ NH_3}$$

A special, but particularly useful, multiple for a balanced chemical equation is Avogadro's number. When multiplied by Avogadro's number, the equation will then represent the number of particles in a sample large enough for you to weigh on a balance. Let us take the balanced equation for the rusting of iron and multiply through by Avogadro's number (Section 8.3). Once again, the equation will still be balanced. But most important, the resulting equation can be expressed in terms of *moles* of atoms, molecules, or formula units because 6.022×10^{23} particles represent 1 mol of particles.

4 Fe	**+**	**3 O_2**	$\longrightarrow$	**2 Fe_2O_3**
4 Fe atoms		3 O_2 molecules		2 Fe_2O_3 formula units
$4 \times 6.022 \times 10^{23}$ Fe atoms		$3 \times 6.022 \times 10^{23}$ O_2 molecules		$2 \times 6.022 \times 10^{23}$ Fe_2O_3 formula units
4 mol Fe atoms	**+**	**3 mol O_2 molecules**	$\longrightarrow$	**2 mol Fe_2O_3 formula units**

The stoichiometric coefficients in the balanced equation can now be interpreted in terms of moles of each reactant and product, not just in terms of the numbers of atoms, molecules, or formula units. Also, the relationships given by the coefficients can be rewritten in terms of moles so that they apply to large quantities of matter. These new relationships are summarized below.

4 mol Fe	require	3 mol O_2
4 mol Fe	yield	2 mol Fe_2O_3
3 mol O_2	require	4 mol Fe
3 mol O_2	yield	2 mol Fe_2O_3

Remember that mol is the abbreviation for mole. When writing equations and performing calculations (i.e., when using the unit with a number), chemists generally use the abbreviation mol.

The stoichiometric coefficients are now interpreted to mean that every 4 mol of Fe requires exactly 3 mol of O_2 to react according to this equation. And, for every 4 mol of Fe or every 3 mol of O_2 that react, exactly 2 mol of Fe_2O_3 are produced.

Molar amounts are easily related to weighable masses (gram amounts) through the molar mass (Section 8.6). In the laboratory, chemists almost always interpret the coefficients of balanced chemical equations as numbers of moles of reactants or products. Thus, you can, and should, read chemical equations in terms of moles. Additional examples are given below.

$$\begin{array}{ccccccccc} Ca & + & 2\,H_2O & \longrightarrow & Ca(OH)_2 & + & H_2 \\ 1\text{ mol }Ca & + & 2\text{ mol }H_2O & \longrightarrow & 1\text{ mol }Ca(OH)_2 & + & 1\text{ mol }H_2 \end{array}$$

$$\begin{array}{ccccc} 2\,KClO_3 & \longrightarrow & 2\,KCl & + & 3\,O_2 \\ 2\text{ mol }KClO_3 & \longrightarrow & 2\text{ mol }KCl & + & 3\text{ mol }O_2 \end{array}$$

sample exercise 1

a. Interpret the following equation in terms of the interacting particles:

$$Ca + 2\,H_2O \rightarrow Ca(OH)_2 + H_2\uparrow$$

b. Write relationships for reactants and products given by the stoichiometric coefficients in this balanced equation.

solution

a. Identify the nature of each reactant and product. That is, ask yourself of what kinds of units or particles are they constructed. The coefficients tell you the relative numbers of the particles.

   Ca = A symbol alone (no coefficient or subscript) denotes a single *atom.*

   H_2O = Combinations of nonmetallic elements are molecular compounds; therefore, this is a *molecule.*

   $Ca(OH)_2$ = Combinations of metals and nonmetals are ionic compounds; therefore, this is a *formula unit* made up of a Ca^{2+} ion and two OH^- ions.

   H_2 = The subscript 2 reminds us that hydrogen is an element that exists as a *diatomic molecule.*

   Thus the interpretation of the equation is one Ca atom plus two H_2O molecules yields one $Ca(OH)_2$ formula unit plus one H_2 molecule.

b. Relationships given by the coefficients are as follows.

ATOMIC AND MOLECULAR INTERPRETATION

1 Ca atom	requires	2 H_2O molecules
1 Ca atom	yields	1 $Ca(OH)_2$ formula unit
1 Ca atom	yields	1 H_2 molecule
2 H_2O molecules	require	1 Ca atom
2 H_2O molecules	yield	1 $Ca(OH)_2$ formula unit
2 H_2O molecules	yield	1 H_2 molecule

MOLAR INTERPRETATION

1 mol Ca	requires	2 mol H_2O
1 mol Ca	yields	1 mol $Ca(OH)_2$
1 mol Ca	yields	1 mol H_2
2 mol H_2O	require	1 mol Ca
2 mol H_2O	yield	1 mol $Ca(OH)_2$
2 mol H_2O	yield	1 mol H_2

sample exercise 2

Prove that the coefficients of the equation $3\ H_2 + N_2 \rightarrow 2\ NH_3$ tell you the molar relationships between reactants and products.

solution

1. The coefficients tell the relationships of reacting particles:

$$3\ H_2 \text{ molecules} + 1\ N_2 \text{ molecule} \rightarrow 2\ NH_3 \text{ molecules}$$

2. Multiplication of all coefficients by Avogadro's number will not disrupt the balancing. Multiplication of both sides of an equation by the same number leaves the equation balanced.

$$3(6.022 \times 10^{23})\ H_2 \text{ molecules} + 1(6.022 \times 10^{23})\ N_2 \text{ molecules} \rightarrow 2(6.022 \times 10^{23})\ NH_3 \text{ molecules}$$

3. Because 1 mol = 6.022×10^{23} molecules, we can also read the equation as

$$3 \text{ mol } H_2 + 1 \text{ mol } N_2 \rightarrow 2 \text{ mol } NH_3$$

Remember that mol is the abbreviation for mole.

••• problem 1

Interpret the following equations in terms of the numbers of moles of reactants and products.

a. $2\ Al(OH)_3 + 3\ H_2SO_4 \rightarrow Al_2(SO_4)_3 + 6\ H_2O$

b. $4\ Li + O_2 \rightarrow 2\ Li_2O$

10.3 conversion factors from balanced chemical equations

To perform stoichiometric calculations, you will use conversion factors obtained from the coefficients in the balanced chemical equation. These conversion factors result directly from the relationships you wrote in Section 10.2, which connect molar amounts of one reactant to molar amounts of another reactant, or molar amounts of any reactant to molar amounts of any product. For example, for the balanced equation $N_2 + 3\ H_2 \rightarrow 2\ NH_3$, you can write the following relationships using the molar interpretation of coefficients.

$$1 \text{ mol } N_2 = 3 \text{ mol } H_2$$
$$1 \text{ mol } N_2 = 2 \text{ mol } NH_3$$
$$3 \text{ mol } H_2 = 2 \text{ mol } NH_3$$

Or in summary,

$$1 \text{ mol } N_2 = 3 \text{ mol } H_2 = 2 \text{ mol } NH_3$$

Note that each relationship has now been written as a mathematical equality in the same manner you wrote equalities in Section 3.3. You should also note that *each of these equalities between molar amounts apply only to the particular reaction described by the balanced chemical equation.* For different reactions involving these same materials, you obtain different equalities. For example, N_2 and H_2 can also react to form hydrazine, N_2H_4, a rocket fuel. This reaction is represented by the equation

$$N_2 + 2\ H_2 \rightarrow N_2H_4$$

For this reaction, you can write the following equalities using the molar interpretation of the coefficients in the balanced equation. Particularly note that these equalities are different than those in the previous example, even when they relate the same elements, N_2 and H_2.

$$1\ \text{mol}\ N_2 = 2\ \text{mol}\ H_2$$
$$1\ \text{mol}\ N_2 = 1\ \text{mol}\ N_2H_4$$
$$2\ \text{mol}\ H_2 = 1\ \text{mol}\ N_2H_4$$

Or in summary,

$$1\ \text{mol}\ N_2 = 2\ \text{mol}\ H_2 = 1\ \text{mol}\ N_2H_4$$

Equalities for the same two elements (or compounds) may be different for two different balanced equations.

Each of the equalities for the two balanced chemical equations above can be made into conversion factors just as you have done previously for other equalities in Section 3.3. For example, for the reaction between N_2 and H_2 to form NH_3,

$$N_2 + 3\ H_2 \rightarrow 2\ NH_3$$

you can write the conversion factors as follows.

$$\frac{1\ \text{mol}\ N_2}{3\ \text{mol}\ H_2} \qquad \frac{1\ \text{mol}\ N_2}{2\ \text{mol}\ NH_3} \qquad \frac{3\ \text{mol}\ H_2}{2\ \text{mol}\ NH_3}$$

You know that you can also write conversion factors that are the reciprocals of these fractions. Which conversion factor you use will depend on the *given* units you wish to cancel when you convert a *given* measurement into a *new* set of units. The conversion factor will have *given* units in its denominator and *new* units in its numerator.

All of these conversion factors are fractions, called **mole ratios**, that relate molar amounts of one reactant to molar amounts of another reactant, or molar amounts of a reactant to molar amounts of a product. Stoichiometric calculations *always* use appropriate *mole ratios* that have been obtained from a balanced chemical equation. Mole ratios are applied just as any other conversion factor in the unit conversion method.

sample exercise 3

What is the conversion factor (or mole ratio) of H_2O to NaOH in the following reaction?

$$2\ Na(s) + 2\ H_2O(l) \rightarrow 2\ NaOH(aq) + H_2 \uparrow$$

Na is more active than hydrogen in the activity series (Table 9.2). As a result, it will replace hydrogen in water to produce H_2 gas (Figure 10.3).

solution

The coefficients give you equalities between molar amounts. In this case, 2 mol of H_2O = 2 mol of NaOH. The conversion factor written from this equality is the mole ratio

$$\frac{2\ \text{mol}\ H_2O}{2\ \text{mol NaOH}}$$

Note that the number 2 in the numerator and denominator has not been canceled. Coefficients in conversion factors are often left uncanceled because it helps you to recheck the conversion factor with the balanced equation directly. Common factors, such as the 2's, will automatically cancel when the conversion factors are used in calculations.

Figure 10.3

Sodium metal replaces hydrogen in water to produce H_2 gas in a vigorous reaction that releases heat.

sample exercise 4

What is the conversion factor (mole ratio) of HF to SnF_2 in the following reaction, which is a means of making tin(II) fluoride (stannous fluoride), the "fluoride" in toothpaste?

$$SnO(s) + 2\ HF(aq) \rightarrow SnF_2(aq) + H_2O(l)$$

solution

The coefficients give you the ratio of the reacting molar amounts. In this case, 2 mol of HF produces 1 mol of SnF_2. The mole ratio, which can be written and used as a conversion factor, is

$$\frac{2\ \text{mol HF}}{1\ \text{mol SnF}_2}$$

problem 2

What is the mole ratio of Na to H_2 in the reaction given in Sample Exercise 3?

problem 3

What is the mole ratio of SnO to HF in the reaction given in Sample Exercise 4?

10.4 mole-to-mole conversions

With a balanced chemical equation and the conversion factors it represents, you are prepared, in principle, for most stoichiometric calculations. Every balanced equation tells you directly the mole ratio (or conversion factor) of any two reactants, of any two products, or of any reactant with any product. Hence, you may directly convert *moles* of any one material *to moles* of any other material in the balanced equation.

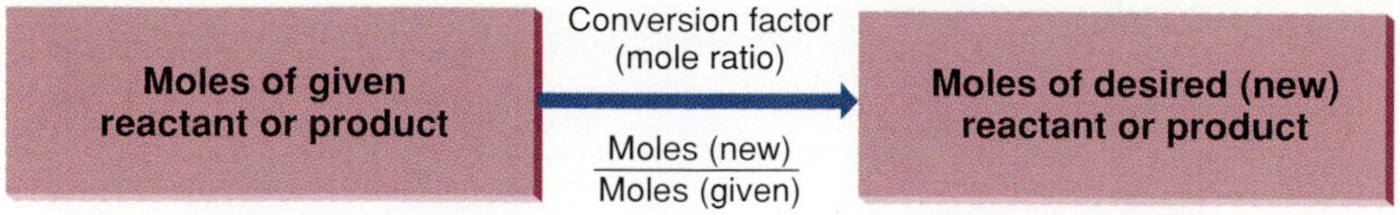

Figure 10.4
Summary of a simple stoichiometric calculation to convert moles of any reactant or product to moles of any other reactant or product. The conversion factor (mole ratio) for the calculation is obtained directly from the coefficients in the balanced chemical equation.

The simplest stoichiometric calculations are summarized in Figure 10.4. You are given a balanced chemical equation and a molar amount of one reactant or product. You are asked to calculate the molar amounts of the other reactants and products. These problems are handled directly by the unit conversion method in the same manner used in Section 3.3. You begin by identifying the *given* quantity and extracting a conversion factor from the balanced equation. These stoichiometric calculations are illustrated in the following sample exercises.

sample exercise 5

The decomposition of $MgCO_3$, a compound used in the paint and ink industries, is represented by the following equation.

$$MgCO_3(s) \xrightarrow{\Delta} MgO(s) + CO_2 \uparrow$$

If 6 mol of $MgCO_3$ decompose at 350°C, how many moles of MgO are produced? Magnesium oxide, produced in this reaction, is used as a nutritional supplement for the deficiency of magnesium. It is also used in making fire bricks because of its high melting point (2,800°C) and in making compounds to prevent boiler scale because of its slightly basic reaction with water (Figure 10.5).

Figure 10.5
At 350°C, 6 mol of $MgCO_3$ decompose to yield 6 mol of MgO. Both compounds are white crystalline solids, but 6 mol of each occupy significantly different volumes.

solution

Follow the steps of the unit conversion method.

Step 1. The *given* quantity is 6 mol of $MgCO_3$, a reactant.

Step 2. The *new* quantity is moles of MgO, a product. You are determining how many moles of this product can be produced from a given quantity of reactant.

Step 3. The equality relating the *given* and *new* quantities is obtained from the coefficients of the balanced chemical equation:

$$1 \text{ mol } MgCO_3 = 1 \text{ mol MgO}$$

Choose the conversion factor with the *new* quantity in the numerator and the *given* quantity in the denominator. Thus,

$$\frac{1 \text{ mol MgO}}{1 \text{ mol } MgCO_3}$$

Step 4. Set up the proper format:

Given × Conversion factor = New

$$6 \cancel{\text{ mol } MgCO_3} \times \frac{1 \text{ mol MgO}}{1 \cancel{\text{ mol } MgCO_3}} = 6 \text{ mol MgO}$$

Mole-to-mole conversions are simple, *one-step* unit conversions, and you are undoubtedly ready to apply the method even to very complex and unfamiliar reactions.

sample exercise 6

How many moles of oxalic acid, $H_2C_2O_4$, are required to react completely with 1.50 mol of aqueous potassium permanganate, $KMnO_4$, in acid solution according to the following equation?

$$2\ KMnO_4 + 6\ HCl + 5\ H_2C_2O_4 \rightarrow 2\ MnCl_2 + 10\ CO_2 + 2\ KCl + 8\ H_2O$$

solution

Step 1. The *given* quantity is 1.50 mol of $KMnO_4$, a reactant.

Step 2. The *new* quantity is moles of $H_2C_2O_4$, another reactant. You are determining how many moles of this reactant are required to react with a given amount of the other reactant.

Step 3. The equality from the balanced equation is

$$2\ \text{mol}\ KMnO_4 = 5\ \text{mol}\ H_2C_2O_4$$

and the conversion factor is

$$\frac{New}{Given} = \frac{5\ \text{mol}\ H_2C_2O_4}{2\ \text{mol}\ KMnO_4}$$

Step 4. The format is

$$1.50\ \cancel{\text{mol}\ KMnO_4} \times \frac{5\ \text{mol}\ H_2C_2O_4}{2\ \cancel{\text{mol}\ KMnO_4}} = \frac{1.50 \times 5\ \text{mol}\ H_2C_2O_4}{2} = 3.75\ \text{mol}\ H_2C_2O_4$$

Notice that the coefficients from the equation (5 and 2) are not measured numbers and may be treated as exact. As a result, they do not limit the number of sig figs in the answer.

Oxalic acid, $H_2C_2O_4$, a reactant in this reaction, is a white, crystalline solid used as a bleach and cleanser. It occurs naturally in many plants and vegetables such as spinach, rhubarb, and sorrel. In quantity, it is poisonous and corrosive to skin and mucous membranes (Figure 10.6).

Figure 10.6
3.75 mol of oxalic acid, $H_2C_2O_4$, are required to react completely in acid solution with the 1.5 mol of potassium permanganate, $KMnO_4$, shown here. Progress of this reaction is shown by the evolution of CO_2 gas from the reaction mixture.

••• problem 4

Write a balanced equation for the reaction between K and Br_2 that forms the sedative KBr and calculate the number of moles of KBr produced from the reaction of 7.50 mol of Br_2.

10.5 gram-to-mole and mole-to-gram conversions

As you know from Section 8.6, it is not possible for you to weigh out moles of any substance without first converting the given number of moles to grams using the molar mass of the substance. If you are to conduct a chemical reaction in the laboratory or in an industrial plant with the molar ratios required by the balanced equation, practical necessity demands that you extend your mole-to-mole calculations to include at least one further conversion factor. You must use the molar mass of your *given* substance to convert either its *number of moles to grams* or its *number of grams to moles.*

If your *given* substance has units of grams and your *new* substance has units of moles, you must first convert the *grams* of *given* substance *to moles* of *given* substance. Once you know the number of moles of the *given* substance, you may continue with a mole-to-mole stoichiometric calculation using the balanced equation as shown in the exercises and problems you worked through in Section 10.4.

Figure 10.7

Summary of a gram-to-mole conversion calculation.

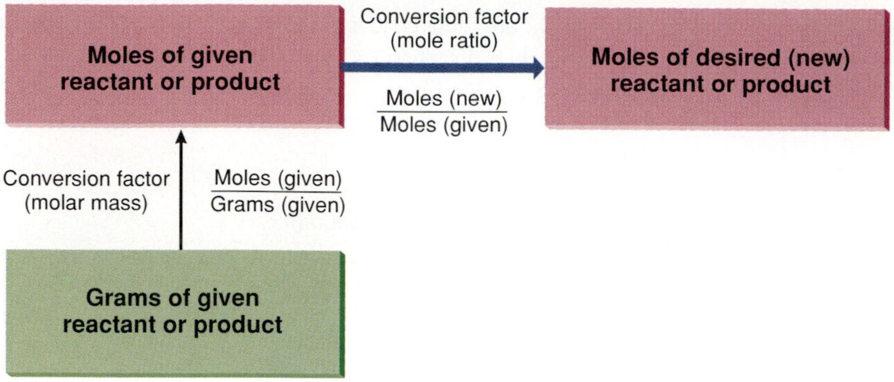

In Sample Exercise 7, you are given grams of the reactant aluminum and asked to calculate the number of moles of the product, Al_2O_3, produced. You must first convert the grams of the reactant, Al, to moles of the reactant, Al. Then you must convert the moles of the reactant, Al, to moles of the product, Al_2O_3. These conversions are summarized in Figure 10.7.

sample exercise 7

According to the equation

$$Fe_2O_3(s) + 2\ Al(s) \rightarrow Al_2O_3(s) + 2\ Fe(s)$$

how many moles of Al_2O_3 are produced by the reaction of 81 g of Al?

solution

Use the unit conversion method. Two conversion factors will be required: (1) the molar mass of the reactant Al and (2) the mole ratio between the product Al_2O_3 and the reactant Al.

Step 1. The *given* quantity and material is 81 g of Al, a reactant.

Step 2. The *new* material is moles of Al_2O_3, a product. You are determining the moles of product formed from a given quantity of reactant.

Step 3. The mole ratio necessary to convert moles of Al to moles of Al_2O_3 is

$$\frac{\textit{New}\text{ material}}{\textit{Given}\text{ material}} = \frac{1\text{ mol }Al_2O_3}{2\text{ mol Al}}$$

To use this factor, grams of Al must be changed to moles of Al through its molar mass obtained from the periodic table.

$$\frac{\textit{New}\text{ unit}}{\textit{Given}\text{ unit}} = \frac{1\text{ mol Al}}{27\text{ g Al}}$$

Step 4. Use a format beginning with the *given* quantity and using conversion factors of the form *new/given* to cancel out units.

Given × **Conversion factors** = **New**

$$81\ \cancel{\text{g Al}} \times \frac{1\ \cancel{\text{mol Al}}}{27\ \cancel{\text{g Al}}} \times \frac{1\text{ mol }Al_2O_3}{2\ \cancel{\text{mol Al}}} = \frac{81 \times 1 \times 1\text{ mol }Al_2O_3}{27 \times 2} = 1.5\text{ mol }Al_2O_3$$

Note that calculations with *multiple conversions* are accomplished with a single setup, similar to unit conversion problems in Section 3.3.

If your *given* substance has units of moles and your *new* substance has units of grams, you must first convert the *given* number of moles (reactant or product) to moles of the new substance (reactant or product) using the mole ratio. Then you must use the molar mass of the new substance to find the number of grams of that substance. Both of these conversions are summarized in Figure 10.8.

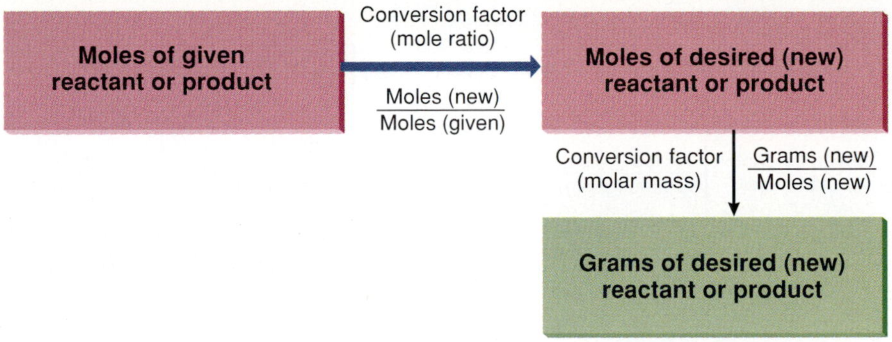

Figure 10.8
Summary of a mole-to-gram conversion calculation.

sample exercise 8

According to the equation

$$2\ KClO_3 \rightarrow 2\ KCl + 3\ O_2$$

how many grams of $KClO_3$ are required to produce 4.50 mol of O_2?

solution

Use the unit conversion method. Two conversion factors will be required: (1) the molar mass of the reactant $KClO_3$ and (2) the mole ratio between the moles of the reactant $KClO_3$ and the product O_2.

Step 1. The *given* quantity and material is 4.50 mol of the product O_2.

Step 2. The *new* material is grams of the reactant $KClO_3$. You are determining the amount of reactant needed to produce a given number of moles of product.

Step 3. The mole ratio from the balanced equation necessary to convert moles of O_2 to moles of $KClO_3$ is

$$\frac{\textit{New}\text{ material}}{\textit{Given}\text{ material}} = \frac{2\text{ mol }KClO_3}{3\text{ mol }O_2}$$

To use this factor and obtain the answer in grams of $KClO_3$, moles of $KClO_3$ must be converted to grams through the molar mass.

K	1	×	39.1	=	39.1
Cl	1	×	35.5	=	35.5
O	3	×	16.0	=	48.0
					122.6

$$1\text{ mol }KClO_3 = 122.6\text{ g }KClO_3$$

Thus, the appropriate conversion factor is

$$\frac{\textit{New}\text{ unit}}{\textit{Given}\text{ unit}} = \frac{122.6\text{ g }KClO_3}{1\text{ mol }KClO_3}$$

Step 4. Use a format beginning with the *given* quantity and using conversion factors of the form *new/given* to cancel out units.

Given × **Conversion factors** = **New**

$$4.50\ \cancel{\text{mol }O_2} \times \frac{2\ \cancel{\text{mol }KClO_3}}{3\ \cancel{\text{mol }O_2}} \times \frac{122.6\text{ g }KClO_3}{1\ \cancel{\text{mol }KClO_3}} = \frac{4.50 \times 2 \times 122.6\text{ g }KClO_3}{3 \times 1} = 368\text{ g }KClO_3$$

••• problem 5

Refer to the balanced equation in Sample Exercise 8 to do this problem. How many grams of KCl are produced when 3.00 mol of $KClO_3$ react?

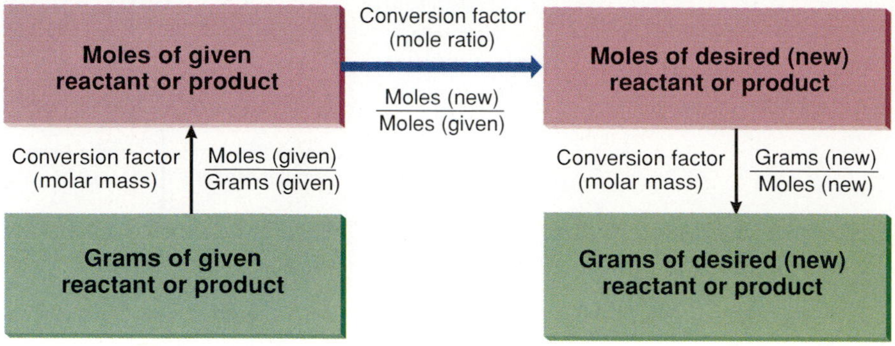

Figure 10.9

Summary of a gram-to-gram stoichiometric conversion calculation.

10.6 gram-to-gram conversions

For many laboratory problems, you will need to know the gram amounts of both substances used in your reaction: the *given* substance and the *new* substance. Typical examples of laboratory problems include the following. For a *given* number of grams of reactant, how many grams of product (*new* quantity) may be produced? Or, for a *given* number of grams of product, how many grams of reactant (*new* quantity) are required?

As in previous stoichiometry problems, *gram-to-gram* stoichiometry problems depend on a balanced chemical equation to give you a mole-ratio conversion factor that relates the materials involved in the conversion. *Once again, the mole ratio will be the central focus and should be determined first.* The other conversion factors will be gram-to-mole or mole-to-gram factors and are used to extend the conversion to gram quantities. *Gram-to-gram* stoichiometric calculations are summarized in Figure 10.9. This figure is a useful summary of all stoichiometry calculations since simpler stoichiometry calculations also follow this outline; they just use that portion of the pathway that applies to them.

sample exercise 9

Sodium iodide is used in making photographic emulsions. Because of iodide's accumulation in the thyroid gland, NaI is used medicinally as a diagnostic aid, treatment (Graves disease) and preventative (Figure 10.10) for thyroid problems. NaI can be made in the laboratory by combining metal and iodine. The balanced equation for this reaction is $2\ Na + I_2 \rightarrow 2\ NaI$. How many grams of I_2 are required to produce 225 g of NaI?

solution

Use the unit conversion method. Three conversion factors will be required: (1) the molar mass of the reactant I_2, (2) the mole ratio between the reactant I_2 and the product NaI, and (3) the molar mass of the product NaI. Determine the mole ratio first.

Step 1. The *given* quantity is 225 g of NaI, a product.

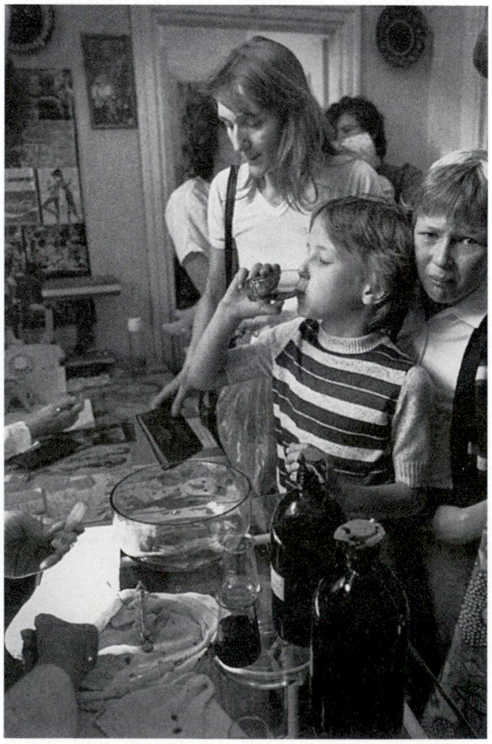

Figure 10.10

A child in a Warsaw clinic drinks a solution of aqueous sodium iodide to displace radioactive iodine which may have been absorbed in his thyroid gland from radioactive fallout. In this case, the radioactive fallout came from a nuclear power plant accident at Chernobyl, Ukraine.

Step 2. The *new* quantity is grams of I_2, a reactant. You are determining how many grams of the reactant I_2 are required to produce a given amount of NaI product.

Step 3. The mole ratio necessary to convert NaI to I_2 is obtained from the *balanced* chemical equation.

$$\frac{\textit{New}\text{ material}}{\textit{Given}\text{ material}} = \frac{1\text{ mol } I_2}{2\text{ mol NaI}}$$

To use this factor, grams of NaI must be changed to moles of NaI through the molar mass.

$$\frac{\textit{New}\text{ unit}}{\textit{Given}\text{ unit}} = \frac{1\text{ mol NaI}}{150.\text{ g NaI}}$$

Atomic weights
Na 23.0
I 127.
150.

To obtain the answer in grams of I_2, moles of I_2 must be converted to grams of I_2 through the molar mass.

$$\frac{\textit{New}\text{ unit}}{\textit{Given}\text{ unit}} = \frac{254\text{ g } I_2}{1\text{ mol } I_2} \qquad (2 \times 127 = 254)$$

Step 4. Use the usual format, beginning with the *given* quantity.

$$225\ \cancel{\text{g NaI}} \times \frac{1\ \cancel{\text{mol NaI}}}{150.\ \cancel{\text{g NaI}}} \times \boxed{\frac{1\ \cancel{\text{mol } I_2}}{2\ \cancel{\text{mol NaI}}}} \times \frac{254\text{ g } I_2}{1\ \cancel{\text{mol } I_2}} = \frac{225 \times 1 \times 1 \times 254\text{ g } I_2}{150. \times 2 \times 1} = 191\text{ g } I_2$$

Notice that recognizing the mole ratio as the central focus in Sample Exercise 9, or any problem, helps in setting up the format. All gram-to-gram conversions have the following "skeleton," which applies the conversion factors in the order outlined in Figure 10.9.

$$\text{Grams } \textbf{(given)} \times \frac{\text{Moles } \textbf{(given)}}{\text{Grams } \textbf{(given)}} \times \boxed{\frac{\text{Moles } \textbf{(new)}}{\text{Moles } \textbf{(given)}}} \times \frac{\text{Grams } \textbf{(new)}}{\text{Moles } \textbf{(new)}} = \text{Grams } \textbf{(new)}$$

Mole ratio

Gram-to-mole conversion factors arranged to cancel out units

---

sample exercise 10

How many grams of Na are required to react completely with 161 g of S to produce sodium sulfide, Na_2S? Na_2S is prepared from its elements, which have been dissolved in liquid ammonia according to the reaction

$$2\text{ Na} + \text{S} \rightarrow Na_2S$$

solution

Work this through as in Sample Exercise 9, recognizing 161 g of S as the *given* quantity and grams of Na as the *new* quantity, *or* try fitting the data into the preceding "skeleton." Begin this by inserting the quantities into the clearly labeled parts of the "skeleton."

$$\underset{\textbf{Given}}{161\text{ g S}} \times \frac{\quad}{\quad} \times \underset{\textbf{Mole ratio}}{\boxed{\frac{2\text{ mol Na}}{1\text{ mol S}}}} \times \frac{\quad}{\quad} = \underset{\textbf{New}}{\text{Grams of Na}}$$

Now the first conversion factor must cancel grams of S. So it will be

$$\frac{1\text{ mol S}}{32.1\text{ g S}} \qquad (1\text{ mol S} = 32.1\text{ g S})$$

The last conversion factor must produce the new unit grams of Na. So it will be

$$\frac{23.0 \text{ g Na}}{1 \text{ mol Na}} \qquad (1 \text{ mol Na} = 23.0 \text{ g Na})$$

Thus, the setup is

$$161 \cancel{\text{g S}} \times \frac{1 \cancel{\text{mol S}}}{32.1 \cancel{\text{g S}}} \times \boxed{\frac{2 \cancel{\text{mol Na}}}{1 \cancel{\text{mol S}}}} \times \frac{23.0 \text{ g Na}}{1 \cancel{\text{mol Na}}} = \frac{161 \times 1 \times 2 \times 23.0 \text{ g Na}}{32.1 \times 1 \times 1} = 231 \text{ g Na}$$

• • • problem 6

How many grams of Na_2S are produced by the reaction described in Sample Exercise 10? Na_2S is a white, crystalline solid that is unstable and may explode upon impact or rapid heating. Nevertheless, it has many manufacturing applications, such as in removing hair from animal hides and producing rubber.

10.7 conversions summarized

All stoichiometric conversions require a balanced equation from which one determines an appropriate mole ratio. Because only the *mole ratio can be used as a conversion factor* relating moles of new material to moles of given material, gram amounts of material must be converted into molar amounts (or vice versa) in the course of stoichiometric calculations. A general plan of attack for all stoichiometry problems follows.

Guidelines for stoichiometry problems

1. Determine the mole ratio necessary to convert the given material into the new material. The mole ratio comes from the coefficients of the balanced equation and will have the form

$$\frac{\text{Number of moles } \textit{new} \text{ material}}{\text{Number of moles } \textit{given} \text{ material}}$$

2. If both the given and new units are stated in moles, then the problem will be a one-step conversion, as seen in Section 10.4.
3. If either or both of the given and new units are stated in grams, then there must be gram-to-mole or mole-to-gram conversion factors obtained from the molar mass, as seen in Sections 10.5 and 10.6 (also Figure 10.9).

10.8 limiting reactant

The balanced equation for the rusting of iron,

$$4 \text{ Fe}(s) + 3 \text{ O}_2(g) \rightarrow 2 \text{ Fe}_2\text{O}_3(s)$$

was interpreted in Section 10.2 to mean that every 4 mol of iron, Fe, requires 3 mol of oxygen, O_2, for the reaction to occur. That is, the coefficients of the balanced equation require that the molar amounts of the two reactants be consumed in a fixed ratio.

If 4 mol of Fe atoms were mixed with 4 mol of O_2 molecules, all 4 mol of the iron could react because there are at least 3 mol of O_2 present. Once all 4 mol of Fe reacted, there would be no Fe left and the reaction would no longer continue. There would be 1 mol of O_2 remaining, however, because only 3 mol of O_2 would have been used up in the reaction. The final mixture would contain iron oxide, Fe_2O_3, and the *excess* 1 mol of O_2, which did not react.

The initial and final amounts of reactants and products for this reaction are summarized below.

Initial Amounts		Change from Reaction		Final Amounts
4 mol Fe	−	4 mol Fe (reacted)	=	0 mol Fe
4 mol O_2	−	3 mol O_2 (reacted)	=	1 mol O_2
0 mol Fe_2O_3	+	2 mol Fe_2O_3 (formed)	=	2 mol Fe_2O_3

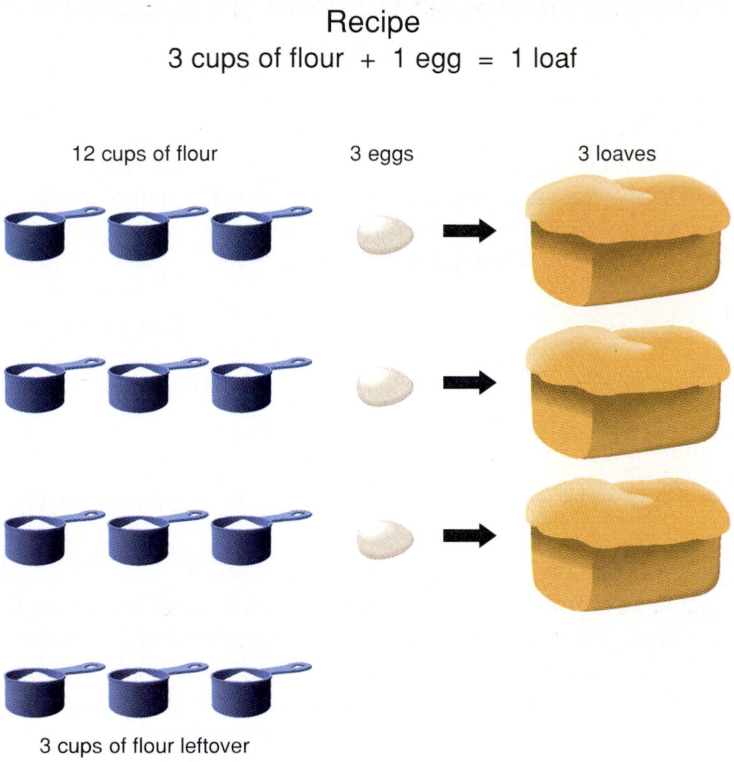

Figure 10.11
The three eggs limit you to making three loaves of bread by the recipe, even though you have enough flour to make four loaves. The excess flour is leftover.

The 4 mol of the reagent Fe *limits* the reaction because this amount is used up while the other reactant, O_2, is still present. Fe is called the **limiting reactant** (or **limiting reagent**) because it *limits and determines* the maximum amount of product that may be obtained. Once the limiting reactant is used up, the reaction stops and product may no longer be formed.

The concept of a limiting reactant is one you are already familiar with in your everyday experience. If you set out to bake some loaves of bread, you need flour and eggs in the fixed ratio given in your recipe. If your recipe for one loaf calls for 3 cups of flour and 1 egg, how many loaves can you make if you have available in your kitchen 3 eggs and 12 cups of flour (Figure 10.11)?

With the given recipe and these amounts available, you can make a maximum of 3 loaves of bread. The 3 eggs *limit* the maximum number of loaves to only 3 (3 ~~eggs~~ × 1 loaf/1 ~~egg~~ = 3 loaves). The 12 cups of flour are sufficient to make 4 loaves (12 ~~cups~~ × 1 loaf/3 ~~cups~~ = 4 loaves), but there are not 4 eggs available for 4 loaves. Thus, the amount of flour available is in excess of what can be used up and eggs are the *limiting* ingredient.

You may summarize the preceding flour, egg, and loaf data in the form of a "balanced" chemical equation:

Reactants	→	**Products**
3 cups flour + 1 egg	→	1 loaf

Available reactants and product obtained:

12 cups flour	+	3 eggs	→	3 loaves	+	3 cups flour leftover
Excess		**Limiting reactant**				(only 9 cups of flour can "react" with 3 eggs)

Whenever you are told the amounts of two or more reactants in a chemical reaction, you must identify the **limiting reactant,** that is, the one that *limits and determines* the maximum amount of product that can be obtained. The other reactant or reactants are in excess. When you are told the amount of only one reactant, that reactant is the limiting reactant and any other reactants are assumed to be present in excess.

In our example of Fe reacting with O_2 to form Fe_2O_3 according to the equation,

$$4\ \text{Fe}(s) + 3\ \text{O}_2(g) \rightarrow 2\ \text{Fe}_2\text{O}_3(s)$$

you had 4 mol of Fe and 4 mol of O_2. The equation tells you through mole ratios the amount of product that you might obtain from the given amount of each reactant. You must use the mole-to-mole conversion for *each* reactant as shown below to find which reactant is the limiting reactant, that is, which reactant limits the reaction and produces the *smaller amount of product,* Fe_2O_3.

From *4 mol of Fe,* you can obtain *2 mol of* Fe_2O_3.

$$\mathbf{4}\ \cancel{\textbf{mol Fe}} \times \frac{2\ \text{mol Fe}_2\text{O}_3}{4\ \cancel{\text{mol Fe}}} = \textbf{2 mol Fe2O3}$$

Limiting reactant Fe **Smaller amount of product**

From *4 mol of* O_2, you can obtain *2.67 mol of* Fe_2O_3.

$$\mathbf{4}\ \cancel{\textbf{mol O}_2} \times \frac{2\ \text{mol Fe}_2\text{O}_3}{3\ \cancel{\text{mol O}_2}} = \mathbf{2.67\ mol\ Fe_2O_3}$$

Reactant O_2 **Larger amount of product**

Fe is the limiting reactant because its given initial amount produces the *smaller amount of product. The limiting reactant is the reactant that limits the amount of product formed.* In this case, Fe, the limiting reactant, determines that a maximum of 2 mol of Fe_2O_3 may be formed with the given amount of reactants.

The limiting reactant is always identified by working with moles, as shown throughout the previous discussion. This means that if you are told the amounts of reactants in grams, you must convert grams to moles for each reactant before attempting to identify the limiting reactant.

General Method For Determining The Limiting Reactant

1. Recognize the necessity of determining the limiting reactant when you are given the amounts of two or more reactants.
2. Convert amounts of reactants given to moles.
3. Calculate the molar amount of product obtained based on the amount of *one* of the reactants as if it were the only reactant amount given. Then calculate the amount of product obtained based on the *other* reactant as if it were the only one given.
4. Choose the reactant that produces the *least* amount of product as the limiting reactant. This amount of product is the maximum amount that can be produced (see discussion of theoretical yield in Section 10.9).

sample exercise 11

A 50. g sample of the base, magnesium hydroxide, $Mg(OH)_2$, is mixed with 70. g of phosphoric acid, H_3PO_4. A neutralization (double-replacement) reaction (Section 9.8) takes place, which is represented by the equation

$$3\ \text{Mg(OH)}_2(s) + 2\ \text{H}_3\text{PO}_4(aq) \rightarrow \text{Mg}_3(\text{PO}_4)_2(s) + 6\ \text{H}_2\text{O}$$

How many grams of $Mg_3(PO_4)_2$ are produced? (The molar mass of $Mg(OH)_2$ = 58.3; the molar mass of H_3PO_4 = 98.0; and the molar mass of $Mg_3(PO_4)_2$ = 262.9.) This reaction would occur in your stomach if you took a dose of milk of magnesia (which contains $Mg(OH)_2$) after drinking a typical carbonated soft drink (which contains H_3PO_4).

solution

Step 1. Because the amounts of *both reactants* are given, you must decide which is the limiting reactant and proceed according to the general method.

Step 2. Use the molar masses given to calculate the moles of each reactant.

$$50.\ \cancel{g\ Mg(OH)_2} \times \frac{1\ mol\ Mg(OH)_2}{58.3\ \cancel{g\ Mg(OH)_2}} = 0.86\ mol\ Mg(OH)_2$$

$$70.\ \cancel{g\ H_3PO_4} \times \frac{1\ mol\ H_3PO_4}{98.0\ \cancel{g\ H_3PO_4}} = 0.71\ mol\ H_3PO_4$$

Step 3. Perform a mole-to-mole conversion on each reactant to determine how much product could be obtained from that reactant.

$$0.86\ \cancel{mol\ Mg(OH)_2} \times \frac{1\ mol\ Mg_3(PO_4)_2}{3\ \cancel{mol\ Mg(OH)_2}} = \mathbf{0.29\ mol\ Mg_3(PO_4)_2}$$

Limiting reactant — Mole ratios from balanced equation — **Smaller amount of product**

$$0.71\ \cancel{mol\ H_3PO_4} \times \frac{1\ mol\ Mg_3(PO_4)_2}{2\ \cancel{mol\ H_3PO_4}} = 0.36\ mol\ Mg_3(PO_4)_2$$

Larger amount of product

Step 4. $Mg(OH)_2$ is the limiting reactant because it is the reactant that produces the smaller amount of product. 0.29 mol of $Mg_3(PO_4)_2$ is the maximum amount of product that can be formed from the given quantities of reactants.

You know the molar amount of product, $Mg_3(PO_4)_2$. To get the gram amount, use the given molar mass.

$$0.29\ \cancel{mol\ Mg_3(PO_4)_2} \times \frac{262.9\ g\ Mg_3(PO_4)_2}{1\ \cancel{mol\ Mg_3(PO_4)_2}} = \mathbf{76\ g\ Mg_3(PO_4)_2}$$

Once the limiting reactant is determined, the amount of product obtained is also known, and the amount of excess reactant left over can be calculated in the manner outlined below and illustrated in Sample Exercise 12.

Guidelines for calculating excess reactant left over after reaction

1. Calculate the number of moles of the excess reactant that must have reacted with the limiting reactant to form the product.
2. Subtract the amount determined in step 1 from the original molar amount present.

   Number of moles originally – Number of moles reacted = Number of moles left over

3. Use a gram-to-mole conversion factor to change moles to grams if needed.

---

sample exercise 12

How many grams of the excess reactant are left over after the reaction described in Sample Exercise 11 is completed?

solution

The excess reactant was H_3PO_4.

Step 1. Calculate the number of moles of H_3PO_4 that is required to react with 0.86 mol of $Mg(OH)_2$.

$$0.86\ \cancel{mol\ Mg(OH)_2} \times \frac{2\ mol\ H_3PO_4}{3\ \cancel{mol\ Mg(OH)_2}} = 0.57\ mol\ H_3PO_4 \text{ reacts with } 0.86\ mol\ Mg(OH)_2$$

Step 2. Subtract the moles that reacted from the original molar amount present of the excess reactant. Originally, there was 0.71 mol of H_3PO_4.

Moles originally	–	Moles reacted	=	Moles left over
0.71 mol	–	0.57 mol	=	0.14 mol H_3PO_4 left over

Step 3. Use the molar mass of H_3PO_4 to change moles to grams.

$$0.14 \cancel{\text{mol } H_3PO_4} \times \frac{98.0 \text{ g } H_3PO_4}{1 \cancel{\text{mol } H_3PO_4}} = 14 \text{ g } H_3PO_4 \text{ left over}$$

---

sample exercise 13

A mixture of 30.0 g of Al and 80.0 g of Fe_2O_3 is prepared. The mixture is heated and a vigorous single-replacement reaction occurs according to the following equation.

$$Fe_2O_3(s) + 2\,Al(s) \rightarrow Al_2O_3(s) + 2\,Fe(l)$$

What is the mass of iron produced? (The atomic weight of Al = 27.0; the formula weight of Fe_2O_3 = 160.; and the atomic weight of Fe = 55.9.)

This reaction of ferric oxide, Fe_2O_3, with aluminum, Al, is called the thermite reaction. So much heat is released that temperatures reach far above the melting point of Fe (1,535°C) and this product is formed as a liquid (Figure 10.12).

Figure 10.12

So much heat is released in the thermite reaction that it produces liquid iron as a product and may be used to weld iron and steel.

solution

Step 1. The given amounts of both reactants signals a limiting reactant problem.

Step 2. Use the molar masses to calculate the moles of each reactant.

$$30.0 \cancel{\text{g Al}} \times \frac{1 \text{ mol Al}}{27.0 \cancel{\text{g Al}}} = 1.11 \text{ mol Al}$$

$$80.0 \cancel{\text{g } Fe_2O_3} \times \frac{1 \text{ mol } Fe_2O_3}{160. \cancel{\text{g } Fe_2O_3}} = 0.500 \text{ mol } Fe_2O_3$$

Step 3. Determine the molar amount of iron that could be obtained from each reactant.

$$1.11 \cancel{\text{mol Al}} \times \frac{2 \text{ mol Fe}}{2 \cancel{\text{mol Al}}} = 1.11 \text{ mol Fe}$$

Larger amount of product

$$0.500 \cancel{\text{mol } Fe_2O_3} \times \frac{2 \text{ mol Fe}}{1 \cancel{\text{mol } Fe_2O_3}} = 1.00 \text{ mol Fe}$$

Limiting reactant — **Smaller amount of product**

Step 4. Fe_2O_3 is the limiting reactant because it produces the smaller amount of product, 1.00 mol of Fe. Notice that Fe_2O_3 is limiting even though we began with a larger mass of Fe_2O_3. *The limiting reactant cannot be determined by simply looking at the given masses of the two reactants.*

The mass of iron product is obtained from the molar amount using the atomic weight of iron.

$$1.00 \cancel{\text{mol Fe}} \times \frac{55.9 \text{ g Fe}}{1 \cancel{\text{mol Fe}}} = 55.9 \text{ g Fe}$$

---

sample exercise 14

Calculate the amount of excess Al in Sample Exercise 13.

solution

Step 1. Determine the amount of Al that must have reacted with the limiting reactant, Fe_2O_3.

$$0.500 \cancel{\text{mol } Fe_2O_3} \times \frac{2 \text{ mol Al}}{1 \cancel{\text{mol } Fe_2O_3}} = 1.00 \text{ mol Al}$$

Step 2. Subtract the amount of the Al that reacted from the original molar amount of the excess reactant. Originally, there were 1.11 mol of Al.

Moles originally	–	Moles reacted	=	Moles left over
1.11 mol	–	1.00 mol	=	0.11 mol Al left over

Step 3. Use the molar mass of Al to change moles to grams.

$$0.11\ \cancel{\text{mol Al}} \times \frac{27.0\ \text{g Al}}{1\ \cancel{\text{mol Al}}} = 3.0\ \text{g Al left over}$$

••• problem 7

Sulfur trioxide, SO_3, is prepared by the action of oxygen, O_2, on sulfur dioxide, SO_2, in the presence of a catalyst such as platinized magnesium sulfate or ferric oxide. What mass of SO_3 is produced when 3.60 g of SO_2 is combined with 2.40 g O_2?

$$2\ SO_2(g) + O_2(g) \xrightarrow{\textit{Catalyst}} 2\ SO_3(g)$$

SO_3 is a reactant used in the manufacture of sulfuric acid (Figure 10.13). It is also a major atmospheric pollutant leading to the formation of acid rain.

This section has focused on chemical reactions that do not have stoichiometric amounts of reactants in the reaction mixture; that is, the reactants have not been mixed with the ratio of moles specified by the balanced chemical equation. Perhaps you are asking yourself why chemists don't simply use correct stoichiometric amounts of all reactants in chemical reactions. Why not use amounts such that all reactants are consumed and none are in excess? In Sample Exercise 11, this would require the reaction of 50. g (0.86 mol) of $Mg(OH)_2$ and 56 g (0.57 mol) of H_3PO_4 to give exactly 76 g (0.29 mol) of $Mg_3(PO_4)_2$.

More than stoichiometric amounts of reactants are often used intentionally because many reactions do not produce the stoichiometric amount of products. Sometimes, reactants are not completely converted into products; that is, the reaction does not go to completion. Once formed, products may interact to re-form reactants, and the reaction reaches a state of balance between reactants and products. This state of balance is called **chemical equilibrium** and will be studied in detail in Chapter 14.

One way to encourage a reaction to go completely to products is to use one reactant in excess. A typical way that chemists decide on the amounts of reactants to use in a reaction is to follow the reasoning outlined in the columns below.

1. Decide the amount of product that is desired.	1. A chemist wants to make 12.0 g of AgF by the reaction $2\ Ag + F_2 \rightarrow 2\ AgF$.
2. Consider the relative availability and cost of the reactants. Usually a chemistry lab has less of the more expensive reactant.	2. Ag is very expensive and there is not much in the laboratory. F_2 is abundantly available.
3. Calculate the correct stoichiometric amount of the less available (more expensive) reactant needed for the desired amount of product.	3. $12.0\ \cancel{\text{g AgF}} \times \frac{1\ \cancel{\text{mol AgF}}}{127\ \cancel{\text{g AgF}}} \times \frac{2\ \cancel{\text{mol Ag}}}{2\ \cancel{\text{mol AgF}}} \times \frac{108\ \text{g Ag}}{1\ \cancel{\text{mol Ag}}} = \frac{12.0 \times 108\ \text{g Ag}}{127} = 10.2\ \text{g Ag}$
4. Use an excess of the more available (cheaper) reactant.	4. Use an excess of F_2; that is, use more than 1.80 g of F_2, which is the correct stoichiometric amount. $12.0\ \cancel{\text{g AgF}} \times \frac{1\ \cancel{\text{mol AgF}}}{127\ \cancel{\text{g AgF}}} \times \frac{1\ \cancel{\text{mol F}_2}}{2\ \cancel{\text{mol AgF}}} \times \frac{38.0\ \text{g F}_2}{1\ \cancel{\text{mol F}_2}} = \frac{12.0 \times 38.0\ \text{g F}_2}{127 \times 2} = 1.80\ \text{g F}_2$

Figure 10.13
About 50 million tons of sulfuric acid are produced industrially in plants such as this one throughout the United States each year. Sulfuric acid is widely used as a laboratory reagent and in the manufacture of fertilizers, explosives, dyestuffs, paper, and glue.

From the preceding reasoning, the chemist would use more than 1.80 g of F_2 with 10.2 g of Ag and be assured of producing 12.0 g of AgF at the minimum cost.

10.9 theoretical yield, actual yield, and percentage yield

The **theoretical yield** is the maximum number of grams of product that can be obtained based on the amount of limiting reactant given and the balanced chemical equation. So, whenever you use stoichiometry to calculate the number of *grams* of product obtained from a specified amount of reactant, you are calculating the theoretical yield. Thus, you already know how to perform the theoretical yield calculation. This is just an added bit of terminology.

When conducting chemical reactions in the laboratory, you rarely can collect the entire theoretical yield of product. Even if you use an excess of a reactant to shift the reaction to completion (Section 10.8), some product is generally "lost" as the reaction is performed. A portion of a product may remain dissolved in solution and not be collected as a solid product, or some solid material may be left behind when the reaction mixture is transferred from one flask to another. The law of conservation of mass is not violated; the lost material can be accounted for. But the amount of product you actually collect is *less than* the theoretical yield. What you actually collect is the **actual yield.** The actual yield can only be determined experimentally; it cannot be calculated.

The **percentage yield** is the ratio of the actual yield to the theoretical yield, converted into a percentage by multiplying by 100. It is a measure of the closeness of the actual yield to the maximum possible yield. For a reaction that proceeds to completion, the percentage yield should approach 100% if your experimental technique has taken every precaution to minimize loss. The percentage yield cannot be predicted; however, it should be reproducible for the same reaction conducted under the same conditions. The reproducibility of the percentage yield indicates the reproducibility of your experimental technique.

$$\frac{\textbf{Actual yield}}{\textbf{Theoretical yield}} \times \textbf{100} = \textbf{Percentage yield}$$

sample exercise 15

Calculate the theoretical yield of AlF_3 obtained from the reaction of 0.56 mol of Al in the reaction

$$2\ Al + 3\ F_2 \rightarrow 2\ AlF_3.$$

Aluminum fluoride, AlF_3, is used in a wide variety of applications: as a flux in metallurgy, in the manufacture of aluminum, and as an inhibitor (agent to reduce the rate) of fermentation.

solution

The mass of AlF_3 is determined from the moles of Al, the only given amount of reactant.

Step 1. The mole ratio is

$$\frac{\text{New}}{\text{Given}} = \frac{2\ \text{mol}\ AlF_3}{2\ \text{mol Al}}$$

Step 2. Because the answer must be in grams of AlF_3, a mole-to-gram conversion will be necessary using the molar mass relation

$$1\ \text{mol}\ AlF_3 = 84\ \text{g}\ AlF_3$$

1 × Al	27 g
3 × F	57 g
	84 g

Step 3. Set up the unit conversion format:

$$0.56\ \cancel{\text{mol Al}} \times \frac{2\ \cancel{\text{mol}\ AlF_3}}{2\ \cancel{\text{mol Al}}} \times \frac{84\ \text{g}\ AlF_3}{1\ \cancel{\text{mol}\ AlF_3}} = 47\ \text{g}\ AlF_3$$

The theoretical yield is 47 g of AlF_3, the maximum amount possible from 0.56 mol of Al.

sample exercise 16

In the reaction described in Sample Exercise 15, suppose that the amount of product actually obtained was 40. g. Calculate the percentage yield.

solution

$$\text{Percentage yield} = \frac{\text{Actual yield}}{\text{Theoretical yield}} \times 100 = \frac{40.\,\not{g}}{47\,\not{g}} \times 100 = 85\%$$

10.10 heat as a reactant or product

Whenever a chemical reaction occurs, there is an energy change that can be measured in the form of heat energy using the units of kilojoules (kJ). Heat can either be released by a chemical reaction or supplied for the reaction to occur. Throughout this text, you have seen examples of reactions in which reactants have been heated. In Section 9.7, a special symbol, Δ, was introduced, and when it is placed over the yield arrow of a balanced chemical equation it indicates the addition of heat to the reaction.

When heat must be supplied for a reaction to occur, heat may be regarded as a reactant. A reaction in which heat is a reactant is called an **endothermic reaction** (*endo* means "in"; *thermic* means "heat"; thus, heat is put in or absorbed). The formation of laughing gas, dinitrogen oxide (N_2O), from N_2 and O_2 is an example of an endothermic reaction. In the example below, the amount of heat needed for the reaction has been written as a reactant with units of kilojoules (kJ).

$$2\,N_2(g) + O_2(g) + \underset{\textbf{Reactant}}{163\text{ kJ}} \longrightarrow 2\,N_2O(g) \qquad (10.1)$$

Just as you have seen many examples of reactions being heated, you have seen many examples of heat released during a reaction. In Figure 10.14, the Bunsen burner is producing heat as a result of the combustion reaction of the fuel, methane, CH_4, with oxygen gas, O_2, which is present in the air. When heat is released during a reaction, the heat is regarded as a product. A reaction in which heat is a product is called an **exothermic reaction** (*exo* means "out"; *thermic* means "heat"; hence, heat comes out or is released). The exothermic reaction for the combustion of methane, natural gas, is shown below with the amount of heat released written as a product with units of kJ.

$$CH_4(g) + 2\,O_2(g) \longrightarrow CO_2(g) + 2\,H_2O(l) + \underset{\textbf{Product}}{891\text{ kJ}} \qquad (10.2)$$

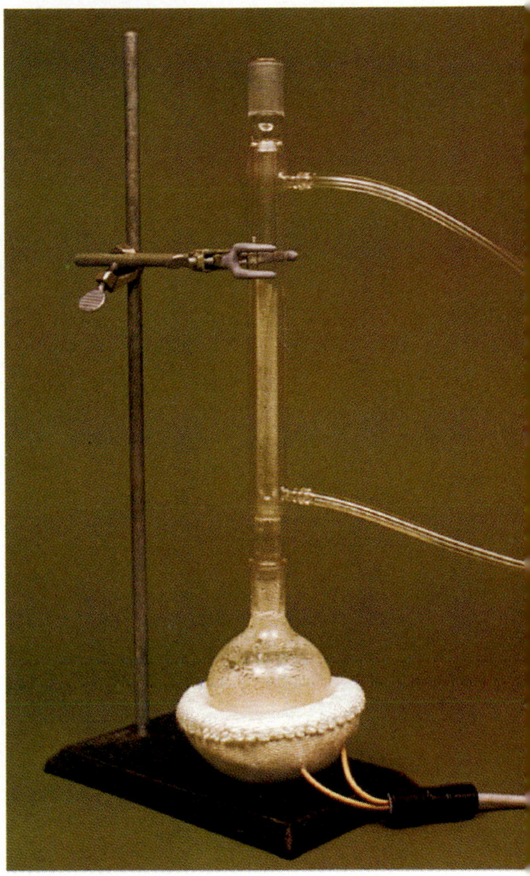

Figure 10.14
Many chemical reactions require the input of heat. Reactions using materials that do not burn are heated with a Bunsen burner. Reactions using combustible materials or combustible vapors are heated with an electric heating mantle.

While chemical reactions always involve a change in energy in the form of heat being absorbed or released, so do changes in the physical state of a pure material. You remember from Section 1.3 that heat must be supplied when a solid melts, a liquid vaporizes, or a solid sublimes. All of these changes in physical state are endothermic processes and can be written with heat as a reactant, that is, on the left side of the equation. They are summarized for water below.

Physical Change in State	Equation
Melting (fusion)	$H_2O(s) + 6.01\text{ kJ} \rightarrow H_2O(l)$
Vaporization	$H_2O(l) + 40.66\text{ kJ} \rightarrow H_2O(g)$
Sublimation	$H_2O(s) + 46.67\text{ kJ} \rightarrow H_2O(g)$

You also remember from Section 1.3 that heat is released when a liquid freezes and when a gas condenses to a liquid or a solid. These processes are exothermic and can be written with heat as a product, that is, on the right side of the equation.

Physical Change in State	Equation
Freezing	$H_2O(l) \rightarrow H_2O(s) + 6.01\text{ kJ}$
Condensation	$H_2O(g) \rightarrow H_2O(l) + 40.66\text{ kJ}$
Condensation	$H_2O(g) \rightarrow H_2O(s) + 46.67\text{ kJ}$

problem 8

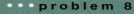

Identify the following reactions as *exo*thermic or *endo*thermic.

a. $2\ PCl_3 + 604\ kJ \rightarrow 2\ P + 3\ Cl_2$

b. $C + O_2 \rightarrow CO_2 + 394\ kJ$

c. $2\ O_3 \rightarrow 3\ O_2 + 285\ kJ$

Because heat can be regarded as a reactant or product in a chemical reaction, you can obtain from the balanced equation mathematical relations and conversion factors that relate the amount of heat supplied or produced to the amounts of reactants used or products formed. The equalities for Equations 10.1 and 10.2 are as follows. From Equation 10.1:

$$2\ \text{mol}\ N_2 = 1\ \text{mol}\ O_2 = 163\ \text{kJ} = 2\ \text{mol}\ N_2O$$

From Equation 10.2:

$$1\ \text{mol}\ CH_4 = 2\ \text{mol}\ O_2 = 1\ \text{mol}\ CO_2 = 2\ \text{mol}\ H_2O = 891\ \text{kJ}$$

Conversion factors can be constructed from these equalities in the usual way, except now the conversion factors are not simply mole ratios relating two materials. They are ratios relating moles of a reactant or product to the amount of heat absorbed or released in the reaction.

Conversion factors for Equations 10.1 and 10.2 as well as their reciprocals may be written. For Equation 10.1:

$$\frac{2\ \text{mol}\ N_2}{163\ \text{kJ}} \qquad \frac{1\ \text{mol}\ O_2}{163\ \text{kJ}} \qquad \frac{2\ \text{mol}\ N_2O}{163\ \text{kJ}}$$

For Equation 10.2:

$$\frac{1\ \text{mol}\ CH_4}{891\ \text{kJ}} \qquad \frac{2\ \text{mol}\ O_2}{891\ \text{kJ}} \qquad \frac{1\ \text{mol}\ CO_2}{891\ \text{kJ}} \qquad \frac{2\ \text{mol}\ H_2O}{891\ \text{kJ}}$$

Using the same general format as for other stoichiometry problems, you can calculate the amount of heat supplied or produced given an amount of some chemical reactant or product.

sample exercise 17

How much heat is released when 8.00 g of CH_4 burns? Burning is the reaction of CH_4 with O_2, shown in Equation 10.2.

solution

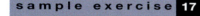

Use the unit conversion method. Two conversion factors are needed: (1) the molar mass of methane and (2) the ratio of moles to the amount of heat from the balanced chemical equation. The equation gives you the equality

$$1\ \text{mol}\ CH_4 = 891\ \text{kJ}$$

The molar mass of CH_4 is 16.0; thus,

$$1\ \text{mol}\ CH_4 = 16.0\ \text{g}$$

Given	×	Conversion factors		=	New
$8.00\ \cancel{\text{g}\ CH_4}$	×	$\frac{1\ \cancel{\text{mol}\ CH_4}}{16.0\ \cancel{\text{g}\ CH_4}}$ ×	$\frac{891\ \text{kJ}}{1\ \cancel{\text{mol}\ CH_4}}$	=	446 kJ

How much heat is required to produce 132 g of laughing gas, N_2O, according to Equation 10.1? N_2O is used as an anesthetic, particularly in dental surgery.

••• problem 9

$$2\ N_2(g) + O_2(g) \rightarrow 2\ N_2O(g)$$

10.11 page from a laboratory notebook

You are now fully prepared to understand a typical page from the laboratory record of a working chemist (Figure 10.15). A chemist working on insecticide research set out to prepare 500. g of DDT ($C_{14}H_9Cl_5$) from chlorobenzene (C_6H_5Cl) and trichloroethanal (C_2HOCl_3). His first task was to write a balanced chemical equation:

$$C_2HOCl_3 + 2\ C_6H_5Cl \rightarrow C_{14}H_9Cl_5 + H_2O$$

From the balanced equation, he determined the necessary amounts of reactants through appropriate *mole ratios.* In this case, he used 295 g of trichloroethanal as the *limiting reactant* and 500. g of chlorobenzene (450. g was needed; the *excess* helped the reaction go to completion) to obtain a theoretical yield of 709 g of DDT. He set his theoretical yield higher than the amount he actually wanted because he knew that the actual yield might be considerably less than the theoretical yield.

When the reaction was complete, he found that his *actual yield* was 510. g, i.e., more than the amount desired. However, the actual yield was less than the 709 g *theoretical yield* that was possible from the reactant amounts. He calculated the *percentage yield* as 71.9%.

DDT would only be prepared in small quantities in the laboratory for use in research projects. Commercially, DDT is no longer sold as an insecticide in the United States because of environmental and health hazards.

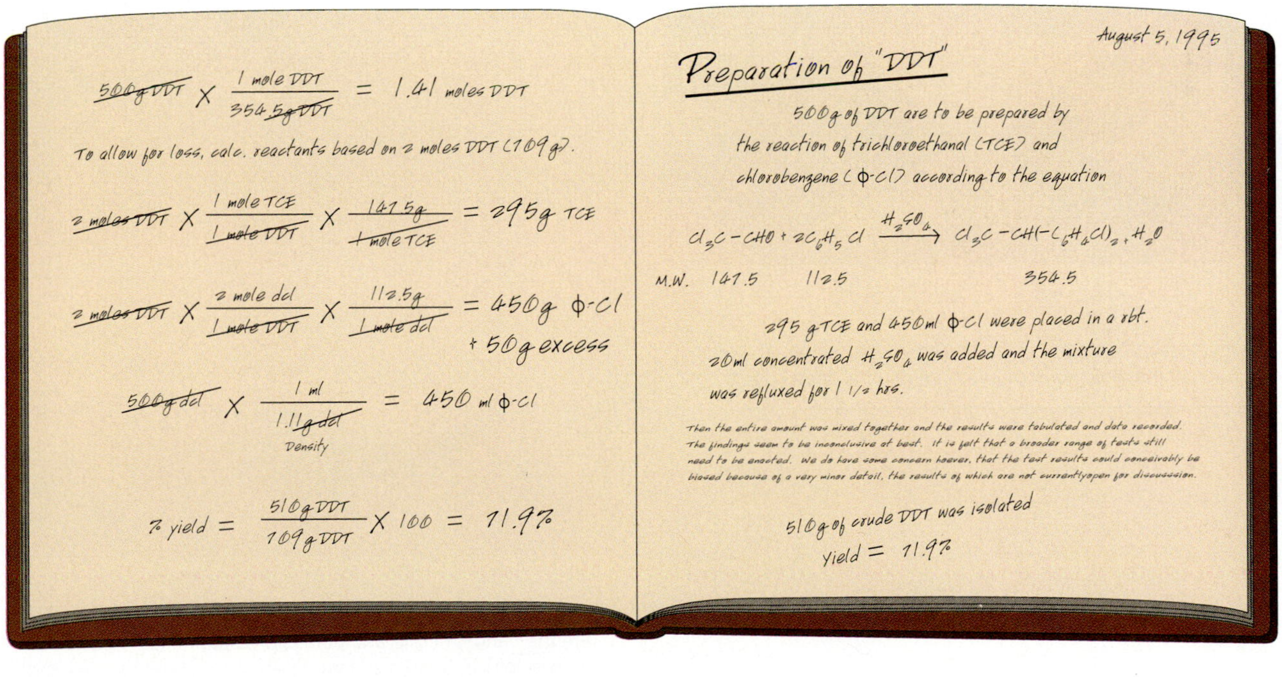

Figure 10.15

Facing pages from the notebook of a working chemist. The formulas in the equations are partial structural formulas that organic chemists sometimes use rather than the molecular formulas that you are used to. Trichloroethanal is abbreviated TCE and chlorobenzene is abbreviated ϕ-Cl.

CHEMLAB

Laboratory exercise 10: stoichiometry

In this experiment, you will apply your understanding of stoichiometric calculations to the reaction in which the very insoluble barium sulfate, $BaSO_4$, is formed by a double-replacement reaction using the reactants barium nitrate, $Ba(NO_3)_2$, and potassium sulfate, K_2SO_4.

$$Ba(NO_3)_2(aq) + K_2SO_4(aq) \rightarrow BaSO_4 \downarrow + 2\ KNO_3(aq)$$

Weigh out exactly 3.92 g of $Ba(NO_3)_2$ into a clean, dry, labeled 250 mL beaker. Add 75 mL of distilled water and several drops of dilute nitric acid and stir to dissolve the barium nitrate (Figure 10.16).

Into a clean, dry 150 mL beaker, weigh out the stoichiometric amount of K_2SO_4 needed to react completely with 3.92 g of $Ba(NO_3)_2$. Add 50. mL of distilled water and stir to dissolve the potassium sulfate. When dissolved, slowly pour the K_2SO_4 solution into the $Ba(NO_3)_2$ solution while stirring. Allow this mixture to sit undisturbed so that the precipitate may settle while you proceed with the rest of the experiment (Figure 10.17).

To separate the $BaSO_4$ precipitate from the reaction mixture by filtration, set up a Buchner funnel in a filter flask and attach the filter flask to a vacuum aspirator to provide suction (Figure 10.18). Place a piece of weighed filter paper in the funnel and moisten it with distilled water. Use only very fine glass, microfiber filter paper such as Whatman® GF/C; otherwise, the very fine precipitate will probably leak through. Filter the precipitate, which should have settled from the nearly clear liquid, called the supernatant, above. Decant (pour off the liquid, leaving the solid behind) most of the supernatant liquid through the filter. Then complete the transfer of the wet precipitate using a rubber policeman (a small rubber spatula attached to the end of a glass rod) to scrape out the beaker. Any $BaSO_4$ precipitate remaining in

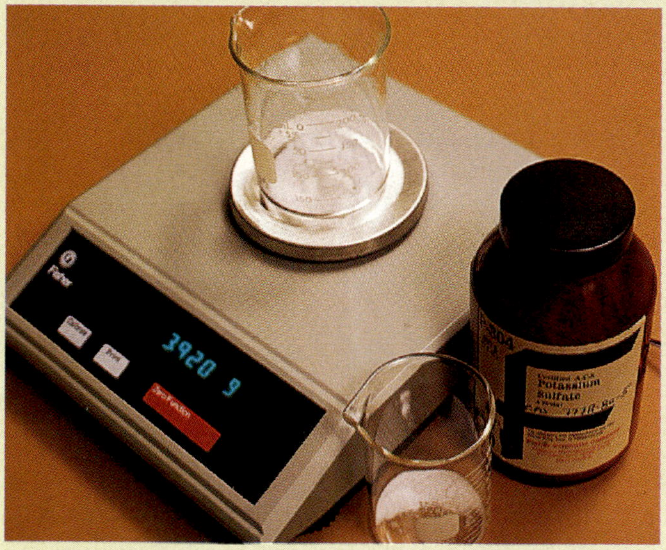

Figure 10.16

The 250 mL beaker contains 3.92 g of barium nitrate; the 150 mL beaker contains the stoichiometric amount of potassium sulfate.

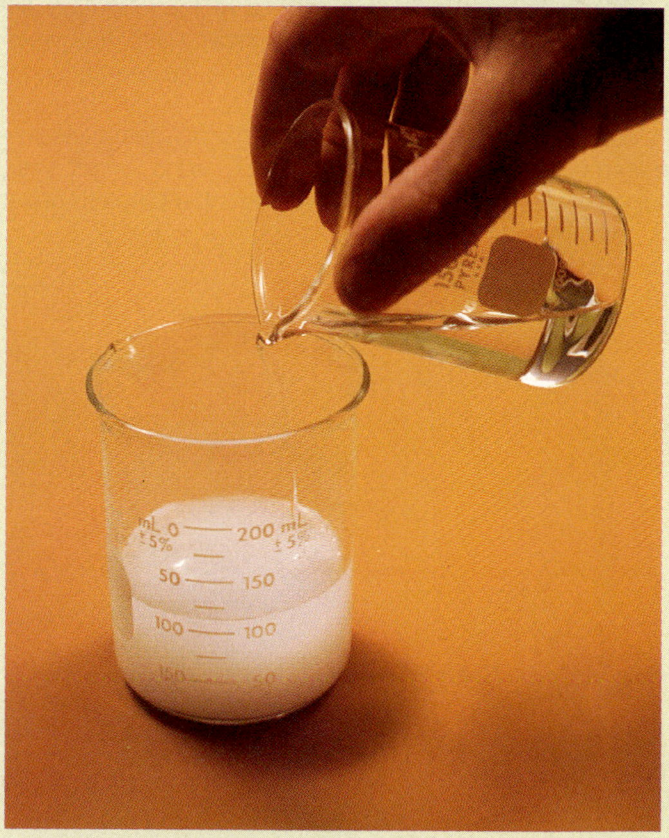

Figure 10.17

When the barium nitrate and potassium sulfate solutions are mixed, a white precipitate, barium sulfate, forms throughout the reaction mixture. The precipitate is so fine that most of it does not settle to the bottom of the reaction beaker; the precipitate appears to be suspended uniformly throughout the solution. Upon standing or after centrifugation, the precipitate will settle to the bottom of the beaker.

the beaker should be carefully washed into the funnel with a jet of distilled water from a wash bottle (Figure 10.19).

Rinse the precipitate with two successive 5 mL portions of cold distilled water and then with two 5 mL portions of acetone, a solvent that dries quickly. Allow air to be pulled through the precipitate by the suction apparatus until the precipitate and filter paper are dry.

Carefully remove the precipitate and filter paper from the Buchner funnel and place them on a tared watch glass. A tared watch glass means that the mass of the watch glass has been automatically subtracted from the reading displayed on your balance. The balance will read 0.000 g when the empty watch glass has been tared. When you place the filter paper with precipitate on the balance, their combined mass will be read directly from the balance. By subtracting the mass of the filter paper from this reading, you will have found the mass of the $BaSO_4$ precipitate formed in your reaction.

Questions:

1. What amount of K_2SO_4 is required to react completely with 3.92 g of $Ba(NO_3)_2$?
2. The filter paper and precipitate weighed 3.458 g. If the filter paper weighed 0.268 g, what is the weight of the $BaSO_4$ precipitate you prepared?
3. What is the theoretical yield you would predict for $BaSO_4$ based on the starting amounts of reactants?
4. What is the percentage yield of $BaSO_4$ from your experiment?
5. The formation of an insoluble precipitate such as $BaSO_4$ is a reaction that goes to completion, so you might expect a percentage yield approaching 100%. If your percentage yield is not 100%, suggest possible ways the precipitate may have been lost.

Figure 10.18

The solid, wet precipitate of $BaSO_4$ is transferred to the funnel by scraping the beaker with a rubber policeman on the end of a glass stirring rod.

Figure 10.19

The precipitate remaining in the reaction beaker is washed into the funnel with a jet of distilled water from a wash bottle.

chapter accomplishments

After completing this chapter, you should be able to define all key terms and do the following.

10.2 Molar interpretation of the balanced equation

- ❑ Interpret a chemical equation in terms of numbers of interacting particles.
- ❑ Interpret the coefficients of a chemical equation as numbers of moles of reactants and products.

10.3 Conversion factors from balanced chemical equations

- ❑ Given a balanced chemical equation, write equalities between molar amounts of reactants and/or products.
- ❑ Construct mole-ratio conversion factors from molar equalities.

10.4 Mole-to-mole conversions

- ❑ Given a balanced equation and a molar amount of one reactant or product, calculate the molar amounts of other substances reacting or produced.

10.5 Gram-to-mole and mole-to-gram conversions

- ❑ Given a balanced equation and a gram amount of one reactant or product, calculate the molar amounts of other substances reacting or produced.
- ❑ Given a balanced equation and a molar amount of one reactant or product, calculate the gram amounts of other substances reacting or produced.

10.6 Gram-to-gram conversions

- ❑ Given a balanced equation and a gram amount of one reactant or product, calculate the gram amounts of other substances reacting or produced.

10.7 Conversions summarized

- ❑ State the importance of the mole ratio in all stoichiometric calculations.

10.8 Limiting reactant

- ❑ Given the amounts of two or more reactants and a chemical equation, identify the limiting reactant.
- ❑ Determine the maximum amount of product that can be obtained based on the limiting reactant.
- ❑ Determine how much of an excess reactant is left over after the completion of a reaction.

10.9 Theoretical yield, actual yield, and percentage yield

- ❑ Given some amount of reactant or reactants, calculate the theoretical yield of product.
- ❑ Given the actual and theoretical yields, calculate the percentage yield.
- ❑ Given the actual yield and enough information to calculate the theoretical yield, calculate the percentage yield.

10.10 Heat as a reactant or product

- ❑ Identify chemical equations as representing either exothermic or endothermic reactions.
- ❑ Given a balanced chemical equation, write equalities between the amount of heat consumed or produced and the molar amount of reactants or products.
- ❑ Given a balanced equation and an amount of one reactant or product, calculate the amount of heat consumed or produced as the reaction proceeds.
- ❑ Given a balanced equation and an amount of heat consumed or produced, calculate the amount of reactant consumed or product produced.
- ❑ Identify changes in the physical state of a pure substance that are endothermic and exothermic.
- ❑ Given an amount of heat consumed or produced in a change of physical state, calculate the amount of pure substance involved.

10.11 Page from a laboratory notebook

- ❑ Recognize the stoichiometric calculations in a laboratory notebook.

key terms

stoichiometry
stoichiometric coefficients
mole ratios
limiting reactant (or limiting reagent)
chemical equilibrium
theoretical yield
actual yield
percentage yield
endothermic reaction
exothermic reaction

10.1 What is stoichiometry?

10. Define in your own words the term *stoichiometry*.
11. What must be true about any equation used in a stoichiometric calculation?

10.2 Molar interpretation of the balanced equation

12. Interpret the following balanced chemical equations in terms of numbers of atoms, molecules, and/or formula units.
    a. $2\ P(s) + 3\ H_2(g) \rightarrow 2\ PH_3(g)$
    b. $HBr(g) + KOH(aq) \rightarrow KBr(aq) + H_2O(l)$
    c. $2\ CO(g) + O_2\ (g) \rightarrow 2\ CO_2(g)$
    d. $C(s) + 2\ Cl_2(g) \rightarrow CCl_4(l)$
    e. $Mg(s) + 2\ HCl(aq) \rightarrow MgCl_2(aq) + H_2 \uparrow$
13. Interpret the equations in Problem 12 in terms of the numbers of moles of reactants and products.
14. Which interpretation, moles (Problem 13) or particles (Problem 12), is used in doing stoichiometry problems?
15. Could an equation be read directly in units of mass? For example, can $C + O_2 \rightarrow CO_2$ be read 1 g of C + 1 g of O_2 yields 1 g of CO_2? Explain your answer.

10.3 Conversion factors from balanced chemical equations

16. Write all of the "equalities" between reactants and products that the balanced equations in Problem 12, parts a and e, indicate.
17. Write the mole ratios corresponding to the equalities you wrote in Problem 16.
18. For the following equation, write all possible mole-ratio conversion factors that relate the number of moles of each product to the number of moles of each reactant.

$$2\ C_2H_6 + 7\ O_2 \rightarrow 4\ CO_2 + 6\ H_2O$$

19. Explain how the law of conservation of mass is satisfied by the use of mole ratios in stoichiometry problems.

10.4 Mole-to-mole conversions

20. Given the balanced equation

$$Na_2Cr_2O_7 + 6\ HI + 4\ H_2SO_4 \rightarrow 3\ I_2 + Cr_2(SO_4)_3 + Na_2SO_4 + 7\ H_2O$$

    a. How many moles of I_2 form when 3.0 mol of HI react?
    b. How many moles of HI react with 3.0 mol of H_2SO_4?
    c. How many moles of water form at the same time that 9.0 mol of I_2 form?
    d. How many moles of HI are required to produce 0.41 mol of H_2O?

21. Given the balanced equation

$$5\ C + 2\ SO_2 \rightarrow CS_2 + 4\ CO$$

    a. How many moles of CS_2 form when 3.00 mol of carbon react?
    b. How many moles of SO_2 are needed to react with 3.00 mol of carbon?
    c. How many moles of CO form at the same time that 7.00 mol of CS_2 form?
    d. How many moles of C are required to produce 1.50 mol of CS_2?

22. The balanced equation for the reaction of methane (CH_4) with oxygen is

$$CH_4 + 2\ O_2 \rightarrow CO_2 + 2\ H_2O$$

    a. How many moles of oxygen are required to react with 5.00 mol of CH_4?
    b. How many moles of oxygen are required to produce 3.50 mol of CO_2?
    c. How many moles of water will be produced from 1.15 mol of CH_4?

23. Write a balanced equation and calculate the following.
    a. How many moles of aluminum oxide can be produced from the reaction of 6.0 mol of aluminum with oxygen? Remember that oxygen exists as a diatomic molecule. *A balanced equation is always necessary for stoichiometry.*
    b. How many moles of oxygen must react with the 6.0 mol of Al?

10.5 Gram-to-mole and mole-to-gram conversions

24. Given the equation

$$2\ Al + 6\ HCl \rightarrow 2\ AlCl_3 + 3\ H_2 \uparrow$$

    a. How many grams of Al are needed to release 0.54 mol of H_2?
    b. How many grams of $AlCl_3$ are obtained from 12 mol of HCl?
    c. How many grams of H_2 are released from 8.0 mol of HCl?

25. Given the balanced equation

$$2\ KOH + H_2SO_4 \rightarrow K_2SO_4 + 2\ H_2O$$

    a. How many moles of H_2SO_4 are needed to make 78 g of K_2SO_4?
    b. How many moles of H_2O are produced from the reaction of 16.9 g of H_2SO_4?
    c. How many moles of KOH are needed to produce 125 g of water?

26. One of the steps in the manufacture of nitric acid is represented by the following reaction.

$$3\ HNO_2 \rightarrow 2\ NO + HNO_3 + H_2O$$

a. How many grams of HNO_2 are required to make 5.0 mol of HNO_3?

b. How many grams of NO are produced along with the 5.0 mol of HNO_3?

27. A propellant used in rocket engines is a mixture of hydrazine (N_2H_4) and hydrogen peroxide (H_2O_2). This mixture reacts spontaneously to yield N_2 and H_2O.

$$N_2H_4 + 2\ H_2O_2 \rightarrow N_2 + 4\ H_2O$$

a. How many grams of hydrazine are needed to react completely with 0.5 mol of H_2O_2?

b. How many grams of N_2 are produced by the combination of ingredients described in part a?

28. Chlorine is prepared by passing an electric current through a solution of sodium chloride. Sodium hydroxide and hydrogen are important by-products of this reaction.

$$2\ NaCl(aq) + 2\ H_2O(l) \xrightarrow{\text{Electricity}} 2\ NaOH(aq) + Cl_2(g) + H_2(g)$$

a. How many grams of chlorine are produced from 1.50 mol of NaCl?

b. If you wish to prepare 9.00 mol of Cl_2, how many grams of NaCl are needed?

c. How many moles of NaOH and H_2 will be produced if you start with the number of grams of NaCl calculated in part b?

10.6 Gram-to-gram conversions

29. Given the equation

$$MnO_2 + 4\ HCl \rightarrow MnCl_2 + Cl_2 + 2\ H_2O$$

a. How many grams of chlorine are formed from 12.5 g of MnO_2?

b. How many grams of HCl are needed to make 3.00 g of $MnCl_2$?

c. How many grams of water form when 18.0 g of HCl react?

30. The reaction that occurs in Clorox®, a bleach that leads to whitening, is

$$NaClO + NaCl + H_2O \rightarrow Cl_2 + 2\ NaOH$$

In a cup of Clorox®, there is approximately 12 g of NaClO.

a. How many moles of Cl_2 are liberated by 12 g of NaClO?

b. How many grams of NaOH are produced?

31. Soda-lime glass is made by the reaction of Na_2CO_3, limestone ($CaCO_3$), and sand (SiO_2). 478g.

$$Na_2CO_3 + CaCO_3 + 6\ SiO_2 \rightarrow \underset{\text{Soda-lime glass}}{Na_2O{\cdot}CaO{\cdot}6\ SiO_2} + 2\ CO_2 \uparrow$$

a. How many grams of sand are needed to make one champagne bottle weighing 618 g?

b. How many moles of CO_2 are released as the glass for the champagne bottle forms?

c. How many moles of limestone are needed to make 0.720 mol of glass?

32. Write a balanced equation and calculate how many grams of chlorine (Cl_2) are needed to produce 153 g of sodium chloride from sodium metal and chlorine gas.

33. Iron reacts with bromine to form iron(III) bromide.

a. How many grams of iron are needed to produce 65.0 g of the compound?

b. How many moles of Br_2 will be needed for this reaction?

34. Aluminum displaces copper from copper(II) sulfate. How many grams of Al are required to displace 34.0 g of Cu?

35. An important source of energy for humans and other living species is the breakdown of energy-rich glucose ($C_6H_{12}O_6$) to carbon dioxide and water. The overall chemical reaction is

$$C_6H_{12}O_6 + 6\ O_2 \rightarrow 6\ CO_2 + 6\ H_2O$$

a. Calculate the number of grams of oxygen required to react with 280. g of $C_6H_{12}O_6$.

b. Calculate the number of grams of water produced from the given amount of glucose in part a.

36. Some organisms such as yeast derive their energy from the breakdown of glucose ($C_6H_{12}O_6$) to carbon dioxide and ethyl alcohol (C_2H_6O), a reaction known as *fermentation.*

$$C_6H_{12}O_6 \rightarrow 2\ C_2H_6O + 2\ CO_2$$

What mass of glucose is needed to form 10.0 g of ethyl alcohol?

37. The combination of nitrogen with hydrogen to form ammonia (NH_3) is an important industrial process.

$$N_2(g) + 3\ H_2\ (g) \rightarrow 2\ NH_3(g)$$

a. How much ammonia can be produced from 2.00 kg of N_2?

b. How much H_2 is necessary to prepare 8.50 kg of NH_3?

10.7 Conversions summarized

38. Which conversion factor is used in all stoichiometry problems?

10.8 Limiting reactant

39. Chloroform ($CHCl_3$) can be made by the reaction of chlorine and methane (CH_4) according to the equation

$$3\ Cl_2 + CH_4 \rightarrow CHCl_3 + 3\ HCl$$

For each of the following combinations of reactants, decide which is the limiting reactant.

a. 3.00 mol of Cl_2 and 3.00 mol of CH_4

b. 5.00 mol of Cl_2 and 1.50 mol of CH_4

c. 0.50 mol of Cl_2 and 0.20 mol of CH_4

40. The "fluoride" in toothpaste is stannous fluoride [tin(II) fluoride], which can be made through the reaction of Sn and HF.

$$Sn + 2\ HF \rightarrow SnF_2 + H_2$$

For each of the following combinations of reactants, decide which is the limiting reactant.

a. 50.0 g of Sn and 1.00 mol of HF

b. 120. g of Sn and 60. g of HF

c. 120. g of Sn and 39 g of HF

41. How many moles of $CHCl_3$ are produced in the reaction in Problem 39, part a?

42. How many grams of SnF_2 are produced in the reaction in Problem 40, part b?

43. How much of the excess reactant is left over after the reaction is complete in Problem 40, part b?

44. How many grams of magnesium nitride are produced when 63 g of Mg and 41 g of N_2 react?

$$3\ Mg + N_2 \rightarrow Mg_3N_2$$

45. How much of the excess reactant is left over after the reaction is complete in Problem 44?

46. Calcium hydroxide reacts with phosphoric acid according to the equation:

$$3\ Ca(OH)_2 + 2\ H_3PO_4 \rightarrow Ca_3(PO_4)_2 + 6\ H_2O$$

a. When 215 g of $Ca(OH)_2$ reacts with 201 g of H_3PO_4, which reactant is limiting according to the above equation?

b. What mass of $Ca_3(PO_4)_2$ can form?

c. How much of the excess reactant is left over at the end of the reaction?

10.9 Theoretical yield, actual yield, and percentage yield

47. What is the theoretical yield of CO_2 based on 115 g of $CaCO_3$ according to the equation in Problem 31?

48. In a laboratory, 34.0 g of Li metal reacts with an excess of Br_2. What is the theoretical yield of LiBr?

49. Old oil paintings are darkened by PbS, which forms by the reaction of the Pb in paint with H_2S in air.

$$Pb(s) + H_2S(g) \rightarrow PbS(s) + H_2 \uparrow$$

Suppose 0.40 g of H_2S comes in contact with 2.0 g of Pb. What is the theoretical yield of PbS?

50. Old oil paintings can be cleaned by the reaction of hydrogen peroxide, H_2O_2, with the dark lead sulfide, PbS.

$$PbS(s) + 4\ H_2O_2(l) \rightarrow PbSO_4(s) + 4\ H_2O(l)$$

a. What is the theoretical yield of $PbSO_4$ based on 0.80 g of H_2O_2?

b. How much PbS will be removed?

51. a. What is meant by the actual yield of a chemical reaction?

b. Can the actual yield be calculated before the reaction is carried out?

52. In doing the reaction described in Problem 48, a chemist obtains 340.0 g of LiBr. What is the percentage yield?

53. An art restorer actually obtains 1.16 g of $PbSO_4$ in the cleaning process described in Problem 50. What is the percentage yield?

54. Chromium metal can be removed from its oxide ore by reaction with carbon.

$$Cr_2O_3 + 3\ C \rightarrow 2\ Cr + 3\ CO$$

When 1.00 kg of Cr_2O_3 reacts with 0.300 kg of C,

a. What is the theoretical yield of Cr?

b. How much of which reactant is left over in excess?

c. The actual yield obtained was 531 g of Cr. What is the percentage yield?

55. Elemental iron can be recovered from iron oxide ore by reaction with carbon monoxide.

$$Fe_2O_3 + 3\ CO \rightarrow 2\ Fe + 3\ CO_2$$

a. When 145 g of iron(III) oxide reacts with 95 g of carbon monoxide, how much iron will form?

b. Assume that 93 g of iron is produced in the reaction in part a. What is the percentage yield of this reaction?

56. A chemist sets out to make nitroglycerin from the reaction of glycerine with nitric acid.

$$\underset{\text{Glycerine}}{C_3H_8O_3} + 3\ HNO_3 \rightarrow \underset{\text{Nitroglycerin}}{C_3H_5O_9N_3} + 3\ H_2O$$

She uses 4.6 g of glycerine and 15 g of HNO_3 and gets 9.8 g of nitroglycerin. What is the percentage yield?

10.10 Heat as a reactant or product

57. a. Give an example of an exothermic process with which you are familiar.

b. Give an example of an endothermic process.

58. Identify the following reactions as exothermic or endothermic.

a. $CaCO_3 + 176\ kJ \rightarrow CaO + CO_2$

b. $MnO_2 + Mn \rightarrow 2\ MnO + 249\ kJ$

c. $C + H_2O + 130\ kJ \rightarrow CO + H_2$

59. The propellant reaction given in Problem 27 is highly exothermic.

$$N_2H_4(l) + 2\ H_2O_2(l) \rightarrow N_2(g) + 4\ H_2O(g) + 640.\ kJ$$

a. How much heat is liberated when 5.00 mol of H_2O_2 react?

b. How much heat is released as 1.50 mol of water form?

60. The air pollutant SO_3 combines with water to form H_2SO_4 according to the reaction below. H_2SO_4 in the atmosphere is observed as acid rain, which is destructive to marble and limestone. The reaction also contributes to thermal pollution (excessive heat in the atmosphere).

$$SO_3(g) + H_2O(l) \rightarrow H_2SO_4(aq) + 130.\ kJ$$

How many grams of H_2SO_4 and how many kilojoules of heat are produced from the reaction of 21 g of SO_3?

hello my name is Jennifer
hello

Stoichiometry
Stoichiometry
Stoichiometry

61. The reaction of H_2SO_4 with marble can be represented as follows.

$$H_2SO_4(aq) + CaCO_3(s) \rightarrow CaSO_4(s) + H_2O(l) + CO_2(g) + 113\ \text{kJ}$$

a. When 26 g of $CaCO_3$ reacts, how much heat is released?

b. Look back at Problem 60. How much $CaCO_3$ is destroyed by 21 g of SO_3?

62. How many moles of carbon dioxide can be released from $MgCO_3$ if 46.9 kJ of energy is supplied?

$$MgCO_3 + 118\ \text{kJ} \rightarrow MgO + CO_2$$

63. When 5.0 g of nitroglycerin blows up, the reaction can be represented by the equation

$$4\ C_3H_5O_9N_3(l) \rightarrow 12\ CO_2\uparrow + 6\ N_2\uparrow + O_2\uparrow + 10\ H_2O(l) + 7217\ \text{kJ}$$

a. How much heat is evolved?

b. Given that 4.18 kJ of heat can raise the temperature of 1 kg of water by 1°C, what will be the final temperature of 200. g of water initially at 40.°C if it is subjected to the heat evolved in part a?

Additional problems

64. Baking soda (sodium hydrogen carbonate) is often added in baking recipes for cookies and bread. When the batter is heated, the hydrogen carbonate decomposes, yielding gaseous carbon dioxide, which causes the cookies or bread to rise.

$$2\ NaHCO_3(s) \xrightarrow{\Delta} Na_2CO_3(s) + CO_2(g) + H_2O(l)$$

How many grams of $NaHCO_3$ are required to yield 30.0 g of CO_2?

65. Ammonia can react with copper(II) oxide to form metallic copper, nitrogen, and water.

a. Write a balanced equation.

b. How many grams of copper can be produced by treating 231 g of copper(II) oxide with 68.4 g of ammonia?

c. The actual yield of copper in the above reaction is 151 g. What is the percentage yield?

66. Ammonia is manufactured from nitrogen and hydrogen by the Haber process.

$$N_2(g) + 3\ H_2(g) \rightarrow 2\ NH_3(l)$$

How many grams of ammonia can be produced from the reaction for 128 g of nitrogen and 42 g of hydrogen?

67. Ethanol, the alcohol in all alcoholic beverages, is formed from glucose in a fermentation process.

$$C_6H_{12}O_6 \xrightarrow{\textit{Enzyme catalyst}} C_2H_5OH + 2\ CO_2\uparrow$$

a. Balance the above equation.

b. What amount of ethanol in grams can be formed from 57.0 g of glucose?

c. What is the volume of ethanol corresponding to the mass you calculated in part b? (The density of ethanol = 0.789 g/mL.)

68. Nitric acid, HNO_3, is produced from ammonia by the Ostwald method, a process that consists of three steps:

$$4\ NH_3(g) + 5\ O_2(g) \rightarrow 4\ NO(g) + 6\ H_2O(g)$$

$$2\ NO(g) + O_2(g) \rightarrow 2\ NO_2(g)$$

$$3\ NO_2(g) + H_2O(l) \rightarrow 2\ HNO_3(aq) + NO(g)$$

How many kilograms of nitric acid can be produced from 8.30×10^4 kg of ammonia? (Hint: You may wish to think in terms of extending the conversion factor method to a sequence of three equations.)

69. Elemental liquid bromine, Br_2, is formed in the displacement reaction:

$$2\ NaBr(aq) + Cl_2(g) \rightarrow Br_2(l) + 2\ NaCl(aq)$$

How many grams of bromine can be formed from the reaction of 30.0 g of NaBr with 25.0 g of Cl_2?

70. Magnesium metal burns in air (such as in road flares) to form magnesium oxide. If the magnesium oxide is treated with water, magnesium hydroxide results.

a. Write balanced equations for the two reactions.

b. How much magnesium oxide can form from the combustion of 4.50 g of magnesium?

c. How many grams of magnesium hydroxide can be produced from this amount of magnesium?

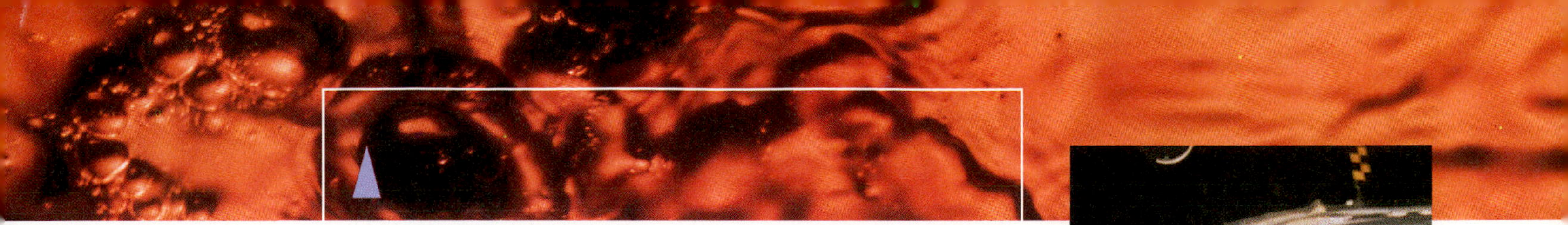

chapter 11

gases

An automobile airbag inflates instantaneously on impact. The pressure of gas within an airbag resists the force of the driver against the steering wheel and greatly reduces the likelihood of serious injury.

11.1 introduction

Throughout Chapter 10, you determined the relative amounts of reactants and products that participate in balanced chemical reactions. From the examples in Chapter 10, you learned how to convert the stoichiometric number of moles of any reactant or product to its corresponding mass so that it could be measured on a balance in the laboratory. When a reactant or product is a gas, it is typically more convenient and simpler to measure its volume rather than its mass. Easily measurable volumes of gases have very small masses that are more difficult to measure accurately. In this chapter, you will learn to extend your mastery of stoichiometry to gaseous reactions where you express the reactant or product gases in terms of their volumes. But first, you must become more familiar with the physical properties of gases and how the ideal gas law prescribes changes in those properties as you vary the temperature, pressure, volume, or number of moles of gas present.

In Chapter 1, you learned properties of solids, liquids, and gases (Table 1.1). Since all gases expand to take the shape of and to fill the volume of their container, gases have no definite shape or volume of their own. You are familiar with how quickly the gases from cooking food spread to fill your home and disperse their aroma to a room remote from the kitchen. Conversely, all gases may be compressed easily; that is, the volume of gases is reduced when the volume of their containers become smaller (Figure 11.1). An automobile piston greatly compresses the gaseous fuel mixture in each of its cylinders.

These two properties of gases—that they are easily *compressed* and readily *expand*—are summarized in the first two statements of the list following. Additional physical properties unique to the gaseous state are also described in the following list; they will be useful in understanding and describing the behavior of gases with the gas laws that are developed throughout this chapter.

Figure 11.1

Pumping air from the atmosphere into an automobile or bicycle tire greatly reduces its volume and illustrates how easily a gas, like air, may be compressed.

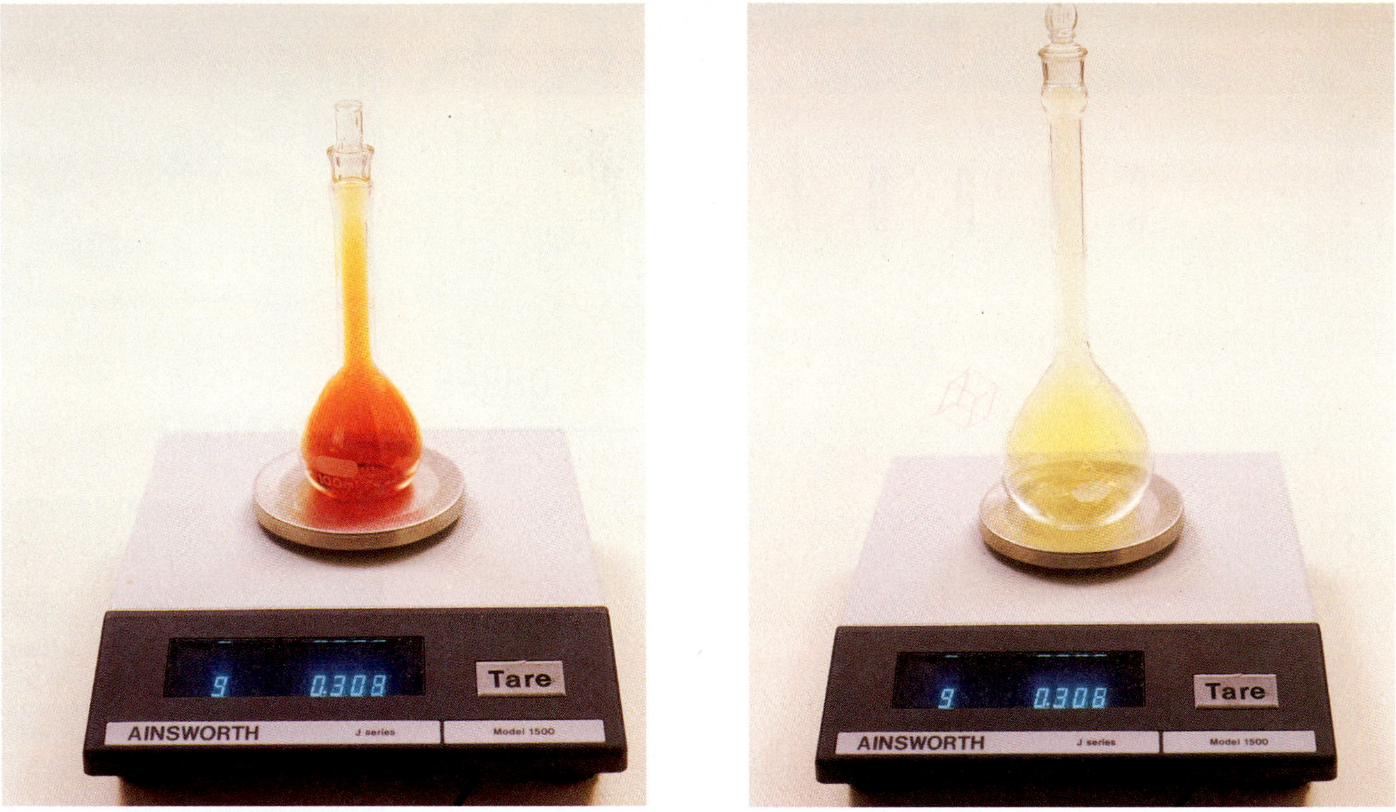

(a) (b)

Figure 11.2
The tared *mass* or "weight" of gas in each volumetric flask is the *same*. The mass of the empty flask has been automatically subtracted (tared) from the total mass on each balance, so the mass readings represent the mass of just the gas in each flask. Since the *volumes* of the gas samples are considerably *different*, their respective *densities* ($d = m/V$) will also be *different*. As you can observe by the more intense color, the gas in the smaller flask (a) has a high density. (b) Conversely the less intense color indicates that the gas in the larger flask has a lower density.

- volume of a gas varies w/ container
- mass remains constant
- density varies as the volume of the container varies

★ characteristics and physical properties of gases

1. *Gases can be easily compressed* by applying pressure to a movable piston in a container with rigid walls (Figure 11.1). Compression implies a *decrease* in volume. By contrast, when you try "pressing" on a container filled with water, you find that liquids are not easily compressed. If you applied pressure to the surface of water in a bottle, the bottle would break before any noticeable decrease in the volume of liquid could be observed. Similarly, solids are not easily compressed and, indeed, are the least compressible of the three states. The compressibility of a material is a physical property that is often listed in handbooks of chemical substances.
2. *Gases expand to fill the entire volume of their container.* Expansion implies an *increase* in volume. If a small container of a gas, such as the vapor from a bottle of perfume or cologne, is opened in a classroom, the gas escapes and soon expands to fill the entire room. The mass of the gas undergoing expansion remains the same, only the volume changes; that is, the same amount of material is spread out over a larger space.
3. *Gases have indefinite densities.* As you learned in Chapter 2, density, a physical property of any material, is the ratio of mass to volume ($d = m/V$). Because the volume of a gas (V) varies with its container while its mass (m) remains constant, the density (d) also varies as the volume of the container varies. The expanding gas described in statement 2 has a lower density when it fills the room than when it was in the small container (Figure 11.2).

   Density is a physical property of each gas and may be used to identify it. However, as you will learn from the gas laws that describe gaseous behavior (Sections 11.3 through 11.9), care must be taken to specify the temperature and pressure at which the density of the gas is determined.

••• problem 1

Originally, 0.96 g of a gas at 20.°C occupies 0.50 L. The gas is allowed to expand to fill 12 L at 20.°C. Calculate the original and final densities.

Figure 11.3
Gases have lower densities than liquids; this is why gas bubbles within a liquid tend to rise.

4. *Gases have a low density* compared with liquids and solids (Table 3.2). Because the volume of a gas is always larger than the volume of the same amount (mass) of liquid, the density of the gas will always be lower than the density of its liquid ($d = m/V$). In general, all gases will be less dense than liquids; this is why gas bubbles within a liquid tend to rise (Figure 11.3).
5. Assuming that they do not react chemically, *gases can diffuse (mix) rapidly through each other* in all directions. Ammonia gas released at the front of a classroom rapidly diffuses through the air, as can be verified by a student sitting in the back. The odor of ammonia is quite pungent and is the ingredient released by smelling salts to revive semiconscious individuals.
6. *Gases exert a pressure on the wall of any container or surface that they touch.* You measure the pressure of air pressing on the wall of your automobile or bicycle tires by attaching a pressure gauge to the tire valve. In overinflated tires, the gas is pressing too hard on the tire walls, making them stiff, less resilient, and giving a "hard" ride. In underinflated tires, the gas is not pressing hard enough on the tire walls, making them soft and reducing your steering control. A special case of air pressing on every surface it touches is that of the atmosphere, the volume of air surrounding the earth. At sea level, air presses against every square inch of our bodies or any other object with a force of 14.7 lb.
7. *Gases expand in volume when they are heated and contract when they are cooled.* The expansion of a material is a physical property characteristic of that material. Liquids and solids also expand and contract with temperature, but not to the same extent as gases. How much a gas expands depends on its temperature and pressure (Figure 11.4).

Figure 11.4
A hot air balloon has a heater to heat the air, which rises into the balloon above. The air expands as it is heated, and the heated air that fills the balloon is less dense than the cooler air outside the balloon. As a result, the balloon rises until the air within it is balanced with the less dense air of the atmosphere at higher altitude. When the air inside the balloon cools, it becomes more dense and the balloon descends.

All of the properties and characteristics of gases listed above can be explained and accounted for by the kinetic molecular theory (Section 11.10). This theory explains the measured properties of gases in terms of the behavior of individual gas molecules. Kinetic theory is a fairly recent development; consequently, it will appear toward the end of this chapter according to its chronological development in the knowledge of gases.

At the beginning of this chapter, you will explore the gas laws, which give the quantitative relationships among the *number of moles*, the *volume*, the *pressure*, and the *temperature* of a gas. To understand how the gas laws were developed, you must know how to measure each of these four gas variables: number of moles, *n*, volume, *V*, pressure, *P*, and temperature, *T*. If any one of these four variables changes for a given sample of gas, then at least one other variable must also change. The ideal gas law in Section 11.9 will summarize all of these possible changes and enable you to calculate the properties of a gas under any conditions. The designated letters *n*, *V*, *P*, and *T* are the standard symbols for the four gas variables.

11.2 measurement of the four gas variables: *n*, *T*, *V*, and *P*

Measurements of number of moles, *n*, temperature, *T*, and volume, *V*, are straightforward and familiar to you from previous experiments. Since pressure, *P*, and its measurement are probably less familiar to you, however, it will be defined and discussed in great detail in this section.

The number of moles, *n*, can be determined directly by measuring the mass of a gas sample. The conversion between mass and moles using the molar mass has been well known to you since Chapter 8. The mass of a gas sample may be determined by measuring the mass of an empty, closed container and subtracting it from the mass of the same container filled with a specific gas.

The temperature, *T*, of a gas can be measured with a thermometer. Laboratory measurements of temperature are usually made using the Celsius (centigrade) scale. You will see in Section 11.5, however, that the relationships among the temperature, volume, and pressure of a gas can be greatly simplified by the introduction of a new temperature scale known as the *Kelvin temperature scale*. To convert a temperature measured in degrees Celsius (°C) to a temperature measured in degrees on the Kelvin scale (K), you simply add 273.15.

$$\text{K} = {}^\circ\text{C} + 273.15$$

In gas law problems, the Kelvin temperature K is represented by *T* and 273.15 is rounded to 3 sig figs, 273, so that the above equation becomes

$$T = {}^\circ\text{C} + 273$$

The volume, *V*, of a gas can be measured most easily by placing the gas in a container of known volume. The unit of volume most often employed in gas measurements is the liter, and many sizes of containers calibrated in fractions or multiples of liters are routinely available in chemical laboratories.

In Section 11.1, you learned that gases exert a pressure on the wall of any container or surface that they touch. You use your intuitive understanding of pressure to describe the amount of air compressed into a tire. Before you may use pressure as a gas variable, however, you must keep in mind a more formal definition. **Pressure**, *P*, is defined as the force exerted per unit area, or

$$P = \frac{\text{Force}}{\text{Area}}$$

Forces are observed in pushes, pulls, and collisions. You can see from the preceding formula that, given the same area, a larger force results in a greater pressure. You can also see from the formula that, given the same force, pressure is greater when the force operates over a smaller area. Notice that force and pressure are not the same thing. Identical forces can lead to very different pressures, depending on the area over which they are distributed (Figure 11.5).

Yeow!
Ouch!
Oh!
Excuse me!
Ouch!
Oh!
Excuse me!

A woman wearing stiletto-type heels (area = 0.50 cm^2) steps on a man's toes. The pressure exerted is

$$P = \frac{125 \text{ lb}}{0.50 \text{ cm}^2} = 250 \text{ lb/cm}^2$$

It really ***hurts!***

The same woman wearing flat heels (area = 4.0 cm^2) steps on the man's toes. The pressure exerted is

$$P = \frac{125 \text{ lb}}{4.0 \text{ cm}^2} = 31 \text{ lb/cm}^2$$

It doesn't hurt nearly as much.

Pressure is eight times greater in the case of the smaller heel, even though the force remains the same.

Figure 11.5

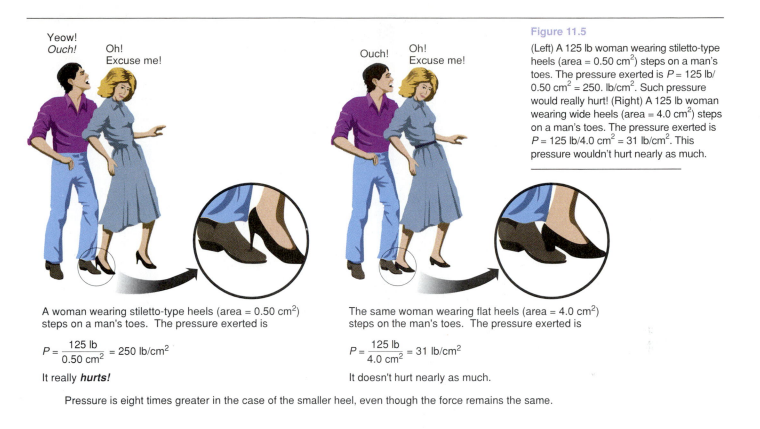

(Left) A 125 lb woman wearing stiletto-type heels (area = 0.50 cm^2) steps on a man's toes. The pressure exerted is P = 125 lb/0.50 cm^2 = 250. lb/cm^2. Such pressure would really hurt! (Right) A 125 lb woman wearing wide heels (area = 4.0 cm^2) steps on a man's toes. The pressure exerted is P = 125 lb/4.0 cm^2 = 31 lb/cm^2. This pressure wouldn't hurt nearly as much.

According to kinetic molecular theory (Section 11.10), all molecules of a gas are constantly moving. The pressure of a gas comes from the force exerted on the inside area of the container by the collisions of moving gas molecules with the walls of the container (Figure 11.6). The pressure of the air around us, which is known as **atmospheric pressure,** is the pressure exerted when molecules in air collide with our bodies, the ground, or other objects on earth. This pressure is approximately 14.7 lb per square inch (lb/in.^{2}). *P*ounds per *s*quare *i*nch can also be abbreviated *psi.*

$$\frac{14.7 \text{ lb}}{1 \text{ in.}^2} = \frac{\text{Force}}{\text{Area}}$$

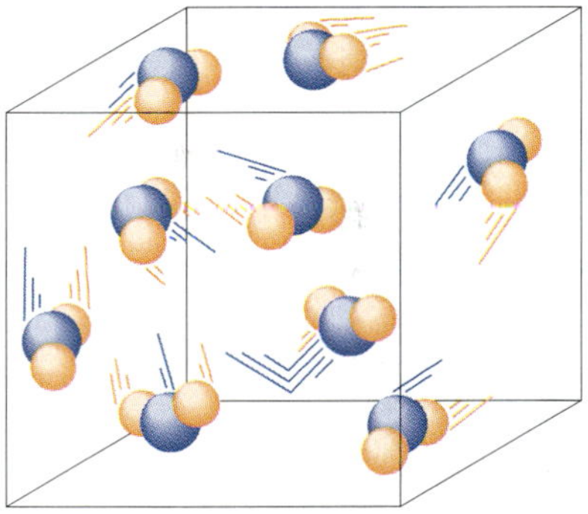

Figure 11.6

The pressure of a gas results from the force of collisions of moving gas molecules exerted on the inside walls of the container.

Air pressure, or atmospheric pressure, can be measured with a barometer (Figure 11.7). A long (about 80 cm), narrow (diameter less than 1 cm) tube, open at one end, is filled with mercury, a dense, liquid metal. The tube is stoppered and inverted into a dish containing mercury. When the stopper is removed, some of the mercury empties out of the tube into the dish, but a column of mercury approximately 76 cm (760 mm) high remains in the tube. Because the tube was originally completely filled with mercury, the space that is left above the mercury column when some of the mercury flowed out of the tube contains *no* air; it is a totally empty space, called a vacuum. A vacuum contains no molecules to collide with the top surface of the mercury column, so there is no force pushing down on the top of the mercury in the tube. There is a force from the weight of air on the surface of the mercury in the dish, however, pushing the mercury upward into the tube. The pressure of the atmosphere is exerted on the mercury in the dish and exactly balances the pressure from the weight of the mercury column in the tube (Figure 11.8).

Figure 11.7

Steps in the construction of a barometer. (a) A stoppered tube full of mercury is inverted in a dish of mercury. (b) The stopper is removed and some mercury falls out of the tube. The remaining height of mercury in the tube above the mercury surface in the dish is about 760 mm. Atmospheric pressure from the weight of air pushing down on the open dish of mercury holds up the column of mercury in the tube.

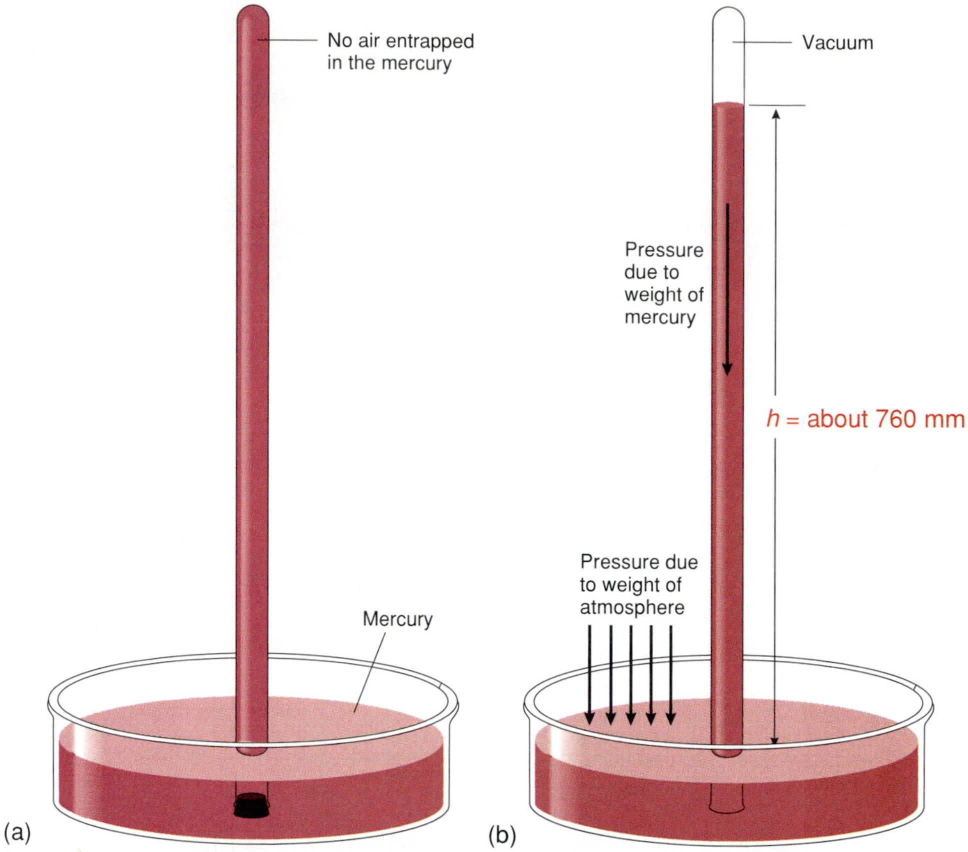

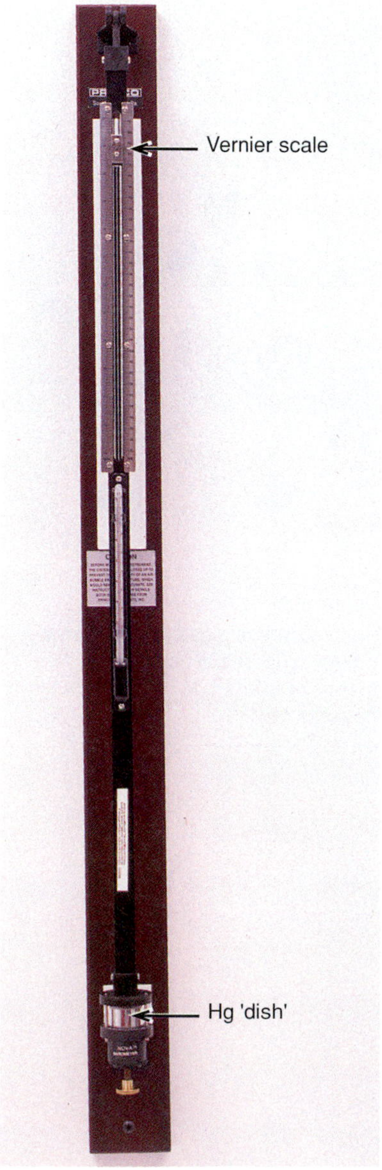

(a)

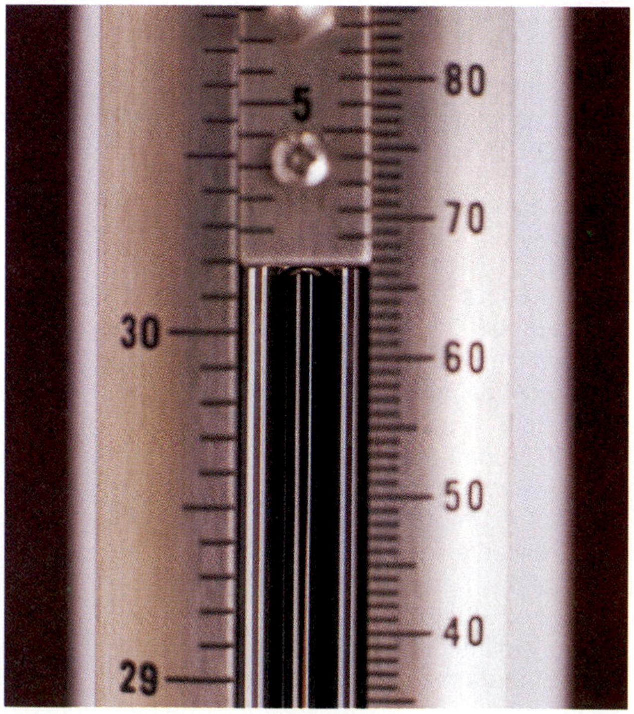

(b)

Figure 11.8

An actual mercury barometer, shown in (a), follows the same construction outlined in Figure 11.7, but the dish of mercury is reduced in size for convenience. An accurate sliding scale (called a vernier scale), shown in (b), is found near the top of the mercury column and enables you to adjust the bottom of the marker to the exact height of the mercury. Pressure readings may be taken to within 0.1 mmHg (0.1 torr). The height of the column of mercury in (b) can be read as 766.7 mmHg from the right scale.

table 11.1 equalities among common units of pressure

1 atm = 760 mmHg	760 mmHg = 760 torr
1 atm = 760 torr	1 mmHg = 1 torr
1 atm = 14.7 lb/in.^{2}	760 mmHg = 14.7 lb/in.^{2}

The height of the mercury in the tube gives us a direct measure of the atmospheric pressure. If the atmospheric pressure increases, the pressure on the surface of mercury in the dish will increase and the height of the mercury in the tube will increase. If the atmospheric pressure decreases, the pressure on the surface of mercury in the dish will decrease and the height of mercury in the tube will decrease. At sea level on an average day, atmospheric pressure can support a column of mercury exactly 760 mm high. This pressure, 760 mmHg, is also called one **atmosphere** (atm). In honor of the Italian scientist, Evangelista Torricelli, who first developed the barometer, the unit of pressure, millimeters of mercury, has been given the name **torr.** So 1 atm = 760 mmHg = 760 torr at 0°C. Table 11.1 illustrates equalities among common units of pressure. From these equalities, you may construct conversion factors to change pressure from one set of units to another.

$$1 \text{ atm} = 760 \text{ mmHg} = 760 \text{ torr}$$

••• problem 2

Weather reporters typically say, "Today's barometric pressure is 30. inches and falling." The 30. in. is the day's atmospheric pressure in inches of mercury. What is the pressure in torr? (The conversion factors for inches to millimeters of mercury are given in Tables 2.2 and 3.1.)

Now that you have reviewed and are familiar with the four gas variables, you are prepared to apply them to chemical and biological problems. So far, you have considered the measurement of pressure only as it applies to a single, pure gas. Many real-life situations involve a mixture of gases or a nearly pure gas that contains small amounts of contaminating gases. Dalton's law in Section 11.3 will allow you to extend your understanding of gas pressure to gaseous mixtures.

11.3 Dalton's law of partial pressures for mixtures of gases

John Dalton (the same scientist who proposed the atomic theory) conducted many experiments to measure the total pressure in a container when two or more separate gases are mixed. One such experiment consisted of taking three identical 1.0 L containers. The first 1.0 L container was filled with gas A at pressure p_A, and the second 1.0 L container was filled with gas B at pressure p_B. The contents of both containers were then totally transferred to the third 1.0 L container and the total pressure measured. Note that the gases did not change volume. The volume of the combined gases was 1.0 L, the same as the volume of each of the original pure gases. The temperature, likewise, remained unchanged.

In each experiment like the one just described, Dalton found that the *pressure in the third container was equal to the sum of the pressures that each gas exerted by itself;* that is,

$$P_{\text{total}} = p_A + p_B$$

Dalton's Law of Partial Pressures

Stated another way, each gas in a mixture exerts a pressure as if it were alone in the container. The pressure of each gas in the mixture is known as the **partial pressure** of that gas. **Dalton's law of partial pressures** may be applied to any

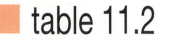

table 11.2 composition of clean, dry air

Gas	Percent by volume
N_2	78.09
O_2	20.94
Ar	0.93
CO_2	0.03
Noble gases other than Ar	0.0024
CH_4	0.00015
Trace amounts of H_2, N_2O, CO, O_3, NH_3, and SO_2	0.0075

number of gases in a mixture and states that *the total pressure of a mixture of gases is equal to the sum of all the partial pressures of each gas in the mixture:*

$$P_{total} = p_1 + p_2 + p_3 + p_4 + \dots \quad (11.1)$$

Partial pressures are commonly represented by lower-case *p*'s.

One common example of a mixture of gases is that of clean, dry air. The composition of air is given in Table 11.2. It consists primarily of the elements nitrogen and oxygen mixed with a small amount of the noble gas argon and trace amounts of numerous other elements and compounds that are naturally occurring. Ozone, O_3, is a small, but extremely important, component of air in the upper atmosphere. It blocks harmful ultraviolet light emitted by the sun and prevents damage to plants and animals on the surface of the earth. As a pollutant at ground level, it is harmful to plants and causes respiratory distress in animals.

Pollutants add many other gases to the mixture of air. Given time, pollutant gases that do not decompose or react with other materials diffuse throughout the atmosphere. Chlorofluorocarbons, very stable chemical compounds used in dry cleaning, refrigeration, and air conditioning, are pollutants that have diffused throughout the atmosphere with harmful consequences. They chemically react with ozone and lower its concentration, eventually leading to "holes," or areas in which the ozone layer is depleted in the upper atmosphere (Figure 11.9). While the manufacture of chlorofluorocarbon compounds is being reduced, their replacement as refrigerants will require research into the properties of substitute materials.

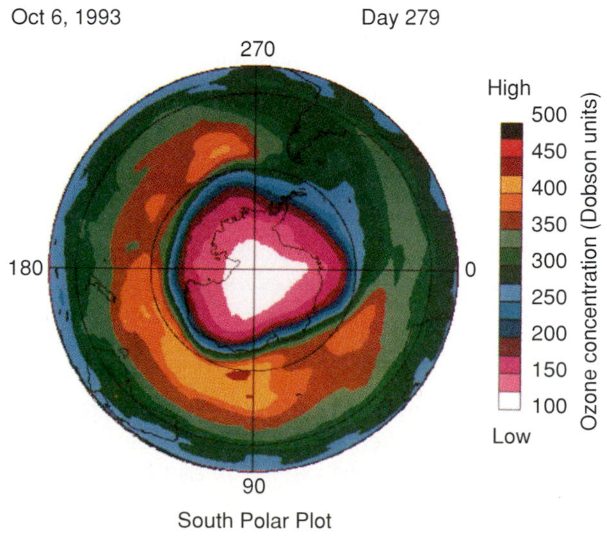

Figure 11.9
A large ozone hole (an area depleted of ozone) in the upper atmosphere has been measured over the South Pole using satellite tracking. The ozone hole appears as a large white area at the center of the picture. The size and location of another ozone hole over the North Pole is under study.

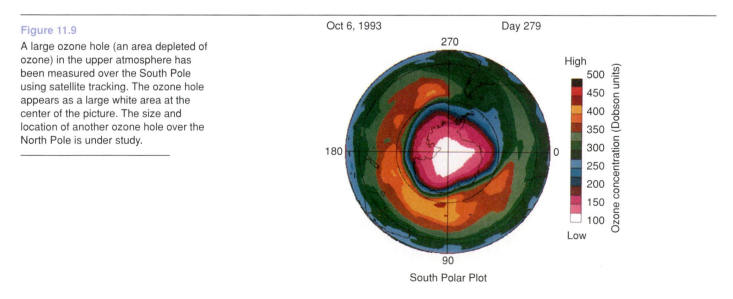

problem 3

A container holds three gases, argon, neon, and krypton, with partial pressures of 1.0 atm, 0.5 atm, and 1.5 atm, respectively. What is the total gas pressure in the container?

Dalton's law is perfectly general and may be applied to any mixture of gases. The remainder of this section will illustrate this law's usefulness with two particularly common applications. First, you will use Dalton's law in the measurement of a gas pressure when that gas is contaminated with water vapor. Second, you will use Dalton's law to understand respiration in multicellular organisms.

Gases such as oxygen, which are only very slightly soluble in water, can be collected by displacing water from a collection bottle (Figure 11.10). In Laboratory Exercise 7, you calculated the percentage of oxygen in potassium chlorate by measuring the difference in mass of the reactant, potassium chlorate, and the product, potassium chloride, according to the decomposition reaction:

$$2\ KClO_3(s) \rightarrow 3\ O_2\uparrow + 2\ KCl(s)$$

The same decomposition reaction is used in Figure 11.10, but the oxygen gas produced is collected by displacing water from a collection bottle. The oxygen obtained is *not* pure, but mixed with water vapor. If Dalton's law is applied, the total pressure of the mixture must equal the sum of the individual pressures of O_2 and H_2O vapor.

$$P_{total} = p_{O_2} + p_{H_2O}$$

If the water level inside the bottle is carefully adjusted so that it is the same as that outside the bottle (as shown in Figure 11.10c), the total gas pressure inside the bottle, P_{total}, must be the same as the atmospheric pressure outside the bottle on the water in the trough, P_{atm}. Then

$$P_{total} = P_{atm} = p_{O_2} + p_{H_2O}$$

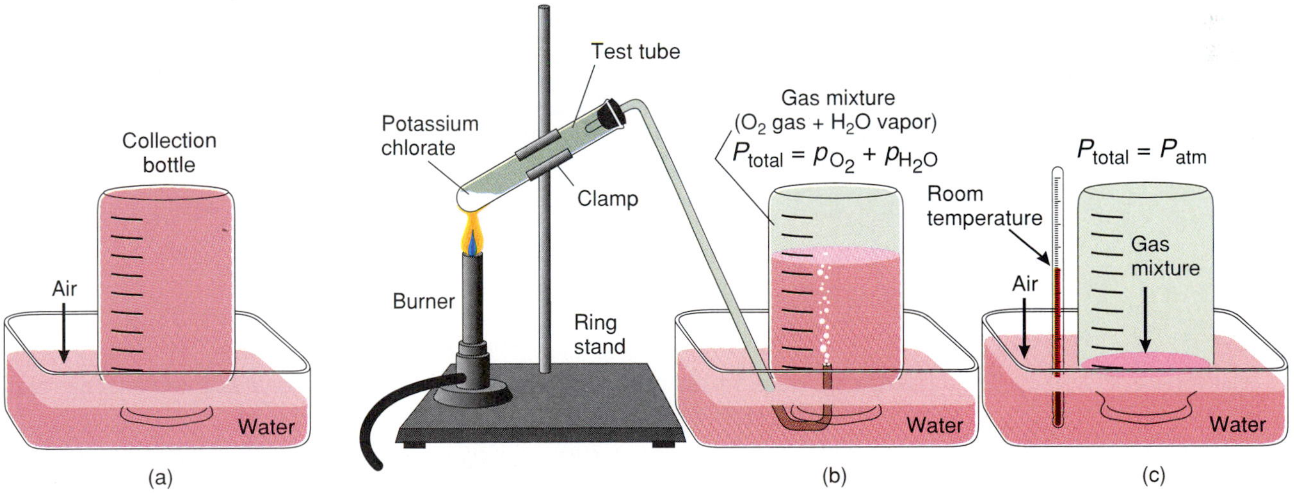

Figure 11.10

O_2 gas produced by the decomposition of $KClO_3$ is collected by displacing water from a collection bottle. (a) The full bottle of water is shown before O_2 gas is introduced. (b) The O_2 gas has displaced some of the water in the collection bottle. (c) The total pressure, P_{total} of the gas mixture in the bottle (O_2 gas + H_2O vapor) is being measured by equalizing the water levels inside and outside the bottle. When the water levels are equal, the total pressure inside the bottle equals the atmospheric pressure outside.

The pressure of O_2 in the collection bottle may be found by subtracting p_{H_2O} from each side of the preceding equation.

$$P_{atm} - p_{H_2O} = p_{O_2}$$

Then the pressure of O_2 is simply the difference between the atmospheric pressure, which you can measure, and the vapor pressure of water, which you can look up.

The measurement of P_{atm} with a barometer is routine in chemical laboratories and has already been described. The vapor pressure of water, p_{H_2O}, depends only on the temperature of the water (Section 12.6) and can be looked up in a table such as the one in Appendix D of this text. Other tables may be found in reference books such as the *CRC Handbook of Chemistry and Physics* and *Lange's Handbook of Chemistry.*

Using Dalton's law of partial pressures, you have found the *pressure* of O_2 gas produced in a specific chemical reaction. If you carefully measure the volume of the water in your collection bottle and subtract it from the total volume of your bottle, you have also found the *volume* of the O_2 gas produced in the reaction. Furthermore, if you read the temperature of a thermometer inserted in the water of the trough, you have also measured the *temperature* of the O_2 gas produced in the reaction.

In this simple experiment, you have now measured three of the four gas variables discussed in Section 11.2. When you learn the ideal gas law in Section 11.9, you will be able to combine the values of temperature, pressure, and volume to solve for the *number of moles* of O_2 gas produced. And, once you know the number of moles of gas produced, you may convert to grams of gas using its molar mass.

In this water-displacement experiment, you have found a method for measuring stoichiometric amounts of gases produced in chemical reactions. Earlier in Section 11.2, we said that mass is not always the easiest quantity to measure when reactions produce gases. Instead, chemists measure the temperature, pressure, and volume of a gas and convert this information to moles or grams using a relationship known as the ideal gas law, which you will see in Section 11.9.

sample exercise 1

A sample of N_2 is collected by water displacement in a setup similar to that in Figure 11.10. The water level inside the bottle is equalized with that in the trough. Barometric pressure is found to be 757 mmHg, and the temperature of the water is 22°C. What is the partial pressure of N_2?

solution

The total gas pressure inside the bottle is

$$P_{total} = p_{N_2} + p_{H_2O}$$

Since the water levels inside and outside the bottle were equalized, the total gas pressure inside the bottle must equal P_{atm}.

$$P_{total} = P_{atm} = p_{N_2} + p_{H_2O}$$

$$p_{N_2} = P_{atm} - p_{H_2O}$$

P_{atm} is given:

$$P_{atm} = 757 \text{ mmHg}$$

So,

$$p_{N_2} = 757 \text{ mmHg} - p_{H_2O}$$

Appendix D indicates that p_{H_2O} at 22°C is 19.8 mmHg.

$$p_{N_2} = 757 \text{ mmHg} - 19.8 \text{ mmHg}$$

$$p_{N_2} = 737 \text{ mmHg} = 737 \text{ torr}$$

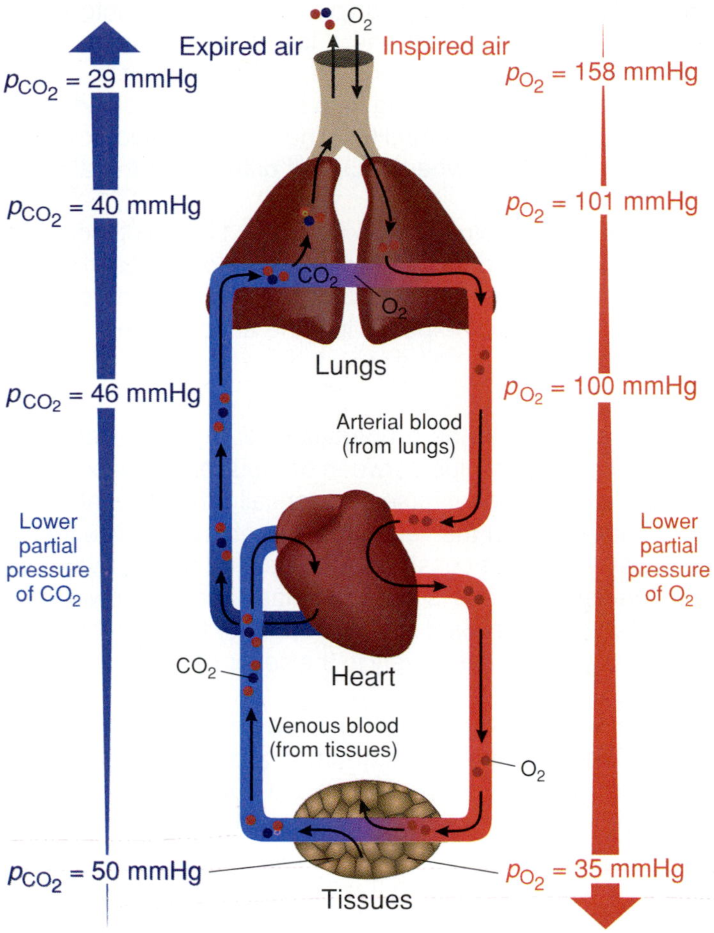

Figure 11.11
Respiration is a cyclic process in which O_2 from the air is transported to body tissues and CO_2 from body tissues is expelled into the air. Each gas flows spontaneously in the direction of lower partial pressure.

••• problem 4

A sample of oxygen is collected by water displacement. What would be the pressure of dry oxygen if the temperature is 18°C and the barometric pressure is 765 mmHg?

The concept of partial pressures also plays an important role in your understanding of respiration in multicellular organisms. Respiration is the physiological process by which oxygen and carbon dioxide are exchanged between the cells, the blood, the lungs, and the outside atmosphere (Figure 11.11).

Gases flow spontaneously from a region of higher partial pressure to a region of lower partial pressure. As Figure 11.11 shows, respiration is a cyclic process that begins when atmospheric O_2 flows into small sacs in the lungs, called alveoli. Because p_{O_2} is greater in the inhaled air (158 mmHg) than in alveolar air (101 mmHg), O_2 is transferred from the atmosphere into the alveoli. Because p_{CO_2} is greater in the alveoli (40 mmHg) than in the atmosphere (0.3 mmHg), CO_2 will be transferred out of the alveoli into the exhaled air.

In a second exchange, O_2 diffuses from the alveoli (p_{O_2} = 101 mmHg) into the circulating venous blood (blood returning to the heart via veins from body tissues), where the partial pressure of O_2 is still lower (p_{O_2} = 45 mmHg). The increase in O_2 in the blood converts it from venous blood, which was deficient in O_2, into arterial blood, which is rich in O_2 and has a p_{O_2} of about 100 mmHg. At the same time that oxygen moves from the lungs to the blood, carbon dioxide migrates from the venous blood (p_{CO_2} = 46 mmHg) into alveolar sacs (p_{CO_2} = 40 mmHg).

Boyle's Law

* P is inversely proportional to V

PV=k(nT)

k=constant

The partial pressure of O_2 in tissue is still lower (about 35 mmHg), and therefore O_2 migrates from the arterial blood (p_{O_2} = 100 mmHg) into the tissues. The deoxygenated blood (venous blood) has a low partial pressure of O_2 (about 45 mmHg) and returns to the lungs, where the cycle is repeated. Meanwhile, CO_2 flows from the tissues (p_{CO_2} = 50 mmHg) into arterial blood (blood flowing from the heart to body tissues) (p_{CO_2} = 40 mmHg). The venous blood flowing back to the lungs has a p_{CO_2} of 46 mmHg, where CO_2 will diffuse into the alveoli and be expelled into the outside air.

By following the red and blue arrows in Figure 11.11, you see that by flowing from a region of higher partial pressure to a region of lower partial pressure, O_2 gas is transported from the air into body tissues to keep them alive and CO_2 gas is removed from body tissues and expelled into the atmosphere.

11.4 Boyle's law

Now that you are familiar with the properties of pure and mixed gases, you are ready to examine the relationships between the four gas variables: pressure, volume, temperature, and number of moles. Historically, these variables were studied two at a time. Values of two of the variables were changed while the values of the remaining variables were kept constant. After several pairs of variables were understood experimentally, the results were combined into one powerful law, the ideal gas law, which you will use in Section 11.9. The first two variables, pressure and volume, were investigated quantitatively in the seventeenth century by the English chemist, Robert Boyle.

You are familiar with the behavior of pressure and volume from your own experiences. When you increase the pressure on a bubble of gas by squeezing it, its volume gets smaller, just as you saw with a tire pump in Figure 11.1. When you lessen the pressure, the volume gets larger again. There appears to be an *inverse* relationship between the volume and the pressure of a gas; raising or lowering one has the opposite (or inverse) effect on the other gas variable.

Boyle's experiments can be illustrated by a gas confined within a cylinder as shown in Figure 11.12. The temperature and the number of moles of gas were held at fixed values; the change in volume of the gas was measured as the external pressure on the gas was varied.

The pressure of the gas inside the cylinder must be the same as the external pressure when the piston is not moving. The pressure exerted upward from the gas is exactly equal to the pressure pushing downward from the weights on the piston. The external pressure is increased by adding more weights to the piston. Note that in Figure 11.12, the gas pressures in the cylinders are 15, 30., and 60. lb/in.^{2}. When the gas pressure has doubled from 15 to 30. lb/in.^{2}, the volume of the gas has decreased to one-half its initial value. When the gas pressure has quadrupled from 15 to 60. lb/in.^{2}, the volume of the gas has decreased to one-fourth its initial value.

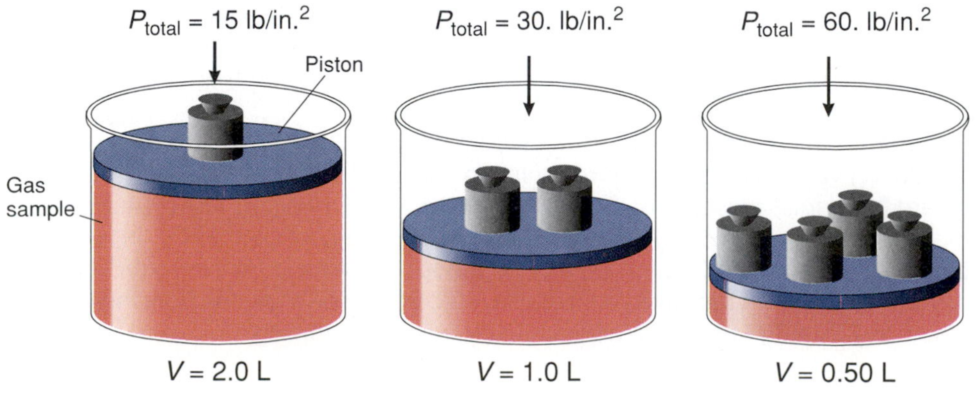

Figure 11.12
The apparatus used to illustrate Boyle's experiments shows that the volume of a gas inside a cylinder decreases as the pressure increases. The pressure of the gas inside the cylinder is equal to the pressure from the weights pressing on the surface of the piston. Pressure increases as more weight is distributed over the same piston area.

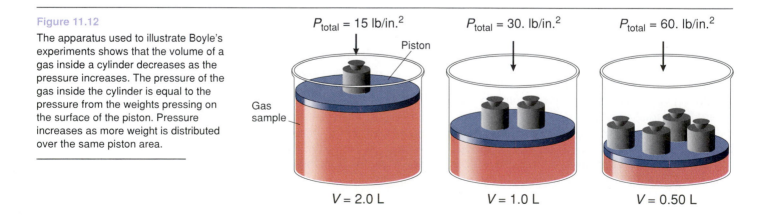

table 11.3 pressure-volume data for fixed moles of gas at a constant temperature

Pressure, $P\left(\frac{\text{lb}}{\text{in.}^2}\right)$	Pressure, P (atm)	Volume, V (L)	PV (L • atm)
15.	1.0	2.0	2.0
30.	2.0	1.0	2.0
45.	3.0	0.67	2.0
60.	4.0	0.50	2.0
75.	5.0	0.40	2.0

Table 11.3 summarizes the data from Boyle's experiments, and three sets of these data are illustrated in Figure 11.12. The measured values of pressure have been converted into units of atmospheres in the second column of Table 11.3 to be consistent with the units used in the remainder of this chapter. Remember, 1 atm = 14.7 lb/in.^{2}. For all measurements given in the table, whenever the volume of the gas is decreased, the pressure of the gas has increased. This inverse relationship between volume and pressure is also shown graphically in Figure 11.13. These results can be stated mathematically by Boyle's law:

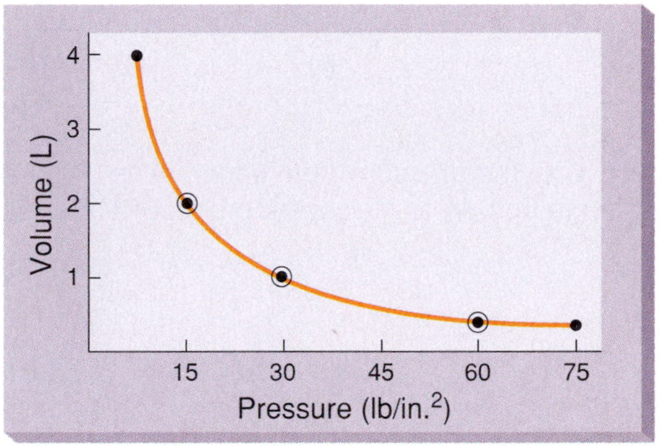

Figure 11.13
The data of volume and pressure measurements from Table 11.3 are graphed to show their inverse relationship. The measurements obtained from the apparatus in Figure 11.12 are indicated by ⊙.

$$V \propto \frac{1}{P} \quad \text{at constant } n \text{ and } T \qquad (11.2)$$

where $\propto$ is the symbol that expresses proportionality. This expression is read, "The volume of a gas is *inversely* proportional to its pressure when temperature and number of moles are held constant." This statement is **Boyle's law.**

A proportionality sign can always be replaced by an equal sign and a constant, thus converting the relationship into an equation (Appendix A, Section A.6). In this case, Boyle's law is written

$$V = \frac{k}{P} \quad \text{at constant } n \text{ and } T \qquad (11.3)$$

where k is a constant, a fixed number. Multiplication of both sides of the equation by P gives

$$PV = k \quad \text{at constant } n \text{ and } T \qquad (11.4)$$

An alternate way of stating Boyle's law is that the product of the pressure and volume of a gas is equal to a constant value when temperature and number of moles are fixed. Experimentally, you can see that this has been confirmed in Table 11.3 by looking at the values of the product PV in the fourth column. For all of the experiments, the product PV has the value of 2.0 L • atm within the error expected for data reported to two significant figures.

Equation 11.4 is particularly useful for calculating changes in either the pressure or volume of a gas. For a fixed quantity of gas at a particular temperature, if

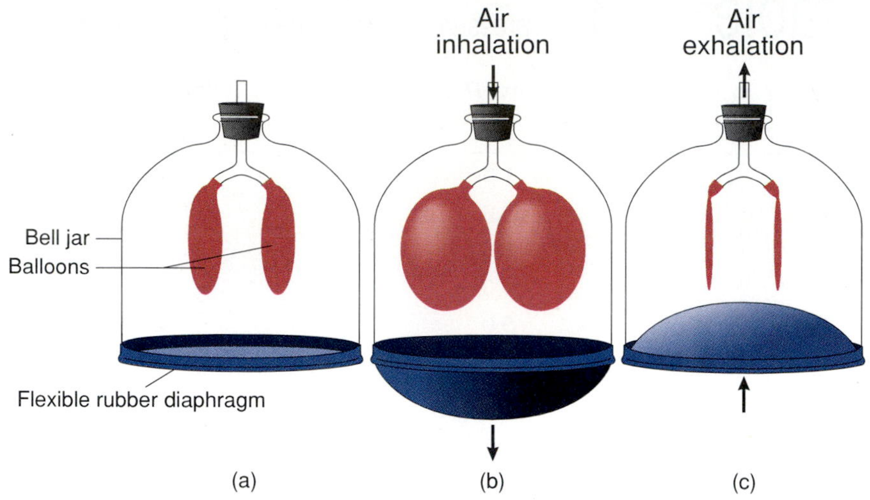

Figure 11.14

Breathing is a direct demonstration of the inverse relationship between volume and pressure. (a) When the rubber diaphragm is pulled down, the volume of the jar increases and the pressure decreases below 1 atm. (b) Outside air at 1 atm pressure rushes into the balloons, corresponding to your inhalation. (c) The diaphragm is released and the volume of the jar decreases. The higher pressure in the jar forces air out of the balloons, corresponding to your exhalation.

you vary the pressure from some value P_1 to a new value P_2, the volume must change from V_1 to V_2, but the product PV will still be equal to the same constant k.

$$P_1V_1 = k \quad \text{at constant } n \text{ and } T$$

and

$$P_2V_2 = k \quad \text{at constant } n \text{ and } T$$

or

$$P_2V_2 = P_1V_1 \quad \text{at constant } n \text{ and } T \tag{11.5}$$

Notice again the variation in P and V but the constancy of PV as shown in Table 11.3.

An application of Boyle's law can be seen in the physiological process of breathing. Breathing is a direct demonstration of the inverse relationship between volume and pressure. Moving your diaphragm downward expands the rib cage and increases the volume V_t of the thoracic cavity (the body cavity surrounding the lungs). Boyle's law tells you that as V_t increases, the pressure in the thoracic cavity, P_t, must decrease, that is, fall below atmospheric pressure. The air outside at atmospheric pressure then rushes into the lungs. Moving the diaphragm upward contracts the rib cage, decreasing the volume V_t of the thoracic cavity. The pressure P_t is increased and air is pushed out of the lungs.

A model that illustrates the breathing process is shown in Figure 11.14, where a sealed bell jar with a flexible, rubber diaphragm on the bottom represents your thoracic cavity, and two balloons attached to a Y-shaped tube represent your lungs. As the rubber diaphragm on the bottom of the jar is pulled down, the volume of the jar increases, just as the volume of your thoracic cavity increases when your diaphragm moves down. As a result, the pressure within the jar decreases below 1 atm and air outside the jar then rushes into the balloons in the same manner as it does into your lungs. When the rubber diaphragm is released, the volume in the jar decreases, like the volume of your thoracic cavity, and pressure increases. As a result, air within the balloons is forced out just as it is forced out of your lungs.

11.5 Charles' law

The two variables, volume and temperature of a gas, were systematically studied by Jacques Charles in 1787, at least 100 years after Boyle had studied pressure and volume. Like the results of Boyle, Charles' measurements are also consistent with your experiences.

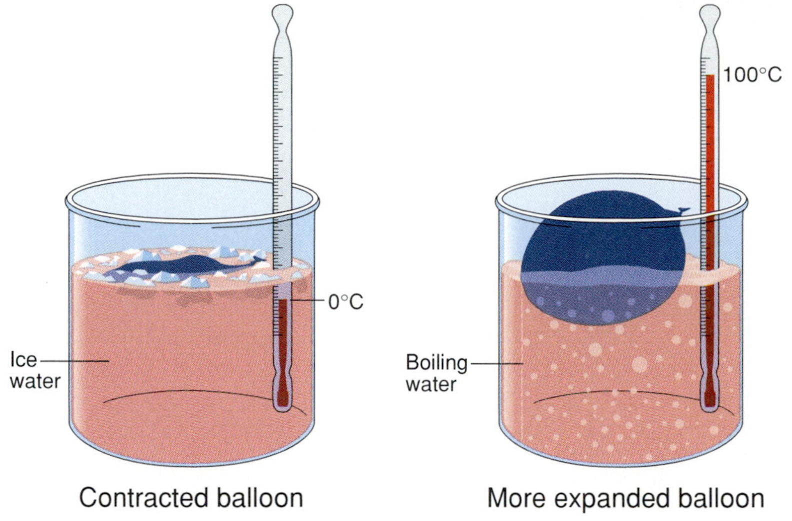

Figure 11.15
The volume of gas in a balloon at 0°C and 100°C shows how the two gas variables (temperature and volume) change when the number of moles is fixed and the pressure is constant at 1 atm. The volume of gas in the balloon increases as the temperature increases.

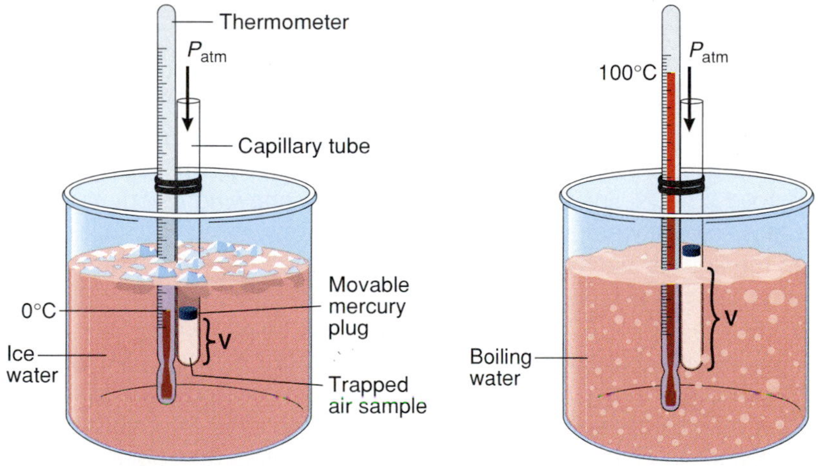

Figure 11.16
The volume of air trapped below the mercury plug expands as the temperature increases. The number of moles of air is fixed and the pressure of the trapped air sample must exactly equal the constant external pressure (the sum of pressures from the atmosphere and from the constant weight of the small mercury plug). The volumes of the trapped air sample at two temperatures are not drawn to scale; the volume increases are exaggerated for visibility.

* V is directly proportional to T

Charles Law

If you take a balloon filled with air and place it in boiling water, the volume of the balloon visibly increases (Figure 11.15). If you take the same balloon and place it in the freezer compartment of your refrigerator, the volume decreases. In both cases, the volume change occurs at a constant pressure of 1 atm (P_{atm}) and with a fixed amount of gas. You may conclude that when the temperature of a gas increases, its volume increases; when the temperature decreases, the volume of the gas decreases. Mathematically, you would say that there is a *direct* relationship between the volume and the temperature of a gas. Raising or lowering one has the *same* effect on the other gas variable.

Figure 11.16 shows an experimental setup similar to that used by Charles. A small plug of liquid mercury is placed in a capillary tube that is open at one end. A fixed mass (therefore, a fixed number of moles) of air is trapped below the mercury. Because the only pressure on the air mass is atmospheric pressure, P_{atm}, plus the pressure from the fixed weight of mercury, the external pressure is constant. The pressure of the trapped air sample, $P_{air\ mass}$, must exactly equal the constant external pressure, so $P_{air\ mass}$ is also constant. The capillary tube is attached to a thermometer, which is used to read the temperature (in °C); in addition, the finely spaced, uniform markings on the thermometer are used to measure the relative volume changes of the trapped air mass.

table 11.4 volume-temperature data for fixed moles of a typical gas at a constant pressure

Volume (L)	Temperature (°C)	Temperature (K)
0.96	–10.0	263
1.00	0.0	273
1.05	10.0	283
1.07	20.0	293
1.11	30.0	303
1.15	40.0	313
1.19	50.0	323
1.22	60.0	333
1.26	70.0	343
1.30	80.0	353
1.34	90.0	363

A typical set of data for Charles' experiment is shown in Table 11.4, where you observe that, as the temperature increases, the volume increases, and as the temperature decreases, the volume decreases. A graph of volume versus temperature gives the solid straight line shown in Figure 11.17, showing clearly that there is a direct relationship between volume and temperature.

When the line for each set of data is extrapolated (shown by dashed lines) to zero volume, all gases have the same temperature value equal to –273.15°C. Or stated in reverse, at –273.15°C, all gases should have zero volume, at least in principle. In practice, however, all gases liquefy before reaching this temperature.

Mathematically, a straight-line relationship between volume and temperature would be simpler if the volume and temperature scales both went through the origin. The extrapolated lines in Figure 11.17 suggest that the best starting point (or zero value) for temperature would be –273.15°C, where the volume of the gas is also zero. You can accomplish this for the volume-temperature relationship by establishing a new temperature scale with the zero of this new scale equal to –273.15°C. This temperature is called absolute zero because it is not possible to achieve lower temperatures than this. This new scale is called the **Kelvin temperature scale** (after Lord Kelvin, the English physicist and mathematician, who constructed it) or the **absolute temperature scale.** The units of this scale are represented by a capital letter K. One does *not* use the degree symbol as in °C and °F. Any Celsius temperature can be converted to Kelvin by using the formula

$$K = {}^\circ C + 273.15$$

The third column of Table 11.4 is a list of the Kelvin temperatures corresponding to each Celsius reading in the second column. As noted in Section 11.2, *T* is used to designate a Kelvin temperature, and for convenience, 273.15 is rounded to 273. Figure 11.18 shows a plot of the same data used in Figure 11.17, but the temperature scale has been converted to Kelvin. You can see clearly how both volume and temperature begin at the origin when the Kelvin temperature scale is used.

• • • problem 5

Convert the following temperatures to Kelvin temperatures.

a. 0°C b. 25°C c. 212°F

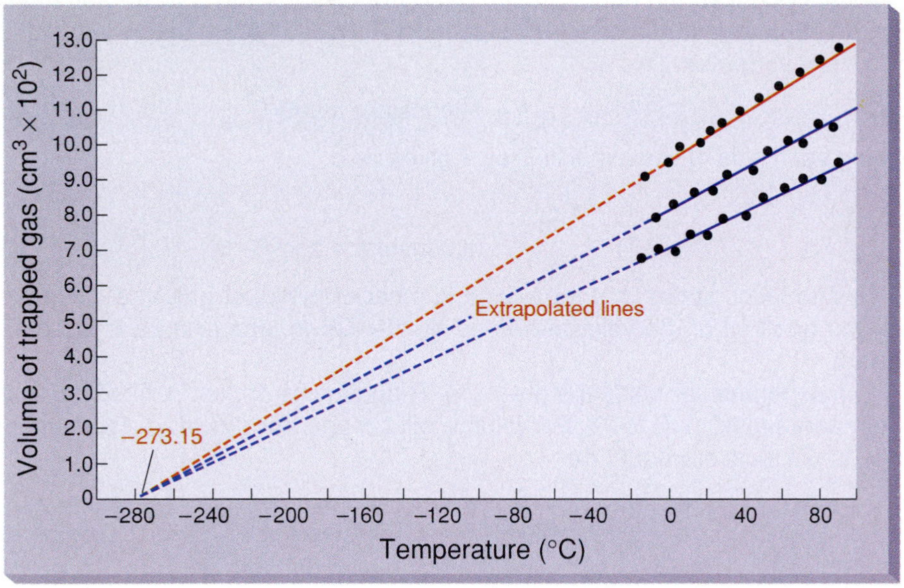

Figure 11.17

A graph of the data from Table 11.4 is shown in red and data from other gases are shown in blue. The linear data illustrate the direct relationship between the volume and temperature of each gas. Extrapolation of the experimental data also shows that the gases would have zero volume at a temperature of −273.15°C.

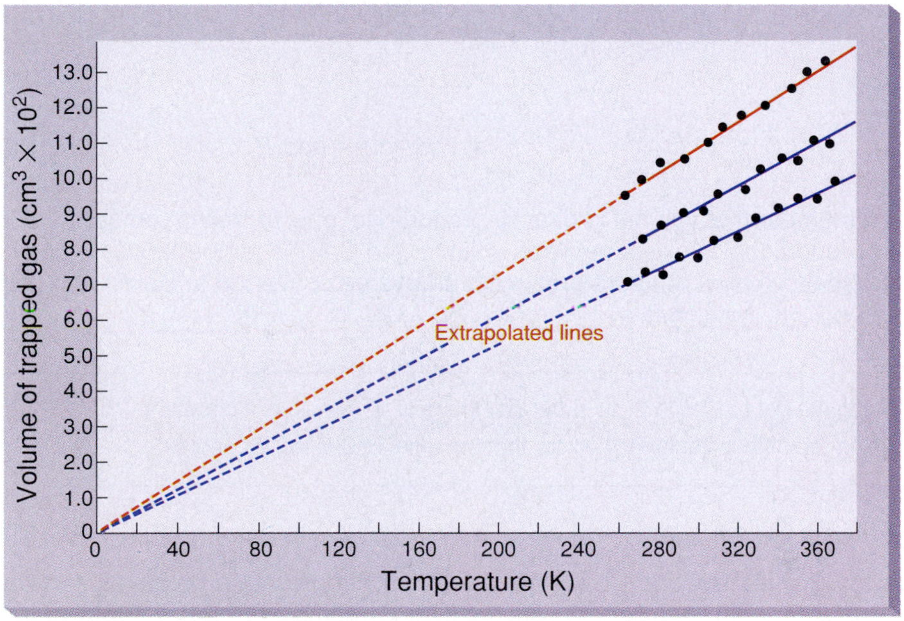

Figure 11.18

A graph of the data from Table 11.4 is shown in red and data from other gases are shown in blue just like Figure 11.17. Here the temperature scale has been converted to Kelvin so that both the volume and temperature scales begin at the same point (origin). The linear data on this scale illustrate a direct proportion between the volume and temperature of each gas.

The direct proportion between volume and temperature shown in Figure 11.18 can be stated mathematically by **Charles' law**:

$$V \propto T \quad \text{at constant } n \text{ and } P \qquad (11.6)$$

This expression is read, "The volume of a gas is *directly* proportional to its Kelvin temperature when pressure and number of moles are held constant." The direct proportion between these two gas variables means that as the temperature is raised, the volume increases *in the same proportion.* That is, if the temperature doubles, the volume will also double. If the temperature is reduced to one-third its initial value, the volume will also be reduced to one-third its initial size. *The direct proportionality between volume and temperature of a gas is true only for the Kelvin temperature scale, not for the Celsius scale.*

You can convert the direct proportionality of Charles' law into an equation by using an equal sign and a constant (Appendix A, Section A.6), just as you did with Boyle's law in Section 11.4:

$$V = mT \quad \text{at constant } n \text{ and } P \tag{11.7}$$

Dividing each side of Equation 11.7 by T gives us

$$\frac{V}{T} = m \quad \text{at constant } n \text{ and } P \tag{11.8}$$

This equation states that for a fixed number of moles of gas at a given pressure, the quotient of the volume divided by the Kelvin temperature is equal to a constant.

If an experimenter holds the pressure, P, and moles of gas, n, fixed but varies the temperature from T_1 to T_2, the volume will change from V_1 to V_2. However, the quotient V/T must still equal m:

$$\frac{V_1}{T_1} = m \quad \text{at constant } n \text{ and } P$$

and

$$\frac{V_2}{T_2} = m \quad \text{at constant } n \text{ and } P$$

or

$$\frac{V_2}{T_2} = \frac{V_1}{T_1} \quad \text{at constant } n \text{ and } P \tag{11.9}$$

Remember that volume is directly proportional only to Kelvin temperature; no direct proportionality exists between volume and Celsius temperature. In gas law calculations, Celsius temperatures must always be converted to Kelvin by adding 273.15 (273 to 3 sig figs).

• • • problem 6

a. Using the data in Table 11.4, show that the ratio V/T (in K) is a constant.

b. Using the data in Table 11.4, show that the ratio $V/°C$ is *not* a constant.

11.6 combined gas laws

Applying Boyle's law and Charles' law to chemical problems becomes simpler if you know that both laws can be combined into a single expression. Then, instead of remembering two separate laws (Equations 11.2 and 11.6), you need to remember only one fundamental expression (Equation 11.10).

$$\text{Boyle's law:} \quad V \propto \frac{1}{P} \quad \text{at constant } n \text{ and } T \tag{11.2}$$

$$\text{Charles' law:} \quad V \propto T \quad \text{at constant } n \text{ and } P \tag{11.6}$$

Therefore, upon combining the two separate laws into one, we obtain

$$\text{Combined law:} \quad V \propto \frac{T}{P} \quad \text{at constant } n \tag{11.10}$$

This expression is read, "For a fixed number of moles of gas, the volume of the gas is directly proportional to its Kelvin temperature and inversely proportional to its pressure." You can write this in an equation form by replacing the proportionality sign with an equal sign and a constant (r).

$$V = \frac{rT}{P} \quad \text{at constant } n \tag{11.11}$$

Multiplying both sides of the equation by P and dividing by T gives

$$\frac{PV}{T} = r \quad \text{at constant } n \tag{11.12}$$

If the pressure, volume, and temperature are changed from P_1, V_1, and T_1 to P_2, T_2, and V_2, then

$$\frac{P_1V_1}{T_1} = r \quad \text{at constant } n$$

and

$$\frac{P_2V_2}{T_2} = r \quad \text{at constant } n$$

or

$$\frac{P_1V_1}{T_1} = \frac{P_2V_2}{T_2} \quad \text{at constant } n \tag{11.13}$$

Equation 11.13 is the most useful equation to solve problems involving a change in any of the three gas variables, pressure, volume, and temperature, when the number of moles of gas is fixed. *The units used on both sides of the equation must be the same.* This equation may be used if any two gas variables change or if all three gas variables change. If P, V, and T all change, the equation must be used directly as stated in Equation 11.13. If any one of the variables does not change, its value will be the same on both sides of the equation and it may be canceled from the equation. For example, if the temperature does not change, that is, $T_1 = T_2$, then equation 11.13 becomes

$$\frac{P_2V_2}{T_1} = \frac{P_1V_1}{T_1} \quad \text{at constant } n$$

becomes

$$P_2V_2 = P_1V_1 \quad \text{at constant } n$$

and Equation 11.13 has been reduced to Boyle's law as shown above. If the pressure does not change, that is, $P_1 = P_2$, then equation 11.13 becomes

$$\frac{P_1V_2}{T_2} = \frac{P_1V_1}{T_1} \quad \text{at constant } n$$

becomes

$$\frac{V_2}{T_2} = \frac{V_1}{T_1} \quad \text{at constant } n$$

and Equation 11.13 has been reduced to Charles' law as shown above. Hence, you need to remember only one relation, Equation 11.13, to solve most gas problems.

sample exercise 2

In the laboratory, 4.00 L of helium gas are trapped in a cylinder at a pressure of 7.00 atm. The pressure is decreased, at a constant temperature, to 2.00 atm. What is the new volume?

solution

1. Write down all given data in the form of initial and final conditions. If necessary, convert all pressures to the same unit, all volumes to the same unit, and all temperatures to Kelvin.

Initial	**Final**
$P_1 = 7.00$ atm	$P_2 = 2.00$ atm
$V_1 = 4.00$ L	$V_2 = ?$

Temperature is constant ($T_1 = T_2$).

2. You recognize this as a problem involving pressure and volume changes at constant temperature and mass. Write down the combined gas law:

$$\frac{P_2V_2}{T_2} = \frac{P_1V_1}{T_1} \tag{11.13}$$

3. Omit terms that are identical on both sides. Since in this problem $T_1 = T_2$, these may be canceled from the equation to give Boyle's law:

$$P_2V_2 = P_1V_1$$

4. Isolate the unknown variable to one side. You are asked to solve for the new volume, so divide both sides by P_2, isolating V_2 on one side:

$$\frac{\cancel{P_2}V_2}{\cancel{P_2}} = \frac{P_1V_1}{P_2}$$

$$V_2 = \frac{P_1V_1}{P_2}$$

5. Substitute values of pressure and volume from the given data, being careful to have all pressures in the same unit and all volumes in the same unit. Perform arithmetic operations and cancel the common units.

$$V_2 = \frac{(7.00\ \cancel{\text{atm}})(4.00\ \text{L})}{(2.00\ \cancel{\text{atm}})}$$

$$V_2 = 14.0\ \text{L}$$

checking the solution

It is a very good idea to check the reasonableness of your answer by thinking through the problem in terms of the qualitative discussions of P, V, and T changes in Sections 11.4 and 11.6. In this problem, because the volume is varying with a change in pressure at constant temperature, a Boyle's law (Section 11.4) relationship should be recognized. Because the pressure is decreasing (from 7 to 2 atm), the volume must increase; i.e., the final volume must be greater than the initial volume of 4 L. The final volume of 14 L is reasonable according to this qualitative analysis.

sample exercise 3

A sealed weather balloon containing 4.00×10^3 m^3 of helium at 20.°C is cooled to –23°C as it rises into the stratosphere. What is the new volume of the balloon in liters?

solution

1. Write down all given data in the form of initial and final conditions, converting units if necessary.

   Hint: 1 m = 10 dm.

Initial	**Final**
$V_1 = (4.00 \times 10^3 \cancel{m^3}) \times \frac{10^3 \cancel{dm^3}}{1 \cancel{m^3}} \times \frac{1\ L}{1 \cancel{dm^3}}$	$V_2 = ?$
$= 4.00 \times 10^6\ L$	
$T_1 = 20.°C + 273 = 293\ K$	$T_2 = -23°C + 273 = 250.\ K$
Pressure is constant ($P_1 = P_2$).	

2. This problem involves volume and temperature changes. The mass of air inside the balloon is fixed and the pressure (P_{atm}) is constant.

$$\frac{P_2 V_2}{T_2} = \frac{P_1 V_1}{T_1} \qquad (11.13)$$

3. Since $P_1 = P_2$, these may be canceled from the equation to give Charles' law:

$$\frac{V_2}{T_2} = \frac{V_1}{T_1}$$

4. Isolate the unknown variable by multiplying both sides by T_2.

$$\frac{V_2 \cancel{T_2}}{\cancel{T_2}} = \frac{V_1 T_2}{T_1}$$

$$V_2 = \frac{V_1 T_2}{T_1}$$

5. Substitute given values and solve for V_2.

$$V_2 = \frac{(4.00 \times 10^6\ L)(250.\ \cancel{K})}{293\ \cancel{K}}$$

$$V_2 = 3.41 \times 10^6\ L$$

checking the solution

The volume is varying because of a change in temperature at constant pressure. This is a Charles' law (Section 11.5) relationship. Because the temperature is decreasing (from 293 to 250. K), the volume must decrease; i.e., the final volume must be less than the initial volume of 4.00×10^6 L. The final volume of 3.41×10^6 L is reasonable.

sample exercise 4

An automobile tire at 20.°C is filled to a pressure of 28 lb/in.^{2} as measured on a tire pressure gauge. The gauge pressure is the pressure *above* the external, atmospheric pressure of 1.0 atm. The actual pressure inside the tire is the gauge pressure plus the atmospheric pressure or 28 lb/in.^{2} + 14.7 lb/in.^{2} = 43 lb/in.^{2}. The tire is driven hard and the temperature increases to 46°C. Assuming that the volume of the tire does not change, calculate the new pressure inside the tire.

solution

1. Write down all given data in the form of initial and final conditions, converting units if necessary.

Initial	Final
$P_1 = \frac{43\ \text{lb}}{\text{in.}^2}$	$P_2 = ?$
$T_1 = 20.°\text{C} + 273 = 293\ \text{K}$	$T_2 = 46°\text{C} + 273 = 319\ \text{K}$

2. This problem involves pressure and temperature changes. Mass and volume are fixed.

$$\frac{P_2V_2}{T_2} = \frac{P_1V_1}{T_1} \qquad (11.13)$$

3. Since $V_1 = V_2$, volume may be canceled from both sides of the equation to give

$$\frac{P_2}{T_2} = \frac{P_1}{T_1}$$

This direct relationship between pressure and temperature at constant volume and mass is known as **Gay-Lussac's law.** While experiments of Joseph Louis Gay-Lussac, the French chemist and physicist, verified this relationship independently, you can see that it is included in Equation 11.13, our composite equation for gas behavior.

4. Isolate the unknown variable by multiplying both sides by T_2.

$$\frac{P_2\cancel{T_2}}{\cancel{T_2}} = \frac{P_1T_2}{T_1}$$

$$P_2 = \frac{P_1T_2}{T_1}$$

5. Substitute given values and solve for P_2.

$$P_2 = \frac{(43\ \text{lb/in.}^2)(319\ \cancel{\text{K}})}{293\ \cancel{\text{K}}}$$

$$P_2 = 47\ \frac{\text{lb}}{\text{in.}^2} \quad \text{the actual new pressure, inside the tire}$$

Since the actual pressure inside the tire equals the gauge pressure plus the atmospheric pressure, the new pressure as measured on a tire gauge would be the actual pressure minus the atmospheric pressure or 47 lb/in.^{2} – 14.7 lb/in.^{2} = 32 lb/in.^{2}.

sample exercise 5

A sample of helium gas has a volume of 1.25 L at –125°C and 5.00 atm. The gas is compressed at 50.0 atm to a volume of 325 mL. What is the final temperature of the helium gas in °C?

solution

1. Write down the given data, remembering to use Kelvin temperatures and the same volume units.

Initial	Final
$P_1 = 5.00$ atm	$P_2 = 50.0$ atm
$V_1 = 1.25$ L	$V_2 = 325$ mL
$T_1 = -125°C + 273 = 148$ K	$T_2 = ?$

The volumes must be expressed in the same unit. Convert 325 mL to liters:

$$V_2 = 325\,\cancel{\text{mL}}\left(\frac{1\text{ L}}{1{,}000\,\cancel{\text{mL}}}\right) = 0.325\text{ L}$$

2. This problem involves pressure, volume, and temperature changes. Moles are fixed.
3. Since P, V, and T all change, there are no terms to be omitted from Equation 11.13.
4. Remember from Section 3.2 that when two fractions are equal,

$$\text{Numerator}_a \times \text{Denominator}_b = \text{Numerator}_b \times \text{Denominator}_a$$

so Equation 11.13,

$$\frac{P_1V_1}{T_1} = \frac{P_2V_2}{T_2}$$

becomes

$$P_1V_1T_2 = P_2V_2T_1$$

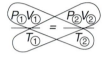

Now isolate the unknown variable, T_2, by dividing both sides by P_1V_1.

$$\frac{\cancel{P_1V_1}T_2}{\cancel{P_1V_1}} = \frac{P_2V_2T_1}{P_1V_1}$$

$$T_2 = \frac{P_2V_2T_1}{P_1V_1}$$

5. Substituting the given values yields

$$T_2 = \frac{(50.0\,\cancel{\text{atm}})(0.325\,\cancel{\text{L}})(148\text{ K})}{(5.00\,\cancel{\text{atm}})(1.25\,\cancel{\text{L}})}$$

$$T_2 = 385\text{ K}$$

The problem asks for the final temperature in °C, so you must convert your answer in K to °C.

$$\text{K} = °\text{C} + 273$$

$$385 = °\text{C} + 273$$

$$385 - 273 = °\text{C}$$

$$112 = °\text{C}$$

---

problem 7

A 100. mL sample of a Freon gas in an automobile air conditioner at 30.0°C and 10.0 atm is expanded rapidly into a volume of 934 mL. Its pressure drops to 1.00 atm. What will be the temperature of the gas after expansion?

Note that a gas is expanded and condensed in a cyclic process for all air conditioning units. Check your answer to see if the final temperature of the gas seems reasonable. Your air conditioner should cool the air surrounding its coils, but not freeze moisture from the air onto the coils.

Figure 11.19

The tubes in (a), (b), and (c) contain different amounts of matter held at constant temperature by the water bath. The pressure of the trapped air sample is constant because it is equal to the constant external pressure (the sum of the atmospheric pressure and the pressure from the weight of the mercury plug). As the number of moles of air in the tube doubles, the volume of the trapped air sample also doubles.

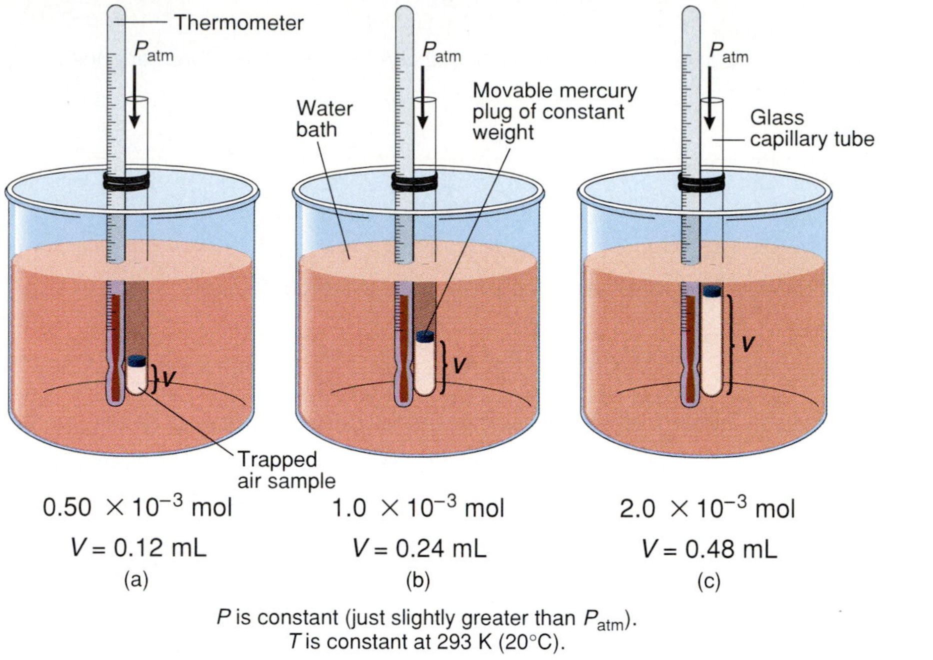

11.7 Avogadro's law

Throughout the foregoing sections, you have worked with a *fixed* amount of gas and observed quantitative variations in three of the four gas variables, P, V, and T. You have learned that the volume of any gas sample is directly proportional to its temperature and inversely proportional to its pressure. Avogadro's experiments found that the volume of any gas sample is also directly proportional to the moles of sample present.

If you take a balloon (at constant T and P) and simply add more gas to it, the volume of the balloon increases. Of course, it is true for any sample of matter that if we add more mass to the sample, it gets bigger (its volume increases). But the remarkable thing about gases is that there is a *direct proportionality* between volume and number of moles that is the same for *all* gases. Such a general proportionality does not hold for all liquids or solids (Chapter 12).

An experimental apparatus that illustrates Avogadro's findings is shown in Figure 11.19. A small capillary tube is strapped to a thermometer in a manner similar to the Charles' law apparatus in Figure 11.16. As you see from the volume changes illustrated, as the number of moles of air within the tube is doubled, the volume of the trapped air sample is doubled.

As you have seen previously in Charles' law, a direct proportion is expressed:

$$V \propto n \qquad \text{at constant } T \text{ and } P \tag{11.14}$$

Writing this as an equation using A as a constant of proportionality yields

$$V = An \tag{11.15}$$

Dividing each side of the equation by n, you get

$$\frac{V}{n} = A \qquad \text{at constant } T \text{ and } P \tag{11.16}$$

Holding the pressure and temperature fixed and varying V from V_1 to V_2 and n from n_1 to n_2 gives

$$\frac{V_1}{n_1} = A \quad \text{at constant } T \text{ and } P$$

and

$$\frac{V_2}{n_2} = A \quad \text{at constant } T \text{ and } P$$

or

$$\frac{V_1}{n_1} = \frac{V_2}{n_2} \quad \text{at constant } T \text{ and } P \qquad (11.17)$$

This equation can be read to give an important result for two equal volumes of any gas. At the same temperature and pressure, if the volume of one gas, V_1, equals the volume of a second gas, V_2, then according to Equation 11.17, the number of moles of the first gas, n_1, must equal the number of moles of the second gas, n_2.

If $V_1 = V_2$, volume may be canceled from each side of Equation 11.17 to give

$$\frac{1}{n_1} = \frac{1}{n_2}$$

After cross-multiplying, you find the important result that

$$n_2 = n_1$$

Equal volumes of all gases at the same temperature and pressure contain the same number of moles.

A very significant statement that extends this result to molecules is **Avogadro's law,** which states that *equal volumes of all gases at the same temperature and pressure contain the same number of molecules.* Because the number of molecules is directly proportional to the number of moles (Section 8.4), the two statements on the equal volumes of gases are identical.

Historically, it is useful to know that Avogadro gained his insight into the relationship between the number of moles and the number of molecules from experiments of Gay-Lussac that measured the volumes of gases that combined in chemical reactions. However, for now it is important to recognize the usefulness of Avogadro's law. It gives the direct proportionality between the volume of any gas sample and the number of moles or molecules it contains. As a result, the fourth gas variable, number of moles, *n,* will be combined in Section 11.9 with the other three gas variables, *T, P,* and *V,* into one overall gas equation, the ideal gas law, which can be used to solve all gas problems. Before constructing the ideal gas law in Section 11.9, however, you must learn about the volume occupied by 1 mol of any gas and the special conditions under which it is measured (Section 11.8).

---

sample exercise 6

At 738 mmHg and 24°C, 0.393 mol of nitrogen gas is contained in 9.87 L. What volume would 0.393 mol of a different gas, argon, occupy at the same pressure and temperature?

solution

The two gases are at the same temperature and pressure. Therefore, 0.393 mol of argon must be contained in the same volume as 0.393 mol of nitrogen. Since the volume of nitrogen is 9.87 L, the volume of argon is 9.87 L.

••• problem 8

At 0°C and 1 atm, 0.76 mol of helium occupies 17 L. How many moles of hydrogen gas would occupy this volume at the same temperature and pressure?

STP = standard - temp. - pressure

11.8 molar gas volume and standard temperature and pressure

You learned in Section 11.1 that measuring the volume of a gas produced or consumed in a chemical reaction is often more convenient than measuring its mass. However, you now understand that a measurement of the volume of a gas is meaningless unless you also know its temperature and pressure. You have seen that the volume of any gas changes as the temperature and pressure change. Consequently, volumes of gases can only be compared under the same condition, that is, at the same temperature and pressure.

Scientists have agreed on a convenient set of reference conditions for comparing all gases. The temperature 273 K (0°C) and the pressure 1 atm (760 mmHg) have been chosen as **standard temperature** and **standard pressure.** The acronym STP is often used to indicate standard temperature and pressure in comparing gases.

STP = 273 K (0°C) and 1 atm (760 mmHg)

Since exactly 760 mmHg is defined as 1 atm, it is not a measured number and does not limit the number of significant figures in a calculation.

Conversion of any gas volume to its volume at standard conditions is straightforward with the combined gas law of Equation 11.13. Sample Exercise 7 illustrates such a conversion.

sample exercise 7

A given mass of oxygen gas produced in a reaction displaced 435 mL of water at 25°C and 740. mmHg. What will be its new volume at STP? Assume that the vapor pressure of water has already been subtracted from the gas pressure given.

solution

1. The initial and final conditions using correct units are

Initial	**Final**
P_1 = 740. mmHg	The phrase "at STP" tells you the final T and P.
V_1 = 435 mL	P_2 = 760. mmHg
T_1 = 25°C + 273 = 298 K	V_2 = ?
	T_2 = 273 K

2. This problem involves pressure, volume, and temperature changes. Moles are fixed.

$$\frac{P_2V_2}{T_2} = \frac{P_1V_1}{T_1} \qquad (11.13)$$

3. Since P, V, and T all change, there are no terms to be canceled from the equation.
4. To isolate the unknown V_2, multiply both sides by T_2/P_2.

$$\frac{\cancel{T_2}}{\cancel{P_2}} \times \frac{\cancel{P_2}V_2}{\cancel{T_2}} = \frac{P_1V_1}{T_1} \times \frac{T_2}{P_2}$$

$$V_2 = \frac{P_1V_1T_2}{T_1P_2}$$

5. Substituting given values yields

$$V_2 = \frac{(740.\ \cancel{\text{mmHg}})(435\ \text{mL})(273\ \cancel{\text{K}})}{(298\ \cancel{\text{K}})(760.\ \cancel{\text{mmHg}})}$$

$$V_2 = 388\ \text{mL}$$

Standard conditions are particularly important in establishing the volume for 1 mol of any gas. The volume of 1 mol of gas is called the **molar gas volume.** *At the special conditions of STP, the molar gas volume has been measured experimentally; 1 mol of any gas occupies a volume of 22.4 L* (Figure 11.20).

All problems involving gases at STP can often be solved most readily by use of the following equality.

1 mol of a gas at STP = 22.4 L

Figure 11.20
The molar volume of any gas at STP is approximately equal to the volume of a cylinder large enough to hold a basketball.

In the next section, you will combine Boyle's, Charles', and Avogadro's laws into the ideal gas law. The relationship given above for the molar volume of any gas will enable you to evaluate the constant of proportionality in the equation for the ideal gas law.

The use of the molar volume as a conversion factor in solving gas problems is illustrated in the sample exercises that follow.

---

sample exercise 8

What is the volume of 11.9 g of nitrogen gas at STP? You recall from Table 11.2 that nitrogen gas is the major constituent of air.

solution

This problem can be solved by the unit conversion method.

1. The given quantity and unit is 11.9 g of N_2 at STP.
2. The quantity to be determined is number of liters at STP.
3. We do not have a conversion factor to convert grams to liters directly, but at STP, we known that 1 mol of N_2 = 22.4 L of N_2 and that the molar mass, obtained from the periodic table, relates moles to grams. In this case, 1 mol of N_2 = 28.0 g of N_2.
4. Set up the proper format.

Given × **Conversion factors** = **New**

$$11.9\ \cancel{g\ N_2} \times \frac{1\ \cancel{mol\ N_2}}{28.0\ \cancel{g\ N_2}} \times \frac{22.4\ L\ N_2}{1\ \cancel{mol\ N_2}} = 9.52\ L\ N_2$$

---

sample exercise 9

What is the density of hydrogen gas at STP?

solution

The *given* information in this case comes from the statement that the gas is at STP. At STP for any gas, 1 mol = 22.4 L. For H_2 gas in particular, 1 mol of H_2 = 2.02 g of H_2.

Density is mass divided by volume. For gases, the units used are grams per liter. At STP, you know both the mass and the volume of a 1 mol sample of H_2 gas, and you can substitute these in the equation for density:

$$d = \frac{2.02\ g}{22.4\ L} = 0.0902\ \frac{g}{L}$$

Note that the density of H_2 gas is lower than other gases because of its small molar mass. Because of its low density, it was used in airships (dirigibles), but not without the danger of combustion in air. News films of the Hindenburg disaster vividly portray this danger. Helium, which is chemically inert but also low in density, has replaced hydrogen for such purposes. Jacques Charles first introduced the use of hydrogen gas in balloons in France in the eighteenth century.

••• problem 9

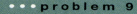

Compare the density of hydrogen gas at STP (from Sample Exercise 9) with the densities of helium and air at STP. (Use 28.8 g for the molar mass of air.)

••• problem 10

Calculate the volume of 7.50 g of hydrogen gas at STP.

11.9 ideal gas law

The volume of any gas has been characterized by Boyle's, Charles', and Avogadro's laws to be inversely proportional to the pressure, *P*, and directly proportional to the temperature, *T*, and the number of moles, *n*.

Boyle's law: $$V = \frac{k}{P} \tag{11.3}$$

Charles' law: $$V = mT \tag{11.7}$$

Avogadro's law: $$V = An \tag{11.15}$$

All of these laws may be combined into a single proportionality that is valid under all conditions, so that no variables are held constant.

$$V \propto \frac{nT}{P} \tag{11.18}$$

Replacing the proportionality with an equal sign and a constant will give the overall gas law.

$$V = \frac{RnT}{P} \tag{11.19}$$

For this equation, however, the proportionality constant is specifically designated as *R* and is called the **universal gas constant**. The value of *R* is evaluated below. When written with the universal gas constant *R*, Equation 11.19 is called the **ideal gas law**, an equation that can be used to solve all gas problems with changes in any or all of the four gas variables. The most common form of the ideal gas law is obtained by multiplying both sides of Equation 11.19 by *P* to give

$$PV = nRT \tag{11.20}$$

A useful way of rewriting the ideal gas law in Equation 11.20 is to divide both sides by *nT* to obtain the quotient *PV/nT*, which isolates and tells you how to evaluate *R*.

$$\frac{PV}{nT} = \frac{\cancel{n}R\cancel{T}}{\cancel{nT}}$$

$$\frac{PV}{nT} = R \tag{11.21}$$

Substituting STP conditions into Equation 11.21 gives a value and unit for *R* that is valid under *all* conditions. Remember, *R* is a constant; it has the same value for any combination of *n, V, T,* and *P* in the same units.

$$R = \frac{(1\ \text{atm})(22.4\ \text{L})}{(1.00\ \text{mol})(273\ \text{K})}$$

$$R = 0.0821\ \frac{\text{L} \cdot \text{atm}}{\text{mol} \cdot \text{K}} \tag{11.22}$$

This value of *R*, with units as expressed in Equation 11.22, will be used for most gas problems in which pressure is given in atmospheres.

Should it be desirable to work with pressure expressed in millimeters of mercury, then R would have a different value and unit as follows.

$$R = \frac{(760 \text{ mmHg})(22.4 \text{ L})}{(1.00 \text{ mol})(273 \text{ K})}$$

$$R = 62.4 \frac{\text{L} \cdot \text{mmHg}}{\text{mol} \cdot \text{K}} \tag{11.23}$$

When P is in millimeters of mercury, use this value of R.

Another value of R that is gaining widespread usage is its value in SI units (Section 2.4). In SI units, $R = 8.314$ J/(mol • K), where joules (J) are units of energy. Unfortunately, to use R in SI units, pressure must be measured in units of pascals and volume must be measured in cubic meters. Because of these new units for pressure and volume, R will not be used in SI units throughout this book. It is mentioned so that you will not be bewildered should you use a more advanced chemistry text in the future. No matter what set of units you use, the ideal gas law will apply as long as the measurements of pressure and volume have units that correspond to your value of R. Temperature is always expressed in K.

Hints on the use of the ideal gas law equation

1. Pay special attention to the units. As always, for gas law problems, T must be in Kelvin. V must be in liters. If P is in atmospheres, use the R value of Equation 11.22.

$$R = 0.0821 \frac{\text{L} \cdot \text{atm}}{\text{mol} \cdot \text{K}} \tag{11.22}$$

If P is in millimeters of mercury, use the R value of Equation 11.23.

$$R = 62.4 \frac{\text{L} \cdot \text{mmHg}}{\text{mol} \cdot \text{K}} \tag{11.23}$$

2. Because determining the number of moles of a sample, given the number of grams, always involves dividing grams of the sample by molar mass (see, for example, Sample Exercise 8), you can substitute the ratio, grams of sample (g)/molar mass, for n in the ideal gas law equation:

$$PV = nRT \tag{11.20}$$

$$PV = \frac{g}{\text{Molar mass}} RT \tag{11.24}$$

Equation 11.24 allows us to determine molar masses experimentally in a manner exemplified later in Sample Exercise 12.

3. The ideal gas equation may be used when there are *changing* conditions of P, V, T, or n by building on the form given in Equation 11.21. (Remember that R is a constant for any set of P, V, T, and n.)

$$\frac{P_1V_1}{n_1T_1} = R \quad \text{and} \quad \frac{P_2V_2}{n_2T_2} = R$$

so

$$\frac{P_2V_2}{n_2T_2} = \frac{P_1V_1}{n_1T_1}$$

If $n_1 = n_2$, the number of moles may be canceled from both sides of this equation. Then this form of the ideal gas law immediately reduces to Equation 11.13.

$$\frac{P_2V_2}{T_2} = \frac{P_1V_1}{T_1} \tag{11.13}$$

Let us examine three applications of the ideal gas law in Sample Exercises 10 to 12.

sample exercise 10

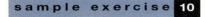

What volume will be occupied by 3.25 mol of oxygen gas at 735 mmHg and 25°C?

solution

You recognize that this problem can be solved by the ideal gas law because you are given three variables (*n, P,* and *T*) and are asked to find the fourth gas variable (*V*). Therefore, you will use

$$PV = nRT \qquad (11.20)$$

1. Isolate the variable you are being asked to find. In this problem, the unknown variable is *V,* so you write

$$V = \frac{nRT}{P}$$

2. Choose the value of *R* that you will use based on the units of *P* given. In this case, because *P* is given in millimeters of mercury, choose the *R* value of Equation 11.23.
3. Perform any necessary conversions of given units to conform to the units of *R.* In this case,

$$25°C = 298\ K \qquad (25°C + 273)$$

4. Substitute the given values with their proper units in the equation shown in step 1. Cancel common units and perform the indicated arithmetic.

$$V = \frac{(3.25\ \cancel{mol})\left(62.4\ \frac{L \cdot \cancel{mmHg}}{\cancel{mol} \cdot \cancel{K}}\right)(298\ \cancel{K})}{735\ \cancel{mmHg}}$$

$$V = 82.2\ L$$

5. Check to be sure that the unit you are left with after canceling units is the proper unit for the unknown variable. In this case, liters is the proper unit for volume.

sample exercise 11

How many moles of argon gas are present in a typical cylinder whose volume is 70.0 L and registers a pressure of 180. atm at 21°C (Figure 11.21)?

solution

Three variables are given (*V, T,* and *P*) and the fourth (*n*) must be determined. Use Equation 11.20.

$$PV = nRT \qquad (11.20)$$

1. Isolate *n:*

$$\frac{PV}{RT} = n$$

2. Choose *R* based on the *P* units given. In this case, *P* is given in atmospheres, so the *R* value of Equation 11.22 is chosen.

3. Convert units as necessary.

$$21°\text{C} = 294\ \text{K} \qquad (21°\text{C} + 273)$$

4. Substitute, cancel, and do the arithmetic.

$$n = \frac{(180.\ \cancel{\text{atm}})(70.0\ \cancel{\text{L}})}{\left(0.0821\ \dfrac{\cancel{\text{L}} \cdot \cancel{\text{atm}}}{\text{mol} \cdot \cancel{\text{K}}}\right)(294\ \cancel{\text{K}})}$$

$$n = 522\ \frac{1}{\dfrac{1}{\text{mol}}} = 522\ \text{mol}$$

Note: The unit of moles belongs in the numerator because

$$\frac{1}{\dfrac{1}{\text{mol}}} = 1 \div \frac{1}{\text{mol}} = 1 \times \frac{\text{mol}}{1} = \text{mol}$$

5. Moles is the proper unit for n.

Figure 11.21
For commercial, industrial, and laboratory use, gases are stored in heavy, metal gas cylinders, usually at a pressure greater than 2,000 psi. When the cylinder cap is removed and a regulator valve is attached, the cylinder must be secured with a strap or chain to prevent the cylinder from falling. If the gas regulator valve were broken because of the tank falling, the high-pressure gas would escape from the small opening in the top of the tank, and the escaping gas would propel the tank dangerously like a rocket. In accidents like this, tanks have gone through thick walls and done extensive damage.

sample exercise 12

In the laboratory, 10.0 g of an unknown gas is found to occupy a volume of 5.60 L at 20.0°C and 740. mmHg. What is the molar mass of this unknown gas?

solution

Whenever we are asked to determine the molar mass of a gas, we use Equation 11.24 (see hint 2).

$$PV = \frac{g}{\text{Molar mass}}\ RT \qquad (11.24)$$

1. Isolate molar mass by multiplying each side by molar mass:

$$PV \times (\text{Molar mass}) = gRT$$

and dividing each side by PV:

$$\text{Molar mass} = \frac{gRT}{PV}$$

2. Choose the R corresponding to millimeters of mercury.
3. Convert temperature to Kelvin.

$$20.0°\text{C} = 293\ \text{K} \qquad (20.0°\text{C} + 273)$$

4. Substitute, cancel, and do the arithmetic.

$$\text{Molar mass} = \frac{(10.0\ \text{g})\left(62.4\ \dfrac{\cancel{\text{L}} \cdot \cancel{\text{mmHg}}}{\text{mol} \cdot \cancel{\text{K}}}\right)(293\ \cancel{\text{K}})}{(740.\ \cancel{\text{mmHg}})(5.60\ \cancel{\text{L}})} = 44.1\ \frac{\text{g}}{\text{mol}}$$

5. Grams per mole is a proper unit for molar mass.

••• problem 11

How many moles of gas are there in a 400. mL aerosol can at a temperature of 20.°C and a pressure of 3.00 atm?

••• problem 12

At STP, 700. mL of a gas weighs 1.452 g. What is the molar mass of the gas?

11.10 kinetic molecular theory of gases

The kinetic molecular theory of gases uses the behavior of individual molecules to account for the physical properties and gas laws that you have learned in this chapter, which result from measurements on a macroscopic (or bulk) scale. The theory describes the hypothesized behavior of individual gas atoms or molecules. This description of the behavior of individual molecules or atoms creates a picture, or model, that helps you to explain the properties of a gas, such as its compressibility or the relationship between its volume and pressure. If gas atoms or molecules behave according to the theory's assumptions listed below, then the properties of gases and the gas laws can be explained in terms of their molecular motion.

1. Gases are made up of small particles (either atoms or molecules) that are constantly moving in random, straight-line motion.
2. The distance between particles is large compared with the size of the particles. A gas is mostly empty space.
3. There are no attractive forces between particles. The particles move independently of each other.
4. The particles collide with each other and with the walls of the container without incurring a loss of energy.
5. The average kinetic energy (the energy of motion described in Section 3.7) of the particles is directly proportional to the absolute (Kelvin) temperature of the gas. When the average kinetic energy increases, the particles move faster.

Table 11.5 shows how specific observations of gas properties may be explained in terms of the motion of individual molecules as described in kinetic molecular theory.

The gas laws discussed throughout this chapter may also be explained in terms of kinetic molecular theory. An explanation for each gas law is given in terms of molecular motion in Table 11.6.

Throughout this chapter, you have learned of the experiments and reasoning that led to the development of the ideal gas law. The law is so named because it deals with **ideal gases,** that is, with gases that have no attractive forces between their particles. All gases behave ideally at pressure less than a few atmospheres and at high temperatures. At low pressure, the molecules of a gas are so far apart that forces between the molecules cannot be felt even if they exist. At high temperatures, the molecules are moving too fast to feel any forces between the molecules.

For discussions and problems in this book, all gases will be considered to be ideal. At high pressure or at low temperature, however, attractive forces that do exist between molecules slightly change the properties of the gases. For most calculations, the values of any of the four gas variables would not be expected to change by more than a few percent from those obtained using the ideal gas law.

11.11 gas stoichiometry

One of the primary purposes in studying the ideal gas law is to apply it to the stoichiometry of chemical equations when gases are included among the reactants and products. In Chapter 10, you learned that the stoichiometric coefficients of a balanced chemical equation give fundamental relationships and conversion factors between the numbers of moles of any reactants or products as summarized in Figure 11.22.

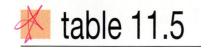

table 11.5 explanations of physical properties based on the molecular theory of gases

Physical property	Explanation in terms of kinetic molecular theory
Gases can be compressed easily.	Because gas particles are far apart, they easily can be squeezed closer together by an outside force.
Gases expand to fill the volume of their container.	Gas particles are constantly moving with no attractive forces between particles, so they will expand until they meet an outside force, namely, the wall of the container.
Gases have a low density.	Because a gas is mostly empty space, there are few particles (low mass) per unit volume.
Gases can diffuse through each other.	Gas particles are constantly moving and are separated by large distances; this leads to freedom for particles of one gas to move through the empty space of another gas.
Gases exert a pressure on container walls.	Moving gas particles collide with container walls, thus exerting a force on every square inch.

table 11.6 explanations of gas laws based on the kinetic molecular theory

Gas law	Explanation in terms of kinetic molecular theory
Boyle's law: $V \propto \frac{1}{P}$ at constant T and n	When the volume of a gas is *decreased,* the particles collide with the container walls more often, leading to a *greater pressure;* when the volume of a gas is *increased,* the particles collide with the walls less often, leading to a *decreased pressure.*
Charles' law: $V \propto T$ at constant P and n	Increased temperature causes particles to move faster, leading to more and "harder" collisions with container walls. Pressure inside the walls is increased until the volume expands to the point where the pressure inside the walls is again equal to the pressure outside.
Avogadro's law: $V \propto n$ at constant P and T	When more molecules are added to a container at constant temperature and pressure, the volume must increase. If the volume did not increase, there would be more molecules colliding with the walls more often and the pressure would increase. Since pressure is constant and may not increase, the volume must increase so that the frequency of molecular collisions with the walls will remain the same.
Dalton's law of partial pressures: $P_{total} = p_1 + p_2 + p_3 + \ldots$	Since the particles move independently of one another, each gas in a mixture will exert a pressure independent of the pressure of the other gases; the total pressure will be the sum of the individual pressures.

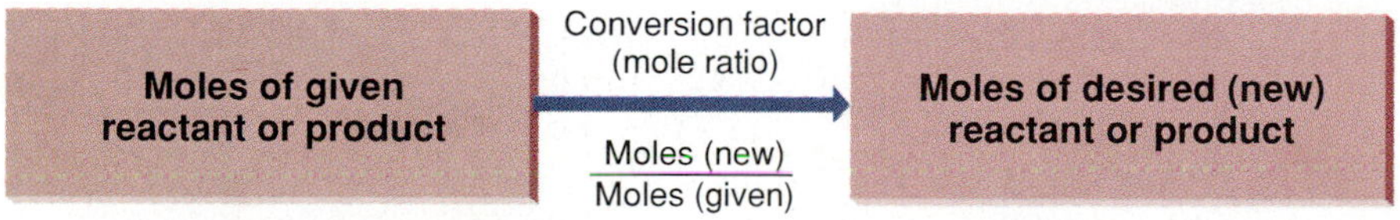

Figure 11.22
Summary of a simple stoichiometric calculation to convert moles of any reactant or product to moles of any other reactant or product. The conversion factor (mole ratio) for the calculation is obtained directly from the coefficients in the balanced chemical equation.

You applied this diagram directly to mole-to-mole calculations and extended it to gram-to-mole and mole-to-gram calculations by including the molar mass as an additional conversion factor for either a reactant or product.

Further extending your knowledge to stoichiometric problems with gases follows the same procedure except that, instead of converting moles to grams or grams to moles, you must convert moles of gas to volume (liters of gas) or volume (liters of gas) to moles. In this section, calculations will be classified into two groups that will modify the procedure you used in Section 10.4. The two groups of sample exercises in this section include (1) *liter-to-liter* calculations, which are similar to mole-to-mole calculations, and (2) *gram-to-liter* or *liter-to-gram* calculations, which are similar to gram-to-mole and mole-to-gram calculations.

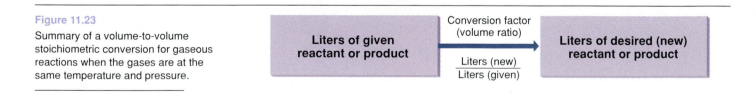

The new conversion factors you need have been discussed in Sections 11.7 to 11.9. To relate mole ratios to volume ratios, you will use Avogadro's law. To convert liters of gas to moles or moles to liters of gas, you will use either the molar volume if the gases are at STP or the ideal gas law if the gases are at other temperatures and pressures.

liter-to-liter calculations

You learned from Avogadro's law (Section 11.7) that equal volumes of gas at the same temperature and pressure contain equal numbers of moles, or, conversely, equal numbers of moles of gas occupy the same volume. This statement may be applied to the moles of gases represented by the stoichiometric coefficients of any balanced chemical equation. From Avogadro's law, you see that the *mole ratios represented by the stoichiometric coefficients of gaseous reactants and products also represent volume ratios* as long as the gases are at the same temperature and pressure. For example, you may read the equation $3\ H_2(g) + N_2(g) \rightarrow 2\ NH_3(g)$ as either "3 mol of hydrogen combine with 1 mol of nitrogen to form 2 mol of ammonia," or "3 volumes of hydrogen combine with 1 volume of nitrogen to form 2 volumes of ammonia," because these reactants and products are gases.

A summary of liter-to-liter calculations is given in Figure 11.23. The flow chart in this figure has been applied in solving Sample Exercise 13.

sample exercise 13

In the combustion of hydrogen gas, how many liters of hydrogen would be required to react completely with 6.00 L of oxygen to form liquid water at a constant temperature and pressure?

The explosive reaction between hydrogen and oxygen gases requires extreme caution in handling a mixture of these gases. In 1937, the German airship, the Hindenburg, which was filled with several million cubic feet of hydrogen gas, exploded and burned at its mooring mast at Lakehurst, New Jersey, killing 35 of the 97 people aboard.

solution

Write the balanced equation for this reaction.

$$2\ H_2(g) + O_2(g) \rightarrow 2\ H_2O(l)$$

This exercise can be completed in a manner similar to that of the stoichiometric mole-to-mole conversions we did in Section 10.4.

1. The given quantity is 6.00 L of O_2.
2. The new quantity is liters of H_2.

Figure 11.23

Summary of a volume-to-volume stoichiometric conversion for gaseous reactions when the gases are at the same temperature and pressure.

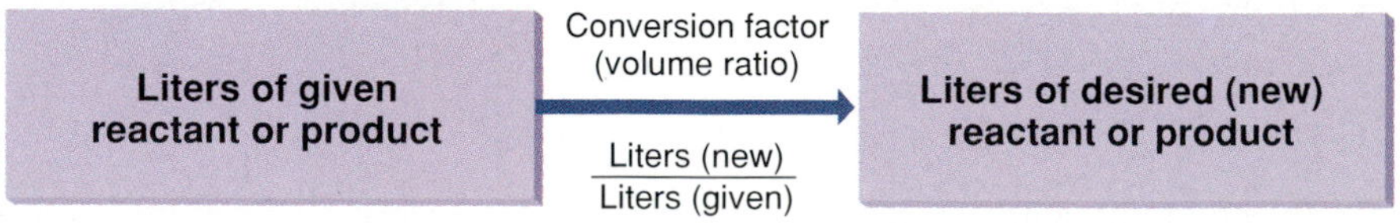

3. Just as the balanced equation gives us the equality

$$2 \text{ mol } H_2 = 1 \text{ mol } O_2$$

and the corresponding conversion factor, so too does the equation relate volumes for gases (Avogadro's law). In this case,

$$2 \text{ L } H_2 = 1 \text{ L } O_2$$

4. The proper format is

Given ¥ Conversion factor = New

$$6.00 \cancel{\text{L } O_2} \times \frac{2 \text{ L } H_2}{1 \cancel{\text{L } O_2}} = 12.0 \text{ L } H_2$$

This is the *volume ratio,* which, because of Avogadro's law, is used similarly to the *mole ratio.*

Note that since H_2O is not present in the gas phase in Sample Exercise 13, Avogadro's law does *not* apply to it and the molar coefficients of either of the gaseous reactants and the liquid water do not represent volume ratios.

gram-to-liter calculations

In many stoichiometric problems involving gases, you are given a mass of some reactant or product and are asked to calculate a volume of a gaseous reactant or product. You may apply the techniques developed in Chapter 10 to calculate the number of moles of the gas. Then, if the reaction is done at STP, simply multiply the number of moles by the conversion factor:

$$\frac{22.4 \text{ L}}{\text{mol}} \qquad \text{(from 1 mol = 22.4 L at STP)} \qquad (11.25)$$

Use of this conversion factor is illustrated in Sample Exercise 15. If the reaction is *not* at STP, use the ideal gas law to calculate the volume corresponding to the number of moles at the *T* and *P* conditions given (Sample Exercise 14). Compare Sample Exercises 14 and 15 to see how each conversion factor is used. A flow chart with both conversions is given in Figure 11.24.

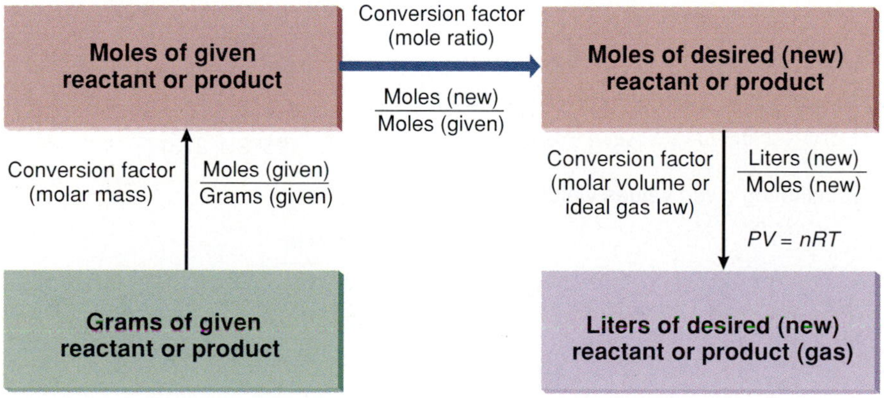

Figure 11.24
Summary of a gram-to-liter stoichiometric conversion is similar to the gram-to-mole and mole-to-gram conversions in Section 10.5.

sample exercise 14

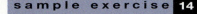

How many liters of oxygen can be produced from the decomposition of 18.5 g of potassium chlorate at 25°C and 750. mmHg according to the following balanced equation?

$$2\ KClO_3(s) \rightarrow 2\ KCl(s) + 3\ O_2(g)$$

In Laboratory Exercise 7, you measured the number of grams of oxygen gas produced in this reaction as a difference in mass between the solid KCl product and the $KClO_3$ reactant. Now you are able to predict the volume of oxygen gas that will be produced and to compare it directly with your measurement of the volume of water the oxygen gas displaces from a collection bottle (Figure 11.10).

solution

Because the given substance, $KClO_3$, is present as a solid, you can employ the usual techniques of *stoichiometry* involving the mole ratio from the balanced equation and then apply your knowledge of the *ideal gas law*.

Stoichiometry (review section 10.7 if necessary)

1. The mole ratio relating the given and new quantities is

$$\frac{\text{New}}{\text{Given}} = \frac{3\ \text{mol}\ O_2}{2\ \text{mol}\ KClO_3}$$

2. Because grams of $KClO_3$ is given, we will need the conversion factor we get from the molar mass of $KClO_3$, 122.6 g:

$$\frac{\text{New unit}}{\text{Given unit}} = \frac{1\ \text{mol}\ KClO_3}{122.6\ \text{g}\ KClO_3}$$

3. The proper format is

$$18.5\ \cancel{\text{g}\ KClO_3} \times \frac{1\ \cancel{\text{mol}\ KClO_3}}{122.6\ \cancel{\text{g}\ KClO_3}} \times \frac{3\ \text{mol}\ O_2}{2\ \cancel{\text{mol}\ KClO_3}} = 0.226\ \text{mol}\ O_2$$

The conditions given are not STP, so we must use the equation $PV = nRT$ to solve for V.

Ideal gas law (review section 11.9 if necessary)

1. Isolate V:

$$V = \frac{nRT}{P}$$

2. R is chosen as 62.4 (L • mmHg)/(mol • K) because P is given in millimeters of mercury.
3. Converting temperature to Kelvin yields

$$25°\text{C} = 298\ \text{K} \qquad (25°\text{C} + 273)$$

4. Substituting given values yields

$$V = \frac{(0.226\ \cancel{\text{mol}}\ O_2)\left(62.4\ \frac{\text{L} \cdot \cancel{\text{mmHg}}}{\cancel{\text{mol}} \cdot \cancel{\text{K}}}\right)(298\ \cancel{\text{K}})}{750.\ \cancel{\text{mmHg}}}$$

$$V = 5.60\ \text{L}\ O_2$$

5. Liters is the proper unit for V.

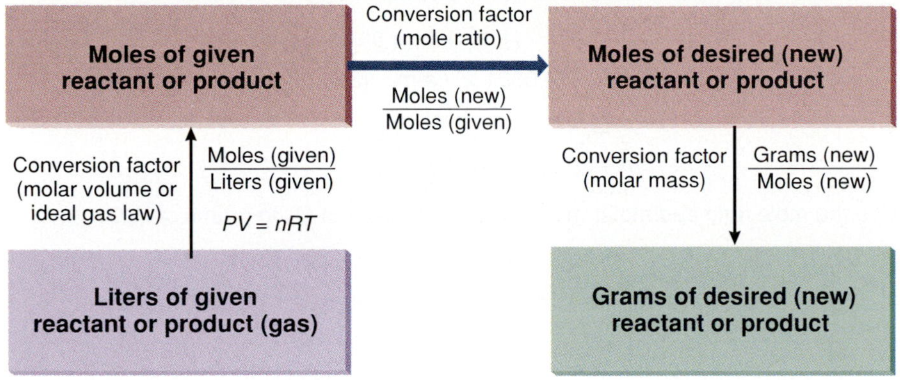

Figure 11.25
Summary of a liter-to-gram stoichiometric conversion is similar to the mole-to-gram and gram-to-mole conversions in Section 10.5.

If you are given a volume of gas and asked to calculate the mass of another substance in a balanced equation, first convert your volume of gas to moles of gas. The problem will then become a typical liter-to-gram conversion as summarized in Figure 11.25. Sample Exercise 15 illustrates this conversion.

sample exercise 15

How many grams of calcium oxide can be produced from the reaction of 6.00 L of oxygen with excess calcium at 0°C and 1.00 atm, according to the following balanced equation?

$$2\ Ca(s) + O_2(g) \rightarrow 2\ CaO(s)$$

solution

If the given 6.00 L of O_2 is converted to moles of O_2, then the problem can be completed according to the techniques learned in Chapter 10. Because the conditions given are STP (0°C, 1 atm), you can use the molar gas volume (Section 11.8) for the liter-to-mole conversion:

$$6.00\ \cancel{L\ O_2} \times \frac{1\ mol\ O_2}{22.4\ \cancel{L\ O_2}} = 0.268\ mol\ O_2$$

$$0.268\ \cancel{mol\ O_2} \times \frac{2\ \cancel{mol\ CaO}}{1\ \cancel{mol\ O_2}} \times \underbrace{\frac{56.1\ g\ CaO}{1\ \cancel{mol\ CaO}}}_{\text{Molar mass}} = 30.1\ g\ CaO$$

If the volume had been given at non-STP conditions, you would use the ideal gas equation ($PV = nRT$) to convert the given volume to moles.

sample exercise 16

How many grams of water can be produced from the reaction of 4.00 L of oxygen with excess hydrogen at 25°C and 1.10 atm, according to the following balanced equation?

$$2\ H_2(g) + O_2(g) \rightarrow 2\ H_2O(l)$$

solution

Use the ideal gas equation in the form $n = PV/RT$ to convert 4.00 L of O_2 to moles of O_2 and then proceed as in any typical stoichiometry problem.

1. Establish the values to be used in the ideal gas equation.

$P = 1.10$ atm

$V = 4.00$ L

$R = 0.0821$ L • atm/(mol • K) because P is in atmospheres

$T = 25°C = 298$ K (25 + 273)

2. Substitute these given values in the ideal gas equation and solve for n.

$$n = \frac{(1.10\ \cancel{atm})(4.00\ \cancel{L})}{\left(0.0821\ \frac{\cancel{L} \cdot \cancel{atm}}{mol \cdot \cancel{K}}\right)(298\ \cancel{K})}$$

$$n = 0.180\ mol\ O_2$$

3. Use the mole ratio and molar mass to convert moles of O_2 to grams of H_2O.

$$0.180\ \cancel{mol\ O_2} \times \frac{2\ \cancel{mol\ H_2O}}{1\ \cancel{mol\ O_2}} \times \frac{18.0\ g\ H_2O}{1\ \cancel{mol\ H_2O}} = 6.48\ g\ H_2O$$

••• problem 13

How many liters of chlorine gas are needed to react with 15.5 g of magnesium to form magnesium chloride at 25°C and 755 mmHg? Magnesium chloride, a white, ionic solid, is used in disinfectants and fire extinguishers.

••• problem 14

How many grams of sulfur tetrafluoride can be formed from the reaction of 4.93 L of fluorine with excess sulfur at 30.°C and 1.00 atm? Sulfur tetrafluoride is a colorless, corrosive gas that attacks glass and reacts violently with water. It is used in preparing other compounds that contain fluorine atoms.

chapter accomplishments

After completing this chapter, you should be able to define all key terms and do the following.

11.1 Introduction

- ❑ State the characteristic properties of gases.
- ❑ State the four measurable gas variables.

11.2 Measurement of the four gas variables: *n, T, V*, and *P*

- ❑ Describe how the number of moles, volume, and temperature of a gas are measured.
- ❑ Describe how pressure changes with increasing (or decreasing) force and increasing (or decreasing) area.
- ❑ Describe how atmospheric pressure is measured.
- ❑ State the molecular basis of gas pressure.
- ❑ State the "equalities" among 1 atm and millimeters of mercury, torr, and pounds per square inch.

11.3 Dalton's law of partial pressures for mixtures of gases

- ❑ State Dalton's law of partial pressures.
- ❑ Given the pressure of each gas in a mixture of gases, calculate the total pressure.
- ❑ Calculate the partial pressure of a dry gas that has been collected by displacement of water, given the atmospheric pressure and temperature.
- ❑ Explain the relationship between partial pressure and gas exchange between two regions of different partial pressures.

11.4 Boyle's law

- ❑ State Boyle's law in words and mathematically.
- ❑ Given an increase or decrease of pressure by a given factor, state the effect on the volume of a fixed amount of gas at constant temperature.

11.5 Charles' law

- ❑ State Charles' law in words and mathematically.
- ❑ Convert a Celsius temperature to Kelvin.
- ❑ Given an increase or decrease of Kelvin temperature by a given factor, state the effect on the volume of a fixed amount of gas at constant pressure.

11.6 Combined gas laws

- ❑ Considering the three variables, volume, temperature, and pressure, and holding one variable constant (for example, P), calculate the effect of a change in the second variable (for example, T) on a given amount of the third variable (for example, V).
- ❑ Considering the three variables, volume, temperature, and pressure, calculate the effect of a change in two variables (for example, P and T) on a given amount of the third variable (for example, V).

11.7 Avogadro's law

- ❑ Given the volumes of two gases at the same temperature and pressure, compare the relative numbers of moles present.

11.8 Molar gas volume and standard temperature and pressure

- ❑ Given the volume of a fixed amount of gas at a given temperature and pressure, calculate the volume at STP.
- ❑ State the volume of 1 mol of any gas at STP.
- ❑ Calculate the volume of a given mass of gas at STP.
- ❑ Calculate the density of a given gas at STP.
- ❑ Calculate the number of moles that corresponds to a given volume of gas at STP.

11.9 Ideal gas law

- ❑ State the ideal gas equation that relates the four gas variables.
- ❑ Choose the proper value of *R,* the ideal gas constant, for a given value of pressure.
- ❑ Given three of the gas variables (for example, *P, T,* and *n*), calculate the fourth (for example, *V*).
- ❑ Solve for the molar mass of a gas, given the mass, volume, temperature, and pressure of the gas.
- ❑ Solve for the mass of a known gas, given the volume, temperature, and pressure of the gas.

11.10 Kinetic molecular theory of gases

- ❑ State the assumptions of the kinetic molecular theory of gases.
- ❑ Explain the characteristic properties of gases, Boyle's law, Charles' law, and Dalton's law of partial pressures in terms of the kinetic theory of gases.
- ❑ State the conditions under which gas ideality (zero attractive forces) is most nearly approached.

11.11 Gas stoichiometry

- ❑ Given a balanced chemical equation for a reaction occurring at constant *T* and *P* and the volume of one gaseous reactant or product, calculate the volume of a second gaseous reactant or product.
- ❑ Given a balanced chemical equation, a given *T* and *P,* and the mass of one reactant or product, calculate the volume of a gaseous reactant or product.
- ❑ Given a balanced chemical equation, a given *T* and *P,* and a volume of a gaseous reactant or product, calculate the mass of another reactant or product.

key terms

pressure
atmospheric pressure
atmosphere
torr
partial pressure
Dalton's law of partial pressures
Boyle's law
Kelvin temperature scale
absolute temperature scale
Charles' law
Gay-Lussac's law
Avogadro's law
standard temperature
standard pressure
molar gas volume
universal gas constant
ideal gas law
ideal gases

problems

11.1 Introduction

15. State three physical properties of gases in general.
16. Describe at least two ways in which the properties of gases differ from those of liquids and solids.

11.2 Measurement of the four gas variables: *n, T, V,* and *P*

17. Describe why a container must be evacuated before it can be used in determining the mass of a gas.
18. Originally 3.57 g of a gas at 20.°C occupies 2.5 L. The gas is allowed to expand to fill 15 L at 20.°C. Calculate the original and final densities.
19. Describe the cause of atmospheric pressure.
20. Why might atmospheric pressure be lower on top of a mountain than at sea level?
21. Explain why all the mercury does not run out of a barometer.
22. Do the following conversions.
    a. 735 mmHg to atmospheres
    b. 1.75×10^{-2} atm to millimeters of mercury
23. The atmospheric pressure in Mexico City is about 580. torr.
    a. What is this pressure in millimeters of mercury?
    b. In atmospheres?

11.3 Dalton's law of partial pressures for mixtures of gases

24. In your own words, state Dalton's law of partial pressures.
25. Oxygen and chlorine gas are mixed in a container with partial pressures of 401 mmHg and 0.639 atm, respectively. What is the total pressure inside the container?
26. Oxygen gas produced from the decomposition of mercury (II) oxide is collected by water displacement on a day when the barometric pressure and temperature are 765 mmHg and 21°C, respectively. What is the partial pressure of the oxygen gas?
27. Two mixtures of gases, *A* and *B,* are separated by a barrier that gases can cross. In mixture *A,* gas 1 has a partial pressure of 80. mmHg and gas 2 has a partial pressure of 138 mmHg. In mixture *B,* p_1 = 98 mmHg and p_2 = 120. mmHg. Will there be any gas flow between *A* and *B,* and if so, in which direction? Explain.

28. If 2.0 L of N_2 at 355 mmHg, 2.0 L of H_2 at 1.9 atm, and 2.0 L of O_2 at 751 torr are all placed in a 2.0 L container, what will be the total pressure inside the container?

11.4 Boyle's law

29. a. State Boyle's law in terms of a proportionality.
    b. State Boyle's law in the form of an equation.
    c. Give a common example that demonstrates Boyle's law.
30. Describe how Boyle's law is related to the breathing process.
31. Consider the gas data in the following volume-pressure graph.

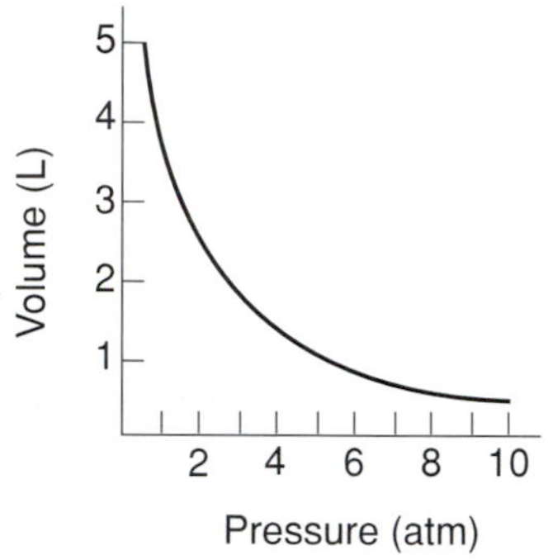

    a. At what pressure will the gas occupy a volume of 3 L?
    b. What is the value of *PV* for the volume in part a?
    c. What is the value of *PV* at a volume of 1 L?
    d. What is the value of *PV* at a volume of 2 L?

11.5 Charles' law

32. a. State Charles' law in the form of a proportionality.
    b. State Charles' law in the form of an equation.
    c. Give a common example that demonstrates Charles' law.
33. How can the thermometer be used to measure the volume of trapped air in Figure 11.19?
34. Convert the following.
    a. 37°C to Kelvin
    b. –221°C to Kelvin
    c. 398 K to degrees Celsius
35. Why does warm air rise above cold air?
36. a. A gas is at a temperature of 25°C. Assuming that the pressure and mass are held constant, will the volume of the gas double if the temperature is increased to 50.°C?
    b. To what temperature must the gas be heated to double its volume?
37. a. Is the volume of a gas directly proportional to the Celsius temperature?
    b. What is the relationship between the volume of a gas and the Celsius temperature?

11.6 Combined gas laws

38. A 1.00 L balloon is released at sea level where the pressure is 760. mmHg. Assuming that the temperature remains constant, what will be the volume of the balloon at an altitude where the pressure is 580. mmHg?
39. At constant pressure, 1.20 L of exhaled gas undergoes a change in temperature from 98°F to 18°C. What is the new volume of the gas?
40. In a laboratory, 273 mL of a gas at 0°C is cooled to –1°C at constant pressure.
    a. What is the new volume of the gas?
    b. By what factor of the original volume has the gas decreased?
    c. If the gas were cooled to –2°C, what would be the new volume?
    d. Now, by what factor of the original volume has the gas decreased?
41. Would your answers in Problem 40 be the same if *P* were not held constant? Explain.
42. The pressure on a gas is tripled at constant temperature.
    a. Will the volume increase or decrease?
    b. By what factor will the volume change?
43. Gasoline vapor is injected into a 4.00 mL cylinder at a temperature and pressure of 21°C and 753 mmHg. The vapor is compressed under a pressure of 9.50 atm to 1.07 mL. What is the final temperature of the gasoline vapor?
44. A 4.00 L balloon of air is contained within a rocket ship at a temperature of 18°C and 1.00 atm. The balloon is released into space at a temperature of –245°C and 2.0 mmHg pressure. What is the new volume of the balloon?
45. An aerosol can at 19°C contains a gas at 2.20 atm. What pressure is exerted by the gas at 95°C? Assume that the can is rigid and thus there is no volume change.
46. A balloon filled with helium has a volume of 2.50 L at 25°C. The balloon is placed in a freezer at –12°C. What is the volume of the balloon in the freezer?
47. If 325 L of a gas at 18°C and 1.00 atm is compressed into a cylinder at 135 atm and 21°C, what is the volume of the cylinder?

11.7 Avogadro's law

48. Do 3.03 g of hydrogen gas molecules and 48.0 g of oxygen gas molecules occupy the same volume at 50°C and 730. mmHg?
49. A 1.50 mol sample of CO_2 gas at 18°C and 1.00 atm occupies a volume of 36 L. What volume would a 1.50 mol sample of SO_2 gas take up at the same temperature and pressure?

11.8 Molar gas volume and standard temperature and pressure

50. A student collects 31.8 mL of argon gas at a temperature of 21°C and a pressure of 755 mmHg. What is the volume of argon at STP?
51. In a laboratory, 75.0 mL of hydrogen gas is collected by water displacement at a temperature of 18°C. The barometric pressure is 751 mmHg. What would be the volume of hydrogen at STP?
52. Why is it necessary to compare volumes of gas only at the same temperature and pressure?
53. Calculate the volume of 3.0 mol of nitrogen gas at STP.

54. Calculate the volume of a cylinder that contains 5.67 g of hydrogen gas at STP.
55. Calculate the density of nitrogen gas at STP.
56. What volume will be occupied by the following weights of each gas at STP?
    a. 15.4 g O_2
    b. 7.56 g HCl
    c. 395 mg CO
    d. 18.1 g NH_3
57. What is the molar mass of a gas with a density of 1.16 g/L at STP?
58. What is the mass of 10.0 L of H_2 at STP?
59. a. Does 1.00 mol of a gas always occupy 22.4 L?
    b. What are the restrictions?

11.9 Ideal gas law

60. Calculate the volume of 4.0 mol of nitrogen gas at 754 mmHg and 85°C.
61. How many moles of propane gas, C_3H_8, will be present in a 2.55 L cylinder if the temperature is 35°C and the pressure is 15.1 atm?
62. Calculate the molar mass of an unknown gas if 987 mL at 18°C and 754 mmHg weighs 6.29 g.
63. How many grams of helium gas are present in a 22.4 L cylinder at 18°C and 100. atm?
64. What pressure will be exerted by 12.5 g of oxygen confined to a volume of 540. mL at 25°C?
65. Calculate the mass of 20.0 L of methane gas (CH_4) at 25°C and 11.0 atm.
66. A 1.65 g sample of gas occupies a volume of 3.00 L at 75°C and 1.05 atm. What is the molar mass of the gas?
67. Calculate the density of methane gas (CH_4) at 45°C and 754 mmHg.
68. A mixture contains 16.0 g of O_2 and 14.0 g of N_2.
    a. What volume will the mixture occupy at STP?
    b. What is the partial pressure of the O_2? Of the N_2?
    c. What is the relationship between the sum of the partial pressures and the total pressure?
    d. The mole fraction of a substance in a mixture is defined as the ratio of the number of moles of that substance divided by the total number of moles in the mixture. Can you find a relationship between the partial pressure of each gas and its mole fraction and the total pressure of the mixture?

11.10 Kinetic molecular theory of gases

69. State the assumptions of the kinetic molecular theory of gases.
70. a. What is meant by an ideal gas?
    b. Under what conditions of temperature and pressure does a real gas most approach ideal behavior?
    c. Under what conditions of temperature and pressure does a real gas deviate most from ideal behavior?
71. Using the kinetic theory of gases, explain why the pressure in an automobile tire increases after the car has been driven at high speed.
72. Using the kinetic theory of gases, explain why an aerosol can should not be heated above the warning temperature listed on the can.
73. Using Charles' law, explain why a given gas decreases in density as its temperature rises.
74. Explain what happens to the particles making up a gas when the temperature of the gas is increased.
75. A sample of nitrogen gas is at 0°C. To what temperature must the gas be heated to double its kinetic energy?

11.11 Gas stoichiometry

76. How many liters of hydrogen chloride gas can be formed from the reaction of 3.5 L of chlorine with excess hydrogen at 20.°C and 1.10 atm?
77. How many liters of oxygen are required to react with 3.0 L of carbon monoxide to form carbon dioxide at STP?
78. Zinc metal reacts with hydrochloric acid to produce zinc chloride and hydrogen gas.
    a. How many liters of hydrogen gas can be produced from 2.0 mol of zinc at STP?
    b. How many liters of hydrogen gas can be produced from 41.3 g of zinc at 18°C and 1.00 atm?
    c. How many grams of zinc are needed to produce 10.0 L of H_2 at STP?
    d. How many grams of zinc are needed to produce 10.0 L of H_2 at 25°C and 755 mmHg?
79. Glucose ($C_6H_{12}O_6$) is metabolized in living systems according to the overall reaction:

$$C_6H_{12}O_6(s) + 6\ O_2(g) \rightarrow 6\ CO_2(g) + 6\ H_2O(l)$$

    What is the volume of CO_2 gas produced from 28.0 g of glucose at 1.00 atm and 37°C?
80. Ammonia can be produced from nitrogen:

$$N_2(g) + 3\ H_2(g) \rightarrow 2\ NH_3(g)$$

    a. What volume of nitrogen is required to react with 1.5 L of hydrogen at 25°C and 1.0 atm?
    b. What volume of NH_3 will be produced from 1.5 L of H_2 at 25°C and 1.0 atm?
    c. If 2.0 L of nitrogen and 3.0 L of hydrogen are mixed together at 25°C and 1.0 atm, which reactant will limit the formation of NH_3 according to the above equation?

Additional problems

81. Describe the physical processes by which the aroma from an open bottle of perfume is perceived by a student who is standing 10 ft from the open bottle.
82. Why might swirling a glass of wine increase the intensity of its aroma in the glass?
83. Where would gas pressure be lower: in outer space or in your classroom? Why?
84. The accurate reporting of a boiling point requires a statement of the pressure at which the boiling point is measured. Why is this necessary?

85. What is the pressure in atmospheres corresponding to a weather report in which the pressure is 29.10 in. Hg?

86. A person with respiratory problems is often provided a face mask connected to a gas source that has a higher concentration of oxygen than air. What gas variable has been affected to provide a greater inhalation of oxygen?

87. Air is predominantly a mixture of oxygen and nitrogen. Can the pressure of air be increased without increasing the oxygen pressure? How?

88. Describe a simple procedure by which the partial pressure of carbon dioxide can be increased in inhaled air.

89. Explain why heating the air in a hot air balloon causes the balloon to rise.

90. Indicate which of the four gas variables are held constant in each of the following gas laws.

    a. Dalton's law of partial pressures

    b. Boyle's law

    c. Charles' law

    d. Gay-Lussac's law

    e. Avogadro's law

91. A sample of methane gas of volume 6.40 L is heated from 35°C to 75°C at constant pressure. What is its final volume?

92. A sample of argon gas in a 1.50 L container is heated from 27°C to 135°C. The initial pressure was 1.25 atm. If the volume is kept fixed, what is the new pressure?

93. The ozone layer in the stratosphere absorbs much of the ultraviolet radiation from the sun. The temperature of the stratosphere is –23°C and the partial pressure of the ozone is 1.4×10^{-7} atm. Calculate the number of moles of ozone in 1.0 L.

94. Calculate the volume that 1.50 mol of ammonia would occupy at 36°C and 655 torr.

95. Calculate the density of water vapor at 100.°C and 1.00 atm. Compare your calculated value to that of liquid water at the same temperature and pressure (0.958 g/mL).

96. A glass vessel contains air at 1.00 atm and 25°C. If the glass would burst at a pressure of 2.00 atm, what is the highest temperature that the glass vessel can be heated to without bursting?

97. The volume of a gas at STP is 475 mL. Calculate its volume at 21.5 atm and 175°C.

98. 5.20 L of a gas has a mass of 7.20 g at 755 torr and 43°C. What is the molar mass of the gas?

99. If a 152 g block of dry ice (the solid form of carbon dioxide) is completely sublimed at 25°C and 1.00 atm, what volume will the carbon dioxide gas occupy?

100. Your lungs have a volume of about 6.5 L. How many moles of oxygen are contained in your lungs at 1.00 atm and 36°C? Note that air is about 21% oxygen by volume.

101. A metal can with a screw cap is partially filled with water, heated to the boiling point of water, and then sealed tightly with the screw cap. The can is then placed in ice water where it is observed to partially collapse. Explain the collapse of the can using the kinetic molecular theory of gases.

102. At which set of conditions would you expect the gas law calculations to be more accurate: low pressure and high temperature, or high pressure and low temperature? Explain.

103. Would you expect Dalton's law of partial pressures to hold more accurately for an ideal gas or a nonideal gas? Explain.

104. The natural gas in your home is mostly methane, CH_4. Methane and air form a combustible mixture that can burn according to the equation:

$$CH_4(g) + 2\ O_2(g) \rightarrow CO_2(g) + 2\ H_2O(g)$$

What volume of air, which is approximately 21% oxygen, is required for the total combustion of 1.0 L of methane if both are at the same temperature and pressure?

105. If ignited, hydrogen and oxygen can react with each other to form water.

    a. What volume of oxygen will react with 1.0 L of hydrogen, assuming both are at 200.°C and 1.0 atm?

    b. What volume of water would be produced at 200.°C and 1.0 atm?

106. How many grams of nitrogen are present in a 2.0 L container of air at 25°C when the partial pressure of N_2 is 1.0 atm?

107. A 1.21 g sample of gas in a sealed 1.16 L container is found to exert a pressure of 1.73 atm at 50.°C. What is the molar mass of this gas?

108. A 3.0 L sample of propane, C_3H_8, is burned in oxygen to form carbon dioxide and water.

    a. How many liters of oxygen are required for the total combustion of the propane?

    b. How many liters of carbon dioxide will form?

109. A 4.00 L flask at 27°C contains 8.80 g of carbon dioxide and 2.00 g of nitrogen. What is the total pressure inside the flask?

110. A 0.395 g sample of gas is contained in a 250 mL container at 0.921 atm and 27°C.

    a. What is the molar mass of this gas?

    b. If the gas has an empirical formula of CH_2, what is its molecular formula?

111. A dent in a table-tennis ball may often be removed by heating the ball in water. Use the kinetic molecular theory of gases to explain this observation.

112. Calculate the volume of O_2 at STP required for the complete combustion of 24 g of methane, CH_4.

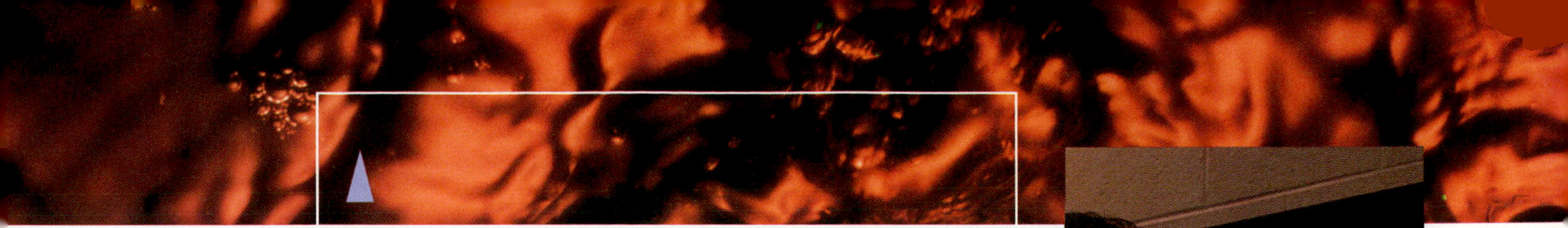

chapter 12

liquids and solids

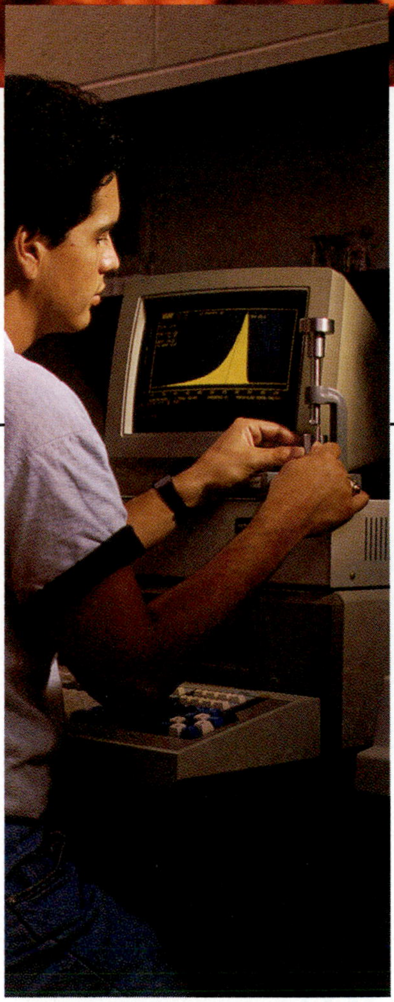

A student is measuring the heat of fusion (the heat of melting) for a pharmaceutical sample. By placing a carefully weighed sample (about 10 mg) in a platinum sample pan and heating it at a specified rate, a differential scanning calorimeter will produce a thermogram (a graph of the heat absorbed by the sample versus time) that has a peak where the sample melts. Computer software in the calorimeter determines the melting point and measures the area under the peak shown on the computer screen, which is a direct measure of the heat of fusion of the sample. The melting point and heat of fusion are indicators of sample purity. Differential scanning calorimeters are routinely used in analytical laboratories, chemical research laboratories, and industrial production plants.

12.1 introduction

In Chapter 11, you learned that the kinetic molecular theory of gases was a model, or picture, of the motion of individual atoms or molecules that explains the gas laws and the properties of gases. In this chapter, you will extend the kinetic model to explain the different properties of liquids and solids.

In Section 6.12, you learned of covalent bonding and the forces that bind atoms to each other *within* molecules. This chapter will expand your understanding of covalent compounds in the liquid and solid state by describing the various types and strengths of forces *between* separated molecules (**intermolecular forces**).

A brief review of the difference in properties between gases and liquids or solids will illustrate the importance of intermolecular forces. In gases, the atoms or molecules are very far apart. Consequently, the forces between them are so weak that they may be ignored. In liquids and solids, the atoms or molecules are much closer together so that the forces between them become significant. As a result of the forces between molecules, liquids and solids have considerably different properties from gases. Solids and liquids are *not easily compressed* because the atoms or molecules repel one another as they are pushed more closely together. Unlike gases, solids and liquids *retain their own volume* and *do not expand to fill the volume of their container.* Forces of attraction between the molecules in solids and liquids resist their spreading apart. Unlike gases, which have indefinite densities, solids and liquids have *characteristic densities* for each physical state that are used to identify the material (Section 1.7). And, because of the much smaller volume occupied by liquids and solids, you expect that their densities ($d = m/V$) are much larger. Like gases, most liquids and solids *expand when heated* and *contract when cooled,* but the change in volume is very small compared to the change in the volume of gases. However, there are exceptions when some solids are heated. Ice (solid water) contracts, it does not expand, when it melts to form the liquid state; correspondingly, liquid water expands when it freezes to form ice. The basis for this exception will be discussed in terms of intermolecular forces in Section 12.4.

A better understanding of the properties of liquids and solids demands a closer look at ionic and intermolecular forces. Since most of this chapter will discuss the properties of molecular solids and liquids, our discussion will begin with forces which exist *between* molecules. But before doing so, you must look at two concepts that play an important role in determining the strength of intermolecular forces, namely, molecular shape and molecular polarity.

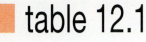

table 12.1 electron-pair geometry and molecular shape predicted by VSEPR theory when both are the same

Sets of electron pairs	Arrangement of sets	Electron-pair geometry and molecular shape	Bond angle	
2 bonding		Linear; atoms are arranged in a line.	180°	
3 bonding	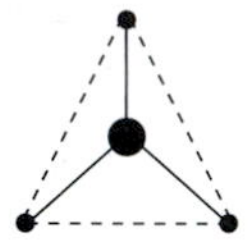	Planar triangular (or trigonal planar); all atoms are in one plane, and a triangle is formed (dashed lines) if the outer atoms are connected.	120°	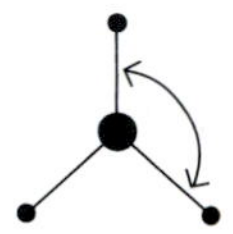
4 bonding	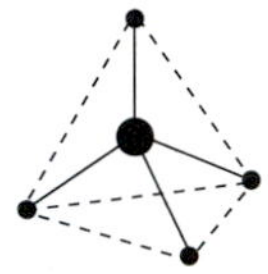	Tetrahedral; a tetrahedron is formed (dashed lines) if the outer atoms are connected (a tetrahedron has four faces, each an equilateral triangle).	109.5°	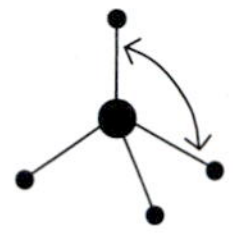

Central atom	●
Outer atom	•
Bonding electron pair	—

12.2 molecular shape

Since many molecular properties depend on the shape of the molecule, it is important to predict the shapes of molecules. You know that all molecules contain atoms that are covalently bonded to each other, and you have learned how to distribute the electrons around the central atom in any molecule to form a Lewis structure (Section 6.13). You also know that most central atoms in molecules obey the octet rule; that is, they share electrons with other atoms until their outer shell contains a noble-gas configuration of eight valence electrons. By simply extending your knowledge of Lewis structures, you will quickly be able to predict the three-dimensional shape of many molecules.

The eight valence electrons around any central atom are divided into pairs of electrons that are either bonding or nonbonding (Section 6.11). The number of bonding electron pairs in any molecule and the number of single, double, or triple bonds are directly determined from your Lewis structure. Also determined from the Lewis structure is the number of nonbonding electron pairs (lone pairs) around the central atom. Your counting of these pairs of electrons and how they are distributed in sets is the basis of **valence shell electron-pair repulsion (VSEPR) theory**. VSEPR theory (pronounced "vesper" for short) allows you to use Lewis structures to predict the three-dimensional arrangement of electrons around the central atom in most molecules.

The fundamental concept in VSEPR theory is the repulsion of sets of electron pairs; to minimize this repulsion, *sets of electron pairs keep as far away from each other as possible.* A set may consist of one pair of electrons (either a single bond or a nonbonding pair), two pairs (a double bond), or three pairs (a triple bond). Table 12.1 shows the geometries associated with the maximum possible separation in space for two, three, and four sets of electron pairs all attached to a central atom. For all cases shown in the table, the sets of electron pairs contain only *bonded pairs.* In these cases, the *geometry of the electron pairs is identical to the geometry of the molecule.*

The geometry that will keep two sets of electron pairs farthest apart is a linear geometry with a bond angle of 180°. For example, Lewis electron structures for carbon dioxide, CO_2, and hydrogen cyanide, HCN, are given below.

$$:\ddot{O}{=}C{=}\ddot{O}: \qquad H{-}C{\equiv}N:$$

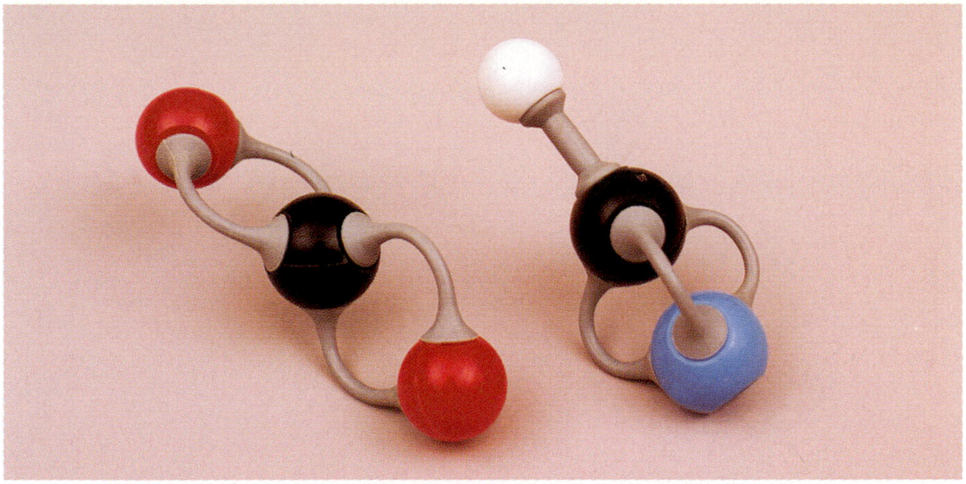

Figure 12.1

Carbon dioxide, CO_2, has two sets of electron pairs, each set containing a double bond. Hydrogen cyanide, HCN, has two sets of electron pairs, one set containing a single bond and the other set containing a triple bond.

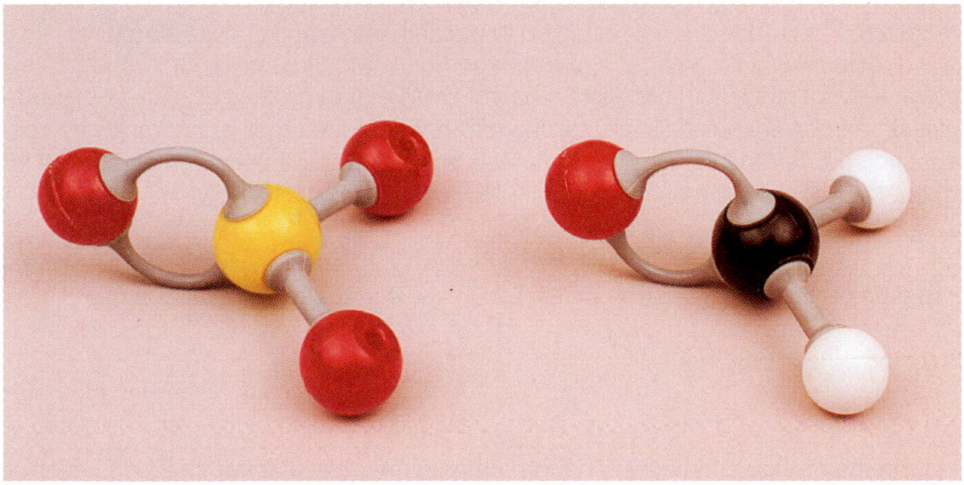

Figure 12.2

Both sulfur trioxide, SO_3, and formaldehyde, CH_2O, have three sets of electron pairs, one set containing a double bond and two sets each containing a single bond.

Note that there are two sets of electron pairs on each central atom (Figure 12.1). In CO_2, the eight valence electrons on the central carbon atom are divided into two sets of double bonds. Each of the sets repels the other to form a linear molecule. In HCN, the eight valence electrons on the central carbon atom are divided into two sets. One set forms a single bond to the H atom and the second set forms a triple bond to the N atom. HCN is also a linear molecule.

The geometry that will keep three sets of electron pairs farthest apart is a planar, triangular geometry with bond angles of 120°. For example, Lewis electron structures for sulfur trioxide, SO_3, and formaldehyde, CH_2O, are given below.

In both molecules, the eight valence electrons on the central atom are divided into three sets (Figure 12.2). One set forms a double bond and two sets each form a single bond. The three sets of bonds are in a plane and directed toward the corners of a triangle. This molecular geometry is called planar triangular (also called trigonal planar).

The geometry that will keep four sets of electron pairs farthest apart is tetrahedral, not a square, as most people first envision. When the four sets of electron pairs

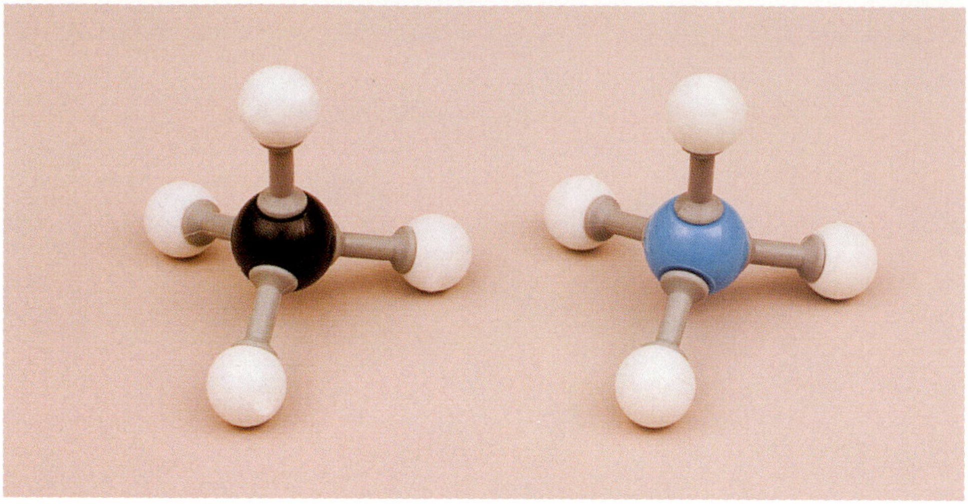

Figure 12.3

Both methane, CH_4, and ammonium ion, NH_4^+, have four sets of electron pairs, each containing a single bond.

repel each other and are not restricted to a plane as drawn in Lewis structures, they are directed toward the corners of a tetrahedron with bond angles of 109.5°. At this point, a set of molecular models or a bag of small marshmallows with a box of toothpicks may be very helpful in visualizing tetrahedral geometry. For example, try constructing the tetrahedral geometry represented by the Lewis electron structures for methane, CH_4, and the ammonium ion, NH_4^+, that are given below.

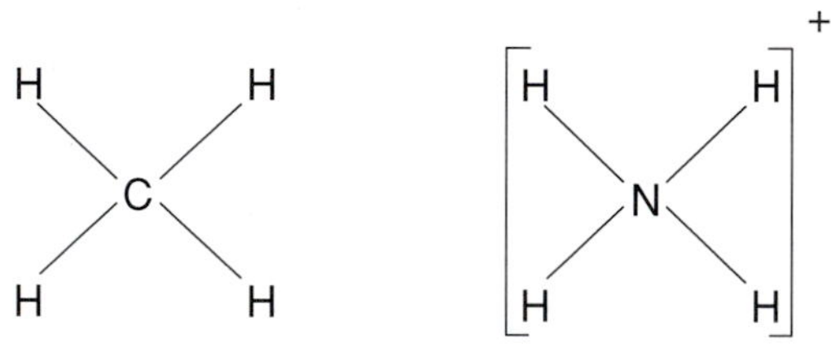

In both the methane molecule and the ammonium molecular ion, the eight valence electrons on the central atom are divided into four single-bond sets (Figure 12.3). The four bonds (electron pairs) are arranged tetrahedrally with the four outer atoms at the corners of a tetrahedron. The molecular geometry of both species is called tetrahedral, the same geometry as the electron pairs around the central atom.

The geometry of the molecule is often different from the geometry of the electron pairs around the central atom, however. *When the molecule contains one or more nonbonding pairs, the geometry of the molecule is different from the geometry of the total number of electron pairs. The arrangement of bonded atoms describes molecular shape, not the arrangement of electron pairs.* The geometry of the molecule (molecular shape) is named from the shape of the bonded atoms only, not from the electron pairs, as summarized in Table 12.2.

For example, the Lewis electron structure for the water molecule, H_2O,

$$H—\ddot{\underset{..}{O}}—H$$

indicates that the central O atom has four sets of electron pairs. Two of the sets each have a singly bonded pair and two of the sets each have a nonbonded pair. The four sets of electrons have a tetrahedral geometry and are directed toward the corners of a tetrahedron just as they were when all of the electron pairs were bonded. However, the geometry of the water molecule is called bent, not tetrahedral, because *the shape of the molecule, as determined by the bonded atoms,* is three atoms *bent away* from a straight line (Figure 12.4).

Figure 12.4

Water, H_2O, has four sets of electrons, two sets each containing a nonbonding pair and two sets each containing a singly bonded pair.

table 12.2 electron-pair geometry and molecular shape predicted by VSEPR theory when they are different

Sets of electron pairs	Arrangement of sets	Electron-pair geometry	Molecular shape	Bond angle
4 (3 bonding + 1 nonbonding)		Tetrahedral	Pyramidal; a pyramid is formed if the outer atoms are connected (the pyramid has four faces that are not equilateral).	109.5° 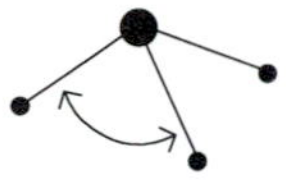
4 (2 bonding + 2 nonbonding)		Tetrahedral	Bent; the atoms are not linear, but rather *bent away* from a straight line.	109.5° 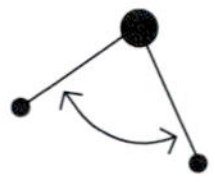

Central atom	●
Outer atom	•
Nonbonding electron pair	
Bonding electron pair	—

Another example of a Lewis electron structure with only one nonbonding pair of electrons is ammonia, NH_3, below.

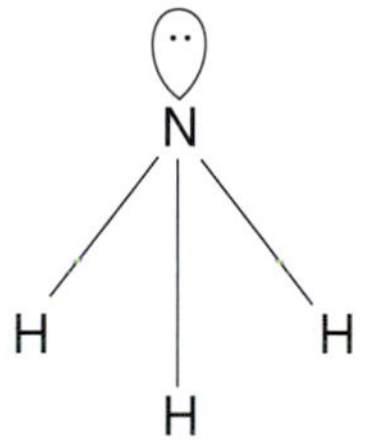

NH_3 has four sets of electron pairs. One set forms a nonbonding pair and three sets each form singly bonded pairs. The geometry of the four electron pairs is still tetrahedral. However, the geometry of the molecule as determined by the N and three H atoms is not tetrahedral, but *pyramidal. The shape of the molecule, as determined by the bonded atoms,* is a pyramid (Figure 12.5).

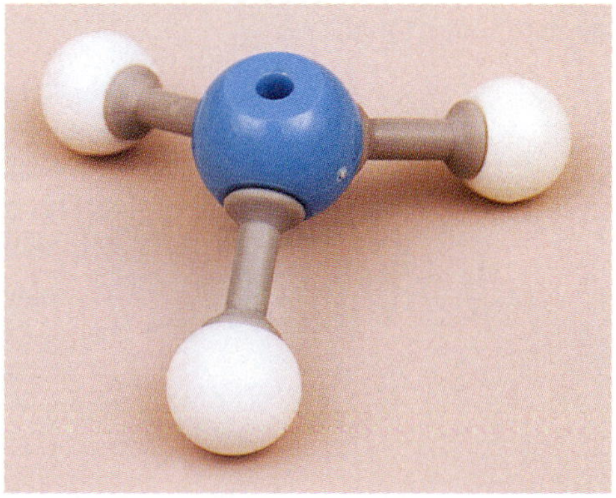

Figure 12.5
Ammonia, NH_3, has four sets of electrons, one set containing a nonbonding pair and three sets each containing a singly bonded pair.

You have seen throughout the previous examples that the electron-pair geometry predicted from VSEPR theory leads directly to your prediction of molecular shape. Only in the case of lone, nonbonding pairs of electrons around the central atom does the electron-pair geometry differ from the geometry of the molecule. Since only the atoms, and not the electron pairs around them, can be determined in experiments, only the atoms are considered in naming the

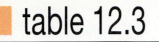

table 12.3 electron-pair geometry and molecular shape

Sets of electron pairs	Number of nonbonding pairs*	Electron-pair geometry	Molecular shape	Examples
2	0	Linear	Linear	CO_2, HCN
3	0	Planar triangular (or trigonal planar)	Planar triangular (or trigonal planar)	H_2CO, SO_3
4	0	Tetrahedral	Tetrahedral	CH_4, NH_4^+
4	1	Tetrahedral	Pyramidal	NH_3
4	2	Tetrahedral	Bent	H_2O

*Notice that when the number of nonbonding pairs is zero, molecular shape is identical to electron-pair geometry.

molecular shape. Table 12.3 summarizes all of the relationships you have seen among the sets of electron pairs, the number of nonbonding electron pairs, electron-pair geometry, and molecular shape.

sample exercise 1

What are the electron-pair geometry and molecular shape of the following compounds?

a. CH_2Cl_2, methylene chloride (dichloromethane), a colorless, nearly nonflammable and nonexplosive liquid, is used as a component in nonflammable paint-remover mixtures and as a solvent in food processing.

b. H_2S, hydrogen sulfide, a flammable, toxic gas with the characteristic odor of rotten eggs, is used in the manufacture of chemicals, in metallurgy, and as a laboratory reagent.

c. $COCl_2$ (C is the central atom), phosgene (carbonyl chloride), a highly toxic, colorless gas, has been used in gas warfare, but is now used in the preparation of polyurethane (plastic).

solution

Write Lewis structures for each molecule according to the guidelines of Section 6.13 and determine the number of sets and the number of nonbonding electron pairs around the central atom. The electron-pair geometry and molecular shape can be determined from these numbers and Table 12.3.

a.

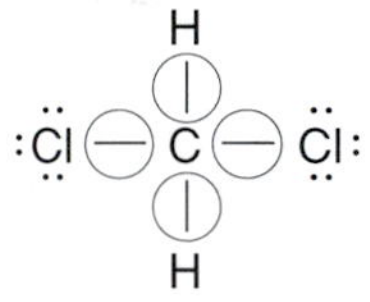

For CH_2Cl_2, there are four sets of electrons around the central carbon, so the set number is four and the electron-pair geometry is tetrahedral. Because there are no nonbonding electrons, the molecular shape is also tetrahedral.

b.

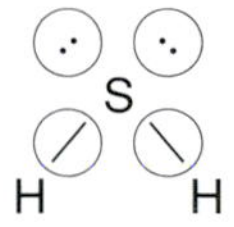

The electron-pair geometry of H_2S is tetrahedral because once again there are four sets of electrons. However, there are only two bonded atoms, so the molecular shape is defined by the shape

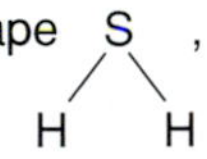

, which is bent.

c.

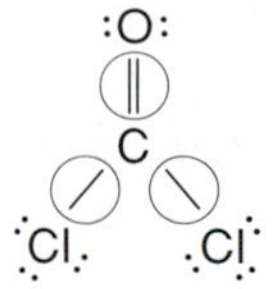

The electron-pair geometry of $COCl_2$ is planar triangular because there are three sets of electrons. Since there are no nonbonding electrons on the central atom, the molecular shape is also planar triangular.

What is the electron-pair geometry and molecular shape of the following compounds?

a. CS_2, carbon disulfide, a very flammable and toxic liquid with a sweet, pleasing odor, is used to manufacture rayon and to dissolve fats, oils, and rubbers.

b. SO_2, sulfur dioxide, a colorless, toxic gas, is used in preserving fruits and vegetables and in bleaching raisins, apricots, and beet sugars.

c. CH_3Br, methyl bromide, a colorless, toxic gas, is used in extracting oils from nuts, seeds, and flowers and as an insect fumigant for mills, warehouses, and ships. Widely used in tropical regions for insect control, methyl bromide contributes to ozone depletion in the upper atmosphere.

••• problem 1

12.3 molecular polarity

Each of the bonds in a molecule may be polar or nonpolar depending on the difference in electronegativity of the atoms sharing electrons (Section 6.15). The greater the difference in electronegativity between two atoms, the more unequally they share electrons and the more polar the bond. You recall from Section 6.15 that a polar bond means that one atom is electron rich and carries a partial negative charge while the other atom is electron deficient and carries a partial positive charge. The separation of partial positive, δ^+, and partial negative, δ^-, charges forms a **bond dipole.** As you shall see, a bond dipole may or may not result in the molecule as a whole having a **molecular dipole.**

polar: strong intermolecular forces

nonpolar: weak intermolecular forces

If a molecular dipole exists, the molecule is said to be a **polar molecule.** Strong intermolecular forces arise between polar molecules. If a molecular dipole does not exist, the molecule is said to be a **nonpolar molecule.** Weak intermolecular forces arise between nonpolar molecules. The strength of forces between molecules is responsible for many important properties discussed later in this chapter.

Whether a molecule is polar and has a permanent molecular dipole depends on both the existence of bond dipoles within the molecule and the molecular shape. In some molecules, the bond dipoles will add together to create a permanent molecular dipole. In such molecules, the bond dipoles arrange themselves such that the center of all of the partial positive charges and center of all of the partial negative charges remain apart.

In Section 6.15, you saw that the simple molecule, HCl, has a bond dipole:

$$^{\delta+}\text{H—Cl}^{\delta-} \quad \text{or} \quad \overset{+\!\!\longrightarrow}{\text{H—Cl}}.$$

The partial positive and partial negative charges remain at opposite ends of the molecule to create a permanent molecular dipole. Because this bond dipole involves the entire molecule, the bond dipole and molecular dipole are the same thing. Therefore, HCl is a polar molecule.

In more complicated molecules, such as the bent H_2O molecule shown below, both bond dipoles point toward the more electronegative O atom.

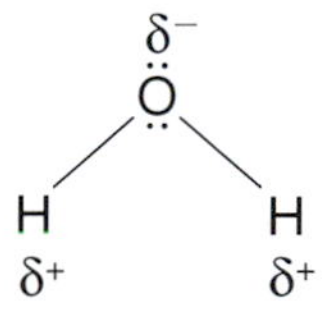

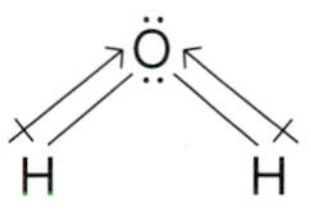

The two bond dipoles add together to create a permanent molecular dipole. The center of the partial negative charges from both dipoles is on the O atom. The center of the partial positive charges from both dipoles is between the two H atoms.

The permanent molecular dipole points from the center of the partial positive charges to the center of the partial negative charges as illustrated below.

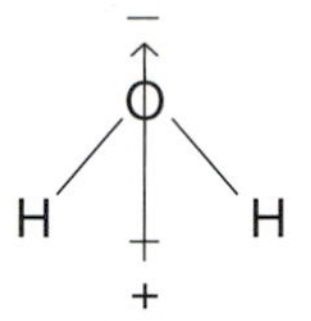

Water molecules are polar because their molecular shape is such that the two bond dipoles add together to create a permanent molecular dipole. That is, there exists a region of higher electron density (the negative end) and a region of lower electron density (the positive end). The experimental measurement of a permanent dipole in water supports the VSEPR prediction of bent geometry.

In some molecules, because of the molecular shape, bond dipoles exert their effects in opposite directions and cancel each other out, leaving the molecule as a whole with no molecular dipole. In that case, the center of all of the partial positive charges and the center of all of the partial negative charges will be at the same point, not separated from each other.

Three general models in which molecular shape results in the cancellation of bond dipoles are given below.

Z–A–Z

1. Molecular shape is linear, with two identical atoms, *Z,* bonded to a central atom *A*. Carbon dioxide is an example:

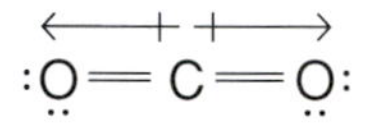

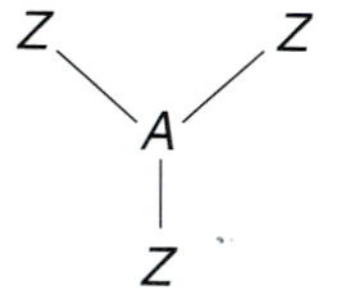

2. Molecular shape is planar triangular, with three identical atoms, *Z,* arranged around a central atom *A*. Boron trifluoride is an example:

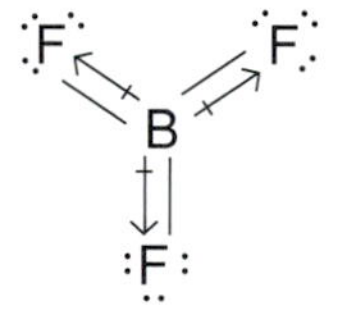

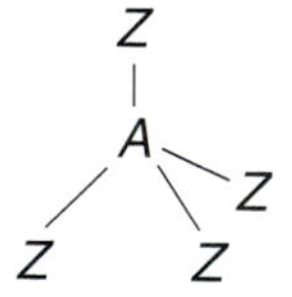

3. Molecular shape is tetrahedral, with four identical atoms, *Z,* tetrahedrally arranged around a central atom *A*. Carbon tetrachloride is an example:

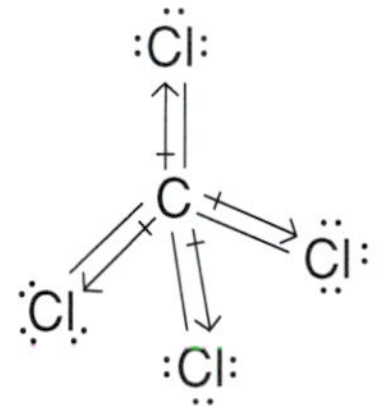

In each of the models above, the molecular dipole points symmetrically away from the central atom. For each bond dipole, the partial positive charge is on the central atom and the partial negative charge is on the more electronegative outer atom. The center of the partial positive charges is always on the central atom. Because of the symmetry of each molecule, the center of the partial negative charges is also on the central atom. As a result, there is no *net* separation of charge. The bond dipoles cancel and molecules represented by the above three models are all *nonpolar.*

An experimental test for molecular polarity can be carried out by placing a compound between two charged electrical plates, as shown in Figure 12.6. Polar molecules will orient themselves in an electric field so that the positive end of the molecule will point to the negative plate and the negative end of the molecule will point to the positive plate. Nonpolar molecules will not line up in an electric field because they have no net separation of charge.

No electric field

(a)

Polar molecules in electric field

(b)

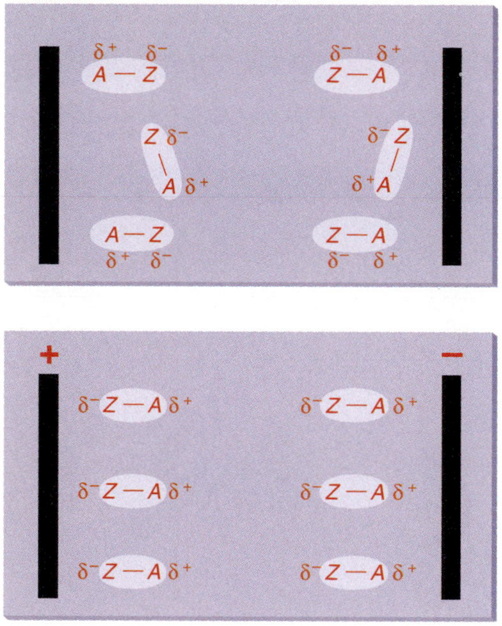

Figure 12.6

(a) Polar molecules accent themselves randomly when no electric field is present. (b) Polar molecules tend to align themselves in an electric field with the positive end of the molecules toward the negative plate and the negative end of the molecules toward the positive plate.

When molecules line up between charged plates, there is a change in the capacitance (the capacity of the plates to hold a charge at a given voltage) that can be observed and measured. The difference in capacitance between the plates when a polar liquid is present and when it is not can be used to evaluate the size of the dipole, called the **dipole moment,** of the liquid. Chemists routinely measure the dipole moment of a liquid using a device called a capacitance bridge.

• • • problem 2

Which of the molecules in Problem 1 have a dipole moment?

12.4 intermolecular forces

Intermolecular forces are the forces that act *between* (*inter* means "between") separated molecules. They are the "glue" that holds separated molecules together in a liquid and solid. The covalent bond you studied in Chapter 6 is an *intramolecular force* that holds the atoms together within a molecule (*intra* means "within"). Intermolecular forces are much weaker than covalent and ionic bonds. For example, the intermolecular forces between liquid water molecules can be broken at the boiling point (100°C) and separate gas molecules (steam) form, but the covalent bonds in the individual water molecules are not broken. Breaking the covalent bonds within water molecules would result in forming elements or new compounds. Temperatures much higher than 100°C are necessary to decompose water molecules into hydrogen and oxygen molecules by breaking the covalent bonds in the water molecules.

Intermolecular forces < Intramolecular forces

H_2O molecules can be broken apart at 100°C

But covalent bonds in individual H_2O molecules require higher temperatures

The three types of intermolecular forces are summarized in Table 12.4. Dipole-dipole forces are stronger than dispersion forces and are the dominant forces between *polar* molecules. Dispersion forces are the only forces between *nonpolar* molecules and explain the weak "glue" that holds nonpolar molecules together in the liquid and solid states. Intermolecular hydrogen bonds are special forces that exist only between a polar, bonded H atom in one molecule and a N, O, or F atom in another molecule. Hydrogen bonds are extremely important in biological systems and in binding the genetic material of all living cells. Each of these forces is discussed in a following section.

table 12.4 types of intermolecular forces

Force	Description	
Dispersion (London) forces	Weak; only force between nonpolar molecules	Increasing Strength ↓
Dipole-dipole forces	Strong; dominant force between polar molecules	
Hydrogen bonds	Strong intermolecular force; only between a H atom in one molecule and an O, N, or F atom in another molecule	

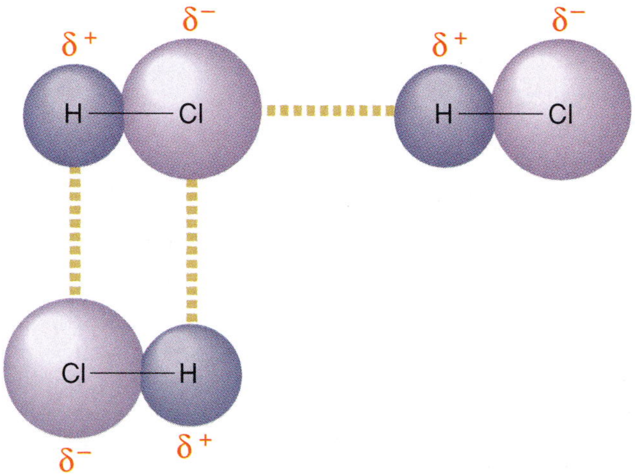

Figure 12.7
Dipole-dipole forces (dashed lines) are the mutual attractions between the positive end of one polar molecule and the negative end of another polar molecule.

dipole-dipole interactions

Dipole-dipole interactions occur between molecules that have a permanent dipole. The positive end of one molecule is attracted to the negative end of another molecule, as shown in Figure 12.7. When many molecules are closely arranged, as they are in liquids and solids, they tend to align themselves. The positive end of each molecule is closest to the negative end of surrounding molecules, and the negative end of each molecule is closest to the positive end of surrounding molecules. Notice the alignment of molecules throughout Figure 12.7.

All polar molecules are attracted to one another through this **dipole-dipole force**. In the gas phase, the molecules are either too far apart or moving too rapidly for this force to be important. You will recall that one of the assumptions of the kinetic molecular theory (Section 11.10) is that gases, particularly at high temperature and low pressure, have no intermolecular interactions and follow the ideal gas law. As the temperature of a gaseous, polar compound is lowered and its pressure is increased, the forces between the polar molecules become more important and result in the gas condensing into a liquid. In liquids and solids, the strength of the dipole-dipole attraction depends on the magnitude of the dipole moment. Nonpolar molecules do not exhibit this interaction because they do not have permanent dipoles.

The *partial* separation of charge (permanent dipole) within a polar molecule results in strong attractive forces between individual molecules. The forces are not as strong as the attraction between oppositely charged ions, however, where the charge has been *fully* separated. As a result, the melting points and boiling points of polar materials are higher than nonpolar materials, but not nearly as high as in ionic compounds. For example, the melting and boiling points of water, a polar compound, are 0°C and 100°C in comparison with 801°C and 1,431°C for ionic NaCl.

dispersion (London) forces

Dispersion forces result from distortions of the electron distributions in nonpolar molecules when the molecules approach one another closely. Electrons in molecules are in constant motion. At any instant, the electron distribution within a molecule may become unsymmetrical (Figure 12.8b), so that an "instantaneous dipole" develops. This instantaneous dipole induces an unsymmetrical charge distribution in a nearby molecule, causing a net attractive force to develop between the two molecules (Figure 12.8c). An instant later, the electron distributions shift, setting up instantaneous dipoles with other molecules. These instantaneous forces between nonpolar molecules are called **dispersion forces** or **London forces** after Fritz London, a U.S. physicist who first offered a mathematical understanding of these interactions.

Dispersion forces exist between all kinds of molecules, whether polar or nonpolar. In polar molecules, the larger dipole-dipole forces dominate and the weak dispersion forces become less significant. In nonpolar molecules, however, where no other forces exist, weak dispersion forces are the only intermolecular forces. They cause nonpolar gases such as He, H_2, and CH_4 to liquefy and solidify at lower temperature and higher pressure. Gaseous nonpolar molecules must be at low temperature and moving slowly before they can be liquefied. They must be at even lower temperatures before they can freeze. Consequently, nonpolar molecules have very low melting points and boiling points. For He, H_2, and CH_4, the melting points are –272.2°C, –259.1°C, and –182.4°C, and the boiling points are –268.9°C, –252.9°C, and –164°C, respectively. Notice how much lower these melting and boiling points are in comparison with those of the polar water molecule.

The strength of dispersion forces depends on the number of electrons in a molecule and the ease with which their distribution can be distorted. Molecules with greater numbers of atoms and hence greater molar masses have larger numbers of electrons. You would expect an increase in dispersion forces as molar mass (and number of electrons) increases. An illustrative example can be seen by comparing the physical states of elements in the halogen series. In order of increasing molar mass, the halogens are fluorine, F_2, chlorine, Cl_2, bromine, Br_2, and iodine, I_2. When compared at room temperature, the least massive molecules, F_2 and Cl_2, are gases, the more massive molecule, Br_2, is a liquid, and the most massive molecule, I_2, is a solid. The dispersion forces between Br_2 molecules must be considerably higher than between F_2 and Cl_2 molecules for Br_2 to exist as a liquid. Likewise, the dispersion forces between I_2 molecules must be even higher for I_2 to have frozen into the solid state at room temperature.

Electron distributions in molecules are more easily disturbed by neighboring molecules if the "surfaces" of the electron clouds can more nearly touch. In Figure 12.9, the dispersion forces between two molecules of different shape but the same molecular formula (and, hence, the same molecular mass) are compared. For any given molecular weight, a molecule like *n*-pentane that has the shape of a spaghetti strand will have greater dispersion forces than a molecule like neopentane that is ball-shaped. Spaghetti-shaped molecules have long surfaces that are close together. Multiple attractions along the elongated surfaces between *n*-pentane molecules result in large dispersion forces and a higher boiling point (36.2°C). Ball-shaped molecules have only a small region between them that is closely spaced. The short interaction surfaces between spherical neopentane molecules result in smaller dispersion forces and a lower boiling point (9.5°C).

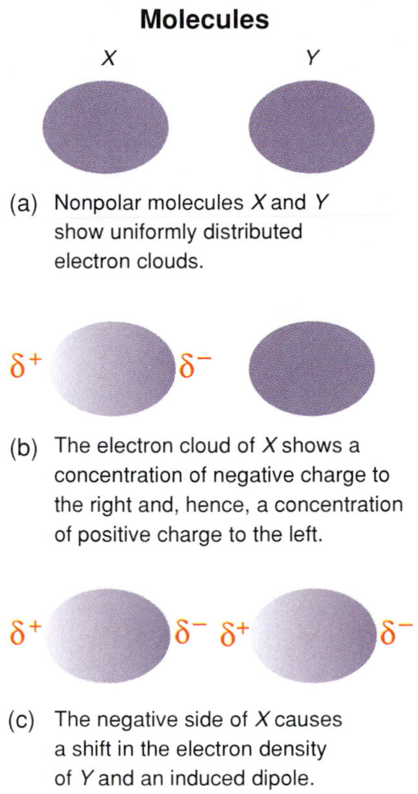

Molecules
X
Y
(a) Nonpolar molecules X and Y show uniformly distributed electron clouds.
δ+
δ−
(b) The electron cloud of X shows a concentration of negative charge to the right and, hence, a concentration of positive charge to the left.
δ+
δ−
δ+
δ−
(c) The negative side of X causes a shift in the electron density of Y and an induced dipole.

Figure 12.8
(a) Separated nonpolar molecules show uniformly distributed electron clouds. (b) Dispersion forces between two nonpolar molecules result from distortions of electron distributions when the molecules approach one another closely. (c) These distortions create instantaneous dipoles that attract the nonpolar molecules to each other.

Figure 12.9

Dispersion forces vary with molecular shape when the molecular mass is the same. *n*-Pentane (BP = 36.2°C) has larger dispersion forces because multiple attractions are possible between the elongated surfaces of two molecules. Neopentane (BP = 9.5°C) has weaker dispersion forces because the spherical shape of the molecules permits only a small region of surface attraction between two molecules.

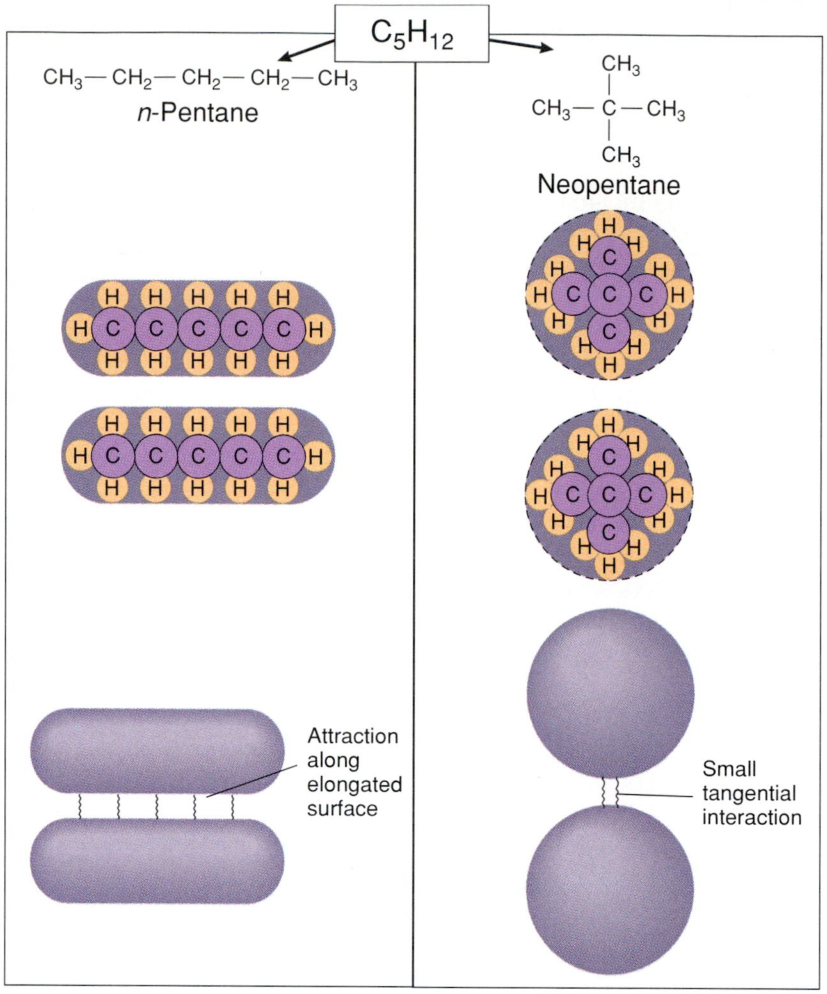

hydrogen bonds

The intermolecular force between a hydrogen atom covalently bonded to N, O, or F in one molecule and a nonbonding pair of electrons on N, O, or F in another molecule is called a **hydrogen bond.** Water, H_2O, is the most commonly encountered molecule that shows hydrogen bonding. The electron pair in the covalent bond between O and H is shifted toward the more electronegative oxygen, leaving the hydrogen partially positive in charge. Because of the small size of the hydrogen atom, its positive charge is concentrated in a small volume, and this leads to a strong intermolecular attraction with a nonbonding pair of electrons on the oxygen of a nearby water molecule. This attraction is known as the hydrogen bond.

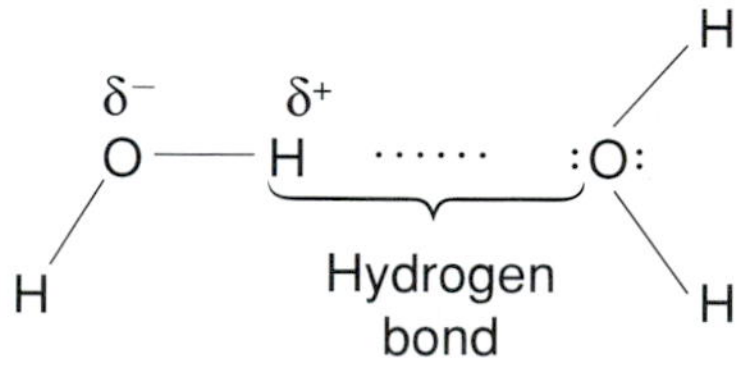

Each oxygen offers two nonbonding pairs of electrons for hydrogen bonding with hydrogens of two nearby water molecules. The hydrogen bond is stronger than other dipole-dipole attractions or dispersion (London) forces, but much weaker than the normal covalent bond.

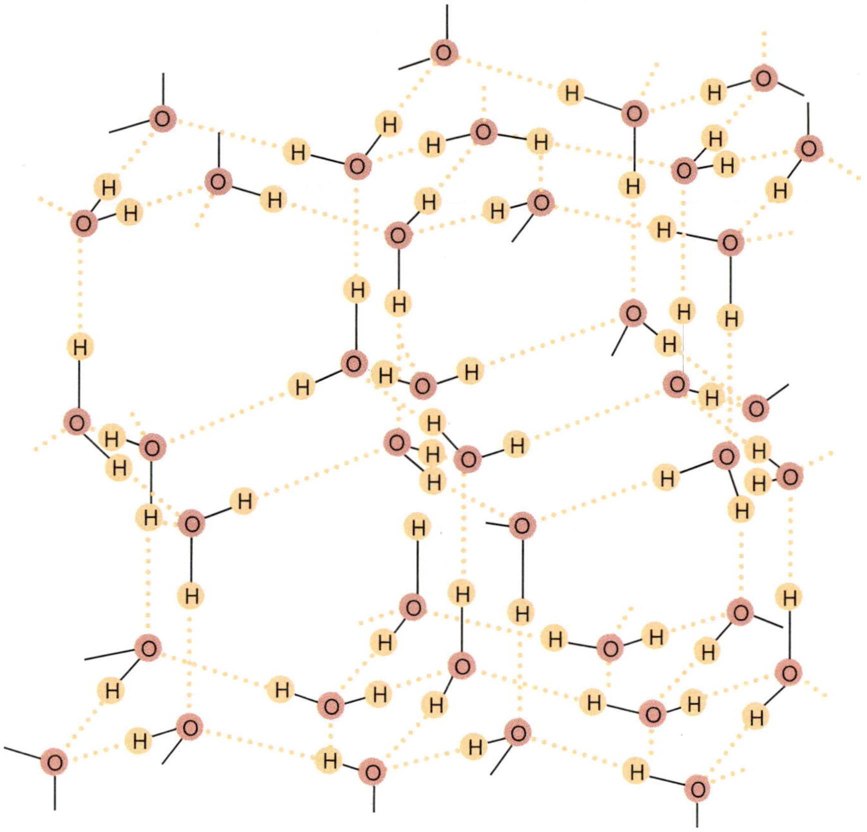

Figure 12.10

The spacious, three-dimensional structure of ice is fixed by hydrogen bonds. Each O atom is bonded to four H atoms by means of two covalent and two hydrogen bonds. The low density of ice is caused by the large amount of empty space in the ice structure.

Many of the special physical properties of water (compared with molecules of similar molecular weight and polarity) arise from the strong hydrogen bonding between water molecules. One special property of water is its ability to climb upward in a narrow tube, a phenomenon known as **capillary action.** Oxygen atoms are a part of the molecular structure of glass tubes or the network of pores in soil. Hydrogen bonds form between water molecules and these oxygen atoms, and this attraction helps to pull the water up the tube or pore. A blood sample (which is an aqueous mixture) is drawn into a capillary tube by this same action.

Another special property of water is its unusually high boiling point compared with other compounds of similar, or even higher, molar mass that lack hydrogen bonding. For example, the polar molecule H_2S with a molar mass of 34 has a boiling point of $-61°C$, which is much lower than the boiling point of water, 100°C.

In Section 12.1, you learned that water was an exception in that its volume decreased, rather than increased, when it melted. Whereas most substances have a smaller volume and correspondingly a greater density in the solid state than in the liquid state, water is unique in that ice is less dense than liquid water from 0°C to 4°C. This property results from an open, spacious network of molecules that is held apart because of hydrogen bonds between the water molecules (Figure 12.10). Each O atom is bonded to four H atoms by means of two covalent and two hydrogen bonds. Because of its lower density, ice floats on top of liquid water rather than sinking. As a result, a body of water freezes from top to bottom, allowing aquatic life to survive below the top ice layer.

Hydrogen bonds are also very important in determining both the three-dimensional structure and the function of large biological molecules such as proteins and nucleic acids. The genetic material of all living cells is deoxyribonucleic acid (DNA), which consists of two complex strands (called sugar (S)-phosphate (P) backbones) that

Figure 12.11

(a) A molecule of deoxyribonucleic acid (DNA) consists of two complex strands (called sugar (S)-phosphate (P) backbones) that are held together to form a double helix by hydrogen bonds (dotted lines). The making and breaking of these hydrogen bonds are required for DNA replication.(b) This model represents an enlarged portion corresponding to the shaded area.

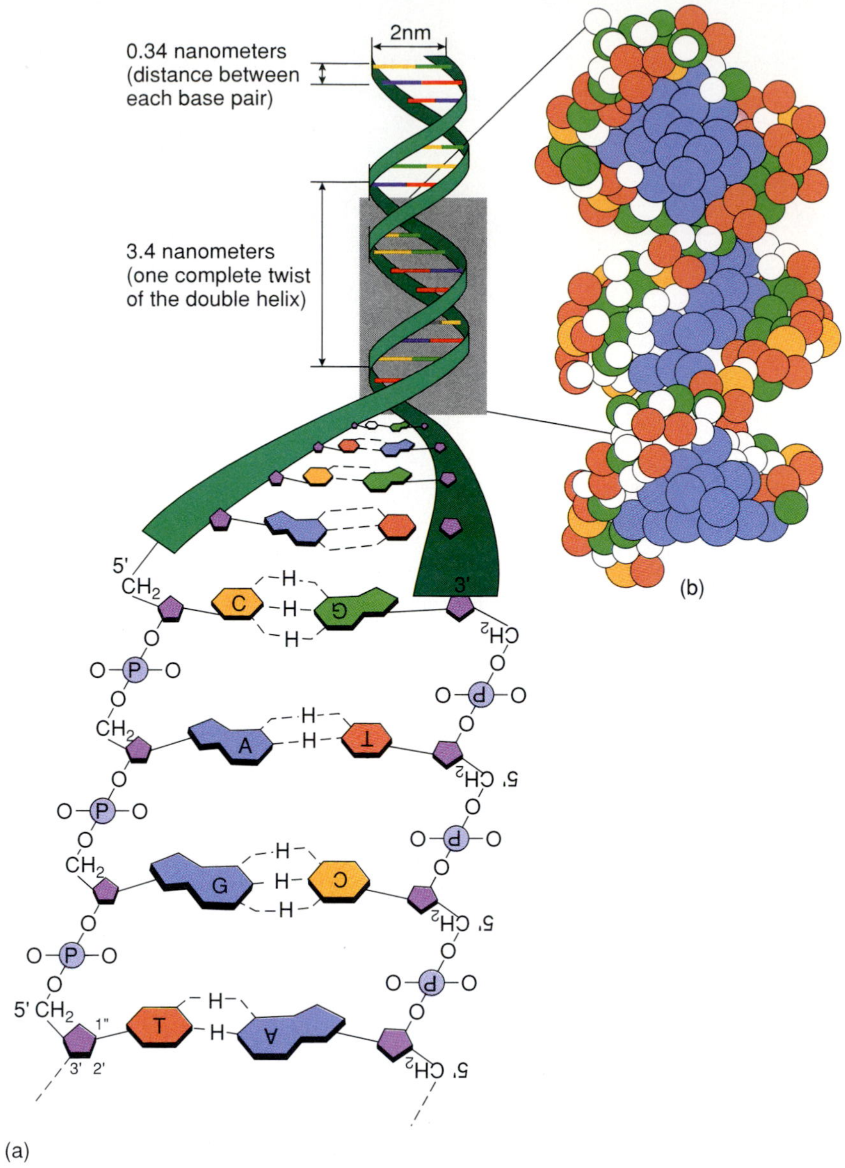

are held together to form a double helix by hydrogen bonds (shown by dotted lines in Figure 12.11). The making and breaking of hydrogen bonds are required for DNA replication (duplication) and for the translation of genetic information into your body's synthesis of proteins. Hydrogen bonds also affect the three-dimensional shape of proteins.

sample exercise 2

Which of the following molecules are capable of forming hydrogen bonds with one another?

a. NH_3, ammonia
b. CH_4, methane (natural gas)
c. HF, hydrogen fluoride
d. H—C(—H)=O, formaldehyde

solution

Molecules with N—H, O—H, or F—H bonds exhibit hydrogen bonding.

a. NH_3, because it has a polar N—H bond, shows hydrogen bonding.

b. CH_4 has only a nonpolar C—H bond and will *not* show hydrogen bonding.

c. HF has a polar F—H bond and will show hydrogen bonding.

d. In this case, although an oxygen with its nonbonding pair of electrons is present, the hydrogen is *not* bonded to N, O, or F, and so hydrogen bonding is not possible.

••• problem 3

Which of the following molecules are capable of forming hydrogen bonds with one another?

a. H_2S

Hydrogen sulfide occurs naturally in coal pits, gas wells, and sulfur springs.

c. 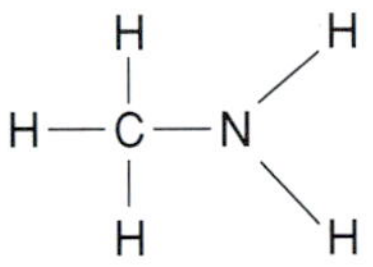

Methylamine is used in tanning leather.

b. 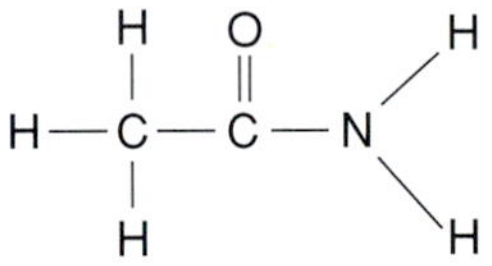

Acetamide is a crystalline solid that, when molten, is used as a solvent for organic compounds.

d. 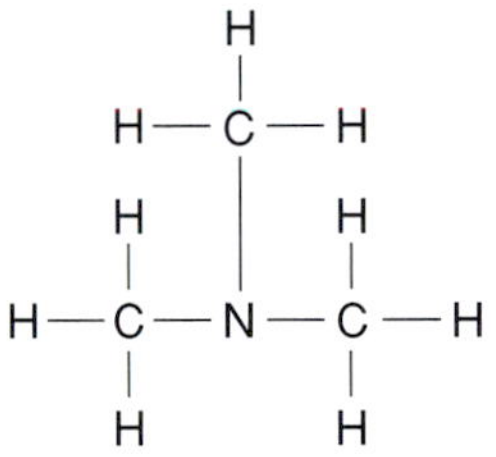

Trimethylamine is used as an insect attractant.

condensation pt. = boiling pt.

12.5 converting gases to liquids and solids

As you learned in Section 12.1, intermolecular forces between gas molecules become more significant as the gas is cooled or its pressure is increased. As the pressure increases, the gas molecules are closer together and undergo more collisions. As the gas molecules are cooled, they travel more slowly and have less kinetic energy to overcome attractive forces between them as they collide. Consequently, intermolecular forces become strong enough to bind the molecules to each other as a liquid.

The temperature at which a gas begins to condense into a liquid could be called the **condensation point**, but it is more common to call it the **boiling point**. For any substance, the change in state from gas to liquid (condensation) occurs at the same temperature as the change in state from liquid to gas (boiling) (Section 1.3). The boiling point of a substance increases with increasing pressure. When the pressure on any gas is increased, the distance between molecules is decreased and less cooling is necessary before intermolecular attractions bind the molecules into a liquid.

The stronger the intermolecular forces that are present, the less cooling or compression is required to cause liquefaction. Examples in Section 12.4 show that polar molecules, which have strong intermolecular forces, condense (or boil) at a higher temperature than nonpolar molecules, which have very weak intermolecular forces. Polar molecules that are capable of forming hydrogen bonds condense (or boil) at still higher temperatures than similarly sized polar molecules that are not capable of forming hydrogen bonds.

sample exercise 3

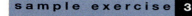

Predict the relative values of the boiling points of the following compounds, based on the intermolecular forces present in each.

a. 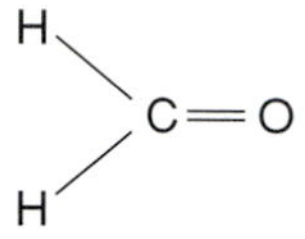

Formaldehyde is widely used in solution as a disinfectant for dwellings, ships, and storage buildings, as a fungicide for plants and vegetables such as oats and wheat, and as an embalming fluid. It is also used to manufacture press board, photographic materials, melamine plastics, and foam insulation.

b. 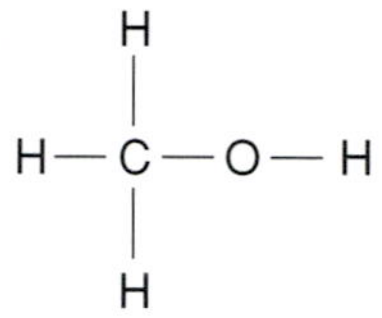

Methanol, also known as methyl alcohol and wood alcohol (because it is obtained from the destructive distillation of wood), is a poisonous alcohol known for causing blindness and even death. It is widely used as a solvent in industry, as a deicer in gasoline, and as antifreeze in windshield washer fluid.

c. H—C≡C—H

Acetylene is a gas that burns in an atmosphere of pure oxygen to produce a large amount of heat. It is commonly used to weld, cut, and solder metals.

solution

All compounds have dispersion forces. Acetylene has only dispersion forces, and we therefore expect it to have the lowest boiling point (–82°C); i.e., it must be cooled the most for the weak attractions to bind the molecules together. The dispersion forces in acetylene and formaldehyde are about equal because of their similar molar masses. However, formaldehyde is polar because of the C═O bond dipole. Dipole-dipole interactions will cause it to liquefy at a higher temperature (–21°C) than acetylene. Methyl alcohol has an O—H bond, which signals the existence of the strongest intermolecular force, the hydrogen bond; its boiling point is 65°C.

• • • problem 4

In each pair, pick the compound with the higher boiling point.

a. NH_3 or NF_3, ammonia or nitrogen trifluoride

b. H_2O or H_2S, water or hydrogen sulfide

c. Cl_2 or Br_2, chlorine or bromine

Just as gas molecules become bound to each other as they are cooled or their pressure is raised, molecules in the liquid state become more rigidly bound into a solid, crystalline state under further cooling or increased pressure. In a liquid, the forces of attraction between molecules are strong enough to hold the molecules together in a definite volume, but not strong enough to hold them together in rigidly fixed positions. Liquid molecules are free to move, but not far enough to become independent of each other. As a result, a liquid conserves its volume, but it may be poured and it flows to assume the shape of its container. Molecules in a solid are so closely held that they are not free to move; they retain their highly ordered arrangement. Solids, in general, do not flow, but retain a fixed shape that resists being compressed into a smaller volume.

table 12.5 observable properties of gases, liquids, and solids and their molecular explanation

State	Property	Molecular explanation
Gases		
	Indefinite volume	Gas molecules are so far apart that forces between the molecules are negligible. There are no forces of attraction to hold the molecules together.
	Indefinite shape	Freely moving molecules travel into all parts of a container regardless of its shape.
	High compressibility	A gas is mostly empty space, so there is nothing to resist squeezing the molecules together more closely.
Liquids		
	Definite volume	Forces of attraction are strong enough to hold liquid molecules together in a definite volume.
	Indefinite shape	Particles in a liquid are free to move (but not independently of each other), so that a liquid flows into the shape of its container.
	Low compressibility	Particles in a liquid lie close together with very little space between them.
Solids		
	Definite volume	Forces of attraction are very strong, holding particles in a definite volume.
	Definite shape	Particles are held in fixed positions by strong attractive forces.
	Very low compressibility	Particles are touching; increasing pressure cannot squeeze particles closer together.

The conversion of a solid to a liquid (melting or fusion) or a liquid to a gas (vaporization or evaporation) requires energy to overcome the attractive forces binding the molecules so that they become farther apart and more freely moving. As you learned in Section 10.10, each of these processes is endothermic and requires the addition of heat energy. Conversely, the freezing of a liquid or the condensation of a gas requires that energy be removed, so that the molecules move slowly enough and are close enough together that the attractive forces between them dominate. You also learned in Section 10.10 that each of these processes is exothermic and releases heat energy. While you learned about the corresponding heat change with a change in physical state in Section 10.10, you should now understand how molecular motion accounts for this behavior.

melting pt. = evaporation pt.

A molecular explanation for the properties of gases, liquids, and solids is summarized in Table 12.5.

12.6 physical properties of liquids

In the previous section, you combined intermolecular forces with the kinetic molecular theory of gases to explain the condensation of gases and the freezing of liquids. Liquids and solids have properties distinctly different from gases, yet they may still be explained by molecular motion and how that motion is modified by the forces between molecules. In this section, specific physical properties of liquids, namely, vapor pressure, boiling point, heat of vaporization, and surface tension, will be discussed in terms of models of molecular motion.

vapor pressure

Liquids left in an open container slowly evaporate or vaporize; i.e., the liquid's molecules escape into the gas phase. A model of liquids can aid in the understanding of this phenomenon. Some molecules near the surface of the liquid have more than

Figure 12.12

When liquid is placed in a closed container, it begins to evaporate (↑). As evaporation proceeds, more gas molecules (shown as dots) accumulate and the pressure increases. Gas molecules also begin to condense (↓). When the rates of evaporation and condensation are equal, the pressure remains constant at a value called the *vapor pressure* of the liquid.

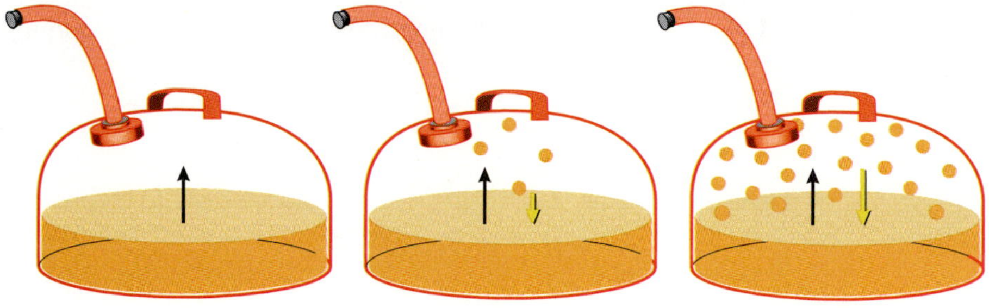

the average amount of kinetic energy, the energy of motion (Section 3.7), which they have gained through collisions with other molecules. These more energetic molecules can overcome the attractive forces of neighboring molecules and escape into the gas phase. Since it is the more energetic molecules that are leaving, the average kinetic energy of the molecules in the liquid decreases, and therefore the temperature of the liquid also decreases. It is for this reason that evaporation is said to be a cooling process.

For example, perspiration evaporates from your skin, absorbing heat from the body. Sweating followed by evaporation is one of the body's cooling mechanisms. Liquids with low intermolecular attractions vaporize more readily at the same temperature than liquids with high intermolecular attractions. For example, ether, a nonpolar laboratory solvent and anesthetic that does not show hydrogen bonding, evaporates much more readily than water, which has the strong hydrogen-bonding intermolecular attractions. However, the evaporation of ether absorbs less heat from the body than does the evaporation of water, because less heat energy is needed to overcome the weaker attractive forces between ether molecules. Alcohol, a substance that evaporates more readily than water, but is polar, has hydrogen bonds, and absorbs more heat from the body than ether, is rubbed on the skin of a patient with a high fever to introduce an additional evaporative cooling mechanism.

When a liquid partially fills a closed container such as gasoline in a can, the molecules that escape into the gas phase (vaporize) find themselves enclosed in a fixed space. As more and more gasoline molecules enter the gas phase, a pressure develops in the can (Figure 12.12). Since the gas-phase molecules cannot escape from the enclosed volume, they eventually strike the liquid surface and a few are "recaptured" (condense) into the liquid phase. After a period of time, the rate of condensation becomes equal to the rate of vaporization. Vaporization and condensation continue, but there is no *net* change in the number of molecules leaving the liquid for the gas phase. The liquid and gas phases are said to be in **dynamic equilibrium,** or balanced. The pressure of the gas phase in equilibrium with its liquid phase is known as the **vapor pressure** of the liquid at a specified temperature.

The value of the vapor pressure depends on the temperature and the nature of the liquid. Nonpolar liquids with low intermolecular forces, such as ethers and gasoline, have high vapor pressures at room temperature. Polar liquids and liquids with hydrogen bonding that have high intermolecular forces, such as water and alcohols, have low vapor pressures at room temperature (Table 12.6).

The vapor pressure does *not* depend on the volume of liquid present. If the surface area of the liquid is increased, the rate of vaporization is initially faster, but there is also a larger liquid surface for gas molecules to strike and be "recaptured," and so the rate of condensation also increases. The point of dynamic equilibrium occurs at the same vapor pressure as with the original surface area.

The volume of the enclosed gas space also does not affect the vapor pressure. A larger enclosure allows more molecules to vaporize before dynamic equilibrium is

table 12.6 vapor pressure of common liquids

Liquid	Polar or nonpolar	Intermolecular forces	Vapor pressure (mmHg) at 25°C
Diethyl ether	Nonpolar	Dispersion	530
Hexane (a component of gasoline)	Nonpolar	Dispersion	154
Isopropyl alcohol (rubbing alcohol)	Polar	Dipole-dipole and hydrogen bonds	43.0
Water	Polar	Dipole-dipole and hydrogen bonds	23.8

established. However, a larger number of gas molecules spread over a larger volume yields the same pressure as a smaller number of gas molecules over a smaller volume.

Temperature, on the other hand, *does* affect the vapor pressure of a liquid. Vapor pressure increases rapidly with increasing temperature (Figure 12.13). According to kinetic molecular theory, energy and temperature are proportional. An increase in temperature gives rise to a corresponding increase in kinetic energy, and hence more molecules near the surface of the liquid are able to overcome the attractive forces of their neighbors and escape into the gas phase. Also at higher temperatures, molecules in the gas phase, having greater kinetic energy, are less likely to be recaptured into the liquid. The result is a greater number of gas molecules constrained within the same volume and hence a higher vapor pressure.

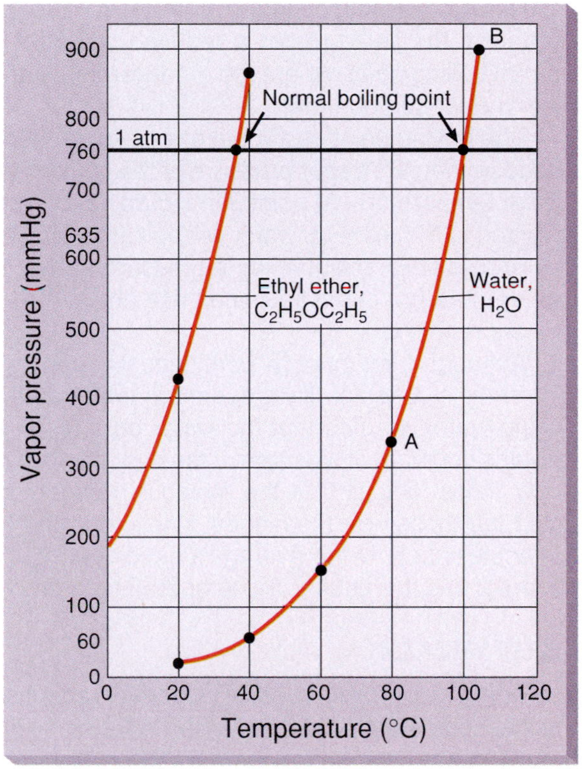

Figure 12.13
The vapor pressure of a liquid increases rapidly with increasing temperature.

Note the danger of raising the temperature of containers with flammable or toxic liquids. The higher vapor pressure at higher temperatures means the vapors may escape more rapidly and threaten safety. Always be sure that stored liquids are well sealed in containers that can withstand an increase in their internal pressure. Improper containers might explode from increased vapor pressure of the liquid inside.

boiling point

A liquid *boils* when its vapor pressure is equal to the external pressure on the surface of the liquid. When this condition exists, we say we are at the boiling point of the liquid. During evaporation, only molecules at the surface escape into the vapor phase, but at the boiling point, some molecules *within* the liquid have sufficient energy to overcome the intermolecular attractive forces of their neighbors, so that bubbles of vapor form within the liquid. The bubbles rise in the liquid, and the vapor is released at the surface. It is the formation of vapor bubbles within the liquid itself that characterizes boiling and distinguishes it from evaporation.

Polar liquids with high intermolecular attractions, such as water, require a relatively high temperature before their vapor pressure equals the external pressure; hence these liquids are found to have a high boiling point. Nonpolar liquids with low intermolecular attractions, such as ethers, have a lower boiling point.

The so-called normal boiling point is the temperature at which the vapor pressure is equal to an external pressure of 1 atm. For example, water boils at 100°C when the external (atmospheric) pressure is 1 atm or 760 mmHg, because the vapor pressure of water is 760 mmHg at 100°C (Figure 12.13).

The boiling point of a liquid can be intentionally reduced by lowering the external pressure (using a water aspirator or vacuum pump). The vapor pressure of the liquid will equal the lower external pressure at a lower temperature, so its boiling point will be reduced. At point A in Figure 12.13, you see that if the external pressure is lowered to 350 mmHg, water will boil at 80°C. One application of this principle is found in the food industry. Water is removed from such substances as coffee by boiling the liquid under a reduced pressure at a temperature lower than the normal boiling point, where the product might decompose (break down into simpler elements or compounds).

The boiling point of a liquid can be increased by raising the external pressure, because then the vapor pressure of the liquid is equal to the external pressure at a higher temperature. At point B in Figure 12.13, you see that if the external pressure is raised to 900 mmHg, water will boil at 103°C. A home pressure cooker works on this principle. By maintaining a pressure above 1 atm inside the pressure cooker, the temperature of the liquid can rise above 100°C, thus allowing the food to cook in a shorter time.

Although heat must be continuously supplied, the temperature of a boiling liquid remains constant. If we add more heat (raise the flame) to an uncovered pot of boiling water, we find that the water boils faster but the temperature of the water remains constant. If we remove the heat, the boiling process slows down and eventually stops. Recall that the molecules escaping from an evaporating or boiling liquid are those with the highest kinetic energy. Therefore, as these higher-energy molecules escape, the average kinetic energy of the remaining liquid molecules is lowered, and this causes the temperature to drop unless heat is added to the liquid.

heat of vaporization

The quantity of heat needed to convert a fixed mass of liquid at a fixed temperature to the gaseous state is known as the **heat of vaporization,** which is designated in handbooks as ΔH_{vap}. You learned about vaporization in Section 1.3 and used the heat of vaporization in Section 10.10 as part of chemical equations that express vaporization (the change from liquid to gaseous state) and condensation (the change from gaseous to liquid state). The SI unit for the heat of vaporization is joules per gram (J/g) or kilojoules per mole (kJ/mol); common units of calories per gram (cal/g) and kilocalories per gram (kcal/g) are still in everyday use. When an amount of heat at least equal to the heat of vaporization is supplied to a boiling liquid, the liquid continues to boil at a constant temperature.

When the heat of vaporization is expressed in kilojoules per mole, it is known as the **molar heat of vaporization.** The molar heat of vaporization gives a measure of the strength of the intermolecular forces in the liquid. The large difference between the molar heat of vaporization for water (40.7 kJ/mol) and that for methane (9.25 kJ/mol) reflects the strong hydrogen-bonding attractions between water molecules in contrast to the weaker dispersion forces between methane molecules.

A greater amount of heat is required to evaporate a mole of water than the same number of molecules of any other common liquid. This property adds to the suitability of water as our body liquid because the evaporation of only a small amount of perspiration leads to significant cooling. The excess heat of an "overheated" person goes into the evaporation process and body temperature is maintained.

When a gas condenses to the liquid state, heat called the **heat of condensation** is given off in an amount exactly equal to the heat of vaporization. The heat of condensation must be removed for the gas to condense to a liquid at a constant temperature. A steam burn is severe because of the large amount of heat that is liberated when the steam vapor condenses to liquid water on your skin. Surgical and laboratory equipment is sterilized in an atmosphere of steam rather than boiling water because steam can be heated to temperatures above 100°C and because large amounts of heat are released when steam condenses. The instrument using pressurized steam for sterilization is called an autoclave.

---

sample exercise 4

How much heat is required to vaporize 2.5 mol of ethyl alcohol at its normal boiling point? The molar heat of vaporization of ethyl alcohol is 39.2 kJ/mol.

solution

The heat needed to convert a liquid to a gas at its boiling point is the heat of vaporization. Since you are given moles of ethyl alcohol, you will use the molar heat of vaporization (39.2 kJ/mol).

$$2.5\,\text{\sout{mol ethyl alcohol}} \times \frac{39.2\text{ kJ}}{1\,\text{\sout{mol ethyl alcohol}}} = 98\text{ kJ}$$

••• problem 5

How much heat is liberated when 22.5 g of steam is condensed to liquid water at 100°C? The molar heat of vaporization of water is 40.7 kJ/mol.

---

surface tension

Unbalanced intermolecular attractions at the surface of a liquid lead to the property of surface tension. All molecules in the interior of a liquid are completely surrounded by other molecules and feel balanced intermolecular attractions. However, molecules at the surface feel an unbalanced force because they are not attracted to other liquid molecules on one side (Figure 12.14). As a result, surface molecules experience a pull into the liquid known as **surface tension.**

Surface tension is especially strong in water, where the hydrogen-bonding attractive force is present between the liquid water molecules. The tension at the surface can balance a steel needle if it is carefully placed on the water and not pushed below the surface. If the needle is pushed below the surface, it will immediately sink because of the greater density of the steel needle compared with water. An insect such as a water strider can walk on water, balanced by surface tension. The strong surface tension in water is also evident by the tendency of water to form droplets. Water beads on the newly polished surface of your

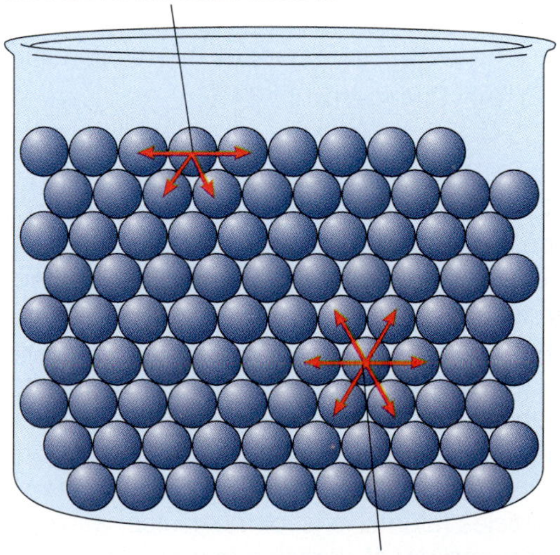
A molecule at the surface is attracted only by molecules below and beside it; there are no molecules above it.
Molecules in the interior of a liquid are attracted equally on all sides by surrounding molecules.

Figure 12.14
Molecules at the surface of a liquid experience an unbalanced attractive force, called surface tension, toward the interior of the liquid.

Figure 12.15
You can test the quality of your car polishing by how well water beads on the surface. The surface tension of water causes surface water molecules to pull together into a droplet.

table 12.7 relationship between intermolecular forces and physical properties of liquids

Type of force	Rate of evaporation	Vapor pressure	Boiling point	Heat of vaporization	Surface tension
Strong intermolecular force	Low	Low	High	High	High
Weak intermolecular force	High	High	Low	Low	Low

car because the primary forces acting on the surface water molecules are the inward attractive forces that pull them together to minimize surface area (Figure 12.15). There are no forces between the water molecules and the polished surface of the car. Water tends to form spherical droplets because a sphere has the lowest surface area for any given volume of water. This spherical shape maximizes the inward attractive forces between the water molecules. When water forms spots on a dirty surface, the surface water molecules may interact with the dirt and distort the spherical shape.

Table 12.7 summarizes the relationship between intermolecular forces and the physical properties of liquids.

12.7 crystalline and amorphous solids

Solids may be classified into two distinct categories based on the arrangement of the particles of which the solid is composed. You know that all solids have a definite volume and shape (Table 12.5). Some solids, known as **crystalline solids**, are made up of a highly regular, repeating pattern of particles, whereas others, known as **amorphous solids**, do not have any regular, repeating pattern in their structure.

You have learned that all crystals of table salt are in the shape of cubes with very smooth surfaces (Figure 6.16). Other materials have crystals of other shapes, but all *crystalline solids* have characteristic angles between their faces (Figure 12.16).

The visible geometry in a crystalline solid arises from the arrangement of the particles (atoms, ions, or molecules) within the solid. The repeated, three-dimensional arrangement of particles in a crystalline solid is known as the **crystal lattice**. The lattice structure gives each material a definite shape that can be seen in the geometry of bulk crystal samples by the characteristic angles between crystal faces. The cubic, crystalline lattice of NaCl is illustrated in Figure 6.16. Other crystal shapes, such as hexagons, reflect their corresponding crystal lattice structures.

Figure 12.16

Crystals of different compounds have different shapes and different angles between their faces, but all crystals of the same compound have a characteristic shape and characteristic angles.

The portion of the crystal lattice that, when repeated in three dimensions, can generate the entire lattice is called the **unit cell.** From a two-dimensional point of view, the unit cell can be thought of as the repeating pattern in a roll of wallpaper. Every crystalline solid has a characteristic crystal lattice and a sharply defined, characteristic melting point.

In contrast to crystalline solids, *amorphous solids* have their particles packed in an irregular manner so they lack any regular overall shape or form. Examples of amorphous solids are glass, plastics like polystyrene, asphalt, tar, and rubber. Amorphous solids can be distinguished from crystalline solids experimentally by the fact that they do not have a sharply defined melting point. When you heat an amorphous solid such as asphalt, it simply softens over a wide range of temperatures, as you have seen on a road, driveway, or parking lot on a hot summer's day.

You learned in Section 6.9 that when a crystalline solid such as NaCl is broken, it retains the same basic geometry of the original sample. The angles between the crystal faces remain the same because of the same underlying lattice structure. Amorphous solids, by contrast, lack an underlying lattice structure, so when they break, their edges are irregular and their surfaces exhibit no reproducible geometry. Consider pieces of broken Styrofoam™, glass, plastic, or rubber.

Crystalline solids are divided into four general classes depending on the particles that occupy the crystal lattice. Ionic crystals are composed of cations and anions, molecular crystals are composed of molecules, covalent crystals are composed of atoms, and metallic crystals are composed of cations surrounded by freely moving valence electrons. Following is a discussion of each class.

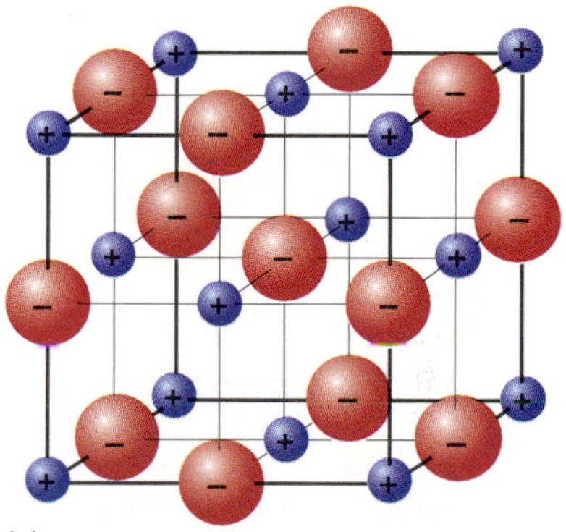

(a)

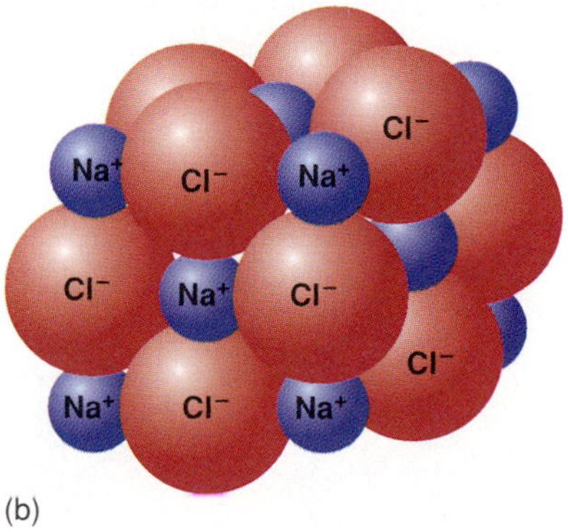

(b)

Figure 12.17

(a) The cubic lattice structure of NaCl depicts the relative sizes of the cations and anions. (b) The ions pack as closely as possible in relation to their respective sizes.

ionic crystals

A crystalline solid in which the lattice is made up of oppositely charged *ions* is said to be an **ionic crystal** (Figure 12.17). All ionic compounds fit into this class (review Section 6.9). Ionic crystals melt at high temperatures because of the very strong electrostatic forces between oppositely charged particles. The strong electrostatic force between each pair of ions is enhanced, as you can see in Figure 12.17, because each Na^+ cation is surrounded by six Cl^- anions and each Cl^- anion is surrounded by six Na^+ cations. As a result, each ion is locked into a rigid arrangement that makes it difficult for the individual ions to slide past one another. Therefore, ionic solids are also very hard and resist deformation.

molecular crystals

Crystalline solids in which *molecules* occupy the lattice points are known as **molecular crystals.** Examples of molecular crystals include ice, in which a H_2O molecule occupies each lattice point, dry ice (solid carbon dioxide), in which a CO_2 molecule is at each lattice point, and solid hydrogen, which has a H_2 molecule at each lattice point. The intermolecular attractive forces in these solids are the dipole-dipole, hydrogen bond, and dispersion forces discussed in Section 12.4. Even the strongest intermolecular forces are considerably weaker than the electrostatic attraction between oppositely charged ions. Because of the weaker attractive forces between molecules, molecular solids possess lower melting points than ionic solids. You have already learned, however, that solids such as ice have higher melting points than hydrogen, because the hydrogen bonding forces present in ice are much stronger than the weak dispersion forces in hydrogen. Because of the weaker attractions between molecules in molecular crystals, molecules may slide past one another. Consequently, molecular crystals tend to be softer than ionic crystals and are more easily deformed.

covalent crystals (covalent network crystals)

Crystalline solids in which the lattice points are occupied by *atoms* covalently bonded to one another are called **covalent crystals** or **covalent network crystals.** Each atom in a covalent crystal is covalently bonded to at least two other atoms, and the bonding between atoms spreads into an extensive three-dimensional network as shown for C atoms in diamond (Figure 4.17). The entire diamond crystal, like other covalent crystals, is actually one large molecule. Because of the strength of covalent bonds holding the atoms together in an extensive network, covalent crystals possess very high melting points and are very hard. Other examples of covalent crystals include quartz (silicon dioxide), shown in Figure 12.18, and Carborundum (silicon carbide), both of which are very hard abrasives and, in ground form, are used in stonecutting and grinding glass.

Figure 12.18

(a) Silicon dioxide, SiO_2, commonly called quartz, is a covalent crystal with every Si atom bonded to four O atoms in a tetrahedral array. (b) These covalent bonds spread through an extensive, three-dimensional network that occupies the entire crystal.

(a) From Uno Kask and J. David Rawn, *General Chemistry.*

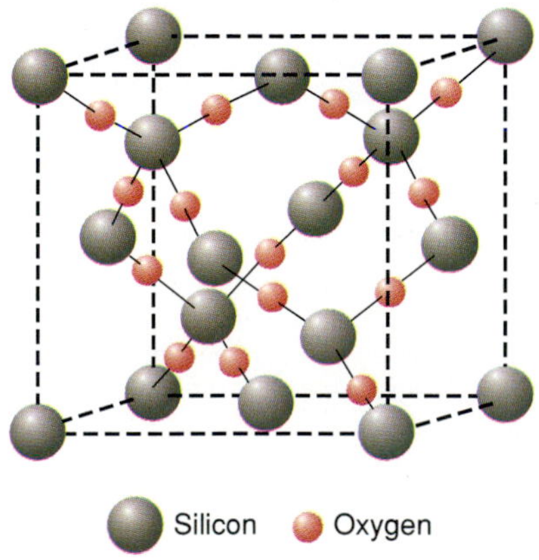

(a)

(b)

metallic crystals

Crystalline solids in which the lattice points are occupied by metallic cations surrounded by a sea of freely moving valence electrons are called **metallic crystals** (Figure 12.19). In these crystals, the valence electrons lost from the metallic atoms are no longer attached to, or remain near, any specific cation; rather, they are "free" to move through the entire lattice. The ability of charged electrons to move freely throughout metallic crystals gives metals their property of excellent electrical conductivity.

The attractive forces in metallic crystals are the attractions between the positive metal cations and the mobile valence electrons. The strength of the attractive forces varies according to the size of the ion and the number of electrons separated from each atom. Mercury's low melting point (–38°C) suggests that it has very weak forces, whereas tungsten (melting point of 3,400°C) clearly has very strong attractive forces. While it is difficult to classify the differences in the melting points of metals in general, a metal such as Li with a smaller cation will have a higher melting point (180.5°C) than Na (97.81°C) or K (63.7°C), which have the same cationic charge as Li, but larger cation radii. Smaller cations will pack into a smaller lattice structure, and the free valence electrons will be closer to the nucleus of each atom and, therefore, more strongly attracted.

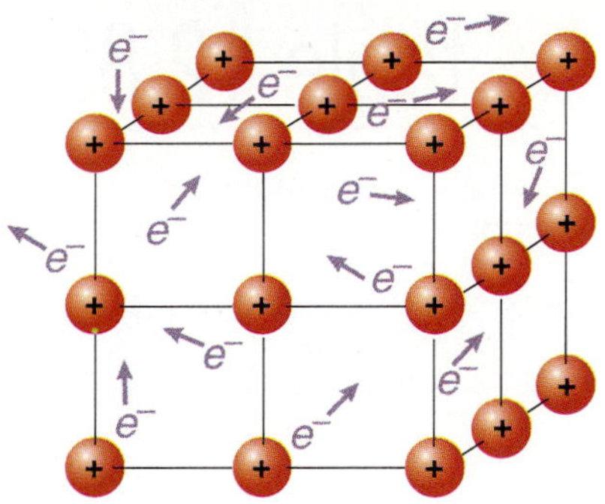

Figure 12.19

In a metallic crystal, metal cations occupy the lattice points. The valence electrons no longer remain near any specific cation, but are free to move throughout the entire lattice. The mobility of the electrons gives metals their property of excellent electrical conductivity.

The properties of pure metals may be modified by adding atoms of other elements to their crystalline lattice structure. A mixture of elements that retains the properties of metals is called an **alloy,** and alloys are extremely important in industrial and engineering applications. The new atoms in an alloy may either replace original metal atoms or may fit into spaces between the original metal atoms, depending on their size. For example, when zinc atoms replace some of the atoms in pure copper, the alloy brass results. When carbon atoms fit into spaces between the atoms of pure iron, the alloy steel results. Since carbon atoms tend to form covalent bonds, the introduction of localized bonding within the metallic iron crystal binds the lattice structure more strongly, so that steel is harder and stronger than iron. By adding elements such as nickel and chromium to iron, its properties are further modified to include resistance to rusting and corrosion. The resulting alloy, stainless steel, is commonly used in household cookware and utensils.

other crystals

A new category of crystalline solids is rapidly being developed by modern research chemists. In Section 4.13, you learned that a new structural form of carbon called fullerene, C_{60}, has only recently been studied in detail. Fullerene contains 60 carbon atoms that are covalently bonded into an extended network, except that the network does not extend indefinitely. Instead, the carbon atom network closes on itself to form a large, cagelike molecule shaped like a hollow soccer ball, with C atoms at each of the 60 points where the seams on a soccer ball intersect. The crystalline solid of C_{60} results from the packing of these soccer-ball-shaped molecules into a lattice structure. The properties of fullerene molecules are between those of a covalent network and a molecular crystal. In addition to fullerene with 60 C atoms, larger fullerene-type molecules, some with hundreds of C atoms, have also been discovered and their properties are being studied.

Another familiar type of crystal that defies classification among the types listed previously is that of "liquid" crystals. **Liquid crystals** are liquid phases that retain some degree of fluidity but also exhibit some of the long-range ordering of crystals. The molecules in liquid crystals are usually characterized by a long, rodlike shape so that the molecules may line up in the same direction as their neighbors, but lack the three-dimensional spatial ordering of true solid crystals.

The alignment of the molecules often depends on the presence of impurity ions, whose charges help to orient the molecules when an electric field is present. When the molecules do line up in an electric field, the arrangement is somewhat similar to a large box of toothpicks, with all of the toothpicks lying in the same

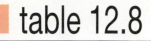

table 12.8 classification of crystalline solids and their properties

Type of crystal	Example	Particles occupying lattice sites	Attractive forces	Melting point	Water solubility	Electrical conductivity
Ionic	Sodium chloride, NaCl	Cations and anions	Ionic bond	High	Often water soluble	Conducts in molten state or in solution
Molecular	Ice, $H_2O(s)$	Molecules	Hydrogen bonds, dipole-dipole and dispersion forces	Low	Polar molecules are water soluble; nonpolar molecules dissolve in organic solvents	Nonconductors in pure state
Covalent network	Diamond, C	Atoms	Covalent bonds	Very high	Not soluble	Usually nonconductors; a few are semiconductors
Metallic	Iron, Fe	Cations	Attraction between the cations and "free" valence electrons	Ranges from low to high	Not soluble	Good conductors in solid or molten state

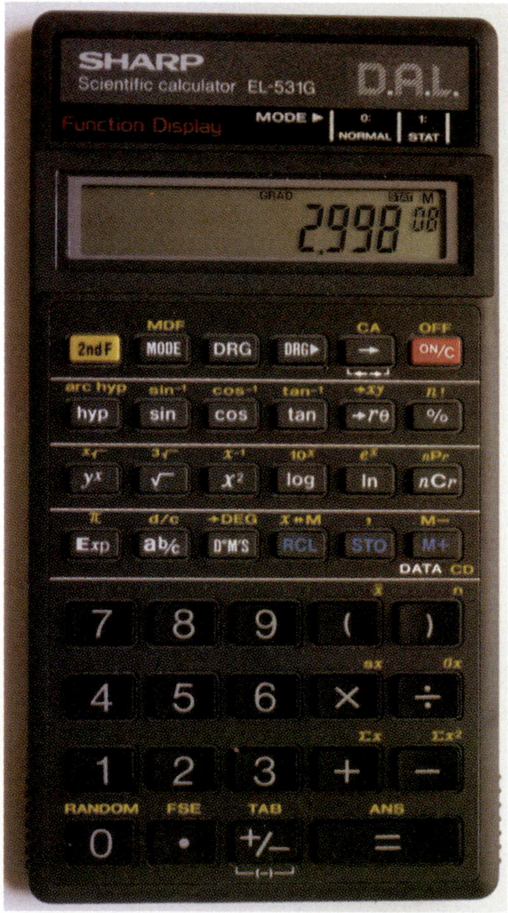

Figure 12.20
By adjusting the orientation of an electric field to liquid crystals, the aligned molecules may be used to form letters and numbers such as those in this scientific calculator.

direction. However, their ends and layers are arranged in a random pattern, not uniform rows. By constantly controlling the orientation of the electric field around the liquid crystals, the aligned molecules may be used to form letters and numbers. While this effect is complicated, it has found widespread application in display devices for electronic calculators and computers (Figure 12.20).

Table 12.8 lists the four basic classifications of crystalline solids with a summary of the properties of each.

12.8 physical properties of solids

In Section 12.7, you combined intermolecular forces with kinetic molecular theory to explain some of the properties of crystalline and amorphous solids. The stronger the intermolecular forces between liquid molecules, the higher the temperature at which that compound will freeze and become a solid; conversely, the higher the temperature the solid will melt (or fuse) to become a liquid. The temperature at which a liquid is converted to a solid is known as the **freezing point** (Section 1.3); it is the same temperature at which the solid melts to become a liquid, which is called the **melting point** or **temperature of fusion**.

In contrast to boiling points (Section 12.5), the melting points of most solids do not change greatly with a change in external pressure.

If you heat a solid that is initially at a temperature below its melting point, the temperature increases until it reaches the melting point. As the solid is heated, the particles in the lattice vibrate more rapidly about their fixed positions, though they remain in their same relative positions as if they were "marching in place." At the melting point, the vibrations become so rapid that the particles begin to move apart and the lattice structure begins to break down. Though the temperature remains constant, heat must be added continuously to the system to continue breaking the lattice down and converting the solid to a liquid. The stronger the intermolecular forces, the more heat must be added to melt any solid (Table 12.8).

The amount of heat needed to convert 1 g of a solid to a liquid at the melting point (temperature of fusion) is called the **heat of fusion** and is designated in handbooks as ΔH_{fus}. You used heats of fusion in Section 10.10 as part of chemical equations that express fusion (the change from solid to liquid state) and freezing (the change from liquid to solid state). When using values for heats of fusion, you should note carefully the unit given them. Sometimes the heat of fusion will represent the heat energy needed to melt 1 g of the compound and will have a unit of joules per gram (J/g); other times, the heat of fusion will represent the heat energy needed to melt 1 mol of the compound and will have a unit of joules per mole (J/mol). The heat needed for the liquefaction of 1 mol of solid is known as the **molar heat of fusion.** The decreasing molar heats of fusion for ionically bonded NaCl (30.1 kJ/mol), hydrogen bonded water (6.02 kJ/mol), and methane (0.92 kJ/mol) with only dispersion forces reflect the lessening strength of intermolecular attraction between the ions or molecules.

During the reverse process of solidification, an amount of heat that is equal in magnitude to the heat of fusion must be removed or liberated. This quantity of heat liberated, which is exactly equivalent to the heat of fusion, may be called the **heat of solidification.** Vineyard owners protect their grapes by using water's high heat of solidification (6.02 kJ/mol). When a nighttime freeze is expected in the fall, they spray a mist of water over the grapevines. Of course, they hope that the temperature will not fall to 0°C, but if it does, the moisture on the surface of the grapes will release heat as it begins to freeze. The heat released may be enough to keep the grapes themselves from freezing. As the water freezes, 6.02 kJ of heat is released per mole of water that freezes, and this may prevent the temperature of the grapes from falling below 0°C.

Although the particles in a solid are more restricted in movement than those in a liquid, some molecules at the surface of a solid may break away directly into the gas phase; consequently, many solids have a measurable vapor pressure and can evaporate directly from the solid to the vapor state without passing through the liquid state. This process is called **sublimation** (Section 1.3). Molecular crystals provide the most likely candidates for sublimation because they generally contain the weakest attractive forces between their particles. You recall from Section 1.3 that mothballs (nonpolar naphthalene molecules), ice (polar, hydrogen-bonded molecules), carbon dioxide (nonpolar molecules), and iodine (nonpolar molecules) were examples of molecular solids that sublime.

12.9 energy requirements to raise temperature and change phase

When a solid is heated, its temperature rises until it reaches the melting point. As you saw in Section 3.8, the heat needed to raise the temperature of any substance from t_1 to t_2 can be calculated as the product of the mass, *m,* the change in temperature, $\Delta t = (t_2 - t_1)$, and the specific heat of the solid, *SH*(*s*).

$$\text{Heat} = m \times \Delta t \times SH(s) \qquad (12.1)$$

Figure 12.21

This graph summarizes how to calculate the amount of heat energy added (released) when the temperature of a substance increases (decreases). Sloped line segments (a, c, and e) represent energy changes in heating a single phase: solid, liquid, or gas. Horizontal line segments (b and d) represent the energy changes required for the substance to change from one physical state to another.

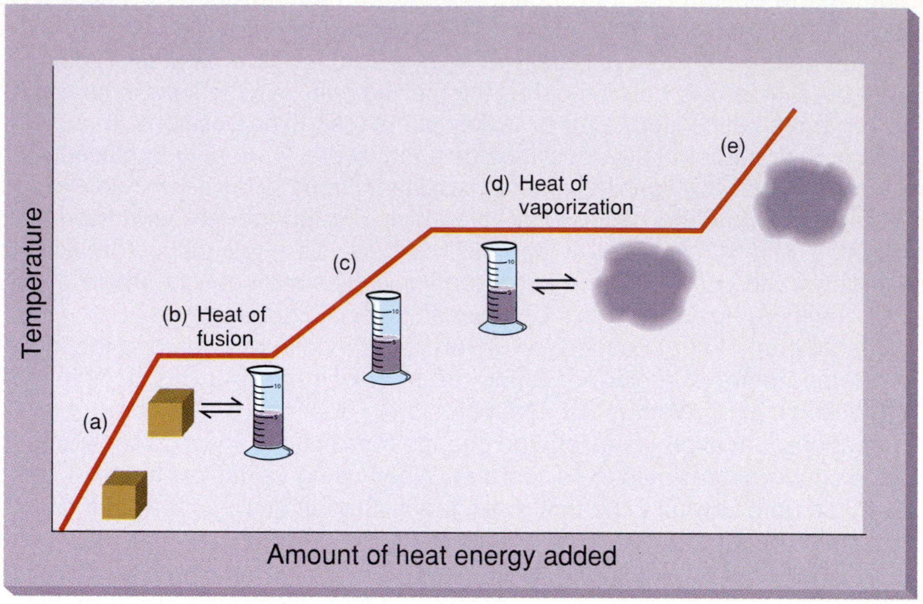

The heat calculated from this formula applies not only to any solid, but also to any liquid or gas as long as the value of the specific heat corresponds to the correct physical state (phase). Once a solid reaches its melting point, its temperature remains constant until the solid is totally converted to a liquid. The heat needed to melt the solid can be calculated as the product of the heat of fusion, ΔH_{fus}, and the mass, *m*, of solid that melts. If the heat of fusion is given in joules per gram, the amount of solid must be in grams; if the heat of fusion is given in joules per mole, the amount of solid must be in moles.

$$\text{Heat} = m \times \Delta H_{fus} \quad (12.2)$$

If heating continues, the temperature of the liquid increases until it reaches the boiling point. Again, the heat needed to raise the temperature of the liquid can be calculated from Equation 12.1 as long as the specific heat has a value corresponding to that of the liquid, $SH(l)$. The temperature then remains constant until the liquid is totally converted to a gas. The heat needed to vaporize the liquid can be calculated as the product of the heat of vaporization, ΔH_{vap}, and the mass, *m*, of liquid that vaporizes. If the heat of vaporization is given in joules per gram, the amount of liquid must be in grams; if the heat of vaporization is given in joules per mole, the amount of liquid must be in moles.

$$\text{Heat} = m \times \Delta H_{vap} \quad (12.3)$$

If the gas continues to be heated, the temperature of the gas increases. Once again, the heat needed to raise the temperature of the gas can be calculated from Equation 12.1 as long as the specific heat has a value corresponding to that of the gas.

Figure 12.21 summarizes the changes in temperature of a pure substance as it is heated, beginning with a solid and continuing to the gaseous state as has just been described. If you are given the specific heat (review Section 3.8) and the heats of fusion and vaporization for a particular substance, you can calculate the amount of heat required for a fixed mass of that substance to undergo any increasing temperature change. Note that cooling a substance follows the same pathway as shown in Figure 12.21, except in reverse. When a substance is cooled, all of the steps liberate or give off heat, rather than absorb it.

sample exercise 5

Given the following values, calculate the amount of heat required to convert 25.0 g of ice initially at –15.0°C to steam at 125°C.

Specific heat of ice = 2.09 J/g°C

Specific heat of water = 4.18 J/g°C

Specific heat of steam = 2.01 J/g°C

ΔH_{fus} = 6.02 kJ/mol = 334 J/g

ΔH_{vap} = 40.7 kJ/mol = 2,260 J/g

solution

1. Calculate the amount of heat required to raise the temperature of ice from –15.0°C to its melting point of 0°C (review Section 3.8).

$$\text{Heat} = \text{Mass} \times \text{Temperature change} \times \text{Specific heat} = m \times \Delta t \times SH(s)$$

$$\text{Heat} = 25.0\,\cancel{\text{g}} \times (0\,\cancel{^\circ\text{C}} - (-15.0\,\cancel{^\circ\text{C}})) \times 2.09\,\frac{\text{J}}{\cancel{\text{g}^\circ\text{C}}} = 784\text{ J} = 0.784\text{ kJ}$$

2. Use the heat of fusion of ice to calculate the amount of heat necessary to melt 25.0 g of ice.

$$\text{Heat} = m \times \Delta H_{fus}$$

$$\text{Heat} = 25.0\,\cancel{\text{g}} \times 334\,\frac{\text{J}}{\cancel{\text{g}}} = 8{,}350\text{ J} = 8.35\text{ kJ}$$

3. In a manner similar to step 1, calculate the heat required to raise the temperature of water from 0°C to its boiling point (100°C).

$$\text{Heat} = m \times \Delta t \times SH(l)$$

$$\text{Heat} = 25.0\,\cancel{\text{g}} \times (100\,\cancel{^\circ\text{C}} - 0\,\cancel{^\circ\text{C}}) \times 4.18\,\frac{\text{J}}{\cancel{\text{g}^\circ\text{C}}} = 10.5 \times 10^3\text{ J} = 10.5\text{ kJ}$$

4. Use the heat of vaporization of water to calculate the amount of heat necessary to vaporize 25.0 g of water.

$$\text{Heat} = m \times \Delta H_{vap}$$

$$\text{Heat} = 25.0\,\cancel{\text{g}} \times 2{,}260\,\frac{\text{J}}{\cancel{\text{g}}} = 56{,}500\text{ J} = 56.5\text{ kJ}$$

5. As in steps 1 and 3, calculate the heat required to heat steam through the indicated temperature change.

$$\text{Heat} = m \times \Delta t \times SH(g)$$

$$\text{Heat} = 25.0\,\cancel{\text{g}} \times (125\,\cancel{^\circ\text{C}} - 100\,\cancel{^\circ\text{C}}) \times 2.01\,\frac{\text{J}}{\cancel{\text{g}^\circ\text{C}}} = 1{,}260\text{ J} = 1.26\text{ kJ}$$

6. The total heat input for steps 1 through 5 is

Step 1	0.784 kJ	Heating ice
Step 2	8.35 kJ	Ice → liquid water
Step 3	10.5 kJ	Heating liquid water
Step 4	56.5 kJ	Liquid water → steam
Step 5	1.26 kJ	Heating steam
Total heat	77.4 kJ	

Notice that the largest portion of the added heat is used in converting liquid water into steam (vaporization).

CHEMLAB

Laboratory exercise 12: heat of fusion of water

The purpose of this experiment is to determine the heat of fusion of water by measuring the amount of heat transferred to melt each gram of ice. The experiment is conducted in a coffee-cup calorimeter, that is, in two Styrofoam™ cups nested and supported inside a beaker (Figure 12.22). A thermometer used to measure the temperature inside the cup should rest in the lip of the beaker. The Styrofoam™ cups provide an insulated container in which heat may be transferred inside the cups, but not to the surroundings.

By placing a measured mass (approximately 100 g) of warm (approximately 35°C) water in the inside cup of the calorimeter and adding a measured mass (approximately 25 g) of crushed ice, the ice will melt and the warm water will cool until a final equilibrium temperature, t_f, is reached. The warmer water transfers heat to the ice, causing it to melt. After all of the ice has melted, the warmer water raises the temperature of the melted ice to its final temperature.

The heat lost by the warmer water may be evaluated according to Equation 12.1, rewritten here as Equation 12.4.

$$\text{Heat lost from warmer water} = m_w \times \Delta t_w \times SH_w \qquad (12.4)$$

where m_w is the mass of the warmer water, Δt_w is the change in temperature of the warmer water from its initial value, t_w (approximately 35°C), to its final value, t_f, and SH_w is the specific heat of liquid water (4.18 J/g°C). All terms in this equation are known. The specific heat of water is given, and the mass of water and its temperature change are experimentally measured.

The heat gained by the ice can be evaluated from two heat terms similar to those in Equations 12.2 and 12.1: the first term is the heat needed to melt the ice and the second term is the heat needed to raise the *melted ice* to its final temperature.

$$\text{Heat gained by ice} = (m_i \times \Delta H_{fus}) + (m_i \times \Delta t_i \times SH_w) \qquad (12.5)$$

where m_i is the mass of the ice, ΔH_{fus} is the heat of fusion of ice in joules per gram, Δt_i is the change in temperature of the ice from its initial temperature of 0°C to its final temperature, t_f, and SH_w is the specific heat of liquid water (4.18 J/g°C). Note that the specific heat of liquid water, not ice, is used in the second term because all of the ice has melted and it is the liquid state that is being heated in the second term. All terms in Equation 12.5 are known except for ΔH_{fus}. The specific heat of water is given, the change in temperature of the ice is t_f, and the mass of the ice was measured. Even the heat gained by the ice is known because it is exactly equal to the heat lost by the warmer water, which was calculated in Equation 12.4. Heat values, however, carry a positive sign to indicate heat absorption and a negative sign to indicate heat release. Consequently, *the heat gained by the ice is exactly equal in magnitude, but opposite in sign, to the heat lost by the water.* Therefore, combining Equations 12.4 and 12.5,

$$\text{Heat gained by ice} = -\text{Heat lost from warmer water} \qquad (12.6)$$

$$(m_i \times \Delta H_{fus}) + (m_i \times \Delta t_i \times SH_w) = -m_w \times \Delta t_w \times SH_w$$

Isolating the heat of fusion of water yields

$$\Delta H_{fus} = \frac{(-m_w \times \Delta t_w \times SH_w) - (m_i \times \Delta t_i \times SH_w)}{m_i}$$

$$\frac{(-m_w \times (t_f - t_i) \times SH_w) - (m_i \times t_f \times SH_w)}{m_i} \qquad (12.7)$$

Because all of the terms on the right side of Equation 12.7 are known, you can evaluate ΔH_{fus} from this experiment having used the heat energy relationships you learned in Section 12.9.

Questions:

1. Find the heat lost when 100.00 g of warm water at 80.0°C is cooled to a final temperature of 48.0°C.
2. Find the heat needed to raise the temperature of 25.00 g of water at 0.0°C to a final temperature, t_f, of 48.0°C.
3. Use Equation 12.7 or combine your calculations above to evaluate the heat of fusion of water. Compare your answer with the value given in Sample Exercise 5.

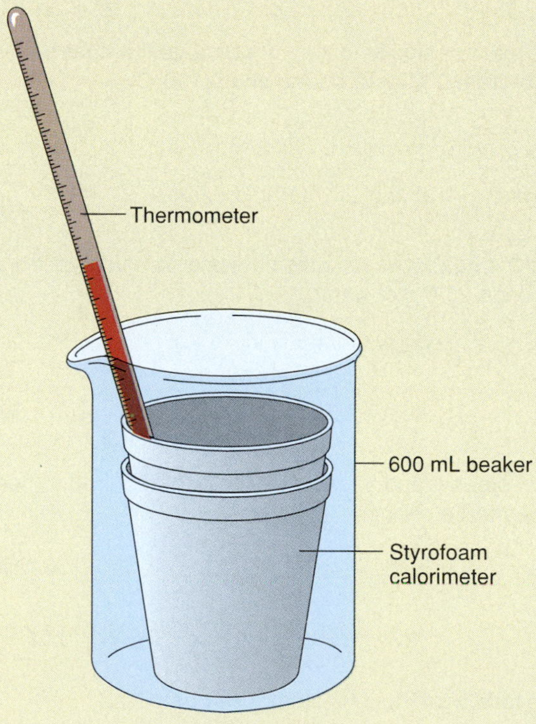

Figure 12.22
A calorimeter consisting of two nested Styrofoam™ cups provides an insulated container in which heat may be transferred inside the cups, but not to the surroundings. A mass of 100 g of warm water is added to melt the 25 g of ice inside the calorimeter. Heat from the warmer water is used to melt the ice and to raise the temperature of the melted ice to its final equilibrium value.

chapter accomplishments

After completing this chapter, you should be able to define all key terms and do the following.

12.1 Introduction

- ❑ Compare the strengths of the intermolecular forces in liquids and solids with the strength of such forces in gases.

12.2 Molecular shape

- ❑ State the essential idea of VSEPR theory.
- ❑ State the geometries of maximum separation for two, three, or four sets of electron pairs.
- ❑ Given the Lewis structure of a molecule, predict the electron-pair geometry.
- ❑ Given the Lewis structure of a molecule, predict the molecular shape.

12.3 Molecular polarity

- ❑ Given a Lewis structure of a molecule and a table of electronegativities, predict whether a molecule will be polar or nonpolar.
- ❑ Describe an experimental test of molecular polarity.

12.4 Intermolecular forces

- ❑ Describe and distinguish among the three intermolecular forces.
- ❑ State the relationship between the strength of the dipole-dipole force and the molecular polarity of a molecule.
- ❑ State the relationship between the strength of dispersion forces and the molecular weight and shape of a molecule.
- ❑ Given the Lewis structure of a molecule, predict the existence of hydrogen bonding, dipole-dipole, and dispersion intermolecular forces.

12.5 Converting gases to liquids and solids

- ❑ State the relationship between the boiling point of a liquid and the strength of the intermolecular forces within the liquid.
- ❑ Explain the observable, distinguishing properties of liquids and solids by using the molecular model of liquids and solids.

12.6 Physical properties of liquids

- ❑ Describe the process by which a constant vapor pressure for a liquid is attained in a closed container.
- ❑ State the relationship between the vapor pressure of a liquid and the strength of the intermolecular forces in the liquid.
- ❑ State the relationship between the temperature of a liquid and its vapor pressure.
- ❑ Describe the relationship between the boiling point of a liquid and the external pressure.
- ❑ Distinguish between evaporation and boiling in a liquid.
- ❑ State the relationship between the boiling point of a liquid and the strength of the intermolecular forces in the liquid.
- ❑ Give the unit in which the heat of vaporization of a liquid is commonly expressed.
- ❑ State the relationship between the molar heat of vaporization and the strength of intermolecular forces within a liquid.
- ❑ Given the heat of vaporization (condensation), calculate the amount of heat needed (released) to vaporize (condense) a given amount of liquid (gas).
- ❑ Explain how the phenomenon of surface tension arises.

12.7 Crystalline and amorphous solids

- ❑ Distinguish between crystalline and amorphous solids.
- ❑ Name the four classes of crystalline solids.
- ❑ Distinguish among the four classes of crystalline solids with respect to the nature of the particles occupying lattice points and the nature of the attractive forces between these particles.
- ❑ State the relationship between the melting point of a solid and the strength of the interparticle attractions within the solid.
- ❑ Distinguish between fullerenes and covalent and molecular crystals.

12.8 Physical properties of solids

- ❑ State the unit in which the heat of fusion is commonly expressed.
- ❑ Given the heat of fusion (solidification), calculate the amount of heat needed (released) to melt (solidify) a given amount of solid (liquid).
- ❑ Give an example of a material that exhibits sublimation.

12.9 Energy requirements to raise temperature and change phase

- ❑ Given appropriate specific heats and the heats of fusion and vaporization, calculate the amount of heat needed for a specified mass of a substance to undergo a given temperature change.

key terms

intermolecular forces
valence shell electron-pair repulsion (VSEPR) theory
molecular shape
bond dipole
molecular dipole
polar molecule
nonpolar molecule
dipole moment
dipole-dipole force
dispersion forces (London forces)
hydrogen bond
capillary action
condensation point
boiling point
dynamic equilibrium
vapor pressure
heat of vaporization
molar heat of vaporization
heat of condensation
surface tension
crystalline solids
amorphous solids
crystal lattice
unit cell
ionic crystal
molecular crystals
covalent (network) crystals
metallic crystals
alloy
liquid crystals
freezing point
melting point (temperature of fusion)
heat of fusion
molar heat of fusion
heat of solidification
sublimation

problems

12.1 Introduction

6. a. In which physical state are the attractive forces between particles the weakest?
   b. Could an ideal gas be condensed to a liquid?

12.2 Molecular shape

7. What is the electron-pair geometry of the following compounds?
   a. CF_4
   b. PH_3
   c. H_2Se
8. What is the molecular shape of each of the compounds in Problem 7?
9. a. Why does moving electron pairs farther apart lower their energy?
   b. In the case of a linear arrangement of electron pairs, what prevents their moving apart to an angle of 200° in a molecular substance?
10. What is the electron-pair geometry and molecular shape of each of the following compounds?
    a. H_2CO
    b. CH_3OH
    c. OF_2

12.3 Molecular polarity

11. a. Write the Lewis structures of CH_2Cl_2 and CCl_4.
    b. Compare the molecular polarities of CH_2Cl_2 and CCl_4.
12. List HF, HCl, HBr, and HI in order of increasing molecular polarity.
13. a. Draw Lewis structures for $BeCl_2$ (assume covalent bonding), H_2S, and H_2O.
    b. Which of the molecules in part a probably has the smallest molecular polarity? Explain.
    c. Which of these molecules has the largest molecular polarity?
14. a. Draw a Lewis structure for NH_3.
    b. Mark the bond dipoles in NH_3.
    c. Mark the molecular dipole in NH_3.
15. Describe how the molecular polarity of a molecule can be measured experimentally.
16. On the basis of molecular shape only, which of the compounds in Problem 8 have a net zero dipole moment?
17. Does any compound in Problem 10 have a zero dipole moment? If so, which one?
18. Identify the polar molecules in the following list.
    a. $CHCl_3$
    b. CH_4
    c. O_2
    d. ClF
    e. 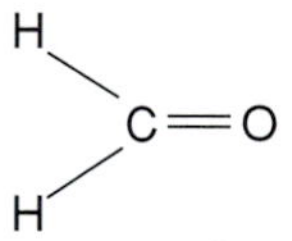
    f. SO_2
    g. CO
    h. PCl_3
19. Which compound do you predict to have the larger dipole moment: NH_3 or PH_3? Explain your answer.

12.4 Intermolecular forces

20. a. Explain the difference between the terms *intermolecular force* and *intramolecular force.*
    b. Give one example of each type of force.
21. a. State the three intermolecular attractive forces and list them in order of increasing strength.
    b. Describe the molecular conditions necessary for the existence of each of these forces in molecules.
22. State the intermolecular attractive forces present in samples of the following substances.
    a. F_2
    b. H_2O
    c. CH_2F_2
    d. HF

23. For each substance in Problem 22, determine which type of attractive force will be *most* significant in determining its physical properties (such as boiling point).

24. Determine the intermolecular attractive forces present in the following substances.
   a. NH_3 b. HCN c. Ne

25. For each substance in Problem 24, determine which type of attractive force will be *most* significant in determining its physical properties (such as boiling point).

26. a. In which types of molecules do we find dispersion forces?
   b. What are the two factors that determine the strength of dispersion forces?

27. Explain why the boiling point of the inert gases decreases with decreasing molar mass.

28. Which of the following molecules are capable of hydrogen bonding between themselves?
   a. $H_3C—O—H$
   b. 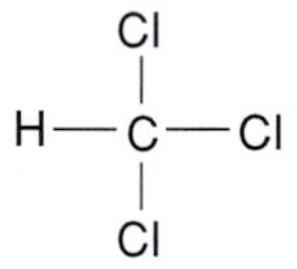
   c. NF_3
   d. H_2S
   e. 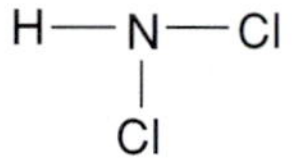
   f. HI

29. Which of these substances have dipole-dipole attractive forces between their molecules?
   a. CF_4
   b. HCl
   c. S═C═S
   d. H—C≡N
   e. BF_3
   f. PF_3
   g. Cl_2
   h. SO_2

30. Compounds *A* and *B* below have the same molecular weight, but the boiling point of *A* (80°C) is considerably higher than *B* (–24°C). Explain.

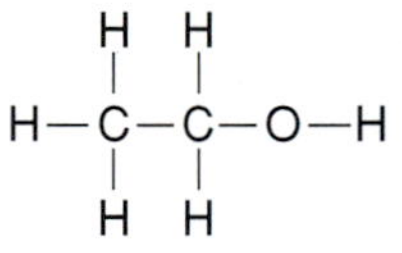
Compound *A*

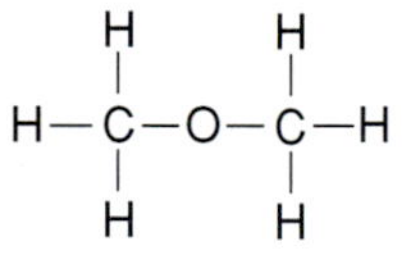
Compound *B*

31. Would you predict C_5H_{12} or $C_{10}H_{22}$ to have the stronger intermolecular forces? Explain.

12.5 Converting gases to liquids and solids

32. Explain how increasing the pressure on a gas allows it to be liquefied at a higher temperature than it could be if the gas were at standard atmospheric pressure.

33. Explain why an increase in intermolecular attractions leads to an increase in the boiling point of a liquid.

34. Predict the order of increasing boiling point for the halogens F_2, Cl_2, Br_2, and I_2. Justify your answer.

35. Arrange the following molecules in order of increasing boiling point.
   a. CH_3F b. CH_4 c. He d. CH_2F_2
   e. 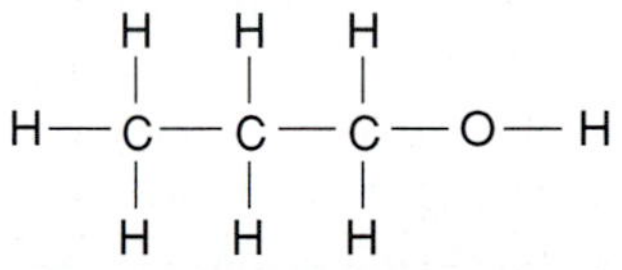

36. Which substance do you predict to have a higher boiling point: H_2O or NaCl? State your reasons.

37. Use the models of liquids and solids developed in Table 12.5 to explain why liquids but not solids can flow.

38. Using the model of a liquid developed in Table 12.5 and the model of a gas developed in Section 11.10, explain why diffusion is much more rapid in a gas than in a liquid.

39. Explain how the models of liquids and solids in Table 12.5 account for the low compressibility of these physical states compared to gases.

12.6 Physical properties of liquids

40. Why do liquids with low intermolecular attractive forces have high vapor pressures?

41. a. Why does the kinetic energy of an isolated liquid decrease as liquid evaporates into the gas phase?
   b. Will the remaining liquid evaporate as fast as the initial liquid?

42. Why is the process by which perspiration evaporates from your skin a cooling process?

43. Indicate which substance in each pair you would expect to have the higher vapor pressure at the same temperature.
   a. CH_4 *or* H_2O
   b. 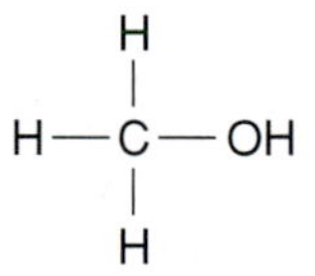 *or* 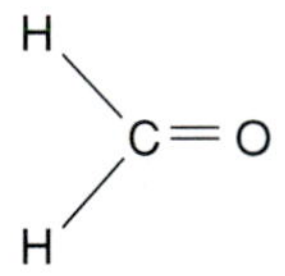
   c. 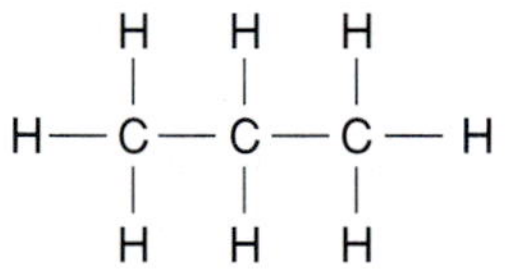 *or* 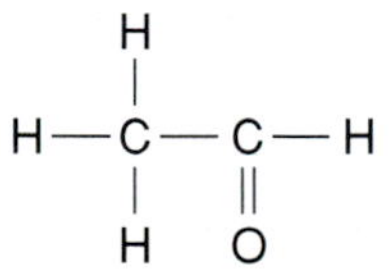

44. a. Explain why increasing the surface area *does* increase the rate of evaporation of a liquid.
   b. Explain why increasing the surface area *does not* increase the vapor pressure of a liquid.

45. Why does increasing the temperature of a liquid increase its vapor pressure?

46. At 20°C, the vapor pressure of substance *X* is 700. mmHg and that of substance *Y* is 335 mmHg.
   a. Which substance has the higher vapor pressure?
   b. Which substance will have the lower normal boiling point?
   c. Which substance has the stronger intermolecular forces?

47. Describe at least two differences between the processes of evaporation and boiling.
48. The vapor pressure of a substance at 25°C is 835 mmHg. If the atmospheric pressure is 758 mmHg, will the substance be a liquid or a gas at 25°C?
49. a. What is meant by the statement that water has a normal boiling point of 100°C?
    b. Can water be made to boil at temperatures other than 100°C? Explain.
50. A newly discovered liquid is very sensitive to heat and, in fact, decomposes at temperatures above 75°C before it can be converted to a gas. Describe how you would distill this liquid without decomposition.
51. a. What name is given to the quantity of heat required to convert a mole of liquid to a gas at its boiling point?
    b. A liquid and a gas can coexist at the boiling point. Which state has the greater amount of energy? Where does the greater amount of energy come from?
    c. Which physical state has the stronger attractive forces at the boiling point?
    d. On a molecular level, explain the molar heat of vaporization.
52. Liquid *A* has a molar heat of vaporization of 72.4 kJ/mol. Liquid *B* has a molar heat of vaporization of 364 kJ/mol.
    a. Which liquid has the larger intermolecular forces?
    b. Which liquid would you predict to have the lower vapor pressure at 25°C?
    c. Which liquid would you predict to have the lower boiling point?
53. The heat of vaporization of water is 2,260 J/g. Calculate how much heat is required to vaporize 31.1 g of H_2O at 100°C.
54. Using the ΔH_{vap} of water from Problem 53, calculate how much heat is liberated when 3.4 mol of steam are condensed to liquid water at 100°C.
55. Predict what effect an increased temperature would have on the surface tension of a liquid.
56. Igniting a warmed dish containing alcohol gives a much more vigorous display of flames than igniting a cold dish. Explain.
57. Cheese boards typically have domed covers. Why should cheese not be left uncovered?
58. At 760 mmHg and 100°C, large bubbles begin to form within a sample of water. What is contained in the bubbles?
59. A typical atmospheric pressure in Mexico City is 580 mmHg.
    a. Use Figure 12.13 to determine the typical boiling point of water in Mexico City.
    b. Will an egg take a longer or shorter period of time to reach the state of hard boiled in Mexico City as compared to at sea level?

12.7 Crystalline and amorphous solids

60. Describe two physical properties that distinguish crystalline solids from amorphous solids.
61. Given an unknown white, powdery solid, how might you determine whether it is crystalline or amorphous?
62. A purplish solid with shiny, flat surfaces is given to you. You determine that its melting point is 33 to 34°C. Is this solid likely to be crystalline or amorphous?
63. Indicate the type of species that occupies the lattice points for each of the following types of crystalline solids.
    a. Ionic b. Molecular c. Covalent d. Metallic
64. a. Give specific examples for each type of crystalline solid.
    b. Identify the particles that would occupy the lattice points in each of your examples.
65. a. Indicate the attractive forces involved in each of the four types of crystalline solids.
    b. Which type of crystalline solid generally has the weakest attractive forces?
66. Ionic solids and metallic solids are made up of charged particles. However, metallic solids conduct electricity, whereas ionic solids do not conduct electricity in the solid state. Explain.
67. Explain why the melting point of sodium chloride (804°C) is so much higher than that of pure hydrogen iodide (–51°C), a compound of higher molecular weight than NaCl.
68. Diamond and graphite are both forms of carbon. Diamond has a covalent crystal structure, whereas graphite has a molecular crystal arrangement. Use this fact to explain why diamond is very hard and graphite is soft.
69. a. Describe how the crystalline structure of fullerene resembles a covalent network.
    b. Describe how the crystalline structure of fullerene resembles a molecular crystal.

12.8 Physical properties of solids

70. a. What name is given to the quantity of heat required to convert a mole of solid to a liquid at its melting point?
    b. A solid and a liquid can coexist at the melting point. Which state has the greater amount of energy? Where does the greater amount of energy come from?
    c. Which physical state has the larger attractive forces at the melting point?
    d. On a molecular basis, what is the function of the molar heat of fusion?
71. The heat of fusion of water is 334 J/g. Calculate the amount of heat liberated by the solidification of 41.1 g of water at 0.0°C.
72. The heat of fusion for a substance is much less than its heat of vaporization. From a knowledge of the strengths of intermolecular forces in each state, explain this statement.

73. The heat required to convert a solid directly to a gas is called the heat of sublimation. Would you predict the heat of sublimation to be larger or smaller than the heat of vaporization? Why?
74. When we detect the odor of a solid, what must we actually be smelling?

12.9 Energy requirements to raise temperature and change phase

75. How many joules are required to convert 46.0 g of solid ethyl alcohol at –153°C to vapor at 79°C? The melting point of the alcohol is –117°C and the boiling point is 79°C. The specific heat of the liquid alcohol is 2.24 J/g°C and the specific heat of the solid alcohol is 0.971 J/g°C. ΔH_{fus} = 104 J/g and ΔH_{vap} = 854 J/g.
76. Use the data in Sample Exercise 5 to calculate the number of kilojoules of heat required to convert 1.50 mol of ice initially at –6.0°C to vapor at 110.°C.
77. A sponge bath using isopropyl alcohol (rubbing alcohol) (molar mass = 60.1 g) can be used to lower the temperature of a person suffering from a high fever. Evaporation of the alcohol consumes heat that comes from the patient. The molar heat of vaporization for isopropyl alcohol is 42.3 kJ/mol (assume that it is the same at room temperature as at the boiling point). How much heat is removed in the evaporation of 150. g of the alcohol?
78. At room temperature (25°C), oxygen is a gas, water is a liquid, and sodium chloride (table salt) is a solid. What do these observations imply about the relative strengths of the intermolecular forces in these substances.
79. Xenon gas condenses at 166 K and its molar heat of vaporization, ΔH_{vap}, is 12.6 kJ/mol. Calculate the quantity of heat evolved when 1.00 kg of Xe condenses.
80. The heat of vaporization of the refrigerant Freon-12, CCl_2F_2, is 155 J/g. Determine the amount of Freon-12 that must be evaporated to freeze a tray of 16 1 oz ice cubes with the water initially at 20.°C (1 oz = 28 g).

Additional problems

81. Use VSEPR theory to predict the molecular shape of each of the following.
    a. AsH_3
    b. OCl_2
    c. H_3O^+
    d. $AlCl_3$
    e. $AlCl_4^-$
82. Explain how it is possible for a molecule to have bond dipoles and yet not have a molecular dipole.
83. Would a molecule such as OCS (C is the central atom) have a higher or lower molecular dipole than CO_2? Explain.
84. Arrange the following substances in order of increasing boiling point: He, C_3H_8, Ne, and C_2H_5OH.
85. The atmospheric pressure on the top of a mountain is 400. mmHg. What would be the boiling point of water on this mountain top?
86. People who live in the colder climates of the United States usually drain the water from unheated plumbing. Explain why this is necessary.
87. Although their molar masses are very similar, the boiling point of ethyl alcohol, C_2H_5OH (79°C), is much higher than that of propane, C_2H_6 (–42°C). Provide a reasonable explanation.
88. N—H, O—H, and F—H bonds can engage in hydrogen bonding while C—H bonds do not display such interactions. Provide a reasonable explanation.
89. Classify each of the following crystalline solids as an ionic crystal, molecular crystal, covalent crystal, or metallic crystal.
    a. Mg
    b. CO_2
    c. SiO_2
    d. Na
    e. NaCl
    f. H_2O
90. A 25 mL sample of a volatile liquid is placed in a 0.5 L container, and another 25 mL sample of the same liquid is placed in a 0.25 L container.
    a. Which container will develop the higher vapor pressure from the volatile liquid?
    b. How would your answer in part a change if 50. mL of the liquid were placed in the 0.5 L container?
    c. Would your answer in part a be different if 50. mL were placed in the 0.25 L container?
91. Explain why a burn from steam at 100.°C can cause greater harm than a burn from water at 100.°C.
92. Calculate the amount of heat needed to convert 66.0 g of ice at –11°C to steam at 111°C.
93. Drinking water does not add or consume any energy from the body; however, obtaining water from the ingestion of snow or ice would cause the body to expend energy. Explain.

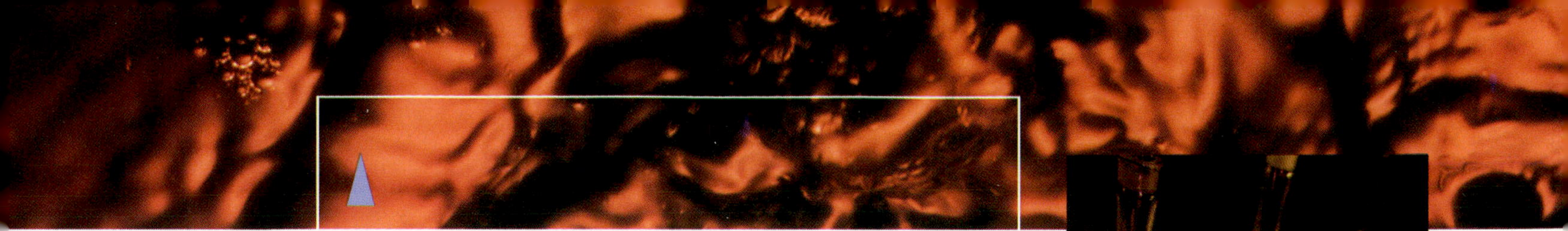

chapter 13

solutions

homogeneous

13.1 introduction

In the previous two chapters, you have studied the properties and molecular behavior of *pure substances* in each of their physical states: solid, liquid, and gas. *Mixtures* of pure substances are called solutions when they combine pure substances homogeneously, or uniformly throughout. A **solution** (**homogeneous mixture**) is a combination of two or more pure substances whose proportions may vary in different mixtures (Section 1.5), but whose composition is constant throughout any given mixture, no matter how finely the mixture is divided. For a solution to be truly homogeneous, the particles of the intermixed substances must be ionic or molecular in size (Section 13.3).

In describing solutions, you will find it useful to identify and name the substance dissolved and the substance doing the dissolving. *The substance dissolved* is called the **solute**. The *substance doing the dissolving* is the **solvent**. In most solutions, the solute is the component that is present in lesser amount and the solvent is the component present in greater amount, so you may envision the solute molecules (or ions) surrounded by the solvent molecules (Figure 13.1). The number of solvent molecules surrounding each solute molecule (or ion) will be the same throughout the solution.

$$\textbf{Solute + Solvent = Solution} \qquad (13.1)$$

Because there are three physical states of matter, and each state, in principle, can be either the solute or solvent in a solution, there are conceivably nine different types of solutions. Table 13.1 gives a common example of each type when it actually exists. In discussing mixtures of gases (Section 11.3), you have already encountered gaseous solutions such as *air,* in which oxygen gas (solute) is dissolved in nitrogen gas (solvent). You have also encountered solid solutions such as the alloy *brass,* in which the metal zinc (solute)

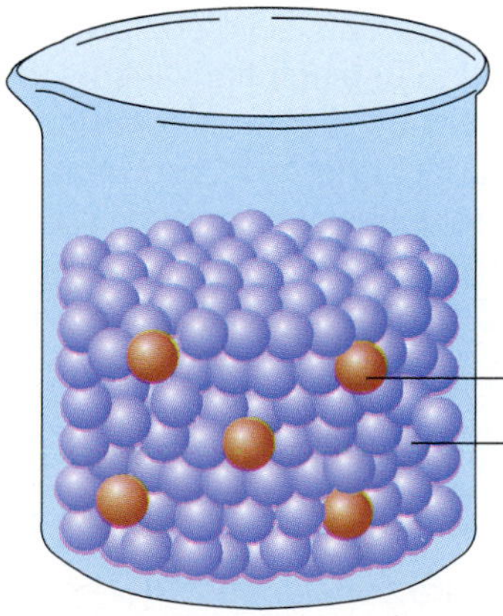

Figure 13.1
A molecular view of solute and solvent molecules in a liquid solution. The composition of any given solution is uniform; the number of solvent molecules surrounding each solute molecule (or ion) does not change throughout the mixture.

Preparing solutions with accurately known concentrations and analyzing solutions for the concentrations of their solute are routine tasks for a chemist. A colorimeter is an instrument that passes a beam of light through a given sample of colored solution. The more light that is absorbed, that is, the less light that passes through the solution, the higher the concentration of solute. A colorimeter is a fast, convenient method for measuring the concentrations of unknown solutions.

In most solutions

solvent = greater
solute = lesser

★ # solvent mol. surrounding solute mol. will be the same throughout.

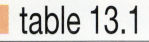

table 13.1 examples of the nine solution types based on the physical states of solute and solvent

Solute	Solvent		
	Gas	**Liquid**	**Solid**
Gas	$O_2(g)$ in $N_2(g)$ (air)	$CO_2(g)$ in $H_2O(l)$ (soda)	$H_2(g)$ in Pd(*s*) (hydrogenation catalyst)*
Liquid	No common example	Alcohol(*l*) in $H_2O(l)$ (martini)	Hg(*l*) in Ag(*s*) (dental fillings or amalgams)
Solid	No common example	NaCl(*s*) in $H_2O(l)$ (salt water or saline solution)	Zn(*s*) in Cu(*s*) (brass)

*This catalyst is used in hydrogenation reactions in organic chemistry.

alloy → brass
metal zinc (solute) is dissolved in metal copper (solvent)

is dissolved in the metal copper (solvent) (Section 12.7). An amalgam is a special alloy containing the liquid element mercury and is an example of a solution in which the liquid metal, Hg (solute), is dissolved in a solid metal (solvent). *Dental fillings* are amalgams of mercury with silver.

While there are no common examples of a solution with either a solid or a liquid dissolved in a gas, several examples are often questioned as possibilities. Students often suggest fog as an example of a liquid dissolved in a gas, because the liquid, as tiny droplets of water, appears to be dissolved uniformly throughout the air. Dust or sooty pollution is often suggested as an example of a solid dissolved in a gas, because the finely powdered earth or soot appears to be dissolved uniformly throughout the air. Both examples are not *truly* solutions, however, because the size of the solute particles is larger than that of individual molecules. Water droplets, no matter how small, contain many water molecules, so they are not truly dissolved in air when they form fog. Dust or soot particles, no matter how small, contain many molecules, so they, too, are not truly dissolved in air. Both fog and polluted air would be classified as **heterogeneous mixtures**, or mixtures that are neither uniform nor homogeneous throughout. No matter how fine the water droplets or soot particles, some part of the fog will contain more water droplets than others and some part of polluted air will contain more soot particles than others.

While solutions may be homogeneous mixtures of substances in any physical state, the most common solutions are those in which the solvent is a liquid. Solutions formed with a pure solid, liquid, or gas dissolved in a liquid such as water are used in your everyday life as well as in the chemical laboratory. *Soda* is a common example of carbon dioxide gas dissolved in water, a *martini* is liquid alcohol dissolved in water, and *salt water* is solid sodium chloride dissolved in water.

You will be required to use solutions with solid or liquid solutes and liquid solvents. Of those, water will be the most common solvent by far. By the end of this chapter, you should understand how to prepare such solutions in the laboratory and how to use these solutions as reagents in chemical reactions. Since solutions may be prepared with varying composition, you must learn how to describe solutions with different proportions of solute and solvent and understand the properties used to describe them.

13.2 properties of solutions with liquid solvents

The general classifications of solutions as mixtures implies the following two statements.

1. *The proportion of pure substances in a solution is variable.* When you add antifreeze such as ethylene glycol or methanol (solute) to the water (solvent) in your car radiator, you may add a small amount or a large amount and still have a solution, that is, a mixture that is homogeneous throughout.
2. *The solute and solvent may be separated by physical means.* You add methanol antifreeze to your radiator during cold weather. After driving your car for several weeks, the methanol component, which has a higher vapor

pressure (Section 12.6) than water, evaporates more readily from the antifreeze solution. Through evaporation, the methanol has been separated from the solvent, water, by physical means (Section 1.5).

The classification of solutions as homogeneous mixtures implies the following three statements.

1. *Solute particles are uniformly distributed among solvent particles throughout a solution.* No matter where you take a sample from the antifreeze solution in your radiator, the proportion of water and ethylene glycol (or water and methanol) will be the same.
2. *Solute particles will not settle out no matter how long a solution is left undisturbed.* Because the solute and solvent particles are intermixed at the molecular level, no particle is significantly heavier than any other. Gravity does not act to separate them. By contrast, soot or dust in the air does settle out because the soot or dust particles, no matter how finely divided, are much more massive than the air molecules surrounding them.
3. *A solution has the same chemical and physical properties in every part.* Because the particles in a solution are intermixed at the molecular level and are uniformly distributed, the intermolecular forces between solute and solvent molecules will be the same throughout the solution. As a result, those physical properties that are related to intermolecular forces, such as polarity, melting point, boiling point, electrical conductivity, vapor pressure, heat of fusion, and heat of vaporization, will be the same for all samples of the solution. All other physical and chemical properties will also be the same throughout the solution. For example, the uniform distribution of solvent and solute molecules will result in a constant mass for any given volume of solution. Consequently, the density of a homogeneous solution ($d = m/V$) will be constant.

particles in a solution are intermixed at the molecular level, and are uniformly distributed, so the intermolecular forces between solute and solvent mol.'s will be the same throughout.

A physical property that helps to identify a liquid solution from other mixtures is its transparency. *A true liquid solution, even though it may be colored, is transparent.* It transmits light so that objects may be viewed distinctly through it. Cola, white wine, and filtered apple juice are examples of colored solutions that are transparent (Figure 13.2).

Figure 13.2
Some commonly encountered solutions whose solvent is water.

table 13.2 solubility of some common ionic substances at two temperatures

		Solubility, $\frac{\text{Maximum grams of solute}}{100 \text{ g } H_2O}$	
Solute	**Name**	**20°C**	**60°C**
NaCl	Sodium chloride	36.0	37.3
KBr	Potassium bromide	65.2	85.5
KNO_3	Potassium nitrate	31.6	110.0
$AgC_2H_3O_2$	Silver acetate	1.04	1.89
$K_2Cr_2O_7$	Potassium dichromate	13.1	50.5
$KMnO_4$	Potassium permanganate	6.4	22.2
$AgNO_3$	Silver nitrate	222	525
$BaSO_4$	Barium sulfate	0.00023	0.00036

Some of the most useful applications of solutions result from how their properties change as the amount of solute varies. *The physical and chemical properties of all solutions change as the proportion of solute and solvent is varied.* For example, as you add antifreeze to your car radiator, the freezing point of the water in your radiator is lowered. The more antifreeze you add, the lower the freezing point of the resulting solution. You adjust the freezing point of the radiator solution to be lower than the lowest expected outside temperature. Since water expands upon freezing (Section 12.4, hydrogen bonds), if allowed to freeze in your engine block, it would crack the block and destroy your engine.

Lowering the freezing point by adding antifreeze is one example of a solution property that changes as the proportion of solute is increased. The proportion of solute in any solution may be expressed in numerous ways. *Solubility* of the solute tells you the *maximum* amount of solute that can dissolve in a given amount of solvent. The *concentration* of the solute tells you the *actual* quantity of solute dissolved in a given amount of solvent. The next several subsections will explain these terms and describe different ways in which concentration may be expressed.

solubility

Solubility is a measure of how much solute can dissolve in a given amount of solvent at a given temperature. Its value may be found from the ratio

$$\textbf{Solubility} = \frac{\textbf{Maximum amount of solute}}{\textbf{Amount of solvent}}$$

and usually has the unit grams of solute per 100. g of solvent (Table 13.2). For example, 36.0 g of sodium chloride, NaCl, will dissolve in 100. g of water at 20°C, while 222 g of silver nitrate, $AgNO_3$, and 2.3×10^{-4} g of barium sulfate, $BaSO_4$, will dissolve in the same mass of water at the same temperature. Qualitatively, solutes are **soluble** when a clearly observable amount of solute dissolves or **insoluble** when no observable amount of solute dissolves. Modifiers such as *slightly* or *very* soluble loosely describe smaller or larger amounts of solute dissolving in a given amount of solvent. NaCl is clearly soluble in water. In comparison, $AgNO_3$ is very soluble and $BaSO_4$ is insoluble.

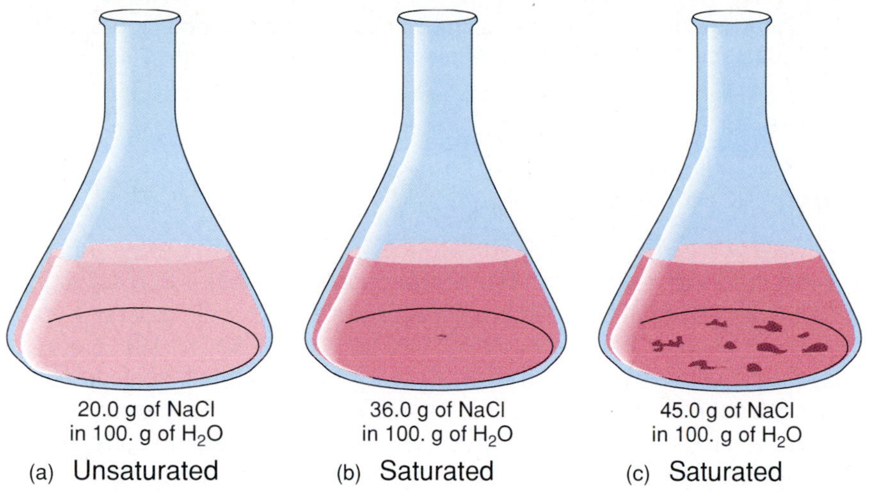

Figure 13.3
Unsaturated and saturated solutions of NaCl at 20°C. (a) 20.0 g is less than the maximum amount of NaCl that can dissolve in 100. g of H_2O. (b) 36.0 g is exactly the solubility limit—a small speck of undissolved solid in equilibrium with the dissolved solute may or may not be visible. (c) 36.0 g of NaCl is in solution; the extra 9.0 g of NaCl appears as undissolved solid on the bottom of the flask.

saturated solution = dynamic equilibrium (solute is dissolving and precipitating)

saturated solutions

A **saturated solution** is one in which no more solute can dissolve in a given amount of solvent at a given temperature. That is, a saturated solution has dissolved in it an amount of solute equal to its solubility at a specified temperature. Table 13.2 lists the solubilities of some common ionic compounds at two temperatures. To prepare a saturated solution of NaCl, you dissolve at least 36.0 g of NaCl in 100. g of water at 20°C. At 20°C, 36.0 g is the maximum amount of NaCl that will dissolve; any excess NaCl will settle to the bottom of the container (Figure 13.3b and c). If the amount of solute dissolved is *less than* the solubility limit, then the solution is **unsaturated** (Figure 13.3a).

Although a saturated solution containing excess solid appears to your eye to be static (exhibiting no activity), the solute particles are actually involved in a *dynamic* (active) *equilibrium* process. The solid solute is dissolving (going into solution), and at the same time, some solute particles are precipitating (coming out of solution) as pure solid. At equilibrium, these two opposing processes are occurring at equal rates, just as you saw for the rates of evaporation and condensation of a liquid in Section 12.6. In the latter case, the liquid and the vapor of a pure substance were the materials in dynamic equilibrium. Here, the equilibrium is between dissolved and undissolved solute. For a solution in which water is the solvent, this equilibrium process may be represented by the equation

$$\textbf{Undissolved solute } (s) \rightleftharpoons \textbf{Dissolved solute } (aq)$$

where (s) indicates a solid and (aq) indicates an *aqueous* or water solution. The two arrows ($\rightleftharpoons$) indicate that the process is reversible and can proceed in either direction between undissolved and dissolved solute. You can usually see any undissolved solute in a saturated solution, except at the solubility limit (for NaCl, this is 36.0 g of NaCl in 100. g of H_2O at 20°C), when the amount of undissolved solute may be too small to be seen (Figure 13.3b).

In Problem 1, the *actual* amount of solute added to 100. g of H_2O has been given for several different solutions. For these solutions, the actual amount of solute in 100. g of H_2O and the solubility of each solute (Table 13.2) have the same units, so they may be compared directly. If the actual amount of solute is less than the solubility, the solution is *unsaturated* because more solute could dissolve. If the actual amount of solute is greater than or equal to the solubility, the solution is *saturated* because the maximum amount of solute is already dissolved.

• • • problem 1

In each of the following solutions, the actual amount of solute added to 100. g of H_2O is given. Consult Table 13.2 and decide whether the following solutions are saturated or unsaturated.

a. 15 g of KBr in 100. g of H_2O at 20°C
b. 15 g of $KMnO_4$ in 100. g of H_2O at 60°C
c. 115 g of $AgNO_3$ in 100. g of H_2O at 20°C

In most laboratory situations, the actual quantity of solute in a solution is not added to exactly 100. g of H_2O. If 25 g of NaCl were added to 50. g of H_2O, you know that the actual amount of solute would have to be 50. g of NaCl if 100. g of H_2O had been used. In your mind you established a ratio. If twice as much solvent were used, twice as much solute must also be used. Similarly, for any arbitrary amounts of solute and solvent, you must establish a ratio so that the amount of solute corresponding to 100. g of H_2O may be found. This procedure is illustrated in Sample Exercise 1.

sample exercise 1

Use Table 13.2 to decide whether the following solutions are saturated or unsaturated.

a. 19 g of NaCl in 50. g of H_2O at 20°C
b. 135 g of KNO_3 in 150. g of H_2O at 60°C
c. 1.5 g of $KMnO_4$ in 20. g of H_2O at 20°C

solution

Because the given solutions use varying masses of H_2O, you must first establish a ratio to find the actual number of grams of solute per 100. g of H_2O in each of the solutions described. This can be done by setting up the *given ratio* (given grams of solute to given grams of H_2O) and setting it equal to the *desired ratio* (x grams of solute to 100. g of H_2O). Then solve the equation for x, the number of grams of solute in 100. g of H_2O.

a. **Given** **Desired**

$$\frac{19 \text{ g NaCl}}{50. \text{ g } H_2O} = \frac{x \text{ g NaCl}}{100. \text{ g } H_2O}$$

$$50x = 1{,}900$$

$$x = 38 \text{ g}$$

At 20°C, 38 g of NaCl per 100. g of H_2O would be a *saturated solution* because 38 is greater than 36, the solubility of NaCl.

b. **Given** **Desired**

$$\frac{135 \text{ g } KNO_3}{150. \text{ g } H_2O} = \frac{x \text{ g } KNO_3}{100. \text{ g } H_2O}$$

$$150.\,x = 13{,}500$$

$$x = 90. \text{ g}$$

At 60°C, 90. g of KNO_3 per 100. g of H_2O is an unsaturated solution because it is less than the solubility of KNO_3, which is 110. g per 100. g of H_2O.

c. **Given** **Desired**

$$\frac{1.5 \text{ g } KMnO_4}{20. \text{ g } H_2O} = \frac{x \text{ g } KMnO_4}{100. \text{ g } H_2O}$$

$$20.\,x = 150$$

$$x = 7.5 \text{ g}$$

At 20°C, 7.5 g of $KMnO_4$ per 100. g of H_2O is a *saturated solution* because it is greater than the solubility of $KMnO_4$, which is 6.4 g per 100. g of H_2O.

dilute and concentrated solutions

The terms *dilute* and *concentrated* are used to classify unsaturated solutions with different proportions of the same solute and solvent. A **dilute solution** has a comparatively small amount of solute; a **concentrated solution,** a relatively large amount. The terms are most useful in comparing two solutions of the *same* solute and solvent. For example, 1 or 2 g of NaCl in 100. g of water at 20°C would be a dilute solution, whereas, 30. g of NaCl in 100. g of water is concentrated because 30. g is approaching the solubility limit (36 g of NaCl per 100. g of water) at that temperature. It is not very useful to use the terms *dilute* and *concentrated* in comparing solutions that contain different solutes or solvents because different solutes have different solubility limits.

problem 2

In comparison with a solution in which 3.0 g of KBr are dissolved in 5.0 g of H_2O, is the solution in part a of Problem 1 dilute or concentrated?

supersaturated solutions

Under carefully controlled conditions, you can prepare a solution that contains a dissolved amount of solute that exceeds the normal solubility limit. Such a solution is described as **supersaturated.** To prepare and maintain a supersaturated solution, you must protect the solution from disturbances. For example, the normal solubility limit of sodium acetate is 46.5 g per 100. g of H_2O at 20°C. If you mix 75 g of sodium acetate and 100. g of H_2O and heat the mixture to 50°C, the solid will dissolve (solubility at 50°C is 83 g of sodium acetate per 100. g of H_2O). If you then slowly and *carefully* cool the solution to 20°C so that no dust particles enter the solution and no shock waves transverse the solution, the 75 g of solute will remain in solution and the solution will be supersaturated. However, if a small "seed crystal" of sodium acetate is added to the supersaturated solution, or if the container is shaken, the solute in excess of the normal solubility limit will suddenly precipitate out of solution. The result will be a saturated solution with the excess solute appearing as a solid precipitate.

A supersaturated solution of sodium acetate is used commercially to produce Magic Heat™ instant hand warmers. A small, sealed plastic pouch containing sodium acetate and water may be heated in boiling water or in a microwave oven to form a sodium acetate solution within the pouch. When the pouch is cooled to room temperature, a supersaturated solution is formed, which may be stored as long as the pouch remains undisturbed. In the commercial product, a small metal disk has been sealed within the pouch. When flexed rapidly, the disk sets up a shock wave within the supersaturated solution and the excess solute suddenly precipitates. Since the precipitation of sodium acetate is an exothermic process (Section 10.10), released heat will warm the pouch as well as your hands (Figure 13.4). The pouch is convenient for use in cold weather and has the advantage of being reusable.

miscible and immiscible solutions

The terms *miscible* and *immiscible* are used in describing mixtures of liquids with liquids. **Miscible liquids** dissolve in each other in all proportions. For example, alcohol and water are totally miscible and form a solution whether the water or the alcohol is the solvent (present in greater amount). In contrast, oil and water are totally **immiscible;** that is, they are completely insoluble in one another regardless of the proportions involved. The miscibility of two liquids depends directly on the similarity of their intermolecular forces (Section 12.4). Molecules such as water and alcohol are both polar and both have hydrogen bonds. They mix completely because the forces between the solute and solvent molecules are very similar to the forces between the solute and solute molecules or between solvent and solvent molecules.

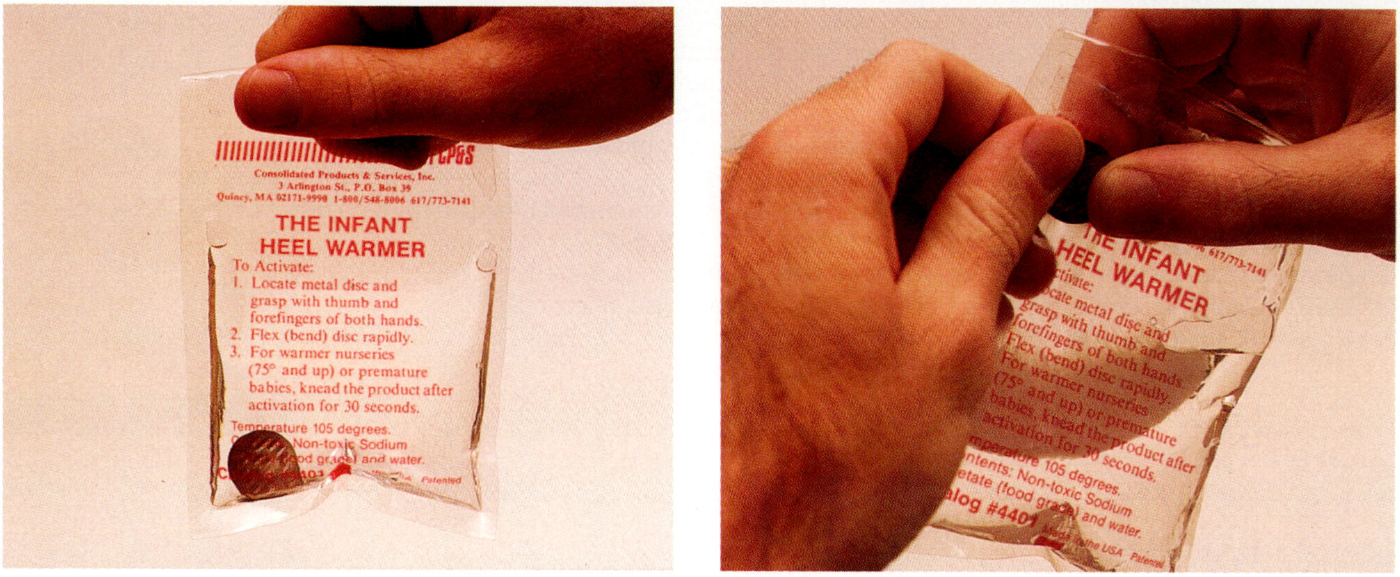

(a) (b)

Figure 13.4

(a) A supersaturated solution of sodium acetate is formed by heating the pouch in boiling water or a microwave oven and cooling it to room temperature. The supersaturated solution may then be stored at room temperature. (b) Flexing a small metal disk within the pouch sets up a shock wave to disturb the supersaturated solution and precipitate the excess solute. Heat released as the precipitate forms may be used to warm your hands in cold weather.

13.3 dissolving ionic solutes in water

It will help you to visualize how a typical ionic compound like NaCl dissolves in water by remembering several ideas that you have already encountered.

1. Ionic compounds are constructed of cations and anions that are held together in orderly arrays in the solid state. Each cation is surrounded by anions and each anion is surrounded by cations. All ions are held rigidly in a lattice structure with characteristic angles and distances between the layers of ions.
2. The attraction between cations and anions in the solid state is a very stable (low-energy) condition.
3. Water molecules are freely moving in the liquid state.
4. Water molecules are polar, with the oxygen atom bearing a partial negative charge and the hydrogen atoms each bearing a partial positive charge.

Figure 13.5a represents solid NaCl just at the moment it has been added to a beaker of water. At the surface of the solid, the moving water molecules collide with the ions. These collisions gradually "chip" away the ions from the crystal. The collisions are particularly effective when the negatively charged end of the water molecule collides with a positively charged ion

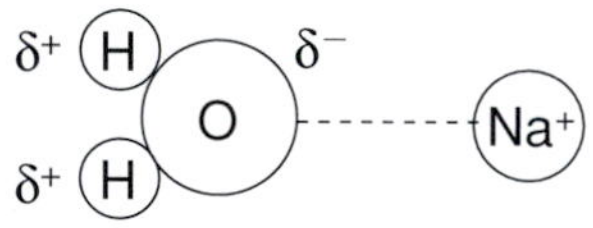

or the positive end of the water molecule collides with a negative ion.

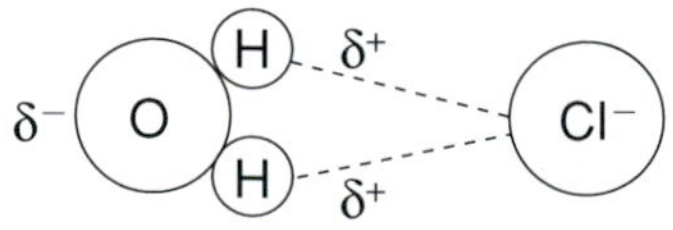

In these cases, the water molecules exert an attraction (ion-dipole force) that helps to pull apart the array of ions.

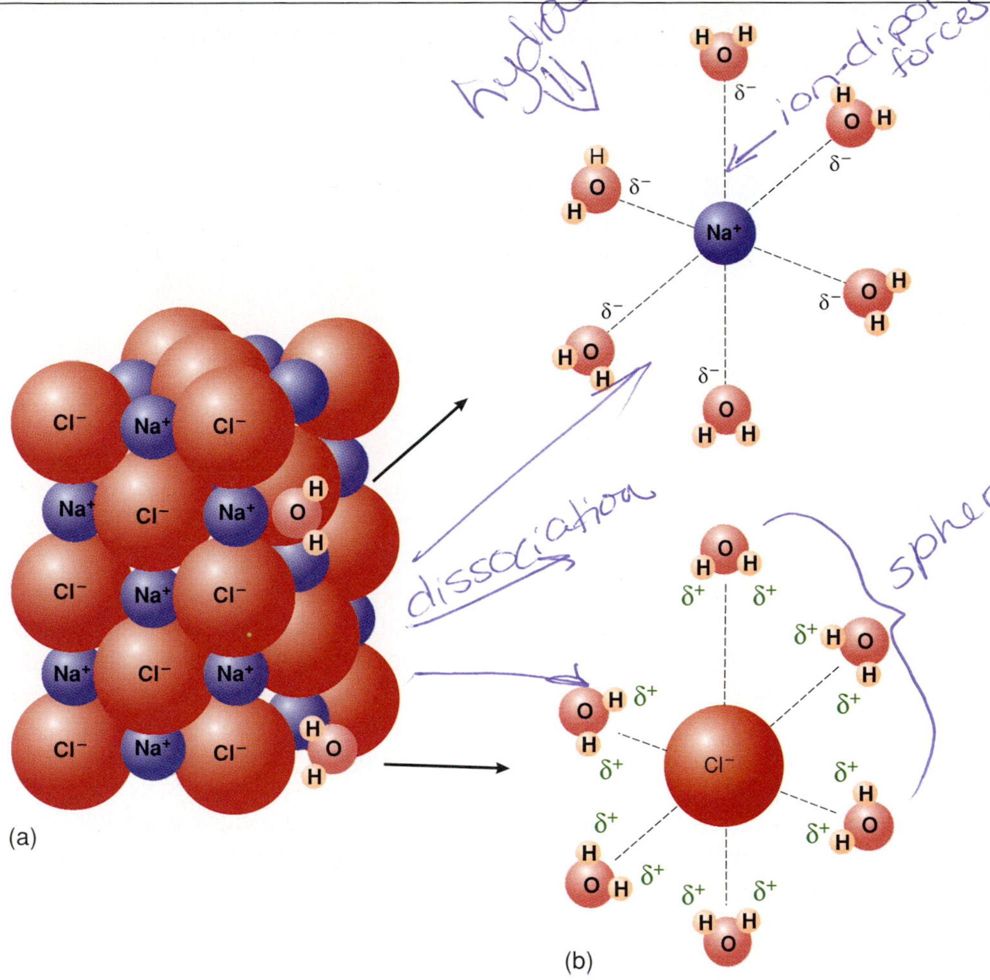

Figure 13.5

(a) Dissolving an ionic solute such as NaCl requires energy to separate the undissolved ions from their lattice structure. (b) Hydration energy is released as dissolved ions are surrounded by solvent water molecules.

When the cations and anions are separated from the crystal, the water molecules surround each of them (Figure 13.5b) and remain bonded to them by ion-dipole forces. The shell of water molecules surrounding each ion in solution is called the **sphere of hydration**. Notice that in the hydration sphere of the anions, the *positively charged ends* of the water dipoles point toward the Cl^-. In the hydration sphere of the cations, the *negatively charged ends* of the water dipoles point toward the Na^+. This process of bonding solvent water molecules to dissolved ions is known as **hydration**. For the general case in which solvent molecules other than water surround dissolved ions, the process is called **solvation**. Regardless of the solvent molecule, all solvation processes release energy because they involve the formation of new bonds. When water is the solvent molecule, the energy released is called **hydration energy**. It is the release of energy by the solvating ions that provides energy to separate the undissolved ions from their lattice structure.

Ions in solution are solvated and move through the solution with their shell of surrounding solvent molecules. As you can see in Figure 13.5, solvated Na^+ and Cl^- ions are much larger than the bare ions.

The process by which ions in the crystal lattice of a solute break apart or dissociate to become independent hydrated ions in solution is called **dissociation**. Dissociation may be represented by a chemical equation of the form

$$NaCl(s) \rightleftharpoons Na^+(aq) + Cl^-(aq)$$

Figure 13.6

Dissolving NaCl is a reversible process. (a) When solid NaCl is first placed in water, only separation of ions from the solid lattice occurs (upward arrow); this process will continue until the solubility limit of NaCl is reached. (b) Once the maximum amount of NaCl has dissolved, the dissolved ions will combine and visibly precipitate out of solution (downward arrow). When equilibrium between the two processes is established, the solution is saturated.

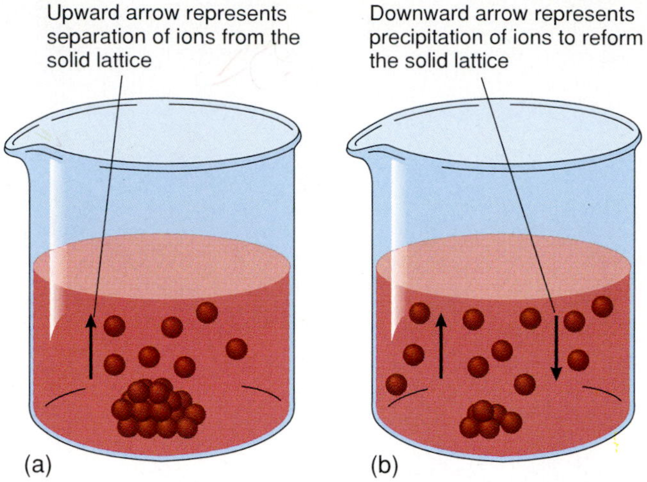

Upward arrow represents separation of ions from the solid lattice
Downward arrow represents precipitation of ions to reform the solid lattice
(a)
(b)

ionic solubility? = is the lattice energy too strong for water mol.'s to pull it apart through ion-dipole attractions?

The double arrow is used because dissociation is a reversible process (Figure 13.6). The solute will continue to dissolve, as shown by the forward arrow and as observed by a decreasing amount of solid NaCl, until the solubility limit of NaCl is reached. As solid NaCl dissolves, its rate of dissolving will decrease until it reaches its solubility limit. As Na^+ and Cl^- ions increase in solution, their rate of precipitation back to solid NaCl increases. When the rates of the forward and reverse processes become equal, the solution is saturated. At that point, you will observe no further decrease in the amount of solid NaCl.

The solubility limit of an ionic compound is reached when there are not enough water molecules for the complete hydration of more dissolved ions. The electrostatic forces of attraction between dissolved anions and cations become stronger than the ion-dipole forces between dissolved ions and solvent water molecules. As a result, dissolved ions associate to form new solid and precipitate from the solution.

Different ionic compounds dissolve to different extents. Table 13.2 shows quite large differences in solubility, ranging from the *very soluble* silver nitrate to the *slightly soluble* silver acetate to the *insoluble* barium sulfate. When ionic compounds dissolve only to the very small extent that $BaSO_4$ does (about 2.3×10^{-4} g per 100 g of H_2O), they are classified as *insoluble.* In the representation of the reversible dissolving process, $BaSO_4(s) \rightleftharpoons Ba^{2+}(aq) + SO_4^{2-}(aq)$, there is *almost no tendency for the ions to separate.* The process from left to right ($\rightarrow$) essentially does not occur. Similarly, since the solubility of ZnS in water at 25°C is only about 3×10^{-11} g per 100 g of H_2O, you find that ZnS is classified as insoluble according to the solubility rules given later in Table 13.4.

One factor that determines just how soluble an ionic compound will be is the magnitude of its hydration energy compared with the energy associated with the attraction between its ions in the crystalline solid. In many ionic solids, the ions are held together too tightly in the crystal for water molecules to pull them away through ion-dipole attractions.

Table 13.4 is a useful summary of the solubilities of several groups of compounds that are classified by the anion or cation that determines their solubility. Knowledge of these solubility rules will give you the ability to predict the solubility of a large number of commonly encountered compounds.

13.4 factors influencing solubility

As you have seen, different solutes dissolve to different extents. The factors that influence the extent to which a given solute dissolves in a given solvent are discussed in the following subsections.

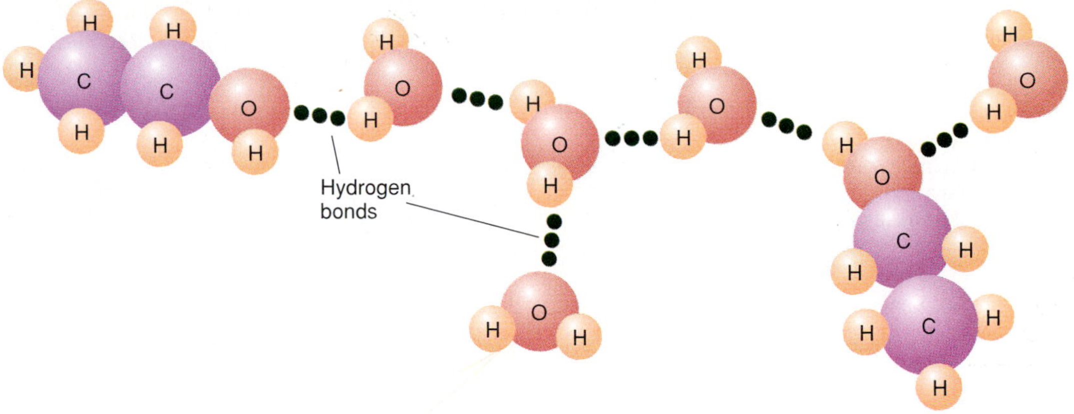

Figure 13.7
The OH end of the ethyl alcohol molecule forms hydrogen bonds with water molecules. The entire alcohol molecule, CH_3CH_2OH, is easily accommodated within the hydrogen-bonding network.

high KE at high T = fast "mixing"

nature of solute and solvent

You have examined the solution process for an ionic compound in water. This process depends on the ionic nature of the solute and the dipole nature of the solvent, which together make interaction and intermingling of particles possible.

Solvation, in general, involves the solvent molecules "sneaking in between" the solute particles. For this reason, substances with similar intermolecular forces tend to dissolve in one another. Substances with dissimilar intermolecular forces do not tend to form solutions. You have already learned that water and ethyl alcohol (grain alcohol) are completely miscible because of the similarity of their intermolecular forces. Both are polar and both form hydrogen bonds. Figure 13.7 shows the hydrogen bonds that enable them to mix intimately.

Gasoline, which can be represented as C_8H_{18}, will not dissolve in either water or alcohol because gasoline is nonpolar. It will, however, dissolve in pentane, C_5H_{12}, or hexane, C_6H_{14}, both of which, like gasoline, display weak dispersion forces (Section 12.4) as their principal intermolecular forces.

This generalization about similar intermolecular forces in solute and solvent leading to solubility is often referred to as the principle of "like dissolves like." Polar solutes dissolve in polar solvents, and nonpolar solutes dissolve in nonpolar solvents (Figure 13.8). Although this principle is a somewhat useful guideline, the solution process does not depend on relative polarities alone. Fortunately, it is easy to test solubility directly in the laboratory.

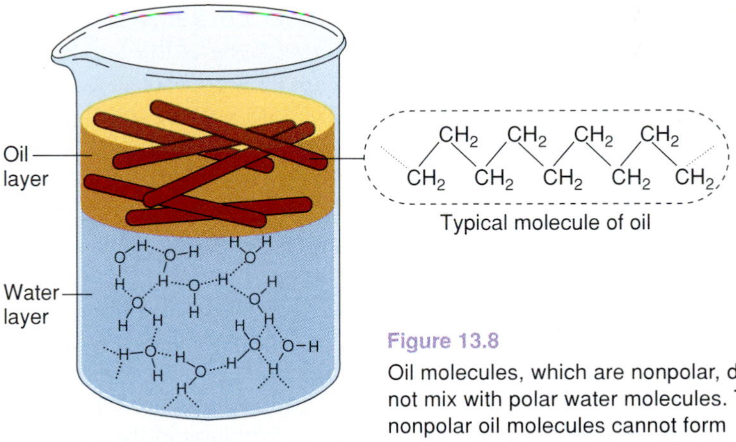

Figure 13.8
Oil molecules, which are nonpolar, do not mix with polar water molecules. The nonpolar oil molecules cannot form equally strong interactions within the hydrogen-bonding network of water molecules. As a result, a layer of oil remains undissolved and, because of its lower density, floats on the surface of water.

temperature

Temperature affects both the speed and the extent of the solution process. Molecules move more quickly with higher kinetic energies at higher temperatures, so the intermingling process of solution formation is speeded up. Sugar dissolves more quickly in hot tea than in iced tea.

In Section 14.6, we will see that temperature has a great effect on equilibrium processes. In the solution process, temperature exerts its effect on the equilibrium between undissolved and dissolved solute. For solid solutes in liquid solvents, the effect is *often* an increase in solubility with an increase in temperature (Figure 13.9). Not only does sugar dissolve faster in hot tea, but also more sugar dissolves in a given amount of hot tea than in the same amount of iced tea. For gaseous solutes

Figure 13.9

Solubility of various compounds in water. Notice that the solubility of all of the compounds, except Li_2SO_4 (a solid) and HCl (a gas), increases as temperature increases. For the two exceptions, solubility decreases as temperature increases. The behavior shown for HCl gas is typical of all gases.

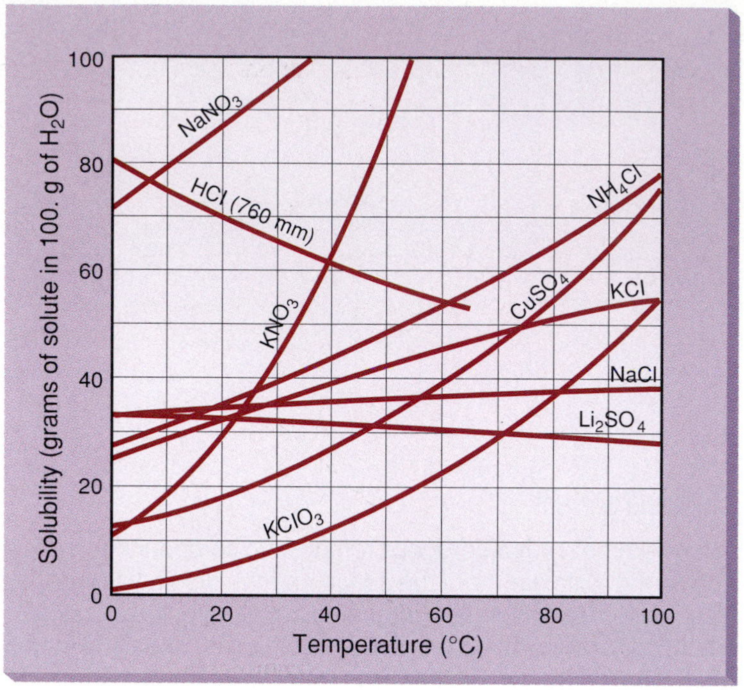

in liquid solvents, however, increasing temperature decreases solubility. Cold soda and cold beer maintain their fizz (gaseous CO_2 solute) more readily than do warm soda and warm beer.

pressure

Pressure has almost no effect on the solubility of solids in liquids. The dynamic equilibrium between solid solute and dissolved ions is not affected by pressure on the surface of the solution. The solubility of a gas in a liquid, on the other hand, is directly proportional to the partial pressure of the gas above the solution. The higher the partial pressure of gas above the surface of the solution, the higher the rate of those gas molecules dissolving and the slower the rate of those dissolved molecules escaping from solution. Carbonated beverages are bottled in such a way that the partial pressure of CO_2 above the solution is greater than 1 atm. This increases the solubility of CO_2 in H_2O. When the bottle is opened, CO_2 gas escapes, thus lowering the partial pressure of CO_2 above the solution. Hence, the solubility of CO_2 in the soda is reduced.

The external pressure on the human body affects the solubility of gases in blood. Deep-sea divers are subjected to high pressures (much greater than 1 atm) at low depths, and this increases the amount of nitrogen (from the air they breathe) that dissolves in their blood. If they rise to the surface (1 atm pressure) too rapidly, the nitrogen gas leaves solution in much the same way as CO_2 gas bubbles leave an opened bottle of soda. Gas bubbles in the bloodstream can cause pain and even death because the bubbles act as embolisms (clots) and block blood circulation. This condition is known as the "bends."

The effect of pressure on gas solubility is put to good use in hyperbaric therapy. Patients are placed in a chamber in which the pressure can be maintained at more than twice atmospheric pressure. This chamber can be used to slowly reduce the pressure on divers thus avoiding the "bends."

Often the hyperbaric chamber is filled with pure oxygen. The increased partial pressure of oxygen leads to greater solubility of oxygen in blood. This is helpful in some cases of hypoxia or oxygen deficiency, such as in carbon monoxide poisoning.

It is also helpful in treating infections, such as gangrene, caused by anaerobic bacteria because increased oxygen is toxic to these bacteria. When X ray radiation of cancer cells is carried out in a hyperbaric chamber under high oxygen pressure, the cancer cells are destroyed more effectively.

particle size and stirring

Because the solution process occurs at the surface of a solid solute, increasing the surface area speeds up the process. Finely pulverized solids offer more surface area per unit mass than big chunks of solid, so there are more opportunities for solute molecules to interact with solvent molecules. Similarly, stirring the solution provides more opportunities for solute molecules to interact with solvent molecules, speeding up the solution process.

While small particle size and vigorous stirring increase the speed of solution formation, they have no effect on the total amount of solute that dissolves. Solubility is set by other factors and does not depend on whether a fine powder quickly dissolves or a large chunk slowly dissolves.

13.5 concentration expressions

It is important that you be able to prepare, use, and analyze solutions in the laboratory and in your everyday lives. Accurate solution concentrations are critical in a medical facility. The concentration of a saline solution injected into a patient's bloodstream must have the same concentration of salt as the blood itself or the injection could be fatal. Research studying the effect of prescription drugs must measure the concentration of each ingredient as it is absorbed into the bloodstream as well as how the concentration in the bloodstream changes with time. An auto mechanic must know what concentration of antifreeze is needed to lower the freezing point of your car's radiator fluid below the coldest outside temperature. Because solutions may be prepared with varying composition, you must learn how to express the proportion of solute and solvent and you must understand the various units used to measure this proportion.

When you state the composition of a solution, you call that statement the *concentration* of the solution. In Section 13.2, you defined concentration as a ratio of masses, grams of solute per 100 g of solvent, which is the same unit used to define solubility. Notice the difference between solubility, which is the *maximum* amount of solute per amount of solvent at a given temperature, and concentration, in which the solute amount may take *any* value up to the solubility limit. The concentration of solute may be expressed in many other units, however, and chemists have established units that are especially convenient for specific purposes. In general, the different units for concentration may be classified into two forms. The **concentration** of a solution is the *amount of solute* present in either some *given amount of solvent* or some *given amount of solution.* Concentration is a ratio in one of two forms:

$$\frac{\text{Amount of solute}}{\text{Amount of solvent}} \quad \text{or more often} \quad \frac{\text{Amount of solute}}{\text{Amount of solution}}$$

Which ratio is used for a particular solution depends on how the solution is prepared. If you weigh the solvent or measure its volume *before* mixing it with the solute, the first ratio is convenient. More often, however, you mix the solute and solvent and then weigh the entire solution or measure its entire volume. In that case, the second ratio is more convenient.

As stated above, the various concentration expressions differ in the *units* used for the amounts of solute and solution. The *molarity* expression used most often by chemists employs the units of moles for the solute and liters for the solution as you will see later in this section. In *percentage concentrations,* the units for the solute and solution may be either grams (weight) or milliliters (volume). This gives rise to three possible percentage expressions, *percentage weight per weight* (% w/w), *percentage weight per volume* (% w/v), and *percentage volume per volume* (% v/v). These three percentage expressions and a fourth, parts per million, as well as molarity (grams per liter) are discussed in the following subsections.

percentage mass/mass (percentage weight/weight, % w/w)

If you were given 100 g of a 10% (by mass) sugar solution, you could easily figure out how much sugar and how much water were mixed together by recognizing that the percentage refers to the *percentage of solute* in the total mass of the solution. Hence, 10% of 100 g of solution = 0.10 × 100 g = 10 g of sugar. The other 90% of the solution is water, hence, 90 g of water (10 g of solute + 90 g of solvent = 100 g of solution). Equation 13.1 may be rewritten as read:

$$\text{Grams of solution} = \text{Grams of solvent} + \text{Grams of solute} \qquad (13.2)$$

Percentage is always part of the whole divided by the whole times 100 (review Section 3.6, which concerns percentage calculations). For calculations involving solutions, you will find the following equations useful. Remember that, by convention, *mass is represented as weight in these expressions.*[1]

$$\% \frac{\text{w}}{\text{w}} \text{ of solute} = \frac{\text{Grams solute}}{\text{Grams solution}} \times 100 \qquad (13.3)$$

$$\% \frac{\text{w}}{\text{w}} \text{ of solute} = \frac{\text{Grams solute}}{\text{Grams solute} + \text{Grams solvent}} \times 100 \qquad (13.4)$$

The mass (weight) percent unit of concentration is independent of temperature. The number of grams of solute or solvent is constant, unaffected by temperature. Since it does not change with temperature, the mass (weight) percent is an appropriate measure of concentration when the same solution is being used over a widely varying temperature range.

sample exercise 2

In a laboratory, 233 g of a NaCl solution was evaporated to dryness; i.e., the water was boiled off. What was left was 28.4 g of solid NaCl.

a. What was the percentage by mass (weight) of NaCl in the solution?
b. How much water was boiled away?

solution

a. Using Equation 13.3,

$$\% \frac{\text{w}}{\text{w}} \text{ of solute} = \frac{28.4 \text{ g NaCl}}{233 \text{ g solution}} \times 100 = 12.2\%$$

b. Using Equation 13.2,

$$233 \text{ g solution} = ? \text{ g water} + 28.4 \text{ g NaCl}$$

$$233 - 28.4 = 204.6 \text{ g water} = 205 \text{ g water (calculation requires 3 sig figs)}$$

sample exercise 3

What is the percentage by mass (weight) of potassium hydroxide in a solution prepared by dissolving 5.0 g of KOH in 55 g of water?

1. Percentage weight/weight, or weight percent, is probably more accurately referred to as percentage mass/mass, or mass percent; however, common usage employs the term *weight* instead of *mass*.

solution

Using Equation 13.4,

$$\% \frac{w}{w} \text{ of solute} = \frac{5.0 \text{ g KOH}}{5.0 \text{ g KOH} + 55 \text{ g } H_2O} \times 100 = 8.3\%$$

To do problems involving the preparation of some given amount of solution of a given percentage, you must use Equation 13.2 and your basic knowledge of what percentage means.

In many problems, it is useful to recognize the ratio of grams of solute per 100 g of solution as a conversion factor. That is, every percentage weight/weight gives you a conversion factor relating grams of solute to grams of solution. For example,

$$\frac{5 \text{ g solute}}{100 \text{ g solution}} \quad \text{for} \quad 5\% \frac{w}{w}$$

The use of percentage weight/weight as a conversion factor is illustrated in Sample Exercise 4.

---

sample exercise 4

How many grams of sugar and water would you use to prepare 700. g of a 7.0% w/w solution?

solution

The grams of sugar needed can be obtained using the unit conversion method.

1. The *given* quantity and unit is 700. g of solution.
2. The *new* unit is grams of sugar.
3. The percentage weight/weight provides the conversion factor 7.0 g of sugar/100 g of solution.
4. Set up the usual format:

Given	×	**Conversion factor**	=	**New quantity and unit**
700. ~~g solution~~	×	7.0 g sugar / 100. ~~g solution~~	=	49 g sugar

The grams of water can be obtained by using Equation 13.2:

$$700. \text{ g solution} = ? \text{ g water} + 49 \text{ g sugar}$$

$$700. - 49 = 651 \text{ g water}$$

---

problem 3

In a laboratory, 12.0 g of $AgNO_3$ is dissolved in 26.0 g of H_2O. What is the percentage by mass (% w/w) of $AgNO_3$ in the solution? Silver nitrate is a key starting material in the preparation of silver halides, which are used in photographic emulsions.

---

problem 4

How would you prepare 500.0 g of a 32.0% w/w solution of KBr in water? Potassium bromide, used in the manufacture of photographic papers, is also a sedative.

percentage mass/volume (percentage weight/volume, % w/v)

Percentage mass/volume (called weight/volume) is the expression used most often in medicine because the most common solutions are those in which the solute is a solid, most easily measured in grams, and the solution is a liquid, most easily measured in volume units, e.g., milliliters.

The numerical value expressed in a mass (weight) per volume percentage (% w/v) is equal to the number of grams of solute present in 100. mL of solution. For example, a 10. % w/v dextrose solution has 10. g of dextrose per 100. mL of solution. A 0.9 % w/v salt solution (physiological saline) contains 0.9 g of salt per 100. mL of solution. This value can be readily calculated through the equation

$$\% \frac{\text{w}}{\text{v}} \text{ of solute} = \frac{\text{Grams of solute}}{\text{Milliliters of solution}} \times 100 \qquad (13.5)$$

Percentage mass/volume is a unit of concentration that is dependent on the temperature of the solution. While the number of grams of solute is a constant, the volume of the solution will expand slightly as the temperature of the solution increases. As a result, the increasing volume of solution will result in a slightly smaller concentration (% w/v) at higher temperature. For accurate concentration, you should always record the temperature at which a percentage mass/volume (% w/v) solution is prepared and include it on the label of your solution.

sample exercise 5

What is the percentage mass/volume (% w/v) of KOH in 60. mL of a solution that contains 5.0 g of KOH?

solution

Using Equation 13.5,

$$\% \frac{\text{w}}{\text{v}} \text{ of KOH} = \frac{5.0 \text{ g KOH}}{60. \text{ mL solution}} \times 100 = 8.3\% \frac{\text{w}}{\text{v}}$$

There is 8.3 g of KOH per 100 mL of solution.

As with mass/mass (called weight/weight) calculations, the given percentage weight/volume can be used as a conversion factor; this is illustrated in the following sample exercise.

sample exercise 6

Physiological saline solution is a 0.90% w/v NaCl solution used for storing red blood cells because it has the same concentration of dissolved particles as human blood. How much salt (in grams) is needed to prepare 250. mL of physiological saline solution?

solution

Use the unit conversion method.

1. The *given* quantity and unit is 250. mL of solution.
2. The *new* unit is grams of NaCl.
3. The percentage mass/volume (% w/v) provides the conversion factor 0.90 g of NaCl/100. mL of solution.
4. Set up the usual format:

Given	×	**Conversion factor**	=	**New quantity and unit**
250. ~~mL solution~~	×	$\frac{0.90 \text{ g NaCl}}{100. \cancel{\text{mL solution}}}$	=	2.3 g NaCl

••• problem 5

How many milliliters of a 7.0% w/v glucose solution would contain 49 g of glucose? Glucose is a main source of energy for living organisms.

percentage volume/volume (% v/v)

This expression is encountered only for solutions that are mixtures of liquids. For example, a water solution of alcohol might be labeled 45% v/v. This indicates 45 mL of alcohol per 100. mL of solution. In general, the numerical value of percentage volume/volume equals the number of milliliters of solute present in 100 mL of solution.

parts per million (ppm)

The concentration unit of parts per million is most useful for very dilute solutions wherein percentage expressions would involve very small numbers. For example, the concentration of fluoride in drinking water is usually reported in units of parts per million because it is so small.

To understand the expression parts per million, it is useful to realize that a statement of percentage as discussed in the previous subsections is actually a statement of parts per hundred. A 5% solution indicates 5 parts of solute per 100 parts of solution. A very dilute solution that has the concentration 5 ppm contains 5 parts of solute per 1 million parts of solution. Drinking water typically contains only 1 mg (0.001 g) of fluoride ion (F^-) per liter (1,000 mL) of solution. In percentage weight/volume, this would be

$$\frac{0.001 \text{ g F}^-}{1{,}000 \text{ mL solution}} \times 100 = 0.0001\ \%\ \frac{\text{w}}{\text{v}}$$

To convert the above ratio from grams per 1,000 mL to grams per 1,000,000 mL, multiply the left term by 1,000/1,000 = 1. The concentration of fluoride ion (F^-) in drinking water in parts per million is 1 ppm, as calculated below.

$$\frac{0.001 \text{ g F}^-}{1{,}000 \text{ mL solution}} \times \frac{1{,}000}{1{,}000} = \frac{1 \text{ g F}^-}{1{,}000{,}000 \text{ mL solution}} =$$

$$\frac{1 \text{ part F}^-}{1 \text{ million parts solution}} = 1 \text{ ppm}$$

You can see that the small value of fluoride ion concentration is much more conveniently expressed in parts per million than in percentage weight/volume. For even more dilute solutions, the expression parts per billion (ppb) is employed.

molarity

Molarity is the concentration unit used most frequently by chemists because solutions expressed in molarity are straightforward to prepare in the laboratory. **Molarity** (M) is defined as moles of solute per liter of solution:

$$\textbf{Molarity } (M) = \frac{\textbf{Moles of solute}}{\textbf{Liter of solution}} \qquad (13.6)$$

Whenever you are told the number of moles of solute or the amount of solute in grams that is dissolved in a specified volume of solution, *the molarity can be calculated by dividing the number of moles of solute by the number of liters of solution.*

The concentration unit of molarity is also dependent on the temperature of the solution. While the number of moles of solute is constant, the volume of the solution will expand slightly as the temperature increases.

Moles of solute are conveniently measured in the laboratory when studying chemical reactions (Chapter 10) because any number of moles of material is always readily related to grams of material through the molar mass (Section 8.5). Flasks with accurately calibrated volumes from 10 mL to 2 L are routinely available in the laboratory. The temperature (usually 20 or 25°C) at which the volume has been calibrated is labeled on the flask. Preparing a solution with a known molarity is no more difficult than transferring a weighed amount of solute to a volumetric flask, adding some solvent and dissolving the solute by swirling, and then filling the flask with solvent to the calibration mark on its neck (Figure 13.10). As you can see in Figure 13.10, the shape of a volumetric flask is designed to facilitate swirling the solute and solvent, a process that speeds the dissolving of solute (Section 13.4). Each volumetric flask also has a tightly fitting, ground glass stopper. Once the flask has been filled carefully to its exact volume, the solution is mixed thoroughly by inverting the stoppered flask several times.

sample exercise 7

Determine the molarity of the following solutions.

a. 4 mol of NaCl are dissolved in enough water to make 2 L of solution.

b. 58.8 g of KOH are dissolved in enough water to make 700.0 mL of solution.

solution

a. Because the units given are moles of solute and liters of solution, the problem is completely straightforward and we can immediately divide:

$$\frac{4 \text{ mol NaCl}}{2 \text{ L solution}} = \frac{2 \text{ mol}}{\text{L}} = 2\,M$$

(Remember *M* is the abbreviation for the unit molarity.)

b. In this case, grams of solute must be converted to moles of solute and milliliters of solution to liters of solution before you divide.

$$58.8 \cancel{\text{g KOH}} \times \frac{1 \text{ mol KOH}}{56.1 \cancel{\text{g KOH}}} = 1.05 \text{ mol KOH}$$

$$700.0 \cancel{\text{mL}} \times \frac{1 \text{ liter}}{1{,}000 \cancel{\text{mL}}} = 0.7000 \text{ L}$$

$$\frac{1.05 \text{ mol KOH}}{0.7000 \text{ L}} = \frac{1.50 \text{ mol}}{\text{L}} = 1.50\,M$$

The distinction between "moles" and "molarity" is very important. The number of moles of material is an absolute amount and can be directly converted to grams. Molarity is a *ratio* that tells the number of moles distributed through a volume of solution. It is convenient to recognize that the total number of moles in a given solution can be obtained by multiplying $M \times V$ (in liters). This relationship comes about by rearranging the equation by which molarity is calculated:

$$\boldsymbol{M} = \frac{\textbf{Moles of solute}}{\textbf{Volume of solution (in liters)}} \qquad (13.6)$$

Multiplying each side of Equation 13.6 by *V* yields

$$\boldsymbol{M} \times \boldsymbol{V} \text{ (in liters)} = \textbf{Moles of solute} \qquad (13.7)$$

Equation 13.7 is extremely important for finding the molarity of a solution that has been diluted. It will provide the basis for our discussion in Section 13.6.

(a) (b) (c) (d)

Figure 13.10

The process of accurately preparing a solution with known molarity is summarized in the following sequence of photographs. (a) Solid solute has been (or is being) transferred to the volumetric flask through a powder funnel. (b) The flask has been only partially filled with solvent so that the solute may be dissolved efficiently by swirling. (c) Filling the flask with solvent to the calibration mark requires adding the final amount of solvent dropwise. The bottom of the meniscus (concave surface of the solution) in the neck of the flask should appear exactly at the calibration mark when your eye is level with the mark. (d) Inverting the solution in the flask several times mixes the solution so that it is uniform throughout.

When you speak of 1 *M*, 2 *M*, and 6 *M* solutions (read "1 molar," "2 molar," and "6 molar"), your statement of molarity provides a conversion factor between moles of solute and liters of solution. For example,

Molarity:	1 *M*	2 *M*	6 *M*
Conversion factor:	$\frac{1 \text{ mol solute}}{1 \text{ L solution}}$	$\frac{2 \text{ mol solute}}{1 \text{ L solution}}$	$\frac{6 \text{ mol solute}}{1 \text{ L solution}}$

Of course, the reciprocals are also conversion factors. Recognizing molarity as a conversion factor enables us to handle many different types of concentration calculations with the unit conversion method.

sample exercise 8

How many moles of KBr, a sedative, are there in 35.8 mL of a 0.172 *M* solution?

solution

Use the unit conversion method, with molarity as a conversion factor.

1. The *given* quantity and unit is 35.8 mL of solution (0.172 *M* is also given).
2. The *new* unit is moles of KBr.
3. The necessary conversion factors are 1 L/1,000 mL (to use molarity, the volume of the solution must be in liters) and 0.172 mol KBr/1 L solution.
4. Set up the usual format:

$$\textbf{Given} \times \textbf{Conversion factor} = \textbf{New quantity and unit}$$

$$35.8\,\cancel{\text{mL}} \times \frac{1\,\cancel{\text{L}}}{1{,}000\,\cancel{\text{mL}}} \times \frac{0.172 \text{ mol KBr}}{1\,\cancel{\text{L}}} = 0.00616 \text{ mol KBr}$$

This problem can also be solved by remembering that

$$M \times V \text{(in liters)} = \text{Moles of solute}$$

$$\frac{0.172 \text{ mol KBr}}{1\,\cancel{\text{L}}} \times 35.8\,\cancel{\text{mL}} \times \frac{1\,\cancel{\text{L}}}{1{,}000\,\cancel{\text{mL}}} = 0.00616 \text{ mol KBr}$$

The unit conversion method is preferred, however, because it has more diverse applications than the $M \times V$ formula.

The use of molar mass as a conversion factor in conjunction with a solution's molarity is illustrated in the following sample exercise.

sample exercise 9

How many grams of NaOH are there in 2.50 L of a 0.343 *M* solution? Solutions of NaOH are used as drain cleaner; such solutions cause fats to react with water to form water-soluble soap.

solution

1. The *given* quantity and unit is 2.50 L of solution (0.343 *M* is also given).
2. The *new* unit is grams of NaOH.
3. The conversion factors are

$$\frac{0.343 \text{ mol NaOH}}{1 \text{ L}} \quad \text{and} \quad \frac{40.0 \text{ g NaOH}}{1 \text{ mol NaOH}}$$

4. Using the unit conversion method,

$$2.50\,\cancel{\text{L}} \times \frac{0.343\,\cancel{\text{mol NaOH}}}{1\,\cancel{\text{L}}} \times \frac{40.0 \text{ g NaOH}}{1\,\cancel{\text{mol NaOH}}} = 34.3 \text{ g NaOH}$$

sample exercise 10

How would you prepare 0.500 L of a 0.140 *M* solution of $CuSO_4$ (molar weight = 160. g)? You have available solid $CuSO_4$, a balance, and a 500 mL volumetric flask.

solution

Using the unit conversion method, determine the number of grams of $CuSO_4$ required to prepare this solution.

1. The *given* quantity and unit is 0.500 L of solution (0.140 *M* is also given).
2. The *new* unit is grams of $CuSO_4$.
3. The conversion factors are

$$\frac{0.140 \text{ mol } CuSO_4}{1 \text{ L}} \quad \text{and} \quad \frac{160. \text{ g } CuSO_4}{1 \text{ mol } CuSO_4}$$

4. The proper format is

$$0.500 \cancel{\text{L}} \times \frac{0.140 \cancel{\text{mol } CuSO_4}}{1 \cancel{\text{L}}} \times \frac{160. \text{ g } CuSO_4}{1 \cancel{\text{mol } CuSO_4}} = 11.2 \text{ g } CuSO_4$$

To prepare the desired solution, weigh out 11.2 g of $CuSO_4$ and transfer this to the 500 mL volumetric flask (500. mL = 0.500 L). Add some distilled water and swirl it to dissolve the solute. When the solute is dissolved, add water exactly to the etched calibration mark, which gives 0.500 L of solution (Figure 13.10).

You know from Section 8.5 that the number of grams of solute may be found from the number of moles by using its molar mass:

$$\text{Moles of solute} \times \text{Molar mass of solute} = \text{Grams of solute}$$

From Equation 13.7, in any given solution, the moles of solute = $M \times V$, so

$$\boldsymbol{M \times V} \text{ (in liters)} \times \textbf{Molar mass of solute} = \textbf{Grams of solute} \qquad (13.8)$$

Equation 13.8 enables you to calculate the number of grams of solute necessary to make a given volume of a solution with any specific molarity. This multiplication was performed in step 4 of Sample Exercise 10. You may acquire further practice using it in Problem 6.

••• problem 6

How many grams of NaOH are needed to make 125 mL of a 2.44 *M* solution?

In summary, all problems involving molarity can be solved by using one or more of the following conversion factors or their reciprocals:

$\frac{\text{Moles of solute}}{\text{1 L solution}}$	$\frac{\text{Molar mass of solute}}{\text{1 mol solute}}$	$\frac{\text{1 L}}{\text{1,000 mL}}$
Molarity, *M*	**Molecular weight or formula weight**	**Volume conversion factor**

A further example that combines all of these is Sample Exercise 11, where you must calculate not the number of grams of solute, but the volume of solution containing a given number of grams. Knowing these conversion factors provides you with a strong foundation for working in the laboratory. You may prepare a solution with known molarity or you may take a solution with a given molarity and measure the volume you would use to transfer a given number of moles or grams of solute.

sample exercise 11

Find the number of milliliters of a 1.32 *M* solution that contains 6.72 g of NaF (molar mass = 42.0 g).

solution

This problem requires all of the previously listed factors for its solution.

1. The *given* quantity and unit is 6.72 g of NaF (1.32 *M* is also given).
2. The *new* unit is milliliters of solution.
3. Set up the conversion factors in the usual form of *new* units/*given* units:

This factor gives moles of NaF. $\frac{1 \text{ mol NaF}}{42.0 \text{ g NaF}}$

This factor gives liters of solution from which milliliters may easily be determined. $\frac{1 \text{ L solution}}{1.32 \text{ mol NaF}}$

4. The proper format is

$$6.72 \cancel{\text{g NaF}} \times \frac{1 \cancel{\text{mol NaF}}}{42.0 \cancel{\text{g NaF}}} \times \frac{1 \cancel{\text{L}} \text{ solution}}{1.32 \cancel{\text{mol NaF}}} \times \frac{1{,}000 \text{ mL}}{1 \cancel{\text{L}}} = 121 \text{ mL solution}$$

Reciprocal of molar mass — **Reciprocal of molarity** — **Volume conversion**

problem 7

How many liters of a 0.643 *M* solution contain 114 g of NaF?

13.6 dilution

An extremely important and convenient method of preparing solutions of some desired molarity is through the process of dilution. Often you have a more concentrated solution available in the laboratory, and, rather than preparing a less concentrated solution from scratch, you may simply dilute the solution you have. **Dilution** involves adding solvent (usually water) to a concentrated solution. The addition of water produces a more dilute solution because the *same amount of solute* is distributed through a larger amount of solvent or solution. For example, if 1 L of water is added to 1 L of a 6 *M* solution, the solution becomes a 3 *M* solution; the volume of the solution has doubled,[2] so the molarity of the solution is half as large.

$$\frac{6 \text{ mol}}{1 \text{ L solution}} + 1 \text{ L water} \rightarrow \frac{6 \text{ mol}}{2 \text{ L solution}}$$

Original 6 *M* solution — **Final 3 *M* solution**

The dilution process may be seen in Figure 13.11, where 1 L of distilled water is added to 1 L of concentrated solution in a graduated cylinder.

Calculations involving dilution problems center on the fact that the *same amount of solute* is present in both the original solution

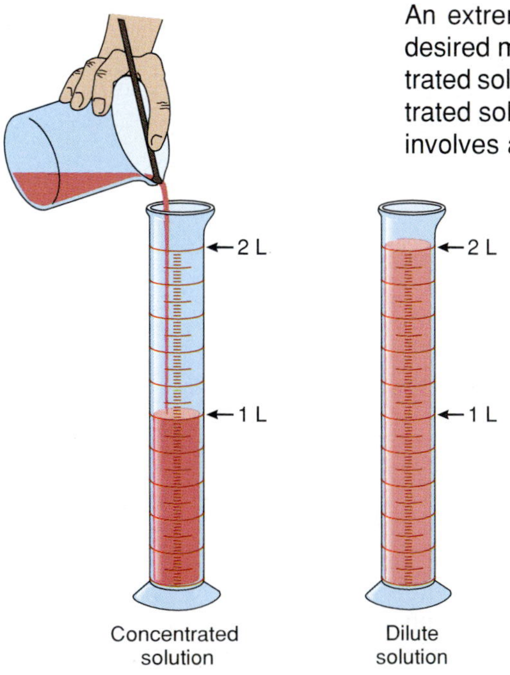

Figure 13.11
The dilution of a concentrated solution by adding water (solvent) results in the same amount of solute being distributed throughout a larger volume of solution.

2. Combining 1 L of solution with 1 L of water does not produce exactly 2 L of solution. However, this assumption is sufficiently accurate to 2 sig figs.

and the final diluted solution; that is, the number of moles of solute in any solution is conserved (unchanged) by dilution.

$$\textbf{Moles of solute}_{\text{original solution}} = \textbf{Moles of solute}_{\text{diluted solution}} \qquad (13.9)$$

But Equation 13.7 states that the number of moles of solute present in any solution can always be determined by multiplying $M \times V$ (in liters); that is,

$$M \times V = \text{Moles of solute} \qquad (13.7)$$

$$\frac{\text{Moles of solute}}{1\cancel{L}} \times 1\cancel{L} = \text{Moles of solute}$$

Therefore, you may substitute $M \times V$ for moles of solute and rewrite Equation 13.9 as

$$\boldsymbol{M_o \times V_o = M_d \times V_d} \qquad (13.10)$$

where *o* stands for the original solution and *d* stands for the diluted solution. You will use Equation 13.10 for all calculations in which solutions are diluted, such as those illustrated in the following sample exercises.

sample exercise 12

What is the molarity of the solution prepared by adding 2.0 L of water to 1.5 L of a 0.50 *M* KOH solution?

solution

Whenever water (solvent) is added to a solution, the problem should be recognized as a dilution problem. Then use Equation 13.10.

Original solution		Diluted solution
$M_o = 0.50\ M$		$M_d = ?$
$V_o = 1.5\ \text{L}$		$V_d = 1.5\ \text{L} + 2.0\ \text{L} = 3.5\ \text{L}$
$M_o \times V_o$	=	$M_d \times V_d$
$0.50\ M \times 1.5\ \text{L}$	=	$M_d \times 3.5\ \text{L}$

Divide each side of the equation by 3.5 L to isolate M_d.

$$M_d = \frac{0.50\ M \times 1.5\cancel{L}}{3.5\cancel{L}}$$

$$M_d = 0.21\ M$$

Because volume appears on both sides of Equation 13.10, the units of volume must be *consistent,* but they do not necessarily have to be liters. In this case, you could have used 1,500 mL and 3,500 mL and obtained the same result. However, one *cannot* use liters on one side and milliliters on the other.

sample exercise 13

What volume of a 0.34 *M* $MgCl_2$ solution is required to make 250. mL of a 0.10 *M* solution by dilution?

solution

The word *dilution* signals the use of Equation 13.10.

Original solution		Diluted solution
$M_o = 0.34\ M$		$M_d = 0.10\ M$
$V_o = ?$		$V_d = 250.\ \text{mL}$
$M_o \times V_o$	=	$M_d \times V_d$
$0.34\ M \times V_o$	=	$0.10\ M \times 250.\ \text{mL}$

Divide each side by 0.34 *M* to isolate V_o

$$V_o = \frac{0.10\ \cancel{M} \times 250.\ \text{mL}}{0.34\ \cancel{M}}$$

$$V_o = 73.5\ \text{mL} = 74\ \text{mL}$$

Proper use of significant figures demands that the answer be reported as 74 mL.

••• problem 8

In a laboratory, 500. mL of water is added to 250. mL of a 0.75 *M* NaOH solution. What is the molarity of the resulting solution?

13.7 electrolytes and nonelectrolytes in aqueous solution

In Section 13.3 and Figure 13.5, you learned that solutions of ionic compounds contain cations and anions that result from dissociation of the original compound. For example, a solution of NaCl is actually Na^+ ions and Cl^- ions intermingled between (and hydrated by) H_2O molecules. Experimentally, it is found that a solution of an ionic compound conducts electricity, and you know that it is the presence of ions in a solution that allows that solution to conduct electricity in the first place. Solutions that conduct electricity are classified as **electrolyte solutions.**

Electric current is a movement of electric charge. In metal wires, electrons carry the charge; in solution, ions carry the charge. As shown in Figure 13.12, if we attach electrodes (metal rods) to the poles of a battery with wires and immerse the electrodes in a solution containing ions, the ions will carry the electric charge from one pole to the other to complete the circuit and make the light glow. In solution, cations and anions move in opposite directions. Cations (being positively charged) are attracted to the negatively charged electrode; anions (being negatively charged) are attracted to the positively charged electrode.

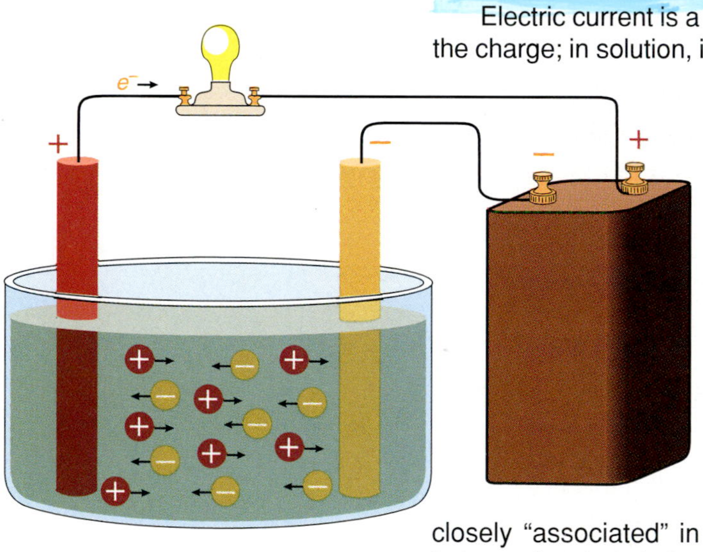

Figure 13.12

In metal wires, electrons carry the charge; in solutions, ions carry the charge. Cations (being positively charged) are attracted to the negatively charged electrode; anions (being negatively charged) are attracted to the positively charged electrode.

If there are no ions in solution, no current flows and the light does not glow. *A solution that conducts electricity must contain ions.* Soluble ionic compounds are **strong electrolytes** because they completely dissociate in water. You recall that dissociation is the name given to the process described in Section 13.3 whereby ions that are closely "associated" in the solid crystal dissociate, or break apart, and become independent in solution. Counting the number or moles of ions in a solution is very important in the stoichiometry of ionic reactions (Section 13.11) as well as for certain properties of solutions that change with the number of ions (Section 13.12). You might wish to review chemical formulas in Sections 8.6 and 8.7 at this point to recall the number of ions present in a given ionic compound.

The formula for an ionic compound tells how many ions are present per formula unit or how many moles of ions there are per mole of compound. For example,

$$KOH(s) \rightarrow K^+(aq) + OH^-(aq)$$

$$Na_2SO_4(s) \rightarrow 2\ Na^+(aq) + SO_4^{2-}(aq)$$

$$Al(NO_3)_3(s) \rightarrow Al^{3+}(aq) + 3\ NO_3^-(aq)$$

The breaking apart or dissociation of an ionic solid into ions is complete and so the above equations have been written with a single arrow in the direction of dissociation. The tendency of the ions to recombine can be ignored unless the solubility limit is exceeded.

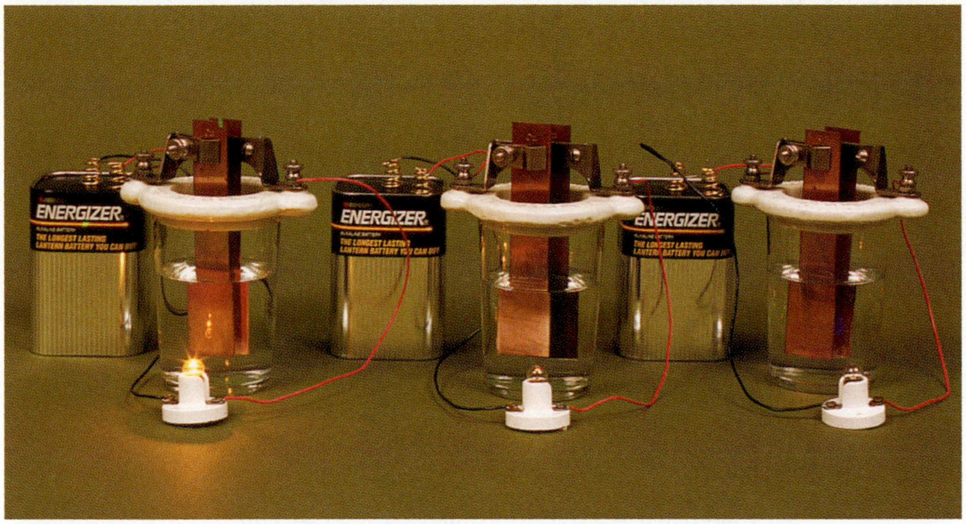

Figure 13.13
Electrical conductivity of a solution depends on the number of ions in that solution. For NaCl, a strong electrolyte, the solution conducts electricity well, and the light bulb burns brightly. For vinegar, a weak electrolyte, the solution conducts electricity poorly and the light bulb is dim. For pure water and a sucrose solution, both nonelectrolytes, no electricity is conducted and the light bulb remains unlit.

Most molecular compounds are *non*electrolytes; their solutions do *not* conduct electricity. Molecular compounds are not composed of ions and, thus, cannot dissociate in water. If there are no ions in solution, then the solution is a nonelectrolyte. However, some molecular compounds react with water in such a way as to produce ions. This process is called **ionization.**

One particularly important class of molecular compounds that undergoes ionization in water is **acids.** Acids are molecular compounds in which hydrogen is bound to an electronegative element. Their water solutions always contain a measurable quantity of H^+ ions, a property that will be discussed fully in Chapter 15. The water solutions of acids are electrolytes because acids react in water to form ions, i.e., *acids ionize in water.* For example, in water, HCl molecules form H^+ ions and Cl^- ions:

$$HCl(g) \xrightarrow{H_2O} H^+(aq) + Cl^-(aq)$$

The tendency of the ions to recombine is so small in this case, it may be ignored. Thus, you can say that HCl (hydrochloric acid) ionizes 100%.

If, by reaction with water, a molecular compound such as HCl produces many ions per mole of compound, it is a *strong* electrolyte. Since every molecule of HCl in solution is ionized, HCl is a strong electrolyte. Hydrochloric acid is also termed a **strong acid** because of its 100% ionization. There are six common strong acids (strong electrolytes): HCl, HBr, HI, HNO_3, H_2SO_4, and $HClO_4$ (Table 13.3). All other common acids are weak. By knowing the six strong acids, you automatically know how to classify thousands of other acids that are weak.

If only a few ions are produced per mole of compound, the compound is a **weak electrolyte** (Figure 13.13). Vinegar is an aqueous solution of acetic acid, $HC_2H_3O_2$. Acetic acid ionizes only to a very limited extent:

$$HC_2H_3O_2 \rightleftharpoons H^+(aq) + C_2H_3O_2^-(aq)$$

This representation ($\rightleftharpoons$) indicates that a few molecules of $HC_2H_3O_2$ will form a few ions in water, but most of the $HC_2H_3O_2$ will remain as un-ionized molecules. Because only a small number of ions is produced, $HC_2H_3O_2$ is a weak electrolyte (Figure 13.13).

Most acids are weak electrolytes and thus weak acids. Learn the list of the six strong acids given in Table 13.3. All other acids, then, are weak. Aside from acids, almost all other *molecular compounds* are nonelectrolytes. One notable exception

table 13.3 strong acids

Formula	Name
HCl	Hydrochloric acid
HBr	Hydrobromic acid
HI	Hydriodic acid
HNO_3	Nitric acid
H_2SO_4	Sulfuric acid
$HClO_4$	Perchloric acid

is ammonia, NH_3, which is a weak electrolyte. When NH_3 reacts with water, it ionizes to form ammonium ions and hydroxide ions according to the equation

$$NH_3(g) + H_2O(l) \rightleftharpoons NH_4^+(aq) + OH^-(aq)$$

Because ammonia produces hydroxide ions when it reacts with water, it is a **base.**

13.8 summary of solute particles in solution

For many chemical reactions conducted in solution, the reactants are solutes. Reaction products may also be solutes, or they may leave the solution if they are either insoluble solids or gases. When you mix baking soda ($NaHCO_3$) with vinegar ($HC_2H_3O_2$), the mixture foams profusely because of the gaseous product, CO_2, rapidly leaving the reaction mixture. When you add lemon juice (citric acid) to milk, it curdles; that is, a solid product of the reaction (milk proteins called alpha- and beta-caseins) precipitates. To describe a reaction in solution accurately, it is necessary to know the nature of the solute particles as well as the solubilities of the reactants and products.

From the preceding discussion about electrolyte solutions, you may classify the nature of solute particles into four specific types, listed as follows.

Type 1. The solute particles of *ionic compounds* are *ions.* An ionic compound in solution should be represented as the sum of its hydrated ions. For example, NaCl in solution is represented as $Na^+(aq) + Cl^-(aq)$ because the ions are separate and independent and surrounded by water molecules as represented by the abbreviation (*aq*).

Type 2. The solute particles of *molecular compounds that are nonelectrolytes* are *molecules.* Sugar is a nonelectrolyte. The solid is composed of the molecules $C_{12}H_{22}O_{11}$. A sugar solution is a mixture of water molecules and sugar molecules. The process of dissolving the nonelectrolyte sugar in water can be represented as

$$C_{12}H_{22}O_{11}(s) \rightarrow C_{12}H_{22}O_{11}(aq)$$

Type 3. The solute particles of *molecular compounds that are strong electrolytes* are *ions.* Therefore, such a compound should be represented as the sum of the ions it forms. HCl, for example, should be expressed as $H^+(aq) + Cl^-(aq)$ if it is in water (aqueous) solution.

Type 4. The solute particles of *molecular compounds that are weak electrolytes* are mixtures of *molecules* and *ions.* Because the number of molecules is much greater than the number of ions, we usually represent weak electrolytes by their molecular formulas. Acetic acid, for example, is represented as $HC_2H_3O_2(aq)$.

13.9 ionic equations

Now that you are equipped with a picture of solute particles in all types of solutions, you can better represent the detailed behavior of chemical reactions in solution. Up to this point, you have written chemical equations using the chemical formulas for each reactant and each product. Now you can write them in terms of the ions that exist in solution. You will see how writing **ionic equations** not only conveys more detail, it also helps you to predict which products are soluble (remain as solutes) and which may leave the solution as insoluble solids (precipitates) or gases.

Consider the example used in Section 9.8, where you represented the single-replacement reaction between magnesium metal and hydrochloric acid to yield magnesium chloride and hydrogen gas as

$$Mg + 2\ HCl \rightarrow MgCl_2 + H_2\uparrow$$

Traditional equation

Figure 13.14
A strip of magnesium metal reacts with a solution of HCl to produce H_2 gas (seen as bubbles) and a solution of $MgCl_2$.

Figure 13.14 shows a strip of Mg metal in an HCl solution. The result is bubbles of H_2 gas and a clear solution of $MgCl_2$. The physical picture of this reaction is much better described by an ionic equation in which you are careful to indicate appropriate solute particles for materials in solution. Thus, the ionic equation for the reaction would be

$$Mg(s) + \underbrace{2\ H^+(aq) + 2\ Cl^-(aq)} \rightarrow \underbrace{Mg^{2+}(aq) + 2\ Cl^-(aq)} + H_2\uparrow$$

Mg(s)	2 H⁺(aq) + 2 Cl⁻(aq)	Mg²⁺(aq) + 2 Cl⁻(aq)	H₂↑
The solid does not dissolve; Mg disappears because it reacts	**See solute type 3 in Section 13.8**	**See solute type 1 in Section 13.8**	**The gas escapes; it is not in solution**

Ionic equation

Notice that this ionic equation also tells you that magnesium and hydrogen were involved in chemical changes, but that chloride ion was unchanged. Mg metal becomes Mg^{2+} ions, and H^+ ions become H_2 gas, but Cl^- ions remain Cl^- ions. Ions that are unchanged in chemical reactions are called **spectator ions**. Spectators do not participate in the reaction; they simply "watch" the other ions react.

You may totally represent the chemical changes that occur in the preceding reaction by a so-called **net ionic equation,** which shows only the reacting species, *not spectators.*

$$Mg(s) + 2\ H^+(aq) \longrightarrow Mg^{2+}(aq) + H_2\uparrow$$

Net ionic equation

Previously, you represented the double-replacement reaction between solutions of NaCl and $AgNO_3$ in the following way.

$$NaCl(aq) + AgNO_3(aq) \longrightarrow AgCl\downarrow + NaNO_3(aq)$$

Traditional equation

The corresponding ionic equation shows clearly the nature of the solute particles in solution.

$$\underbrace{Na^+(aq) + Cl^-(aq) + Ag^+(aq) + NO_3^-(aq)}_{\text{See solute type 1 in Section 13.8}} \rightarrow \underbrace{AgCl\downarrow}_{\text{Precipitate; not in solution}} + \underbrace{Na^+(aq) + NO_3^-(aq)}_{\text{See solute type 1 in Section 13.8}}$$

Ionic equation

The *net* ionic equation is written by identifying the spectator ions and realizing that, because they appear on both sides of the equation, they can be canceled out. The spectators are Na^+ ions and NO_3^- ions. Thus,

$$Cl^-(aq) + Ag^+(aq) \longrightarrow AgCl\downarrow$$

Net ionic equation

The net ionic equation for the double-replacement reaction between NaCl and $AgNO_3$ focuses your attention on the actual chemical change that has occurred. It also tells you that other soluble chlorides, such as LiCl or KCl, reacting with $AgNO_3$ would also give the same net ionic equation. The Li^+ ion from reactant LiCl or the K^+ ion from reactant KCl would be spectator ions, just as the Na^+ ion is when NaCl is the reactant. As a result, you see that the net ionic equation describes the chemical change that will occur between a great variety of compounds, all of which contribute the same reactant ions but different spectator ions.

Once you learn which ions combine to form insoluble solids in the following section, you will be able to predict the products for a great number of solute reactions.

13.10 using solubility rules

For the reaction of Mg with HCl, you saw in Figure 13.14 that the product $MgCl_2$ is soluble because no precipitate appears in the bottom of the reaction flask. Only the escaping bubbles of H_2 gas disturb the appearance of the original solution. In the other preceding example, it was noted that the product AgCl was insoluble, a property that is directly observable by the appearance of AgCl precipitate in the bottom of the reaction flask (Figure 13.15).

Figure 13.15

When a solution of $AgNO_3$ is mixed with a solution of NaCl, a white precipitate, AgCl, is formed and settles to the bottom of the reaction flask.

Usually, you will not be told whether products of a chemical reaction are soluble and you will not have experimental observations directly available to you. However, you can use the solubility rules in Table 13.4 to decide whether a wide variety of ionic compounds are soluble. These solubility rules are applied to the solute reactions in Sample Exercises 14 to 16, and you may gain additional practice solving Problems 9 and 10, which follow.

table 13.4 solubility rules*

Soluble ions	Rules
Group IA, NH_4^+	1. All ionic compounds in which the cation is a Group IA element or NH_4^+ are *soluble.*
NO_3^-	2. All nitrates are *soluble.*
Cl^-, Br^-, I^-	3. All chlorides, bromides, and iodides are *soluble,* except for Ag*X* and PbX_2† (where *X* stands for Cl, Br, or I).
SO_4^{2-}	4. All sulfates are *soluble* except for $CaSO_4$,† $SrSO_4$, $BaSO_4$, $PbSO_4$, and Ag_2SO_4.†
Insoluble ions	
S^{2-}	5. All sulfides are *insoluble* except for those of the Group IA and IIA elements and NH_4^+.
OH^-	6. All hydroxides are *insoluble* except for those of the Group IA elements, calcium, strontium, and barium from Group IIA, and NH_4^+.
	7. All other compounds are *insoluble.*

*Mercury compounds are not covered by these rules as written because of the complexity of the mercury(I) ion, which is beyond the scope of this book.
†Actually slightly soluble

sample exercise 14

Classify the following compounds as soluble or insoluble.

a. $MgSO_4$, magnesium sulfate
b. CuS, copper(II) sulfide
c. $CaCO_3$, calcium carbonate
d. $Fe(NO_3)_3$, iron(III) nitrate
e. K_2CO_3, potassium carbonate
f. $PbSO_4$, lead(II) sulfate

solution

1. Consult the solubility rules in Table 13.4.
2. Classify each compound with respect to its cation and anion components.
3. Apply the appropriate rule to each compound.

Cation	Anion	Rule
a. Mg^{2+}	SO_4^{2-}	4. All sulfates are *soluble;* this is not an exception.
b. Cu^{2+}	S^{2-}	5. All sulfides are *insoluble;* this is not an exception.
c. Ca^{2+}	CO_3^{2-}	7. This compound does not fit any of the first six rules; such compounds are *insoluble.*
d. Fe^{3+}	NO_3^-	2. All nitrates are *soluble.*
e. K^+	CO_3^{2-}	1. K^+ is in Group IA; it is *soluble.*
f. Pb^{2+}	SO_4^{2-}	4. $PbSO_4$ is *insoluble.*

sample exercise 15

How should the compounds in Sample Exercise 14 be represented in an ionic equation if they are products of a reaction?

solution

Soluble ionic compounds form ions in solution. Insoluble compounds are shown with the precipitate arrow (↓). Thus,

a. $Mg^{2+}(aq) + SO_4^{2-}(aq)$
b. CuS ↓
c. $CaCO_3$ ↓
d. $Fe^{3+}(aq) + 3\ NO_3^-(aq)$
e. $2\ K^+(aq) + CO_3^{2-}(aq)$
f. $PbSO_4$ ↓

Net ionic equations for double-displacement reactions between ionic compounds can be written by the stepwise procedure shown for the reaction yielding solid AgCl in Section 13.9.

Guidelines for writing net ionic equations

1. Write the *traditional* equation.
2. Write the *ionic* equation by dissociating the ions of all soluble ionic compounds.
3. Eliminate spectator ions to obtain the *net ionic equation.*

However, you can shorten this procedure and go directly from the *traditional* to the *net ionic equation* by simply *combining the ions that form the product that precipitates.* All other ions, which do not form a precipitate, will cancel out because they are spectators.

sample exercise 16

Write a net ionic equation for the reaction between solutions of Na_3PO_4 and $Fe_2(SO_4)_3$.

solution

1. Write the traditional equation by switching the cation-anion partners to predict products.

$$Na_3PO_4 + Fe_2(SO_4)_3 \rightarrow Na_2SO_4 + FePO_4$$

Then balance the equation.

$$2\ Na_3PO_4 + Fe_2(SO_4)_3 \rightarrow 3\ Na_2SO_4 + 2\ FePO_4$$

2. Evaluate the solubility of the products. Na_2SO_4 is soluble (rule 1 from Table 13.4); $FePO_4$ is insoluble (rule 7 from Table 13.4).
3. The net ionic equation is the combination of ions producing the precipitate:

$$Fe^{3+}(aq) + PO_4^{3-}(aq) \rightarrow FePO_4 \downarrow$$

••• problem 9

What is the net ionic equation for the combination of solutions of K_2SO_4 and $Ba(NO_3)_2$?

••• problem 10

Is there a net ionic equation for the combination of solutions of $CuCl_2$ and KNO_3? Explain.

13.11 solution stoichiometry

Because chemical reactions often occur in solution, it is necessary and convenient to be able to perform stoichiometric conversions involving volumes of solutions of known molarity. It may help you to review the summary of stoichiometric conversions given in Section 10.7 and in Figure 10.9 before continuing with this section.

The key to solving solution stoichiometry problems is to remember that, for any solution,

$$\mathbf{M} \times \mathbf{V}\ \text{(in liters)} = \textbf{Moles of solute} \qquad (13.7)$$

That is, M is a conversion factor relating moles of solute to liters of solution. Thus, if you are given the molarity (M) and volume (V) of some reactant or product, you can easily determine the number of moles of that reactant or product and then

CHEMLAB

Laboratory exercise 13: testing for the presence of ions in solution

Identifying the constituents in a material is called qualitative analysis; determining the amount of one or more constituents is called quantitative analysis. In this experiment, you will examine some qualitative analytical tests for specific ions.

Each ion has unique properties and undergoes certain characteristic chemical reactions. Often, specific ions can be identified by examining the results of certain test reactions performed with the solution. In any test, you will be looking for a chemical change that is evidenced by

1. The formation of a precipitate
2. A color change in the solution
3. The formation of gas (bubbles)

With your knowledge of solubility rules and your understanding of ionic equations, you should be able to suggest simple laboratory tests to determine whether specific ions are present in unknown solution samples.

Questions:

1. Using the following reagents, which are available in a typical chemical laboratory, suggest a solubility test to determine whether each of the following ions is present in unknown solution samples. Assume that only one unknown ion is present in each sample.
2. Write the traditional chemical equation, the ionic equation, and the net ionic equation that describes each of your tests. If more than one test is used to confirm the presence of an ion, then write chemical equations using all appropriate reagents.

Reagents: 0.1 *M* $AgNO_3$, 0.1 *M* H_2SO_4, and 0.1 *M* NaCl

Ions to be tested for: Ca^{2+}, Ba^{2+}, Ag^+, Pb^{2+}, Cl^-, I^-, and SO_4^{2-}

proceed to carry out the stoichiometry problem, as you did in Chapter 10. *Since stoichiometry problems are usually presented with amounts of the original reactants and products, and not their ionic counterparts, you should use the traditional form of the complete balanced chemical equation, not the net ionic equation, for these problems.*

sample exercise 17

How many milliliters of a 0.52 *M* $AgNO_3$ solution are needed to react completely with 250. mL of a 0.36 *M* $MgBr_2$ solution according to the following equation?

$$2\ AgNO_3 + MgBr_2 \rightarrow 2\ AgBr + Mg(NO_3)_2$$

solution

1. The *given* quantity and unit is 250. mL of 0.36 *M* $MgBr_2$.
2. The *new* unit is milliliters of 0.52 *M* $AgNO_3$.
3. The central focus in all stoichiometry problems is the mole ratio obtained from the balanced equation, which relates *new* material to *given* material. In this case, the central focus is

$$\frac{2\ \text{mol}\ AgNO_3}{1\ \text{mol}\ MgBr_2}$$

4. To use this factor, recall that $M \times V$ (in liters) will give moles:

$$\underbrace{250.\ \cancel{mL} \times \frac{1\ \cancel{L}}{1{,}000\ \cancel{mL}}}_{\text{Volume (in liters)}} \times \underbrace{\frac{0.36\ \text{mol MgBr}_2}{1\ \cancel{L}}}_{\text{Molarity of reactant solution}} = \underbrace{0.090\ \text{mol MgBr}_2}_{\text{Moles of reactant}}$$

5. The setup is now

$$\underbrace{0.090\ \cancel{\text{mol MgBr}_2}}_{\text{Moles of reactant}} \times \frac{2\ \cancel{\text{mol AgNO}_3}}{1\ \cancel{\text{mol MgBr}_2}} \times \underbrace{\frac{1\ \cancel{L}}{0.52\ \cancel{\text{mol AgNO}_3}}}_{\text{Reciprocal of molarity of reactant solution}} \times \frac{1{,}000\ \text{mL}}{1\ \cancel{L}} = \underbrace{3.5 \times 10^2\ \text{mL of } 0.52\ M\ \text{AgNO}_3}_{\text{Volume of reactant solution}}$$

Notice again the use of molarity (M) as the conversion factor between moles and volume (in liters).

---

sample exercise 18

How many grams of AgCl will precipitate from 34 mL of a 0.24 M $AgNO_3$ solution when excess NaCl solution is added? The reaction is represented by the equation

$$AgNO_3 + NaCl \rightarrow AgCl \downarrow + NaNO_3$$

solution

1. The *given* quantity and unit is 34 mL of 0.24 M $AgNO_3$.
2. The *new* unit is grams of AgCl.
3. The central focus will be

$$\frac{1\ \text{mol AgCl}}{1\ \text{mol AgNO}_3}$$

4. Moles of $AgNO_3$ can be calculated from the V and M of $AgNO_3$.
5. The formula weight for AgCl (143.3) relates grams and moles.
6. Integrating all of the aforementioned information and beginning as always with the given, you can express the setup as follows.

$$\underbrace{34\ \cancel{mL} \times \frac{1\ \cancel{L}}{1{,}000\ \cancel{mL}}}_{\text{Volume of reactant solution}} \times \underbrace{\frac{0.24\ \cancel{\text{mol AgNO}_3}}{1\ \cancel{L}}}_{\text{Molarity of reactant solution}} \times \frac{1\ \cancel{\text{mol AgCl}}}{1\ \cancel{\text{mol AgNO}_3}} \times \frac{143.3\ \text{g AgCl}}{1\ \cancel{\text{mol AgCl}}} = 1.2\ \text{g AgCl}$$

13.12 colligative properties of solutions

In Section 12.6, you studied the physical properties of pure liquids, such as vapor pressure, boiling point, and freezing point, and how these properties are related to the strength of intermolecular forces. The properties of pure solvents are always altered in predictable ways by the presence of dissolved solutes. For example, boiling points of solutions are typically higher than the boiling point of the pure solvent, and solution freezing points are typically lower. The example of lowering the freezing point of your car radiator by adding ethylene glycol or methanol as antifreeze was discussed in Section 13.5. The boiling point of the solution in your car radiator would also increase because of the solute added. The addition of antifreeze not only protects your radiator from freezing in winter, but also protects it from boiling over in the summer. Both of

table 13.5 colligative properties of solutions

Vapor pressure lowering
Boiling point elevation
Freezing point depression
Osmotic pressure

these effects stem from the fact that the vapor pressure of a solution (ethylene glycol in water) is lower than that of the pure solvent (water). Furthermore, solutions demonstrate a property, **osmotic pressure,** that is not present in a pure liquid. Osmotic pressure is a measure of pressure forcing the solvent, water, to flow across a membrane.

The numerical values of vapor pressure lowering, boiling point elevation, freezing point depression, and osmotic pressure depend *only* on the concentration of solute particles in solution and not on the identity of the solute particles. Such properties, which do not depend on the nature of the dissolved species, but only on the *number of dissolved particles,* are called **colligative properties.** The greater the number of particles in solution, the lower its vapor pressure, the lower its freezing point, the higher its boiling point, and the higher its osmotic pressure.

Ionic compounds and ionizable molecular compounds produce more moles of ions in solution than were present in 1 mol of the original compound. For example, 1 mol of the ionic compound NaCl produces 2 mol of ions in aqueous solution, that is, 1 mol of Na^+ cations and 1 mol of Cl^- anions. Nonelectrolytes such as ethylene glycol that do not produce ions in solution, however, dissolve in solution without any increase in the number of particles. A mole of ethylene glycol produces a mole of ethylene glycol molecules in aqueous solution. While all particles in solution affect the solution's properties, ionic solutes with more particles in solution result in a greater change in the colligative properties of solutions than do nonelectrolytes.

Table 13.5 lists four colligative properties of solutions. The first three, vapor pressure lowering, boiling point elevation, and freezing point depression are discussed in the following subsections. Osmotic pressure is discussed separately in Section 13.13.

vapor pressure lowering

In Section 12.6, you learned that the vapor pressure of a solvent is the pressure of its gas phase in equilibrium with its liquid phase. Because thermal energy helps liquid molecules escape into the gas phase, vapor pressure increases rapidly as temperature increases. The graph of vapor pressure versus temperature in Figure 13.16 shows just how rapidly the vapor pressure of a pure solvent, water, increases as the temperature increases from 0°C to 100°C. In addition, a second curve, below that for pure water, shows how the vapor pressure of an aqueous solution is always *lower* than the vapor pressure of pure water.

The lowering of vapor pressure by the presence of solute particles is readily explained by remembering that the development of vapor pressure depends on the ability of molecules to escape from the surface of a liquid. As Figure 13.16 implies, in a solution, solute particles literally get in the way of solvent molecules and interfere with their escape. The greater the number of solute particles, the greater the interference effect, and therefore, the greater is the lowering of vapor pressure.

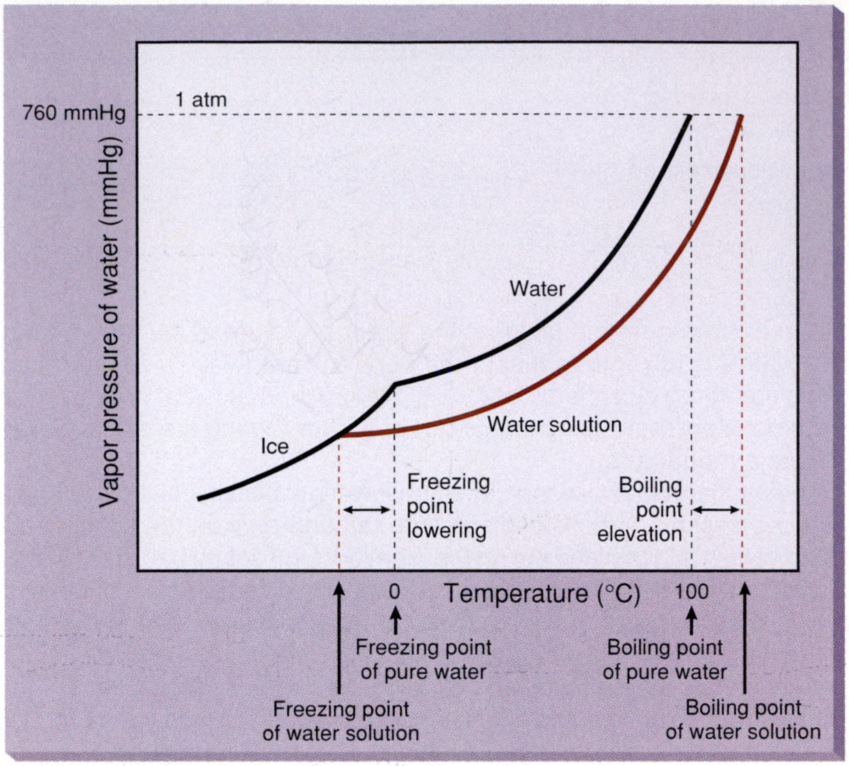

Figure 13.16
The black line shows how rapidly the vapor pressure of pure water increases with increasing temperature. The colored line shows how much lower the vapor pressure of an aqueous solution is compared to that of pure water at any temperature. Because the normal boiling point is the temperature at which the vapor pressure equals 760 mmHg, solutions boil at higher temperatures than pure water. Any aqueous solution freezes when the vapor pressure of the solution equals the vapor pressure of ice. The intersection of the black and colored curves indicates how much the freezing point is lowered.

boiling point elevation

Lowering of vapor pressure leads directly to the elevation of the normal boiling point of a solution compared with that of the pure solvent. Recall that the normal boiling point is defined as the temperature at which the vapor pressure is equal to an external pressure of 760 mmHg (Section 12.6). Because the vapor pressure of a solution is lower at all temperatures than that of the pure solvent, a higher temperature is required to reach a vapor pressure of 760 mmHg. This observation is shown graphically in Figure 13.16.

Adding salt to a pot of boiling water leads to a slightly increased boiling temperature because the salt lowers the vapor pressure of the solution and a higher temperature is necessary to reestablish the vapor pressure of the solution equal to the atmospheric pressure. Adding more salt to the solution will increase the boiling temperature still further. Mole for mole, salt has twice the effect on a solution's boiling point as sugar because sugar is a molecular solid and produces 1 mol of sugar molecules in solution, whereas salt produces 2 mol of ions (1 mol of Na^+ and 1 mol of Cl^-) per mole of compound. Variations in colligative properties depend directly on numbers of particles.

freezing point depression

A liquid freezes at the temperature at which the vapor pressure of the liquid equals the vapor pressure of the solid state of the substance. The vapor pressure of liquid water and the vapor pressure of ice are equal at 0°C. Figure 13.16 shows that the lowering of the vapor pressure curve for an aqueous solution results in the intersection of that curve with the solid-state curve (for pure water) at a lower temperature. This means that solutions freeze at lower temperatures than pure solvents.

This colligative property of freezing point depression is put to practical uses in cold winter weather. Besides adding antifreeze to your automobile radiator to lower

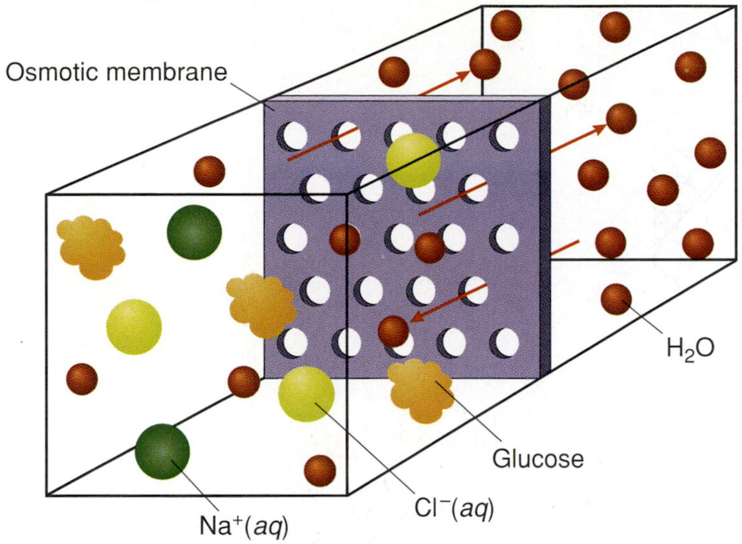

Figure 13.17
An osmotic membrane allows the passage of H_2O molecules in both directions, but prevents the passage of solvated ions and glucose molecules.

its freezing point, you also spread salt on your steps and sidewalk to prevent water from freezing. The aqueous solution formed when the salt dissolves will freeze at many degrees below 0°C. As a result, the outside temperature must fall much lower before ice will form.

The same principle is employed to melt ice on roadways. Remember, melting and freezing occur at the same temperature. Whereas pure water freezes or ice melts at 0°C, a solution undergoes these transitions at lower temperatures. Thus, salt sprinkled on icy roads forms a solution with the ice that melts at a lower temperature. The magnitude of freezing point depression, like all colligative properties, depends on number of particles in solution. Using $CaCl_2$ rather than NaCl on ice produces 3 mol of particles (1 mol of Ca^{2+} and 2 mol of Cl^-) per mole of compound rather than 2 mol of particles from NaCl (1 mol of Na^+ and 1 mol of Cl^-). On a molar basis, $CaCl_2$ is 1.5 times more effective at melting ice on streets and roadways than NaCl.

13.13 osmotic pressure

Osmotic pressure is a property that you may not have encountered previously. It involves a pressure directing the flow of water (solvent) across any membrane. Because of its practical importance in living systems, your familiarity with osmotic pressure could have important consequences on your own safety and the safety of your environment.

If you receive an intravenous feeding, the concentration of ions in the injected solution is critically important; if the concentration is incorrect, osmotic pressure could cause water to flow into or out of your blood cells so that they rupture and lead to your death. When you spread salt on your steps and sidewalk, the concentration of ions in the aqueous solution may spread to neighboring plants and trees. Just as osmotic pressure across your blood cells could lead to their rupture, osmotic pressure across the roots of plants and trees could likewise lead to their rupture and destruction.

Osmotic pressure depends on the phenomenon of *osmosis* and the concept of particle flow through an osmotic membrane. An osmotic membrane can be thought of as a thin sheet with small holes that allow the passage of water and other small molecules, the dimensions of which are not significantly larger than water. Ions such as Na^+ or Cl^- that are solvated with water molecules (Figure 13.5) cannot pass through an osmotic membrane; neither can particles larger than a water molecule, such as a sugar molecule (Figure 13.17).

Figure 13.18

(a) A dilute and a concentrated glucose solution are each placed in a compartment separated from each other by an osmotic membrane. Only water molecules can pass through the membrane. The glucose molecules block the passage of water molecules by deflecting them. The more glucose, the more interference. The net flow of water molecules is from the dilute glucose solution where the concentration of water is higher into the more concentrated glucose solution where the concentration of water is lower. (b) The passage of water results in increased height in the right column. The pressure due to the extra weight of the column of solution on the right is a direct measure of the osmotic pressure across the membrane.

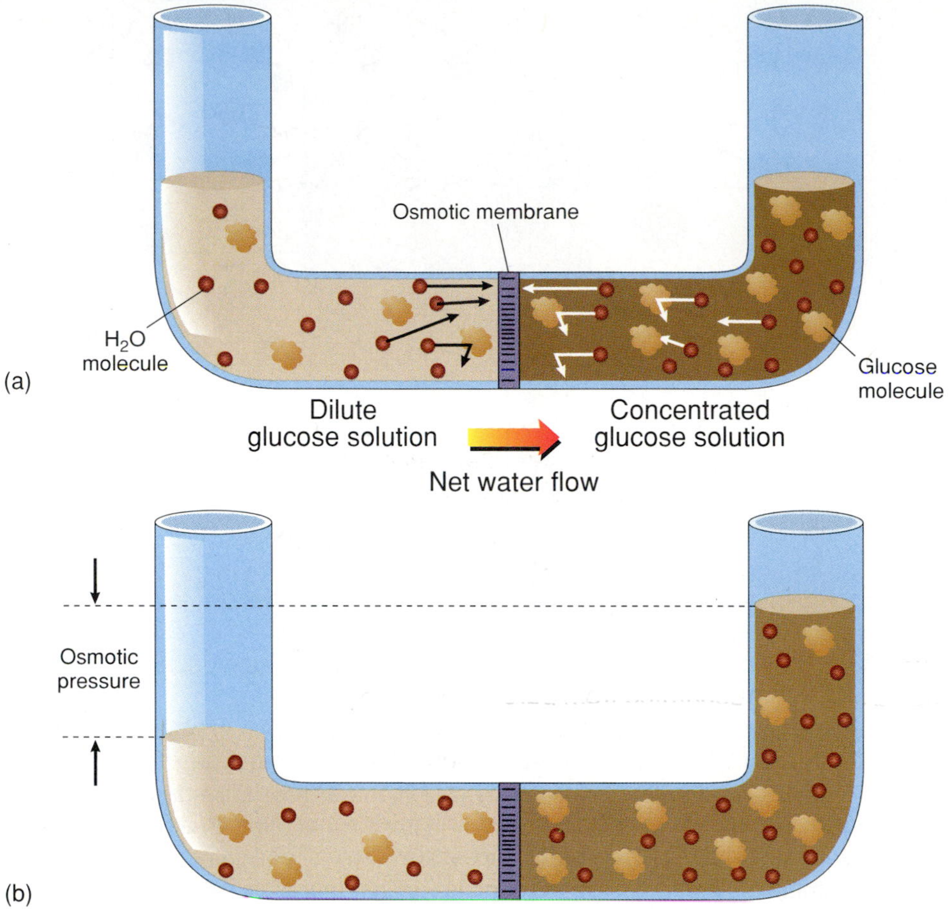

osmosis

Osmosis is defined as the flow of water across an osmotic membrane from a more dilute solution (or pure water) into a more concentrated solution. This represents flow from a region where the water concentration is high to a region where the water concentration is lower. It is important to realize that water flows across the osmotic membrane in both directions, as shown in Figure 13.17. However, there is a *net* flow of water from the solution with a larger relative number of water molecules (dilute solution or pure water) to the one with relatively fewer water molecules (concentrated solution).

Once again, we can use the idea of solute particles "getting in the way" to explain this net flow of water. Figure 13.18a shows solute particles blocking some water molecules from passing through an osmotic membrane. This blockage is less effective in a dilute solution than in a concentrated solution because there are fewer solute particles in a dilute solution. The net flow of water from the dilute to the concentrated chamber increases the volume of the more concentrated solution, as shown in Figure 13.18b. The net movement of water cannot continue indefinitely because the increased height of the more concentrated solution causes a downward pressure from the added weight of liquid, which eventually balances the opposing tendency of water to flow into the more concentrated solution. When this balance of forces occurs, there is no further net movement of water. At this point, the concentrated solution has become more dilute and the dilute solution more concentrated. Another way of saying this is that the concentration gradient has

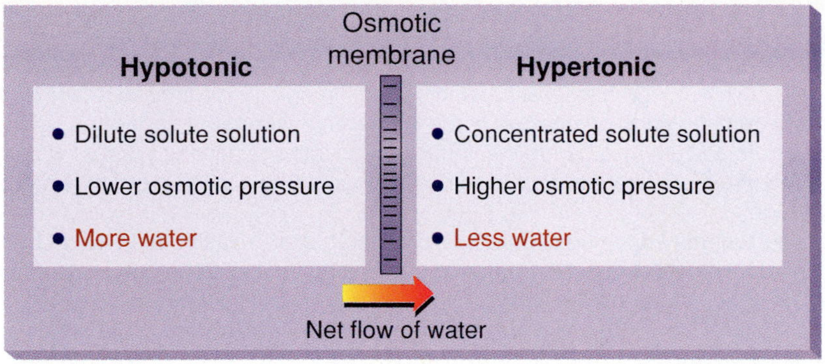

Figure 13.19

The greater the concentration of solute particles, the larger the solution's osmotic pressure and the greater the tendency for water to flow into it.

been diminished. A **concentration gradient** is a *difference* in concentration between solutions. The greater the difference or gradient, the greater the potential for a net flow of water in the direction shown below.

Dilute solution (higher concentration of water)	→	**Concentrated solution** (lower concentration of water)

osmotic pressure

When a solution is separated from pure water by an osmotic membrane, the pressure that develops from the excess height of the solution side as the water flows in is called the **osmotic pressure** of the solution. The osmotic pressure of a solution can also be defined as the external pressure that must be applied to prevent a net flow of water across a membrane from pure water to the solution. Osmotic pressure measures the tendency for water to flow from pure water across an osmotic membrane into a solution. This tendency depends only on the concentration of solute particles in the solution; that is, it is a colligative property. The greater the concentration of solute particles, the larger the solution's osmotic pressure and the greater the tendency for water to flow into it.

It is the difference in osmotic pressure between two solutions with different solute concentrations that causes the net flow of water that we've been discussing. In comparing solutions, a dilute solution with a low osmotic pressure is said to be **hypotonic** compared with a more concentrated solution with a higher osmotic pressure. In turn, the concentrated solution is called **hypertonic** compared with the dilute solution (Figure 13.19). Solutions of exactly the same particle concentration have identical osmotic pressures and are said to be **isotonic.**

osmosis in foods and agriculture

You witness the effects of osmosis in many foods. Prunes swell up if placed in water (a hypotonic medium compared to the prune), cucumbers shrink and become pickles in brine (a hypertonic medium compared to the cucumber), eggplants give off their water when sprinkled with salt, and hams are dehydrated and, therefore, preserved by "cooking" them in a brine solution. The movement of water in each case is through the membrane that surrounds each living cell. This membrane functions as an osmotic membrane, allowing water, but not larger molecules, to pass. In agriculture, osmosis helps one understand how water flows from the ground root system (hypotonic) to the concentrated sap cells (hypertonic) in the upper regions of a tree. The maximum height of trees is limited by the osmotic pressure established across the ground root system.

CHEMQUEST

Separating two or more *insoluble* substances by filtration is accomplished both in the laboratory and in natural processes. In the laboratory, filter paper separates solid particles from a liquid by allowing the liquid to pass through; similarly, in the watershed of your reservoir, acres of soil trap solid contaminants and allow the groundwater to flow through.

Solutes, too, may often be separated from the solvent in which they are dissolved. For example, if you dissolve ink from your pen in a solvent such as ether and place the solution in a glass tube packed with grains of silica gel, the solution will flow downward through the tube. As it does, the ink solute will separate into distinct bands of color that will move down the silica gel column at different rates. Each band of color corresponds to a different dye in the ink.

In the laboratory, this technique of separating solutes from solution using a column packed with solid grains or powder is called chromatography, a name derived from the colored bands that are often observed. The reason that the dyes in ink are separated into different colored bands is that each dye is attracted to (or adsorbed on) the surface of the solid particles, but each dye is adsorbed with a different strength of attraction. As the solution flows downward, the dyes less strongly adsorbed on the solid move more rapidly with the flow of solvent; the dyes strongly adsorbed on the solid move more slowly.

As a result of physical adsorption, the watershed of your reservoir may separate dissolved contaminants from your water supply as well as insoluble material. Polar and slightly polar molecules may be more strongly attracted to the solid particles than they are to the highly polar, solvent water molecules.

At home, you may have extended this technique of separation further by installing a water filter on your drinking water faucet. Many home water filters consist of multiple stages: one or more for the filtration of insoluble particles and another for the separation of dissolved impurities such as pesticides, cleaning agents, and detergents.

Research what materials are used in home water filters and report on the most common impurities they are likely to remove from your drinking water. Discuss whether you would expect them to remove toxic, heavy metal ions that may also be dissolved in your water supply.

chapter accomplishments

After completing this chapter, you should be able to define all key terms and do the following.

13.1 Introduction

- ❑ Name and distinguish between the two components of a solution.
- ❑ Give a specific example of each of the seven known types of solutions shown in Table 13.1.

13.2 Properties of solutions with liquid solvents

- ❑ Distinguish between the terms *solubility, soluble,* and *insoluble.*
- ❑ Distinguish between saturated, unsaturated, and supersaturated solutions.
- ❑ Distinguish between dilute and concentrated solutions.
- ❑ Indicate the type of mixtures to which the terms *miscible* and *immiscible* are applied.
- ❑ Given solubility data, state whether a solution is saturated or unsaturated.

13.3 Dissolving ionic solutes in water

- ❑ Describe the formation of a water solution of an ionic solid.

13.4 Factors influencing solubility

- ❑ State the factors that affect the extent of solubility of a given solute in a given solvent.
- ❑ State the factors that affect only the speed by which a given solute dissolves in a given solvent.
- ❑ Given the Lewis structures of two molecular compounds, predict whether one will dissolve in the other.
- ❑ Describe the relationship between the partial pressure of a gas and its solubility in a liquid.

13.5 Concentration expressions

- ❑ Distinguish between the terms *concentration* and *solubility.*
- ❑ Given the grams of solute and the grams of solvent or solution, calculate the percentage mass/mass (% w/w) of the solution.

- ❑ Calculate the number of grams of solute and solvent needed to prepare a given mass of solution of some given mass percent (% w/w).
- ❑ Given the grams of solute and the volume of solution, calculate the percentage mass/volume (% w/v) of the solution.
- ❑ Calculate the number of grams of solute needed to prepare a given volume of solution of some given percentage mass/volume (% w/v).
- ❑ Given the volume and molarity of a solution, calculate the number of moles of solute in that solution.
- ❑ Given the volume and molarity of a solution, calculate the number of grams of specified solute in that solution.
- ❑ Given a specific solute, describe how you would prepare a given volume of a solution having some specified molarity.
- ❑ Calculate the volume of a solution with some specified molarity that would contain a given mass of a specified solute.

13.6 Dilution

- ❑ Given the volume and molarity of an original solution, calculate the molarity of a new solution prepared by adding a given amount of solvent to the original solution.
- ❑ Calculate the volume of a concentrated solution of given molarity that must be diluted to prepare a specified volume of a new solution of lower molarity.

13.7 Electrolytes and nonelectrolytes in aqueous solution

- ❑ Give specific examples of strong and weak electrolytes and nonelectrolytes.
- ❑ Distinguish between the processes of dissociation and ionization.
- ❑ Give examples of strong and weak acids.

13.8 Summary of solute particles in solution

- ❑ Describe the basic particles found in solutions of ionic compounds and in solutions of molecular electrolytes and nonelectrolytes.

13.9 Ionic equations

- ❑ Given a traditional equation, write a net ionic equation for a reaction in which a gas or precipitate is formed.

13.10 Using solubility rules

- ❑ Use the solubility rules in Table 13.4 to predict whether a given compound is soluble.

13.11 Solution stoichiometry

- ❑ Given a balanced chemical equation and the volume and molarity of one reactant, calculate the volume (or molarity) of a second reactant given the molarity (or volume) of the second reactant.
- ❑ Given a balanced chemical equation and the volume and molarity of one reactant combining with an excess of other reactants, calculate the mass of a product that can be produced.

13.12 Colligative properties of solutions

- ❑ Name four colligative properties.
- ❑ Describe the effect of added solute on the vapor pressure, boiling point, and freezing point of a liquid solvent.

13.13 Osmotic pressure

- ❑ Given the concentrations of two solutions separated by an osmotic membrane, tell the direction in which the net flow of water will occur.
- ❑ Distinguish among isotonic, hypotonic, and hypertonic solutions.

key terms

solution (homogeneous mixture)
solute
solvent
heterogeneous mixtures
solubility
soluble
insoluble
saturated solution
unsaturated solution
dilute solution
concentrated solution
supersaturated solution
miscible liquids
immiscible
sphere of hydration
hydration
solvation
hydration energy
dissociation
concentration
molarity
dilution
electrolyte solutions
strong electrolytes
ionization
acids
strong acid
weak electrolyte
base
ionic equations
spectator ions
net ionic equation
osmotic pressure
colligative properties
osmosis
concentration gradient
hypotonic
hypertonic
isotonic

problems

13.1 Introduction

11. Define in your own words what is meant by a solution.
12. A compound, like a solution, is homogeneous. Describe how a solution differs from a compound.
13. Give specific examples other than those in Table 13.1 for each of the three types of solutions in which the solvent is a liquid.
14. Are all solutions mixtures? Are all mixtures solutions? Explain.
15. Which of the following examples would you classify as solutions?
    a. Helium gas dispersed in air
    b. Scotch whiskey
    c. Urine
    d. Fog
    e. Soda water
    f. Bronze (tin homogeneously dispersed in copper)
    g. Blood
    h. Milk
16. Indicate a possible solute and a possible solvent for each of the examples you classified as a solution in Problem 15.
17. Why is dusty air not considered a solution?

13.2 Properties of solutions with liquid solvents

18. a. Describe how you would experimentally determine the solubility of a solute in a given solvent.
    b. Describe how you would experimentally determine whether a given solution is saturated or unsaturated.
19. a. Does a saturated solution necessarily have to be a concentrated solution? Explain.
    b. Give a specific example from Table 13.2 of a solution that is saturated, yet dilute.
20. Use Table 13.2 to decide whether the following solutions are saturated or unsaturated.
    a. 18 g of NaCl in 100. g of H_2O at 20°C
    b. 12 g of $KMnO_4$ in 100. g of H_2O at 60°C
    c. 1.7 g of $AgC_2H_3O_2$ in 100. g of H_2O at 60°C
    d. 71 g of KBr in 100. g of H_2O at 20°C
21. Consult Table 13.2 and decide whether the following solutions are saturated or unsaturated.
    a. 17 g of KBr in 25 g of H_2O at 60°C
    b. 7.5 g $KMnO_4$ in 50. g of H_2O at 60°C
    c. 50. g of NaCl in 150. g of H_2O at 60°C
    d. 41 g of KNO_3 in 125 g of H_2O at 20°C
22. Other than by adding more solute, describe how a dilute solution of NaCl can be made more concentrated.
23. Describe how you would prepare a supersaturated solution of potassium nitrate in water.
24. You are given three solutions: one is unsaturated, one is saturated, and one is supersaturated. Describe how you would experimentally distinguish among these three solutions.
25. a. To which type of mixture do the terms miscible and immiscible apply?
    b. Give an example of two miscible substances other than alcohol and water.
    c. Give an example of two immiscible substances other than oil and water.

13.3 Dissolving ionic solutes in water

26. a. What attractive forces must be broken in the dissolving of an ionic solid?
    b. What new attractive forces are formed in the dissolving of an ionic solid?
    c. Explain why NaCl does not dissolve in a nonpolar solvent such as gasoline.

13.4 Factors influencing solubility

27. a. How does the nature of a solute affect whether it will dissolve in a given solvent?
    b. What type of solutes will dissolve best in water solution?
    c. What type of solutes will dissolve best in gasoline?
28. Explain why a solute such as sugar dissolves faster in hot water than in cold water.
29. a. Explain why a bottle of soda goes "flat" if left open to the air.
    b. Would keeping the bottle warm help the soda to stay fizzier longer? Explain.
30. a. What factor enables a finely ground solute to dissolve faster than the same solute in the form of chunks?
    b. Give two practical examples of solutes that are finely ground before dissolving them in a solvent.
31. a. Describe how stirring enables a solute to dissolve more quickly.
    b. Does stirring affect the solubility of a solute in a given solvent?
32. a. Write out Lewis structures for methyl alcohol, CH_3OH, and hydrogen chloride, HCl.
    b. Predict whether HCl will dissolve in CH_3OH. Explain your answer.

13.5 Concentration expressions

33. Describe and illustrate by example the difference between the terms solubility and concentration.

34. A solution was prepared by dissolving 18.3 g of KNO_3 in 75.0 g of water at 20°C. What is the percentage mass/mass (% w/w) of this solution?

35. In a laboratory, 179 g of a silver nitrate solution is evaporated to dryness. After all of the water is evaporated, 39.6 g of silver nitrate remains.
    a. What is the mass percent (% w/w) of $AgNO_3$ in the original solution?
    b. How much water was boiled off?

36. An antifreeze solution was prepared by dissolving 250. g of methyl alcohol in 650. g of water. What is the concentration of this solution in mass percent (% w/w)?

37. Describe how you would prepare 250. g of a 5.00% w/w $BaCl_2$ solution.

38. How many grams each of sugar and water would you use to prepare 20.0 g of a 0.90% w/w solution?

39. Using Table 13.2, calculate the mass percent (% w/w) of a saturated, aqueous potassium nitrate solution at 20°C.

40. Using Table 13.2, calculate the mass percent (% w/w) of a saturated, aqueous $KMnO_4$ solution at 60°C.

41. What mass of sugar and water would be needed to prepare 90.0 g of a 15.0% w/w sugar solution?

42. 75.0 mL of solution was prepared by dissolving 13.8 g of KNO_3 in water at 20°C. What is the percentage mass/volume (% w/v) of this solution?

43. In a laboratory, 179 mL of a silver nitrate solution is evaporated to dryness. After all of the water is evaporated, 69.3 g of silver nitrate remains. What is the percentage mass/volume (% w/v) of $AgNO_3$ in the original solution?

44. How many grams of sugar would you use to prepare 20.0 mL of a 0.50% w/v solution?

45. How many milliliters of a 4.0% w/v $Ba(NO_3)_2$ solution would contain 1.25 g of barium nitrate?

46. An antifreeze solution was prepared by dissolving 250. mL of methyl alcohol in 700. mL of water. What is the concentration of this solution in percentage volume/volume?

47. Which solution is more concentrated: a 0.01% w/v glucose solution or a 50 ppm solution?

48. How much NaCl would be needed to prepare 800. mL of a 0.90% w/v physiological saline solution?

49. Describe the differences in the expressions of the concentration of a solution by percentage mass/volume (% w/v) and by molarity.

50. Calculate the molarity of a solution that contains 2.50 mol of KNO_3 dissolved in 5.00 L.

51. How many moles of NaCl are present in 100. mL of a 0.125 *M* solution?

52. How many grams of glucose, $C_6H_{12}O_6$, are present in 1.50 L of a 0.400 *M* solution?

53. Given the equipment available in your laboratory, describe how you would prepare 0.500 L of a 0.900 *M* NaCl solution.

54. Find the number of grams of NaOH present in 375 mL of a 0.300 *M* solution.

55. How many grams of solute are needed to prepare 100.0 mL of a 0.250 *M* silver nitrate solution?

56. Calculate the molarity of an HCl solution with a density of 1.18 g/mL and a mass percent of 36% HCl (% w/w).

57. A solution of $CaCl_2$, of density 1.05 g/mL, contains 6.00% w/w $CaCl_2$. Calculate the molarity of the solution.

58. In an experiment, 0.175 mol of H_2SO_4 is needed for a reaction. The H_2SO_4 solution available is 18.0 *M*. How many milliliters of this solution must be measured out?

59. A student needs to measure out 15.2 g of $NaHCO_3$. The only $NaHCO_3$ present in the laboratory is in a 0.100 *M* solution. How many milliliters of the solution must the student measure out to have the required number of grams?

13.6 Dilution

60. What is the molarity of 50.0 mL of a 0.50 *M* NaOH solution after it has been diluted to 300. mL?

61. If 300.0 mL of water is added to 200.0 mL of a 0.500 *M* Na_2SO_4 solution, what is the molarity of the resulting solution?

62. What volume of a 1.25 *M* KNO_3 solution is required to make 1.00 L of a 0.100 *M* solution by dilution?

63. To what volume must you dilute 80.0 mL of 3.0 *M* $CuSO_4$ to have a 0.50 *M* solution?

64. A student needs to prepare 200.0 mL of a 4.0 *M* HCl solution. The student has available a 12.0 *M* HCl solution. What volume of the concentrated solution must be measured out to prepare the 4.0 *M* solution by dilution?

13.7 Electrolytes and nonelectrolytes in aqueous solution

65. State the evidence for the existence of ions in a solution of NaCl.

66. Describe how you would experimentally show that HCl gas ionizes when dissolved in water.

67. Hydrogen bromide is a covalent substance that acts as a strong electrolyte in aqueous solution; $HC_2H_3O_2$ is a covalent substance that acts as a weak electrolyte in aqueous solution. Explain how one covalent substance can be a strong electrolyte and the other a weak electrolyte.

68. Using NaCl and HCl as specific examples, describe the differences between the processes of dissociation and ionization.

69. An aqueous solution of HCl conducts electricity. However, a solution of HCl in benzene does not conduct electricity. Explain what is happening in the two solutions.

70. Solid sodium chloride, although made up of ions, does not conduct electricity. Molten sodium chloride is, however, a strong electrolyte. What condition other than the presence of ions is necessary for the conduction of electricity?

71. Give the names and formulas of three strong acids and three weak acids.

13.8 Summary of solute particles in solution

72. What solute particles will be found in aqueous solutions of the following?
    a. KBr
    b. $Ca(OH)_2$
    c. HNO_3
    d. $HC_2H_3O_2$
    e. H_2SO_3
    f. Glucose, $C_6H_{12}O_6$
    g. Methyl alcohol, CH_3OH
    h. $(NH_4)_3PO_4$

13.9 Ionic equations

73. Write balanced net ionic equations for the following reactions.
    a. $K_2SO_4(aq) + Ba(NO_3)_2(aq) \rightarrow KNO_3(aq) + BaSO_4 \downarrow$
    b. $Zn(s) + H_2SO_4(aq) \rightarrow ZnSO_4(aq) + H_2 \uparrow$
    c. $NaCl(aq) + AgNO_3(aq) \rightarrow AgCl \downarrow + NaNO_3(aq)$
    d. $(NH_4)_2S(aq) + Pb(NO_3)_2(aq) \rightarrow NH_4NO_3(aq) + PbS \downarrow$
    e. $FeCl_3(aq) + LiOH(aq) \rightarrow LiCl(aq) + Fe(OH)_3 \downarrow$
    f. $HCl(aq) + Na_2CO_3(aq) \rightarrow NaCl(aq) + H_2O + CO_2 \uparrow$

13.10 Using solubility rules

74. Using Table 13.4, indicate which of the following compounds are soluble in water.
    a. K_2SO_4
    b. $CaCl_2$
    c. Na_2S
    d. $Mg_3(PO_4)_2$
    e. $Al(OH)_3$
    f. CsOH
    g. $NH_4C_2H_3O_2$
75. What substances would be formed in solution from the following solutes?
    a. Na_2SO_4
    b. FeS
    c. KOH
    d. $Ca_3(PO_4)_2$
    e. $Mg(NO_3)_2$
    f. $(NH_4)_2CO_3$

13.11 Solution stoichiometry

76. How many milliliters of a 0.100 M H_2SO_4 solution are needed to react completely with 150. mL of a 0.250 M NaOH solution according to the following equation?
    $$H_2SO_4(aq) + 2\ NaOH(aq) \rightarrow Na_2SO_4(aq) + 2\ H_2O(l)$$
77. Find the molarity of a nitric acid solution if 17.5 mL is required to react with 13.9 mL of a 0.755 M NaOH solution according to the following equation.
    $$NaOH(aq) + HNO_3(aq) \rightarrow NaNO_3(aq) + H_2O(l)$$
78. What volume of a 1.00 M K_2CrO_4 solution is needed to react with 50.0 mL of a 0.600 M $BaCl_2$ solution according to the following equation?
    $$BaCl_2(aq) + K_2CrO_4(aq) \rightarrow BaCrO_4 \downarrow + 2\ KCl(aq)$$
79. According to the equation in Problem 78, what mass of $BaCrO_4$ can be obtained from the reaction of 50.0 mL of 0.600 M $BaCl_2$?
80. Find the molarity of a silver nitrate solution if 65.4 mL is required to react completely with 24.9 mL of a 0.500 M NaCl solution according to the following equation.
    $$AgNO_3(aq) + NaCl(aq) \rightarrow AgCl \downarrow + NaNO_3(aq)$$
81. What mass of silver chloride will precipitate from the reaction in Problem 80?

13.12 Colligative properties of solutions

82. Name four colligative properties.
83. Comparing pure water and a 1 M aqueous glucose solution, which has the higher value of each of the following?
    a. Vapor pressure
    b. Boiling point
    c. Freezing point
    d. Osmotic pressure
84. With respect to the same properties, how would a 1 M NaCl solution compare with the two solutions mentioned in Problem 83?

13.13 Osmotic pressure

85. Within a container, compartments A and B are separated by an osmotic membrane. Given the following concentrations of solutions in the compartments, indicate whether there will be a net flow of water and, if so, in which direction.
    a. A contains pure water; B contains 1% w/v glucose.
    b. A contains 1 M glucose; B contains 1 M NaCl.
    c. A contains 1 M NaCl; B contains 0.5 M KBr.
    d. A contains 0.5 M KBr; B contains 1 M glucose.
86. Red blood cells are safely stored in 0.9% w/v NaCl solution because this solution is isotonic with the interior of the cells. What does this mean?
87. Describe each of the following solutions as hypertonic, hypotonic, or isotonic with respect to 0.9% w/v NaCl (0.15 M NaCl).
    a. 2% w/v NaCl
    b. 0.15 M glucose
    c. 0.35 M glucose
    d. 0.12 M $CaCl_2$
    e. 0.30 M glucose

Additional problems

88. Suppose you are given a solution of sodium chloride in water. Can you separate the sodium chloride from the water by filtering the solution? If not, explain how you could separate the sodium chloride.
89. Which solution would have the higher osmotic pressure: 10% w/v NaCl or 10% w/v KCl? Explain.
90. A 300. mL sample of 10% v/v aqueous methyl alcohol contains how many milliliters of methyl alcohol?
91. A cup of coffee may contain as much as 300. mg of caffeine, $C_8H_{10}O_2$. Calculate the molarity of caffeine in one cup of coffee (4 cups = 0.946 L).

92. Wine is approximately 12% v/v ethyl alcohol.
   a. Calculate the volume of alcohol in 750. mL of wine.
   b. Calculate the mass of alcohol in 750. mL of wine (density of alcohol = 0.789 g/mL).
   c. Calculate the molarity of ethyl alcohol, C_2H_6O, in wine.
93. A cucumber placed in concentrated brine (salt water) shrivels into a pickle. Explain.
94. Nitric acid as it is commercially available in the laboratory has a concentration of 16.0 *M*. The density of this solution is 1.42 g/mL.
   a. Calculate the % w/w of nitric acid in this solution.
   b. How many milliliters of this concentrated acid are needed to prepare 250. g of a solution that is 10.0% w/w HNO_3?
95. Vinegar is an aqueous solution of acetic acid. To determine the concentration of a vinegar solution, a student found that 75.0 mL of the solution reacted with 4.00 g of NaOH according to the equation: $HC_2H_3O_2(aq) + NaOH(s) \rightarrow NaC_2H_3O_2(aq) + H_2O(l)$. What is the molarity of the vinegar solution?
96. The LD_{50} of a toxic substance, typically written as (milligrams of substance)/(kilogram of body weight), is the amount of the substance that is found to cause death of 50% of the test animals. The LD_{50} for nicotine ($C_{10}H_{14}N_2$) orally administered to mice is 0.23 mg/kg. How many milliliters of a 0.15 *M* solution of nicotine constitutes an LD_{50} dose for a 250. g mouse?
97. A 29.5 g sample of NaCl is dissolved in 100. mL of an aqueous solution. Calculate the molarity of each ion in this solution.
98. A 50.0 mL sample of a 0.20 *M* $AgNO_3$ solution is mixed with 50.0 mL of a 0.10 *M* $MgCl_2$ solution. Calculate the molarity of each ion in the resulting mixture.
99. Methanol, CH_3OH, is completely soluble in water. However, methane, CH_4, is insoluble in water. Explain.
100. In each case, predict which solution will produce the brightest light in the conductance apparatus shown in Figure 13.12.
   a. 0.1 *M* NaCl versus 0.2 *M* glucose
   b. 0.1 *M* NaCl versus 0.1 *M* Na_3PO_4
   c. 0.1 *M* HCl versus 0.1 *M* acetic acid
101. A 0.90% w/v NaCl solution is isotonic with red blood cells. Calculate the concentration of a solution of Epsom salts ($MgSO_4 \cdot 7H_2O$) that would be isotonic with red blood cells.
102. Methanol, CH_3OH, or ethylene glycol, CH_2OHCH_2OH, can be used as an antifreeze in automobile radiators. If equal weights of each were added to the radiators of two cars, which one would result in the greater freezing point depression?
103. Provide an explanation for each of the following.
   a. Salt is added to ice in an ice cream machine to freeze the ice cream.
   b. Sea water has a lower freezing point than fresh water.
104. Commercial hydrogen peroxide solutions are 3.0% w/w H_2O_2 in water. The density of such a solution is about 1.0 g/mL. Calculate the molarity of the solution.
105. A student has a solution containing sodium nitrate and silver nitrate. What reagent might be used to remove Ag^+ ions from the solution?
106. Sodium chloride is often employed to melt ice on roadways or walkways in the winter. However, the leaching of the salt into the ground and neighboring root systems can affect the osmotic movement of water into trees. Explain.
107. In 1990, the bottled water, Perrier®, was found to contain benzene, C_6H_6, in a concentration of 15 ppb.
   a. Calculate the % w/v of benzene in this solution.
   b. Calculate the molarity of benzene in the solution.

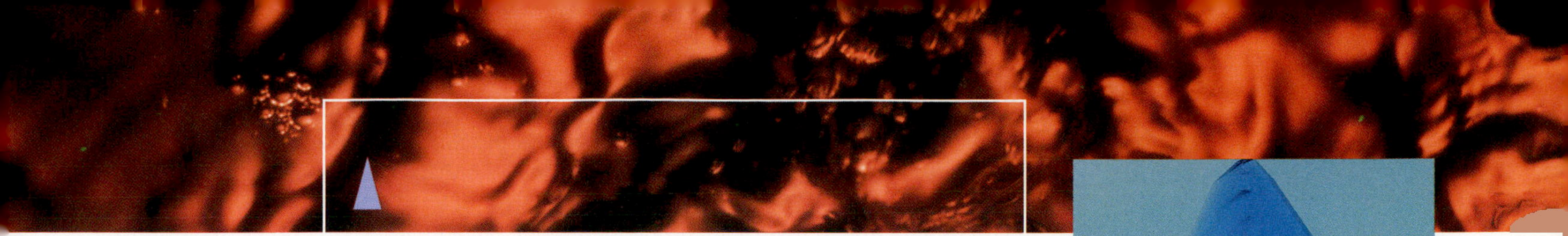

chapter 14

chemical equilibrium

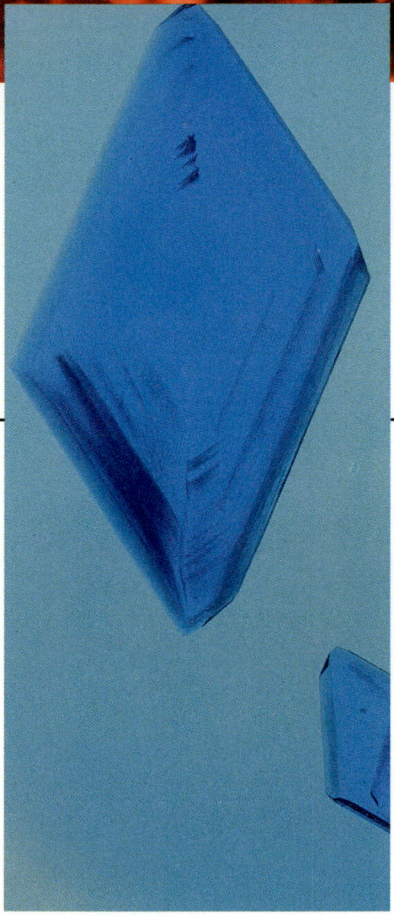

Blue $CuSO_4$ crystals shown above are both dissolving and re-forming at the same rate. Dynamic processes show no net change at equilibrium.

14.1 introduction

Throughout the earlier chapters of this book, you considered chemical reactions that go to completion. When the limiting reagent has been completely consumed, the reaction can proceed no further. Such is the case when you burn a compound in the presence of excess oxygen; all of the material is consumed and carbon monoxide gas is released. Reactions go to completion when a gaseous product or precipitate is formed from reactants in solution. The baking of a cake also releases carbon dioxide gas and continues until the limiting ingredient is consumed. The deposition of insoluble compounds in a water pipe results from the precipitation reaction between dissolved ions in the water (Figure 14.1).

(a)

(b)

(c)

Figure 14.1
(a) Combustion, (b) cake baking, and (c) the clogging of water pipes are the result of reactions going to completion.

In many chemical reactions, however, the products can react to re-form the original reactants. Reactions in which the products can themselves react to re-form reactants are called **reversible reactions.** The formation of stalactites and stalagmites results from a reversible reaction in which calcium ions, Ca^{2+}, and hydrogen carbonate ions, HCO_3^-, in the groundwater react to form the precipitate calcium carbonate, $CaCO_3$, according to the reaction,

$$Ca^{2+}(aq) + 2\ HCO_3^-(aq) \rightleftharpoons CaCO_3(s) + CO_2(aq) + H_2O(l) \qquad (14.1)$$

Stalactites and stalagmites are built from the insoluble $CaCO_3$ deposited from this reaction. Whether the precipitate, $CaCO_3$, forms to build stalactites or stalagmites depends on the conditions of the reaction. When the dissolved product, CO_2, escapes from the solution as a gas, as it can in underground caves, $CaCO_3$ precipitates from the solution and the reaction proceeds to completion. However, when the partial pressure of CO_2 above the solution is high, as it is in soil, the CO_2 formed in Equation 14.1 cannot escape from the solution; instead, the products $CaCO_3(s)$, $CO_2(aq)$, and $H_2O(l)$ recombine to form the reactant ions. In this case, the reactant ions remain dissolved in solution and no net change occurs (Figure 14.2).

The conditions under which a reversible reaction occurs determine just how far the reaction will proceed. Under some conditions, the reaction will completely form products, as in the formation of stalactites and stalagmites. Under other conditions, the reaction will almost completely favor reactants, as in groundwater with dissolved calcium and hydrogen carbonate ions. Under intermediate conditions, the same reaction will yield some product, but will reach an equilibrium mixture with both products and reactants present.

Because there is no limiting reagent in a reversible reaction, you do not know how much product will be formed when the reaction mixture reaches equilibrium. As a result, the yield of product from the equilibrium mixture cannot be determined from the stoichiometry of the balanced chemical equation as you have done in previous examples (Section 10.9).

A major purpose of this chapter will be to learn just how much product may be produced in a reversible reaction. How far a reaction will proceed under given conditions is of critical importance in laboratory and industrial synthesis. The Haber process, which is used to synthesize 120 million tons of ammonia annually from the elements nitrogen and hydrogen, is a reversible reaction (Section 9.4). The conditions that maximize the yield of ammonia are of great economic importance.

Figure 14.2

Whether calcium carbonate, $CaCO_3$, precipitates from groundwater to build stalactites or stalagmites depends on the conditions of a reversible reaction. Calcium ions, Ca^{2+}, and hydrogen carbonate ions, HCO_3^-, precipitate as solid calcium carbonate, $CaCO_3$, and the product carbon dioxide, CO_2, may escape as a gas from the groundwater in underground caves. When the partial pressure of CO_2 surrounding the groundwater is high, as it is in soil, the product CO_2, recombines with water and calcium carbonate to form the reactant ions. As a result, no precipitate of calcium carbonate forms.

14.2 reversible reactions

In addition to the precipitation reaction of calcium carbonate in forming stalactites and stalagmites, you have seen other examples of *reversible* processes. For example, in Section 12.6, you encountered the two opposing processes, vaporization and condensation, that occur whenever a liquid is in a closed container. The liquid evaporates at a constant rate (speed) for any given temperature. As the pressure of the vapor increases above the liquid, the rate of condensation increases until the rate of condensation equals the rate of vaporization, and equilibrium has been reached. At equilibrium, there is *no net change* in the number of molecules in the gas phase, and, thus, the pressure reaches a constant value called the vapor pressure (Section 12.6).

It is important to visualize the correct kinetic molecular model of molecules in any equilibrium process. When measuring the vapor pressure of a liquid, the liquid and vapor molecules are not *static* (inactive). Rather, liquid molecules are continually escaping into the vapor and vapor molecules are continually condensing into the liquid, although the processes are difficult to see because they are occurring at the same rate. In Section 12.6, you learned that whenever the two opposing processes coexist at the same rate, there is no net change and you have a state of **dynamic (active) equilibrium.**

A dynamic equilibrium also exists between dissolved and undissolved solute, such as copper sulfate or sugar, in a *saturated* solution (Figure 14.3).

$$CuSO_4(s) \rightleftharpoons Cu^{2+}(aq) + SO_4^{2-}(aq)$$

$$\text{Sugar}(s) \rightleftharpoons \text{Sugar}(aq)$$

The two arrows ($\rightleftharpoons$) indicate that the process is reversible and can proceed in either direction. When the solution is saturated, the rate at which the solid $CuSO_4$ dissociates into ions is equal to the rate at which ions precipitate from solution. Likewise, the rate at which solid sugar molecules dissolve is equal to the rate at which the molecules precipitate from solution.

Keeping in mind this concept of two opposing processes, one that is exactly the reverse of the other, let us consider another example of a reversible chemical reaction.

If you place 0.1 mol of N_2O_4, dinitrogen tetroxide, a colorless gas, into a closed flask and heat it to 100°C, you observe the formation of a reddish-brown gas. On chemical analysis, this gas proves to be NO_2, nitrogen dioxide, the same constituent visible in photochemical smog. From the color change in your reaction tube, you have visible evidence that the reaction $N_2O_4 \rightarrow 2\ NO_2$ is occurring. The intensity of the reddish-brown color increases rapidly in the beginning, but after a period of time, equilibrium is established and the color remains at a maximum intensity. The contents of the container at this point are found to contain both NO_2 and N_2O_4 (Figure 14.4).

If you reverse the experiment and start with 0.1 mol of NO_2 at the same temperature, the reddish-brown color begins to diminish. This is evidence that the reaction $2\ NO_2 \rightarrow N_2O_4$ is occurring. The reddish-brown color decreases as NO_2 is consumed and colorless N_2O_4 is produced. After a period of time, equilibrium is established and the reddish-brown color reaches a minimum intensity, and does not diminish further.

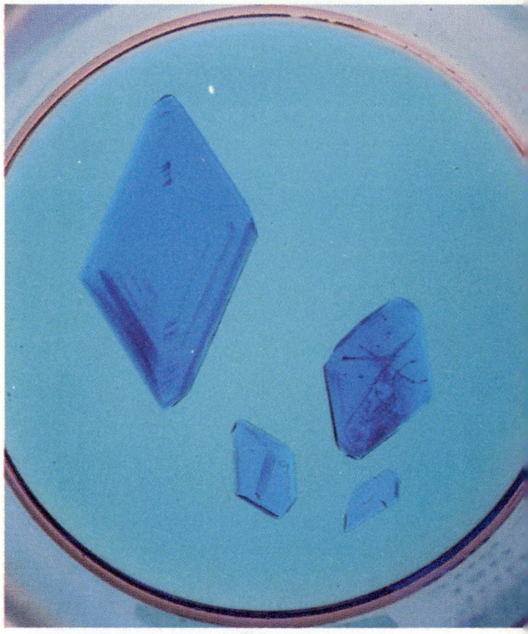

Figure 14.3

Even though a saturated solution of copper sulfate or sugar shows no net change, molecules of the solid are continually dissolving and dissolved ions are continually precipitating in a dynamic equilibrium.

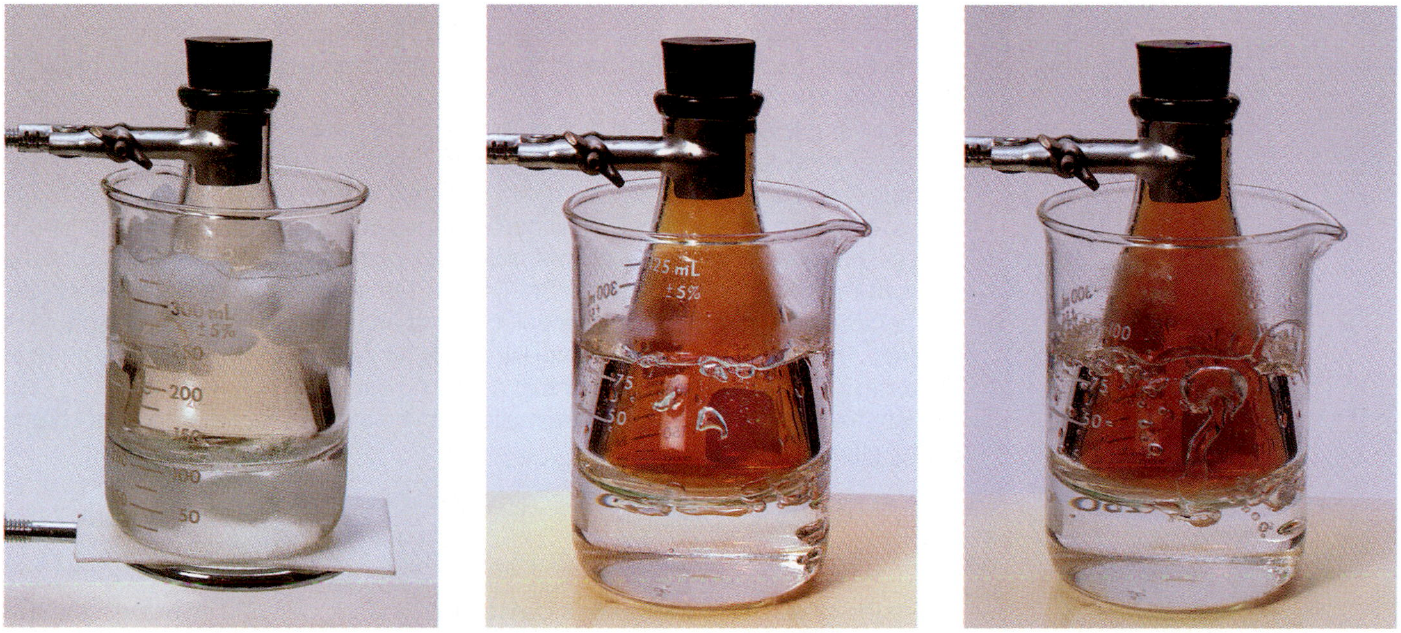

(a) (b) (c)

Figure 14.4

Starting with pure colorless dinitrogen tetroxide, N_2O_4 (a) produces reddish-brown nitrogen dioxide (NO_2) (b) upon heating. The concentration of NO_2 increases until an equilibrium mixture of N_2O_4 and NO_2 is obtained (c).

You may conclude from these experiments that the reaction is reversible because of the following reasons.

1. If you start with N_2O_4, NO_2 forms, and if you start with NO_2, N_2O_4 forms.
2. The reactant is not totally converted to product in either experiment. Evidence for this lies in the color of the mixture. If you begin with colorless N_2O_4, a reddish-brown color as dark as pure NO_2 will never develop. If you begin with NO_2, a totally colorless gas indicative of pure N_2O_4 will not be produced.

The reaction between N_2O_4 (dinitrogen tetroxide) and NO_2 (nitrogen dioxide) can be written like Equation 14.1 with double arrows to indicate a reversible reaction:

$$N_2O_4(g) \rightleftharpoons 2\ NO_2(g) \qquad (14.2)$$

You remember that once any reversible reaction reaches equilibrium, the rate of the forward reaction and the rate of the reverse reaction are equal. You will see in Section 14.3 that the balance between forward and reverse rates leads to an important equilibrium relationship that tells you how much product will be present in the equilibrium mixture. But before you can understand a reaction equilibrium in detail, you need to expand mathematically just what is meant by the rate of a chemical reaction.

14.3 rates of reaction

The **reaction rate,** or speed, of a chemical reaction is a measure of the change in concentration of either a reactant or product in a given amount of time. It is the change in concentration per change in time, just as travel rate (miles or kilometers per hour) is the change in distance per change in time. The more reactant that is used up or the more product that is formed per second, the greater the rate of the chemical reaction.

The rate of decomposition of dinitrogen tetroxide (Equation 14.2) can be determined by measuring the increase in reddish-brown color per unit time. The increase in color is directly related to the increase in NO_2 concentration. Similarly, the rate of the reverse reaction can be determined by measuring the decrease in reddish-brown color per unit time.

For a particular reaction at a given temperature, answering the question of how fast or slow the reaction rate is depends on the concentration (moles per liter) of the reacting species. Before a reaction can occur, the reacting particles must collide with each other. Just as you have envisioned changes in *physical state* using the kinetic molecular theory (Section 11.10), *chemical reactions* may also be understood on the basis of the motion of molecules. The *greater the concentrations of reactants, the greater the number of collisions* per unit time. An analogy can be made to a game of billiards. The greater the number of billiard balls on the table, the greater the number of possible and probable collisions with a moving ball.

Chemists have shown experimentally that you can express the rate of a chemical reaction as

$$\text{Rate} \propto (\text{Concentration of reacting species})^c$$

or

$$\text{Rate} = k[\text{Reacting species}]^c \qquad (14.3)$$

where the brackets [ ] are read as concentration. A proportionality expression can be converted to an equation through the use of a proportionality constant (Appendix A.6). In this case, the proportionality constant k is called a **rate constant,** which relates the rate of reaction to the concentration of reacting species. Note also that the concentration of reacting species is raised to the power c, an exponent that, in general, must be determined from experiment. For elementary, single-step chemical reactions, however, the exponent c of a reactant has the same value as the

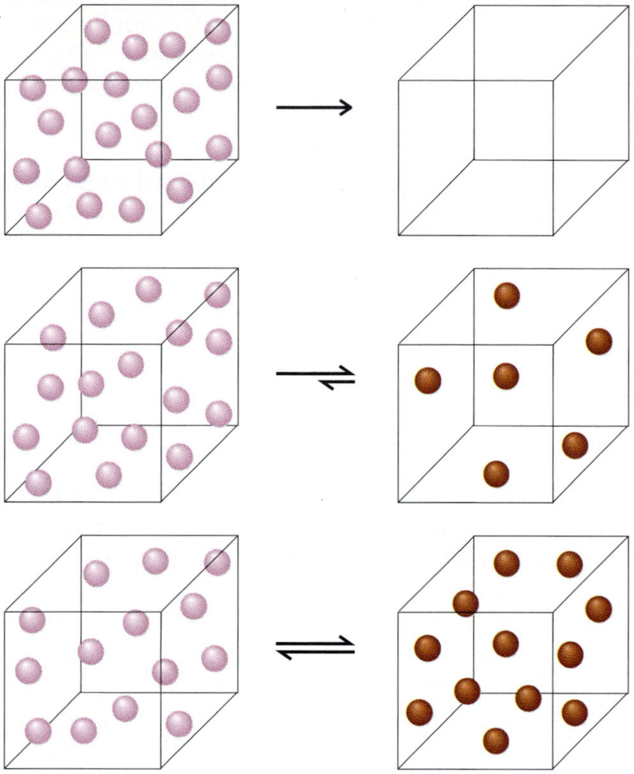

Figure 14.5

The decomposition of $N_2O_4 \rightleftharpoons 2\ NO_2$ is an equilibrium reaction illustrated by the following sequence. The lengths of the arrows indicate the rates of the forward and reverse reactions at each stage. (a) At first, only pure colorless N_2O_4 is present and the forward reaction occurs. (b) As reddish-brown NO_2 forms, the reverse reaction occurs, but slowly because of the small concentration of NO_2. As the reaction proceeds, the rate of the forward reaction decreases because the concentration of N_2O_4 has decreased, and the rate of the reverse reaction increases because the concentration of NO_2 has increased. (c) At equilibrium, the rates of the forward and reverse reactions are equal; $[N_2O_4]$ and $[NO_2]$ remain fixed at their equilibrium values.

stoichiometric coefficient of that reactant in the balanced chemical equation. Equation 14.3 predicts that the greater the concentration of reactant species and the larger the exponent *c,* the greater the reaction rate.

Consider the reaction represented by Equation 14.2

$$N_2O_4(g) \rightleftharpoons 2\ NO_2(g)$$

in terms of the effect of concentration on reaction rate. The rate of the forward reaction depends on the concentration of N_2O_4 (written as $[N_2O_4]$), and the rate of the reverse reaction depends on the concentration of NO_2 (written as $[NO_2]$). Each of the concentrations is raised to an experimentally determined power that, for this reversible reaction, corresponds to its stoichiometric coefficient in the balanced equation. The rate of the forward reaction is proportional to the concentration of dinitrogen tetroxide raised to the first power ($[N_2O_4]^1$), and the rate of the reverse reaction is proportional to the concentration of nitrogen dioxide raised to the second power ($[NO_2]^2$).

If you begin with pure N_2O_4, then initially the concentration of N_2O_4 is large. There are many collisions of N_2O_4 molecules, so the rate of decomposition to NO_2 is fast. As time passes, $[N_2O_4]$ decreases, the chances for collisions decrease, and, consequently, the rate of decomposition decreases. Meanwhile, $[NO_2]$ is increasing, and, thereby, the rate of the reverse reaction is increasing. At some point, the rate of the reverse reaction becomes equal to the rate of the forward reaction, and the concentrations of reactants and products no longer change. This process is illustrated in Figure 14.5.

When the rate of the forward reaction equals the rate of the reverse reaction, the reaction is at **chemical equilibrium,** and the concentrations of reactants and products at equilibrium are called **equilibrium concentrations.** It is important to remember that the equilibrium condition is a dynamic (active) condition.

At equilibrium,

$$\text{Rate}_{\text{forward}} = \text{Rate}_{\text{reverse}} \qquad (14.4)$$

This may be rewritten using Equation 14.3 in terms of the rate constants for the forward and reverse reactions and the corresponding equilibrium concentrations of reacting species each raised to its appropriate power, $[N_2O_4]_e$ and $[NO_2]_e^2$. Note that the subscript *e* denotes explicitly the concentrations *at equilibrium.*

$$k_{\text{forward}}\,[N_2O_4]_e = k_{\text{reverse}}\,[NO_2]_e^2 \qquad (14.5)$$

Isolating the equilibrium concentrations on the right of the equation by dividing both sides by $[N_2O_4]_e$ and k_{reverse} yields

$$\frac{k_{\text{forward}}\,\cancel{[N_2O_4]_e}}{k_{\text{reverse}}\,\cancel{[N_2O_4]_e}} = \frac{\cancel{k_{\text{reverse}}}\,[NO_2]_e^2}{\cancel{k_{\text{reverse}}}\,[N_2O_4]_e}$$

or, upon canceling like terms,

$$\frac{k_{\text{forward}}}{k_{\text{reverse}}} = \frac{[NO_2]_e^2}{[N_2O_4]_e} = \text{constant} = K_{eq} \qquad (14.6)$$

Equation 14.6 is a specific example of the most important concept of chemical equilibrium. Since the forward and reverse rate constants are indeed *constants* and do not change, their ratio on the left side of Equation 14.6 has a constant value called K_{eq}. K_{eq} is called the **equilibrium constant** for the reaction. *It states that when the decomposition reaction of N_2O_4 reaches equilibrium at a given temperature, the ratio of product to reactant concentrations is always the same.*

In the next section, an equilibrium constant will be defined for any chemical reaction. The values of equilibrium constants are particularly important for reversible reactions because they tell you just how far the forward reaction proceeds before it stops, that is, before it reaches equilibrium.

14.4 equilibrium constant

Now that you have seen an equilibrium constant for a specific chemical reaction, it will be straightforward to generalize the concept to any reversible reaction in which *A* and *B* combine in a single step to give *C* and *D,* or *C* and *D* combine to form *A* and *B.*

$$A + B \rightleftharpoons C + D$$

Applying Equation 14.3 to this reaction, where *c* is equal to 1, you can write individual expressions for the rates of the *forward* and *reverse* reactions at any instant in time:

$$\text{Rate} = k\,[\text{Reacting species}] \qquad (14.3)$$

$$\text{Rate}_{\text{forward}} = k_{\text{forward}}[A]\,[B] \qquad (14.7)$$

$$\text{Rate}_{\text{reverse}} = k_{\text{reverse}}[C]\,[D] \qquad (14.8)$$

In the above equations, the proportionality constant *k* is the *rate constant* for either the forward or reverse reaction, and it is fixed in value for a *particular reaction at a given temperature.* The bracket symbol [ ] throughout the remainder of this text means that the concentration of a substance is expressed in molarity (moles per liter). For example, [*A*] is read as "the concentration of *A* in moles per liter."

At the condition of equilibrium, remember that

$$\text{Rate}_{\text{forward}} = \text{Rate}_{\text{reverse}} \qquad (14.4)$$

Therefore, because the left-hand terms of Equations 14.7 and 14.8 are equal, the right-hand terms must also be equal.

$$k_{\text{forward}}[A]_e[B]_e = k_{\text{reverse}}[C]_e[D]_e$$

Rewriting this to isolate the constants to one side gives

$$\frac{k_{\text{forward}}}{k_{\text{reverse}}} = \frac{[C]_e[D]_e}{[A]_e[B]_e} \quad \text{only at equilibrium}$$

Because k_{forward} and k_{reverse} are both constants, their ratio is also a constant, called the *equilibrium constant,* and it is often symbolized by K_{eq}.

$$\frac{k_{\text{forward}}}{k_{\text{reverse}}} = K_{eq} = \frac{[C]_e[D]_e}{[A]_e[B]_e} \qquad (14.9)$$

Equilibrium constant (K_{eq}) **Equilibrium expression** ($\frac{[C]_e[D]_e}{[A]_e[B]_e}$)

Each reversible reaction has a unique K_{eq} that has a fixed value for a particular temperature. The concentrations in the equilibrium expression are the equilibrium concentrations. In Section 14.5, you will see how the magnitude of K_{eq} offers valuable information about the amounts of products and reactants present at equilibrium in a reversible reaction.

The coefficients of an equation for a reversible reaction must be reflected in the equilibrium expression. The numerator of the equilibrium expression contains each product concentration *raised to a power equal to the coefficient of that product in the balanced equation.* The denominator contains each reactant concentration also *raised to a power equal to the coefficient of that reactant in the balanced equation.* In Equation 14.9, the powers of all reactants and products were understood to be 1. For the general balanced equation with individual coefficients,

Coefficients (a, b, c, d)

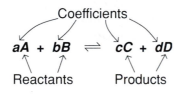

$$aA + bB \rightleftharpoons cC + dD$$

Reactants (A, B) Products (C, D)

the equilibrium expression is written as

$$K_{eq} = \frac{[C]^c[D]^d}{[A]^a[B]^b} \qquad (14.10)$$

where all the concentrations are understood to be at equilibrium.

---

sample exercise 1

Write an equilibrium-constant expression for the reversible reaction $N_2O_4 \rightleftharpoons 2\ NO_2$.

solution

1. In the numerator, write the concentration of each product raised to a power equal to its coefficient in the balanced equation.

$$K_{eq} = \frac{[NO_2]^2}{?}$$

2. In the denominator, write the concentration of each reactant raised to a power equal to its coefficient in the balanced equation.

$$K_{eq} = \frac{[NO_2]^2}{[N_2O_4]}$$

---

sample exercise 2

When the color in the reaction vessel remained the same, indicating that equilibrium had been established for the gas mixture reaction, a chemist found that the concentration of N_2O_4 in the vessel was 0.040 mol/L and the concentration of NO_2 was 0.12 mol/L. Calculate the value of the equilibrium constant for the reaction $N_2O_4(g) \rightleftharpoons 2\ NO_2(g)$.

solution

Use the equilibrium-constant expression determined in Sample Exercise 1.

$$K_{eq} = \frac{[NO_2]^2}{[N_2O_4]}$$

Fill in the appropriate numerical concentrations and perform the mathematical operations indicated.

$$K_{eq} = \frac{(0.12)^2}{0.040} = \frac{0.0144}{0.040} = 0.36$$

The equilibrium constant may have units associated with it depending on the particular equilibrium. However, units of K_{eq} are customarily ignored.

••• problem 1

Write an equilibrium-constant expression for each of the following equilibria. Describe the physical changes that you would observe as the mixture of reactants reaches equilibrium with products in each reversible reaction. H_2 and N_2 are colorless gases, and I_2 is a purple vapor. HI is an acrid, colorless gas that attacks rubber and is used to manufacture hydriodic acid. NH_3 is a colorless gas with a pungent odor; it is used in fertilizers and in the manufacture of nitric acid.

a. $H_2(g) + I_2(g) \rightleftharpoons 2\ HI(g)$

b. $N_2(g) + 3\ H_2(g) \rightleftharpoons 2\ NH_3(g)$

Guidelines for writing K_{eq}

Certain rules have been established for writing correct equilibrium expressions.

1. An equilibrium-constant expression can only be written for a balanced chemical equation.
2. At a particular temperature, each reaction has a unique equilibrium-constant value. The value of the equilibrium constant changes with temperature.
3. The equilibrium expression contains the concentration terms for *gases and substances in solution only.*
4. The concentration term for a pure liquid or solid ([liquid] or [solid]) does not appear in the normal equilibrium expression.

The concentration of a pure liquid or solid is a constant; there is no solute present, so the number of moles per liter cannot change. For example, if we consider pure water, for which we know the density to be 1.0 g/mL, we can calculate the concentration in moles per liter to be 55 *M*.

$$\frac{1.0\ \text{g}}{\text{mL}} \times \frac{1{,}000}{1{,}000} = \frac{1{,}000\ \text{g}}{1{,}000\ \text{mL}} = \frac{1{,}000\ \cancel{\text{g}}}{1\ \text{L}} \times \frac{1\ \text{mol}}{18\ \cancel{\text{g}}} = \frac{55\ \text{mol}}{\text{L}} = 55\ M$$

A sample of pure water always has this concentration. *All pure solids or liquids have constant concentrations.*

As a demonstration of rule 4, consider the following equilibrium in which pure liquid water appears as a product.

$$CO_2(g) + H_2(g) \rightleftharpoons CO(g) + H_2O(l)$$

Using Equation 14.10, you may write

$$K = \frac{[CO]\,[H_2O]}{[CO_2]\,[H_2]}$$

But $[H_2O]$ has a constant value, so if you divide each side of this equation by $[H_2O]$, the left side will still be a constant, but with a different value:

$$\frac{K}{[H_2O]} = K_{eq} = \frac{[CO]}{[CO_2]\,[H_2]}$$

The value of the new constant is defined as the equilibrium constant for the given reaction. As a matter of practicality, simply omitting the concentration of a pure solid or liquid when you write an equilibrium expression has the desired effect of incorporating these concentrations into the value of K_{eq}.

Equilibria involving pure solids and pure liquids are called **heterogeneous equilibria.** You may treat them in the same manner as **homogeneous equilibria,** those not involving pure liquids and pure solids, as long as you follow the rules outlined earlier.

An important application of heterogeneous equilibrium in which the concentration of a pure solid is omitted from the equilibrium expression is that of a saturated aqueous solution of an insoluble ionic compound and its dissolved ions. For example, $CaCO_3$, calcium carbonate, is an insoluble ionic compound that forms stalactites and stalagmites. In a saturated aqueous solution, only a few ions will dissolve according to the equilibrium,

$$CaCO_3(s) \rightleftharpoons Ca^{2+}(aq) + CO_3^{2-}(aq)$$

The expression for the equilibrium constant will include only the product of concentrations of the dissolved ions, not the concentration of the pure, solid calcium carbonate.

$$K_{eq} = [Ca^{2+}]\,[CO_3^{2-}]$$

The equilibrium constant for the equilibrium between a solid and its dissolved ions has a special name, the **solubility product constant,** abbreviated K_{sp}. Therefore, the equilibrium expression for a saturated calcium carbonate solution may be written as

$$K_{sp} = [Ca^{2+}]\,[CO_3^{2-}]$$

---

sample exercise 3

Write the equilibrium-constant expression for the reaction of methyl chloride with chlorine gas to produce carbon tetrachloride, a cleaning solvent, and hydrogen chloride gas.

$$CH_3Cl(g) + 3\,Cl_2(g) \rightleftharpoons CCl_4(l) + 3\,HCl(g)$$

solution

1. Put the product concentrations raised to the power of their coefficients in the numerator. Do not include CCl_4 because it is a pure liquid.

$$K_{eq} = \frac{[HCl]^3}{?}$$

2. The denominator contains the reactant concentrations raised to the power of their coefficients in the balanced equation.

$$K_{eq} = \frac{[HCl]^3}{[CH_3Cl]\,[Cl_2]^3}$$

---

sample exercise 4

An equilibrium exists in a saturated aqueous solution of silver chloride between the solid silver chloride and the dissolved silver and chloride ions.

$$AgCl(s) \rightleftharpoons Ag^+(aq) + Cl^-(aq)$$

Write the equilibrium expression.

solution

Solid silver chloride is omitted from the equilibrium expression, leaving terms only in the numerator. The equilibrium constant for the equilibrium between a solid and its dissolved ions is known as the *solubility product constant,* abbreviated K_{sp}.

$$K_{sp} = [Ag^+]\,[Cl^-]$$

sample exercise 5

Write the equilibrium expression for the reaction of ammonia with oxygen to form nitrogen monoxide (nitric oxide) and water vapor.

$$4\ NH_3(g) + 5\ O_2(g) \rightleftharpoons 4\ NO(g) + 6\ H_2O(g)$$

solution

1. Determine the terms in the numerator.

$$K_{eq} = \frac{[NO]^4\,[H_2O]^6}{?}$$

The concentration of water is included in K_{eq} because water is present as a gas in this reaction.

2. Determine the terms in the denominator.

$$K_{eq} = \frac{[NO]^4\,[H_2O]^6}{[NH_3]^4\,[O_2]^5}$$

NO is used in the bleaching of rayon. However, immediately upon contact with air, it is converted to highly poisonous nitrogen dioxide (NO_2), nitrogen tetroxide (N_2O_4), or both. Adequate ventilation and gas masks are mandatory when handling even small amounts in the laboratory (Figure 14.6).

• • • problem 2

Write the equilibrium-constant expression for each of the following equilibria.

a. $CO(g) + 2\ H_2(g) \rightleftharpoons CH_3OH(l)$

b. $NH_4Cl(s) \rightleftharpoons NH_3(g) + HCl(g)$

c. $2\ Fe(s) + 3\ H_2O(g) \rightleftharpoons Fe_2O_3(s) + 3\ H_2(g)$

• • • problem 3

Write the solubility-product-constant expression for the equilibria in the following saturated solutions.

a. $CaCO_3(s) \rightleftharpoons Ca^{2+}(aq) + CO_3^{2-}(aq)$

b. $PbCl_2(s) \rightleftharpoons Pb^{2+}(aq) + 2\ Cl^-(aq)$

c. $Ba_3(PO_4)_2(s) \rightleftharpoons 3\ Ba^{2+}(aq) + 2\ PO_4^{3-}(aq)$

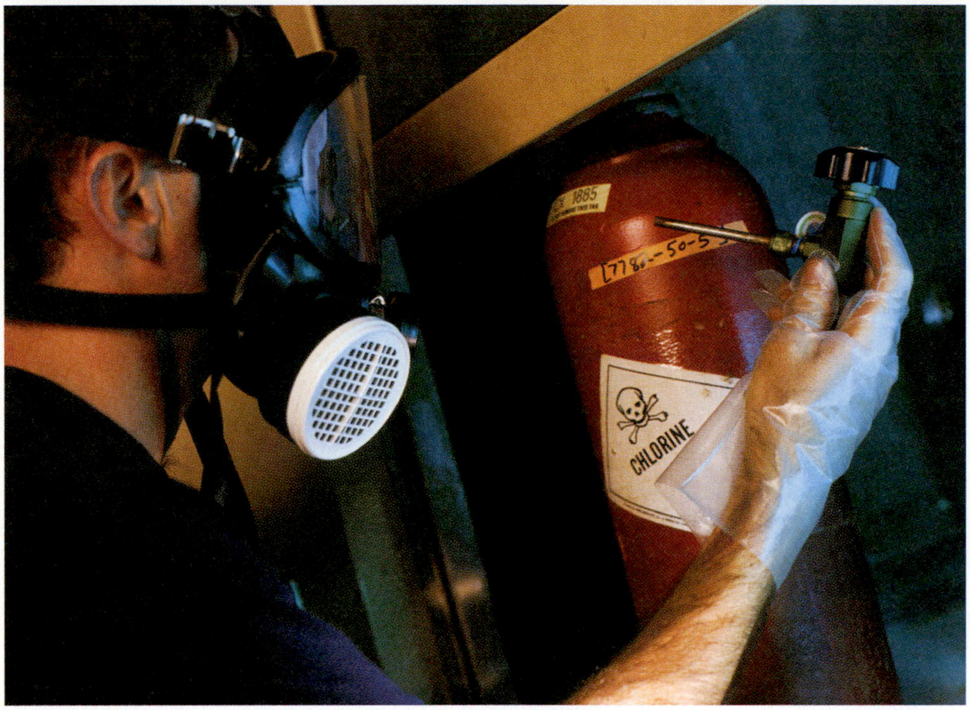

Figure 14.6

Adequate ventilation from a chemical hood and use of a gas mask are mandatory when handling even small amounts of toxic gases such as chlorine, Cl_2, in the laboratory.

14.5 interpreting the value of K_{eq}

Because the equilibrium constant gives you a measure of the concentration of product $[P]$ compared with the concentration of reactant $[R]$ at equilibrium, the value of K_{eq} tells you how far a reaction has proceeded toward product at the equilibrium point. Consider the simple example

$$R \rightleftharpoons P$$

for which the equilibrium expression is

$$K_{eq} = \frac{[P]}{[R]}$$

If the equilibrium constant is large ($>1 \times 10^2$), then the equilibrium system is made up of mostly product, and we say that the reaction has proceeded far to the right. If $[P]/[R] = 100$, then clearly the numerator $[P]$ is considerably greater than the denominator $[R]$.

If the equilibrium constant is small ($<1 \times 10^{-2}$), the equilibrium condition consists mainly of reactant and the equilibrium is said to be far to the left. If $[P]/[R] = 0.01$, then clearly the denominator $[R]$ is considerably greater than the numerator $[P]$.

For reactions in which the numerical value of K_{eq} is between 1×10^{-2} and 1×10^2, the equilibrium state is a mixture containing significant quantities of both products and reactants. Table 14.1 relates K_{eq} values to product and reactant concentrations at equilibrium, and Sample Exercise 6 offers some examples.

sample exercise 6

Given the K_{eq} value, describe qualitatively the equilibrium composition for each of the following reactions.

a. $N_2O_4(g) \rightleftharpoons 2\ NO_2(g)$ — K_{eq} (at 100°C) = 3.6×10^{-1}
b. $N_2(g) + 3\ H_2(g) \rightleftharpoons 2\ NH_3(g)$ — K_{eq} (at 25°C) = 5×10^8
c. $H_2(g) + Cl_2(g) \rightleftharpoons 2\ HCl(g)$ — K_{eq} (at 25°C) = 2.4×10^{33}
d. $Cl_2(g) \rightleftharpoons 2\ Cl(g)$ — K_{eq} (at 1,000°C) = 1.2×10^{-6}
e. $N_2(g) + O_2(g) \rightleftharpoons 2\ NO(g)$ — K_{eq} (at 25°C) = 1×10^{-30}

table 14.1 comparison of K_{eq} values and reactant and product concentrations at equilibrium

K_{eq}	Description of equilibrium condition
1×10^{30} 1×10^{20} 1×10^{10}	Essentially all products at equilibrium
1×10^{2} 1 1×10^{-2}	Significant amounts of products and reactants at equilibrium
1×10^{-10} 1×10^{-20} 1×10^{-30}	Essentially all reactants at equilibrium; reaction has not occurred to any significant extent

solution

a. By the guidelines given in Table 14.1, the equilibrium composition will consist of significant quantities of product and reactant because K_{eq} is between 1×10^{-2} and 1×10^{2}.

b. The equilibrium system will consist mostly of product.

c. The equilibrium state will contain essentially only product. The reaction is said to have gone to completion.

d. There will be a much greater amount of reactant than product in the equilibrium state.

e. Essentially only reactants will remain at equilibrium. The reaction does not proceed as written; the reverse reaction goes to completion.

problem 4

Given the K_{eq} value and the guidelines in Table 14.1, describe qualitatively the equilibrium composition for each of the following reactions.

a. $I_2(g) + Cl_2(g) \rightleftharpoons 2\ ICl(g)$ $\quad K_{eq} = 2.0 \times 10^{5}$

b. $SO_2(g) + NO_2(g) \rightleftharpoons SO_3(g) + NO(g)$ $\quad K_{eq} = 2.5 \times 10^{-1}$

c. $ZnO(s) + H_2(g) \rightleftharpoons Zn(s) + H_2O(g)$ $\quad K_{eq} = 2.0 \times 10^{-16}$

14.6 Le Chatelier's principle

In Section 14.1, you learned that the formation of stalactites and stalagmites was governed by the precipitation of calcium carbonate, $CaCO_3$, from ions dissolved in groundwater according to the equation

$$Ca^{2+}(aq) + 2\ HCO_3^-(aq) \rightleftharpoons CaCO_3(s) + CO_2(aq) + H_2O(l) \qquad (14.1)$$

Whether the precipitate, $CaCO_3$, forms to build the stalactites and stalagmites depends on the conditions of the reaction. When the dissolved product, CO_2, escapes from the solution as a gas, as it can in underground caves, $CaCO_3$ precipitates from the solution and the reaction proceeds to completion. However, when the partial pressure of CO_2 above the solution is high, as it is in soil, the CO_2 formed in Equation 14.1 cannot escape from the solution; instead, the products $CaCO_3(s)$, $CO_2(aq)$, and $H_2O(l)$ recombine to form the reactant ions. In this case, the reactant ions remain dissolved in solution and no net change occurs.

In the discussion of stalactites and stalagmites, you saw that it is possible to shift the point at which equilibrium is achieved (called the **position of equilibrium**) when the pressure of CO_2 is varied. The equilibrium concentrations of products and reactants may also be shifted by altering the concentration or temperature of the equilibrium mixture. This means that you can sometimes shift a reaction in such a way as to produce more of a desired product.

Shifts in equilibrium may be predicted by **Le Chatelier's** (lah shah-tuh-lyay´ 's) **principle,** which states that *if a stress is applied to a system in equilibrium, the equilibrium will shift to relieve that stress.* Let us examine how this principle affects the yield of ammonia (NH_3) in the reaction

$$N_2(g) + 3\,H_2(g) \rightleftharpoons 2\,NH_3(g)$$

The starting point in the following discussion is always the equilibrium condition; that is, you start with equilibrium concentrations of N_2, H_2, and NH_3.

effect of concentration

Stressing the equilibrium by increasing the concentration of either reactant, N_2 or H_2, will cause the equilibrium to shift to the product side away from the stress of added reactant concentration. The equilibrium will reestablish at a higher concentration of ammonia and a lower concentration of the reactant that was not added. Therefore, a higher yield of ammonia is obtained. For example, *stress by adding* N_2 *to the equilibrium*

$$\underset{\textbf{Add}}{N_2} + 3\,H_2 \rightleftharpoons 2\,NH_3$$

results in Shift → to consume N_2 and relieve stress

Such shifting to the right also consumes H_2 and produces more NH_3.

If the equilibrium is disturbed by removing ammonia, the system will react to this stress of lowered product concentration by shifting to the product side. The result will be an increase in the yield of ammonia. More N_2 and H_2 will be consumed. *Stress by removing* NH_3 *from the equilibrium*

$$N_2 + 3\,H_2 \rightleftharpoons \underset{\textbf{Remove}}{2\,NH_3}$$

results in Shift → to replace NH_3 and relieve stress

Such shifting to the right also consumes N_2 and H_2.

If the equilibrium is disturbed by adding ammonia, the system will react to this stress of increased product concentration by shifting to the reactant side. The result will be an increase in the equilibrium concentrations of N_2 and H_2. *Stressing by adding* NH_3 *to the equilibrium*

$$N_2 + 3\,H_2 \rightleftharpoons \underset{\textbf{Add}}{2\,NH_3}$$

results in Shift ← to consume NH_3 and relieve stress

Such shifting to the left also increases $[N_2]$ and $[H_2]$.

effect of temperature

You have seen in Section 10.10 that heat can be treated as if it were a product in an exothermic reaction or a reactant in an endothermic reaction. Increasing the temperature is then thought of as providing more heat for the reaction, and decreasing the temperature as providing less heat for the reaction.

The reaction of hydrogen and nitrogen to produce ammonia is exothermic; that is, heat is released by the reaction and may be written as a product:

$$N_2 + 3\ H_2 \rightleftharpoons 2\ NH_3 + \textbf{heat}$$

Increasing the temperature of this reaction causes the equilibrium to shift to the reactant side, thereby decreasing the yield of ammonia. On the other hand, the reaction responds to the stress of decreased temperature by shifting the equilibrium to the product side, thus increasing the yield of ammonia. *Stressing the equilibrium by reducing heat* (lowering the temperature of the reaction),

$$N_2 + 3\ H_2 \rightleftharpoons 2\ NH_3 + \text{heat}$$

Remove;
lower temperature

results in Shift → to produce heat

Such shifting to the right also produces NH_3.

Stressing the equilibrium by adding heat (increasing the temperature of the reaction),

$$N_2 + 3\ H_2 \rightleftharpoons 2\ NH_3 + \text{heat}$$

Add;
raise temperature

results in Shift ← to consume heat

Such shifting to the left also consumes NH_3.

effect of external pressure

A change in external pressure affects the equilibrium of only those reactions that undergo a change in volume in proceeding from reactant to product. In practice, this means that pressure changes affect those reactions that show a change in the number of moles of gaseous materials. The stress of an increased pressure (decreased V) is relieved by a shift in the equilibrium to the side having fewer gas molecules. A decrease in pressure (increased V) shifts the equilibrium to the side having a greater number of gas molecules. Again, consider the ammonia example,

$$\begin{array}{ccccc} N_2(g) & + & 3\ H_2(g) & \rightleftharpoons & 2\ NH_3(g) \\ \textbf{1 mol } \mathbf{N_2} & + & \textbf{3 mol } \mathbf{H_2} & & \textbf{2 mol } \mathbf{NH_3} \end{array}$$

Here, 4 mol of gaseous reactants are converted into 2 mol of gaseous products. An increase in pressure P will shift the equilibrium to increase the yield of ammonia. Decreased P has the opposite effect. *Stressing the equilibrium*

$$N_2(g) + 3\ H_2(g) \rightleftharpoons 2\ NH_3(g)$$

by increasing P and decreasing V

results in Shift → to produce fewer moles of gas and relieve P

Such shifting to the right also produces NH_3.

Stressing the equilibrium by decreasing P and increasing V

results in Shift ← to produce more moles of gas and restore P

Such shifting to the left also consumes NH_3.

Table 14.2 provides a summary of the ammonia example. Le Chatelier's principle indicates that the best equilibrium yield of ammonia should be obtained from using an excess concentration of one of the reactants at a low temperature and a high pressure. Fritz Haber applied those ideas to the commercial synthesis of ammonia, a very important chemical in the manufacture of farm fertilizers.

2:20

table 14.2 effect of concentration, temperature, and pressure on the equilibrium position of $N_2(g) + 3\ H_2(g) \rightleftharpoons 2\ NH_3(g) + \text{heat}$

Change	Equilibrium shift	Resulting increase in: Reactants	Products
Increase $[N_2]$	$\rightarrow$		X
Increase $[H_2]$	$\rightarrow$		X
Increase $[NH_3]$	$\leftarrow$	X	
Decrease $[N_2]$	$\leftarrow$	X	
Decrease $[H_2]$	$\leftarrow$	X	
Decrease $[NH_3]$	$\rightarrow$		X
Increase temperature	$\leftarrow$	X	
Decrease temperature	$\rightarrow$		X
Increase pressure	$\rightarrow$		X
Decrease pressure	$\leftarrow$	X	

sample exercise 7

State the effect, if any, of the following changes on the equilibrium position for the reaction of methane (natural gas) with chlorine to yield hydrogen chloride and methyl chloride, a colorless gas used as a refrigerant and a local anesthetic.

$$CH_4(g) + Cl_2(g) \rightleftharpoons HCl(g) + CH_3Cl(g) + 110.\ kJ$$

a. Increasing the concentration of CH_4
b. Removing CH_3Cl as it is formed
c. Increasing the reaction temperature
d. Decreasing the total pressure

solution

a. The stress of added CH_4 is relieved by a shift in the equilibrium to products ($\rightarrow$).
b. The effect of removing CH_3Cl is to shift the equilibrium to products ($\rightarrow$) to produce more CH_3Cl.
c. The effect of increasing the reaction temperature is to shift the reaction toward absorbing heat ($\leftarrow$), i.e., back to reactants.
d. Because there is no change in the number of moles of gas (2 mol of gaseous reactants, 2 mol of gaseous products), a pressure change will have no effect on the equilibrium position.

sample exercise 8

What is the effect of adding chloride ions to a saturated silver chloride solution governed by the following equilibrium?

$$AgCl(s) \rightleftharpoons Ag^+(aq) + Cl^-(aq)$$

solution

Increasing the concentration of chloride ions will shift the equilibrium to $AgCl(s)$ so that more $AgCl(s)$ will precipitate and the concentration of Ag^+ will decrease in solution.

CHEMLAB

Laboratory exercise 14: equilibrium

It is possible to affect the position of equilibrium and, therefore, the equilibrium concentrations by altering the conditions of concentration, temperature, or pressure in an equilibrium system. In this experiment, you will observe the effects of concentration changes on several equilibria.

Acid-base indicators are weak electrolyte dyes that undergo color changes as their ionization equilibrium shifts. The symbol HInd represents the acid form of an indicator that ionizes according to the equilibrium

HInd	$\rightleftharpoons$	H^+	+	Ind^-
Molecular form of indicator dye (un-ionized)				**Ionized indicator**

In general, the molecular form of an indicator dye is a different color from its anion.

In the case of the indicator bromthymol blue, the molecular form is yellow and the anion is blue.

HInd	$\rightleftharpoons$	H^+	+	Ind^-
Yellow		**Colorless**		**Blue**

Questions:

1. Add 3 drops of bromthymol blue to 5 mL of distilled water in each of three test tubes. Predict the color of the solution that you will observe when it is held in front of a piece of white paper. Remember that bromthymol blue is a weak electrolyte, so that only a small number of ions will be present in the equilibrium solution.
2. Add 3 drops of 1 *M* HCl to one of the test tubes from Step 1. Predict the color of the solution. Remember, HCl is a strong acid, so it will be totally dissociated into H^+ and Cl^- ions when you add it to your test tube.
3. Add 3 drops of 1 *M* NaOH to another of the test tubes from Step 1. Predict the color of the solution. Remember, NaOH is a strong base, and the OH^- ions it produces will neutralize any H^+ ions present.
4. Figure 14.7 is a photograph of an ammonia fountain. The flask initially contained only ammonia gas. When water was added to the flask from the dropper, the ammonia dissolved, thereby reducing the pressure in the flask. Water from the beaker was pushed into the flask by atmospheric pressure, forming a stream or fountain. The water in the beaker has bromthymol blue indicator in it. Use your understanding of indicator equilibrium to explain the different color of the solutions in the beaker and in the flask.

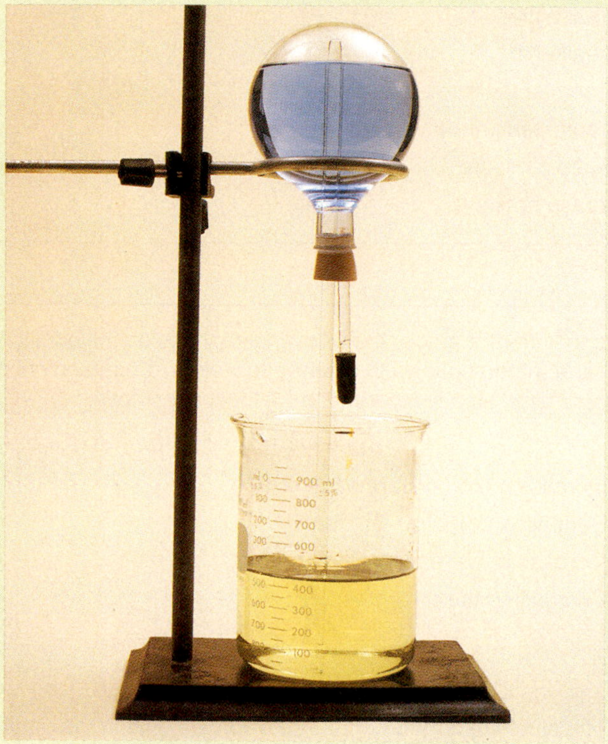

Figure 14.7
Shifting of the indicator equilibrium of bromthymol blue is responsible for the differing colors in the beaker and flask of this ammonia fountain.

problem 5

State the effect, if any, of the following changes on the equilibrium position for the reaction of acetylene with hydrogen to yield ethylene.

$$C_2H_2(g) + H_2(g) \rightleftharpoons C_2H_4(g) + 172\ \text{kJ}$$

Acetylene is used as fuel in welding; ethylene is used commercially to accelerate the ripening of various fruits.

a. Increasing the concentration of H_2
b. Adding C_2H_4
c. Increasing the total pressure
d. Decreasing the reaction temperature

••• problem 6

For the equilibrium

$$BaSO_4(s) \rightleftharpoons Ba^{2+}(aq) + SO_4^{2-}(aq)$$

state the effect of the following on the equilibrium condition.

a. Addition of Ba^{2+} ions
b. Removal of SO_4^{2-} ions
c. Addition of $BaSO_4(s)$

Barium sulfate is widely used in medical and commercial applications. It is used as a diagnostic tool in x-rays; it is also used to manufacture photographic papers and polymeric fibers and to modify pigments in wallpaper.

14.7 predicting the occurrence of reactions

In Section 13.9, you saw how writing *ionic equations* not only conveys more detail about the chemical reaction, it also helps you to predict which products of the reaction are soluble (remain as solutes) and which may leave the solution as insoluble solids (precipitates) or gases. By returning to a consideration of ionic equations and the occurrence of reactions in solutions, you may appreciate another application of the principles of equilibrium.

In general, a chemical reaction between ions in aqueous solution occurs only if one of the products is

1. A gas
2. A precipitate (insoluble solid)
3. H_2O (or some other molecular nonelectrolyte)

Otherwise, the combination of reactants is merely a physical mixing together.

Single- and double-replacement reactions (Section 9.8) occur only in solution. You have already learned to predict the occurrence of a single-replacement reaction, such as those examples given below, based on the activity series of metals given in Table 9.2. Since Mg is higher in the activity series than both H and Cu, it will replace the less reactive material in compounds.

$$Mg(s) + 2\,HCl(aq) \rightarrow MgCl_2(aq) + H_2\uparrow$$

Traditional equation

$$Mg(s) + 2\,H^+(aq) + 2\,Cl^-(aq) \rightarrow Mg^{2+}(aq) + 2\,Cl^-(aq) + H_2\uparrow$$

Ionic equation

$$Mg(s) + 2\,H^+(aq) \rightarrow Mg^{2+}(aq) + H_2\uparrow$$

Net ionic equation

$$Mg(s) + CuSO_4(aq) \rightarrow MgSO_4(aq) + Cu\downarrow$$

Traditional equation

$$Mg(s) + Cu^{2+}(aq) + SO_4^{2-}(aq) \rightarrow Mg^{2+}(aq) + SO_4^{2-}(aq) + Cu\downarrow$$

Ionic equation

$$Mg(s) + Cu^{2+}(aq) \rightarrow Mg^{2+}(aq) + Cu\downarrow$$

Net ionic equation

Thus, Mg replaces the H of $HCl(aq)$ in the first example to yield the gas hydrogen; Mg replaces the Cu of $CuSO_4(aq)$ in the second example to yield the precipitate $Cu(s)$. Notice that these reactions always produce either a gas or an insoluble solid. Ions that have been canceled are spectator ions.

The mixing together of solutions of *ionic compounds* may or may not result in double-displacement reactions, depending on the solubilities of the predicted products. Previous examples (Section 9.8) have always been of the kind where a precipitate formed, and hence there was a reaction. For example, in Sample Exercise 9 in Section 9.8, we completed the equation for the reaction between $MgCl_2$ and K_2CO_3:

$$MgCl_2(aq) + K_2CO_3(aq) \rightarrow MgCO_3 \downarrow + 2\ KCl(aq)$$

Traditional equation

Using Table 13.4, we see that of the above ionic compounds, only $MgCO_3$ is insoluble, so we can write this equation:

$$Mg^{2+}(aq) + 2\ Cl^-(aq) + 2\ K^+(aq) + CO_3^{2-}(aq) \rightarrow MgCO_3 \downarrow + 2\ K^+(aq) + 2\ Cl^-(aq)$$

Ionic equation

K^+ and Cl^- ions are spectators, so

$$Mg^{2+}(aq) + CO_3^{2-}(aq) \rightarrow MgCO_3 \downarrow$$

Net ionic equation

To predict the occurrence of a double-displacement reaction between soluble ionic compounds, *evaluate the solubilities* of the predicted products. A visible reaction occurs if one product is insoluble.

---

sample exercise 9

Predict whether reactions occur upon combination of the following solutions.

a. $CuCl_2(aq) + AgNO_3(aq)$

b. $CuCl_2(aq) + KNO_3(aq)$

solution

Predict the products by the method of switching cation-anion partners, as described in Section 9.8. Then evaluate the solubilities of these products using the rules listed in Table 13.4.

a. $$CuCl_2(aq) + 2\ AgNO_3(aq) \rightarrow Cu(NO_3)_2(aq) + 2\ AgCl \downarrow$$

$Cu(NO_3)_2(aq)$: **Soluble, rule 2**; $AgCl$: **Insoluble, rule 3**

Traditional equation

Since AgCl is insoluble, the reaction occurs. By writing the ionic and net ionic equations, you can see explicitly the chemical change that occurs.

$$Cu^{2+}(aq) + 2\ Cl^-(aq) + 2\ Ag^+(aq) + 2\ NO_3^-(aq) \rightarrow Cu^{2+}(aq) + 2\ NO_3^-(aq) + 2\ AgCl \downarrow$$

Ionic equation

$$Ag^+(aq) + Cl^-(aq) \rightarrow AgCl \downarrow$$

Net ionic equation

b. $$CuCl_2(aq) + 2\ KNO_3(aq) \rightarrow Cu(NO_3)_2(aq) + 2\ KCl(aq)$$

$Cu(NO_3)_2(aq)$: **Soluble, rule 2**; $KCl(aq)$: **Soluble, rule 1**

Since both predicted products are soluble, there is *no* reaction, and you would represent this fact by writing

$$CuCl_2(aq) + KNO_3(aq) \rightarrow NR$$

If you write the ionic equation for this reaction, all chemical species will appear as ions in solution. Since all of the ions are spectators, they will all cancel, and there will be *no net ionic equation.* This is another way of stating that no reaction occurs.

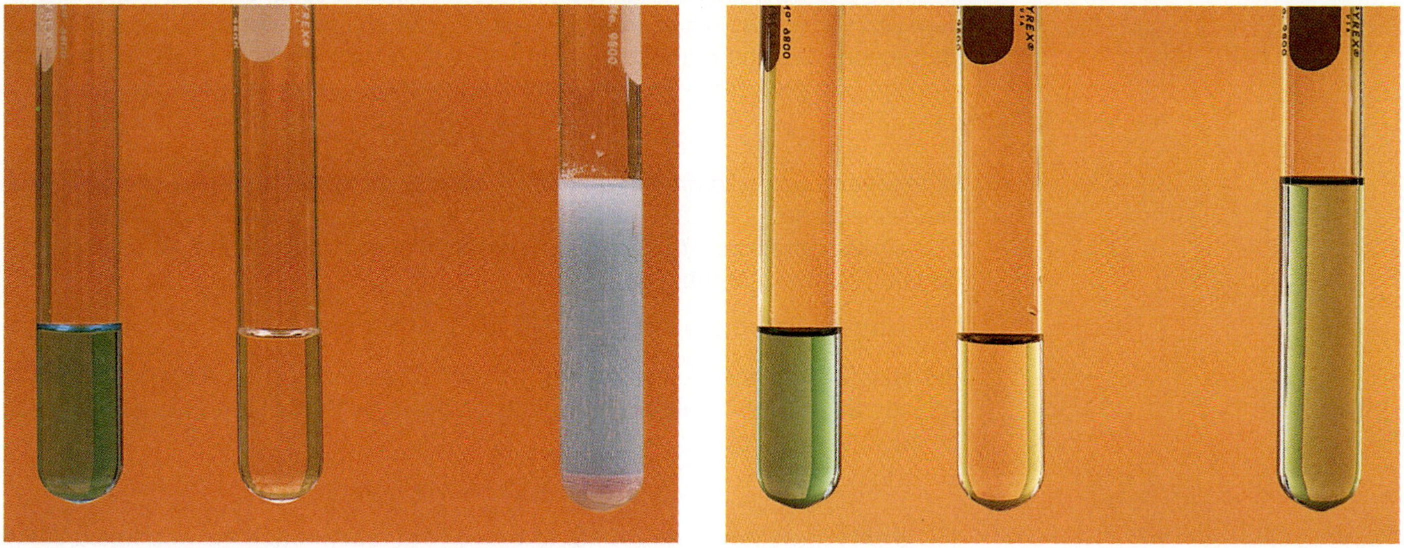

(a) (b)

Figure 14.8
(a) $CuCl_2(aq) + AgNO_3(aq)$ when mixed show the formation of the white precipitate, AgCl. (b) $CuCl_2(aq) + KNO_3(aq)$ when mixed show no net reaction.

If one product of a double-replacement reaction between soluble ionic reactants is a precipitate, as in Sample Exercise 9a, then a visible reaction occurs. If, on the other hand, both predicted products are soluble, as in Sample Exercise 9b, then the result of the combination is a mixture of ions in solution (Figure 14.8).

The other double-replacement reaction mentioned in Section 9.8 was that between a *strong acid* and a *strong base* (soluble metal hydroxide). For example,

$$HCl(aq) + NaOH(aq) \rightarrow NaCl(aq) + H_2O(l)$$

Traditional equation

These reactions occur because water, a liquid, is produced.

Writing this in ionic form by properly writing the solute particles,

$$H^+(aq) + Cl^-(aq) + Na^+(aq) + OH^-(aq) \rightarrow Na^+(aq) + Cl^-(aq) + H_2O(l)$$

Ionic equation

Since the Cl^- ions and Na^+ ions are spectators, they cancel to give

$$H^+(aq) + OH^-(aq) \rightarrow H_2O(l)$$

Net ionic equation

Whenever there is a reaction between a *strong* acid and a *strong* base, the anion of the acid and the cation of the base are always spectators. Therefore, the net ionic equation for this particular kind of reaction is always

$$H^+(aq) + OH^-(aq) \rightarrow H_2O(l)$$

14.8 going to completion

Reversible reactions may go to completion, and not to an equilibrium mixture, when (1) a gas is evolved in an open system, (2) a solid precipitates, or (3) a molecular nonelectrolyte such as water is formed from its ions. For example, if you consider the reversible reaction

$$Mg(s) + 2\,HCl(aq) \rightleftharpoons MgCl_2(aq) + H_2\uparrow$$

in a *tightly closed* container, it would be possible to establish the equilibrium described by the given equation. But in an *open* test tube, the H_2 gas escapes. The escaping gas represents the removal of the product H_2 from the equilibrium mixture.

According to Le Chatelier's principle, there would be a continual shift in the equilibrium position to the right to relieve the stress from the removal of H_2. This continual shift means that the reaction goes to completion from left to right. To indicate the reaction going to completion, you may use the single-arrow designation:

$$Mg(s) + 2\ HCl(aq) \rightarrow MgCl_2(aq) + H_2 \uparrow$$

Whenever a gas is formed in an open system, the reaction goes to completion as the gas escapes.

Whenever a precipitate forms, the reaction goes to completion. Reconsider the double-replacement reaction

$$Na^+(aq) + Cl^-(aq) + Ag^+(aq) + NO_3^-(aq) \rightleftharpoons Na^+(aq) + NO_3^-(aq) + AgCl \downarrow$$

Precipitate formation means that $Ag^+(aq)$ and $Cl^-(aq)$ ions are being removed from solution as $AgCl(s)$ is formed. This is best shown by the net ionic equation for the preceding reaction.

$$Ag^+(aq) + Cl^-(aq) \rightarrow AgCl(s)$$

K_{eq} is very large (typically 1×10^7 to 1×10^{50}) for such reactions and has a value of 6.3×10^9 for this specific reaction. The large value of K_{eq} indicates that only products are significant at equilibrium (Table 14.1) and that the reaction has gone to completion from left to right.

For the reaction of a *strong* acid with a *strong* base, the net ionic equation is always

$$H^+(aq) + OH^-(aq) \rightleftharpoons H_2O(l)$$

The K_{eq} for the formation of $H_2O(l)$ from H^+ and OH^- ions is 1×10^{14}. Once again, the very large value of K_{eq} indicates that the reaction goes to completion. *Whenever a molecular nonelectrolyte is formed from its ions, K_{eq} is large and the equilibrium lies far to the right.*

A combination of the above effects is needed to explain the more complicated equilibrium introduced in Section 14.1 involving the formation of stalactites and stalagmites. *Neither* the precipitation of $CaCO_3(s)$ *nor* the formation of $H_2O(l)$ according to Equation 14.1

$$Ca^{2+}(aq) + 2\ HCO_3^-(aq) \rightleftharpoons CaCO_3(s) + CO_2(aq) + H_2O(l) \qquad (14.1)$$

causes the forward reaction to go to completion unless the dissolved, gaseous CO_2, shown as $CO_2(aq)$, is allowed to escape from the reaction mixture. If the partial pressure of CO_2 gas around the solution is high, the dissolved gas cannot escape; instead, the higher pressure of CO_2 will actually cause more CO_2 gas to dissolve. The addition of the product, $CO_2(aq)$, will shift the equilibrium in the reverse direction (by Le Chatelier's principle) and prevent the precipitation of $CaCO_3(s)$. On the other hand, if the partial pressure of CO_2 gas around the solution is low, as it is in caves, the dissolved gas, shown as $CO_2(aq)$, can escape and the forward reaction will go to completion to form stalactites and stalagmites.

sample exercise 10

Explain why each of the following reactions goes to completion.

a. $3\ KOH(aq) + Al(NO_3)_3(aq) \rightarrow 3\ KNO_3(aq) + Al(OH)_3 \downarrow$

b. $CaC_2(s) + 2\ HCl(aq) \rightarrow CaCl_2(aq) + C_2H_2 \uparrow$

c. $2\ KOH(aq) + H_2SO_4(aq) \rightarrow K_2SO_4(aq) + 2\ H_2O(l)$

solution

a. Large K_{eq} values are generally associated with precipitate formation. In this case, K_{eq} for the formation of solid $Al(OH)_3$ is 2×10^{23}, and consequently, the reaction is driven to completion.

b. Acetylene (C_2H_2) gas escapes, and the equilibrium shifts to replenish it.

c. This is the reaction between a strong base (KOH) and a strong acid (H_2SO_4) to form water. K_{eq} for this reaction is 1×10^{14}.

---

chapter accomplishments

After completing this chapter, you should be able to define all key terms and do the following.

14.2 Reversible reactions

- ❑ Relate the terms *reversible reaction* and *dynamic equilibrium.*
- ❑ Qualitatively describe the changes in initial concentrations of reactants and products in a reversible reaction.

14.3 Rates of reaction

- ❑ State a relationship between reactant concentration and reaction rate.
- ❑ State the relationship between the reaction rates of forward and reverse reactions and chemical equilibrium.

14.4 Equilibrium constant

- ❑ Explain how the equilibrium-constant expression is derived from the forward and reverse rate expressions for one-step reactions.
- ❑ Write the equilibrium-constant expression for a reaction, given a balanced chemical equation.
- ❑ Determine the value of the equilibrium-constant expression, given the equilibrium concentrations.
- ❑ State the convention for the treatment of pure solids and pure liquids in equilibrium-constant expressions.
- ❑ Given a balanced chemical equation, write the equilibrium-constant expression using the common convention regarding pure liquids and solids.
- ❑ Write the solubility-product-constant expression for the equilibrium in a saturated solution of some specified ionic solute.

14.5 Interpreting the value of K_{eq}

- ❑ Given a value of K_{eq}, qualitatively describe the equilibrium composition of a given reaction.

14.6 Le Chatelier's principle

- ❑ Given a chemical equation, describe the effect on the equilibrium concentrations of a change in concentration of one species.
- ❑ Given an exothermic or endothermic chemical reaction, describe the effect of a change in temperature on the equilibrium concentrations.
- ❑ Given a chemical equation involving one or more gases, describe the effect on the equilibrium concentrations of a change in total pressure on the reaction.
- ❑ Given a saturated aqueous solution of an ionic compound, describe the effect of an increase in concentration of one ion on the solubility of the other ion.

14.7 Predicting the occurrence of reactions

- ❑ State the conditions necessary for the occurrence of reactions in solution.
- ❑ Given two reactants in solution, predict whether a double-displacement reaction will occur.
- ❑ Write an ionic equation and a net ionic equation for the reaction of a strong acid with a strong base.

14.8 Going to completion

- ❑ State the three conditions under which a reaction can be said to go to completion.

key terms

reversible reactions
dynamic (active) equilibrium
reaction rate
rate constant
chemical equilibrium
equilibrium concentrations
equilibrium constant
heterogeneous equilibria
homogeneous equilibria
solubility product constant
position of equilibrium
Le Chatelier's principle

problems

14.2 Reversible reactions

7. Describe two nonchemical examples of a dynamic equilibrium.
8. A dynamic equilibrium exists according to the following equation.

$$A(g) + B(g) \rightleftharpoons C(g) + D(g)$$

A is a yellowish colored gas, *D* is a bluish colored gas, and *B* and *C* are colorless gases.

   a. Predict the observable changes that occur when 0.1 mol of *A* and 0.1 mol of *B* are placed in an empty container.
   b. Predict the observable changes that occur if the reaction is begun with 0.1 mol of *C* and 0.1 mol of *D*.
   c. Describe the appearance of the system if the reaction is left at equilibrium.
9. a. Describe how the evaporation of water in a closed container leads to a dynamic equilibrium condition.
   b. Would the evaporation of water in an open container lead to a dynamic equilibrium? Explain.
10. Explain how the concentration of reactants and products stay constant at the state of dynamic equilibrium.
11. What is the difference between *equal* product and reactant concentrations and *constant* product and reactant concentrations?

14.3 Rates of reaction

12. a. Given the single-step reaction $A + B \rightarrow C$, explain why the rate of the reaction increases if the concentration of *A* increases.
    b. Explain what would happen to the rate of the reaction if the concentration of *A* were decreased.
13. What statement can be made concerning the rates of the forward and reverse reactions at the equilibrium condition?

14.4 Equilibrium constant

14. Write an equilibrium-constant expression for each of the following equilibria.
    a. $H_2(g) + Cl_2(g) \rightleftharpoons 2\ HCl(g)$
    b. $PCl_5(g) \rightleftharpoons PCl_3(g) + Cl_2(g)$
    c. $CO(g) + H_2O(g) \rightleftharpoons CO_2(g) + H_2(g)$
    d. $4\ NH_3(g) + 5\ O_2(g) \rightleftharpoons 4\ NO(g) + 6\ H_2O(g)$
    e. $3\ O_2(g) \rightleftharpoons 2\ O_3(g)$
    f. $CH_4(g) + 2\ O_2(g) \rightleftharpoons CO_2(g) + 2\ H_2O(g)$
    g. $4\ HCl(g) + O_2(g) \rightleftharpoons 2\ Cl_2(g) + 2\ H_2O(g)$
15. Write an equilibrium-constant expression for a reaction in which the rate of the forward reaction is given by $\text{Rate}_{forward} = k[A]^2$ and the rate of the reverse reaction is given by $\text{Rate}_{reverse} = k'[B]$.
16. What is the equilibrium-constant expression for each of the following?
    a. $H_2(g) + Br_2(g) \rightleftharpoons 2\ HBr(g)$
    b. $CH_4(g) + Cl_2(g) \rightleftharpoons CH_3Cl(g) + HCl(g)$
    c. $2\ NO_2(g) \rightleftharpoons N_2(g) + 2\ O_2(g)$
    d. $4\ NH_3(g) + 3\ O_2(g) \rightleftharpoons 2\ N_2(g) + 6\ H_2O(g)$
17. a. Write the equilibrium-constant expression for the following equilibrium.

$$2\ HI(g) \rightleftharpoons H_2(g) + I_2(g)$$

    b. Both HI and H_2 are colorless gases, whereas $I_2(g)$ is purple. How might you know when equilibrium in the above reaction has been reached?
    c. A chemist analyzed an equilibrium mixture of HI, H_2, and I_2. She found the concentrations to be [HI] = 0.27 *M*, $[H_2]$ = 0.86 *M*, and $[I_2]$ = 0.86 *M*. Determine the numerical value of K_{eq}.
18. Write an equilibrium-constant expression for a reaction in which the rate of the forward reaction is given by $\text{Rate}_{forward} = k[A]^2[B]$ and the rate of the reverse reaction is given by $\text{Rate}_{reverse} = k'[C][D]$.
19. Write an equilibrium-constant expression for each of the following equilibria.
    a. $C(s) + H_2O(g) \rightleftharpoons CO(g) + H_2(g)$
    b. $4\ H_2O(g) + 3\ Fe(s) \rightleftharpoons Fe_3O_4(s) + 4\ H_2(g)$
    c. $2\ H_2O(l) \rightleftharpoons 2\ H_2(g) + O_2(g)$
    d. $MnO_2(s) + 4\ HCl(aq) \rightleftharpoons MnCl_2(aq) + Cl_2(g) + 2\ H_2O(l)$
    e. $H_2CO_3(aq) \rightleftharpoons CO_2(g) + H_2O(l)$
20. What is the equilibrium-constant expression for each of the following?
    a. $PCl_5(s) \rightleftharpoons PCl_3(l) + Cl_2(g)$
    b. $4\ Al(s) + 3\ O_2(g) \rightleftharpoons 2\ Al_2O_3(s)$
    c. $H_2O(l) \rightleftharpoons H_2O(g)$
    d. $4\ HNO_3(l) \rightleftharpoons 4\ NO_2(g) + 2\ H_2O(g) + O_2(g)$
    e. $H_2SO_3(l) \rightleftharpoons SO_2(g) + H_2O(l)$
21. Write the equilibrium-constant expression for each of the following physiological equilibria.
    a. γ-D-glucose(*aq*) $\rightleftharpoons$ β-D-glucose(*g*)
    b. $CH_3OH(aq) + CH_3COOH(aq) \rightleftharpoons CH_3COOCH_3(aq) + H_2O(l)$
    c. Hemoglobin(*aq*) + oxygen(*g*) $\rightleftharpoons$ oxyhemoglobin(*aq*)
    d. Double-strand DNA $\rightleftharpoons$ 2 single-strand DNA

22. Write the solubility-product-constant expressions for the equilibria in the following saturated solutions.
   a. $AgF(s) \rightleftharpoons Ag^+(aq) + F^-(aq)$
   b. $PbBr_2(s) \rightleftharpoons Pb^{2+}(aq) + 2\ Br^-(aq)$
   c. $Ca_3(PO_4)_2(s) \rightleftharpoons 3\ Ca^{2+}(aq) + 2\ PO_4^{3-}(aq)$
   d. $BaSO_4(s) \rightleftharpoons Ba^{2+}(aq) + SO_4^{2-}(aq)$
   e. $As_2S_3(s) \rightleftharpoons 2\ As^{3+}(aq) + 3\ S^{2-}(aq)$
23. What is the solubility-product-constant expression for the equilibrium in each of the following?
   a. $AgBr(s) \rightleftharpoons Ag^+(aq) + Br^-(aq)$
   b. $Cu(OH)_2(s) \rightleftharpoons Cu^{2+}(aq) + 2\ OH^-(aq)$
   c. $Al(OH)_3(s) \rightleftharpoons Al^{3+}(aq) + 3\ OH^-(aq)$
   d. $Mg_3(PO_4)_2(s) \rightleftharpoons 3\ Mg^{2+}(aq) + 2\ PO_4^{3-}(aq)$
   e. $Bi_2S_3(s) \rightleftharpoons 2\ Bi^{3+}(aq) + 3\ S^{2-}(aq)$
24. Calculate the value of K_{eq} at 127°C for the reaction

$$Ni(CO)_4(g) \rightleftharpoons Ni(s) + 4\ CO(g)$$

given the following set of equilibrium concentrations at this temperature.

$[Ni(CO)_4] = 1.80$ mol/L
$[CO] = 0.800$ mol/L

14.5 Interpreting the value of K_{eq}

25. Use Table 14.1 and the given K_{eq} to describe qualitatively the equilibrium composition of each of the following reactions.

a.
$2\ NO_2(g) \rightleftharpoons N_2(g) + 2\ O_2(g)$ $\quad K_{eq}$ (at 25°C) = 6.7×10^{16}

b.
$CH_3OH(g) \rightleftharpoons CO(g) + 2\ H_2(g)$ $\quad K_{eq}$ (at 100°C) = 7.37×10^{-8}

c.
$N_2(g) + O_2(g) \rightleftharpoons 2\ NO(g)$ $\quad K_{eq}$ (at 2,400°C) = 3.4×10^{-3}

d.
$2\ SiO(g) \rightleftharpoons 2\ Si(l) + O_2(g)$ $\quad K_{eq}$ (at 1,727°C) = 9.62×10^{-13}

26. In each of the following cases, use Table 14.1 and the given K_{eq} to describe qualitatively the equilibrium composition.

a.
$2\ NO(g) \rightleftharpoons N_2(g) + O_2(g)$ $\quad K_{eq}$ (at 25°C) = 2.2×10^{30}

b.
$PCl_5(g) \rightleftharpoons PCl_3(g) + Cl_2(g)$ $\quad K_{eq}$ (at 127°C) = 1.19×10^{-2}

c.
$2\ HCl(g) \rightleftharpoons H_2(g) + Cl_2(g)$ $\quad K_{eq}$ (at 2,727°C) = 1.27×10^{-4}

d.
$2\ Na_2O(s) \rightleftharpoons 4\ Na(l) + O_2(g)$ $\quad K_{eq}$ (at 427°C) = 1.84×10^{-25}

27. From the following solubility-product-constant expressions, pick (1) the reaction that is most complete from left to right as written and (2) the reaction that is least complete from left to right as written.

a.
$AgCl(s) \rightleftharpoons Ag^+(aq) + Cl^-(aq)$ $\quad K_{sp}$ (at 25°C) = 1.7×10^{-10}

b.
$PbSO_4(s) \rightleftharpoons Pb^{2+}(aq) + SO_4^{2-}(aq)$ $\quad K_{sp}$ (at 25°C) = 1.3×10^{-8}

c.
$Fe(OH)_3(s) \rightleftharpoons Fe^{3+}(aq) + 3\ OH^-(aq)$ $\quad K_{sp}$ (at 25°C) = 6×10^{-38}

28. In each of the following physiological reactions, use Table 14.1 and the given K_{eq} to describe qualitatively the equilibrium composition.

a.
$ATP + H_2O \rightleftharpoons ADP + HPO_4^{2-}$ $\quad K_{eq}$ (at 25°C) = 1.3×10^5

b.
Arginine + $HPO_4^{2-} \rightleftharpoons$ arginine-phosphate + H_2O
K_{eq} (at 25°C) = 7.9×10^{-6}

c.
Glucose-6-phosphate + arginine $\rightleftharpoons$
glucose + arginine-phosphate $\quad K_{eq}$ (at 37°C) = 1.5×10^{-3}

d.
ATP + CH_3COOH + coenzyme A $\rightleftharpoons$
AMP + pyrophosphate + acetylcoenzyme A
K_{eq} (at 37°C) = 1.0

14.6 Le Chatelier's principle

29. State the effect, if any, of the following changes on the equilibrium position for the reaction

$$4\ NH_3(g) + 3\ O_2(g) \rightleftharpoons 2\ N_2(g) + 6\ H_2O(g) + 1{,}531\ kJ$$

   a. Increasing the concentration of oxygen
   b. Adding N_2 to the equilibrium mixture
   c. Removing H_2O as it is formed
   d. Increasing the reaction temperature
   e. Decreasing the total pressure
30. State the effect, if any, of the following changes on the equilibrium position for the reaction

$$2\ SO_3(g) + 197\ kJ \rightleftharpoons 2\ SO_2(g) + O_2(g)$$

   a. Increasing the concentration of SO_3
   b. Adding O_2 to the equilibrium mixture
   c. Removing O_2 as it is formed
   d. Increasing the reaction temperature
   e. Decreasing the total pressure
31. Consider the equilibrium

$$C(s) + H_2O(g) \rightleftharpoons CO(g) + H_2(g) + \text{heat}$$

   a. What should be done to the total pressure of the system to maximize the equilibrium concentration of H_2?
   b. What should be done to the pressure of the system to maximize the equilibrium concentration of H_2O?
32. Consider the equilibrium

$$46.0\ kJ + 2\ F_2(g) + O_2(g) \rightleftharpoons 2\ OF_2(g)$$

   a. Should the temperature be increased or decreased to maximize the equilibrium yield of product?
   b. Should the pressure be increased or decreased to maximize the equilibrium yield of the product?

33. For the equilibrium

$$Ca_3(PO_4)_2(s) \rightleftharpoons 3\ Ca^{2+}(aq) + 2\ PO_4^{3-}(aq)$$

state the effect of the following on the equilibrium condition.

a. Adding Ca^{2+} ions

b. Adding PO_4^{3-} ions

c. Removing PO_4^{3-} ions

34. Describe the effect on the dynamic equilibrium

$$AgCl(s) \rightleftharpoons Ag^+(aq) + Cl^-(aq)$$

of adding NaCl to a saturated solution of AgCl.

35. Will $BaSO_4$ ($K_{sp} = 1.5 \times 10^{-9}$) be more soluble in sulfuric acid solution or in pure water? Explain.

36. The equilibrium between carbonic acid (H_2CO_3) and bicarbonate ion is very important in the regulation of the acid-base concentration in blood. State the effect, if any, of the following changes on the position of the equilibrium

$$H_2CO_3(aq) \rightleftharpoons H^+(aq) + HCO_3^-(aq)$$

a. Adding H_2CO_3 to the equilibrium mixture

b. Removing H^+ as it is formed

c. Adding H^+ to the equilibrium mixture

d. Adding HCO_3^- to the equilibrium mixture

e. Removing H_2CO_3 as it is formed

37. Reconsider Problem 36c. What happens to the bicarbonate (HCO_3^-) concentration upon addition of H^+ and the consequent shift?

38. An equilibrium that exists in solutions of carbon dioxide such as soda and beer is

$$H_2CO_3(aq) \rightleftharpoons CO_2(g) + H_2O(l)$$

Use Le Chatelier's principle to explain why an open bottle of beer goes flat.

14.7 Predicting the occurrence of reactions

39. Predict whether the following double-displacement reactions will occur. For each reaction that occurs, write a balanced net ionic equation.

a. $NaBr(aq) + AgNO_3(aq) \rightarrow$

b. $HNO_3(aq) + KOH(aq) \rightarrow$

c. $NH_4Cl(aq) + MgBr_2(aq) \rightarrow$

d. $Ba(OH)_2(aq) + H_2SO_4(aq) \rightarrow$

e. $Ca(NO_3)_2(aq) + K_3PO_4(aq) \rightarrow$

14.8 Going to completion

40. List the three types of products that drive an equilibrium all the way to the product side.

41. How does the formation of a gas drive a reaction to completion?

Additional problems

42. Calculate the value of the equilibrium constant for the equilibrium

$$I_2(g) + H_2(g) \rightleftharpoons 2\ HI(g)$$

given the experimentally determined equilibrium concentrations of $[I_2] = 0.42$ mol/L, $[H_2] = 0.025$ mol/L, and $[HI] = 0.76$ mol/L.

43. An important modern chemical problem is the liquefaction of coal because it is still relatively abundant, whereas oil is a dwindling resource. The first step is heating the coal with steam to produce carbon monoxide.

$$C(s) + H_2O(g) \rightleftharpoons CO(g) + H_2(g) \quad \Delta H = 2{,}290\ \text{kJ/mol}$$

The carbon monoxide can be hydrogenated to form the key substance methyl alcohol.

$$CO(g) + 2\ H_2(g) \rightleftharpoons CH_3OH(g) \quad \Delta H = 2{,}240\ \text{kJ/mol}$$

Use Le Chatelier's principle to suggest conditions that maximize the yield of CH_3OH from $CO(g)$ and $H_2(g)$.

44. Consider the following equilibrium.

$$C\ (\text{graphite}) \rightleftharpoons C\ (\text{diamond}) \qquad \Delta H = 33\ \text{kJ/mol}$$

The densities of diamond and graphite are 3.52 g/mL and 2.25 g/mL, respectively. Is the formation of diamond from graphite favored by

a. High or low temperature?

b. High or low pressure?

45. Phosgene, a toxic gas, decomposes according to the reaction

$$COCl_2(g) \rightleftharpoons CO(g) + Cl_2(g)$$

A sample of phosgene is heated in a closed vessel to 527°C and the reaction is allowed to come to equilibrium. At 527°C, the equilibrium concentrations of the products and reactant are found to be $[CO]_e = 0.0456\ M$, $[Cl_2]_e = 0.0456\ M$, and $[COCl_2]_e = 0.449\ M$. Calculate the equilibrium constant for this reaction at 527°C.

46. When heated, phosphorous pentachloride decomposes according to the following equation.

$$PCl_5(g) \rightleftharpoons PCl_3(g) + Cl_2(g)$$

A sample of PCl_5 is added to a 1.0 L reaction vessel to a concentration of 1.10 M. The vessel is heated to 200.°C and equilibrium is attained. The concentration of PCl_5 at equilibrium is found to be 0.33 M. Calculate the equilibrium constant for this reaction at 200.°C.

47. The equilibrium between N_2O_4 and NO_2 according to the equation

$$N_2O_4(g) \rightleftharpoons 2\ NO_2(g)$$

has an equilibrium constant $K_{eq} = 0.20$ at 100°C. If the equilibrium concentration of N_2O_4 is found to be 0.630 *M*, calculate the equilibrium concentration of NO_2.

48. For the equilibrium

$$2\ NO(g) + Br_2(g) \rightleftharpoons 2\ NOBr(g)$$

use Le Chatelier's principle to help you predict the effect, if any, on the equilibrium concentrations of NOBr and NO of an increase in the concentration of Br_2.

49. State the effect, if any, of the following changes on the equilibrium position for the reaction

$$C(s) + 2\ H_2(g) \rightleftharpoons CH_4(g) + 75\ \text{kJ/mol}$$

 a. Increasing the reaction temperature
 b. Increasing the concentration of hydrogen
 c. Decreasing the concentration of CH_4
 d. Increasing the volume of the reaction vessel

50. Consider the following reaction

$$H_2(g) + I_2(g) \rightleftharpoons 2\ HI(g)$$

for which the equilibrium constant at 430.°C is $K_{eq} = 54.3$. During the course of this reaction in a 3.50 L vessel, the following amounts of H_2, I_2, and HI were present: 0.0218 mol of H_2, 0.0145 mol of I_2, and 0.0783 mol of HI.

 a. Is this system now at equilibrium?
 b. If it is not at equilibrium, which way will the reaction proceed?

51. Ammonium carbamate, a solid of formula $NH_4CO_2NH_2(s)$, decomposes to ammonia and carbon dioxide according to the equation

$$NH_4CO_2NH_2(s) \rightleftharpoons 2\ NH_3(g) + CO_2(g)$$

At 40.°C, the total combined concentration of NH_3 and CO_2 is 0.363 mol/L. Calculate the equilibrium constant for this reaction at 40°C.

chapter 15

acids and bases

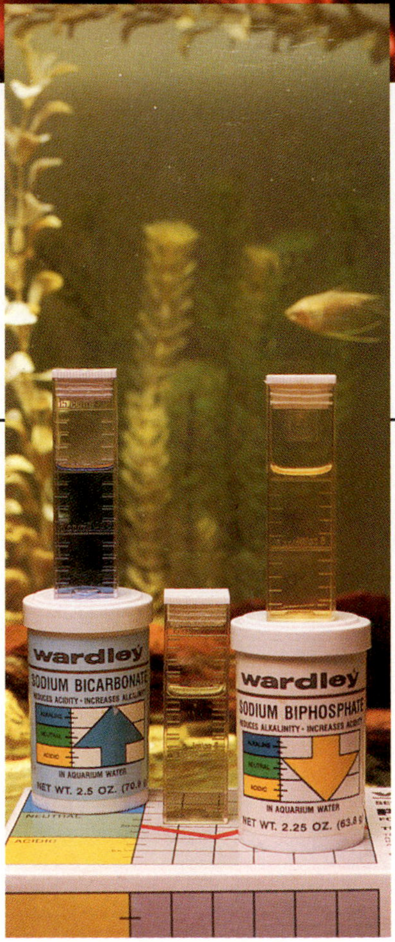

A sample of water from an aquarium is being tested with color-indicator solution, which uses the indicator equilibria discussed in Laboratory Exercise 14. The color of the aquarium water (with indicator added) tells you the acidity, or pH, of the water which is critical to the health of species living in the aquarium.

15.1 introduction

You have daily contact with acids and bases. Acids are necessary for digestion of proteins in the stomach (hydrochloric acid); they are found in vinegar (acetic acid) dressing on salads, in fruits such as lemons (citric acid), in battery acid (sulfuric acid), and in the pain-relieving drugs aspirin (acetylsalicylic acid) and ibuprofen (isobutylphenylpropionic acid). Bases occur in drain and oven cleaners (sodium hydroxide), window cleaner (ammonia), soaps and detergents (mixtures of various bases), and antacids (aluminum hydroxide, magnesium hydroxide, and sodium bicarbonate). Many products used in personal hygiene such as skin cleanser, cosmetics, and shampoos are advertised as "acid balanced," meaning that the degree of acidity or basicity is compatible with their physiological use. Figure 15.1 depicts some familiar acid and base products.

Figure 15.1

Acids and bases are commonly used. Acids: vitamin C (ascorbic acid), aspirin (acetylsalicylic acid), vinegar (acetic acid). Bases: milk of magnesia (magnesium hydroxide), baking soda (sodium bicarbonate), Drano (sodium hydroxide).

table 15.1 properties of acids and bases

Acids	Turn litmus, an indicator dye, red (remember aci*d* and re*d* end with the same sound) Have a pH less than 7 Taste sour (not a recommended general test) React with bases in a neutralization reaction Oxoacids are produced by the reaction of water and a nonmetallic oxide
Bases	Turn litmus, an indicator dye, blue (remember *b* for blue, *b* for base) Have a pH greater than 7 Taste bitter (not a recommended general test) Feel slippery (soapy) React with acids in a neutralization reaction Metal hydroxide bases are produced by the reaction of water and a metallic oxide

Figure 15.2
Red litmus paper turns blue to indicate that soap or ammonia is basic; blue litmus paper turns red to indicate that lemon juice is acidic.

Properties of acids and bases in aqueous solution are listed in Table 15.1. While acids and bases may exist in solvents other than water, aqueous acids and bases will be your primary concern throughout this chapter. You have learned in Section 13.7 that acids ionize in water to produce $H^+(aq)$ ions. In Section 15.2, you will learn that bases ionize to produce $OH^-(aq)$ ions. As a result, the properties of aqueous acids and bases can be attributed to an excess of either $H^+(aq)$ or $OH^-(aq)$ in solution. The excess concentration of $H^+(aq)$ in an aqueous solution is called its **acidity**, and the excess concentration of $OH^-(aq)$ is called its **basicity** or **alkalinity**.

One of the fundamental tests to identify an acidic or basic solution is to observe the color of the indicator dye, litmus, when exposed to the solution. Excess $H^+(aq)$ ions shift the litmus equilibrium to red, so *acid solutions turn blue litmus paper red.* Excess $OH^-(aq)$ ions shift the litmus equilibrium to blue, so *basic solutions turn red litmus paper blue* (Figure 15.2). To help you remember this identification test, all of the tables in this chapter have *acids* shown in *red* type and *bases* shown in *blue* type. A mnemonic that may help you remember the same information is that both *b*ase and *b*lue begin with the same sound and both aci*d* and re*d* end with the same sound.

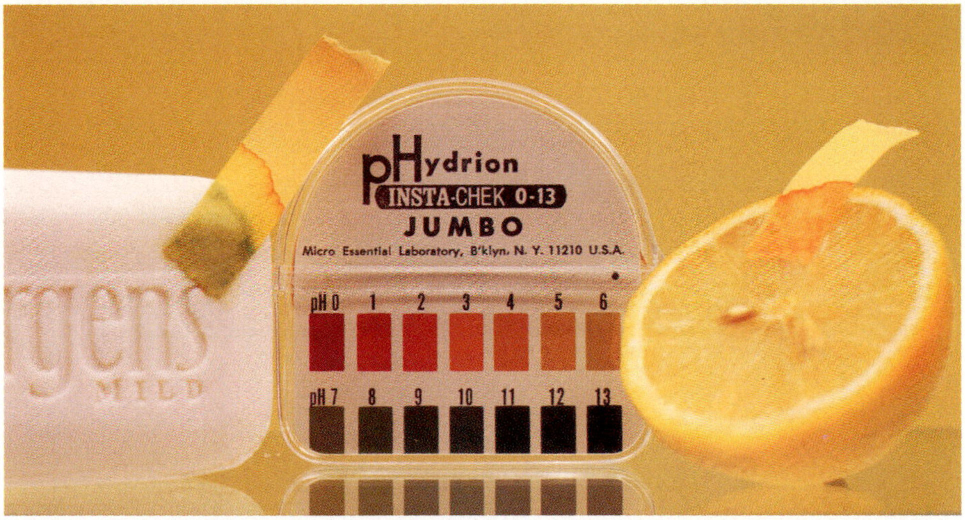

Figure 15.3
pH paper turns color to indicate whether a solution is acidic or basic. By comparing the color of the paper after being immersed in the solution to that of a color chart, you can quickly determine the pH, or $H^+(aq)$ ion concentration, in the solution. Note that the pH of lemon juice is less than 7, so it is acidic; the pH of soap or ammonia is greater than 7, so it is basic.

Another laboratory test to identify an acidic or basic solution is to measure the pH (pronounced as it's spelled, "p-H"). You will learn in Section 15.6 that pH is a measure of the concentration of $H^+(aq)$ ions in solution. Acidic solutions have a larger concentration of $H^+(aq)$ ions and a smaller concentration of $OH^-(aq)$ ions, whereas basic solutions have a smaller concentration of $H^+(aq)$ ions and a larger concentration of $OH^-(aq)$ ions.

Two common methods of measuring the pH of a solution are to use pH paper or a pH meter. pH paper, or universal indicator paper, is like litmus paper in that it is impregnated with indicator dyes that change color with changes in $H^+(aq)$ ion concentration. Each dye on universal indicator paper only changes color for a specific, narrow range of $H^+(aq)$ ions. When the pH paper has been immersed in an unknown solution, it will change color. By matching this color with the closest color on the chart provided with the pH paper, you can quickly determine the pH of the solution far more accurately than with litmus paper (Figure 15.3). A pH meter is an electronic instrument with an electrode that is sensitive to the concentration of $H^+(aq)$ ions in solution. By immersing the electrode directly in your unknown solution, a calibrated scale will give you a direct pH reading that is more accurate than using pH paper (Figure 15.4).

Figure 15.4
With its electrode immersed in a solution, a pH meter directly displays the value of the solution's pH. The pH of the Coca-Cola measured here is 2.28, indicating that it is acidic.

Neutralization of a strong acid with a strong base is an example of a double-replacement reaction discussed in Section 9.8. Later in Section 14.7, you also learned that all strong acid-base neutralizations may be reduced to the same net ionic equation,

$$H^+(aq) + OH^-(aq) \rightarrow H_2O(l)$$

table 15.2 common laboratory acids and bases

Name		Formula	Strength	Use or occurrence
Acids	**Hydrobromic acid**	HBr	Strong	Manufacture of chemicals
	Hydrochloric acid	HCl	Strong	Gastric juice
	Hydriodic acid	HI	Strong	Manufacture of iodine compounds
	Nitric acid	HNO_3	Strong	Lab test for proteins
	Perchloric acid	$HClO_4$	Strong	Manufacture of explosives
	Sulfuric acid	H_2SO_4	Strong	Battery acid and most widely used acid in industrial processes
	Acetic acid	$HC_2H_3O_2$	Weak	Vinegar
	Boric acid	H_3BO_3	Weak	Eyewash
	Carbonic acid	H_2CO_3	Weak	Carbonated water
	Hydrocyanic acid	HCN	Weak	Extermination of rodents
	Hydrofluoric acid	HF	Weak	Etching of glass and other industrial uses
	Phosphoric acid	H_3PO_4	Medium	Soft drink additive
Bases	**Barium hydroxide**	$Ba(OH)_2$	Strong	Manufacture of glass
	Calcium hydroxide	$Ca(OH)_2$	Strong	Cement component
	Lithium hydroxide	LiOH	Strong	Photographic developer
	Potassium hydroxide	KOH	Strong	Manufacture of liquid soap
	Sodium hydroxide	NaOH	Strong	Drain and oven cleaners
	Strontium hydroxide	$Sr(OH)_2$	Strong	Refining of sugar
	Aluminum hydroxide	$Al(OH)_3$	Strong*	Antiperspirant
	Ammonia	NH_3	Weak	Respiratory stimulant
	Magnesium hydroxide	$Mg(OH)_2$	Strong*	Milk of magnesia
	Zinc hydroxide	$Zn(OH)_2$	Strong*	Surgical dressing

*These bases are strong in that to whatever extent they dissolve, they dissociate. However, they are at best only slightly soluble.

This equation specifies that equal numbers of moles of H^+ from an acid and OH^- from a base participate in the reaction. Whenever a given number of moles of H^+ is added, the same number of moles of OH^- will be neutralized, or vice versa. If you know the molarity of an acid (base) solution, you may add it slowly to *a given volume of base* (acid) of unknown concentration until the base (acid) is completely neutralized and you will have mixed equal numbers of moles of H^+ and OH^-. Since you now know *the number of moles of OH^- in a given volume of base,* you may calculate the molarity of the base solution using the definition of molarity,

$$M = \frac{\text{Moles of solute}}{\text{Liters of solution}}$$

In Section 15.9, you will learn how to use this process, called **titration,** in greater detail.

Although we will discuss properties of acids and bases in more detail as this chapter unfolds, you can begin to recognize various foods as acids if they taste sour, such as lemon juice and vinegar, or as bases if they taste bitter, such as chicory, tonic water (quinine), unsweetened chocolate, and milk of magnesia. You will also recognize various solutions as bases if they feel slippery, such as detergents, soap, milk of magnesia, and drain cleaners.

Table 15.2 lists the names, formulas, and strengths of some common acids and bases and cites their occurrence or use in food, medicine, agriculture, or industry. The name of the acid actually applies to the aqueous solution. Hydrochloric acid, for

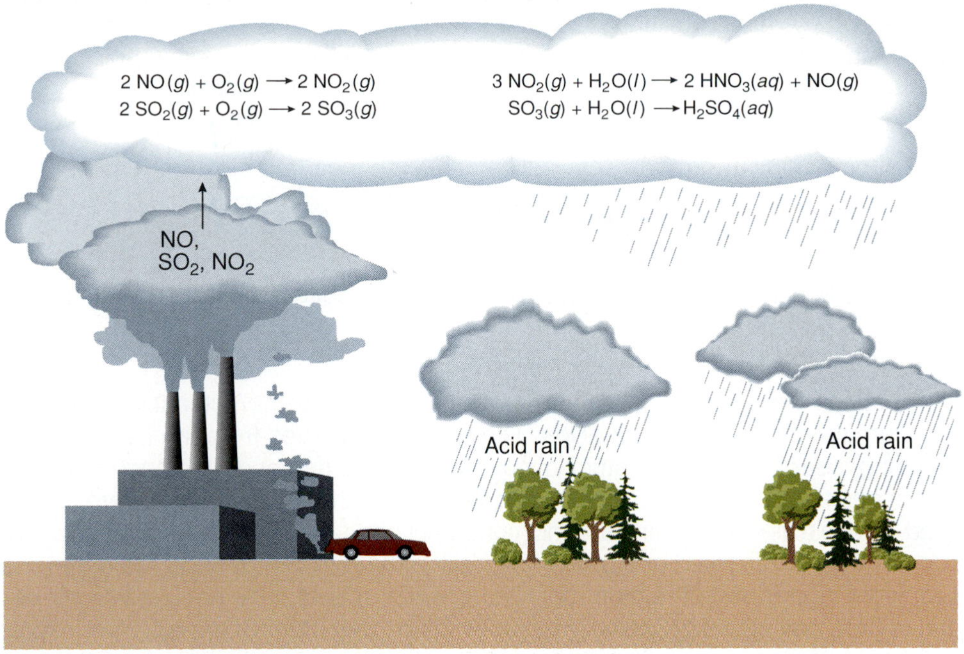

Figure 15.5
Nitrogen oxides and sulfur dioxide are emitted by industrial smokestacks and automotive exhausts. These oxides react with oxygen and water in the air to form nitric and sulfuric acid. This is the origin of acid rain.

example, is formed by dissolving hydrogen chloride, a gas, in water. Notice that, with the exception of ammonia, all of the bases in the table are ionic compounds, which is to say they are combinations of metal ions and hydroxide ions.

You recall from Section 13.7 that electrolytes are designated as strong if they dissociate completely in solution. Strong acids ionize completely in aqueous solution to yield $H^+(aq)$ ions, and strong bases ionize completely to yield $OH^-(aq)$ ions. In Table 15.2, the six strong acids you memorized in Table 13.4 are listed first under acids and should provide a convenient reference as you continue through this chapter. You recall that all acids other than these six are weak acids, so you may automatically classify other acids as weak. Also in Table 15.2, the six commonly used strong bases are listed first under bases. While other bases may be strong in the sense that they ionize completely in aqueous solution, their solubility is so low that the concentration of $OH^-(aq)$ ions in solution is extremely small.

To function effectively in a modern society requires some awareness of the physical and chemical properties of acidic and basic solutions. An understanding of sweet (alkaline) and sour (acidic) soils, acid indigestion and how to buffer (oppose) excess acidity in your stomach, and acid rain requires a knowledge of acidity and basicity. Economically, acids and bases are of tremendous importance simply in terms of the amounts used in industrial processes. Eight billion pounds of sulfuric acid alone are used in storage batteries, in metallurgy, and in the manufacture of plastics, fertilizers, and detergents each year.

Acid rain has become an issue of international importance as industrial atmospheric pollutants such as sulfur dioxide (SO_2) travel across national boundaries to react with water and oxygen in the air and form rain and snow containing sulfuric acid (Figure 15.5). The change in the acidity of rain and snow because of industrial pollution has been measured by comparing the pH of current rain and snow samples with the pH of samples obtained inside glaciers, where they were deposited many years ago. The acidity of rain and snow samples has been increasing during the past twenty years, thus increasing the acidity of some lakes and streams where they collect. There appears to be a direct relationship between this increased acidity and the decrease in fish population of such lakes. For example, more than 200 high-elevation lakes in the Adirondack Mountains of New York are now totally devoid of fish due to the acid rain falling there.

Since you now have a knowledge of solutions and ionic equations (Sections 13.8 and 13.9) and equilibrium concepts (Chapter 14), you are prepared to proceed with a detailed study of acids and bases. An important starting point is that you remember the aqueous ions responsible for acidity and basicity.

An acidic solution has more H^+(*aq*) ions
(hydrogen ions or protons, that is, H atoms with their one electron removed).

A basic solution has more OH^-(*aq*) ions
(hydroxide ions).

Let us see how these ions contribute to the common definitions of acids and bases.

15.2 Arrhenius definitions of acids and bases

Historically, acids and bases were classified according to characteristic properties, for example, the sour taste of acids. The word *acid* comes from the Latin *acidus,* meaning "sour." As scientific study of acids and bases proceeded through the nineteenth and twentieth centuries, various definitions of these substances were formulated. These different definitions do not contradict one another. The newer definitions, such as that of Bronsted and Lowry (Section 15.3), are broader and more general definitions that include the older definitions, such as that of Arrhenius. We will examine both definitions because sometimes one or the other offers a clearer explanation of some observed phenomenon.

As early as 1884, Svante Arrhenius, a Swedish chemist who received the Nobel Prize in 1903, defined acids and bases in terms of the ions they release in aqueous solution:

Acids release H^+ ions into water.

Bases release OH^- ions into water.

Considering that metal hydroxides (MOH) are bases, solutions of these compounds must produce OH^- ions. These ionic compounds are made up of metal cations and hydroxide anions in the solid state. When ionic compounds dissolve, the ions separate (dissociate) and are solvated by water (Section 13.3). You may represent this **dissociation** by using a general equation for the solubility equilibrium in a saturated base solution:

$$\mathbf{MOH}(s) \rightleftharpoons \mathbf{M^+}(aq) + \mathbf{OH^-}(aq)$$

••• problem 1

Use balanced chemical equations to show how KOH(*s*), potassium hydroxide, and $Ca(OH)_2$(*s*), calcium hydroxide, release OH^- ions into solution. In addition to its use in the manufacture of liquid soap (Table 15.2), potassium hydroxide is used to treat cotton yarns or fabrics in order to increase their strength, luster, and affinity for dyes. In addition to its use in cement (Table 15.2), calcium hydroxide is used in preserving eggs and in manufacturing lubricants and pesticides.

Acids are *molecular* compounds that undergo ionization in aqueous solution; indeed, as you saw in Section 13.8, they are the principal examples of *molecular* electrolytes. **Ionization** is the reaction of a *molecular* compound with water to produce ions. In the case of the ionization of molecular compounds of the general formula HA, one of the ions produced is H^+; therefore, HA is classified as an **Arrhenius acid.**

$$HA(g \text{ or } l) \rightleftharpoons H^+(aq) + A^-(aq)$$

This reaction is more correctly represented as

$$\mathbf{HA}\ (g \text{ or } l) + \mathbf{H_2O}(l) \rightleftharpoons \underset{\textbf{Hydronium ion}}{\mathbf{H_3O^+}(aq)} + \mathbf{A^-}(aq)$$

which shows that HA ionizes by reaction with H_2O. For simplicity, we often just write $H^+(aq)$, as before. However, you should remember that an H^+ ion in aqueous solution will always form a coordinate covalent bond (Section 6.14) with a lone pair of electrons on a water molecule's oxygen. This arrangement is the **hydronium ion.**

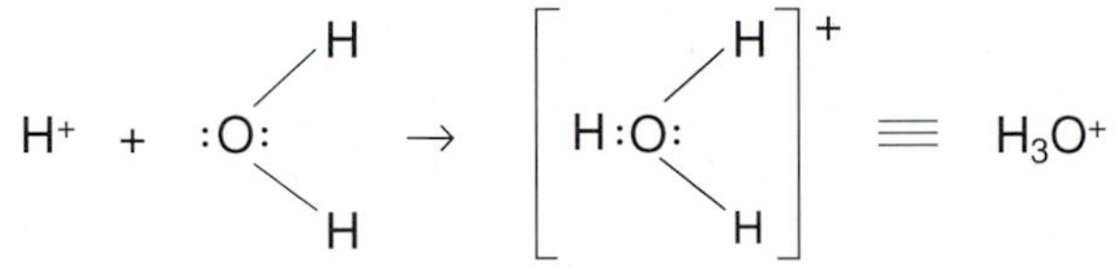

Hydronium ion

15.3 Bronsted-Lowry definitions of acids and bases

If you examine Table 15.2 again, you will find that all of the listed acids and bases easily fit the Arrhenius definitions except for ammonia. That is, the bases are of the form MOH and the acids are of the form HA. Because of NH_3 and some other materials with acidic or basic properties that do not obviously fit the Arrhenius definitions, Johannes Nicolaus Bronsted, a Danish chemist, and Thomas Martin Lowry, an English chemist, independently offered new definitions in 1923:

Acids are H^+ (proton) donors.

Bases are H^+ (proton) acceptors.

In some respects, the Bronsted-Lowry definition of an acid is identical to the Arrhenius definition. An acid donates H^+ or releases H^+ into the solution. However, in the Bronsted-Lowry definition of a base, for compounds of the type MOH, M^+ is regarded strictly as a spectator ion. The true basic species is the hydroxide anion (OH^-), because it is OH^- that will quite readily accept H^+ (a proton) to form water.

$$\underset{\textbf{Hydroxide}}{OH^-} + \underset{\textbf{Proton}}{H^+} \rightarrow \underset{\textbf{Water}}{H_2O}$$

Hydroxide ion (OH^-) is clearly a base because it accepts H^+. MOH is a base by the Arrhenius definition because in water it dissociates to produce the basic OH^- ions. However, it is customary to ignore M^+ when writing Bronsted-Lowry acid-base expressions that involve metal hydroxide bases.

The Bronsted-Lowry definition of bases allows us to understand ammonia's (NH_3) basicity by recognizing that the lone electron pair on nitrogen is available for bonding with H^+. Thus, the NH_3 molecule can accept H^+.

$$\ddot{N}H_3 + H^+ \rightleftharpoons \left[\overset{H}{\ddot{N}}H_3\right]^+ \equiv NH_4^+$$

For an aqueous solution of NH_3, water is the source of H^+.

$$H_3N\!: + H\!:\!\ddot{\underset{..}{O}}\!-\!H \rightleftharpoons \overset{+}{H_3N\!:\!H} + :\!\ddot{\underset{..}{O}}\!-\!H^-$$

This equation shows us that NH_3 is a base in the Arrhenius sense also because the reaction of NH_3 with H_2O releases OH^- ions in solution. Ammonia is a weak base because the above equilibrium lies to the left.

Notice in the ammonia equilibrium

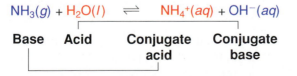

$$\underset{\textbf{Base}}{NH_3(g)} + \underset{\textbf{Acid}}{H_2O(l)} \rightleftharpoons \underset{\textbf{Conjugate acid}}{NH_4^+(aq)} + \underset{\textbf{Conjugate base}}{OH^-(aq)}$$

that water acts as a Bronsted-Lowry acid because it donates H^+ to NH_3. When we examine the preceding equilibrium from right to left ($\leftarrow$), you observe that the ammonium ion (NH_4^+) acts as an acid (i.e., it gives up H^+) and the hydroxide ion (OH^-) acts as a base (i.e., it accepts H^+). The NH_4^+ ion is said to be the *conjugate acid* of the base NH_3. The hydroxide ion (OH^-) is the *conjugate base* of the acid H_2O. Every acid has a **conjugate base,** the species remaining after the acid has donated a hydrogen ion. Every base has a **conjugate acid,** the species formed when the base has accepted a hydrogen ion. According to Bronsted-Lowry definitions, every acid-base reaction may be expressed in the form,

$$\underset{\text{Acid}}{HA} + \underset{\text{Base}}{B} \rightleftharpoons \underset{\text{Conjugate acid}}{HB^+} + \underset{\text{Conjugate base}}{A^-}$$

Water, which we saw act as a Bronsted-Lowry acid in the preceding example, can also act as a Bronsted-Lowry base. Water acts as a base when it accepts H^+ to form hydronium ions. For example, when hydrogen chloride gas is dissolved in water, the following reaction occurs.

$$\underset{\text{Acid}}{HCl(g)} + \underset{\text{Base}}{H_2O} \rightleftharpoons \underset{\text{Conjugate acid}}{H_3O^+(aq)} + \underset{\text{Conjugate base}}{Cl^-(aq)}$$

Here, HCl is an acid because it donates a hydrogen ion. H_2O is a base in this reaction because it accepts a hydrogen ion. H_3O^+ is the conjugate acid of the base H_2O, and Cl^- is the conjugate base of the acid HCl. A substance such as H_2O that can act as either an acid or a base is called **amphoteric.** Aluminum is a common metal that is amphoteric and readily forms an amphoteric oxide, Al_2O_3, which reacts with both acids and bases.

sample exercise 1

Identify the acid-conjugate base and base-conjugate acid pairs for the equilibrium

$$HCl + NH_3 \rightleftharpoons NH_4^+ + Cl^-$$

solution

You recognize HCl as an acid by consulting Table 15.2 or by noticing that it gives up H^+ as the reaction proceeds from left to right. Cl^- is the conjugate base, the species left when H^+ is released from the acid HCl.

You recognize NH_3 as a base by consulting Table 15.2 or by noticing that it accepts H^+ as the reaction proceeds from left to right. NH_4^+ is the conjugate acid, the species formed when the base NH_3 accepts H^+.

••• problem 2

Label the acid, base, conjugate acid, and conjugate base for the equilibrium

$$OH^- + HNO_3 \rightleftharpoons NO_3^- + H_2O$$

About 9 million tons of nitric acid produced each year in the United States are used in the manufacture of fertilizers, dyes, explosives, and many other chemicals.

Clearly, experience tells you that water does not display the characteristics of either an acid or a base as they are described in Table 15.1. Water does not turn blue litmus paper red nor red litmus paper blue. It does not taste sour or bitter, and it has a pH of 7, indicating that it is neither acid nor base. This brings us to the important point that any substance that donates H^+ can be classified *theoretically* as an acid. However, unless the concentration of H^+ produced reaches some lower limit (10^{-6} *M*), we do not witness the properties in Table 15.1. It is easy to recognize the acids and bases listed in Table 15.2 by their properties because they produce sufficiently large concentrations of H^+ or OH^- in solution. Let us now examine the idea of acid and base strength and see how it is related to the concentration of H^+ or OH^- in solution.

Strong acid must ionize completely in order to be considered strong.

15.4 acid and base strength, K_a and K_b

In Section 13.7, you saw that a strong electrolyte is one that dissociates or separates completely into ions in aqueous solution. Similarly, the criterion for a strong acid is that it ionizes completely and produces many ions, specifically $H^+(aq)$ ions (or, more correctly, H_3O^+ ions). The six strong acids you encountered in Section 13.7 and reviewed at the top of Table 15.2 essentially ionize 100%. For example, because the equilibrium

$$HCl(aq) \rightleftharpoons H^+(aq) + Cl^-(aq)$$

lies so far to the right, you may write

$$HCl(aq) \rightarrow H^+(aq) + Cl^-(aq)$$

Note that only those species blocked in color exist in the solution. All six strong acids may be written with an ionization equation similar to that for HCl.

Weak acids, on the other hand, ionize to only a limited extent. For example, acetic acid ionizes less than 1% to form hydrogen ions and the polyatomic acetate anion.

$$HC_2H_3O_2(aq) \rightleftharpoons H^+(aq) + C_2H_3O_2^-(aq)$$

Note that the intensity of color of each chemical species indicates its relative concentration in the solution. Because the equilibrium lies to the left and only a small number of H^+ ions (H_3O^+ ions) are produced, acetic acid is a weak acid. The equilibrium constant corresponding to the ionization of an acid is called the **ionization constant** and is designated K_a. For $HA \rightleftharpoons H^+ + A^-$,

$$K_a = \frac{[H^+][A^-]}{[HA]}$$

Table 15.3 lists the relative strengths of selected acids and bases arranged according to their K_a and K_b values. Notice that the ordering of K_a values places the strongest acids at the top of the table, so acid strength increases from bottom to top. The strength of conjugate bases is automatically arranged in reverse order, with the strongest conjugate base (largest K_b) at the bottom of the table. Base strength increases from top to bottom.

As with other equilibrium constants, a large value for K_a indicates that the equilibrium lies far to the right, and a small value indicates that the equilibrium position is to the left. Thus, strong acids have high K_a values and weak acids, low ones. In particular, compare the K_a values of HCl (1×10^7, strong) and acetic acid (1.8×10^{-5}, weak).

Table 15.3 also shows that if acids have more than one ionizable H^+, we consider the equilibria for removing one H^+ at a time. For example, H_2SO_4 is a **diprotic acid,** which means it has two ionizable hydrogens (yielding two protons). We represent the equilibria for their ionization as

$$H_2SO_4(aq) \rightleftharpoons H^+(aq) + HSO_4^-(aq)$$

$$HSO_4^-(aq) \rightleftharpoons H^+(aq) + SO_4^{2-}(aq)$$

There is a greater tendency for the first H^+ to ionize than the second. H_2SO_4 is a stronger acid than HSO_4^-. In general, there is always a greater tendency for the first H^+ to ionize in any acid with more than one ionizable H^+. This principle also applies to **triprotic acids** such as phosphoric acid, H_3PO_4 (Figure 15.6), or boric acid, H_3BO_3.

table 15.3 relative strengths of acids and bases

K_a	Acid name	Acid formula		Base formula	Base name	K_b
1×10^{10}	Perchloric	$HClO_4$	$\rightleftharpoons$	$H^+ + ClO_4^-$	Perchlorate ion	1×10^{-24}
	Hydriodic	HI	$\rightleftharpoons$	$H^+ + I^-$	Iodide ion	
	Hydrobromic	HBr	$\rightleftharpoons$	$H^+ + Br^-$	Bromide ion	
1×10^{7}	Hydrochloric	HCl	$\rightleftharpoons$	$H^+ + Cl^-$	Chloride ion	1.4×10^{-21}
	Nitric	HNO_3	$\rightleftharpoons$	$H^+ + NO_3^-$	Nitrate ion	
	Sulfuric	H_2SO_4	$\rightleftharpoons$	$H^+ + HSO_4^-$	Hydrogen sulfate ion	
	Hydronium ion	H_3O^+	$\rightleftharpoons$	$H^+ + H_2O$	Water	
1.2×10^{-2}	Hydrogen sulfate ion	HSO_4^-	$\rightleftharpoons$	$H^+ + SO_4^{2-}$	Sulfate ion	
	Phosphoric	H_3PO_4	$\rightleftharpoons$	$H^+ + H_2PO_4^-$	Dihydrogen phosphate ion	
	Hydrofluoric	HF	$\rightleftharpoons$	$H^+ + F^-$	Fluoride ion	1.4×10^{-11}
1.8×10^{-5}	Acetic	$HC_2H_3O_2$	$\rightleftharpoons$	$H^+ + C_2H_3O_2^-$	Acetate ion	
	Carbonic	H_2CO_3	$\rightleftharpoons$	$H^+ + HCO_3^-$	Bicarbonate ion	2.4×10^{-8}
	Hydrosulfuric	H_2S	$\rightleftharpoons$	$H^+ + HS^-$	Hydrogen sulfide ion	
	Dihydrogen phosphate ion	$H_2PO_4^-$	$\rightleftharpoons$	$H^+ + HPO_4^{2-}$	Hydrogen phosphate ion	
	Boric	H_3BO_3	$\rightleftharpoons$	$H^+ + H_2BO_3^-$	Dihydrogen borate ion	
	Ammonium ion	NH_4^+	$\rightleftharpoons$	$H^+ + NH_3$	Ammonia	1.8×10^{-5}
4.0×10^{-10}	Hydrocyanic	HCN	$\rightleftharpoons$	$H^+ + CN^-$	Cyanide ion	
	Bicarbonate ion	HCO_3^-	$\rightleftharpoons$	$H^+ + CO_3^{2-}$	Carbonate ion	
	Hydrogen phosphate ion	HPO_4^{2-}	$\rightleftharpoons$	$H^+ + PO_4^{3-}$	Phosphate ion	5.9×10^{-3}
1.0×10^{-15}	Hydrogen sulfide ion	HS^-	$\rightleftharpoons$	$H^+ + S^{2-}$	Sulfide ion	
	Water	H_2O	$\rightleftharpoons$	$H^+ + OH^-$	Hydroxide ion	5.5×10^{1}

Figure 15.6

In low concentration, phosphoric acid is used in carbonated beverages, particularly cola drinks, where it imparts a sour flavor.

Base strength follows the same principle as electrolyte strength or acid strength; i.e., extensive dissociation or ionization producing many ions means that the base is strong, and little dissociation or ionization yielding few ions means it is weak. In the case of bases, it is the hydroxide ion (OH^-) concentration that is of interest. Therefore, soluble metal hydroxides, such as the Group IA hydroxides LiOH, NaOH, and KOH (Table 15.2), are strong bases because these ionic compounds dissociate 100% as they dissolve.

Magnesium hydroxide is classified as a strong base because all formula units that do enter solution totally dissociate.

$$Mg(OH)_2(s) \rightleftharpoons Mg^{2+}(aq) + 2\ OH^-(aq)$$

However, Group IIA hydroxides, such as $Mg(OH)_2$, have limited solubilities in aqueous solution. In contrast to soluble strong bases, few formula units of magnesium hydroxide dissolve, and a saturated solution of magnesium hydroxide contains only a few hydroxide ions. Consequently, unlike a soluble metal hydroxide base such as NaOH (lye), $Mg(OH)_2$ solutions can be ingested. They are used to treat acidic stomach conditions and peptic ulcers by neutralization reaction with stomach acid (Section 15.8). All bases, strong or weak, neutralize acids, as was pointed out in Table 15.1 and will be discussed in Section 15.8.

The equilibrium point for

$$NH_3(aq) + H_2O(l) \rightleftharpoons NH_4^+(aq) + OH^-(aq)$$

lies to the left. Therefore, because there is little ionization and only a small number of OH^- ions is produced, ammonia is a weak base. The equilibrium constant corresponding to this reaction is the so-called K_b (base ionization constant) of NH_3 and equals 1.8×10^{-5}.

$$K_b = \frac{[NH_4^+][OH^-]}{[NH_3]} = 1.8 \times 10^{-5}$$

You recall from Section 14.4 that pure solids and liquids are excluded from equilibrium-constant expressions. As a result, water is not indicated in the expression for K_b shown above.

Notice that this equilibrium constant gives a measure not only of the extent to which OH^- ions are produced (the Arrhenius criterion for basicity), but also of how well the base NH_3 *accepts* H^+ from water (the Bronsted-Lowry criterion).

The values of K_b in Table 15.3 give you a measure of base strength. In this case, the measure of base strength is in terms of the Bronsted-Lowry test of how well the base accepts H^+. Recall that there is a conjugate base corresponding to every acid.

$$\underset{\textbf{Acid}}{HA} \rightleftharpoons \underbrace{H^+ + A^-}_{\textbf{Conjugate base}}$$

Therefore, because Table 15.3 ranks the tendency of acids to give up H^+ (tendency for the reaction to proceed →), it also tells us the tendency of conjugate bases to accept H^+ (tendency for the reaction to proceed ←).

$$HA \underset{\text{Base reaction}}{\overset{\text{Acid reaction}}{\rightleftharpoons}} H^+ + A^-$$

Strong acids have weak conjugate bases. For example, because HCl completely ionizes to H^+ and Cl^-, Cl^- has essentially no tendency to accept H^+ and therefore is a very weak base. Conversely, *very weak acids have strong conjugate bases.* Water has only a very slight tendency to ionize.

$$HOH \rightleftharpoons H^+ + OH^-$$

The conjugate base of water is the strong base OH^-.

problem 3

Arrange the following list of acids in order of *decreasing acid strength* (use Table 15.3).

HNO_3, nitric acid, an important industrial acid used in fertilizer manufacture
H_3BO_3, boric acid, used as an antiseptic and eyewash
HF, hydrofluoric acid, used to etch glass
H_2CO_3, carbonic acid, formed by dissolving $CO_2(g)$ in water, is used in carbonated beverages
H_2O, water

problem 4

Arrange the conjugate bases of the acids in Problem 3 in order of *increasing base strength.*

15.5 ionization of water, K_w

Although water is a molecular compound, it always contains a very small concentration of ions. In viewing water as an acid (HA), we say that it undergoes the ionization reaction

$$\underset{\textbf{(HA)}}{HOH(l)} \rightleftharpoons H^+(aq) + \underset{\textbf{(A}^-\textbf{)}}{OH^-(aq)}$$

As for any aqueous acid, this equilibrium is more correctly shown as

$$\underset{\textbf{Acid}}{HOH(l)} + \underset{\textbf{Base}}{H_2O(l)} \rightleftharpoons \underset{\textbf{Acid}}{H_3O^+(aq)} + \underset{\textbf{Base}}{OH^-(aq)}$$

This representation shows you that the equilibrium is possible because of the amphoteric nature of water. The value of K_a for water in Table 15.3 further tells you that the equilibrium lies far to the left because H_3O^+ and OH^- are the stronger acid-base pair. Therefore, in pure water, the concentrations of hydronium and hydroxide ions are very small.

As we proceed in our discussion, we will use the simpler expression for the ionization equilibrium of water, that is,

$$HOH(l) \rightleftharpoons H^+(aq) + OH^-(aq)$$

The equilibrium-constant expression for this reaction is

$$K_{eq} = [H^+][OH^-]$$

(Once again, you remember that pure liquids and solids do not appear in K_{eq} expressions, so $[H_2O]$ has not been included as the denominator.) Because this equilibrium constant is so important in acid-base chemistry, it is given the special symbol K_w. Also, the product of the concentrations, $[H^+][OH^-]$, is sometimes called the **ion product of water.**

$$K_w = [H^+][OH^-]$$

Since pure water is *neutral,* that is, it has no net charge, the concentration of H^+ must equal the concentration of OH^-. The concentration of each ion has been measured to be 1.0×10^{-7} *M*. Thus, the value of K_w can be found by substitution. Since

$$[H^+] = [OH^-] = 1.0 \times 10^{-7}\ \frac{\text{mol}}{\text{L}}$$

values for each of these concentrations may be substituted into the expression for K_w at 20.°C.

$$\begin{aligned} K_w &= [H^+][OH^-] \\ &= (1.0 \times 10^{-7})(1.0 \times 10^{-7}) \\ &= 1.0 \times 10^{-14} \end{aligned}$$

The units of K_w, as with most equilibrium constants, are customarily omitted.

The addition of acids to water will raise the concentration of H^+ above 1.0×10^{-7} *M*. The addition of bases will raise the concentration of OH^- above 1.0×10^{-7} *M*. The *product of the concentrations* of H^+ and OH^- ions in water, however, must remain constant and equal to K_w. This is because the water equilibrium $HOH(l) \rightleftharpoons H^+(aq) + OH^-(aq)$ is present in all aqueous solutions and the equilibrium-constant expression,

$$K_w = [H^+][OH^-] = 1.0 \times 10^{-14}$$

for the equilibrium

$$HOH(l) \rightleftharpoons H^+(aq) + OH^-(aq)$$

must always be satisfied.

By applying Le Chatelier's principle to the above equilibrium, you see that when H^+ is added, *the equilibrium will shift to the left to use up some of the excess* H^+; as a result, $[OH^-]$ decreases. You also see that when OH^- is added to the above equilibrium, *the equilibrium will shift to the left to use up some of the excess* OH^-; as a result, $[H^+]$ decreases.

In summary,

Neutral solution:

$[H^+] = [OH^-] = 1.0 \times 10^{-7}$ *M*

Acidic solution:

$[H^+] > 1.0 \times 10^{-7}$ *M*
$[OH^-] < 1.0 \times 10^{-7}$ *M*

Basic solution:

$[OH^-] > 1.0 \times 10^{-7}$ *M*
$[H^+] < 1.0 \times 10^{-7}$ *M*

---

sample exercise 2

An aqueous solution is found to have $[H^+] = 1.0 \times 10^{-4}$ *M*. Is this solution acidic, basic, or neutral?

solution

Since 1×10^{-4} *M* is greater than 1×10^{-7} *M*, the solution is acidic.

---

sample exercise 3

If you are given $[H^+]$ or $[OH^-]$ and asked to determine $[OH^-]$ or $[H^+]$, you will need to use the fact that $K_w = [H^+][OH^-] = 1.0 \times 10^{-14}$. In Sample Exercise 2, $[H^+]$ is given as 1.0×10^{-4} *M* and $[OH^-]$ is unknown. What is $[OH^-]$?

solution

Rearranging the K_w expression,

$$K_w = [H^+][OH^-] = 1.0 \times 10^{-14}$$

you may solve for the concentration of OH^-.

$$[OH^-] = \frac{K_w}{[H^+]} = \frac{1.0 \times 10^{-14}}{[H^+]}$$

$$= 1.0 \times 10^{-14}/1.0 \times 10^{-4}$$

$$= 1.0 \times 10^{-10}\ M$$

Consistent with this being an acidic solution, $[OH^-]$ turns out to be less than 1.0×10^{-7} *M*.

••• problem 5

What is $[OH^-]$ in a solution in which $[H^+] = 3.2 \times 10^{-9}$ *M*? Is this solution acidic, basic, or neutral?

15.6 calculating pH from $[H^+]$

Concentration values of H^+ and OH^- mentioned in the previous section were somewhat unwieldy exponential numbers. They may vary over many powers of 10. To be able to express the acidity or basicity of aqueous solutions without the use of exponential numbers, chemists devised the pH scale of acidities. The pH of a solution is defined as

$$\mathbf{pH = -log\ [H^+]}$$

which is read as, "pH equals the negative logarithm of the hydrogen ion concentration in moles per liter." pH is very easy to calculate for solutions in which $[H^+]$ is an exact power of 10, that is, $1 \times$ some power of 10. In this case, the value of the pH is just the absolute value of the exponent.

$$[H^+] = 1.0 \times 10^{-x}$$

$$pH = x$$

The logarithm of 1.0×10^{-x} is $-x$. The negative logarithm is $-(-x) = x$.

sample exercise 4

Determine the pH of the following solutions and specify if the solution is acidic, basic, or neutral.

a. $[H^+] = 1.0 \times 10^{-7}$
b. $[H^+] = 1.0 \times 10^{-11}$
c. $[H^+] = 0.001$

solution

a. pH = 7. The logarithm of 1.0×10^{-7} is –7; the negative log is –(–7) = 7. This solution is neutral because $[H^+] = 1.0 \times 10^{-7}$ *M* and the pH = 7.
b. pH = 11. The solution is basic because $[H^+] < 1.0 \times 10^{-7}$ *M* and the pH > 7.
c. First write 0.001 in scientific notation: $0.001 = 1 \times 10^{-3}$. Then, as before, the pH = 3. This solution is acidic because $[H^+] > 1.0 \times 10^{-7}$ *M* and the pH < 7.

••• problem 6

Given the following $[H^+]$ values, determine the pH of each solution.

a. $[H^+] = 1.0 \times 10^{-5}$
b. $[H^+] = 0.000010$

The pH characteristics of neutral, acidic, and basic solutions may be summarized as follows.

Neutral solution:	Acidic solution:	Basic solution:
$[H^+] = 1.0 \times 10^{-7}$ *M*	$[H^+] > 1.0 \times 10^{-7}$ *M*	$[H^+] < 1.0 \times 10^{-7}$
pH = 7.00	pH < 7.00	pH > 7.00

The pH scale is usually shown from values of 0 to 14, with values less than 7 indicating an acidic solution and values greater than 7, a basic solution. This range

(a) (b) (c)

Figure 15.7
(a) Scientific calculator with 4.0.
(b) Scientific calculator with entry value of 4.0×10^{-3} and with a pen pointing to the LOG key. (c) Scientific calculator with –2.39794 E 00 displayed.

of pH values spans the concentrations of H^+ ions that are measurable with pH paper or a standard pH meter. In acidic solutions with $[H^+] > 1$ *M,* the pH has a negative value, and in basic solutions with $[OH^-] > 1$ *M,* the pH has a value above 14. Calculations beyond the range of pH between 0 and 14 will not be encountered in this text.

For most solutions, however, the hydrogen ion concentration is *not* an exact power of 10. That is, in most solutions, $[H^+]$ is not equal to 1.0×10^{-x}, but rather $[H^+] = N \times 10^{-x}$, where *N* is some number other than 1. To determine the pH in this case, you must be able to use the logarithm key on your scientific calculator. As an example, the key sequence to evaluate the pH for a solution in which $[H^+] = 4.0 \times 10^{-3}$ *M* is summarized below and shown in Figure 15.7.

Procedure for calculating pH

(This procedure is ordered for a Texas Instruments calculator. Other calculators may invert the order of steps 2 and 3.)

1. Clear the calculator by pressing the CLEAR key.
2. Enter the number equal to $[H^+] = 4.0 \times 10^{-3}$ or 0.0040, and then strike the = (or ENTER) key.

   Remember, to enter a number in scientific notation, you must enter the 4.0, strike the EE key to indicate exponential notation, and then enter the exponent as –3, using the ± key to enter a negative sign for the exponent.
3. Press the LOG key. After striking the LOG key, the logarithm of $[H^+]$, which is –2.39794 E 00, should be displayed on your calculator.

   Be careful that you use the LOG key for logarithm base 10 and not the LN key for logarithm base e.
4. Change the sign of the displayed answer and record this to the correct number of significant figures.

   Since $[H^+]$ was given to 2 sig figs, the pH should also have 2 sig figs. However, the whole number to the left of the decimal is not significant in the value of pH; it simply indicates the nearest power of 10. *Only the figures to the right of the decimal point are significant in a value of pH.* To 2 sig figs, the pH = 2.40.

sample exercise 5

Determine the pH of a solution in which $[H^+] = 2.3 \times 10^{-5}$ *M*.

solution

1. Recognize that this H^+ concentration is of the form

$$[H^+] = N \times 10^{-x}$$

2. Take the log of 2.3×10^{-5} on your scientific calculator.

$$\log 2.3 \times 10^{-5} = -4.638272164 \text{ E } 00$$

3. Change the sign of your answer and report to 2 sig figs.

$$pH = -\log[H^+] = -(-4.64) = 4.64$$

sample exercise 6

Determine the pH of a solution in which $[H^+] = 0.032$ *M*.

solution

1. Write the concentration in scientific notation to conform to

$$[H^+] = N \times 10^{-x}$$

$$[H^+] = 3.2 \times 10^{-2}$$

2. Take the log of 3.2×10^{-2} on your scientific calculator.

$$\log 3.2 \times 10^{-2} = -1.494850022 \text{ E } 00$$

3. Change the sign of your answer and report to 2 sig figs.

$$pH = -\log[H^+] = -(-1.49) = 1.49$$

If the $[H^+]$ of a solution is not given, but the hydroxide ion concentration, $[OH^-]$, is, you can also determine the pH of the solution. Substituting the given $[OH^-]$ into the expression for K_w, you first calculate $[H^+]$ and then pH, as shown in Sample Exercise 7.

sample exercise 7

Determine the pH of a solution in which $[OH^-] = 5.0 \times 10^{-5}$ *M*.

solution

Calculate $[H^+]$ using K_w (see Sample Exercise 3).

$$K_w = [H^+][OH^-] = 1.0 \times 10^{-14}$$

$$[H^+][5.0 \times 10^{-5}] = 1.0 \times 10^{-14}$$

$$[H^+] = \frac{1.0 \times 10^{-14}}{5.0 \times 10^{-5}} = 2.0 \times 10^{-10}\ M$$

Now determine the pH of a solution in which $[H^+] = 2.0 \times 10^{-10}$ *M* (see Sample Exercises 5 and 6).

$$\log [H^+] = -9.69897 \text{ E } 00 = -9.70$$

$$pH = -\log [H^+] = -(-9.70) = 9.70$$

••• problem 7

Determine the pH of the following solutions with the given concentrations.

a. $[H^+] = 6.8 \times 10^{-6}$ *M*
b. $[OH^-] = 7.1 \times 10^{-1}$ *M*

table 15.4 normal pH range for some bodily fluids

Fluid	Normal pH range
Liver bile	7.4–8.0
Blood	7.35–7.45
Gastric juice	1.0–2.0
Pancreatic juice	7.0–8.0
Saliva	6.4–7.0
Sweat	4.5–7.5
Tears	7.0–7.4
Urine	4.5–7.5

table 15.5 pH range for optimum plant growth

Plant	Required pH range
Apples	5.0–6.5
Cherries	6.0–7.5
Strawberries	5.0–6.5
Peas	6.0–7.5
Snap beans	6.0–7.5
Tomatoes	5.5–7.5
Tulips	6.0–7.0
Roses	6.0–8.0
Azaleas	4.5–6.0
Hydrangeas	5.0–6.0

15.7 measurement of pH

Most bodily fluids such as blood and urine have a pH range indicative of normal health (Table 15.4). Continued deviation from a particular range usually indicates some pathological condition and therefore offers a means of diagnosis. To provide this medical information, medical laboratories require a simple and quick method of measuring pH.

Plants grow best in soil with a specific pH range and do not show maximum growth or fruit production outside this range (Table 15.5). Farmers must be able to measure the pH of moist soil in which each crop is planted so they can adjust it if necessary.

Many chemical reactions yield different products depending on the pH, so laboratory workers must also have a way of quickly and conveniently measuring pH.

As you learned in Section 15.1 and in Lab Experiment 14, the pH of a colorless or lightly colored solution is often determined by the use of an **indicator**, a dye that changes color as pH changes. Frequently, the dye is impregnated on paper to which a test solution can easily be applied. Indicators are weak acids that have one color in the acid form (symbolized by HIn) and another color in the conjugate-base form (In^-).

$$\underset{\text{One color at low pH}}{HIn} \rightleftharpoons H^+ + \underset{\text{Different color at higher pH}}{In^-}$$

table 15.6 approximate pH of some common materials

pH	Material
0	Battery acid
1	Stomach acid
2	Lemon juice, lime juice
3	Vinegar, wine, soft drinks, beer, orange juice, pickles
4	Tomatoes, grapes
5	Black coffee, rainwater
6	Urine, milk, saliva
7	Pure water, blood
8	Sea water
9	Clorox, phosphate detergent
10	Soap, milk of magnesia
11	Household ammonia
12	Hair remover
13	Oven cleaner

When H^+ is removed from the equilibrium (by reaction with some base), Le Chatelier's principle predicts that there will be a shift in the position of equilibrium toward the In^- side (to the right). When the concentration of In^- exceeds that of HIn, the color is that of the high pH range. For the dye named **litmus,** a commonly used indicator, this is blue. If the solution is now made acidic (that is, H^+ is added), the equilibrium shifts towards the HIn side, giving the color of the low pH range, which, for litmus, is red (Figure 15.2).

$$\underset{\text{Red}}{\text{HLit}} \rightleftharpoons H^+ + \underset{\text{Blue}}{\text{Lit}^-}$$

Litmus changes color over the range of pH 5 to 8, but other indicators change at other pH values. Each changes over a unique pH range depending on its acid ionization constant.

The use of litmus paper, which is red at any pH less than 5 and blue at any pH greater than 8, enables us to determine only whether a solution is acidic or basic, as you have learned in Section 15.1. A universal pH paper can be prepared by impregnating paper with several dyes, each of which change color at some narrow and different pH range (Figure 15.3). The pH of an unknown solution is then determined by comparing the color developed on pH paper with that on a chart relating color to pH.

A pH meter (Figure 15.4) is used to determine pH values more accurately and to determine those of highly colored solutions, which might obscure the color of an indicator dye. The pH meter works by using an electrode sensitive to the concentration of H^+ ions in solution. Table 15.6 lists the pH values of some commonly encountered materials.

15.8 reactions of acids and bases

Probably the most common and important reaction of acids and bases is their reaction with one another, the **neutralization reaction,** which we will discuss first. You are already familiar with the neutralization of a strong acid with a strong base (Section 9.8) and its net ionic equation (Sections 14.7 and 15.1). Since any acid and base will undergo a neutralization reaction, it is important to extend your understanding of neutralization reactions to acids and bases that are weak. We will begin by reviewing strong acid-strong base neutralization reactions and introducing some new terminology.

strong acid-strong base reactions

When acids and bases react with one another, ionic compounds called salts form. Table salt, NaCl, is only one example of the class of compounds known as salts. **Salts** are ionic compounds formed from the reaction of an acid with a base: the cation of the salt comes from the base, and the anion comes from the acid.

If the base reacting in the neutralization reaction is a hydroxide base, water is also formed. Consider, once again, the neutralization reaction between HCl(*aq*) and NaOH(*aq*):

HCl(*aq*)	+	NaOH(*aq*)	→	NaCl(*aq*)	+	H_2O
Acid		**Base**		**Salt**		**Water**

The salt gets its cation (Na^+) from the base and its anion (Cl^-) from the acid. Water forms from the hydroxide of the base and the H^+ of the acid.

For a strong acid-strong base reactant pair, the reaction can be viewed as a double replacement (Section 9.8), in which cations and anions switch partners.

$$H^+\ Cl^- + Na^+\ OH^- \rightarrow Na^+\ Cl^- + H{-}OH$$

Other examples of strong acid-strong base neutralization reactions are

$HNO_3(aq)$	+	$LiOH(aq)$	→	$LiNO_3(aq)$	+	$H_2O(l)$
$2\ HCl(aq)$	+	$Ca(OH)_2(aq)$	→	$CaCl_2(aq)$	+	$2\ H_2O(l)$
$H_2SO_4(aq)$	+	$2\ NaOH(aq)$	→	$Na_2SO_4(aq)$	+	$2\ H_2O(l)$
Acid		**Base**		**Salt**		**Water**

Because the strong acid is completely ionized and the strong base completely dissociates in aqueous solution, the following ionic equations (Section 13.9) apply,

HA(*aq*)	+	MOH(*aq*)	→	MA(*aq*)	+	$H_2O(l)$
Strong acid		**Strong base**		**Salt (base cation + acid anion)**		

Traditional equation

This equation may be rewritten as an ionic equation to emphasize ionized and dissociated components.

$$H^+(aq) + \cancel{A^-}(aq) + \cancel{M^+}(aq) + OH^-(aq) \rightarrow \cancel{M^+}(aq) + \cancel{A^-}(aq) + H_2O(l)$$

Full ionic equation

As you know, the net ionic equation for neutralization of any strong acid-strong base combination is

$$H^+(aq) + OH^-(aq) \rightarrow H_2O(l)$$

Net ionic equation

The pH at neutralization of a strong acid-strong base pair is 7.0. At neutralization, there is no excess of either $H^+(aq)$ or $OH^-(aq)$ ions. As determined from K_w in Section 15.5, the presence of water at neutralization requires that $[H^+] = [OH^-] = 1.0 \times 10^{-7}\ M$.

In terms of Bronsted-Lowry theory, this strong acid-strong base reaction is a proton transfer reaction in which the equilibrium lies far to the right.

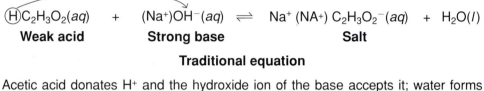

$HCl(aq)$	+	$OH^-(aq)$	$\rightleftharpoons$	$Cl^-(aq)$	+	$HOH(l)$
Strong acid		**Strong base**		**Weak base**		**Weak acid**

A single-headed arrow is typically used for the strong acid-strong base reaction because the equilibrium is so far to the right.

$$HCl(aq) + OH^-(aq) \rightarrow Cl^-(aq) + HOH(l)$$

Notice again that in this Bronsted-Lowry equation, NaOH is represented simply as OH^- because Na^+ is merely a spectator.

weak acid-strong base reactions

Consider the reaction between a weak acid (acetic acid) and a strong base (sodium hydroxide) to yield a salt (sodium acetate) and water.

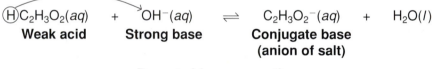

$HC_2H_3O_2(aq)$	+	$(Na^+)OH^-(aq)$	$\rightleftharpoons$	Na^+ (NA+) $C_2H_3O_2^-(aq)$	+	$H_2O(l)$
Weak acid		**Strong base**		**Salt**		

Traditional equation

Acetic acid donates H^+ and the hydroxide ion of the base accepts it; water forms because the base is a hydroxide. The salt, sodium acetate, forms in the usual manner, its cation from the base and its anion from the acid. The neutralization reactions of weak acids or weak bases are best considered in terms of Bronsted-Lowry theory because the weak species is not fully ionized in solution The sodium ion is not shown because spectator ions are generally not represented in Bronsted-Lowry reactions.

$HC_2H_3O_2(aq)$	+	$OH^-(aq)$	$\rightleftharpoons$	$C_2H_3O_2^-(aq)$	+	$H_2O(l)$
Weak acid		**Strong base**		**Conjugate base (anion of salt)**		

Bronsted-Lowry equation

weak acid-weak base reactions

Now consider the neutralization reaction of a weak acid (acetic acid) and a weak base (ammonia) to yield a salt (ammonium acetate).

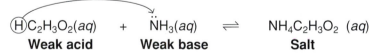

$HC_2H_3O_2(aq)$	+	$\ddot{N}H_3(aq)$	$\rightleftharpoons$	$NH_4C_2H_3O_2\ (aq)$
Weak acid		**Weak base**		**Salt**

Again, a salt forms wherein the cation comes from the base and the anion from the acid. No water forms because the base is not a hydroxide base.

strong acid-weak base reactions

The reaction between a strong acid such as HCl (hydrochloric acid) and a weak base such as NH_3 (ammonia) to yield a salt (ammonium chloride) gives a similar result.

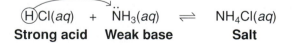

$HCl(aq)$	+	$\ddot{N}H_3(aq)$	$\rightleftharpoons$	$NH_4Cl(aq)$
Strong acid		**Weak base**		**Salt**

applications of neutralization reactions

Neutralization reactions help relieve indigestion. Food overindulgence or stressful situations can lead to an oversecretion of hydrochloric acid in the stomach, for which an antacid is commonly used. $Al(OH)_3$ and $Mg(OH)_2$ are the two most common hydroxide-base antacids. They function by neutralizing some of the excess H^+ ions in the stomach.

$$2\ HCl(aq) + Mg(OH)_2(s) \rightarrow MgCl_2(aq) + 2\ H_2O(l)$$

$$3\ HCl(aq) + Al(OH)_3(s) \rightarrow AlCl_3(aq) + 3\ H_2O(l)$$

Other antacid medications employ the reactions of acids with basic bicarbonate and carbonate anions. Acids stronger than H_2CO_3 (Table 15.3) react with carbonates and bicarbonates to form carbonic acid, H_2CO_3, most of which (because of the instability of H_2CO_3) decomposes to $CO_2(g)$ and $H_2O(l)$. You can write an equation for the reaction between hydrochloric acid and sodium bicarbonate that also shows the carbonic acid equilibrium leading to water and carbon dioxide gas.

$$\underset{}{HCl(aq)} + \underset{\text{Sodium hydrogen carbonate}}{NaHCO_3(aq)} \rightarrow NaCl(aq) + \underset{\text{Carbonic acid}}{H_2CO_3(aq)}$$

Traditional equation

Keeping in mind that $Na^+(aq)$ and $Cl^-(aq)$ are spectator ions, the net ionic equation may be written,

$$H^+(aq) + HCO_3^-(aq) \rightarrow H_2CO_3(aq)$$

Net ionic equation

However, because of its instability, carbonic acid decomposes to $CO_2(g)$ and $H_2O(l)$, which, because of the escape of the carbon dioxide gas, shifts the equilibrium far to the right.

$$H_2CO_3(aq) \rightarrow H_2O(l) + CO_2(g)$$

The overall net ionic equation for the reaction of an acid with a bicarbonate is

$$H^+(aq) + HCO_3^-(aq) \rightarrow H_2O(l) + CO_2(g)$$

Similarly, for a carbonate, the reaction would be

$$2\ H^+(aq) + CO_3^{2-}(aq) \rightarrow H_2O + CO_2(g)$$

Many popular antacids remove excess H^+ through a bicarbonate (Alka-Seltzer, Brioschi) or carbonate (TUMS, Pepto-Bismol) ingredient. They make you burp because CO_2 gas is released in the stomach.

An interesting application of an acid with a carbonate is that of limestone (calcium carbonate) dissolving in the presence of acid. Many buildings, their ornamentation, and statues are made from limestone or marble, which are principally calcium carbonate, $CaCO_3$. The surfaces of these buildings and statues undergo deterioration over a period of many years in the presence of acid rain, which contains dilute sulfuric acid. The chemical equation representing this change is

$$CaCO_3(s) + H_2SO_4(aq) \rightarrow CaSO_4(s) + H_2CO_3(aq) \rightarrow CaSO_4(s) + H_2O(l) + CO_2(g)$$

Because of the escape of carbon dioxide gas from the aqueous solution on the surface of the building and the formation of liquid water, the equilibrium lies far to the right. As a result, the surface of the marble, which is primarily calcium carbonate, when exposed to sulfuric acid, is converted into calcium sulfate, $CaSO_4$. Calcium sulfate is more soluble in water than the original calcium carbonate, so it is washed away more readily. The greater the exposure to acid rain, the more surface that is susceptible to erosion (Figure 15.8).

Figure 15.8
When exposed to sulfuric acid, the surface of marble (calcium carbonate) is changed to calcium sulfate, which is more soluble in water than calcium carbonate. As a result, acid rain, over a period of many years, slowly dissolves the surfaces of buildings and statues.

Acid-base reactions are also important in baking. In baking powder, sodium bicarbonate (baking soda) is combined with acidic ingredients. When wet, this mixture releases carbon dioxide, causing the rising of such baked goods as biscuits and certain breads.

reactions of acids with active metals

Another example of an acid reaction is the reaction of an acid with some metals. Acids react with Group IA, IIA, and IIIA metals to form hydrogen gas and a salt of the metallic ion. For example,

$2\ Al(s)$	+	$6\ HCl(aq)$	$\longrightarrow$	$2\ AlCl_3(aq)$	+	$3\ H_2(g)$
Metal		**Acid**		**Salt of the metal**		**Hydrogen gas**

Some transition metals, such as iron and zinc, react in a similar way, whereas others, such as gold, silver, copper, and mercury, do not react with acids to form hydrogen gas (Section 9.8 and Table 9.2).

Acidic foods such as wine or tomatoes should not be brought into contact with cast-iron or carbon-steel utensils because the acid in these foods will react with the iron. In stainless steel, iron is alloyed in such a way that it no longer reacts with acid. There is also increasing concern that carbonic acid, H_2CO_3 ($CO_2 + H_2O$), in carbonated beverages reacts with aluminum soda cans, albeit slightly, to produce undesirable concentrations of aluminum ions in the soda. A great deal of research is being conducted into limiting the concentration of $H^+(aq)$ ions for beverages in aluminum cans. One approach is to use a buffer (Sections 15.10 and 15.11) to hold the pH of such a solution at a higher (less acidic) value.

15.9 determination of acid-base concentrations: titration

In Section 15.1, you were introduced to the procedure of titration in which neutralization reactions (Section 15.8) can be used to measure the unknown concentration of an acid or base if you are given a base or acid of known concentration. Consider again the *traditional equation* for the neutralization reaction between HCl and NaOH.

$$HCl(aq) + NaOH(aq) \rightarrow NaCl(aq) + H_2O(l)$$

Or, as an *ionic equation,*

$$H^+(aq) + Cl^-(aq) + Na^+(aq) + OH^-(aq) \rightarrow Na^+(aq) + Cl^-(aq) + H_2O(l)$$

At the point of neutralization, called the **equivalence point,** the number of moles of H^+ from the acid exactly equals the number of moles of OH^- from the base. This is particularly apparent from the *net ionic equation* for the reaction of any strong acid and strong base.

$$H^+(aq) + OH^-(aq) \rightarrow H_2O(l)$$

In doing titration calculations, you must always remember that *at the equivalence point,*

Moles of H^+ = Moles of OH^-

In doing a titration experiment, a known volume of an unknown acid or base is placed in a flask with an indicator. The indicator must be one that undergoes a color change at the equivalence point. In practice, the indicator will rarely change color exactly at the equivalence point, but rather at a close pH called the **endpoint.** A buret is then used for the addition of a base to the unknown acid, or for the addition of an acid to the unknown base, until the endpoint is reached. The buret is calibrated so that the volume of solution it delivers through a small bore in the stopcock may be very accurately measured. The stopcock may be rotated so that the solution can be delivered at whatever rate the user chooses; it may be delivered in a steady stream or one drop at a time.

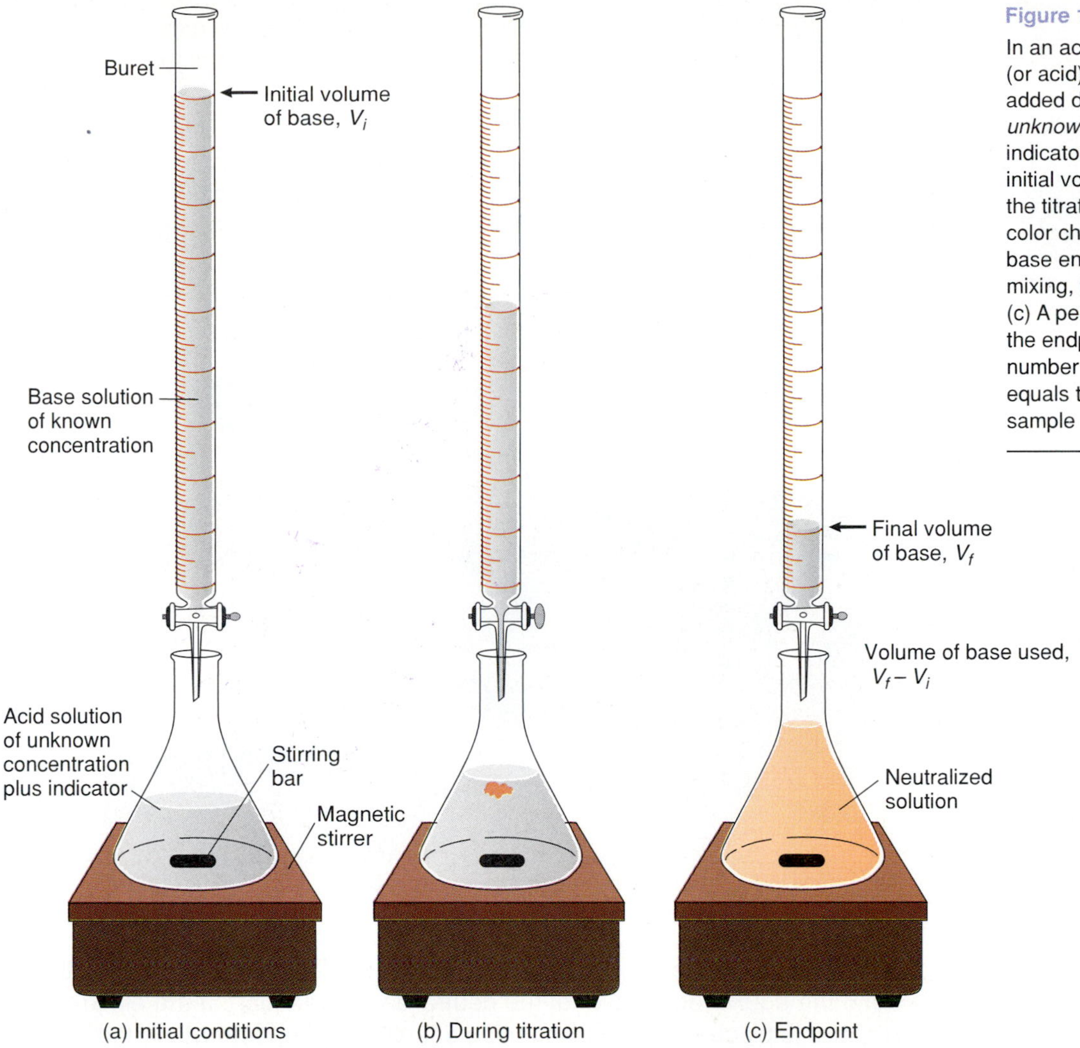

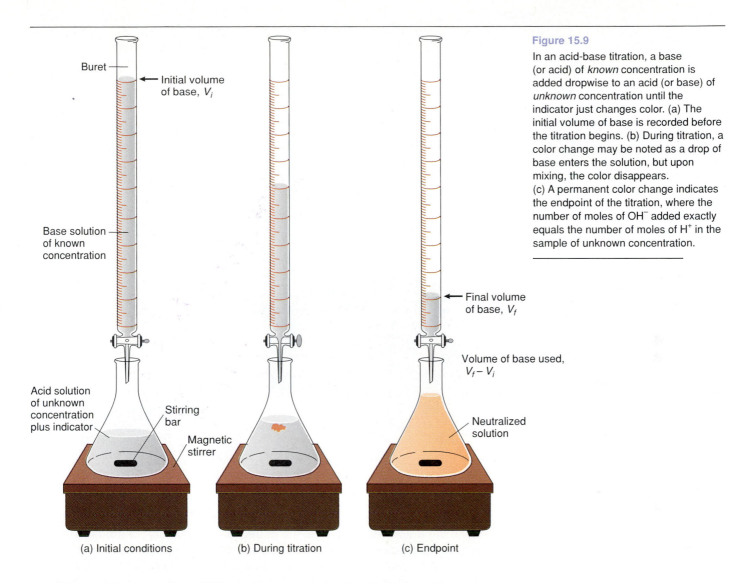

Figure 15.9
In an acid-base titration, a base (or acid) of *known* concentration is added dropwise to an acid (or base) of *unknown* concentration until the indicator just changes color. (a) The initial volume of base is recorded before the titration begins. (b) During titration, a color change may be noted as a drop of base enters the solution, but upon mixing, the color disappears. (c) A permanent color change indicates the endpoint of the titration, where the number of moles of OH^- added exactly equals the number of moles of H^+ in the sample of unknown concentration.

Figure 15.9 shows the addition of a measured volume of base of known concentration until the known volume of unknown acid is neutralized, as witnessed by the change in indicator color at the endpoint. When adding base, the solution in the flask must be constantly swirled or stirred with a magnetic stirring bar to provide thorough mixing. As the endpoint is approached, the solution will begin to show the indicator color, but fade as the solution is mixed. From that point on, base should be added dropwise and mixed between each drop.

When an indicator that changes color at a pH near the equivalence point of the reaction (in this case, a pH of 7) is chosen and careful technique employed, the equivalence point and the endpoint will be nearly identical. At the equivalence point, the number of moles of OH^- added exactly equals the number of moles of H^+ in the sample of unknown concentration. From the change in volume of base in the buret, you know the volume of base added and can calculate the number of moles of OH^- added, which is also equal to the number of moles of H^+ neutralized.

Calculation of the number of moles of added OH^- from the known base concentration, *M,* and volume, *V,* is done by remembering from Section 13.5 that $M \times V$ (in liters) = moles of OH^-, and at the equivalence point, moles of OH^- = moles of H^+. Then the molarity of H^+ is calculated by dividing this number of moles by the volume (in liters) of the unknown acid. Titration calculations always involve a calculation of the number of moles of H^+ or OH^- neutralized and then a calculation of the molarity of H^+ or OH^-. An example of these calculations is given in

Figure 15.10

Instead of using an indicator to change color at the equivalence point, the electrode of a pH meter is placed in the acid solution of unknown concentration. The electrode monitors the increase in pH as the neutralization progresses. When the pH reaches a value of 7.0, the equivalence point has been reached for the neutralization of a strong acid with a strong base. As the pH approaches 7.0, the base should be added dropwise from the buret.

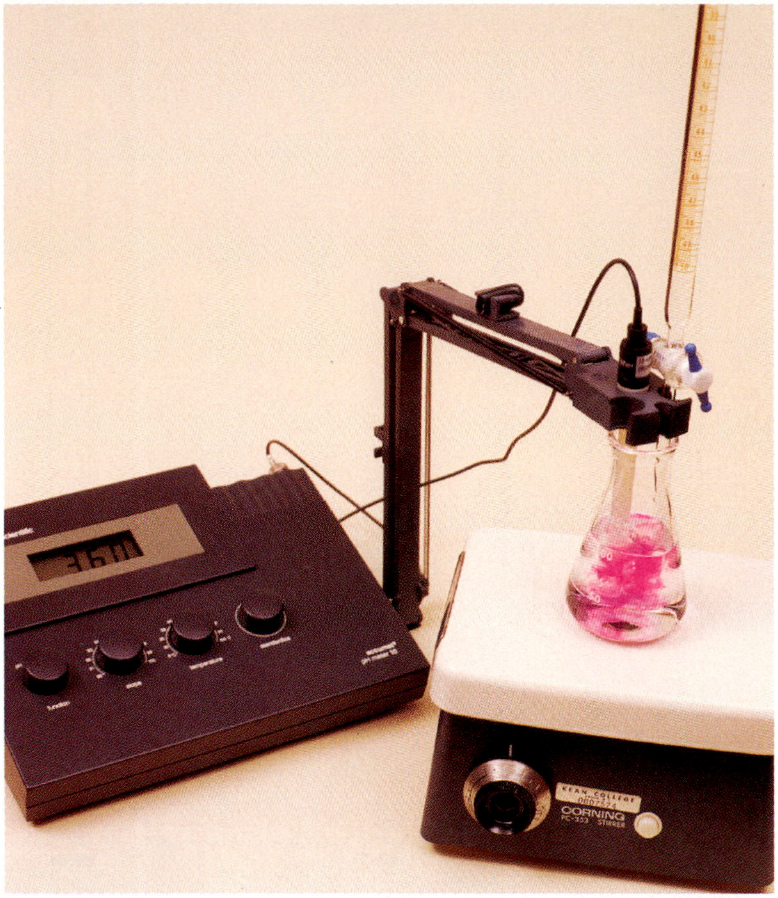

Sample Exercise 8, where a pH meter has been used, instead of an indicator, to determine the equivalence point of the neutralization reaction. If the electrode of the pH meter is placed in the acid solution of unknown concentration and base is added from the buret in the manner described above, the pH of the mixture will increase until it reaches a value of 7.0 at the equivalence point (Figure 15.10).

sample exercise 8

Using a pH meter, a chemist finds that 42.50 mL of a 0.100 *M* NaOH solution is required to titrate (neutralize) 31.00 mL of a hydrochloric acid solution of unknown concentration. What is the molarity of the hydrochloric acid?

solution

1. Calculate the moles of added OH^- from the equation *M* (of OH^-) × *V* (in liters) = moles of OH^-.

$$\frac{0.100 \text{ mol OH}^-}{1 \not{L}} \times 0.04250 \not{L} = 0.00425 \text{ mol OH}^-$$

(0.100 *M* NaOH is 0.100 *M* in OH^-; 42.50 mL × 1 L / 1,000 mL = 0.04250 L.)

2. At the equivalence point, moles of OH^- = moles of H^+. Therefore, 0.00425 mol of H^+ was neutralized.
3. Calculate the acid molarity by dividing moles by liters of solution. There is 0.00425 mol of H^+ in 0.00425 mol of HCl.

$$31.00\ \cancel{mL}\ \text{HCl} \times \frac{1\ \text{L}}{1{,}000\ \cancel{mL}} = 0.03100\ \text{L HCl solution}$$

$$M \text{ of HCl} = \frac{\text{Moles of HCl}}{\text{Liters of solution}}$$

$$M \text{ of HCl} = \frac{0.00425\ \text{mol HCl}}{0.03100\ \text{L HCl}} = 0.137\ M$$

In doing titration calculations, you must sometimes take into account the fact that a given acid or base can provide more than one H^+ or OH^- ion per molecule or formula unit. Such cases occur for diprotic and triprotic acids, for example, H_2SO_4 and H_3PO_4, and for metal hydroxides with more than one hydroxide anion, for example, $Mg(OH)_2$. The subscript of the hydrogen ion or hydroxide ion relates the number of moles of that ion to the number of moles of acid or base.

For example, when H_2SO_4 ionizes according to the equation

$$H_2SO_4(aq) \rightarrow 2\ H^+(aq) + SO_4^{2-}(aq)$$

the coefficients in the balanced chemical equation tell you that 2 mol of H^+ are formed from each mole of H_2SO_4 that ionizes. Therefore, if your H_2SO_4 solution were prepared with 1 mol in each liter (1 *M*), then upon complete ionization there would be 2 mol of H^+ ions in each liter. The concentration of H^+ would be 2 *M*. In general, acid-base neutralizations give complete ionization; and you may multiply the acid molarity by the subscript of hydrogen to determine the molarity of H^+. Similarly, you may multiply the base molarity by the hydroxide subscript to determine the molarity of OH^-. For example, 0.01 *M* $Ca(OH)_2$ is 0.02 *M* in OH^- because, according to the balanced ionization equation

$$Ca(OH)_2(aq) \rightarrow Ca^{2+}(aq) + 2\ OH^-(aq)$$

there are 2 mol of OH^- per mole of $Ca(OH)_2$.

A similar relationship may be used for the concentration of ions that results from the dissociation of any strong electrolyte.

••• problem 8

Assuming complete ionization for the acids and complete dissociation for the bases, calculate the molarity of H^+ or OH^- in each of the following solutions.

a. 0.20 *M* H_2SO_4
b. 0.45 *M* HBr
c. 0.3 *M* KOH
d. 0.75 *M* $Ba(OH)_2$
e. 0.50 *M* H_3PO_4

sample exercise 9

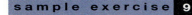

A student neutralizes 29.10 mL of a potassium hydroxide solution with 15.30 mL of a 0.500 *M* H_2SO_4 solution. What is the molarity of the KOH solution? In this titration, the acid, which has the known concentration, was placed in the buret, and the base was placed in the flask.

solution

Note that in this case, an acid of *known* concentration is used to titrate a base of *unknown* concentration. The principles are the same as those of the reverse type of titration.

1. As before, you want to calculate the moles of added known, in this case H^+, from *M* (of H^+) × *V* (in liters) = moles of H^+. A 0.500 *M* H_2SO_4 solution is 1.00 *M* in H^+ because there are 2 mol of H^+ per 1 mol of H_2SO_4.

$$\frac{0.500\ \cancel{\text{mol } H_2SO_4}}{\text{L}} \times \frac{2 \text{ mol } H^+}{1\ \cancel{\text{mol } H_2SO_4}} = \frac{1.00 \text{ mol } H^+}{\text{L}}$$

$$15.30\ \cancel{\text{mL}} \text{ solution} \times \frac{1 \text{ L}}{1{,}000\ \cancel{\text{mL}}} = 0.01530 \text{ L solution}$$

$$M \times V = \frac{1.00 \text{ mol } H^+}{\cancel{\text{L}}} \times 0.01530\ \cancel{\text{L}} = 0.0153 \text{ mol } H^+$$

2. At the equivalence point, moles of H^+ = moles of OH^-. Therefore, 0.0153 mol of OH^- was neutralized.
3. Calculate the base molarity by dividing moles of base by liters of solution. There is 0.0153 mol of OH^- in 0.0153 mol of KOH.

$$29.10\ \cancel{\text{mL}} \text{ KOH} \times \frac{1 \text{ L}}{1{,}000\ \cancel{\text{mL}}} = 0.02910 \text{ L KOH}$$

$$M = \frac{0.0153 \text{ mol KOH}}{0.02910 \text{ L KOH}} = 0.526\ M \text{ KOH}$$

sample exercise 10

A student neutralizes 14.40 mL of a H_2SO_4 solution with 35.20 mL of 0.200 *M* NaOH solution. What is the molarity of the H_2SO_4 solution? The base, NaOH, which has a known concentration, was placed in the buret, and the acid, H_2SO_4, with unknown concentration, was placed in the flask for titration.

solution

1. Calculate the moles of added known.

$$M \text{ (of } OH^-) \times V \text{ (in liters)} = \text{moles of } OH^-$$

$$\frac{0.200 \text{ mol } OH^-}{\cancel{\text{L}}} \times 0.03520\ \cancel{\text{L}} = 0.00704 \text{ mol } OH^-$$

2. At the equivalence point, moles of OH^- = moles of H^+. Therefore, 0.00704 mol of H^+ was neutralized.
3. Calculate the acid molarity by dividing moles of acid by liters of solution. First, determine the moles of acid from the moles of H^+.

$$0.00704\ \cancel{\text{mol of } H^+} \times \frac{1 \text{ mol } H_2SO_4}{2\ \cancel{\text{mol } H^+}} = 0.00352 \text{ mol } H_2SO_4$$

$$14.40\ \cancel{\text{mL}} \text{ solution} \times \frac{1 \text{ L}}{1{,}000\ \cancel{\text{mL}}} = 0.01440 \text{ L solution}$$

$$M = \frac{0.00352 \text{ mol } H_2SO_4}{0.01440 \text{ L solution}} = 0.244\ M\ H_2SO_4$$

15.10 what is a buffer?

A buffer is a chemical system that resists change in pH and is formed by mixing a weak acid and its conjugate base. Buffers are extremely important in regulating the pH in laboratory studies, commercial manufacturing, and biological systems.

As Table 15.4 shows, the pH values of various bodily fluids can be very different, for example, a pH of 1 to 2 for gastric juice versus a pH of 7 to 8 for pancreatic juice. However, the normal pH range for any particular fluid is quite narrow. This is especially true for blood plasma, where the pH in a healthy individual must remain between 7.35 and 7.45. Should the blood pH fall to 7.2, oxygenated hemoglobin (HbO_2), the carrier of O_2 from the lungs to all cells in the body, releases its oxygen and does not carry oxygen to the cells. As a result, the cells are deprived of oxygen and the body dies.

The body uses a system of buffers to maintain the pH of bodily fluids within their proper narrow ranges. A **buffer** is a weak acid-weak base pair that, by reacting with added amounts of a base or acid, can resist large changes in the buffer solution's pH. For example, if 1 mL of 1.0 *M* HCl is added to 100 mL of pure water, the pH of the water plummets 5 units from 7.0 to 2.0. If the same 1 mL of 1.0 *M* HCl is added to 100 mL of bicarbonate buffer (we will discuss the composition of this buffer in the next section), the pH of the buffer solution decreases from 7.4 to 6.9, a difference of only 0.5 pH units (Figure 15.11).

In general, buffers are composed of weak conjugate acid-base pairs: either (1) a weak acid (e.g., acetic acid) and is conjugate base (the acetate ion) or (2) a weak base (e.g., ammonia) and its conjugate acid (the ammonium ion). Many bodily fluid buffers are of the weak acid-conjugate base type, and so we will concentrate our attention on these buffers, using the acetic acid-acetate ion system as a general example.

preparation of a buffer

Before we examine how buffers work, let us see first how to prepare a buffer of the weak acid-conjugate base type. A solution of acetic acid is not a buffer by itself; it does contain both the molecular (un-ionized) acid ($HC_2H_3O_2$) and its conjugate base, the anion $C_2H_3O_2^-$, as formed in the equilibrium

$$\underset{\textbf{Acid}}{HC_2H_3O_2(aq)} \rightleftharpoons H^+(aq) + \underset{\textbf{Conjugate base}}{C_2H_3O_2^-(aq)} \tag{15.1}$$

However, in this solution of the weak acid, the anion (base) concentration is very small because the equilibrium lies to the left. You can increase the anion concentration by adding to the solution an acetate salt such as sodium acetate, which completely dissociates according to the equation

$$NaC_2H_3O_2(aq) \rightarrow Na^+(aq) + C_2H_3O_2^-(aq)$$

The solution you have now formed contains the weak acid, its conjugate base (the acid anion), H^+ ions, and Na^+ ions. Because it contains both the weak acid and its conjugate base in a concentration higher than that from the acid ionization alone, the mixture is a buffer.

A buffer is made by combining a solution of known concentration of a weak acid with a solution of known concentration of the salt (anion) of that acid, as illustrated above. The H^+ concentration, and hence the pH, of the solution depends on the ratio of the acid to the anion. This can be seen easily by an examination of the equilibrium-constant expression for the acetic acid ionization shown in Equation 15.1.

$$K_a = \frac{[H^+][C_2H_3O_2^-]}{[HC_2H_3O_2]} \tag{15.2}$$

Isolating $[H^+]$ by multiplying both sides by $[HC_2H_3O_2]$ and dividing by $[C_2H_3O_2^-]$ yields

$$\frac{K_a[HC_2H_3O_2]}{[C_2H_3O_2^-]} = [H^+]$$

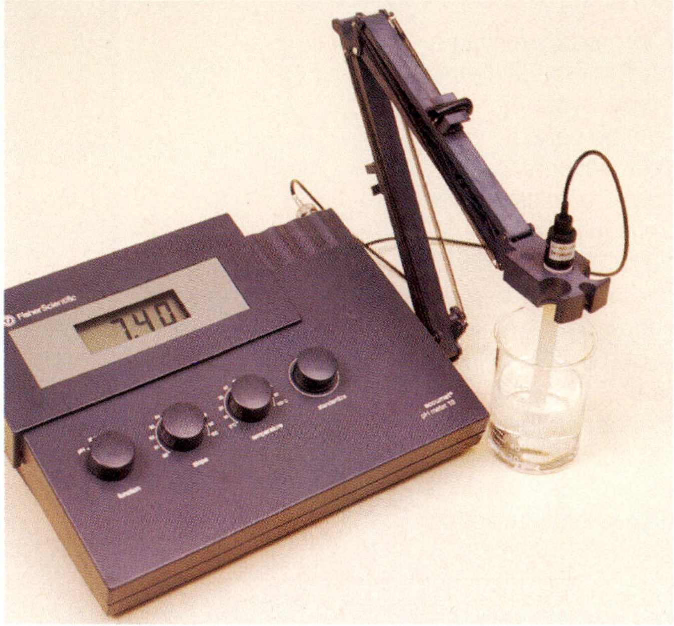

(a)

(b)

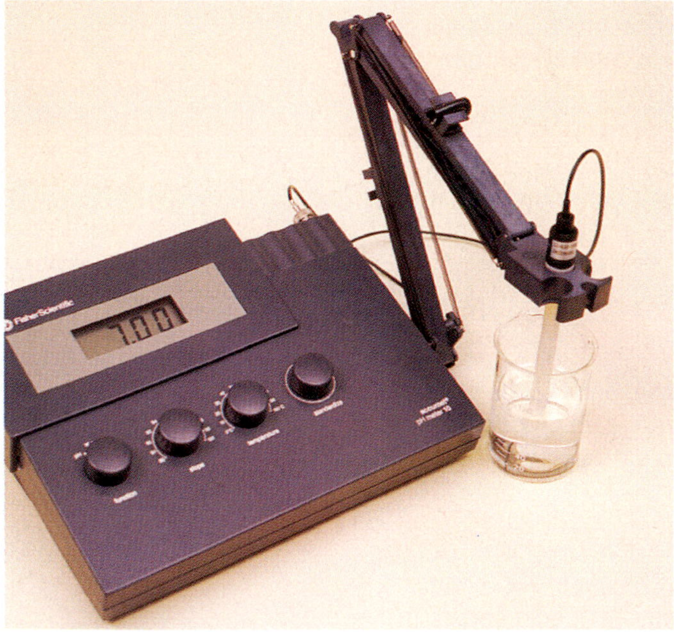

(c)

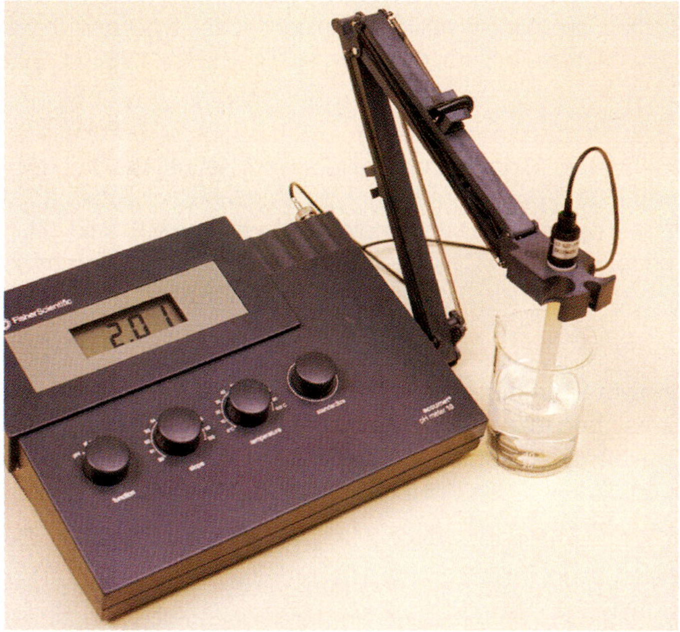

(d)

Figure 15.11

(a) This buffer resists a change in pH when 1 mL of strong acid, HCl, is added to solution. (b) The pH only decreases from 7.4 to 6.9. With no buffer present, the pH falls dramatically from a value of 7.0 (c) to 2.0 (d) when 1 mL of the acid is added to distilled water in (c).

Because K_a is a constant, the $[H^+]$, and therefore the pH, changes only as the ratio of the concentrations of weak acid and its conjugate base change.

how a buffer functions

To understand how a buffer resists a change in pH, let us examine the buffer equilibrium and stress the system by adding either acid (H^+) or base (OH^-).

Suppose that you have made a buffer solution that is 0.1 M $NaC_2H_3O_2$ and 0.1 M $HC_2H_3O_2$. The $[H^+]$ must be very small because it can only be formed by the ionization of the weak acid.

$$HC_2H_3O_2(aq) \rightleftharpoons C_2H_3O_2^-(aq) + H^+(aq)$$

If the equilibrium is *stressed* by adding H^+ ions,

$$HC_2H_3O_2(aq) \rightleftharpoons C_2H_3O_2^-(aq) + H^+(aq)$$

Add H⁺

the equilibrium shifts to the left to relieve the stress by consuming most of the added $H^+(aq)$. Acetate ion is also consumed and acetic acid is formed. However, since the concentrations of acid and acetate ion were initially large, the ratio after equilibrium is reestablished is not greatly affected, and the hydrogen ion concentration after equilibrium is reestablished is nearly the same as it was before acid was added. As a result, the pH of the solution is nearly unchanged.

If the equilibrium is *stressed* by adding OH^- ions,

$$HC_2H_3O_2(aq) \rightleftharpoons C_2H_3O_2^-(aq) + H^+(aq)$$

Add OH⁻

The added OH^- reacts with H^+ to form water,

$$OH^-(aq) + H^+(aq) \rightarrow H_2O$$

thereby removing H^+ from the equilibrium and decreasing $[H^+(aq)]$. The original acetic acid equilibrium shifts to the right to relieve the stress by forming $H^+(aq)$. The concentration of acetic acid is also decreased and that of acetate ion is increased. However, the new ratio of acid to anion is relatively unaffected, and the hydrogen ion concentration after equilibrium is reestablished is nearly the same as it was before hydroxide ions were added. Once again, the pH is nearly unchanged.

The rearranged equilibrium-constant expression in which $[H^+]$ is isolated readily shows the hydrogen ion concentration of a buffer solution as a function of acid and anion concentrations.

$$[H^+] = \frac{K_a[HC_2H_3O_2]}{[C_2H_3O_2^-]} \qquad (15.3)$$

The application of some simple math to this expression yields an equation for the pH of a buffer solution. By taking the negative logarithm of both sides of Equation 15.3, you obtain

$$-\log [H^+] = -\log K_a - \log \frac{[HC_2H_3O_2]}{[C_2H_3O_2^-]}$$

Since $pH = -\log [H^+]$ and $pK_a = -\log K_a$, you can substitute these into the above equation to obtain the expression

$$pH = pK_a - \log \frac{[HC_2H_3O_2]}{[C_2H_3O_2^-]}$$

which may be rearranged to

$$pH = pK_a + \log \frac{[C_2H_3O_2^-]}{[HC_2H_3O_2]} \qquad (15.4)$$

Equation 15.4 can be written for any weak acid in the form

$$pH = pK_a + \log \frac{[A^-]}{[HA]} \qquad (15.5)$$

This equation is known as the **Henderson-Hasselbach equation** for the biochemists who developed it.

The Henderson-Hasselbach equation is extremely convenient when you prepare a buffer in which the concentrations of weak acid and conjugate base are the same. In this case, the ratio of concentrations in Equation 15.5 equals 1, and the log of 1 equals 0. So the Henderson-Hasselbach equation reduces to $pH = pK_a$ and no calculations are needed. The pH of a buffer made with equal concentrations of weak acid and conjugate base is simply equal to the value of the pK_a of the weak acid.

The pH of a buffer may be calculated if you know the identity of the acid, and thereby its pK_a, and the ratio of concentrations of anion to weak acid. If the pK_a or K_a of the acid is not given, the pK_a may be calculated from the K_a obtained in Table 15.3 or tables found in a chemical handbook. You will see examples of pH calculations using Equation 15.5 in Section 15.11.

Another important observation that becomes immediately obvious from the Henderson-Hasselbach equation is the fact that *an aqueous buffer solution, when diluted with more water, retains the same pH* and, therefore, the same $[H^+]$. From Equation 15.5, you can see that the ratio of concentrations, $[A^-]/[HA]$, does not change when the buffer solution is diluted. If the volume of the solution were doubled by adding solvent, both $[A^-]$ and [HA] would be reduced to one-half their initial value. However, the *ratio* of the new concentrations would be unchanged. As a result, the pH of the diluted solution, as calculated with Equation 15.5, would remain the same.

15.11 blood buffers

Now that you understand how buffers work, you can appreciate the body's mechanisms for protecting blood from dramatic pH changes. Three interconnected systems maintain the pH in blood: (1) blood buffers, which actually serve to neutralize added hydrogen and hydroxide ions that form from the body's metabolic reactions, (2) the lungs, which are involved in the balance of carbon dioxide and thereby maintain the concentration of carbonic acid in blood, and (3) the kidneys, which excrete hydrogen ions and bicarbonate ions from the blood into the urine. In this section, we will concentrate on the role of blood buffers and witness the supporting roles of the other two systems.

There are three major body buffers: (1) the H_2CO_3/HCO_3^- buffer (carbonic acid, hydrogen carbonate), (2) the $H_2PO_4^-/HPO_4^{2-}$ buffer (dihydrogen phosphate, hydrogen phosphate), and (3) protein buffers. The carbonic acid-hydrogen carbonate buffer is the major buffer in blood. As we have seen, carbonic acid is an unstable weak acid that in aqueous solution, in this case blood, is always in equilibrium with $CO_2(aq)$.

$$H_2CO_3(aq) \rightleftharpoons CO_2(aq) + H_2O(l) \qquad (15.6)$$

The position of this equilibrium is to the right. Dissolved $CO_2(aq)$ is also in equilibrium with $CO_2(g)$ in the lungs.

$$\underset{\textbf{Blood}}{CO_2(aq)} \rightleftharpoons \underset{\textbf{Lungs}}{CO_2(g)} \qquad (15.7)$$

A moment's thought in applying Le Chatelier's principle should convince you that the concentration of $H_2CO_3(aq)$ can be directly affected by that of $CO_2(g)$.

Carbonic acid is also in equilibrium with the other half of the buffer mixture, the bicarbonate anion, through the acid dissociation

$$H_2CO_3(aq) \rightleftharpoons HCO_3^-(aq) + H^+(aq) \qquad (15.8)$$

The operation of the buffering action here is completely analogous to the acetic acid-acetate ion system discussed in Section 15.10. Acidic by-products (H^+) from metabolic cycles can be neutralized by $HCO_3^-(aq)$, forming $H_2CO_3(aq)$ (equilibrium of Equation 15.8 shifts left). When in excess, H_2CO_3 is removed from the body as $CO_2(g)$ (equilibria of Equations 15.6 and 15.7 shift right). Excess base is

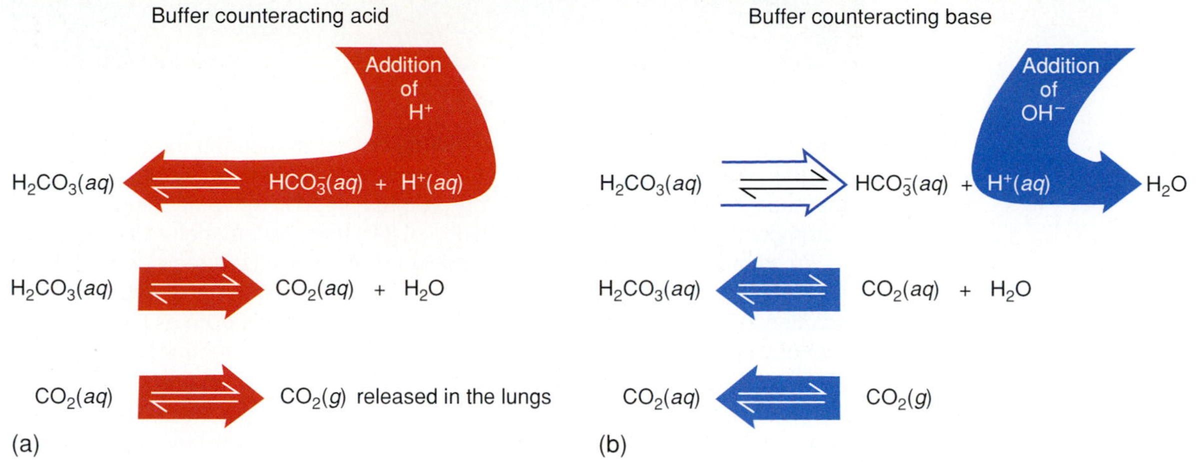

Figure 15.12
Action of the carbonic acid-bicarbonate buffer upon addition of acid or base. (a) Addition of H^+ shifts the equilibrium to the left, producing H_2CO_3. This new H_2CO_3 constitutes a stress on the second equilibrium, shifting it to the right. $CO_2(aq)$ is thus produced, and this shifts the third equilibrium to the right. (b) Addition of OH^- consumes H^+ and shifts the first equilibrium to the right, thus diminishing H_2CO_3. To compensate, the second equilibrium shifts left to replace H_2CO_3. The third equilibrium shifts left to replace $CO_2(aq)$.

neutralized by $H_2CO_3(aq)$, forming $HCO_3^-(aq)$ (equilibrium of Equation 15.8 shifts right). Figure 15.12 summarizes the equilibrium shifts in this buffer system caused by additions of acid or base.

The concentrations of $HCO_3^-(aq)$ and $H_2CO_3(aq)$ in the blood of a healthy individual are 2.5×10^{-2} *M* and 1.25×10^{-3} *M*, respectively. The pK_a of carbonic acid is 6.1. Using these data and the Henderson-Hasselbach relationship in Equation 15.5, you obtain

$$\mathrm{pH} = \mathrm{p}K_a + \log \frac{[HCO_3^-]}{[H_2CO_3]}$$

$$\mathrm{pH} = 6.1 + \log \frac{2.5 \times 10^{-2}}{1.25 \times 10^{-3}}$$

$$\mathrm{pH} = 6.1 + \log 20.$$

$$\mathrm{pH} = 6.1 + 1.3 = 7.4$$

The pH we should expect to find in normal blood is 7.4.

The capacity of a particular buffer to resist changes in pH, called the **buffering capacity,** depends on the ratio of anion to acid remaining fairly constant. For the carbonic acid-bicarbonate buffer, this means a relative constancy in the concentration ratio, $[HCO_3^-]/[H_2CO_3]$, of about 20/1.

A key factor in the maintenance of this 20/1 ratio is the coupling of the H_2CO_3 concentration to the partial pressure of CO_2 in the lungs (Figure 11.14), as shown in Equations 15.6 and 15.7. The CO_2 in our lungs provides a ready and essentially limitless source of more H_2CO_3, so that a shortage is unlikely. Furthermore, excess H_2CO_3 can be disposed of by shifting the equilibria of Equations 15.6 and 15.7 to the right in the direction of $CO_2(g)$. Increased $CO_2(g)$ exhalation by faster and deeper breathing accomplishes this.

The concentration of HCO_3^- is regulated by the kidneys. If the HCO_3^- concentration drops, the kidneys remove H^+ from the blood and the H^+ concentration in urine increases. Removal of H^+ shifts the equilibrium of Equation 15.8 to the right, thus replenishing $[HCO_3^-]$. As we discussed, the H_2CO_3 lost by this shift can be replaced by CO_2. The kidneys promote the excretion of excess concentrations of HCO_3^-.

CHEMLAB

Laboratory exercise 15: estimating pH of solutions

You have learned that two convenient methods for measuring the pH of solutions are the use of universal indicator paper or the use of a pH meter. While each of these methods is available in the laboratory, you may readily estimate the pH of many common solutions using the knowledge of $[H^+]$ that you have already mastered.

To estimate the pH of a solution, consider whether it is formed from a strong acid, a strong base, a weak acid, a weak base, or a combination of acid and base. If the solution is that of a strong acid, you may calculate the pH from its $[H^+]$. If the solution is that of a strong base, you may calculate the pH by finding the $[H^+]$ using K_w. If a weak acid and its conjugate base are present, the solution is a buffer and you may calculate the pH using the Henderson-Hasselbach equation. If only a weak acid or a weak base is present, you do not know how to calculate the pH directly, but may indicate whether the pH is less than or greater than 7.0.

Question:

1. Estimate the pH of the following aqueous solutions.
   a. Vinegar, which is 6% acetic acid, $HC_2H_3O_2$
   b. 0.02 *M* $Ba(OH)_2$, barium hydroxide
   c. 0.50 *M* HNO_3, nitric acid
   d. 0.25 *M* solution of table sugar, a molecular compound
   e. A vitamin C (ascorbic acid) tablet dissolved in a glass of water
   f. A glass of carbonated beverage to which has been added 1/4 teaspoon of sodium hydrogen carbonate, $NaHCO_3$
   g. A solution formed by mixing 50.0 mL of a 0.10 *M* acetic acid solution with 50.0 mL of a 0.20 *M* $NaC_2H_3O_2$ solution

The carbonic acid-bicarbonate system in blood has a high buffering capacity because of the coupled equilibria between H_2CO_3 and the unlimited supply of CO_2 in the lungs and the regulation of $[HCO_3^-]$ by the kidneys. If both $[HCO_3^-]$ and $[H_2CO_3]$ are maintained at nearly constant values, then the ratio $[HCO_3^-]/[H_2CO_3]$ remains constant and so does the pH of the blood.

Another buffer system is the dihydrogen phosphate-hydrogen phosphate ($H_2PO_4^-/HPO_4^{2-}$) weak acid-conjugate base system. This is the major buffer operating within fluids of living cells. This buffer functions through the equilibrium

$$\underset{\substack{\textbf{Weak acid}\\ \textbf{neutralizes}\\ \textbf{added base}}}{H_2PO_4^-(aq)} \rightleftharpoons \underset{\substack{\textbf{Weak base}\\ \textbf{neutralizes}\\ \textbf{added acid}}}{HPO_4^{2-}(aq)} + H^+(aq)$$

problem 9

The pK_a of $H_2PO_4^-$ is 7.21; for normal cell concentrations, $[HPO_4^{2-}] = 2.4 \times 10^{-3}$ *M* and $[H_2PO_4^-] = 1.5 \times 10^{-3}$ *M*. Use the Henderson-Hasselbach equation,

$$pH = pK_a + \log \frac{[HPO_4^{2-}]}{[H_2PO_4^-]}$$

to show that a pH of 7.4 is maintained within cells by this phosphate buffer system.

chapter accomplishments

After completing this chapter, you should be able to define all key terms and do the following.

15.1 Introduction

- ❑ List some common acids and bases and their applications in everyday life.
- ❑ State the physical and chemical properties that distinguish acids from bases.

15.2 Arrhenius definitions of acids and bases

- ❑ State the Arrhenius definitions of acids and bases.
- ❑ Recognize acids and bases according to the Arrhenius definitions.
- ❑ State the ions present in an aqueous solution of a given metallic hydroxide or a given acid.

15.3 Bronsted-Lowry definitions of acids and bases

- ❑ State the Bronsted-Lowry definitions of acids and bases.
- ❑ Recognize acid-conjugate base pairs and base-conjugate acid pairs in a given equation.
- ❑ State the definition and give an example of an amphoteric substance.

15.4 Acid and base strength, K_a and K_b

- ❑ Describe the relationship between the classification of acids and bases as strong or weak and the extent of their ionization.
- ❑ Describe what is meant by acid and base ionization constants.
- ❑ State the relationship between the magnitude of the acid ionization constant and the strength of an acid and its conjugate base, and between the base ionization constant and the strength of a base and its conjugate acid.
- ❑ Given a table of acid and base ionization constants, arrange a list of acids and bases in order of increasing or decreasing acid and base strength.

15.5 Ionization of water, K_w

- ❑ Write the equilibrium-constant expression corresponding to K_w.
- ❑ State the value of $[H^+]$ and $[OH^-]$ in pure water.
- ❑ State the value of K_w.
- ❑ Given a value of $[H^+]$ or $[OH^-]$, state whether the solution is acidic or basic.
- ❑ Given a value for $[H^+]$, calculate $[OH^-]$, or given a value of $[OH^-]$, calculate $[H^+]$.

15.6 Calculating pH from $[H^+]$

- ❑ Given the pH of a solution, state whether the solution is acidic, basic, or neutral.
- ❑ Calculate the pH of a solution given the value of $[H^+]$ or $[OH^-]$.

15.7 Measurement of pH

- ❑ State two methods by which the pH of a solution can be determined experimentally.

15.8 Reactions of acids and bases

- ❑ Write an equation for the neutralization reaction between a given acid and base.
- ❑ Write an equation for the reaction between a given acid and a bicarbonate or carbonate salt.

15.9 Determination of acid-base concentrations: titration

- ❑ Describe how a titration is performed.
- ❑ Given the volume of one solution (acidic or basic) and the volume and molarity of a second solution (basic or acidic) needed to titrate the first, calculate the molarity of the first solution.

15.10 What is a buffer?

- ❑ State the general composition of buffers.
- ❑ Describe how a buffer of the weak acid-conjugate base type is prepared.
- ❑ State what the $[H^+]$ concentration of a given buffer depends on.
- ❑ Describe the changes in equilibrium that occur when a small amount of acid is added to a given weak acid-conjugate base buffer.
- ❑ Given concentrations of a weak acid and its conjugate base and an appropriate pK_a, use the Henderson-Hasselbach equation to calculate the pH of the weak acid-conjugate base buffer.

15.11 Blood buffers

- ❑ State the three systems that regulate the pH in blood.
- ❑ State the three equilibria that influence the concentration of H_2CO_3 in blood plasma.
- ❑ Describe how excess acid and excess base are neutralized by the carbonic acid-bicarbonate buffer.
- ❑ Describe why the carbonic acid-bicarbonate system in blood has a high buffering capacity.

key terms

acidity
basicity (alkalinity)
titration
dissociation
ionization
Arrhenius acid
hydronium ion
conjugate base
conjugate acid
amphoteric
ionization constant
diprotic acid
triprotic acids
ion product of water
indicator
litmus
neutralization reaction
salts
equivalence point
endpoint
buffer
Henderson-Hasselbach equation
buffering capacity

problems

15.1 Introduction

10. Describe a procedure for distinguishing between an acidic and a basic substance.
11. Give the names of two acids and two bases found in commercial products in your household.

15.2 Arrhenius definitions of acids and bases

12. a. Give two examples of an Arrhenius acid.
    b. Give two examples of an Arrhenius base.
13. Show by an equation how the ionization of HNO_3 in aqueous solution leads to the formation of the hydronium ion.
14. Write equations to show the dissociation or ionization of the following compounds in water.
    a. LiOH
    b. H_2SO_4
    c. $Ca(OH)_2$

15.3 Bronsted-Lowry definitions of acids and bases

15. a. Distinguish between a Bronsted-Lowry acid and a Bronsted-Lowry base.
    b. Distinguish between a Bronsted-Lowry acid and an Arrhenius acid.
    c. Distinguish between a Bronsted-Lowry base and an Arrhenius base.
16. Show by equations how the ion HSO_4^- can act as an amphoteric substance.
17. Show by an equation how the ion NH_4^+ can act as a Bronsted-Lowry acid.
18. Identify the acid-conjugate base and base-conjugate acid pairs in the following reactions.
    a. $HBr + H_2O \rightleftharpoons H_3O^+ + Br^-$
    b. $NH_3 + H_3O^+ \rightleftharpoons NH_4^+ + H_2O$
    c. $HSO_4^- + OH^- \rightleftharpoons SO_4^{2-} + H_2O$
    d. $HCO_3^- + H_3O^+ \rightleftharpoons H_2CO_3 + H_2O$
    e. $NH_3 + H_3PO_4 \rightleftharpoons NH_4^+ + H_2PO_4^-$
    f. $HS^- + H_2O \rightleftharpoons H_2S + OH^-$
19. Give the conjugate bases of the following substances.
    a. H_2O
    b. HF
    c. $H_2PO_4^-$
    d. NH_4^+
    e. H_2SO_4
    f. NH_3
20. Give the conjugate acids of the following substances.
    a. H_2O
    b. F^-
    c. $H_2PO_4^-$
    d. NH_3
    e. HSO_4^-
    f. OH^-
21. Which of the following can act as an amphoteric substance: H_2CO_3, HCO_3^-, or CO_3^{2-}? Write equations to demonstrate your answer.
22. Give the formula of the species formed when:
    a. HCl acts as a Bronsted acid
    b. H_2O acts as a Bronsted base
    c. F^- acts as a Bronsted base
    d. NH_4^+ acts as a Bronsted acid
23. Identify the Bronsted-Lowry acid and the Bronsted-Lowry base in each case and complete the following equations by transferring a proton from the acid to the base.
    a. $NH_4^+ + NH_3 \rightleftharpoons$?
    b. $H_2SO_4 + H_2O \rightleftharpoons$?
    c. $HCl + CN^- \rightleftharpoons$?
    d. $H_2CO_3 + NO_3^- \rightleftharpoons$?
    e. $HSO_4^- + OH^- \rightleftharpoons$?
    f. $HSO_4^- + NO_3^- \rightleftharpoons$?

15.4 Acid and base strength, K_a and K_b

24. Arrange the following substances in order of increasing acid strength: H_3BO_3, HSO_4^-, $HClO_4$, H_2O, and HCN.

25. Arrange the conjugate bases of the acids in Problem 24 in order of increasing base strength.
26. a. Write an equilibrium-constant expression for the acid ionization of $HC_2H_3O_2$.
    b. What quantities would have to be determined experimentally to evaluate this equilibrium constant?
27. a. Write an equilibrium-constant expression for the reaction of HSO_4^- as a Bronsted-Lowry base in aqueous solution.
    b. What quantities would have to be determined experimentally to evaluate this equilibrium constant?
28. What is the difference between the terms *dilute* and *concentrated* and *weak* and *strong* with respect to basicity and acidity?
29. a. Write all steps showing *complete* ionization of H_2CO_3.
    b. Write all steps showing *complete* ionization of H_3PO_4.

15.5 Ionization of water, K_w

30. a. Write an equilibrium-constant expression for the reaction of H_2O as a Bronsted-Lowry acid in pure water.
    b. Write an equilibrium-constant expression for the reaction of H_2O as a Bronsted-Lowry base in pure water.
    c. What is the relationship between the equilibrium constants in parts a and b?
    d. What are the numerical values of these equilibrium constants?
31. Indicate whether the following solutions are acidic, basic, or neutral.
    a. $[H^+] = 1.1 \times 10^{-3}\ M$
    b. $[H^+] = 6.0 \times 10^{-11}\ M$
    c. $[OH^-] = 2.3 \times 10^{-11}\ M$
    d. $[OH^-] = 5.0 \times 10^{-7}\ M$
    e. $[H^+] = 0.00013\ M$
    f. $[H^+] = 0.0000001\ M$
32. Calculate the hydrogen ion concentration in an aqueous solution in which the hydroxide ion concentration is
    a. $2.5 \times 10^{-9}\ M$
    b. $3.0 \times 10^{-6}\ M$
    c. $0.0015\ M$
33. Classify each of the solutions in Problem 32 as acidic, basic, or neutral.
34. Calculate the hydroxide ion concentration in an aqueous solution in which the hydrogen ion concentration is
    a. $5.0 \times 10^{-7}\ M$
    b. $2.4 \times 10^{-3}\ M$
    c. $0.000095\ M$
35. Classify each of the solutions in Problem 34 as acidic, basic, or neutral.

15.6 Calculating pH from $[H^+]$

36. Determine the pH of the following solutions and indicate whether each is acidic, basic, or neutral.
    a. $[H^+] = 1.0 \times 10^{-7}\ M$
    b. $[H^+] = 1.0 \times 10^{-2}\ M$
    c. $[H^+] = 1.0 \times 10^{-10}\ M$
    d. $[H^+] = 0.0001\ M$
37. Determine the pH of the following solutions and indicate whether each is acidic, basic, or neutral.
    a. $[H^+] = 5.0 \times 10^{-3}\ M$
    b. $[H^+] = 9.7 \times 10^{-8}\ M$
    c. $[H^+] = 0.047\ M$
    d. $[H^+] = 0.91\ M$
    e. $[OH^-] = 3.4 \times 10^{-6}\ M$
    f. $[OH^-] = 1.2 \times 10^{-9}\ M$
    g. $[OH^-] = 0.031\ M$
    h. $[OH^-] = 0.000020\ M$
38. Indicate whether the following solutions are acidic, basic, or neutral.
    a. Sea water: pH = 8.1
    b. Vinegar: $[H^+] = 1.6 \times 10^{-3}\ M$
    c. Coffee: pH = 5.42
    d. Orange juice: $[H^+] = 2.0 \times 10^{-4}\ M$
    e. Blood: pH = 7.4
    f. Soda water: pH = 2.8
    g. Stomach fluid: pH = 1.8
    h. Ammonia water: $[H^+] = 1.9 \times 10^{-11}\ M$
39. A solution is prepared by dissolving 25.0 g of NaOH in water and then adding water sufficient to bring the volume to 1.00 L in a volumetric flask. What are $[OH^-]$, $[H^+]$, and the pH?
40. A solution is prepared by diluting 10.0 mL of a 6.00 *M* HCl solution to 0.500 L. Calculate $[H^+]$, $[OH^-]$, and the pH.

15.7 Measurement of pH

41. Describe how a pH meter could be used to indicate the equivalence point of a titration reaction between a strong acid and a strong base.
42. Describe the relationship between an indicator equilibrium, Le Chatelier's principle, and the color change observed in acidic and basic solution.

15.8 Reactions of acids and bases

43. Complete and balance the following neutralization reactions.
    a. $KOH + HCl \rightarrow ?$
    b. $Ba(OH)_2 + HNO_3 \rightarrow ?$
    c. $NaOH + H_3PO_4 \rightarrow ?$
    d. $Ca(OH)_2 + H_2SO_4 \rightarrow ?$
    e. $Ca(OH)_2 + H_3PO_4 \rightarrow ?$

44. Indicate the acid and base needed to prepare the following salts.

a. $NaNO_3$
b. K_2SO_4
c. $CaCl_2$
d. $AlPO_4$
e. $Ca(C_2H_3O_2)_2$
f. $(NH_4)_2SO_4$

45. Complete and balance the following equations.

a. $HNO_3 + KHCO_3 \rightarrow ?$
b. $HCl + Li_2CO_3 \rightarrow ?$
c. $Zn + HCl \rightarrow ?$
d. $Na_2CO_3 + HC_2H_3O_2 \rightarrow ?$
e. $Mg + H_2SO_4 \rightarrow ?$

46. Write a balanced net ionic equation for each of the following reactions.

a. $HCl + Ca(OH)_2 \rightarrow CaCl_2 + H_2O$
b. $H_2SO_4 + KOH \rightarrow K_2SO_4 + H_2O$
c. $H_3PO_4 + NaOH \rightarrow Na_3PO_4 + H_2O$
d. $H_3PO_4 + Ca(OH)_2 \rightarrow Ca_3(PO_4)_2 + H_2O$

15.9 Determination of acid-base concentrations: titration

47. A student found that 68.30 mL of an HCl solution was required to completely neutralize 31.75 mL of 0.150 *M* NaOH. What is the molarity of the acid solution?

48. A chemist finds that 39.52 mL of a calcium hydroxide solution is required to titrate 18.90 mL of a 0.200 *M* H_3PO_4 solution. What is the molarity of the base solution?

$$3\ Ca(OH)_2 + 2\ H_3PO_4 \rightarrow Ca_3(PO_4)_2 + 6\ H_2O$$

49. How many milliliters of 0.250 *M* H_2SO_4 are required to neutralize 25.10 mL of 0.100 *M* NaOH?

$$2\ NaOH + H_2SO_4 \rightarrow Na_2SO_4 + 2\ H_2O$$

50. How many milliliters of 0.150 *M* $HClO_4$ are needed to titrate 35.0 mL of 0.215 *M* LiOH?

51. Explain why a 0.30 *M* H_2SO_4 solution is 0.60 *M* in H^+.

52. In a laboratory, 57.3 g of NaOH are dissolved in water and the solution is diluted to 2.50 L. Then 30.0 mL of this NaOH solution is used to titrate 42.5 mL of a HCl solution. What is the molarity of the HCl solution?

53. In an experiment, 35.0 mL of 1.00 *M* HCl are required to titrate a Drano® solution (active basic ingredient is NaOH). How many moles of NaOH are present in the Drano® solution?

54. In another experiment, 10.0 g of vinegar (active acidic ingredient is acetic acid, $HC_2H_3O_2$) are titrated with 65.40 mL of 0.150 *M* NaOH.

$$HC_2H_3O_2 + NaOH \rightarrow NaC_2H_3O_2 + H_2O$$

a. How many moles of $HC_2H_3O_2$ are present in 10.0 g of vinegar?
b. How many grams of $HC_2H_3O_2$ are present in 10.0 g of vinegar?
c. What is the percentage by weight of acetic acid in the vinegar solution?

55. The active sour ingredient in vinegar is acetic acid, $HC_2H_3O_2$. To test the acid molarity of a vinegar solution, a laboratory technician titrated 24.50 mL of vinegar with 0.4789 *M* NaOH. She found that it took 46.09 mL of the NaOH solution to completely neutralize the vinegar. What is the acid molarity of the vinegar? Assume that the only acid present is $HC_2H_3O_2$.

56. Gastric juice, found in the stomach, contains hydrochloric acid. A 30.10 mL sample of gastric juice is titrated with 0.1050 *M* NaOH. The chemist finds that 17.20 mL of the NaOH solution is needed to neutralize the gastric juice acid. What is the molarity of acid in gastric juice? Assume that HCl is the only substance in gastric juice that reacts with NaOH.

15.10 What is a buffer?

57. Define a buffer solution.

58. What substance would you need to add to hydrosulfuric acid, H_2S, to prepare a buffer solution?

59. Using the appropriate equilibrium equations, describe how a hydrosulfuric acid-sodium hydrogen sulfide mixture can act as a buffer.

60. Describe how the $[H^+]$ of a given buffer varies when the acid/anion concentration ratio is

a. Doubled
b. Decreased to one-third of its original value.

61. Qualitatively describe the changes in the concentrations of $[H^+]$, $[H_2CO_3]$, and $[HCO_3^-]$ when a small amount of hydrochloric acid is added to a carbonic acid-bicarbonate buffer.

62. Calculate the pH of an acetic acid-acetate ion buffer in which $[HC_2H_3O_2] = 0.15\ M$ and $[C_2H_3O_2^-] = 0.10\ M$ (pK_a for acetic acid is 4.7).

15.11 Blood buffers

63. Calculate the pH of the blood sample in which the bicarbonate concentration is twice the carbonic acid concentration (pK_a for H_2CO_3 in blood plasma is 6.1).

64. Explain how the $[H_2CO_3]$ in blood plasma would vary if
    a. The concentration of $CO_2(aq)$ increased
    b. The partial pressure of $CO_2(g)$ in the lungs increased
    c. The concentration of bicarbonate decreased
65. What would be the pH of blood if H_2CO_3 and HCO_3^- were present in equal concentrations?
66. a. Describe the role of the lungs in regulating blood pH.
    b. Describe the role of the kidneys in regulating blood pH.
67. Although the concentration of H_2CO_3 in blood plasma is only one-twentieth of the HCO_3^- concentration, blood has a high buffering capacity toward base. Explain.

Additional problems

68. State two pieces of evidence indicating that water can be only very slightly ionized.
69. Why must aqueous solutions with a high OH^- concentration have a low $[H^+]$?
70. Write an ionization equation for $HClO_4$ in water.
71. Provide a reasonable explanation of why the second ionization of a diprotic acid has a smaller equilibrium constant than the first. (Hint: Consider the species from which H^+ is being separated in each case.)
72. How many milliliters of 0.426 *M* NaOH are needed to neutralize 43.8 mL of 0.242 *M* HNO_3?
73. Hydroxide ion, OH^-, is the strongest base that can be present *in water.* Explain.
74. How many grams of aluminum can react completely with 67.0 mL of 1.0 *M* HCl?
75. Write the formula of the conjugate acid of each of the following.
    a. Br^-
    b. H_2O
    c. CO_3^{2-}
76. Write the formula of the conjugate base of each of the following.
    a. NH_3
    b. H_2O
    c. NH_4^+
77. A glass of beer was found to have a pH of 5.0. Is the beer an acidic, neutral, or basic solution? Explain.
78. The acid ionization constant for acid A is 6.6×10^{-8} while that of acid B is 9.9×10^{-5}. Which is the stronger acid? Explain.
79. Determine the pH of an aqueous solution having the following concentrations.
    a. $[H^+] = 1.0 \times 10^{-8}\ M$
    b. $[H^+] = 3.2 \times 10^{-4}\ M$
    c. $[OH^-] = 3.2 \times 10^{-4}\ M$
80. The acid ionization constant of hydrofluoric acid is 6.8×10^{-4}. Calculate the pK_a.
81. Calculate the pH of a buffered solution containing 0.17 *M* acetic acid and 0.20 *M* sodium acetate.
82. The pH of an aqueous solution is 9.0. Calculate $[H^+]$ and $[OH^-]$.
83. Identify the conjugate acid-base pairs in the following equilibria:
    a. $CH_3NH_2(aq) + H_2O(l) \rightleftharpoons CH_3NH_3^+(aq) + OH^-(aq)$
       Methylamine **Methyl ammonium ion**
    b. $H_2O(l) + H_3PO_4(aq) \rightleftharpoons H_3O^+(aq) + H_2PO_4^-(aq)$
84. $Mg(OH)_2$ is a typical active ingredient in antacids. Assuming that stomach acid is about 0.10 *M* in HCl, calculate the volume of stomach acid in milliliters that can be neutralized by 1.00 g of $Mg(OH)_2$.
85. A chemist finds that 32.1 mL of 0.300 *M* NaOH are required to completely react with 25.0 mL of an aqueous solution of H_3PO_4. Determine the molarity of the H_3PO_4 solution.
86. An important buffer in intracellular fluids involves the $H_2PO_4^-/HPO_4^{2-}$ buffer.

$$H_2PO_4^-(aq) + H_2O(l) \rightleftharpoons H_3O^+(aq) + HPO_4^{2-}(aq)$$

    Determine the pH of a buffered solution that initially contains 0.040 *M* NaH_2PO_4 and 0.040 *M* Na_2HPO_4.
87. Compare the pH of a 0.030 *M* HCl solution with that of a 0.030 *M* H_2SO_4 solution.
88. Estimate the pH of a buffered solution that initially contains 0.15 *M* NH_3 and 0.35 *M* NH_4Cl.
89. The most acidic rainfall ever measured fell in Scotland in 1974. The pH was 2.4.
    a. Calculate $[H^+]$ and $[OH^-]$.
    b. How many times greater was $[H^+]$ compared to $[OH^-]$?

chapter 16

oxidation-reduction

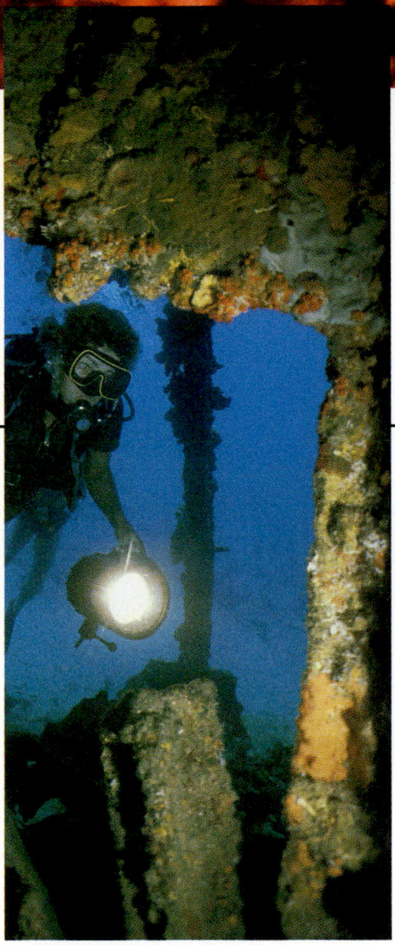

Chemical reactions in which electrons are transferred from one substance to another (oxidation-reduction reactions) are responsible both for the electricity used to power this diver's lantern and for the corrosion that has resulted on the hull of this ship.

16.1 introduction

Electrical energy has enormous practical importance whether it originates from a power company, a local generator, or a chemical or solar battery. The conversion of chemical energy into electrical energy is feasible from chemical reactions that transfer electrons from one substance to another. One substance gains the electrons that have been lost from another. For electrical energy, the transferred electrons must be channeled through a circuit to power your automobile headlights, your flashlight, your digital watch, or your rechargeable handheld vacuum cleaner (Figure 16.1). Yet, not all chemical reactions that transfer electrons are used to produce electrical energy. Some such reactions are useful in the metabolism of your food so that your body has the energy it needs. Combustion reactions, such as those in your automobile engine, furnace, and stove, are used to produce heat energy for transportation, warmth, and cooking. Corrosion reactions, such as those responsible for iron rusting, silver tarnishing, and copper turning green, may be more

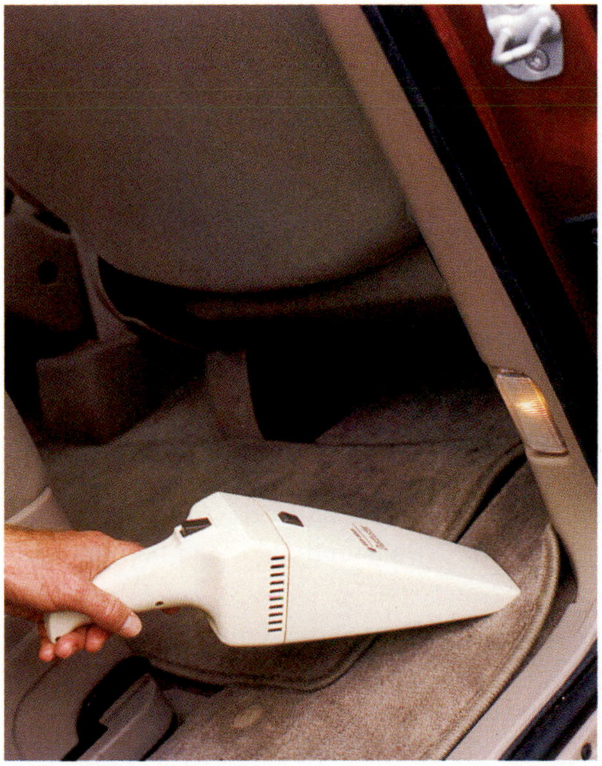

Figure 16.1

Electrochemical reactions in a rechargeable battery power the handheld vacuum cleaner being used to clean this car. Similarly, a lead storage battery, also rechargeable, powers the car's entrance light. Electrochemical reactions convert chemical energy into useful electrical energy, and vice versa.

Figure 16.2
Corrosion reactions responsible for the rusting of the hull of this ship require the constant inspection and maintenance of all structures made from steel.

bothersome than useful. Billions of dollars are spent annually to prevent or repair corrosive damage to buildings, ships, bridges, and cars (Figure 16.2).

Reactions that transfer electrons from one substance to another are classified as **oxidation-reduction reactions,** or **redox reactions** for short. Oxidation was originally associated with the gain of oxygen and reduction with the gain of hydrogen because many common redox reactions can be readily identified based on these observations. However, the general and broader definition focuses on electron transfer. To identify redox reactions and to understand how they may be used, you must learn to keep track of electrons as they rearrange themselves from reactants to products.

16.2 electron-transfer reactions

When chemical reactions transfer electrons, the electron structures of at least two elements have been changed. The element that has lost electrons has lost some negative charge; as a result, it has become less negative or more positively charged. The element that has gained electrons has become more negatively (or less positively) charged.

The transfer of electrons in a redox reaction is particularly obvious in the case of the formation of an ionic compound such as NaCl. Each sodium atom becomes a positively charged cation by transferring its valence electron to a chlorine atom, and each chlorine atom becomes a negatively charged anion by acquiring the valence electron from the sodium. You can represent the transfer of electrons in terms of the loss of electrons by sodium and the gain of electrons by chlorine in two separate reactions:

$Na \rightarrow Na^+ + e^-$	Sodium loses an electron; its charge increases.	oxidation
$Cl_2 + 2\,e^- \rightarrow 2\,Cl^-$	Chlorine gains electrons; its charge is reduced.	reduction

Oxidation is the name given to the process in which there is a *loss of electrons.* The element undergoing oxidation always *increases its charge,* as you see in the example with sodium; each neutral sodium atom acquires a charge of 1+. **Reduction** is the name given to the process in which there is a *gain of electrons.* The element undergoing reduction always *reduces its charge,* as you see in the example with chlorine; each neutral chlorine atom acquires a charge of 1–. The two processes, oxidation and reduction, always occur simultaneously and in such a way that electrons are conserved. That is,

Number of **electrons lost** = Number of **electrons gained**
(in oxidation) (in reduction)

To conserve electrons transferred, two sodium atoms must each be losing one electron for every one chlorine molecule (with its two chlorine atoms) that gains electrons. Once again, oxidation and reduction may each be represented by a separate, balanced chemical equation, called a **half-reaction,** because the sum of the two half-reactions gives the total reaction with which you are familiar.

$2\,Na$	$\rightarrow$	$2\,Na^+ + 2\,e^-$	**Balanced oxidation half-reaction**
$Cl_2 + 2\,e^-$	$\rightarrow$	$2\,Cl^-$	**Balanced reduction half-reaction**
$2\,Na + Cl_2 + 2\,e^-$	$\rightarrow$	$2\,Na^+ + 2\,Cl^- + 2\,e^-$	
$2\,Na + Cl_2$	$\rightarrow$	$2\,Na^+ + 2\,Cl^-$	**Overall reaction**

Every electron-transfer reaction is the sum of an oxidation half-reaction in which electrons are lost and a reduction half-reaction in which electrons are gained. Notice that the electrons cancel out because there are equal numbers gained (left side) and lost (right side). Electron-transfer reactions are called redox reactions because they are the sum of *red*uction and *ox*idation half-reactions.

Almost all reactions in which atoms or molecules form ions and/or ions form atoms or molecules are redox or electron-transfer reactions. In other words, if a species undergoes a change in charge, there must have been an electron-transfer or redox reaction. The overall reaction may still be written as either a traditional equation, an ionic equation, or a net ionic equation (Section 13.9) depending on which is more convenient for the user.

Figure 16.3

Notice how readily the net ionic equation focuses your attention on the chemistry you observe in this redox reaction: $Ca(s) + 2\,H^+(aq) \rightarrow Ca^{2+}(aq) + H_2(g)$. Calcium metal oxidizes to form aqueous ions; hydrogen ions are reduced to form hydrogen gas, which appears vigorously as bubbles.

16.3 writing half-reactions

The net ionic equation (Section 13.9) for a redox reaction shows the species participating in the two half-reactions most readily. For example, consider again (Section 9.8) the reaction between an active metal and an acid to produce an ionic compound and hydrogen gas. For the reaction of calcium metal and hydrobromic acid, the *traditional equation* is

$$Ca(s) + 2\,HBr(aq) \rightarrow CaBr_2(aq) + H_2(g)$$

Presented as an *ionic equation,* it is

$$Ca(s) + 2\,H^+(aq) + 2\,Br^-(aq) \rightarrow Ca^{2+}(aq) + 2\,Br^-(aq) + H_2(g)$$

Simplified as a *net ionic equation,* it becomes

$$Ca(s) + 2\,H^+(aq) \rightarrow Ca^{2+}(aq) + H_2(g)$$

Notice how the net ionic equation helps you to recognize immediately the oxidation and reduction half-reactions. The two half-reactions here are the oxidation of calcium atoms to Ca^{2+} ions and the reduction of H^+ ions to hydrogen molecules:

$Ca(s) \rightarrow Ca^{2+}(aq) + 2\,e^-$ **Oxidation**

$2\,H^+(aq) + 2\,e^- \rightarrow H_2(g)$ **Reduction**

The calcium half-reaction is recognized as an oxidation because each metal atom loses two electrons, increases its charge from 0 to 2+, and becomes a cation. Notice also that two electrons are required on the right of the calcium half-reaction to balance the equation with respect to charge. Alternatively, you may identify the reduction half-reaction. Each of the two hydrogen ions must gain an electron, reduce its charge from 1+ to 0, and combine to form an H_2 molecule (Figure 16.3).

sample exercise 1

Identify the following half-reactions as oxidations or reductions.

a. $Li^+ + e^- \rightarrow Li$
b. $2\,I^- \rightarrow I_2 + 2\,e^-$
c. $H_2 \rightarrow 2\,H^+ + 2\,e^-$
d. $Fe^{3+} + e^- \rightarrow Fe^{2+}$

solution

a. *Reduction*. The lithium ion gains one e^-, which appears on the left; its charge is reduced from 1+ to 0.
b. *Oxidation*. Each iodide ion loses an e^-, which appears as 2 e^- on the right; the charge on each iodide ion increases from 1− to 0.
c. *Oxidation*. Each hydrogen molecule (with two atoms) loses two e^-, which appear as 2 e^- on the right; the charge on each hydrogen *atom* increases from 0 to 1+.
d. *Reduction*. The iron ion gains one e^-, which appears on the left; the charge on Fe is reduced from 3+ to 2+.

problem 1

Identify the following half-reactions as oxidations or reductions.

a. $Cr \rightarrow Cr^{3+} + 3\,e^-$

b. $Cl_2 + 2\,e^- \rightarrow 2\,Cl^-$

c. $S^{2-} \rightarrow S + 2\,e^-$

d. $Hg^{2+} + 2\,e^- \rightarrow Hg$

The material that is oxidized in any redox reaction is also referred to as the **reducing agent** because it provides the electrons that are gained by the material that is reduced. In the example of calcium in acid, the calcium metal is being oxidized (losing electrons) and is thereby acting as a reducing agent because it provides the electrons that reduce H^+. Similarly, the material that is reduced is referred to as the **oxidizing agent** because it accepts the electrons lost in oxidation. Therefore, H^+ is an oxidizing agent for Ca, from which it gains electrons and is thereby reduced. Remember that

Material **oxidized** = **Reducing agent**
(electrons lost)

Material **reduced** = **Oxidizing agent**
(electrons gained)

sample exercise 2

Given the following ionic equation, write the oxidation and reduction half-reactions and identify the oxidizing and reducing agents.

$$Zn(s) + Cu^{2+}(aq) + SO_4^{2-}(aq) \rightarrow Zn^{2+}(aq) + SO_4^{2-}(aq) + Cu(s)$$

solution

1. Write the *net* ionic equation; that is, cross out the spectator ions.

$$Zn(s) + Cu^{2+}(aq) \rightarrow Zn^{2+}(aq) + Cu(s)$$

Now only the materials that have been changed appear in the equation. Once again, notice how readily the chemistry is apparent. Zinc metal dissolves as it is oxidized to Zn^{2+} ions, and copper ions precipitate as they are reduced to Cu.

2. Zn(*s*) must lose electrons (be oxidized) to become Zn^{2+}. Therefore, the oxidation half-reaction is

$$Zn(s) \rightarrow Zn^{2+}(aq) + 2\,e^-$$

Electrons on the right side show electron loss or oxidation; the charge on the zinc atom has increased from 0 to 2+.

3. $Cu^{2+}(aq)$ must gain electrons (be reduced) to become Cu(*s*). Therefore, the reduction half-reaction is

$$Cu^{2+}(aq) + 2\,e^- \rightarrow Cu(s)$$

Electrons on the left side show electron gain or reduction; the charge on the copper ion has been reduced from 2+ to 0.

4. The material oxidized = the reducing agent, which is Zn(*s*). The material reduced = the oxidizing agent, which is $Cu^{2+}(aq)$.

• • • problem 2

Given the following ionic equation, write the oxidation and reduction half-reactions and identify the oxidizing and reducing agents.

$$Ni(s) + 2\ H^+(aq) + 2\ I^-(aq) \rightarrow Ni^{2+}(aq) + 2\ I^-(aq) + H_2\uparrow$$

Nickel is widely used in alloys with iron to produce stainless steels that do not rust. Like other metals above hydrogen in the activity series (Table 9.2), it will replace hydrogen from acid solution; that is, it will undergo oxidation in acid solution and dissolve as shown in the above chemical equation.

As you have seen in the examples above, it is often easy to pick out the substances oxidized and reduced in a net ionic equation. The investigator looks for those ions or atoms that undergo a change in charge during the course of the reaction, keeping in mind that a neutral atom has a charge of zero.

This readily observable change of charge does not always occur during oxidation or reduction reactions. For example, the sulfite ion (SO_3^{2-}) may be oxidized to the sulfate ion (SO_4^{2-}). In this case, there is no obvious change in charge to tell us that the process is oxidation. The fact that one oxygen is gained by sulfur indicates the oxidation of sulfur. *The gain of oxygen by a substance was originally used as a definition of oxidation.* Additional examples that follow this definition include the combustion of carbon with oxygen to form carbon dioxide.

$$C(s) + O_2(g) \rightarrow CO_2(g)$$

and the formation of an oxide from a metal such as aluminum,

$$4\ Al(s) + 3\ O_2(g) \rightarrow 2\ Al_2O_3(s)$$

However, this definition of oxidation is limited to reactions with oxygen and is not always applicable to other chemical reactions. Even when it is applicable, such as to those reactions written above, you do not know how to determine how many electrons have been lost or gained. Oxidation numbers, which are defined in the next section, provide a general method for recognizing oxidized and reduced species and determining the number of electrons gained or lost. These concepts will prove very useful in the balancing of even complex chemical reactions.

16.4 oxidation numbers

Every atom, whether in the elemental state or combined in a molecule or ion, can be assigned an oxidation number based on a set of rules determined by convention. The **oxidation number** of an atom is a hypothetical charge assigned by simple rules. The oxidation number does not always have a physical meaning, but does correspond to the charge attributed to known cations, anions, and neutral elements. In general, it is a bookkeeping device you will find very useful for locating electron transfers during all redox reactions and for balancing the corresponding chemical equations.

The oxidation number of a monatomic ion is identical to the charge on the ion. By convention, we have indicated the charge on an ion by its magnitude followed by its sign, for example, Mg^{2+}. The oxidation number will be distinguished from this by writing first the sign, then the magnitude. For example, for the magnesium ion, the oxidation number is +2; for the oxide ion, O^{2-}, the oxidation number is –2. Free atoms are electrically neutral and are therefore assigned an oxidation number of 0. *The assignment of oxidation numbers to covalently bonded atoms is based on the assumption that the bonding electrons belong completely to the more electronegative of the two bonded atoms.* Read through the following rules and see how they are applied in subsequent examples.

Rules for assigning oxidation numbers

1. *The oxidation number of an atom in its elemental state is 0.* For example, the oxidation number of the potassium atom K is 0, and the oxidation number of the atom H in H_2 is 0.
2. *The oxidation number of any monatomic ion is equal to the charge on the ion.* For example, for Al^{3+}, the charge is 3+ and the oxidation number is +3. For O^{2-}, the charge is 2– and the oxidation number is –2.
3. *The oxygen (–2) rule: In almost all its common compounds, oxygen has the oxidation number –2.* (Hydrogen peroxide, H_2O_2, and its derivatives are notable exceptions. In these cases, the oxidation number is –1.)
4. *The hydrogen (+1) rule: In almost all compounds with other nonmetals, hydrogen has the oxidation number +1.* When hydrogen is combined with a metallic cation to form a metal hydride, as NaH, MgH_2, and AlH_3, the H atom is assigned the oxidation number –1, but these compounds are encountered infrequently.
5. *The sum of the oxidation numbers of atoms in a neutral compound must be 0.*
6. *The sum of the oxidation numbers of atoms in a polyatomic ion must equal the charge on the ion.*

Usually the last two rules are used in conjunction with one or more of the earlier ones. For example, to assign oxidation numbers to the atoms in sodium dihydrogen phosphate, NaH_2PO_4, a component of baking powder, you will use *rules 2, 3, 4, and 6.* This is an ionic compound composed of Na^+ ions and $H_2PO_4^-$ ions.

The oxidation number of Na is +1 (*rule 2*).
The oxidation number of H is +1 (*rule 4*).
The oxidation number of O is –2 (*rule 3*).

Rule 6 will enable us to ascertain the oxidation number of phosphorus by summing up oxidation numbers. The sum of the oxidation numbers of two hydrogens and four oxygens plus the oxidation number of phosphorus must equal –1, the charge on the polyatomic ion, $H_2PO_4^-$. Thus,

$$\begin{aligned} 2(+1) + P + 4(-2) &= -1 \\ 2 + P + (-8) &= -1 \\ P &= -1 + 8 - 2 \\ P &= +5 \end{aligned}$$

The oxidation number of P is +5 according to *rule 6.* Check by *rule 5:*

$$\begin{array}{cccc} Na & H_2 & P & O_4 \\ \downarrow & \downarrow & \downarrow & \downarrow \\ +1 & 2(+1) & +5 & 4(-2) \end{array} = 1 + 2 + 5 - 8 = 0$$

sample exercise 3

Determine the oxidation number of the following atoms.

a. Carbon in CO_2
b. Sulfur in SO_4^{2-}
c. Manganese in $KMnO_4$
d. Sulfur in SO_3^{2-}

solution

a. *Rule 3* tells you that the sum of the oxidation numbers of oxygen in CO_2 would be –4, that is, 2(–2). By *rule 5,* the oxidation number of carbon must be such that the sum for the molecule is zero. Hence, the oxidation number of carbon must be +4, that is, +4 – 4 = 0.

b. In SO_4^{2-}, each oxygen has the oxidation number –2 (*rule 3*). The sum of the oxidation numbers of sulfur and oxygen must equal the charge on the sulfate ion, –2 according to *rule 6.*

$$\begin{aligned} S + 4\,O &= -2 \\ S + 4(-2) &= -2 \\ S + (-8) &= -2 \\ S &= -2 + 8 \\ S &= +6 \end{aligned}$$

The oxidation number of sulfur in SO_4^{2-} is +6.

c. The compound $KMnO_4$ is made up of the K^+ ion and the MnO_4^- ion. The K^+ ion has an oxidation number of +1 (*rule 2*), and each oxygen, –2 (*rule 3*). The sum of the oxidation numbers of potassium, manganese, and oxygen must equal 0 for the neutral compound $KMnO_4$. Therefore,

$$\begin{aligned} K + Mn + 4\,O &= 0 \\ +1 + Mn + 4(-2) &= 0 \\ +1 + Mn + (-8) &= 0 \\ Mn &= +8 - 1 \\ Mn &= +7 \end{aligned}$$

The oxidation number of Mn in $KMnO_4$ is +7.

d. In SO_3^{2-}, each oxygen has the oxidation number –2 (*rule 3*). The sum of the oxidation numbers of sulfur and oxygen must equal the charge on the sulfite ion, –2 according to *rule 6.* Therefore,

$$\begin{aligned} S + 3\,O &= -2 \\ S + 3(-2) &= -2 \\ S + (-6) &= -2 \\ S &= -2 + 6 \\ S &= +4 \end{aligned}$$

The oxidation number of sulfur in SO_3^{2-} is +4.

---

problem 3

Determine the oxidation number of the following atoms.

a. Nitrogen in N_2O_5

b. Nitrogen in NO_3^-

c. Nitrogen in NO_2

The concept of oxidation number gives easily applied, general working definitions of oxidation and reduction. Oxidation represents an increase in charge when ionic species are present, and reduction represents a reduction in charge. In general, when ionic species are not necessarily present,

Oxidation = increase in oxidation number.

Reduction = decrease in oxidation number.

Look back at all previous examples of oxidation and reduction reactions to see that this is the case.

Now you have a general method of recognizing oxidation or reduction even when there is no *obvious* change in charge or transfer of electrons. Notice that you can recognize $SO_3^{2-} \rightarrow SO_4^{2-}$ as the oxidation of S because the oxidation number of sulfur increases from +4 to +6 (Sample Exercise 3, parts b and d). You can also recognize $C + O_2 \rightarrow CO_2$ as the oxidation of C and the reduction of O. The oxidation number of carbon increases from 0 to +4, and the oxidation number of O decreases from 0 to –2. Similarly, you recognize $4\ Al + 3\ O_2 \rightarrow 2\ Al_2O_3$ as the oxidation of Al (oxidation number increases from 0 to +3) and the reduction of O (oxidation number decreases from 0 to –2). A substance gaining oxygen is always undergoing oxidation.

The magnitude of the change in oxidation number tells you the numbers of electrons lost during oxidation or gained during reduction. Since the number of electrons transferred in a redox reaction must be conserved, the overall reaction cannot be balanced unless the number of electrons lost in oxidation equals the number of electrons gained in reduction. You will use this principle to balance redox reactions in the following section. Before undertaking the balancing procedure, however, you should be comfortable identifying oxidation and reduction in a more complicated reaction, such as that in Sample Exercise 4.

sample exercise 4

In the following equation (which is not balanced), determine the substances oxidized and reduced and indicate the oxidizing and reducing agents.

$$Cu(s) + HNO_3(aq) \rightarrow Cu(NO_3)_2(aq) + NO_2(g) + H_2O(l)$$

solution

1. Assign an oxidation number to each atom in every species in the reaction, and write the number below the atom.

Cu	+	HNO_3	→	$Cu(NO_3)_2$	+	NO_2	+	H_2O
0		+1 +5 –2		+2 +5 –2		+4 –2		+1 –2
Rule 1		**Rules 4, 6, 3**		**Rules 2, 6, 3**		**Rules 6, 3**		**Rules 4, 3**

2. Look for an increase in oxidation number for some atom. This will be the oxidation half-reaction. Copper has gone from 0 to +2:

$$Cu \rightarrow Cu^{2+}$$

Because Cu is oxidized, it is the reducing agent.

3. A decrease in oxidation number occurs for some of the nitrogen (+5 to +4). Because N remains bound to oxygen throughout, we include oxygen in the half-reaction:

$$NO_3^- \rightarrow NO_2$$

Because nitrate is reduced, it is the oxidizing agent.

• • • problem 4

In the following equation (which is unbalanced), determine the substances oxidized and reduced and indicate the oxidizing and reducing agents.

$$FeCl_3(aq) + H_2S(aq) \rightarrow FeCl_2(aq) + S(s) + HCl(aq)$$

16.5 balancing redox reactions in acidic and neutral solutions

In Chapter 9, you balanced chemical equations by inspection. However, those equations had been carefully chosen to be suitable; there are many equations that would be tedious or very difficult to balance by a trial-and-error inspection process. A powerful and widely applicable balancing procedure is called the **half-reaction method.** It uses to advantage the fact that in redox reactions, the number of electrons gained in the reduction half-reaction must be equal to the number of electrons lost in the oxidation half-reaction. You arrive at a balanced overall equation by *adding together two half-reactions, each of which has been balanced, so that the number of electrons lost equals the number of electrons gained.*

In Section 16.2, you already used the idea of adding together balanced half-reactions so that electrons lost equal electrons gained for the combination of Na and Cl_2. Even though this was a simple example, the method works even for complex equations.

Before establishing guidelines and exploring more complex examples, consider one more simple example in greater detail. Balance the equation $Al(s) + Cl_2(g) \rightarrow Al^{3+}(aq) + Cl^-(aq)$ by the half-reaction method, remembering that both atoms and charges must be balanced. For this reaction, the balanced oxidation half-reaction is

$$Al \rightarrow Al^{3+} + 3\,e^-$$ **Balanced oxidation half-reaction**

and the balanced reduction half-reaction is

$$Cl_2 + 2\,e^- \rightarrow 2\,Cl^-$$ **Balanced reduction half-reaction**

(Review Sections 16.2 and 16.3 if this is not clear to you.)

Just because each of the half-reactions is individually balanced, however, does not mean that they may be added together directly. In the half-reactions as written, each Al is losing three e^-, while each Cl_2 is gaining two e^-. To achieve the condition that *electrons lost equal electrons gained,* you must multiply the balanced oxidation half-reaction by 2 and the balanced reduction half-reaction by 3. Then, the oxidation reaction will show a loss of six e^- while the reduction reaction will show a gain of exactly six e^-.

$$2(Al \rightarrow Al^{3+} + 3\,e^-) = 2\,Al \rightarrow 2\,Al^{3+} + 6\,e^-$$ **6 e^- lost in oxidation**

$$3(Cl_2 + 2\,e^- \rightarrow 2\,Cl^-) = 3\,Cl_2 + 6\,e^- \rightarrow 6\,Cl^-$$ **6 e^- gained in reduction**

Now, addition of the two balanced and multiplied half-reactions will produce an *overall equation that is balanced.*

$2\,Al$	$\rightarrow$	$2\,Al^{3+} + 6\,e^-$
$3\,Cl_2 + 6\,e^-$	$\rightarrow$	$6\,Cl^-$
$2\,Al + 3\,Cl_2 + \cancel{6\,e^-}$	$\rightarrow$	$2\,Al^{3+} + 6\,Cl^- + \cancel{6\,e^-}$
$2\,Al + 3\,Cl_2$	$\rightarrow$	$2\,Al^{3+} + 6\,Cl^-$ **Balanced overall equation**

The equation is balanced with respect to both mass and charge. Notice that the electrons cancel out because equal numbers of electrons are lost and gained. In this simple example, it was necessary to use only guidelines 5 and 6 from the following list because of the simplicity of the half-reactions and the fact that we were starting with an ionic rather than a traditional equation. Additional guidelines will be necessary to help you balance individual half-reactions when oxygen atoms are present and when reactions occur in either acidic or basic aqueous solutions.

These guidelines are summarized below and applied to an acidic aqueous solution in Sample Exercise 5. You will gain confidence with these guidelines as you follow the sample exercises and work the problems within and at the end of this chapter.

Guidelines for balancing redox equations by the half-reaction method

1. *Write the net ionic equation.*
   From the given traditional equation, first write an ionic equation and cross out spectator ions to establish the unbalanced net ionic equation (Section 13.9).
2. *Assign oxidation numbers to each atom or ion* (Section 16.4).
3. *Write the separate oxidation and reduction half-reactions.*
   Use the changes in oxidation number to determine the substances undergoing oxidation and reduction in the net ionic equation (Section 16.5).
4. *Balance each half-reaction with respect to mass and charge.*
   a. Balance all elements except hydrogen and oxygen.
   b. Balance oxygen by using water (H_2O) as a source of oxygen on whichever side of the equation is deficient in oxygen.
   c. Balance hydrogen by using $H^+(aq)$ ions on whichever side of the equation is deficient in hydrogen. (This procedure applies to neutral and acidic solutions. Basic solutions require an alternate procedure; see Section 16.6.)
   d. Add the charges on each side of each half-reaction.
   e. Balance the charges by adding electrons to the side with the more positive total charge, so that the charges are equal on each side of each half-reaction.
5. *Find the multiplying factors that ensure that electrons lost in oxidation equal electrons gained in reduction.*
   Multiply each half-reaction by the smallest whole number that allows the total number of electrons lost in oxidation to equal the total number gained in reduction. Remember that the multiplying factors multiply all elements in the half-reactions as well as the electrons.
6. *Add the balanced and multiplied oxidation and reduction half-reactions together; cancel electrons and common species.*
   The electrons cancel out because the number of electrons lost in oxidation has been made equal to the number gained in reduction. The total number of electrons transferred is conserved.
7. *Check.*
   Make sure that atoms of each element are balanced and that the total electric charge is the same on both sides of the equation.

sample exercise 5

Write a balanced ionic equation for the following oxidation-reduction reaction.

$$HNO_3(aq) + HI(aq) \rightarrow NO(g) + I_2(aq) + H_2O(l)$$

Nitrate ion is a strong oxidizing agent that oxidizes iodide ion to iodine in acidic solution.

solution

Step1. *Write the net ionic equation* (review Section 13.9 if necessary).

$$H^+(aq) + NO_3^-(aq) + H^+(aq) + I^-(aq) \rightarrow NO(g) + I_2(aq) + H_2O(l)$$

Full ionic equation

$$H^+(aq) + NO_3^-(aq) + I^-(aq) \rightarrow NO(g) + I_2(aq) + H_2O(l)$$

Net ionic equation

(There are no spectator ions, but H^+ need only appear once.)

Step 2. *Assign oxidation numbers and identify species undergoing a change.*

$$\underset{+1}{H^+} + \underset{+5\ -2}{NO_3^-} + \underset{-1}{I^-} \rightarrow \underset{+2\ -2}{NO} + \underset{0}{I_2} + \underset{+1\ -2}{H_2O}$$

Step 3. *Write the separate oxidation and reduction half-reactions.*

$$\underset{-1}{I^-} \rightarrow \underset{0}{I_2}$$ **Oxidation** (increase in oxidation number)

$$\underset{+5}{NO_3^-} \rightarrow \underset{+2}{NO}$$ **Reduction** (decrease in oxidation number)

Step 4. *Balance mass and charge.*

a. *Balance all elements except H and O.*

$2\,I^- \rightarrow I_2$ This balances I.

$NO_3^- \rightarrow NO$ N is balanced.

b. *Balance O* in the reduction half-reaction by using H_2O as a source of O.

$$NO_3^- \rightarrow NO + 2\,H_2O$$

Two oxygens (hence two H_2O molecules) are required on the right to balance oxygen on the left.

c. *Balance H* by using H^+.

$$4\,H^+ + NO_3^- \rightarrow NO + 2\,H_2O$$

d. *Add the charges on each side of each half-reaction.*

$$\underset{2-}{2\,I^-} \rightarrow \underset{0}{I_2}$$

$$\underbrace{4\,H^+ + NO_3^-}_{3+} \rightarrow \underbrace{NO + 2\,H_2O}_{0}$$

e. *Balance the charges* by adding an appropriate number of electrons (negative charges) to the side with the larger positive charge.

$2\,I^- \rightarrow I_2 + 2\,e^-$ **Balanced oxidation half-reaction**

$4\,H^+ + NO_3^- + 3\,e^- \rightarrow NO + 2\,H_2O$ **Balanced reduction half-reaction**

Even though each half-reaction is now balanced with respect to mass and charge, they cannot be added together directly.

Step 5. *Find the multiplying factors* that will ensure that *electrons lost equal electrons gained.* In this case, the least common multiple of two e^- and three e^- is six e^-, so the multiplying factors are 3 for the oxidation half-reaction and 2 for the reduction.

$$3(2\,I^- \rightarrow I_2 + 2\,e^-)$$
$$2(4\,H^+ + NO_3^- + 3\,e^- \rightarrow NO + 2\,H_2O)$$

Step 6. *Add the balanced half-reactions together so that electrons are conserved.*

$6\,I^- \rightarrow 3\,I_2 + 6\,e^-$ **6 e^- lost in oxidation**

$8\,H^+ + 2\,NO_3^- + 6\,e^- \rightarrow 2\,NO + 4\,H_2O$ **6 e^- gained in reduction**

$8\,H^+ + 2\,NO_3^- + 6\,I^- + \cancel{6\,e^-} \rightarrow 3\,I_2 + 2\,NO + 4\,H_2O + \cancel{6\,e^-}$

$8\,H^+(aq) + 2\,NO_3^-(aq) + 6\,I^-(aq) \rightarrow 3\,I_2(aq) + 2\,NO(g) + 4\,H_2O(l)$

Balanced net ionic equation

Step 7. *Check* for balance of atoms.

Atoms of Reactants		**Atoms of Products**
8 H, 2 N, 6 O, 6 I	=	8 H, 2 N, 6 O, 6 I

Check for balance of charges.

Charges of Reactants		**Charges of Products**
8(1+) + 2(1–) + 6(1–)		0 + 0 + 0
0	=	0

The traditional balanced equation can be written by realizing that the H^+ ions must have come from the aqueous acids, $HI(aq)$ and $HNO_3(aq)$. Associate six of the H^+ ions with six I^- ions to form six HI; associate two H^+ ions with the two NO_3^- ions to form two HNO_3.

$$6\ HI(aq) + 2\ HNO_3(aq) \rightarrow 3\ I_2(aq) + 2\ NO(g) + 4\ H_2O(l)$$

Balanced traditional equation

---

sample exercise 6

When chloride ions (Cl^-) are added to an acidic solution of permanganate (MnO_4^-), the color of the solution changes from the deep-purple color of permanganate to a greenish-yellow color from chlorine (Cl_2) (Figure 16.4). Balance the following equation representing this change by using the half-reaction method.

$$KMnO_4(aq) + HCl(aq) \rightarrow MnCl_2(aq) + Cl_2(g) + KCl(aq) + H_2O(l)$$

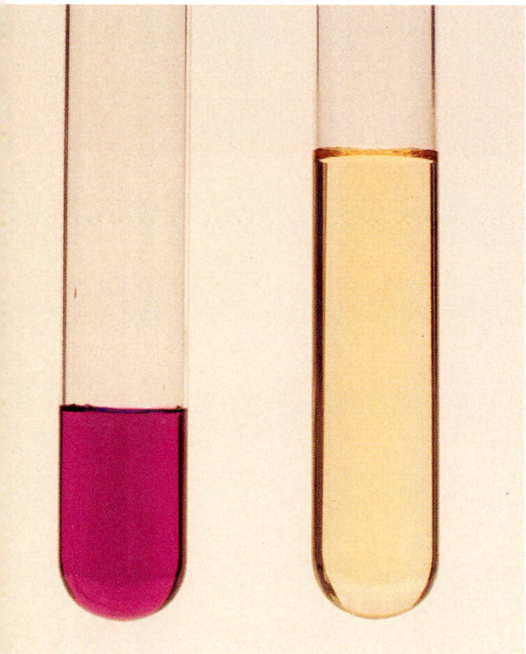

Figure 16.4
When chloride ions (Cl^-) are added to an acidic solution of permanganate (MnO_4^-) (left), the color of the solution changes from the deep-purple color of permanganate to the greenish-yellow color of chlorine gas (Cl_2) (right).

solution

Step 1. *Write the net ionic equation.*

$$K^+(aq) + MnO_4^-(aq) + H^+(aq) + Cl^-(aq) \rightarrow$$
$$Mn^{2+}(aq) + 2\ Cl^-(aq) + Cl_2(g) + K^+(aq) + Cl^-(aq) + H_2O(l)$$

Full ionic equation

$$MnO_4^-(aq) + H^+(aq) + Cl^-(aq) \rightarrow Mn^{2+}(aq) + Cl_2(g) + H_2O(l)$$

Net ionic equation

(The K^+ ion undergoes no change. Cl^- may be omitted from the right side of the equation. Although some reactant Cl^- is changed to Cl_2, product Cl^- ions are present only as spectator ions to balance the charge of K^+ ions.)

Step 2. *Assign oxidation numbers and identify species undergoing a change.*

$$\underset{+7\ -2}{MnO_4^-} + \underset{+1}{H^+} + \underset{-1}{Cl^-} \rightarrow \underset{+2}{Mn^{2+}} + \underset{0}{Cl_2} + \underset{+1\ -2}{H_2O}$$

Step 3. *Write the separate oxidation and reduction half-reactions.*

Cl^-	→	Cl_2	**Oxidation** (increase in oxidation number)
–1		0	
MnO_4^-	→	Mn^{2+}	**Reduction** (decrease in oxidation number)
+7		+2	

Step 4. *Balance mass and charge.*

a. *Balance all elements except H and O.*

$2\ Cl^-$	→	Cl_2	This balances Cl.
MnO_4^-	→	Mn^{2+}	Mn is balanced.

b. *Balance O* by using H_2O as a source of O.

$$MnO_4^- \rightarrow Mn^{2+} + 4\ H_2O$$

c. *Balance H* by using H^+.

$$MnO_4^- + 8\ H^+ \rightarrow Mn^{2+} + 4\ H_2O$$

d. *Add the charges on each side of each half-reaction.*

$$\underset{2-}{2\ Cl^-} \rightarrow \underset{0}{Cl_2}$$

$$\underbrace{MnO_4^- + 8\ H^+}_{7+} \rightarrow \underbrace{Mn^{2+} + 4\ H_2O}_{2+}$$

e. *Balance the charges* by adding an appropriate number of electrons to the side with the larger positive charge.

$2\ Cl^- \rightarrow Cl_2 + 2\ e^-$	**Balanced oxidation half-reaction**
$MnO_4^- + 8\ H^+ + 5\ e^- \rightarrow Mn^{2+} + 4\ H_2O$	**Balanced reduction half-reaction**

Even though each half-reaction is now balanced with respect to mass and charge, they cannot be added together directly.

Step 5. *Find the multiplying factors* that will ensure that *electrons lost equal electrons gained.* In this case, the least common multiple of two e^- and five e^- is ten e^-, so the factors are 5 for the oxidation half-reaction and 2 for the reduction half-reaction.

$$5(2\ Cl^- \rightarrow Cl_2 + 2\ e^-)$$

$$2(MnO_4^- + 8\ H^+ + 5\ e^- \rightarrow Mn^{2+} + 4\ H_2O)$$

Step 6. *Add the balanced half-reactions together so that electrons are conserved.*

$10\ Cl^-$	$\rightarrow$	$5\ Cl_2 + 10\ e^-$	**10 e^- lost in oxidation**
$2\ MnO_4^- + 16\ H^+ + 10\ e^-$	$\rightarrow$	$2\ Mn^{2+} + 8\ H_2O$	**10 e^- gained in reduction**
$2\ MnO_4^- + 16\ H^+ + 10\ Cl^- + \cancel{10\ e^-}$	$\rightarrow$	$2\ Mn^{2+} + 5\ Cl_2 + 8\ H_2O + \cancel{10\ e^-}$	
$2\ MnO_4^-(aq) + 16\ H^+(aq) + 10\ Cl^-(aq)$	$\rightarrow$	$2\ Mn^{2+}(aq) + 5\ Cl_2(g) + 8\ H_2O(l)$	

Balanced net ionic equation

Step 7. *Check* for balance of atoms.

Atoms of Reactants		**Atoms of Products**
2 Mn, 8 O, 16 H, 10 Cl	=	2 Mn, 8 O, 16 H, 10 Cl

Check for balance of charges.

Charges of Reactants		**Charges of Products**
2(1–) + 16(1+) + 10(1–)		2(2+)
4+	=	4+

The traditional equation can be constructed by using the coefficients we have determined for the ions in the balanced net ionic equation and then completing the original equation by inspection. Thus,

$$2\ MnO_4^-(aq) + 16\ H^+(aq) + 10\ Cl^-(aq) \rightarrow 2\ Mn^{2+}(aq) + 5\ Cl_2\,(g) + 8\ H_2O(l)$$

Balanced net ionic equation

$$2\ KMnO_4(aq) + 16\ HCl(aq) \longrightarrow 2\ MnCl_2(aq) + 5\ Cl_2(g) + \_\_\ KCl(aq) + 8\ H_2O(l)$$

Coefficient of MnO_4^-	The ionic equation shows the apparent contradiction of 16 H^+ and 10 Cl^-; this comes about because 10 Cl^- are oxidized to Cl_2, while some remain unchanged; in such a case, choose the larger coefficient.	Coefficient of Mn^{2+}	Coefficient of Cl_2	Coefficient of H_2O

Traditional equation with ionic coefficients

Balancing K by inspection will complete the equation.

$$2\ KMnO_4(aq) + 16\ HCl(aq) \longrightarrow 2\ MnCl_2(aq) + 5\ Cl_2(g) + 2\ KCl(aq) + 8\ H_2O(l)$$

$$2\ K\ \ 2\ Mn\ \ 8\ O \qquad 16\ H\ \ 16\ Cl \longrightarrow 2\ Mn\ \ 4 + 10 + 2 = 16\ Cl \qquad 2\ K \qquad 16\ H\ \ 8\ O$$

••• problem 5

Balance the following equation.

$$Pb(s) + PbO_2(s) + H_2SO_4(aq) \rightarrow PbSO_4(aq) + H_2O(l)$$

This reaction provides the electrical energy in a storage battery to start your car.

In the preceding examples, the solutions have been acidic, so the use of H^+ for balancing was quite legitimate and understandable. Under neutral conditions, we may still introduce H^+ for balancing purposes in guideline 4. Any artificially added H^+ will be canceled out when the half-reactions are added together. Sample Exercise 7 illustrates neutral conditions under which the H^+ used to balance each half-reaction cancels out when the half-reactions are added to obtain the overall balanced equation.

sample exercise 7

Balance the following equation for the combustion of ethane using the half-reaction method.

$$C_2H_6(g) + O_2(g) \rightarrow CO_2(g) + H_2O(g)$$

Ethane gas, C_2H_6, is used as a fuel; it is a component of "bottled gas."

solution

Step 1. *Write the net ionic equation.* Because there are no ions, the net equation in this example is the same as the traditional equation.

$$C_2H_6(g) + O_2(g) \rightarrow CO_2(g) + H_2O(g)$$

Step 2. *Assign oxidation numbers and identify the species undergoing change.*

$$\underset{-3\ +1}{C_2H_6} + \underset{0}{O_2} \longrightarrow \underset{+4\ -2}{CO_2(g)} + \underset{+1\ -2}{H_2O}$$

C and O undergo changes in oxidation number.

Step 3. *Write separate oxidation and reduction half-reactions.*

$$\underset{-3}{C_2H_6} \rightarrow \underset{+4}{CO_2} \qquad \textbf{Oxidation (increase in oxidation number)}$$

$$\underset{0}{O_2} \rightarrow \underset{-2}{H_2O} \qquad \textbf{Reduction (decrease in oxidation number)}$$

Step 4. *Balance mass and charge.*

a. *Balance C.*

$$C_2H_6 \rightarrow 2\ CO_2$$

b. *Balance O* by using H_2O as a source.

$$C_2H_6 + 4\ H_2O \rightarrow 2\ CO_2$$
$$O_2 \rightarrow 2\ H_2O$$

c. *Balance H by using* H^+.

$$C_2H_6 + 4\ H_2O \rightarrow 2\ CO_2 + 14\ H^+$$
$$O_2 + 4\ H^+ \rightarrow 2\ H_2O$$

d. *Add the charges on each side of each half-reaction.*

$$\underbrace{C_2H_6 + 4\ H_2O}_{0} \rightarrow \underbrace{2\ CO_2 + 14\ H^+}_{14+}$$

$$\underbrace{O_2 + 4\ H^+}_{4+} \rightarrow \underbrace{2\ H_2O}_{0}$$

e. *Balance the charges* by using electrons.

$C_2H_6 + 4\ H_2O \rightarrow 2\ CO_2 + 14\ H^+ + 14\ e^-$ **Balanced oxidation half-reaction**

$O_2 + 4\ H^+ + 4\ e^- \rightarrow 2\ H_2O$ **Balanced reduction half-reaction**

Even though each half-reaction is balanced, they cannot be added together directly.

Step 5. *Find the multiplying factors* that will ensure *electrons lost equal electrons gained.* The least common multiple of 14 and 4 is 28.

$$2(C_2H_6 + 4\ H_2O \rightarrow 2\ CO_2 + 14\ H^+ + 14\ e^-)$$
$$7(O_2 + 4\ H^+ + 4\ e^- \rightarrow 2\ H_2O)$$

Step 6. *Add the balanced half-reactions together such that electrons are conserved,* canceling (subtracting) common species.

$2\ C_2H_6 + 8\ H_2O \rightarrow 4\ CO_2 + 28\ H^+ + 28\ e^-$ **28 e^- lost in oxidation**

$7\ O_2 + 28\ H^+ + 28\ e^- \rightarrow 14\ H_2O$ **28 e^- gained in reduction**

$$2\ C_2H_6 + 7\ O_2 + 8\ H_2O + \cancel{28\ H^+} + \cancel{28\ e^-} \rightarrow 4\ CO_2 + 14\ H_2O + \cancel{28\ H^+} + \cancel{28\ e^-}$$
$$-\ 8\ H_2O \qquad\qquad -\ 8\ H_2O$$

$$2\ C_2H_6 + 7\ O_2 \rightarrow 4\ CO_2 + 6\ H_2O$$

Balanced traditional equation

Because there were no ions in the original equation, you have obtained the balanced traditional equation directly.

Step 7. *Check* for balance of atoms.

Atoms of Reactants		Atoms of Products
4 C, 12 H, 14 O	=	4 C, 12 H, 14 O

Check for balance of charges.

Charge of Reactants		Charge of Products
0	=	0

16.6 balancing redox reactions in basic solution

All of the examples of redox reactions in Section 16.5 were either under neutral conditions or in acidic solution. You know that in all acidic aqueous solutions, $[H^+]$ is greater than $[OH^-]$ (Section 15.5), so it was reasonable to use H^+ ions (which are already present in the reaction mixture) to help balance the oxidation and reduction half-reactions. In basic aqueous solutions, $[OH^-]$ is greater than $[H^+]$. Consequently, it is the OH^- ions (which are already present in the reaction mixture) that must be used to balance redox reactions in basic solution.

To balance equations for redox reactions occurring in basic solution, you must modify guidelines 4b and 4c, which deal with balancing oxygen and hydrogen, respectively. This modification is necessary because in a basic solution, the sources of oxygen and hydrogen will be OH^- and H_2O, not H_2O and H^+ as in acidic and neutral solutions. All other guidelines remain the same. The new guidelines for basic solutions are designated 4b and 4c and appear below:

4b. *Balance each oxygen atom* needed in a half-reaction by adding two hydroxide ions (OH^-) to the side needing oxygen and one water molecule (H_2O) to the other side of the half-reaction.

4c. *Balance each hydrogen atom* by adding one water molecule (H_2O) to the side needing hydrogen and one hydroxide ion (OH^-) to the other side of the half-reaction.

Sample Exercise 8 applies these new guidelines to the oxidation of iodide, I^-, by permanganate, MnO_4^-. The appearance of the hydroxide ion (OH^-) on either side of the equation signals a basic solution.

---

sample exercise 8

Balance the following equation.

$$KI(aq) + KMnO_4(aq) + H_2O(l) \rightarrow MnO_2(s) + I_2(aq) + KOH(aq)$$

Comparing this equation with that in Sample Exercise 6, you can see that permanganate acts as an oxidizing agent in both basic and acidic solution.

solution

Step 1. *Write the net ionic equation.*

$$K^+(aq) + I^-(aq) + MnO_4^-(aq) + H_2O(l) \rightarrow MnO_2(s) + I_2(aq) + K^+(aq) + OH^-(aq)$$

Full ionic equation

$$I^-(aq) + MnO_4^-(aq) + H_2O(l) \rightarrow MnO_2(s) + I_2(aq) + OH^-(aq)$$

Net ionic equation

Step 2. *Assign oxidation numbers and identify the species undergoing change.*

$$\underset{-1}{I^-} + \underset{+7\ \ -2}{MnO_4^-} + \underset{+1\ \ -2}{H_2O} \rightarrow \underset{+4\ \ -2}{MnO_2} + \underset{0}{I_2} + \underset{-2\ \ +1}{OH^-}$$

Step 3. *Write the separate oxidation and reduction half-reactions.*

$$\underset{-1}{I^-} \rightarrow \underset{0}{I_2} \qquad \textbf{Oxidation}$$

$$\underset{+7}{MnO_4^-} \rightarrow \underset{+4}{MnO_2} \qquad \textbf{Reduction}$$

Step 4. *Balance mass and charge.*

a. *Balance the elements other than H and O.*

$$2\,I^- \rightarrow I_2$$
$$MnO_4^- \rightarrow MnO_2$$

b. *Balance O* by using OH^- and H_2O as noted in guideline 4b.

$$MnO_4^- \rightarrow MnO_2$$

Two oxygen atoms are needed on the right side, so we add four hydroxide ions (two for each oxygen needed). Four hydroxide ions (OH^-) on the right demands two water molecules on the left.

$$MnO_4^- + 2\,H_2O \rightarrow MnO_2 + 4\,OH^-$$

c. *Balance the H atoms.* In this case, they are already balanced (four on each side).

d. *Add the charges on each side of each half-reaction.*

$$\underset{2-}{2\,I^-} \rightarrow \underset{0}{I_2}$$

$$\underbrace{MnO_4^- + 2\,H_2O}_{1-} \rightarrow \underbrace{MnO_2 + 4\,OH^-}_{4-}$$

e. *Balance the charges* by using electrons.

$$2\,I^- \rightarrow I_2 + 2\,e^-$$ **Balanced oxidation half-reaction**

$$MnO_4^- + 2\,H_2O + 3\,e^- \rightarrow MnO_2 + 4\,OH^-$$ **Balanced reduction half-reaction**

Even though each of the half-reactions is balanced, they may not be added together directly.

Step 5. *Find the multiplying factors* to ensure *electrons lost equal electrons gained.*

$$3(2\,I^- \rightarrow I_2 + 2\,e^-)$$
$$2(MnO_4^- + 2\,H_2O + 3\,e^- \rightarrow MnO_2 + 4\,OH^-)$$

Step 6. *Add the half-reactions together so that electrons are conserved,* canceling common species.

$$6\,I^- \rightarrow 3\,I_2 + 6\,e^-$$ **6 e^- lost in oxidation**

$$2\,MnO_4 + 4\,H_2O + 6\,e^- \rightarrow 2\,MnO_2 + 8\,OH^-$$ **6 e^- gained in reduction**

$$2\,MnO_4^- + 4\,H_2O + 6\,I^- + \cancel{6\,e^-} \rightarrow 3\,I_2 + 2\,MnO_2 + 8\,OH^- + \cancel{6\,e^-}$$

$$2\,MnO_4^-(aq) + 4\,H_2O(l) + 6\,I^-(aq) \rightarrow 3\,I_2(aq) + 2\,MnO_2(s) + 8\,OH^-(aq)$$

Balanced net ionic equation

Step 7. *Check* for balance of atoms.

Atoms of Reactants		Atoms of Products
2 Mn, 12 O, 8 H, 6 I	=	2 Mn, 12 O, 8 H, 6 I

Check for balance of charges.

Atoms of Reactants		Atoms of Products
2(1–) + 6(1–)		8(1–)
8–	=	8–

The original equation can be balanced by using the coefficients of the ions from the balanced net ionic equation; thus,

$$6\,KI(aq) + 2\,KMnO_4^-(aq) + 4\,H_2O(l) \rightarrow 2\,MnO_2(s) + 3\,I_2(aq) + 8\,KOH(aq)$$

8 K 6 I   2 Mn   12 O   8 H   2 Mn   12 O   6 I   8 K   8 H

Balanced traditional equation

problem 6

Balance the following equation.

$$Zn(s) + MnO_2(s) + H_2O(l) \rightarrow Zn(OH)_2(s) + Mn_2O_3(s)$$

This reaction is employed in the cylindrical, alkaline batteries commonly used in toys, audio devices, and cameras.

16.7 alternate method for balancing redox reactions

The half-reaction method is superior to other balancing methods because it offers the maximum amount of information about balancing all elements involved in the reaction. In the foregoing examples, nearly all of the coefficients of the final balanced equations were obtained directly by following the guidelines on pages 500 and 506.

An alternate balancing method is the **oxidation-number method,** which is applied directly to the traditional equation. If you understand the principles of oxidation and reduction, the oxidation-number method may provide you with a shorter method for balancing equations. However, the transfer of electrons is obscured, so if you need to identify the transfer of electrons explicitly, you will have to resort to the half-reaction method. In addition, the stoichiometric coefficients are obtained directly only for the species actually oxidized or reduced. Coefficients for other reactants and products must be obtained by inspection.

There are only three guidelines for the oxidation-number method. Their brevity is particularly appealing after the lengthy guidelines given for the half-reaction method. Should you have any difficulty in applying them to complex redox equations, you may always return to the half-reaction method with confidence.

Guidelines for balancing redox reactions by the oxidation-number method

1. *Assign oxidation numbers to each type of atom and identify the oxidized and reduced elements.*
2. *Find the multiplying factors that ensure that the increase in oxidation number equals the decrease in oxidation number.* Use these factors as coefficients of the oxidized and reduced elements. Clear fractional coefficients if they occur.
3. *Balance the rest of the equation by inspection* (Section 9.4).

sample exercise 9

Balance the following equation by the oxidation-number method.

$$KMnO_4 + HCl \rightarrow MnCl_2 + Cl_2 + KCl + H_2O$$

solution

Step 1. *Assign oxidation numbers* to each type of atom and identify the oxidized and reduced elements.

$$\underset{+1\ +7\ -2}{KMnO_4} + \underset{+1\ -1}{HCl} \rightarrow \underset{+2\ -1}{MnCl_2} + \underset{0}{Cl_2} + \underset{+1\ -1}{KCl} + \underset{+1\ -2}{H_2O}$$

Unbalanced traditional equation

$\underset{+7}{Mn} \rightarrow \underset{+2}{Mn}$ Reduction (decrease of 5 in oxidation number)

$\underset{-1}{Cl} \rightarrow \underset{0}{Cl}$ Oxidation (increase of 1 in oxidation number for each Cl)

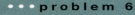

Step 2. *Find the multiplying factor* that ensures that the decrease in oxidation number equals the increase in oxidation number. In this case, the element oxidized must be multiplied by 5, so you can use the following coefficients.

$$KMnO_4 + 5\ HCl \rightarrow MnCl_2 + 5/2\ Cl_2 + KCl + H_2O$$

(5 Cl) under Cl_2; ↑ under KCl: No change as this Cl has not undergone oxidation or reduction

Clear the fractional coefficient of Cl_2 by multiplying through by 2.

$$2\ KMnO_4 + 10\ HCl \rightarrow 2\ MnCl_2 + 5\ Cl_2 + 2\ KCl + 2\ H_2O$$

Step 3. *Balance by inspection* (Section 9.4).

Atoms of Reactants		**Atoms of Products**
2 K, 2 Mn, 8 O, 10 H, 10 Cl		2 Mn, 4 Cl, 10 Cl, 2 K, 2 Cl, 4 H, 2 O
Unbalanced:	10 Cl	16 Cl
	8 O	2 O
	10 H	4 H

Both K and Mn are balanced, but O, H, and Cl are not. To balance the 16 Cl atoms on the right side, increase the coefficient of HCl on the left side to 16.

$$2\ KMnO_4 + 16\ HCl \rightarrow 2\ MnCl_2 + 5\ Cl_2 + 2\ KCl + 2\ H_2O$$

Now K, Mn, and Cl are balanced, but O and H are not.

Atoms of Reactants		**Atoms of Products**
2 K, 2 Mn, 16 Cl, 8 O, 16 H		2 K, 2 Mn, 16 Cl, 2 O, 4 H
Unbalanced:	8 O	2 O
	16 H	4H

To balance the eight O atoms on the left side, raise the coefficient of H_2O on the right side to 8. This simultaneously increases the number of H atoms on the right side to 16, which is the same as the number of H atoms on the left side. In this example, both H and O can be balanced simultaneously by changing the coefficient of H_2O to 8.

$$2\ KMnO_4 + 16\ HCl \rightarrow 2\ MnCl_2 + 5\ Cl_2 + 2\ KCl + 8\ H_2O$$

Balanced traditional equation

For a comparison of balancing methods, refer to Sample Exercise 6, in which this same equation was balanced by the half-reaction method.

---

sample exercise 10

Balance the following equation using the oxidation-number method.

$$Fe + H_2SO_4 \rightarrow Fe_2(SO_4)_3 + SO_2 + H_2O$$

The iron(III) sulfate formed in this reaction is employed as a coagulant in water sewage treatment.

solution

Step 1. *Assign oxidation numbers* to each type of atom and identify the oxidized and reduced elements.

$$\underset{0}{Fe} + \underset{+1\ +6\ -2}{H_2SO_4} \rightarrow \underset{+3\ +6\ -2}{Fe_2(SO_4)_3} + \underset{+4\ -2}{SO_2} + \underset{+1\ -2}{H_2O}$$

Unbalanced traditional equation

$\underset{0}{Fe} \rightarrow \underset{+3}{Fe}$ Oxidation (increase of 3 in oxidation number)

$\underset{+6}{S} \rightarrow \underset{+4}{S}$ Reduction (decrease of 2 in oxidation number)

Step 2. *Find the multiplying factors* that ensure that the decrease in oxidation number equals the increase in oxidation number. The least common multiple of 3 and 2 is 6; thus, you use the coefficient 2 for Fe($2 \times 3 = 6$) and 3 for S($3 \times 2 = 6$).

$$2\ Fe + 3\ H_2SO_4 \rightarrow Fe_2(SO_4)_3 + 3\ SO_2 + H_2O$$

Step 3. *Balance by inspection* (Section 9.4).

Atoms of Reactants		**Atoms of Products**
2 Fe, 3 S, 6 H, 12 O		2 Fe, 6 S, 2 H, 19 O
Unbalanced:	3 S	6 S
	12 O	19 O
	6 H	2 H

Only Fe is balanced; S, H, and O are not. To balance the six S atoms on the right side, increase the coefficient of H_2SO_4 on the left to 6; this increases the number of O atoms to 24 and the number of H atoms to 12 on the left side.

Atoms of Reactants		**Atoms of Products**
2 Fe, 6 S, 12 H, 24 O		2 Fe, 6 S, 2 H, 19 O
Unbalanced:		
	24 O	19 O
	12 H	2 H

$2\ Fe$	+	$6\ H_2SO_4$	$\rightarrow$	$Fe_2(SO_4)_3$	+	$3\ SO_2$	+	H_2O
2 Fe		12 H 6 S 24 O		2 Fe 3 S 12 O		3 S 6 O		2 H 1 O

Now Fe and S are balanced, but H and O are not. Balance the 12 H atoms on the left by raising the coefficient of H_2O on the right to 6. Then the O atoms will also be balanced with 24 O atoms on each side.

$$2\ Fe + 6\ H_2SO_4 \rightarrow Fe_2(SO_4)_3 + 3\ SO_2 + 6\ H_2O$$

Balanced traditional equation

---

problem 7

Balance the following equation using the oxidation-number method.

$$HNO_3 + I_2 \rightarrow NO_2 + HIO_3 + H_2O$$

16.8 activity series revisited

In Section 16.9, you will see that such common devices as batteries and such common industrial processes as chromeplating are based on electron-transfer (redox) reactions. Before considering these applications, it will be useful to revisit and expand the concept of the activity series of metals (Table 9.2), which was presented in Section 9.8.

Table 16.1 is an enlarged, more informative version of Table 9.2. You see that what is meant by a "more active metal" is a metal that more readily loses or donates electrons. A more active metal is one that is more easily oxidized. The more easily oxidized metals are the stronger reducing agents because they so readily donate the electrons necessary to reduce other substances. Likewise, metallic cations that more readily accept electrons (and are more easily reduced) are stronger oxidizing agents.

In Section 9.8 (Figure 9.12), you were only able to say that zinc metal would replace copper cations from their compounds because Zn appears above Cu in the activity series. Now you can see that the reaction occurs because Zn is more readily oxidized (a stronger reducing agent) than Cu, and Cu^{2+} is more readily reduced (a stronger oxidizing agent) than Zn^{2+} (Figure 16.5).

$$Zn(s) + Cu^{2+}(aq) + SO_4^{2-}(aq) \rightarrow Zn^{2+}(aq) + SO_4^{2-}(aq) + Cu(s)$$

$$Zn \rightarrow Zn^{2+} + 2\ e^- \qquad \text{Oxidation (electrons lost)}$$

$$Cu^{2+} + 2\ e^- \rightarrow Cu \qquad \text{Reduction (electrons gained)}$$

table 16.1 activity series of metals (reduction half-reactions)

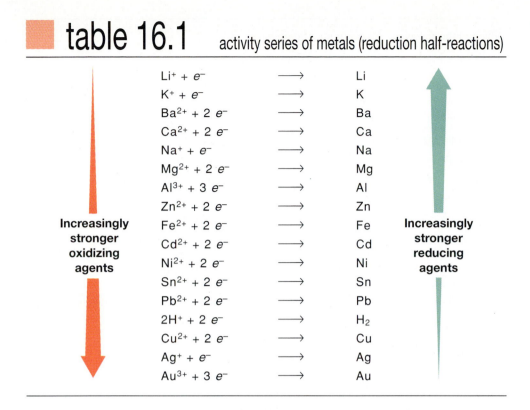

	$Li^+ + e^-$	$\longrightarrow$	Li	
	$K^+ + e^-$	$\longrightarrow$	K	
	$Ba^{2+} + 2\ e^-$	$\longrightarrow$	Ba	
	$Ca^{2+} + 2\ e^-$	$\longrightarrow$	Ca	
	$Na^+ + e^-$	$\longrightarrow$	Na	
	$Mg^{2+} + 2\ e^-$	$\longrightarrow$	Mg	
	$Al^{3+} + 3\ e^-$	$\longrightarrow$	Al	
	$Zn^{2+} + 2\ e^-$	$\longrightarrow$	Zn	
Increasingly stronger oxidizing agents ↓	$Fe^{2+} + 2\ e^-$	$\longrightarrow$	Fe	**Increasingly stronger reducing agents** ↑
	$Cd^{2+} + 2\ e^-$	$\longrightarrow$	Cd	
	$Ni^{2+} + 2\ e^-$	$\longrightarrow$	Ni	
	$Sn^{2+} + 2\ e^-$	$\longrightarrow$	Sn	
	$Pb^{2+} + 2\ e^-$	$\longrightarrow$	Pb	
	$2H^+ + 2\ e^-$	$\longrightarrow$	H_2	
	$Cu^{2+} + 2\ e^-$	$\longrightarrow$	Cu	
	$Ag^+ + e^-$	$\longrightarrow$	Ag	
	$Au^{3+} + 3\ e^-$	$\longrightarrow$	Au	

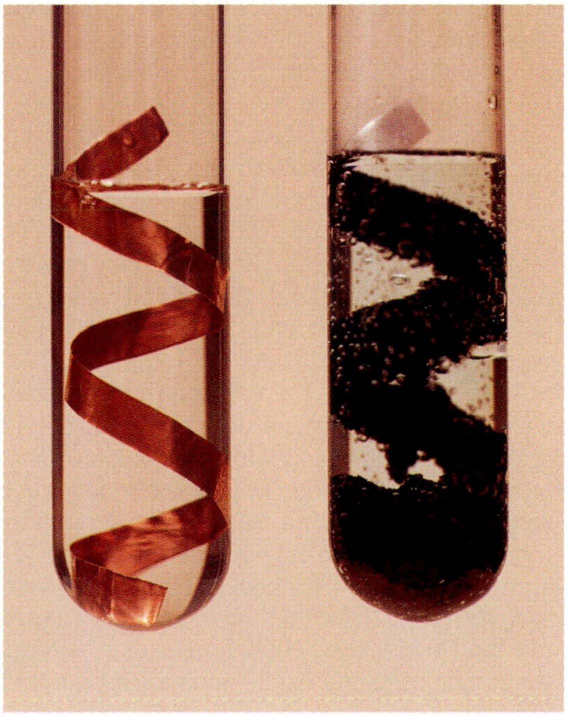

Figure 16.5

The test tube on the left contains a strip of copper metal in a colorless solution of zinc sulfate. The shiny, untarnished appearance of the metal and the clarity of the solution suggests that there is no reaction between the materials. The test tube on the right contains a strip of zinc metal in copper(II) sulfate solution. The submerged portion of the zinc metal has darkened as it reacts with the solution. The solid at the bottom of the test tube is Cu metal that has been produced in the reaction and precipitated out of solution.

In Figure 16.5, you see that the reaction written above (zinc metal replacing copper cations from their compounds) will occur **spontaneously**; that is, it will occur by itself without external assistance. In contrast, the reverse reaction of copper metal replacing zinc cations from their compounds does not occur spontaneously and no change would be visible.

Redox reactions occur spontaneously when the oxidation half-reaction of a relatively stronger reducing agent (Zn in the previous example) is coupled with the reduction half-reaction of a relatively stronger oxidizing agent (Cu^{2+} in the example). Notice that all of the half-reactions shown in Table 16.1 are reduction half-reactions. Metals higher in the table are more easily oxidized and are consequently reactants with stronger reducing strength; ions lower in the table are more easily reduced and are consequently reactants with stronger oxidizing strength. Since Zn is higher in the table than Cu, Zn will be oxidized and copper (ions) will be reduced spontaneously.

sample exercise 11

Which of the following will occur spontaneously?

a. $Pb(s) + Mg(NO_3)_2(aq) \rightarrow Pb(NO_3)_2(aq) + Mg(s)$

b. $Mg(s) + Pb(NO_3)_2(aq) \rightarrow Mg(NO_3)_2(aq) + Pb(s)$

solution

Consider the half-reactions in each case.

a. $Pb \rightarrow Pb^{2+} + 2\,e^-$ — **Oxidation** (electrons lost)
$Mg^{2+} + 2\,e^- \rightarrow Mg$ — **Reduction** (electrons gained)

b. $Mg \rightarrow Mg^{2+} + 2\,e^-$ — **Oxidation** (electrons lost)
$Pb^{2+} + 2\,e^- \rightarrow Pb$ — **Reduction** (electrons gained)

A spontaneous reaction will occur when the stronger reducing agent is oxidized and the stronger oxidizing agent is reduced. Consultation with Table 16.1 reveals that Mg is the stronger reducing agent and Pb^{2+} is the stronger oxidizing agent. Hence, reaction b is spontaneous.

16.9 electrochemistry: redox and energy

voltaic cells

Spontaneous redox reactions of the type just described transfer electrons from the reducing agent to the oxidizing agent. This transfer of electrons does not have to occur directly, however. If the oxidizing and reducing agents are separated from each other, the transfer of electrons can be channeled through a wire to produce a flow of electrons (electricity). Such an apparatus is known scientifically as a **voltaic cell,** or more commonly, as a battery. The electricity produced is a form of energy that may be used to do work, such as run a motor, power a calculator, or power a flashlight.

Figure 16.6 shows two possible ways of making a voltaic cell from the spontaneous reaction between Zn and Cu^{2+}. A voltaic cell cannot be made by simply putting zinc metal into a copper sulfate solution because, in that case, electrons would be directly transferred from Zn to Cu^{2+}. The electrons would not need to flow through a wire and, thus, would not be available for doing useful work. Rather, the reducing agent (Zn, the electron donor) must be separated from the oxidizing agent (Cu^{2+}, the electron acceptor) so the electron flow can be directed through an external circuit to do useful work.

In Figure 16.6a, the oxidation half-reaction, $Zn(s) \rightarrow Zn^{2+}(aq) + 2\,e^-$, takes place in a porous cup; the solid zinc metal electrode (called the anode) dissolves to form zinc cations in a zinc sulfate solution.[1] The electrons lost in the oxidation flow through the wire to the copper electrode (called the cathode), where they reduce copper cations from the copper sulfate solution at the surface of the copper electrode. The reduction half-reaction, $Cu^{2+}(aq) + 2\,e^- \rightarrow Cu(s)$, takes place in the larger beaker; aqueous copper cations precipitate as solid copper onto the copper cathode. Notice how the Zn electrode is eaten away as Zn is oxidized to Zn^{2+} and how the Cu electrode is built up as Cu^{2+} is reduced to more solid Cu.

To complete the circuit, ions must be able to flow between the two solutions. The porous walls of the inner container shown in Figure 16.6a allow this. The pores, or microscopic holes, in the wall of the inner container allow ions in solution to pass through the wall over a period of time, just like water will pass very slowly through

1. All solutions referred to in this section are assumed to be standard concentrations of 1 *M* unless otherwise specified.

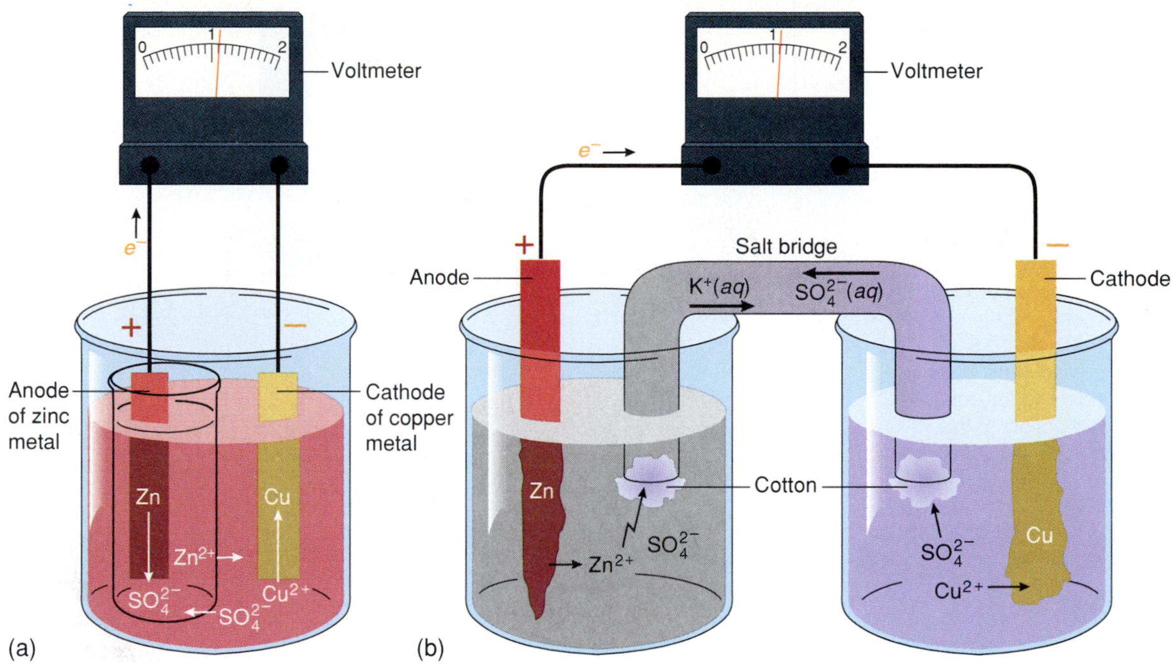

Figure 16.6

Two possible designs for a voltaic cell using zinc and copper electrodes immersed in zinc sulfate and copper sulfate solutions, respectively. The same spontaneous redox reaction, $Zn(s) + Cu^{2+}(aq) \rightarrow Zn^{2+}(aq) + Cu(s)$, occurs in both designs, and the voltmeter shows that both designs produce the same voltage. Notice how the Zn electrode is eaten away as Zn is oxidized to Zn^{2+} and the Cu electrode is built up as Cu^{2+} is reduced to solid Cu. In each case, electrons are transferred through the external wire from the Zn anode to the Cu cathode. The difference is how ions may flow between the two separated solutions to complete the circuit. In (a), ions flow through the wall of a porous cup; in (b), ions flow through a salt bridge.

the walls of a clay plant pot. *Anions* (in this case SO_4^{2-}) *flow toward the anode,* and *cations* (in this case Cu^{2+}) *flow toward the cathode.* This flow maintains the electrical neutrality of the solution.

In Figure 16.6b, the oxidation and reduction half-reactions have been separated in their own beakers. The electrons lost in the oxidation at the zinc anode also pass through the wire to the copper cathode, where reduction occurs. To complete the circuit in this case, the necessary flow of ions between the two solutions is established through a **salt bridge.** A salt bridge is simply a tube containing a salt in a gel. The gel prevents the mixing of the different solutions, but allows ions to flow. In this example, K_2SO_4 is the salt made into a gel; sulfate ions are already present in the solution and potassium ions will not react with any other species present.

Before proceeding, you should review and summarize the names of reactions occurring at the two electrodes of an electrochemical cell. By definition, *oxidation occurs at the anode, reduction at the cathode.* The names given to ions arise because of their migratory tendencies toward the two electrodes. Negative ions are called *an*ions because they always move toward the *an*ode. Positive ions are called *cat*ions because they always move toward the *cat*hode. In summary,

Electrode	Reaction	Ions Moving toward Electrode
Anode	Oxidation	Anions
Cathode	Reduction	Cations

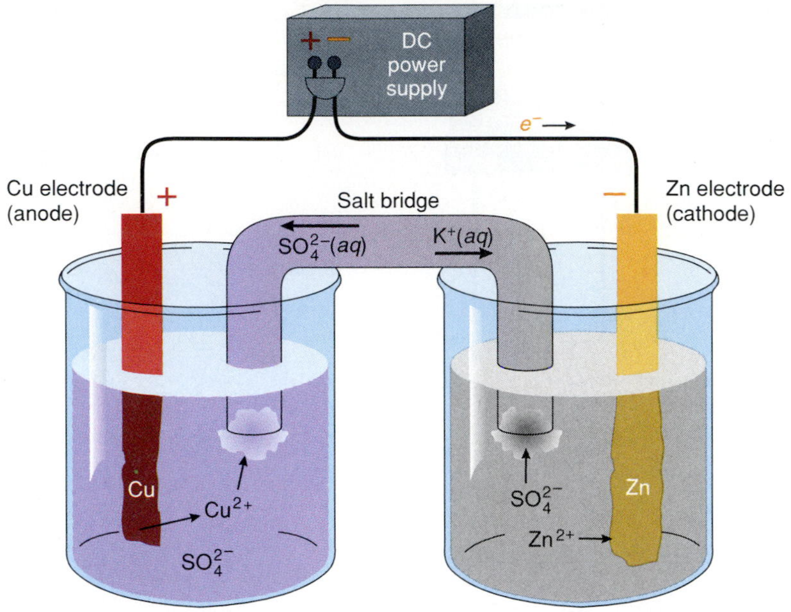

Figure 16.7

In this electrolytic cell, the electric current (from the dc power supply) is applied so that electrons are fed to the Zn electrode. These electrons reduce the aqueous Zn^{2+} ions to Zn metal; as a result, the Zn electrode is built up from the deposit. At the same time, Cu is oxidized to aqueous Cu^{2+} ions; as a result, the Cu electrode diminishes in size. The overall reaction, $Cu(s) + Zn^{2+}(aq) \rightarrow Cu^{2+}(aq) + Zn(s)$, is the reverse of that illustrated in Figure 16.6. In Figure 16.6, which represents a voltaic cell, a voltmeter indicated the voltage produced from the reaction. In this figure, a power source must supply energy to force the reaction to occur.

sample exercise 12

Describe how you would construct a voltaic cell employing the spontaneous reaction discussed in Sample Exercise 11: $Mg(s) + Pb(NO_3)_2(aq) \rightarrow Mg(NO_3)_2(aq) + Pb(s)$.

solution

Refer to Figure 16.6. The desired cell would have a magnesium electrode submerged in a $Mg(NO_3)_2$ solution and a lead electrode submerged in a $Pb(NO_3)_2$ solution. The anode is Mg and the cathode is Pb. The circuit is completed by using a salt bridge (or the arrangement with a porous cup shown in Figure 16.6a) and a wire between the Pb and Mg electrodes.

electrolytic cells

*Non*spontaneous reactions can be forced to occur by the application of an electric current. For example, in Figure 16.7, by introducing a source of electric current, the electron flow in the Zn-Cu system has been reversed. By using a power supply to pump electrons into the circuit, the zinc electrode becomes the cathode and zinc ions are reduced there. Copper metal is oxidized at the copper anode. This figure depicts an **electrolytic cell,** a cell in which a nonspontaneous reaction is made to occur by the application of electric energy (the process of **electrolysis**).

The combination of Na metal and Cl_2 gas to form NaCl is a spontaneous reaction. Conversely, the decomposition of the compound NaCl is nonspontaneous. In fact, the reverse of any spontaneous reaction is always nonspontaneous. However, the nonspontaneous reaction can be made to occur by passing an electric current through molten NaCl (Figure 16.8). The Na^+ cations are attracted to and reduced at the cathode; the Cl^- anions are attracted to and oxidized at the anode. The NaCl must be molten (liquid) so that the ions can flow and complete the circuit. Solid NaCl cannot be electrolyzed because the ions cannot move in the solid state.

electroplating

A common practical application of electrolysis is **electroplating.** Figure 16.9 shows the silver plating of a fork. The fork cathode and silver anode are immersed in a

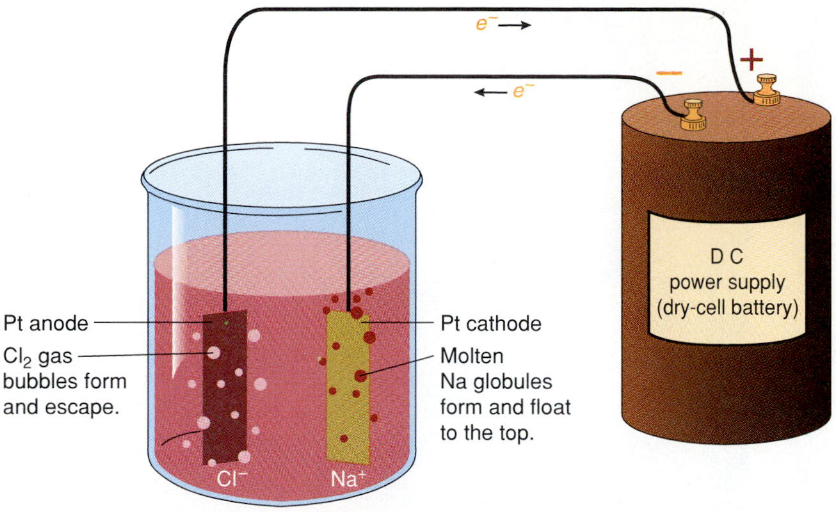

Figure 16.8

This apparatus shows the electrolysis of molten sodium chloride. A nonspontaneous reaction for the decomposition of NaCl takes place because energy added from a dc power supply forces the reaction to occur.

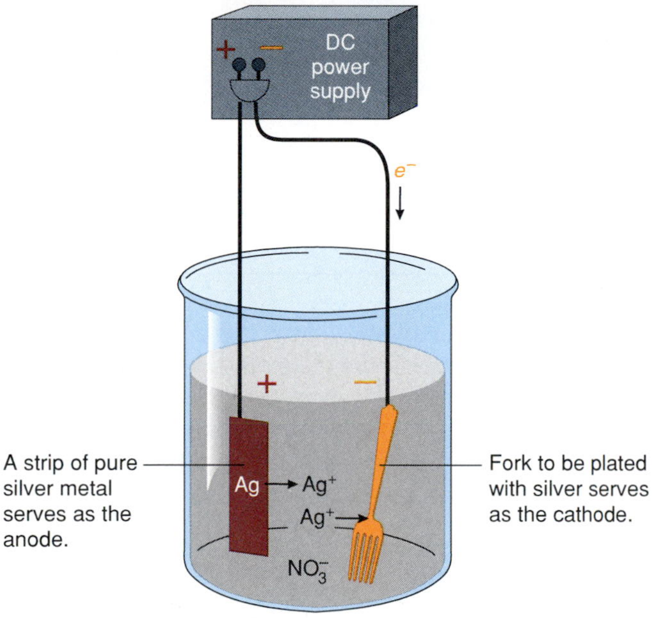

Figure 16.9

Silver metal is electroplated on a fork by reducing Ag^+ ions in solution. These Ag^+ ions in solution are replenished by the oxidation of the silver anode.

silver nitrate solution. Silver is plated onto the fork cathode as a result of the reduction of Ag^+ ions in solution. Silver ions in solution are replenished through the oxidation of silver at the anode.

$Ag^+ + e^- \rightarrow Ag$ **Cathode: reduction**

$Ag \rightarrow Ag^+ + e^-$ **Anode: oxidation**

Chrome trim on antique automobiles and home appliances is electroplated (chromeplated) by a similar procedure. In the case of chromeplating, the cathode is the metallic piece to be plated and the anode is chromium metal. Both are placed in an electrolyte solution of $Cr(NO_3)_3$ or $CrCl_3$. As with all nonspontaneous redox reactions, chromeplating requires the input of a considerable amount of energy, so the cost of chromeplating an automobile bumper or appliance has become economically prohibitive. Now, plastic materials replace the coverings on automobile bumpers and steam irons whenever possible.

Figure 16.10

Aluminum is produced at high temperature by the Hall process in an electrolytic cell that reduces bauxite (Al_2O_3), the principal ore of aluminum, to aluminum metal.

Figure 16.11

Since recycling aluminum requires less than one-tenth the energy of reducing its ore to metal, it is of great benefit, both economically and in the conservation of energy resources.

Another important application of electrolysis is the industrial production of aluminum. Since aluminum is an extremely versatile metal with low density, high strength, and excellent electrical conductivity, it is used widely in the construction of airplanes and in high-voltage transmission lines. Aluminum is produced at high temperature by the Hall process in an electrolytic cell that reduces bauxite (Al_2O_3), the principal ore of aluminum, to aluminum metal according to the equation,

$$\underset{\textbf{Bauxite}}{2\,Al_2O_3} \rightarrow 4\,Al(l) + 3\,O_2(g)$$

Reducing bauxite electrolytically to produce aluminum metal requires an unusually high input of energy, making the production of aluminum costly, both in terms of its market value and in terms of the energy resources it consumes (Figure 16.10). Since recycling aluminum requires less than one-tenth the energy of reducing its ore, it is of great benefit, both economically and in the conservation of energy resources (Figure 16.11).

sample exercise 13

A cooking utensil is to be copperplated. Describe the cathode and anode and write equations for the half-reactions occurring at the electrodes.

solution

In all electroplating procedures, the object to be plated is the cathode and the metal to be deposited is the anode. In this case then, the cooking utensil is the cathode and copper metal is the anode. The *cathode reaction is always the reduction* of the metal ion to the metal, and the *anode reaction is always the oxidation* of the metal to its cation. In this case,

$$Cu^{2+} + 2\,e^- \rightarrow Cu \qquad \textbf{Cathode: } \text{reduction}$$

$$Cu \rightarrow Cu^{2+} + 2\,e^- \qquad \textbf{Anode: } \text{oxidation}$$

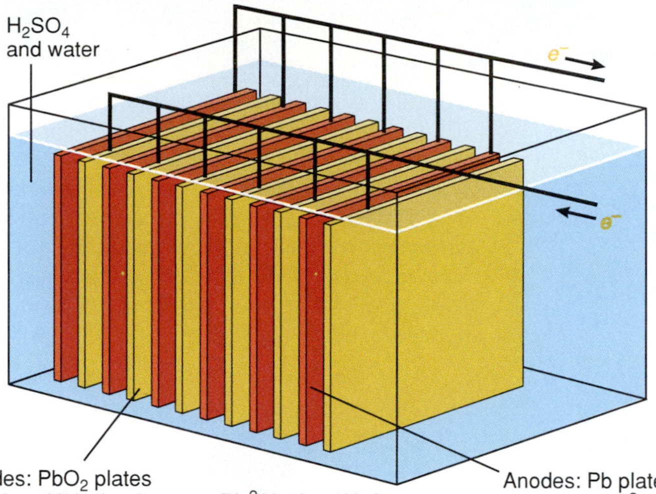

Figure 16.12
The lead storage battery used in your automobile produces a current when electrons released in the oxidation half-reaction at the Pb plates are transferred to the PbO_2 plates, where reduction occurs. Each voltaic cell produces 2 V. Your car battery is composed of six individual cells connected together to produce the standard 12 V.

lead storage batteries

Lead-acid batteries found in automobiles are the most widely used batteries; each year sales amount to about 10 billion dollars. The workings of your car battery are based on the principles of both voltaic and electrolytic cells. When your battery is used to start your car or power your headlights, cassette player, or automatic seat belt, it is operating as a voltaic cell with a spontaneous cell reaction.

$$Pb(s) + PbO_2(s) + 4\,H^+(aq) + 2\,SO_4^{2-}(aq) \rightarrow 2\,PbSO_4(s) + 2\,H_2O(l)$$

The half-reactions are

$$Pb(s) + SO_4^{2-}(aq) \rightarrow PbSO_4(s) + 2\,e^- \qquad \textbf{Anode: oxidation}$$

$$PbO_2(s) + 4\,H^+(aq) + SO_4^{2-}(aq) + 2\,e^- \rightarrow PbSO_4(s) + 2\,H_2O(l) \qquad \textbf{Cathode: reduction}$$

As you can see from the half-reactions, the battery has a lead (Pb) anode and a lead dioxide (PbO_2) cathode that are both immersed in an aqueous solution of sulfuric acid, which acts as the electrolyte. If you measure the voltage from this voltaic cell, you would find that it is only 2 V. Six of these voltaic cells are connected together to produce 12 V in a standard automobile battery (Figure 16.12).

When the same battery is being charged by your car's generator, the generator provides an electric current that reverses the battery reaction, causing the battery to act as an electrolytic cell. If the battery operated only as a voltaic cell, it would soon "run down" as the electrode plates are eaten away and sulfuric acid consumed. Reversing the cell reaction consumes the white, solid lead sulfate, $PbSO_4$, that has accumulated in the cell; it also rebuilds the anode by depositing Pb and rebuilds the cathode by depositing PbO_2. Caution should be used to prevent sparks and to use adequate ventilation when recharging a dead battery. During recharging, some water is decomposed (electrolyzed) into hydrogen and oxygen gas, a potentially explosive mixture (Figure 16.13).

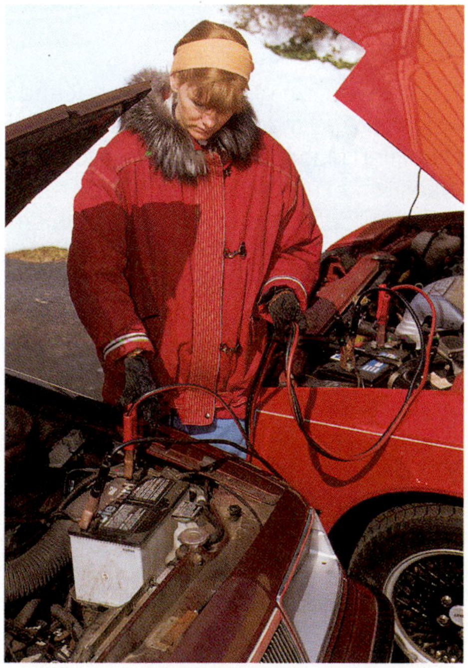

Figure 16.13
Caution should be used to prevent sparks and to provide adequate ventilation when recharging a dead battery. During recharging, some water is decomposed (electrolyzed) into hydrogen and oxygen gas, a potentially explosive mixture.

CHEMLAB

Laboratory exercise 16: electrochemical cells

In connection with single-replacement reactions, you have seen that some metals are more active than others. The basis of the activity series (Table 16.1) is the relative ease of oxidation of various metal atoms to cations. When the oxidation half-reaction of a more easily oxidized metal (stronger reducing agent) is coupled with the reduction half-reaction of a more easily reduced metal cation (stronger oxidizing agent), a spontaneous reaction occurs.

Questions:

1. Consult the activity series to find the relative activities of aluminum and silver. Describe how you would construct a voltaic cell with electrodes made of silver and aluminum. Describe the experimental setup of the cell, the composition of the cathode and anode, the solution(s) you would use, the direction of current flow, and a method for testing your cell.

   The nickel-cadmium cell (nicad cell) is commonly used as a rechargeable storage battery in calculators, shavers, and portable power tools (Figure 16.14). In such a cell, pure Ni is not used as the electrode; rather, hydrated nickel oxide hydroxide, NiOOH(*s*), is used. You may assume that hydrated nickel oxide hydroxide follows the same relative activity as pure Ni. Both nickel and cadmium metals form insoluble hydroxides with the metal having an oxidation number of +2.

2. Consult the activity series to find the relative activities of cadmium and nickel. Using basic conditions for the cell reaction, write half-cell reactions and the overall cell reaction for a nicad voltaic cell. Describe how you would construct a nicad cell. Also describe the composition of the anode and cathode, the solution(s) you would use, the direction of current flow, and a method for testing your cell. Hint: What is the oxidation number of Ni in NiOOH(*s*)?

Figure 16.14

The nickel-cadmium cell ("nicad" cell) is commonly used as a rechargeable storage battery in calculators, shavers, and portable power tools.

16.10 redox in the body

The conversion of the food we eat and the oxygen we breathe into energy for body processes is mediated by redox reactions (Figure 16.15). Figure 16.16 shows a simplified diagram of the respiratory chain that is the central focus of energy-transfer reactions in the body. The overall result of this very complicated stepwise process is the production of water from its elements, a highly exothermic reaction. The advantage of the net reaction is that it supplies a great deal of heat energy, but it does so in small steps so that it may be safely and efficiently used by the body.

Biological oxidation begins at the left side of Figure 16.16 with the oxidation of two H atoms to yield two H^+ ions and two electrons. The electrons continue to be transferred through a chain of Fe^{2+}/Fe^{3+} redox reactions until they reduce O atoms at the far right side of Figure 16.16. The two products of the respiratory chain, H^+ and O^{2-}, combine in the overall reaction to produce energy for the body according to the equation,

$$2\,H^+ + O^{2-} \rightarrow H_2O + \text{Energy}$$

In each step of the chain involving cytochromes, which are iron-containing biological molecules, iron ions are either oxidized or reduced. For example, the electrons

Figure 16.15

The conversion of the food you eat and the oxygen you breathe into energy for body processes is mediated by redox reactions.

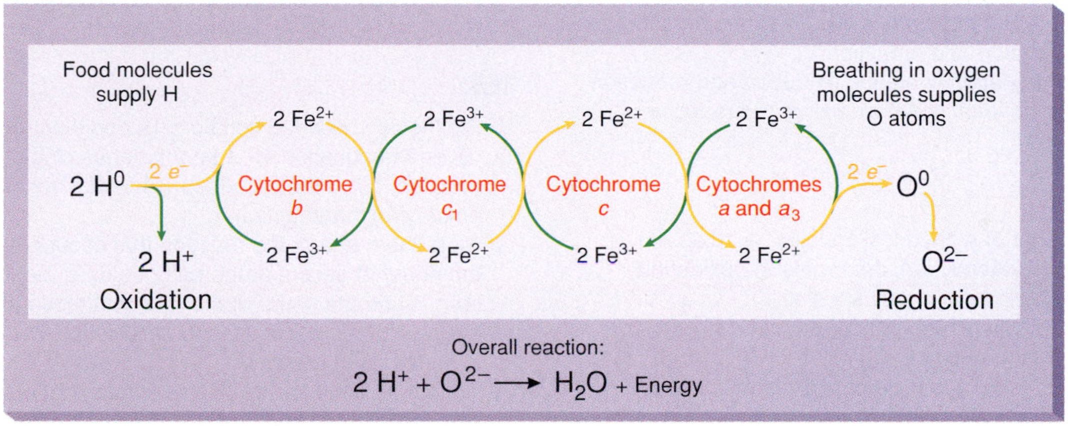

Figure 16.16

In the respiratory chain, oxidation and reduction occur simultaneously along the chain as electrons are lost and gained. Through the mediation of the cytochromes, the electrons, lost by hydrogen as it is oxidized, reduce the oxygen. The overall process is the exothermic production of water.

lost by two H atoms reduce two Fe^{3+} ions to Fe^{2+} in cytochrome *b*. Cytochromes *a* and a_3 lose electrons to oxygen, and in the process, Fe^{2+} ions are oxidized to Fe^{3+} ions.

As you have seen in Figure 16.16, redox reactions, particularly biochemical examples, are often represented by diagrams with intersecting arrows that may be confusing at first glance. It may help you to understand this arrow representation by representing the simple redox reaction,

$$Zn + Cu^{2+} \rightarrow Zn^{2+} + Cu$$

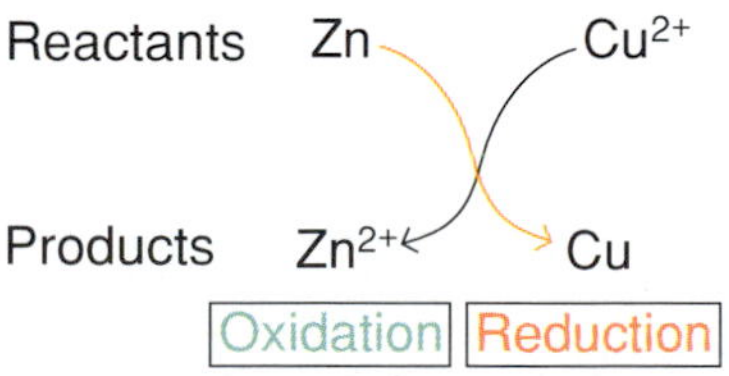

The oxidation of zinc is shown by the left downward arrow and the reduction of copper cations is shown by the right downward arrow. The two electrons transferred in the overall process pass from Zn in the upper left to Cu in the lower right along the yellow line. The point of intersection of arrows represents the electron transfer and consequent oxidation-reduction reaction.

The same type of electron transfer occurs at each of the intersections for the cytochrome molecules along the respiratory chain. For example, in Figure 16.6, one set of intersecting arrows depicts an electron being transferred from Fe^{2+} of cytochrome *b* to Fe^{3+} of cytochrome c_1. Fe^{2+} (cytochrome *b*) is oxidized to Fe^{3+} (cytochrome *b*), while Fe^{3+} (cytochrome c_1) is reduced to Fe^{2+} (cytochrome c_1). The overall path of electron transfer along the entire chain is illustrated by the colored line.

chapter accomplishments

After completing this chapter, you should be able to define all key terms and do the following.

16.2 Electron-transfer reactions

- ❑ State the relationship between electrons gained and electrons lost during any chemical reaction.
- ❑ Given an oxidation half-reaction and a reduction half-reaction, add them to obtain a balanced chemical equation.

16.3 Writing half-reactions

- ❑ Recognize half-reactions as oxidation or reduction.
- ❑ Define oxidizing and reducing agents in terms of materials undergoing oxidation and reduction.
- ❑ Given an ionic equation, write the oxidation and reduction half-reactions and identify the oxidizing and reducing agents.

16.4 Oxidation numbers

- ❑ Given the formula of a chemical species, such as a compound or polyatomic ion, determine the oxidation number of each element in that species.
- ❑ Distinguish between oxidation and reduction in terms of changes in oxidation number.
- ❑ Given a chemical equation, indicate the substances oxidized and reduced and identify the oxidizing and reducing agents.

16.5 Balancing redox reactions in acidic and neutral solutions

- ❑ Given an unbalanced equation for an oxidation-reduction reaction in acidic or neutral solution, write a balanced ionic equation using the half-reaction method.

16.6 Balancing redox reactions in basic solution

- ❑ Given an unbalanced equation for an oxidation-reduction reaction in basic solution, write a balanced ionic equation using the half-reaction method.

16.7 Alternate method for balancing redox reactions

- ❑ Given an unbalanced equation for an oxidation-reduction reaction, write a balanced equation using the oxidation-number method.

16.8 Activity series revisited

- ❑ State the relationships among the activity of a metal, the ease with which it is oxidized, and its strength as a reducing agent.
- ❑ Given an activity series, designate reactions as spontaneous or nonspontaneous based on the strengths of the oxidizing and reducing agents.

16.9 Electrochemistry: redox and energy

- ❑ Distinguish between voltaic cells and electrolytic cells.
- ❑ Given the equation for a spontaneous single-replacement reaction, describe the construction of a voltaic cell employing that reaction.
- ❑ State the name of the reaction that occurs at the anode and the one that occurs at the cathode in all electrochemical cells. State the name of the ions that move toward the anode and the name of those that move toward the cathode in electrochemical cells.
- ❑ Write equations for the half-reactions that occur during the electrolysis of a given molten ionic compound.
- ❑ Given a metal to be electroplated, write equations for the oxidation and reduction half-reactions that occur during the process.

16.10 Redox in the body

- ❑ Recognize oxidation-reduction reactions in the intersecting-arrow representation.

key terms

oxidation-reduction (redox)
oxidation
reduction
half-reaction
reducing agent
oxidizing agent
oxidation number
half-reaction method
oxidation-number method
spontaneously
voltaic cell
salt bridge
electrolytic (electrolysis) cell
electroplating

16.2 Electron-transfer reactions

8. Explain why oxidation-reduction reactions can be thought of as electron-transfer reactions.
9. Oxidation and reduction occur simultaneously in a chemical reaction. Explain why this must be true.
10. a. How many electrons are lost in the half-reaction, $Mg \rightarrow Mg^{2+} + 2\,e^-$?
    b. How many electrons are gained in the half-reaction, $Br_2 + 2\,e^- \rightarrow 2\,Br^-$?
    c. Add the preceding two half-reactions together.
11. a. If we simply add the following half-reactions, will the electrons lost equal the electrons gained?
    $Fe \rightarrow Fe^{3+} + 3\,e^-$
    $Br_2 + 2\,e^- \rightarrow 2\,Br^-$
    b. How can we add the two half-reactions so that the electron change will cancel out in the sum?

16.3 Writing half-reactions

12. Identify the following half-reactions as oxidation or reduction.
    a. $K^+ + e^- \rightarrow K$
    b. $Zn \rightarrow Zn^{2+} + 2\,e^-$
    c. $MnO_4^- + 8\,H^+ + 5\,e^- \rightarrow Mn^{2+} + 4\,H_2O$
    d. $3\,SO_2 + 6\,H_2O \rightarrow 3\,HSO_4^- + 9\,H^+ + 6\,e^-$
13. a. Give a definition of oxidizing and reducing agents.
    b. For each of the half-reactions in Problem 12, indicate whether the reactant is an oxidizing or reducing agent.
14. Identify the following half-reactions as oxidation or reduction and identify the reactant as an oxidizing or reducing agent.
    a. $Br_2 + 2\,e^- \rightarrow 2\,Br^-$
    b. $Pb^{2+}(aq) + 2\,e^- \rightarrow Pb(s)$
    c. $SO_3^{2-} + H_2O \rightarrow SO_4^{2-} + 2\,H^+ + 2\,e^-$
    d. $Fe^{2+} \rightarrow Fe^{3+} + e^-$
15. For each of the following ionic equations, write the oxidation and reduction half-reactions and identify the oxidizing and reducing agents.
    a. $2\,Fe^{3+}(aq) + Sn^{2+}(aq) \rightarrow 2\,Fe^{2+}(aq) + Sn^{4+}(aq)$
    b. $Zn(s) + 2\,H^+(aq) \rightarrow Zn^{2+}(aq) + H_2(g)$
    c. $2\,Fe^{3+}(aq) + 3\,Zn(s) \rightarrow 2\,Fe(s) + 3\,Zn^{2+}(aq)$
16. For each of the following unbalanced ionic equations, write the oxidation and reduction half-reactions and identify the oxidizing and reducing agents.
    a. $Al(s) + Pb^{2+}(aq) \rightarrow Al^{3+}(aq) + Pb(s)$
    b. $SO_4^{2-}(aq) + Zn(s) \rightarrow Zn^{2+}(aq) + SO_2(g)$
    c. $I^-(aq) + NO_3^-(aq) \rightarrow I_2(s) + NO(g)$

16.4 Oxidation numbers

17. Determine the oxidation numbers of the following elements.
    a. Each element in K_2SO_4
    b. Cl in ClO_3^-
    c. Cl in Cl_2
    d. N in NH_4^+
    e. N in NH_3
    f. P in H_3PO_4
    g. C in CH_4
18. Determine the oxidation number of N in each of the following.
    a. N_2
    b. NO
    c. NO_2
    d. N_2O
    e. N_2O_4
    f. NO_2^-
    g. NO_3^-
    h. NH_3
19. Determine the oxidation number of Cl in each of the following.
    a. Cl_2
    b. Cl_2O
    c. Cl_2O_3
    d. Cl_2O_7
    e. Cl_2O_5
20. State the oxidation number of each element in the following compounds.
    a. Na_2SO_4
    b. $KMnO_4$
    c. H_2SO_4
    d. C_2H_6
21. Give the formulas of species having the following oxidation numbers for hydrogen.
    a. +1
    b. −1
    c. 0
22. Give a definition of oxidation and reduction in terms of oxidation-number changes.
23. In the following unbalanced equations, determine the substances oxidized and reduced and identify the oxidizing and reducing agents.
    a. $HNO_3 + HI \rightarrow NO + I_2 + H_2O$
    b. $Bi(OH)_3 + K_2SnO_2 \rightarrow Bi + K_2SnO_3 + H_2O$
    c. $I_2O_5 + CO \rightarrow I_2 + CO_2$
    d. $HCl + Mg \rightarrow MgCl_2 + H_2$
    e. $H_2 + O_2 \rightarrow H_2O$
24. For each of the following equations, determine the substances oxidized and reduced and identify the oxidizing and reducing agents.
    a. $NaNO_3 + Pb \rightarrow NaNO_2 + PbO$
    b. $Na_2SO_4 + 4\,C \rightarrow Na_2S + 4\,CO$
    c. $NH_4NO_2 \rightarrow N_2 + 2\,H_2O$
    d. $As_2O_3 + Cl_2 + H_2O \rightarrow H_3AsO_4 + HCl$

16.5 Balancing redox reactions in acidic and neutral solutions

25. Balance the following half-reactions (solutions are acidic or neutral).
    a. $NO(g) \rightarrow NO_3^-(aq)$
    b. $MnO_4^-(aq) \rightarrow Mn^{2+}(aq)$
    c. $Cr_2O_7^{2-} \rightarrow Cr^{3+}$
    d. $C_2O_4^{2-} \rightarrow CO_2$
26. Write a balanced net ionic equation (if it exists) for the following oxidation-reduction reactions (solutions are either acidic or neutral).
    a. $Zn(s) + HCl(aq) \rightarrow H_2(g) + ZnCl_2(aq)$
    b. $MnO_2(s) + HBr(aq) \rightarrow MnBr_2(aq) + Br_2(aq) + H_2O(l)$
    c. $HNO_3(aq) + I_2(s) \rightarrow NO_2(g) + H_2O(l) + HIO_3(aq)$
    d. $K_3AsO_4(aq) + HI(aq) \rightarrow K_3AsO_3(aq) + I_2(aq)$
    e. $H_2S(aq) + HNO_3(aq) \rightarrow S(s) + NO(g) + H_2O(l)$
    f. $KClO_3(aq) + HI(aq) \rightarrow KCl(aq) + I_2(aq)$
    g. $Cu(s) + H_2SO_4(aq) \rightarrow CuSO_4(aq) + SO_2(g) + H_2O(l)$
    h. $FeSO_4(aq) + KMnO_4(aq) + H_2SO_4(aq) \rightarrow Fe_2(SO_4)_3(aq) + MnSO_4(aq) + H_2O(l) + K_2SO_4$
    i. $H_2(g) + FeCl_3(aq) \rightarrow HCl(aq) + FeCl_2(aq)$
    j. $C_3H_8(g) + O_2(g) \rightarrow CO_2(g) + H_2O(l)$
    k. $C_6H_{12}O_6(aq) + O_2(g) \rightarrow CO_2(g) + H_2O(l)$
27. Write a balanced net ionic equation for the following oxidation-reduction reactions (solutions are either acidic or neutral).
    a. $Ag + HNO_3 \rightarrow AgNO_3 + NO + H_2O$
    b. $HNO_2 + HI \rightarrow NO + I_2 + H_2O$
    c. $Pb(NO_3)_2 + S + H_2O \rightarrow Pb + H_2SO_3 + HNO_3$
    d. $H_2O_2 + HI \rightarrow I_2 + H_2O$

16.6 Balancing redox reactions in basic solution

28. Balance the following half-reactions (solutions are basic).
    a. $Zn \rightarrow Zn(OH)_4^{2-}$
    b. $Sn \rightarrow HSnO_2^-$
    c. $BrO_4^- \rightarrow Br^-$
    d. $CN^- \rightarrow CNO^-$
29. Write a balanced net ionic equation for the following oxidation-reduction reactions (solutions are basic).
    a. $Zn(s) + KMnO_4(aq) \rightarrow Zn(OH)_2(aq) + MnO_2(s) + KOH(aq)$
    b. $Cd(s) + NiOOH(aq) \rightarrow Cd(OH)_2(aq) + Ni(OH)_2(aq)$
    c. $K_2S(aq) + I_2(aq) + KOH(aq) \rightarrow K_2SO_4(aq) + KI(aq)$
    d. $NaOH(aq) + Br_2(aq) \rightarrow NaBrO_3(aq) + NaBr(aq)$
    e. $Bi(OH)_3(aq) + K_2SnO_2(aq) \rightarrow K_2SnO_3(aq) + Bi(s)$
    f. $Al(s) + KNO_3(aq) \rightarrow NH_3(aq) + KAlO_2(aq)$
30. Write a balanced net ionic equation for the following oxidation-reduction reactions (solutions are basic).
    a. $Cr(OH)_3 + NaClO + NaOH \rightarrow Na_2CrO_4 + NaCl + H_2O$
    b. $KMnO_4 + NaClO_2 \rightarrow MnO_2 + NaClO_4$
    c. $Cr^{3+} + ClO^- \rightarrow CrO_4^{2-} + Cl^-$
    d. $CoCl_2 + Na_2O_2 \rightarrow Co(OH)_3 + NaCl$

16.7 Alternate method for balancing redox reactions

31. Balance the following reactions by the oxidation-number method.
    a. $Cu(s) + HNO_3(aq) \rightarrow Cu(NO_3)_2(aq) + NO(g) + H_2O(l)$
    b. $K_2Cr_2O_7(aq) + H_2O(l) + S(s) \rightarrow SO_2(g) + KOH(aq) + Cr_2O_3(aq)$
    c. $H_2O(l) + P_4(s) + HOCl(aq) \rightarrow H_3PO_4(aq) + HCl(aq)$
    d. $PbO_2(aq) + HI(aq) \rightarrow PbI_2(aq) + I_2(aq) + H_2O(l)$
    e. $HNO_3(aq) + H_2S(aq) \rightarrow NO(g) + S(s) + H_2O(l)$
32. Balance the following reactions by the oxidation-number method.
    a. $Br_2 + H_2O + SO_2 \rightarrow HBr + H_2SO_4$
    b. $KClO + H_2 \rightarrow KCl + H_2O$
    c. $SnSO_4 + FeSO_4 \rightarrow Sn + Fe_2(SO_4)_3$
    d. $HNO_2 + HI \rightarrow NO + I_2 + H_2O$

16.8 Activity series revisited

33. Use Table 16.1 to predict which of the following reactions will occur spontaneously.
    a. $2\,Al(s) + 3\,CuSO_4(aq) \rightarrow 3\,Cu(s) + Al_2(SO_4)_3(aq)$
    b. $3\,Ag(s) + Al(NO_3)_3(aq) \rightarrow Al(s) + 3\,AgNO_3(aq)$
    c. $Fe(s) + ZnSO_4(aq) \rightarrow Zn(s) + FeSO_4(aq)$
    d. $Fe(s) + 2\,AgNO_3(aq) \rightarrow 2\,Ag(s) + Fe(NO_3)_2(aq)$
34. Write the half-reactions for the spontaneous reactions you identified in Problem 33.
35. a. Use Table 16.1 to give the name of a metal that is a better reducing agent than Mg.
    b. Use the table to give the names of two metals that will not spontaneously react with HCl.
36. A gold ring can be immersed in a sodium chloride solution or even a concentrated hydrochloric acid solution without damage to the ring. Explain how Table 16.1 enables you to predict these facts.

16.9 Electrochemistry: redox and energy

37. Describe the construction of voltaic cells that might be made employing all appropriate reactions in Problem 33.
38. What is the function of a salt bridge in voltaic and electrolytic cells?
39. A student constructs a battery by immersing an aluminum strip in $Al(NO_3)_3$ solution and a tin strip in $Sn(NO_3)_2$ solution. A salt bridge is used, and the setup is wired in a manner similar to the apparatus in Figure 16.6b.
    a. Which metal is the anode and which is the cathode?
    b. Write the half-reactions and the total cell reaction.
    c. Which electrode becomes heavier as the battery runs?
40. A common flashlight battery works through the coupling of the two half-reactions, $Zn(s) \rightarrow Zn^{2+}(aq) + 2\,e^-$

    $MnO_2(s) + NH_4^+(aq) + e^- \rightarrow MnO(OH)(s) + NH_3(aq)$
    a. Which reaction occurs at the anode and which at the cathode?
    b. Of what material is the anode made?

41. Which reactions in Problem 33 could be induced to occur in an electrolytic cell?

42. The method by which aluminum metal is claimed from its ore involves the electrolysis of molten Al_2O_3.

a. Write the two half-reactions that occur during electrolysis.

b. Does aluminum metal form at the anode or cathode?

43. Given an ordinary metal bracelet, a strip of gold, and a solution of $Au(NO_3)_3$, describe how you would plate the bracelet with gold.

44. Write the anode and cathode reactions that would occur during the goldplating described in Problem 43.

45. The half-reactions shown in Figure 16.12 are those that occur spontaneously and allow the battery to supply current.

a. Which plate is the anode and which is the cathode?

b. Write equations for the half-reactions that occur when a car's generator recharges a battery.

c. During recharging, which plate is the anode and which is the cathode?

16.10 Redox in the body

46. a. Write the half-reactions for the following equation.

$$Mg(s) + Pb(NO_3)_2(aq) \rightarrow Mg(NO_3)_2(aq) + Pb(s)$$

b. Write the half-reactions determined in part a using the intersecting-arrow representation.

47. In the following representation, CoQ is coenzyme Q, an important intermediary in the respiratory chain.

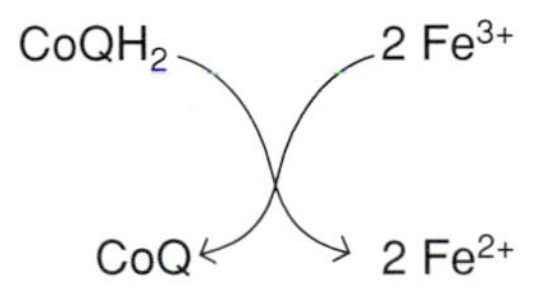

a. What is the oxidizing agent in this reaction? What is the reducing agent?

b. Write the traditional ionic equation for the above intersecting-arrow representation.

Additional problems

48. Sodium metal reacts with bromine to form sodium bromide. Identify the oxidation reaction, reduction reaction, oxidizing agent, and reducing agent.

49. Identify the following changes as oxidation or reduction.

a. $Zn^{2+} \rightarrow Zn$

b. $2\ I^- \rightarrow I_2$

c. $Sn^{2+} \rightarrow Sn^{4+}$

50. The miniature, button-shaped batteries found in watches, calculators, and cameras derive their energy from the following reaction, which takes place in basic medium. Balance the equation.

$$Zn(s) + Ag_2O(s) \rightarrow ZnO(s) + Ag(s)$$

51. What is the function of a reducing agent? Explain why metals tend to be good reducing agents.

52. Copper(II) sulfide dissolves readily in nitric acid, but only slowly in hydrochloric acid. The nitrate ion, but not the chloride ion, can act as a powerful oxidizing agent. Balance the following equation for the reaction with nitric acid.

$$CuS(s) + HNO_3(aq) \rightarrow Cu(NO_3)_2(aq) + S(s) + NO(g) + H_2O(l)$$

53. One of the breath analyzer tests for alcohol depends on the color change exhibited by dichromate, $Cr_2O_7^{2-}$ (bright orange), being reduced to Cr^{3+} (green). Balance the following equation for this reaction.

$$HCl(aq) + K_2Cr_2O_7(aq) + C_2H_5OH(l) \rightarrow CrCl_3(aq) + CO_2(g) + H_2O(l) + KCl(aq)$$

54. Aqua regia, sometimes called royal water, is a mixture of concentrated nitric and hydrochloric acids that dissolves the noble metals, gold and platinum. Balance the following equation for the dissolution of gold.

$$Au(s) + HNO_3(aq) + HCl(aq) \rightarrow HAuCl_4(aq) + NO_2(g)$$

55. An ordinary dry-cell or flashlight battery functions in a basic medium and derives its electrical energy from the following reaction. Balance the equation. Note that ammonium hydroxide, $NH_4OH(aq)$, is not stable and decomposes to $NH_3(aq) + H_2O(l)$.

$$Zn(s) + NH_4Cl(aq) + MnO_2(s) \rightarrow ZnCl_2(aq) + NH_3(aq) + Mn_2O_3(s) + H_2O(l)$$

56. Consider the following two half-reactions.

$$e^- + \text{Oxidizing agent}_1 \rightarrow \text{Reducing agent}_1$$

$$e^- + \text{Oxidizing agent}_2 \rightarrow \text{Reducing agent}_2$$

You are informed that reducing agent_1 is stronger than reducing agent_2.

a. Will the following reaction occur spontaneously or will the reverse occur spontaneously?

$$\text{Oxidizing agent}_1 + \text{Reducing agent}_2 \rightarrow \text{Reducing agent}_1 + \text{Oxidizing agent}_2$$

b. With respect to acid-base strength, what similarity exists between the above reaction and acid-base, conjugate acid-base reactions?

57. a. Use Table 16.1 to choose a reducing agent that will convert Cd^{2+} to Cd, but not Mg^{2+} to Mg.

b. Use the table to choose a metal that will react with $CaCl_2$ to form calcium metal.

58. Silver tarnishes by forming silver sulfide. A method of removing the tarnish is to place the silver object in an aluminum pan that has been polished to remove its aluminum oxide coating. The pan is filled with water and a small amount of table salt is added. Soon the silver is bright and shiny, free of silver sulfide coating. Write a balanced chemical equation that explains the chemistry taking place.

59. The amount of sulfurous acid present in rainwater (from the dissolved pollutant SO_2) can be determined by an oxidation titration with permanganate.

$$H_2SO_3(aq) + KMnO_4(aq) \rightarrow K_2SO_4(aq) + MnO(s)$$

a. Balance the equation.

b. Calculate the number of moles of H_2SO_3 in a 50.0 mL sample of rainwater if 8.12 mL of a 0.00850 M $KMnO_4$ solution is used in the titration.

c. Calculate the number of grams of SO_2 present in the 50.0 mL sample of rainwater.

chapter 17

nuclear chemistry

This technician is shown maneuvering a cylindrical casket of fruit above an accelerator pool used for the study of the irradiation of food.

17.1 introduction

The chemistry you have learned from this book involves changes in the electron structure of atoms, ions, and molecules. The number and/or arrangement of valence electrons in atoms change as chemical reactions occur, but the number of protons and neutrons in each nucleus remains unaltered. However, the nuclei of some isotopes are unstable and undergo changes in their numbers of protons and/or neutrons during a process called radioactive decay, or **radioactivity**. Such isotopes are called **radioisotopes**.

Isotopes of the same element have the same atomic number, but different mass numbers; that is, they have the same number of protons but a different number of neutrons in their nuclei (Section 4.5). The tendency of an isotope to undergo radioactive decay is directly related to the stability of its nucleus with extra neutrons. In Section 17.3, you will learn in detail the various ways in which radioactive decay may occur.

Radioactive decay is characteristically different from traditional chemical reactions. Instead of a rearrangement of valence electrons of the elements, nuclear reactions involve a rearrangement of nuclear structure, that is, the protons and neutrons in the nuclei. As a result, new elements or new isotopes of the same element may be formed. If electrons are involved, they are not the valence electrons of the atoms, but electrons that have resulted from a restructuring of the nucleus. Furthermore, particles or radiation (Section 17.3) may be emitted to conserve mass and energy during a nuclear reaction.

Nuclear chemistry has many valuable applications in medicine, agriculture, and home safety devices. It is also used as an energy resource to generate electricity in power stations and satellites, and it has scientific importance in studying the age of archaeological artifacts and rates of chemical reactions. However, radioactive materials can be hazardous and must be used under carefully monitored conditions (Figure 17.1). Radioactive waste products may also be hazardous for extremely long periods of time and pose serious questions about their safe disposal. Radioactive wastes are classified as those of low level and high level. In the United States, high-level wastes are stored at a special site in Hanford, Washington. In the future, a site for the disposal of low-level waste generated within each state must be found within the state.

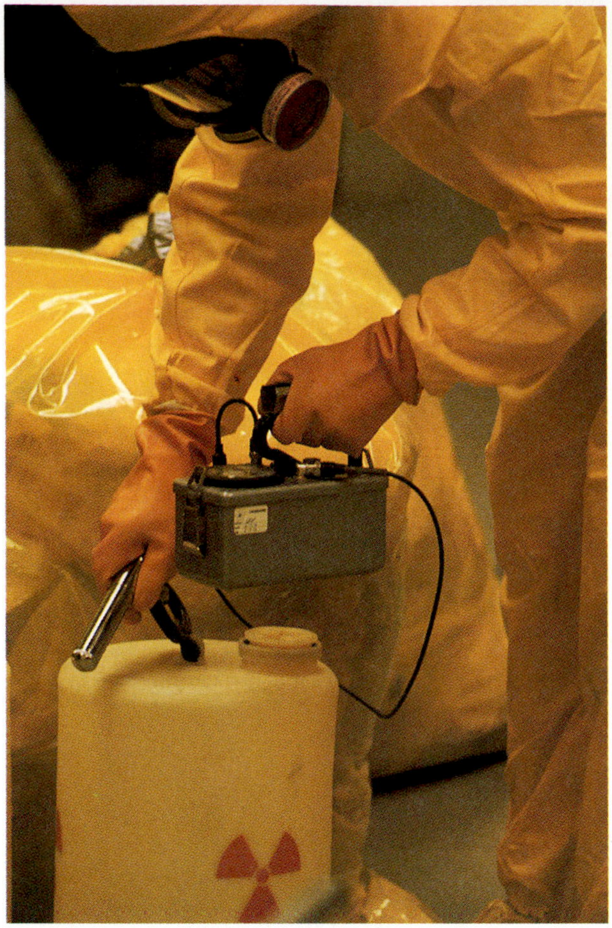

(a)

(b)

Figure 17.1

(a) Radioactive materials are hazardous and must be used under carefully monitored conditions. (b) Film badges are often worn by workers to monitor the amount of radiation to which they are exposed. The principle is derived from Becquerel's observation, that radiation emitted by radioactive elements, exposes undeveloped film. The worker shown is wearing a bell-like radioactive contamination detector near nuclear waste disposal.

17.2 the nucleus and radioactive decay

In Sections 4.5 to 4.7, you learned that the *atomic number* of an element is the number of protons in each nucleus of atoms of that element. The *mass number* is the sum of the number of protons and the number of neutrons in an atom. Consequently,

Number of neutrons = Mass number – Atomic number

Information about the content of a nucleus is conveniently given by a symbol of the form

$$^{\text{mass number}}_{\text{atomic number}}\text{Sy}$$

for example, $^{12}_{6}C$, $^{90}_{38}Sr$, and $^{235}_{92}U$. Because the atomic number is identical for all atoms of a particular element, it is frequently omitted from these symbols, for example, ^{12}C, ^{90}Sr, and ^{235}U. Should an atomic number be needed, the symbol (C, Sr, or U) enables us to find it in the periodic table. An alternative symbolism is carbon-12, strontium-90, and uranium-235, in which the mass number follows the name of the element.

Along with the symbol, the mass number of an atom must always be specified to uniquely identify the isotope. Whereas the number of protons in atoms of a particular element is invariable (and is given from the atomic symbol alone), the number of

Figure 17.2

The isotopes uranium-235 and uranium-238 are separated because their mass differences result in different rates of gaseous effusion of their compounds, $^{235}UF_6$ and $^{238}UF_6$. This photo shows individual units in the separation apparatus. About 2,000 stages are needed to enrich ^{235}U to 99% purity.

neutrons may vary. As you recalled in Section 17.1, atoms with the same atomic number (same number of protons) but different mass numbers (different numbers of neutrons) are called *isotopes.* For example, the most abundant isotope of uranium is $^{238}_{92}U$; it has 92 p^+ and 146 ^1_0n, where p^+ and ^1_0n represent protons and neutrons, respectively. The isotope of uranium that is most useful as a nuclear fuel is ^{235}U (92 p^+, 143 ^1_0n). The two isotopes of uranium are physically separated from each other by differences in the rates of flow of $^{235}UF_6$ and $^{238}UF_6$ (uranium hexafluoride) through the microscopic holes of a porous material (gaseous effusion). The rate of effusion of a gas is inversely proportional to the square root of its mass, so $^{235}UF_6$, with its lighter mass, would effuse slightly more rapidly than the heavier $^{238}UF_6$. After several thousand separations, $^{235}UF_6$ can be obtained with a purity of about 99% (Figure 17.2).

All isotopes with 84 or more protons are unstable and exhibit radioactive decay. Stable isotopes with 20 or fewer protons generally have a neutron/proton ratio, $^1_0n/p^+$, of about 1.0 or slightly greater. Isotopes with 21 to 83 protons are stable with a $^1_0n/p^+$ ratio up to about 1.5. For elements in general, the stability of an isotope is most reliably related to the ratio of neutrons to protons in the nucleus. If the nucleus formed in the decay process is stable, no further radioactive decay occurs. If the new nucleus is not stable, however, additional decay will occur until a stable nucleus is produced.

As an example of stable and unstable isotopes, consider the three isotopes of hydrogen, ^1_1H, ^2_1H, called deuterium, and ^3_1H, called tritium, each containing one proton in its nucleus. Since the $^1_0n/p^+$ ratios of these isotopes have values 0, 1, and 2, only the third isotope, tritium, is predicted to be unstable and is found to be the only isotope of hydrogen to undergo radioactive decay.

The most abundant isotope of carbon, $^{12}_6C$, has no tendency to undergo such a nuclear change. Its $^1_0n/p^+$ ratio is 1. Carbon is typical of the common elements; the most abundant isotopes of the common elements are not radioactive, but usually a small amount of some radioisotope (generally of higher mass) does exist. For example, carbon-14, $^{14}_6C$, is radioactive with six protons and eight neutrons in each nucleus. Its $^1_0n/p^+$ ratio has a value of 1.3.

Upon radioactive decay, each nucleus of carbon changes in such a way that it acquires an extra proton by converting the carbon atom (with six protons) to a nitrogen atom (with seven protons). The number of valence electrons has not changed, but the identity of the element has changed because of a new nuclear structure.

$$^{14}_{6}C \rightarrow ^{14}_{7}N \qquad (17.1)$$

Note that the mass number is conserved in this process. Since the number of protons in the nucleus increases from six to seven, the number of neutrons must correspondingly decrease from eight to seven. The net change seems to be a neutron being converted into a proton. However, nuclear reactions, like chemical reactions, must display a conservation of charge. A neutron (with a 0 charge) may produce a proton (with a 1+ charge) if, in addition, it produces a particle (with a 1– charge) to balance the charge. As a result, Equation 17.1 is not a balanced nuclear reaction because charge has not been conserved. However, if you add to the right side a particle with the same mass and charge as an electron, $^{0}_{-1}e$, with mass number = 0 and "atomic number" = –1,

$$^{14}_{6}C \rightarrow ^{14}_{7}N + ^{0}_{-1}e \qquad (17.2)$$

you have an equation for radioactive decay that conserves both mass number and charge. Careful inspection of Equation 17.2 shows that the sum of the mass numbers equals 14 on both sides and the sum of the atomic numbers equals 6 on both sides. This observation suggests two guidelines that will become important in writing all nuclear equations (Section 17.5).

Guidelines for writing nuclear equations

1. *The sum of the mass numbers on each side of the equation must be equal.*
2. *The sum of the atomic numbers on each side of the equation must be equal.*

A more detailed description of nuclear reactions will be given in Section 17.5. This simple example of the radioactive decay of carbon-14 serves to illustrate that nuclear reactions, like traditional chemical reactions, must conserve both mass number and charge.

While only one isotope of carbon is radioactive, all isotopes of the element uranium (atomic number 92) are radioactive. Nuclei such as those of uranium that undergo nuclear reactions do so in the same exact manner regardless of the nature of their surrounding electron structure. That is, the nuclei of free uranium metal atoms, U, which have the electron structure characteristic of that element, undergo nuclear reactions in the same way as nuclei of chemically combined, and thus electronically altered, uranium atoms, such as those in UF_6. As a result, in considering nuclear chemistry, we will completely ignore the electrons of the elements and look only at the content of the nucleus—the protons and neutrons.

• • • problem 1

Consider the isotopes $^{21}_{11}Na$, $^{27}_{13}Al$, ^{23}Na, ^{79}Br, ^{206}Pb, and ^{241}Am and indicate the numbers of protons and neutrons in the nuclei of each.

17.3 radioactivity

Just as traditional chemical reactions are accompanied by energy changes, nuclear reactions are also accompanied by energy changes, but of much greater magnitude. The spontaneous emission of energy or particles carrying energy that accompanies

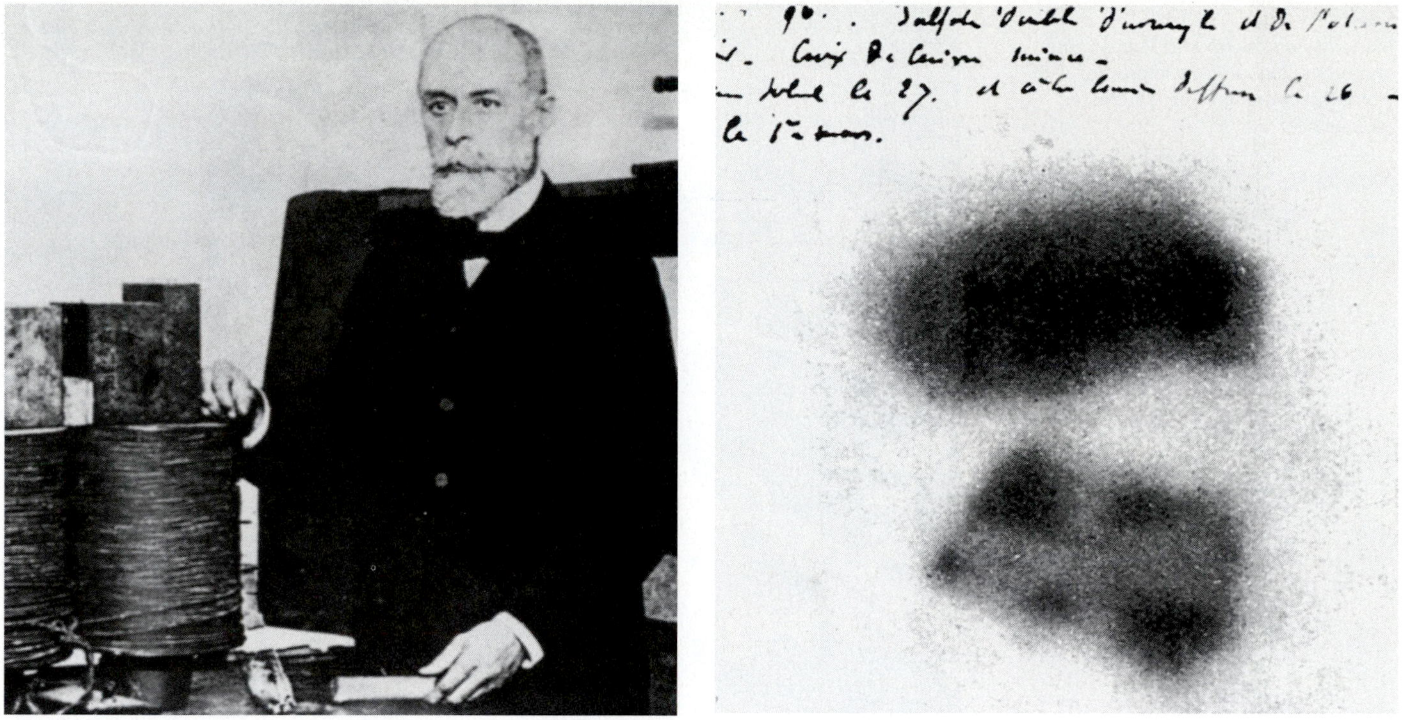

(a) (b)

Figure 17.3
(a) Observations by Becquerel led him to conclude that the nuclei of specific atoms, such as uranium, underwent changes that produced radiation (energy). Shown with Becquerel is his original photographic plate (b) exposed to uranium.

nuclear change is called *radioactivity.* It was the accidental detection of the emission of energy from a uranium salt that led ultimately to knowledge of nuclear reactions. In 1896 in France, Henri Becquerel happened to place a sample of a uranium salt on top of a photographic plate wrapped in opaque, black paper in a dark drawer. Several days later, he developed the photographic plate and found the image of the outlines of the uranium crystal on the plate. He concluded that it was the uranium sample that gave off radiant energy since the photographic plate had been subjected to no other source of radiation. Becquerel soon found that both uranium metal and uranium salts produce this phenomenon, and the amount of radiation was directly proportional to the quantity of uranium *only.* The radiant energy was produced in exactly the same manner by the uranium atoms alone, regardless of the nature of their surrounding electron structure. He therefore concluded that it was changes in the uranium nuclei that were totally responsible for the radiation (Figure 17.3).

Also in France, Marie (Sklodowska) and Pierre Curie took up the study of the mysterious emissions from uranium, and it was they who coined the term *radioactive* to describe elements that spontaneously give off radiation. In the course of their studies of radioactivity, they discovered two previously unknown elements, polonium (Po) (named for Marie's homeland, Poland) and radium (Ra). In 1903, the Curies shared the Nobel Prize in physics with Henri Becquerel for their discoveries (Figure 17.4).

Meanwhile, in England, Ernest Rutherford was also studying radiation, and he was able to establish two distinct types of radiation energy, which he called *alpha* (α) and *beta* (β) *radiation.* (Alpha and beta are the first two letters of the Greek alphabet.) His experiments also showed that both of these two types of nuclear radiation involve the emission of particles from the nucleus.

The existence of a third type of radiation, also named by Rutherford, does not involve particles. This *gamma* (γ) *radiation* is named for the third letter of the Greek alphabet. An experimental setup that identifies the three types of radiation is shown

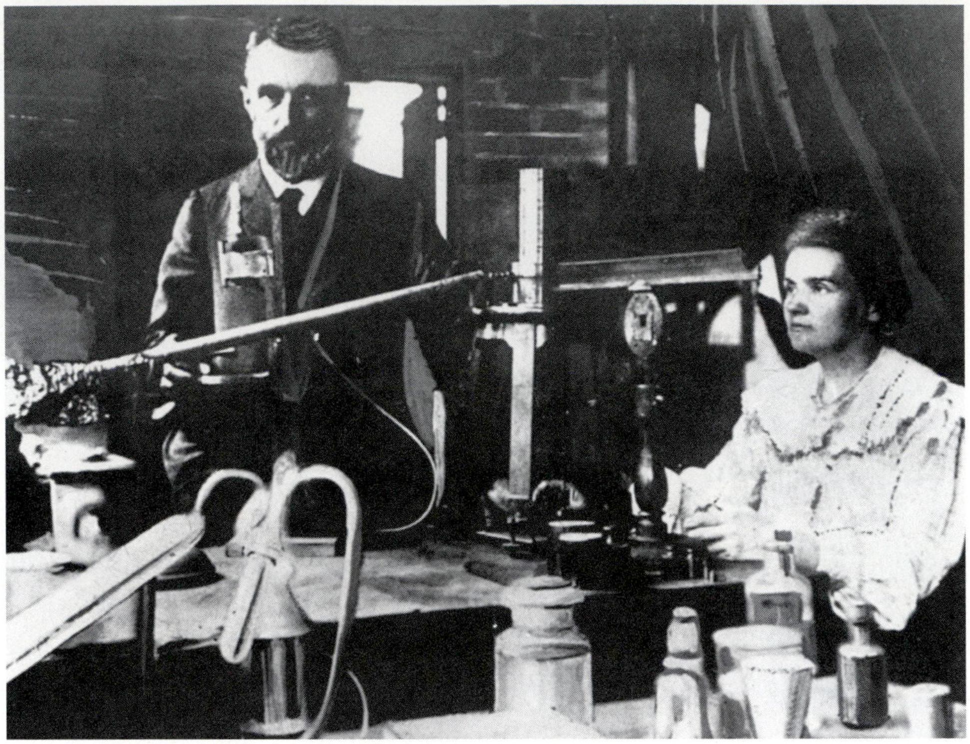

Figure 17.4

Marie (Sklodowska) and Pierre Curie shared the Nobel Prize in physics with Henri Becquerel for their studies of radioactivity.

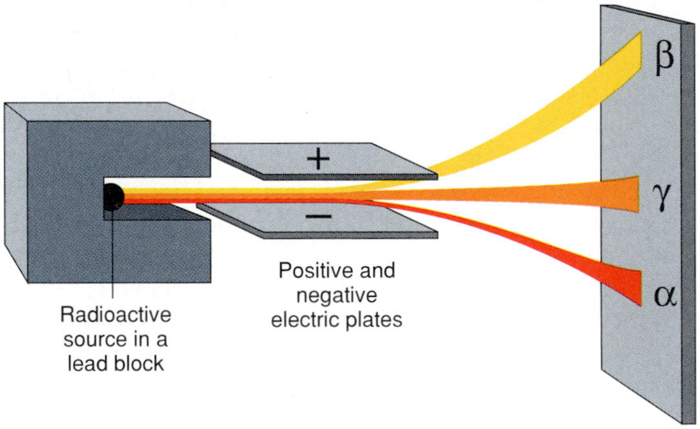

Figure 17.5

Passing a beam of radiation between electric plates will result in one or more of the following observations. If the radiation is unaffected by the electric plates, it is *gamma* (γ) radiation; if the radiation is bent toward the positive plate, it is *beta* (β) radiation; if the radiation is bent toward the negative plate, it is *alpha* (α) radiation. Any source of radiation will have its own characteristic combination of one or more of these types of radiation.

in Figure 17.5. When a single beam of nuclear radiation is passed between electric or magnetic plates, the original beam splits into three separate beams. One beam, α, bends toward the negative plate, indicating that this beam carries a positive charge. Another beam, β, bends toward the positive plate, indicating that it carries a negative charge. The third beam, γ, is uncharged because it is unaffected by the electric plates and passes along the same line as the original emission. The additional observation that β-radiation is deflected more than α-radiation (Figure 17.5) indicates that β-radiation has less mass than α-radiation. These combined observations lead directly to the properties of α-, β-, and γ-radiation discussed in the following section.

table 17.1 characteristics of the three types of nuclear radiation

Type of radiation	Symbols	Composition	Charge	Relative mass (amu)	Penetrating power
Alpha	α, ${}^{4}_{2}\text{He}$	Identical to the He nucleus	2+	4	Low
Beta	β, ${}^{0}_{-1}e$	Identical to the electron	1–	$\frac{1}{1,840}$	Moderate
Gamma	γ, ${}^{0}_{0}\gamma$	High energy	0	0	High

17.4 properties of α-, β-, and γ-radiation

The experiment shown in Figure 17.5 clearly shows that α-radiation is positive because it is attracted toward the negative plate, β-radiation is negative because it is attracted toward the positive plate, and γ-radiation is electrically neutral since it is attracted to neither plate. The magnitudes of these charges were also determined by experiment and found to be $\alpha = 2+$, $\beta = 1-$, and $\gamma = 0$, using the same relative scale as for subatomic particles.

Since α-radiation has twice the charge of β-radiation, you would expect that it would be attracted more strongly toward the oppositely charged plate than β-radiation. However, the experiment in Figure 17.5 shows that β-radiation is attracted to the oppositely charge plate much more strongly. Basic laws of physics have been used to explain that the mass of α-radiation is much larger than the mass of β-radiation. Table 17.1 indicates that the mass of each α-radiation particle is 4/(1/1,840) or 7,360 times larger than the mass of each β-radiation particle.

In carrying energy away from a radioactive nucleus, radiation may penetrate other matter, such as human tissue, and cause tissue damage that may be health threatening. Workers handling radioactive materials must wear film badges to indicate their level of exposure to radiation (Figure 17.1). The importance of wearing these badges cannot be overemphasized.

The properties of mass, charge, composition, and penetrating power for each type of radiation are summarized in Table 17.1.

alpha (α) radiation

Alpha (α) radiation involves the emission of particles from a radioactive nucleus. From a determination of the relative charge (2+) and relative mass (4 amu) of α-particles, scientists were able to conclude that α-radiation consists of particles that are identical to helium nuclei. α-particles are symbolized by ${}^{4}_{2}\text{He}$.

$$\alpha\text{-Particle} \equiv {}^{4}_{2}\text{He (He nucleus)}$$

Because α-particles are the heaviest of the three types of radiation, they are the slowest moving and least penetrating of the three types. They are readily stopped by a few pieces of paper or a layer of human skin. Therefore, α-radiation is usually not hazardous to living organisms unless swallowed or inhaled. Figure 17.6 shows the relative penetrating powers of the three radiation forms.

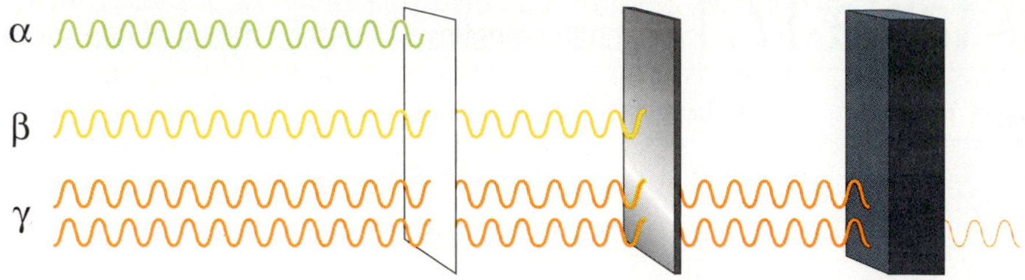

Figure 17.6
The penetrating powers of different types of radiation vary greatly. Alpha particles are stopped by paper alone. Beta particles are stopped by a wood block or aluminum plate. Gamma rays are very penetrating and a small amount may even pass through a lead block 5 cm thick.

beta (β) radiation

Beta radiation also involves the emission of particles from a radioactive nucleus. From a determination of the relative charge (1–) and relative mass (1/1,840 amu) of β-particles, scientists were able to conclude that **beta (β)radiation** consists of particles, traveling at nearly the speed of light, that are identical to electrons. β-particles are symbolized by ${}_{-1}^{0}e$.

$$\beta\text{-Particle} \equiv {}_{-1}^{0}e \text{ (electron)}$$

Because β-particles are smaller and faster moving, they have about 100 times the penetrating power of α-particles. An aluminum plate, a block of wood, or heavy protective clothing is necessary to stop β-radiation. Although most β-radiation is not sufficiently energetic to reach the internal organs of the body, it goes deep enough within the outer layers of skin to cause damage (similar to severe sunburn) and represents a special hazard to eyes.

gamma (γ) radiation

Gamma (γ) radiation is pure energy; *γ-rays are not particles.* Gamma radiation has neither mass nor change. γ-radiation is symbolized by ${}_{0}^{0}\gamma$.

$$\gamma\text{-Radiation} \equiv {}_{0}^{0}\gamma$$

This high-energy radiation, which is similar to X-rays or visible light, travels at the speed of light and can be stopped only by a block of lead or other dense material. Gamma radiation easily penetrates the skin and can cause severe internal damage. You have an idea of the properties of X-rays and know about their ability to travel through space (or the body) from your contact with them in medical and dental diagnosis. Your notions are sufficient for an understanding of the concepts being presented.

17.5 nuclear reactions

You know from the decay of carbon-14 discussed in Section 17.2 that the emission of radiation accompanies a change in the nucleus of the emitting radioactive atom. The change in the nucleus often produces an atom of a new element; in this case, ${}_{6}^{14}C$ is converted to ${}_{7}^{14}N$. To conserve charge, a particle that is identical to an electron is emitted; this particle you now know is called β-radiation or a β-particle because it originates in the nucleus. The balanced nuclear reaction is written as

$${}_{6}^{14}C \rightarrow {}_{7}^{14}N + {}_{-1}^{0}e$$

where ${}_{-1}^{0}e$ indicates the emission of a β-particle.

The change from one element to another is called **transmutation.** Because radioactive elements disappear as they emit radiation and are transmuted into other elements, they are said to disintegrate or *decay.* In Section 17.2, you used conservation of mass number and charge to help characterize the β-particle that was emitted from radioactive carbon-14. From now on, if you are given the type of radiation emitted by any radioactive nucleus, you may use the known composition (mass number and charge) of α- and β-particles to predict the identity of the new element formed from the emitting element as it decays. Using symbols, you can record this conversion, or nuclear reaction, by a nuclear equation of the form

Emitting element → Emitted particle(s) + New element

For each type of radiation emitted, you will apply the guidelines you learned for writing nuclear reactions in Section 17.2 to help you identify the new element formed.

Guidelines for writing nuclear equations

1. *The sum of the mass numbers on each side of the equation must be equal.*
2. *The sum of the atomic numbers on each side of the equation must be equal.*

α-emission

Uranium-238 is an α-emitter. This means $^{238}_{92}U$ nuclei lose α-particles, which you know to be composed of two protons and two neutrons each, that is, helium nuclei, $^{4}_{2}He$. The nuclear equation for this event begins with the uranium-238 emitter and shows the emitted α-particle represented by its nuclear symbol, $^{4}_{2}He$.

$$^{238}_{92}U \rightarrow {}^{4}_{2}He + ?$$

Because of the loss of two protons in the α-particle, the nucleus after emission contains only 90 protons (92 – 2); the atomic number of the new element is 90 (rule 2). Because the number of protons characterizes an atom, this nucleus must be the element thorium, which you determine by looking up atomic number 90 in the periodic table. The mass number of the new Th atom must be 4 less than the original ^{238}U atom because an α-particle, with a mass number of 4, has been lost; that is, the mass number of the new Th isotope must be 238 – 4 = 234 (rule 1). The nuclear equation is completed by showing the symbol for the new thorium isotope.

$$^{238}_{92}U \rightarrow {}^{4}_{2}He + {}^{234}_{90}Th$$

In this, as in all nuclear equations, there is a balance (that is, their sums are equal) of both atomic numbers (92 = 2 + 90) and mass numbers (238 = 4 + 234) on the two sides of the arrow, which is like an equal sign. Mass number is conserved; all protons and neutrons can be accounted for. Similarly, charge is also conserved because there is no change in the atomic number. The decay of uranium-238 is illustrated in Figure 17.7.

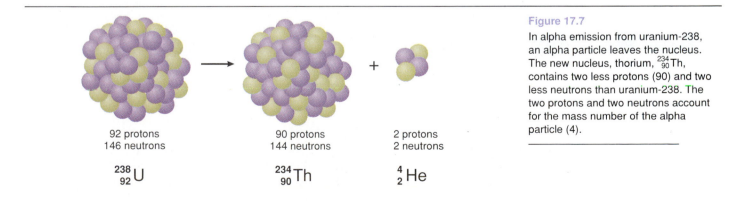

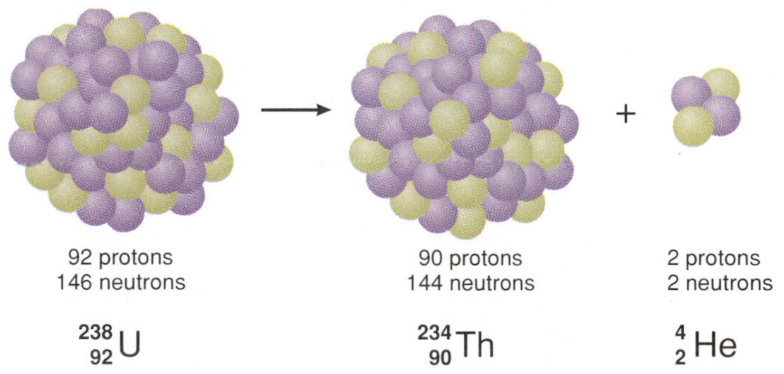

Figure 17.7
In alpha emission from uranium-238, an alpha particle leaves the nucleus. The new nucleus, thorium, $^{234}_{90}Th$, contains two less protons (90) and two less neutrons than uranium-238. The two protons and two neutrons account for the mass number of the alpha particle (4).

Another α-emitter is polonium-218, ${}^{218}_{84}Po$. The new element formed upon α-emission is lead-214. You can quickly verify the following equation by checking that the sums of the mass numbers and atomic numbers are equal on both sides of the equation.

$$ {}^{218}_{84}Po \rightarrow \overbrace{\underbrace{{}^{4}_{2}He + {}^{214}_{82}Pb}_{84}}^{218} $$

---

sample exercise 1

Write the nuclear equation for the change that occurs in radium-226 when it emits an α-particle. Radium-226 was the first radioisotope used to treat cancer.

solution

1. Write the symbol of the emitter, including atomic number and mass number, on the left side of the equation. In this case, the mass number is given, and you must find radium's atomic number in the periodic table.

$$ {}^{226}_{88}Ra \rightarrow $$

2. Write the symbol for the α-particle, which shows that it is a helium nucleus, on the right side of the equation.

$$ {}^{226}_{88}Ra \rightarrow {}^{4}_{2}He + ? $$

3. Complete the equation by writing a symbol for an isotope that has an atomic number 2 less than the original isotope (88 – 2 = 86) and a mass number 4 less than the original (226 – 4 = 222). Use the periodic table to ascertain that atomic number 86 identifies radon, Rn.

$$ {}^{226}_{88}Ra \rightarrow {}^{4}_{2}He + {}^{222}_{86}Rn $$

4. Check the equation to see that the mass numbers and atomic numbers are balanced, that is, that the totals on each side of the arrow are equal.

Mass numbers: 226 = 4 + 222
226 = 226

Atomic numbers: 88 = 2 + 86
88 = 88

Radium-226 is found in granite rocks and concrete building materials. The isotope of radon, radon-222, produced in the above reaction is a radioactive gas and will decay to yet another material. The decay of radium-226 in building materials to yield radioactive radon-222 is a potential health hazard if the gas is allowed to accumulate inside buildings. Considerable effort is now made to monitor the accumulation of radon-222 within buildings and to exhaust it if necessary.

••• problem 2

Write a nuclear equation for α-emission from plutonium-239, ${}^{239}_{94}Pu$. Plutonium-239 is used as a fuel, like uranium-235, in nuclear (breeder) reactors.

β-emission

You know that β-particles are electrons that originate in the nucleus and are not related to the valence electrons of an atom or ion. You also know that an electron may be emitted from a nucleus when a neutron is transformed into a proton; the proton remains in the nucleus and an electron leaves. The overall result is that the number of protons in the nucleus increases by 1 (atomic number increases by 1),

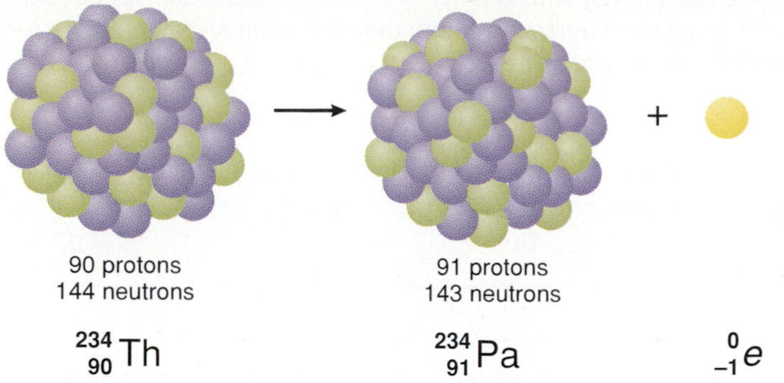

Figure 17.8
In beta emission from thorium-234, a neutron decomposes into a proton and an electron, and the electron leaves the nucleus as a β-particle. The new nucleus, protactinium, Pa, contains one more proton (91) and one less neutron (143).

the number of neutrons decreases by 1, and the mass number remains the same, since the sum of the number of protons and neutrons is not altered.

The first example you saw of a β-emitter was carbon-14. Another example is thorium-234, $^{234}_{90}Th$. One of thorium's 144 neutrons is transformed into a proton and an electron, and the electron is emitted as β-radiation (a β-particle). The original nucleus contained 144 neutrons and 90 protons, and the new nucleus contains 143 neutrons and 91 protons. The element with atomic number 91 is protactinium, Pa. The mass number of the new element is 234 because 143 neutrons and 91 protons have that mass. This reaction is illustrated in Figure 17.8 and can be summarized by the nuclear equation

$$^{234}_{90}Th \rightarrow {}^{0}_{-1}e + {}^{234}_{91}Pa$$

The use of the symbol $^{0}_{-1}e$ for the β-particle (instead of the traditional electron symbol e^-) enables us to check that the sums of the mass numbers and atomic numbers are the same on both sides of the balanced equation.

Mass numbers: $234 = 0 + 234$	Atomic numbers: $90 = -1 + 91$
$234 = 234$	$90 = 90$

Another β-emitter is bismuth-210, $^{210}_{83}Bi$. The new nucleus formed has 84 protons and the same mass number.

$$^{210}_{83}Bi \rightarrow {}^{0}_{-1}e + {}^{210}_{84}Po$$

Once again, you may verify that the sums of the mass numbers and atomic numbers are equal on both sides of this equation.

sample exercise 2

Write the nuclear equation for the change that occurs in cobalt-60 when it emits a β-particle. Cobalt-60 is used as a source of machine-generated radiation used in radiation therapy. In addition to emitting β-particles, cobalt-60 emits γ-radiation, which penetrates tissues and can be focused to destroy tumors.

solution

1. Write the symbol of the emitter, including atomic number and mass number, on the left side of the equation. From the periodic table, the atomic number of cobalt is 27. The mass number 60 is given.

$$^{60}_{27}Co \rightarrow$$

2. Write the symbol for the β-particle, which shows that it is an electron, on the right side of the equation.

$$^{60}_{27}Co \rightarrow {}^{0}_{-1}e + ?$$

3. Complete the equation by writing a symbol for an isotope that has the same mass number and an atomic number 1 greater than the original isotope (27 – (–1) = 28); atomic number 28 indicates Ni.

$$^{60}_{27}Co \rightarrow {}^{0}_{-1}e + {}^{60}_{28}Ni$$

4. Check that the sums of the mass numbers and atomic numbers are balanced.

Mass numbers: 60 = 0 + 60   Atomic numbers: 27 = –1 + 28
60 = 60   27 = 27

---

problem 3

Write a nuclear equation for β-emission from actinium-227, $^{227}_{89}Ac$. Actinium-227 is a long-lived isotope with a half-life of 21.8 years; it has been formed in gram amounts from the β-decay of radium-227.

γ-radiation

Virtually all α- and β-emissions are accompanied by γ-emission. In Sample Exercise 2, the presence of γ-radiation was mentioned, but not included in the equation for the radioactive decay of cobalt-60. It is the high-energy γ-radiation that is used to destroy cancerous tumor tissue. *Because γ-radiation produces no change in nuclear contents, it is usually ignored in writing nuclear equations.* You may explicitly include the fact that γ-emission accompanies β-emission in the nuclear equation for cobalt-60 decaying to nickel-60, but notice that the symbol $^{0}_{0}\gamma$ does not alter the balanced equation.

$$^{60}_{27}Co \rightarrow {}^{0}_{-1}e + {}^{60}_{28}Ni + {}^{0}_{0}\gamma$$

new elements through bombardment

Radioactive decay is a natural process that occurs spontaneously. However, it is also possible to induce nuclear reactions by bombarding stable nuclei with nuclear-sized "bullets." For example, if a high-speed neutron is shot at the stable isotope of aluminum-27, $^{27}_{13}Al$, a radioactive aluminum isotope, aluminum-28, $^{28}_{13}Al$, is produced. The equation is balanced because the sums of the mass numbers and atomic numbers are equal on both sides.

$$^{27}_{13}Al + {}^{1}_{0}n \rightarrow {}^{28}_{13}Al$$ **Neutron bombardment**

In this case, the energy of the collision has simply caused neutron capture; that is, the neutron is taken into the aluminum-27 nucleus to form a new isotope, aluminum-28. The unstable aluminum-28, $^{28}_{13}Al$, is a β-emitter and decays to yield nonradioactive silicon-28 according to the balanced equation

$$^{28}_{13}Al \rightarrow {}^{0}_{-1}e + {}^{28}_{14}Si$$ **Decay**

Neutron bombardment also occurs naturally and produces other results besides simple capture. Carbon-14 exists in the atmosphere because of high-speed collisions between the common isotope of nitrogen and neutrons from high-energy cosmic rays in the upper atmosphere.

$$^{14}_{7}N + {}^{1}_{0}n \rightarrow {}^{14}_{6}C + {}^{1}_{1}H$$

In this case, the bombardment knocks a proton ($^{1}_{1}H$) out of the nitrogen-14 nucleus as the neutron enters. This nuclear equation is also balanced because the sums of the mass numbers and atomic numbers are equal on both sides of the equation.

Subatomic particles other than neutrons can be involved in bombardment. For example, bombardment of ${}^{14}_{7}N$ by an α-particle produces a nonradioactive isotope of oxygen, oxygen-17, according to the balanced nuclear equation

$${}^{14}_{7}N + {}^{4}_{2}He \rightarrow {}^{17}_{8}O + {}^{1}_{1}H$$

Notice that all nuclear equations must be balanced; that is, the sums of the mass numbers and the atomic numbers are identical on both sides of the equation.

---

sample exercise 3

Complete the following nuclear equations, which represent bombardment reactions.

a. ${}^{35}_{17}Cl + {}^{1}_{0}n \rightarrow {}^{34}_{16}S + ?$
b. ${}^{23}_{11}Na + ? \rightarrow {}^{23}_{12}Mg + {}^{1}_{0}n$
c. ${}^{238}_{92}U + {}^{4}_{2}He \rightarrow ? + {}^{1}_{0}n$

solution

The sums of the mass numbers and the atomic numbers must be identical on each side of a nuclear equation. By establishing this equality, the identity of the unknown species can be determined.

a. Because the sum of the mass numbers on the left side is 36, the mass number of the unknown must be 2 (36 = 34 + 2). Similarly, the atomic number of the unknown must be 1 (17 = 16 + 1). Atomic number 1 characterizes H, but the mass number of 2 identifies the isotope as deuterium.

$${}^{35}_{17}Cl + {}^{1}_{0}n \rightarrow {}^{34}_{16}S + {}^{2}_{1}H$$

b. The sum of the mass numbers on the right is 24, so the mass number of the unknown must be 1. The atomic number must also be 1, identifying the bombarding nucleus as hydrogen.

$${}^{23}_{11}Na + {}^{1}_{1}H \rightarrow {}^{23}_{12}Mg + {}^{1}_{0}n$$

c. Because the sum of the mass numbers on the left is 242, the mass number of the unknown must be 241. Similarly, the atomic number must be 94. Plutonium is the element with atomic number 94.

$${}^{238}_{92}U + {}^{4}_{2}He \rightarrow {}^{241}_{94}Pu + {}^{1}_{0}n$$

Perhaps the most exciting application of bombardment reactions is in the preparation of elements that do not occur naturally. Indeed, most of the radioisotopes routinely used in medicine do not occur naturally and are made artificially. The name *technetium,* given to the first element made artificially, is derived from a Greek word meaning "artificial."

Elements with atomic numbers greater than 92 are called **trans-uranium elements.** These elements do not occur naturally, but have been synthesized by bombardment reactions. For example, in Sample Exercise 3, you saw the synthesis of ${}^{241}_{94}Pu$ by α-bombardment of ${}^{238}_{92}U$. Other bombardment reactions lead to other trans-uranium elements. Also, the decay of these elements, all of which are radioactive, often leads to new elements. For example, the β-decay of ${}^{241}_{94}Pu$ produces the radioactive isotope americium-241, used in smoke detectors (Figure 17.16).

$${}^{241}_{94}Pu \xrightarrow{\text{Decay}} {}^{241}_{95}Am + {}^{0}_{-1}e$$

$$^{238}_{92}U \longrightarrow ^{206}_{82}Pb + 8\ ^{4}_{2}He + 6\ ^{0}_{-1}e$$

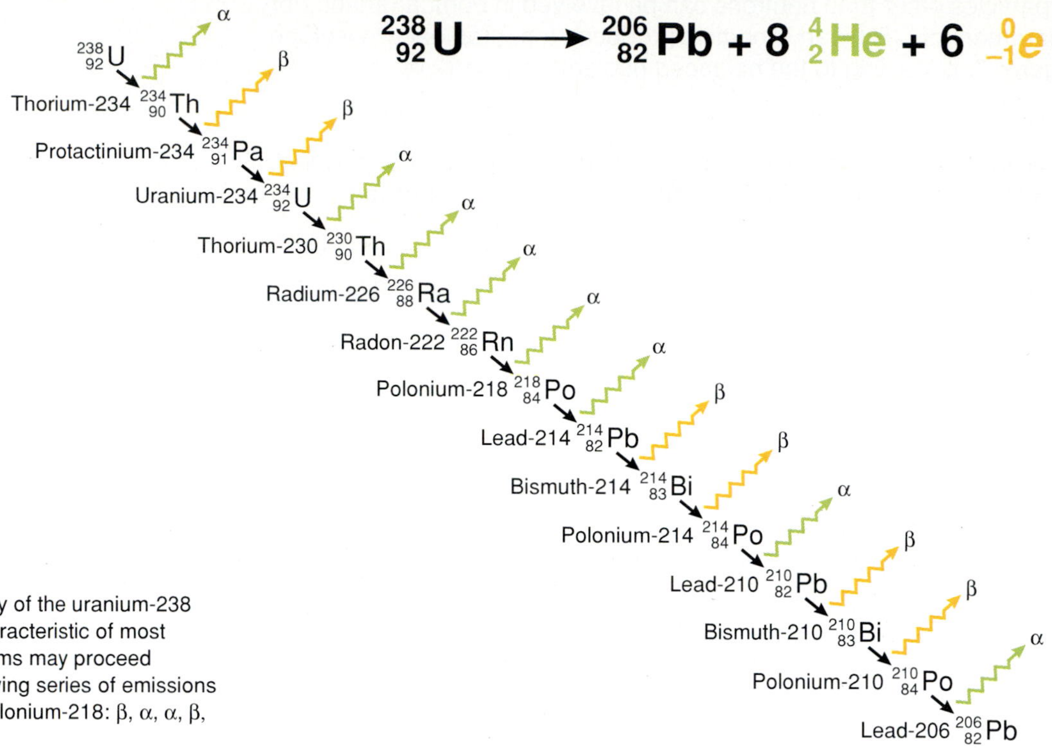

Figure 17.9

The main pathway of the uranium-238 decay series, characteristic of most atoms. Some atoms may proceed through the following series of emissions beginning with polonium-218: β, α, α, β, β, α, β.

problem 4

What radioisotope is produced when curium-242, $^{242}_{96}Cm$, is bombarded by an α-particle, causing a neutron to be ejected?

$$^{242}_{96}Cm + ^{4}_{2}He \rightarrow ^{1}_{0}n + ?$$

The element curium was first made in the United States in 1944 by a research group at the University of California–Berkeley. The element is named for Marie and Pierre Curie, who were among the first explorers of nuclear chemistry.

17.6 radioactive decay series

Because radioactive elements often disappear as they emit radiation and are changed into other elements, they are said to **disintegrate** or **decay.** Very often the product of radioactive decay is also radioactive, and so it will decay also. For example, you know that $^{238}_{92}U$ decays to $^{234}_{90}Th$ by α-emission. You have also seen that $^{234}_{90}Th$ is radioactive and decays to $^{234}_{91}Pa$ by β-emission. It turns out that $^{234}_{91}Pa$ is also radioactive and decays by β-emission to yield $^{234}_{92}U$.

$$^{234}_{91}Pa \rightarrow ^{0}_{-1}e + ^{234}_{92}U$$

$^{234}_{92}U$ is also radioactive, as are all isotopes of uranium (Section 17.2).

These constitute the first steps in a radioactive decay series, which is summarized in Figure 17.9. When decay of a "parent" nucleus produces a radioactive "daughter" nucleus, the daughter will also decay, and so the series continues. A radioactive decay series stops when a stable (*non*radioactive) isotope is produced. In the example given in Figure 17.9, the series stops when nonradioactive $^{206}_{82}Pb$ is formed.

17.7 ionizing radiation

Nuclear radiation, like x-radiation, is also referred to as **ionizing radiation.** This is so because the interaction with α-, β-, γ-radiation or X-rays often causes ionization of an atom or molecule. Just as light energy may excite electrons in an atom or

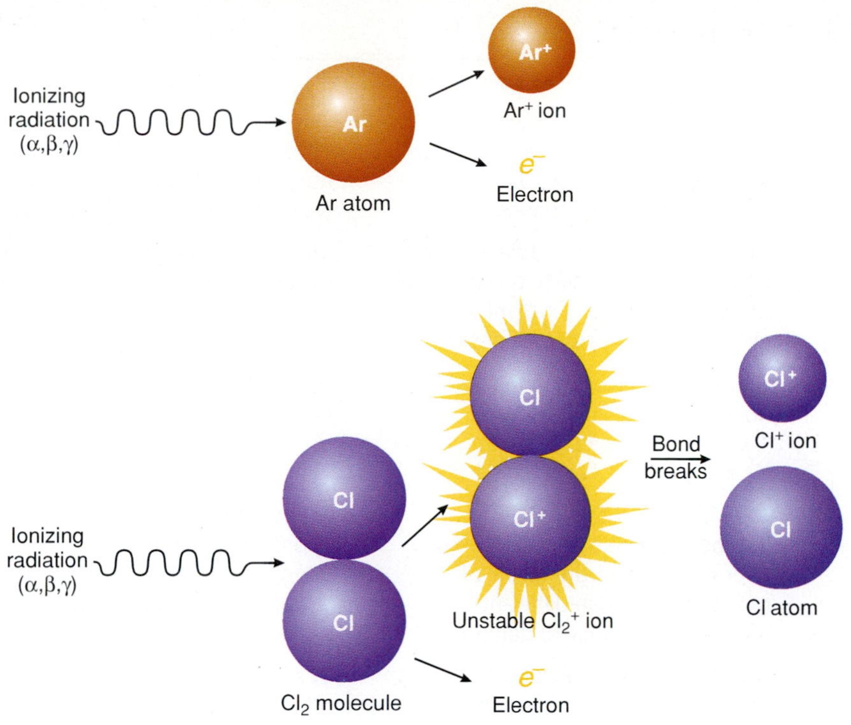

Figure 17.10

Radiant energy knocks electrons out of atoms or molecules to yield ions. Subsequent bond breakage often occurs in the molecular ions formed. Examples of the ionization of atomic argon and molecular chlorine are illustrated. Argon gas is often used in radiation detection devices.

molecule, when α-, β-, γ-, or x-radiation collides with an atom or molecule, some of the energy of the radiation is transferred to the particle that has been hit. At the very least, electrons in the atom or molecule are promoted to higher-energy, less stable states. Often electrons are kicked out of atoms or molecules, producing ions or breaking bonds (Figure 17.10).

Within the cells of the body, ionization effects can be devastating. Body chemistry depends on the presence of specific substances participating in chemical reactions. Alteration of these substances through ionization interferes with necessary cell reactions and leads to other undesirable reactions. The result may be total disruption of cell activity within the affected cells.

The ionizing power of nuclear radiation is inversely related to its penetrating power; that is, γ-radiation is most penetrating but produces the fewest ions per unit volume. On the other hand, the least penetrating radiation, α, produces the largest number of ions, and β-radiation is moderate in penetrating power and ionizing ability. The penetrating powers of α-, β-, and γ-radiation are summarized in Table 17.2. Many of the units of measurement for the power or biological effectiveness of a radioactive source relate to ionizing ability. See Table 17.3 for the definition of units that are used to measure radiation.

Radiation intensity decreases rapidly with increasing distance from the source of radiation. Doubling your distance from a radiation source ensures that the radiation

table 17.2 ionizing power and penetrating power of radiation

Type of radiation	Ionizing power	Penetrating power
α	High	Low
β	Moderate	Moderate
γ	Low	High

table 17.3 units of radiation measurement

Unit	What it measures	Description	Example
Curie (Ci)	Activity	An amount of radioactive material that undergoes 3.7×10^{10} disintegrations per second	^{60}Co γ-ray sources delivered to hospitals for radiation therapy are typically in the millicurie range.
Roentgen (R)	Intensity of exposure	An amount of x- or γ-radiation that produces 1 electrostatic unit of charge (20 billion ion pairs) in 1 mL of dry air at 0°C and 1 atm of pressure	The output from an X-ray machine is described in terms of roentgens.
Radiation absorbed dose (rad)	Absorbed dose	An amount of radiation that, when absorbed by tissue, delivers 1×10^{-5} J per gram of tissue; for X- and γ-rays, 1 R delivers 1 rad	Radiation therapists quote dosages in rads.
Relative biological effectiveness (rbe)	Biological effectiveness	A quality factor that accounts for the fact that the same amount of exposure to different forms of radiation produces different amounts of biological damage	X-, γ-, and β-rays all have approximately the same effectiveness and are given an rbe of 1; α-rays are much more damaging and are rated at 10 rbe.
Roentgen equivalent for human (rem)	Dose equivalent weighted for different kinds of radiation	An amount of radiation that produces the same damage to tissue as 1 R of x-radiation (dose in rems = dose in rads $\times$ rbe); dosages in rems are additive; thus, cumulative effects are expressed by this unit	This unit is used as a preferred dosage statement because it compensates for the differences in effectiveness among the forms of radiation.

you feel will be one-fourth the original amount; at triple the distance, the radiation intensity is one-ninth the original amount. Quantitatively, this is known as the **inverse square law:** Intensity, *I,* is inversely proportional to the square of the distance, *d,* from the source.

$$I \propto \frac{1}{d^2}$$

Reducing the intensity of radiation with distance offers the simplest protection. Placing a shield that is impenetrable to the radiation between you and the source is another effective safety measure. Distance and shielding are the principal safety measures practiced in radiation laboratories, clinics, and power plants.

Many of the devices for detecting and measuring radioactive emissions depend on the fact that nuclear radiation produces ions. For example, consider the cloud chamber shown in Figure 17.11. This chamber contains air supersaturated with water vapor. As radiation passes through this air and ionizes molecules in the air, water vapor condenses on the ions and forms a visible path of microscopic water droplets. The radiation leaves an observable trail as it travels through the cloud chamber, similar to the observable vapor trail a jet leaves behind as it travels through the atmosphere. Radiation trails in a cloud chamber may be photographed for measurement and further study.

Probably the best known device for detecting and measuring radiation is the Geiger-Mueller tube or the Geiger counter (Figure 17.12). Entering radiation ionizes gas within the tube, and an electric current is produced as electrons go to the anode (positive electrode) and argon cations go to the cathode (negative electrode). The

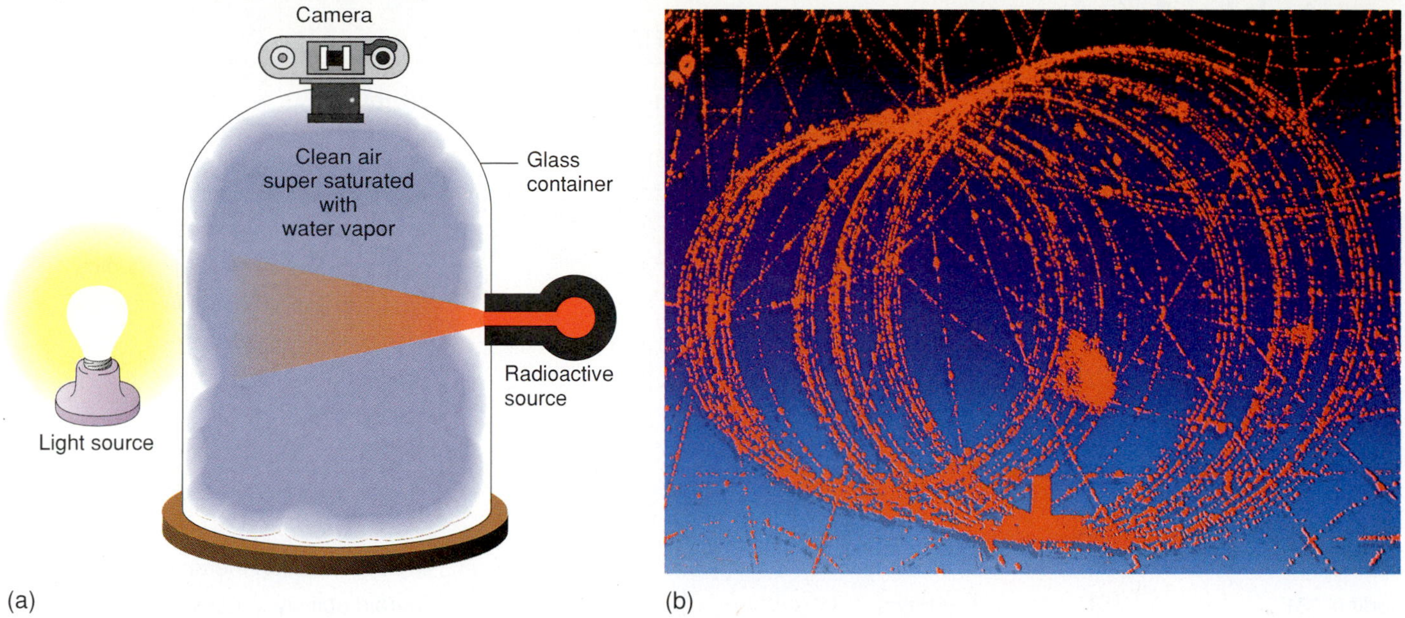

Figure 17.11

(a) The cloud chamber is supersaturated with water vapor. Radioactive emissions ionize the air through which they pass and cause the condensation of visible water droplets that leave a trail behind the traveling emitted ray(s). (b) The visible tracks left by radiation passing through a cloud chamber may be photographed for detailed measurement and further study. This track left by a proton is spiral because the charged particle is moving in a strong external magnetic field.

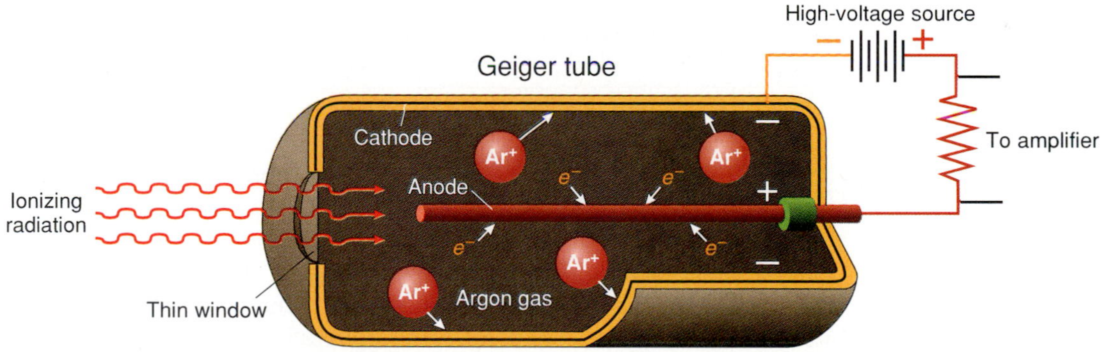

Figure 17.12

Ionizing radiation enters the Geiger tube through the thin window at the left of the apparatus. As the radiation passes through the gas inside the tube, it ionizes argon atoms along its path. The Ar^+ cations are attracted to the negatively charged walls. The electrons are attracted to the central rod, which is positively charged. The flow of charge produces an electric current in the circuit that is amplified and heard as a click.

amount of current is related to the energy and intensity of the radiation. Frequently, the electric current is used to produce audible clicks, the speed and intensity of which increase with increasing radiation. The audible clicks provide a convenient method to monitor the level of radiation without having to watch the instrument.

17.8 half-life, $t_{1/2}$

The rate at which radioactive elements decay varies greatly. Rate of radioactive decay is conveniently measured in terms of an element's **half-life,** symbolized $t_{1/2}$ and defined as the time required for the decay of one-half of a radioactive sample. For example, the half-life of chromium-51, ^{51}Cr, a radioisotope used to study blood volumes, is 28 days. This means that if you have 100 mg of ^{51}Cr today, then 28 days from today you will have only 50 mg because half the sample (50 mg) will

table 17.4 half-lives and uses of selected radioisotopes

Name	Symbol	Half-life	Radiation emitted	Usefulness
Carbon-14	$^{14}_{6}C$	5,720 years	β	Radioactive dating and labeling
Sodium-24	$^{24}_{11}Na$	15 hours	β, γ	Blood circulation studies
Phosphorus-32	$^{32}_{15}P$	15 hours	β	Nucleic acid label for metabolic studies
Sulfur-35	$^{35}_{16}S$	88 days	β	Protein label for metabolic studies
Potassium-42	$^{42}_{19}K$	12 hours	β	Plant and animal nutrition studies
Calcium-45	$^{45}_{20}Ca$	165 days	β	Animal nutrition studies
Iron-59	$^{59}_{26}Fe$	45 days	β	Red blood cell studies
Cobalt-60	$^{60}_{27}Co$	5.3 years	β, γ	Radiation therapy
Strontium-90	$^{90}_{38}Sr$	28.1 years	β	Medical treatment
Yttrium-90	$^{90}_{39}Y$	64 hours	β	Pituitary implant radiation therapy
Technetium-99	$^{99}_{43}Tc$	6 hours	γ	Brain scans
Iodine-123	$^{123}_{53}I$	13 hours	γ	Thyroid radiation therapy
Iodine-131	$^{131}_{53}I$	8 days	β, γ	Thyroid activity studies
Uranium-235	$^{235}_{92}U$	710 million years	α, γ	Nuclear reactors
Plutonium-239	$^{239}_{94}Pu$	24,400 years	β	Fuel in breeder reactors; waste product of some reactors

have decayed. During a second span of 28 days, half the 50 mg will decay so that only 25 mg will remain, and the amount of chromium-51 remaining will continue to decrease by half every 28 days.

$$100\text{ mg } ^{51}\text{Cr} \xrightarrow{28\text{ days}} 50\text{ mg } ^{51}\text{Cr} \xrightarrow{28\text{ days}} 25\text{ mg } ^{51}\text{Cr} \xrightarrow{28\text{ days}} 12.5\text{ mg } ^{51}\text{Cr} \xrightarrow{28\text{ days}} \ldots$$

The listing of half-lives of other commonly used radioisotopes in Table 17.4 points out the wide variation in half-life, from a few hours to millions of years. Some isotopes have half-lives of only fractions of a second. Most isotopes used in medical studies have half-lives of a few hours or days so that the radiation will exist long enough to be measured in the body but will decrease rapidly to low levels. On the other hand, plutonium-239, a waste product of some nuclear reactors, has a half-life of 24,400 years. With such an enormous half-life, safe storage or disposal presents formidable problems for future generations to solve.

Figure 17.13 shows the meaning of half-life graphically and illustrates the rates of radioactive decay for two isotopes with widely different half-lives. Each time a half-life period elapses, half the amount of material present at the beginning of that period disintegrates. After 1 half-life, half the original sample remains; after 2 half-lives, $(1/2)^2 = 1/4$ of the original sample remains; after n half-lives, the fraction of original sample remaining is $(1/2)^n$.

Figure 17.13a shows the general principle of half-life for any sample. In Figure 17.13b and 17.13c, examples of the decay of specific amounts of two radioisotopes are represented. Comparison of Figure 17.13b and c points out the fact that long-lived radioisotopes such as ^{238}U present a radioactive-waste storage problem because they "hang around" almost forever, whereas short-lived radioisotopes "dispose" of themselves. In these examples, a sample of iodine-131 would diminish from 8 g to 0.25 g in only 40 days, whereas 22.5 billion years would be required for the decay of uranium-238 to this extent. Radioactive waste from nuclear power plants consists mostly of long-lived uranium isotopes that must be shielded from the environment for thousands of centuries before the uranium decays sufficiently to be harmless.

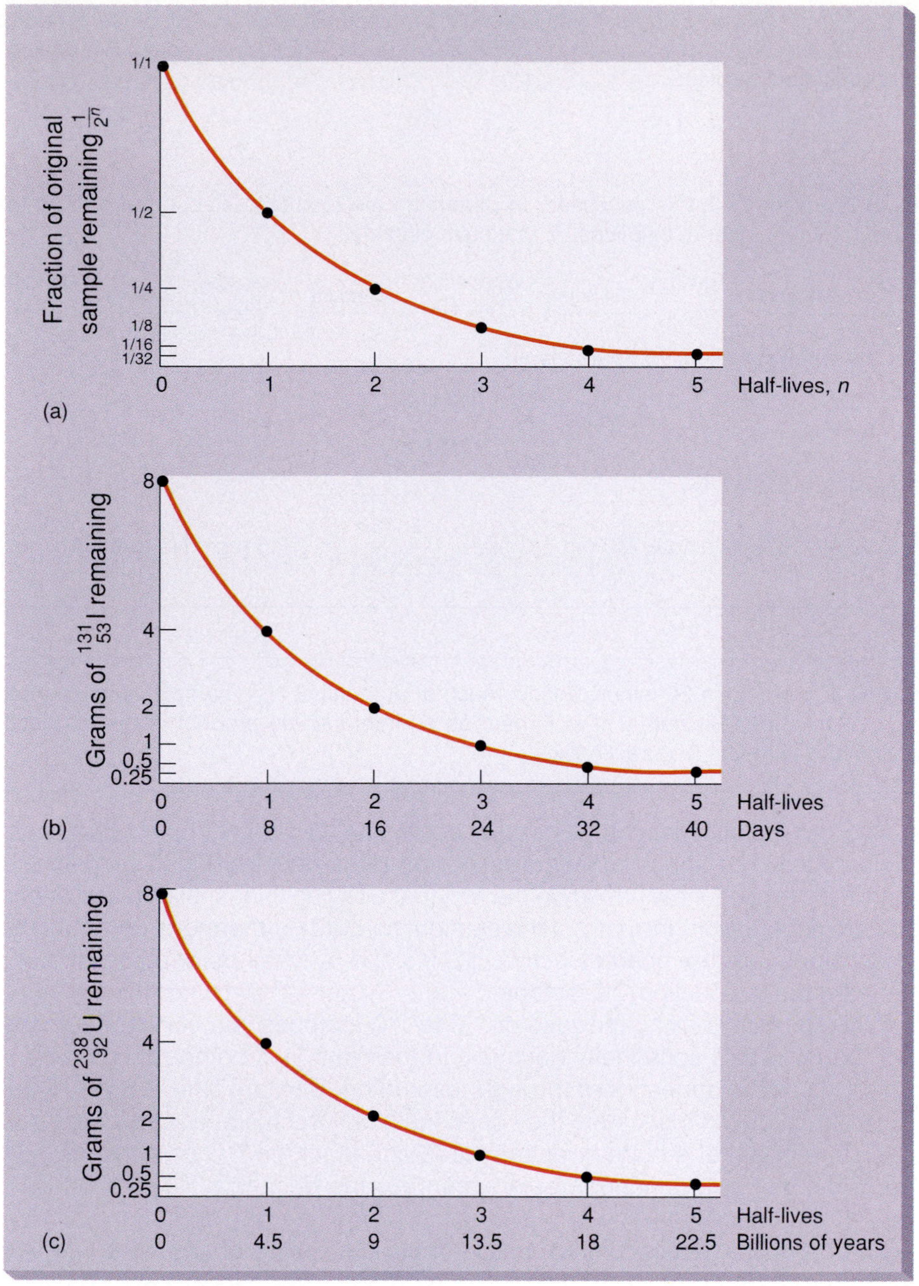

Figure 17.13

(a) The decay curve for any radioactive isotope is expressed in terms of the half-life of the isotope. During each half-life interval, half the material present at the beginning of the interval decays. The fraction of original sample remaining after n half-lives is $(\frac{1}{2})^n$. (b) The decay curve for an 8 g sample of iodine-131 (half-life of 8 days). (c) The decay curve for an 8 g sample of uranium-238 (half-life of 4.5 billion years).

Calculations relating amounts of radioactive material and $t_{1/2}$ can be done using the format shown earlier for the decay of ^{51}Cr, or more compactly, we can recognize that the graphical displays in Figure 17.13 give us an equation for this relationship.

$$\textbf{Quantity remaining after } n \textbf{ half-lives} = \textbf{Original quantity} \times (1/2)^n \qquad (17.3)$$

For ^{131}I, if you begin with 8.0 g, after 5 half-lives ($n = 5$), 0.25 g remains, as the graph and equation indicate.

$$\text{Quantity remaining after 5 half-lives} = 8.0 \text{ g} \times (1/2)^5$$

$$= 8.0 \text{ g} \times \frac{1}{2 \times 2 \times 2 \times 2 \times 2}$$

$$= 0.25 \text{ g}$$

sample exercise 4

The half-life of ^{24}Na is 15 hours. If you have a 240 mg sample of this radioisotope at noon on a Monday, how many milligrams will remain at 3 P.M. on Thursday? Sodium-24 is used in blood circulation studies.

solution

To use Equation 17.3, it is necessary to determine the total time elapsed and, from that, determine the number of half-lives, *n,* that have elapsed.

$$\text{Monday noon} \xrightarrow{24\text{ hours}} \text{Tuesday noon} \xrightarrow{24\text{ hours}} \text{Wednesday noon} \xrightarrow{24\text{ hours}} \text{Thursday noon} \xrightarrow{3\text{ hours}} \text{Thursday 3 P.M.}$$

Over the times specified, 75 hours elapse.

$$75\text{ hours} \times \frac{1\text{ half-life}}{15\text{ hours}} = 5.0\text{ half-lives}$$

Now, using Equation 17.3,

$$\text{Quantity remaining} = 240\text{ mg} \times \frac{1}{2 \times 2 \times 2 \times 2 \times 2} = 7.5\text{ mg }^{24}\text{Na remain}$$

problem 5

The half-life of ^{214}Bi is 20 minutes. How much of an original 16 g sample remains after 2 hours 20 minutes? Bismuth-214 is formed as an intermediate product in the radioactive decay of uranium-234 (Figure 17.9).

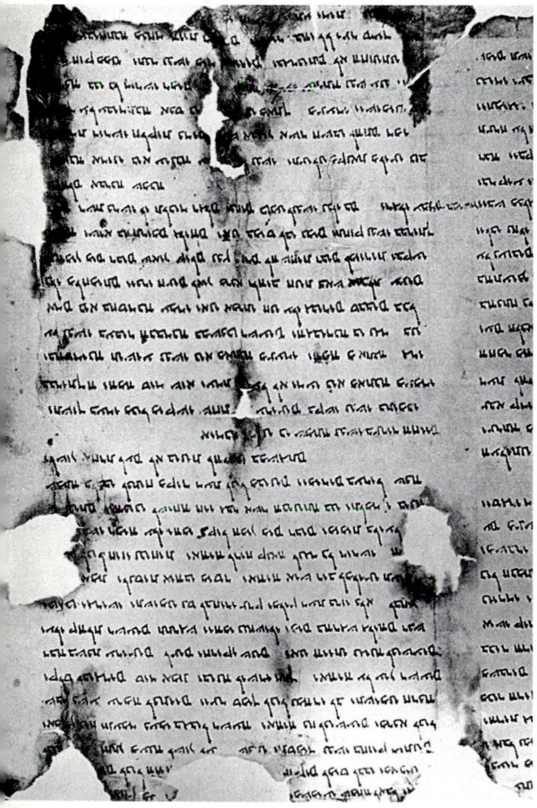

Figure 17.14

The Dead Sea Scrolls were found by carbon-14 dating to be about 1,900 years old.

The half-life concept can be used to determine the age of objects that contain radioisotopes. The age of museum relics can often be determined by carbon-14 dating. This procedure is based on the fact that all living things, plants and animals alike, are composed principally of the element carbon. Furthermore, whereas most carbon atoms are the nonradioactive $^{12}_{6}C$ isotope, a small percentage of carbon atoms are the radioactive $^{14}_{6}C$ isotope.

While a plant is alive, the ratio of ^{14}C to ^{12}C isotopes is a constant because, even though ^{14}C is continually decaying in the plant, it is being replaced by ^{14}C from $^{14}CO_2$ in the atmosphere through respiration. Similarly, the ^{14}C-to-^{12}C ratio is constant in animals because they continually eat fresh plants and inhale some $^{14}CO_2$. The death of a plant or animal does not affect the ^{12}C content. However, there is no longer a continual replacement for the decaying ^{14}C, and thus, the ratio $^{14}C/^{12}C$ diminishes.

The Dead Sea Scrolls contain partial texts of some of the books of the Old Testament and were found in caves near the edge of the Dead Sea in 1947. They were dated by examining the $^{14}C/^{12}C$ ratio in their paper made from papyrus. It was found that the ratio in these scrolls was only 79.5% of the ratio in a living plant. From this and the known $t_{1/2}$ of carbon-14 (5,720 years), archaeologists calculated that the scrolls were approximately 1,900 years old (Figure 17.14).

Using carbon-14, dating for several thousands of years may be determined with reasonable accuracy; over that period of time, there remains a significant amount of the original carbon-14 sample and its change may be measured. Similar dating procedures are applied with other radioisotopes. When using isotopes with much longer half-lives than carbon-14, dating may determine the age of much older samples. For example, rocks can be dated by measuring the remaining amount of uranium-238 ($t_{1/2}$ = 4.5 billion years) or rubidium-87 ($t_{1/2}$ = 500 billion years) content.

sample exercise 5

A fragment of animal-skin clothing from an archaeological dig was found to have a $^{14}C/^{12}C$ ratio only 0.25 times that of the ratio in a living animal's fur. The half-life of carbon-14 is 5,720 years. How old is the clothing?

solution

Let the original $^{14}C/^{12}C$ ratio be represented by *r*. Then every time a half-life elapses, the ratio will be cut in half because the amount of carbon-14 will be halved.

$$r \xrightarrow{\text{5,720 years}} (1/2)\,r \xrightarrow{\text{5,720 years}} (1/4)\,r$$

The given ratio is 0.25*r* or (1/4)*r*. Therefore, two half-lives, or 11,440 years (2 × 5,720), must have elapsed since the animal skin was part of a living animal.

problem 6

In another archaeological dig, a wooden cart displayed a $^{14}C/^{12}C$ ratio 0.5 times that in living trees. How old is the cart?

17.9 physiological effects of radiation

Ionizing radiation is most damaging to the *nuclei* of living cells. The cell nucleus contains the "blueprints" for producing more identical cells. Because the cell nucleus directs cell division and replication, cells that are dividing most rapidly are the first to show the effects of radiation. Because the principal genetic material (DNA) resides in the cell nucleus, radiation damage to nuclei during DNA replication or cell division may produce damage that is passed on to progeny cells. The cells most susceptible to radiation damage are those in the lymphatic system, bone marrow, intestinal tract, reproductive organs, and lens of the eye.

The body can tolerate exposure to small amounts of ionizing radiation without apparent symptoms. Thus, background radiation (from the soil and outer space) and medical x-rays produce no noticeable harm. However, small doses may be cumulative, and it is important for persons who work near sources of radiation to monitor their exposure. A common device employed for this purpose is the film badge, which makes use of the fact that photographic film detects radiation (Figure 17.3b). Film badges are developed periodically to determine the extent of exposure.

One source of background radiation in homes is the gas radon. Specialists fear it may be appearing in many homes in amounts greater than previously believed. The source of the radon gas is the soil or rocks on which a house is built or the actual building materials such as stone or concrete.

Rocks, especially granite, contain minute amounts of radium, Ra-226, a solid that, as such, stays in the rock. However, Ra-226 emits α-particles and is thereby converted to radon, Rn-222, a radioactive gas that enters the air (Sample Exercise 1). Outdoors, radon atoms are quickly dispersed and are harmless. Indoors, on the other hand, they may accumulate and be inhaled by the occupants. The problem is worsened by the fact that gaseous Rn-222 decays to polonium, Po-218, which is itself radioactive and a solid that is not exhaled (Figure 17.9).

In some homes, background radiation from radon may provide more unwanted exposure to radiation in the course of a year than the medical x-rays a person may have in a lifetime. Increased incidence of lung cancer from this exposure is especially likely, since gaseous radon is a source of very damaging alpha radiation that can be inhaled.

Even in the absence of noticeable symptoms of illness, excessive low-level exposure could lead to sterility or birth defects because reproductive cells and fetal

table 17.5 effects on humans of short-term, whole-body radiation exposure

Dose (rems)	Effects
0–25	No detectable clinical effects
25–100	Slight short-term reduction in number of some blood cells; disabling sickness not common
100–200	Nausea, fatigue, vomiting if dose is greater than 125 rem, longer-term reduction in number of some blood cells
200–300	Nausea and vomiting first day of exposure, up to a two-week latent period followed by appetite loss, general malaise, sore throat, pallor, diarrhea, and moderate emaciation; recovery in about three months unless complicated by infection or injury
300–600	Nausea, vomiting, and diarrhea in first few hours, up to a one-week latent period followed by loss of appetite, fever, and general malaise in the second week, followed by hemorrhage, inflammation of mouth and throat, diarrhea, and emaciation; some deaths by about six months
≥600	Nausea, vomiting, and diarrhea in first few hours, rapid emaciation and death as early as second week; eventual death of nearly 100% of victims

Source: E. L. Saenger, Ed., *Medical Aspects of Radiation Accidents.* Washington, DC: United States Atomic Energy Commission, 1963.

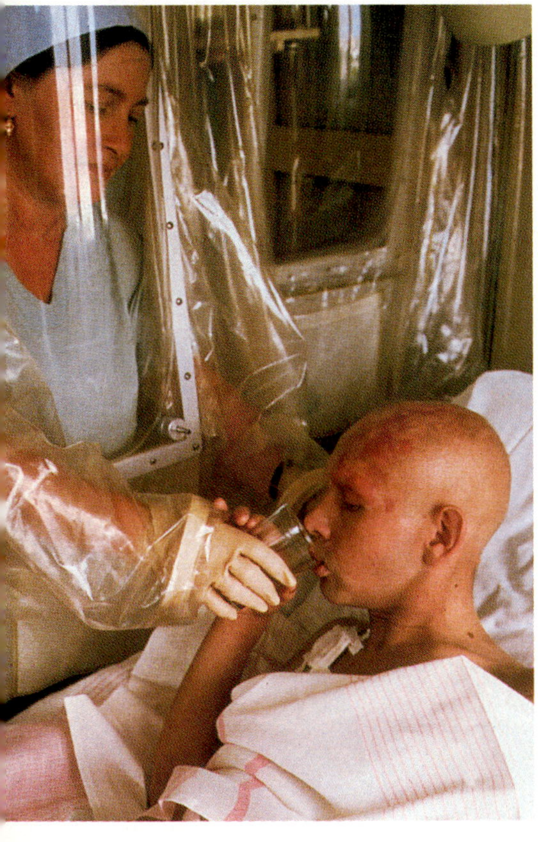

Figure 17.15
A victim of the 1986 Chernobyl radiation disaster in the former USSR exhibits severe symptoms of high-dosage radiation exposure.

tissue are especially sensitive to radiation. The first clinical symptom for higher levels of exposure is a drop in white blood cell count. White blood cells have short life spans; therefore, damage to tissue producing these cells shows up rapidly. A reduction in white blood cell count increases susceptibility to infection because a person's natural resistance is lowered. Red blood cells are also affected, and anemia may result.

Higher doses of radiation cause symptoms such as nausea, vomiting, and diarrhea because of damage to cells in the intestinal tract. The highest doses may produce burns to the skin, clouding of the eye lens (cataracts), and frequently death because of damage to so many essential bodily functions. If a person does survive massive exposure to radiation, the likelihood of developing cancer, particularly leukemia or blood cancer, is greatly increased (Figure 17.15). Table 17.5 lists some of the effects on humans of short-term radiation exposure.

17.10 uses of radioisotopes

Radioisotopes are very useful in chemistry, biology, industry, and especially medicine because of their following properties.

1. They are easily detected even in minute amounts.
2. The chemical reactivity of radioisotopes is identical to that of nonradioactive isotopes of the same element.
3. Radiation damages cells, particularly those that divide rapidly.

The first two properties are used in medical diagnosis and in chemical and industrial applications. Some medical treatments employ the latter two properties. Other properties desirable in radioisotopes that are used in medicine are (1) a fairly short half-life, so that exposure is not long term, and (2) high energy so that they are easily detected.

medical diagnosis with radioactive labeling

Radioisotopes react and combine with other elements just like nonradioactive isotopes. You have learned that all living matter is made of both nonradioactive carbon-12 and radioactive carbon-14 in proportion to their natural abundances. The compound sodium iodide (NaI), which can be used to study thyroid activity, is ordinarily composed of the nonradioactive isotopes ^{23}Na and ^{127}I. However, ^{23}Na can just as readily combine with the radioactive isotope ^{131}I. This sodium iodide ($^{23}Na^{131}I$) is a radioactively **labeled compound.** If it is introduced into the body, such a labeled compound containing a radioactive isotope can be readily detected and followed, that is, traced throughout the body. **Radioisotope tracing** is the basis of medical diagnostic techniques.

The rate at which a thyroid gland absorbs ^{131}I from a $Na^{131}I$ solution ingested by a patient can be readily monitored by a Geiger counter. This rate of absorption indicates whether the thyroid is working normally, is underactive, or is overactive.

A particularly useful radioisotope with several applications is technetium-99, $^{99}_{43}Tc$. For example, labeled sodium pertechnetate ($NaTcO_4$), a combination of Na^+ ions and TcO_4^- ions, is often used for brain studies. Ordinarily, an ion such as TcO_4^- cannot pass through the biological barrier that separates brain cells from most of the constituents of the blood. However, certain tumors or other abnormalities breach this barrier, permitting radioactive TcO_4^- to enter brain tissue. The resulting accumulation of technetium-99 in brain tissue indicates to doctors that brain abnormalities are present.

Technetium-99 combined in other forms has an affinity for tissue types other than brain tissue. Technetium pyrophosphate, $Tc_2P_2O_7$, selectively collects in bone tissue; a ^{99}Tc-sulfur combination is taken up preferentially by cells of the liver, spleen, and bone marrow. Studies of the lung and kidneys are also possible using technetium radiopharmaceuticals.

Other radioisotopes are also used for medical diagnosis. Iron-59 is used to study the formation of red blood cells because the compound hemoglobin in red blood cells contains iron. Sodium-24 is commonly used to study blood circulation. A compound containing ^{24}Na is injected into the bloodstream, and the course of radioactive sodium throughout the body is followed with a Geiger counter.

medical treatment

Radioisotopes have been used for medical treatment since the discovery of radium. Malignant tissues are irradiated with γ-radiation, usually from a cobalt-60 source. Ionizing radiation is more damaging to fast-growing cancer cells than it is to normal tissues, so it is possible to kill the cancer cells while leaving most normal cells unharmed. The γ-rays are directed as much as possible toward the tumor, while normal tissue is shielded.

Another treatment technique is to use radioisotopes that are absorbed only by certain kinds of tissues. Radiation from ^{131}I can be used to treat cancerous thyroid glands because iodine is absorbed only by thyroid tissue and accumulates there. ^{123}I may also be used and has two advantages over ^{131}I: its half-life is shorter (13 hours) and it emits only higher-energy, more penetrating γ-radiation.

Radioactive iodine is also sometimes used to destroy healthy (nonmalignant) thyroid tissue. In Graves' disease, the thyroid is dangerously overactive. One treatment involves ingestion of a pill containing ^{131}I. The radioactive iodine accumulates in the thyroid gland, emits radiation, and destroys some thyroid tissue, thus diminishing the activity of the gland.

Leukemia can be treated by phosphorus-32, which becomes part of bone structure. Bones normally incorporate phosphorus into their makeup from dietary sources. Phosphorus-32 in bone tissue emits beta radiation. The excess white blood cells in the leukemia patient's bone marrow are thereby exposed to radiation and many are killed.

An application of radiation that has been approved more extensively for commercial use is the sterilization of packaged foods by irradiation, shown in the introductory photo to this chapter. The radiation kills microorganisms responsible for the rotting of food. Sterilization of food by irradiation is 50 times more energy efficient than thermal sterilization. Refrigeration requirements may be eliminated, and the shelf life of such foods is greatly enhanced. For example, Gulf shrimp have been treated with cobalt-60 gamma radiation to allow nonfrozen shipment elsewhere. However, considerable controversy still exists as to the conceivably harmful changes that the ionizing radiation may incur in food. Ionizing radiation may produce chemical changes in irradiated foods that can oxidize an oxidizable substance and reduce a reducible substance. The safety of irradiated foods must be subject to scrupulous testing and evaluation.

chemical uses of radioactive labeling

Many of the important concepts and theories of chemistry can be demonstrated or proved through the use of radioactive tagging or radioactive labeling. For example, it can readily be demonstrated that there is indeed a dynamic equilibrium present in a saturated solution of lead(II) chloride, $PbCl_2$, by the use of radioactive ^{212}Pb. The experiment would proceed as follows.

1. Begin with a saturated solution of unlabeled $PbCl_2$ in which we assume the following equilibrium exists.

$$PbCl_2(s) \rightleftharpoons \underset{\text{Nonradioactive}}{^{206}Pb^{2+}(aq)} + 2\,Cl^-(aq)$$

2. Add a few milliliters of tagged $^{212}Pb(NO_3)_2$ solution, a strong electrolyte that supplies aqueous, radioactive lead ions. The solution equilibrium can now be represented by

$$PbCl_2(s) \rightleftharpoons {}^{206}Pb^{2+}(aq) + {}^{212}Pb^{2+}(aq) + 2\,Cl^-(aq) + 2\,NO_3^-(aq)$$

3. If there is truly a dynamic equilibrium, ^{212}Pb will appear in the solid after a period of time. In forming $PbCl_2$, Cl^- does not care whether it teams up with $^{206}Pb^{2+}$ or $^{212}Pb^{2+}$. If there is no equilibrium reaction, the solid will not acquire radioactivity. Since the solid *does* become radioactive as solid $^{212}PbCl_2$ forms, the exchange between solid and ions in solution is demonstrated and can be represented by

$$^{206}PbCl_2(s) + {}^{212}PbCl_2(s) \rightleftharpoons {}^{206}Pb^{2+}(aq) + {}^{212}Pb^{2+}(aq) + 2\,Cl^-(aq) + 2\,NO_3^-(aq)$$

Another example of the use of radioactive labels can be seen in a study of photosynthesis.

$$6\,CO_2 + 6\,H_2{}^*O \xrightarrow{\text{Photosynthesis}} \underset{\text{Glucose}}{C_6H_{12}O_6} + 6\,{}^*O_2$$

The asterisk (*) indicates that the oxygen of water is radioactively tagged, whereas the oxygen of CO_2 is not. The appearance of the tag only in the product O_2 demonstrates that all of the oxygen in glucose comes from CO_2 and all of the oxygen in water becomes free oxygen gas.

Figure 17.16
The radioactive decay of americium-241 ionizes air molecules to produce a current in a smoke detector. When small particles of smoke are present, the current is reduced. The reduced current triggers an audible alarm to warn occupants of fire.

industrial uses of radioisotopes

The harnessing of nuclear energy to generate electric power is probably the most significant industrial use of radioisotopes and will be discussed in the next section. Tagging or labeling techniques, i.e., the replacement of minute numbers of inactive atoms by radioisotopes, are used for wear and corrosion tests. Smoke detection devices contain small amounts of radioactive materials; the detection is based on the interaction of smoke particles and radiation, which reduces an electric current and sounds the alarm (Figure 17.16).

CHEMLAB

Laboratory exercise 17: using radioactive isotopes to study chemical processes

In Section 17.10, you learned of the invaluable information that may be learned about chemical processes by using radioactive labeling. For example, you saw how it could be unequivocally demonstrated that there is a dynamic equilibrium present in a saturated solution of lead(II) chloride. The equilibrium between solid lead(II) chloride and dissolved $Pb^{2+}(aq)$ ions was shown to be dynamic by adding tagged (radioactive) $^{212}Pb^{2+}(aq)$ ions from $^{212}Pb(NO_3)_2$, a soluble salt of lead(II) nitrate. The tagged, aqueous $^{212}Pb^{2+}(aq)$ ions that became incorporated into the solid lead(II) chloride could be directly measured by filtering the solid from the solution and measuring its radioactivity. Besides a few beakers, a saturated solution of lead(II) chloride, and a solution of tagged lead(II) nitrate, the only equipment you would need for this experiment would be a filter funnel, filter paper, a lead safety shield, and a Geiger counter to measure the radiation.

Now that you understand the principle of tagging in chemical experiments and are familiar with chemical reactions, you should be able to design your own experiments to verify some hypothesized chemical processes. For each case below, design an unambiguous experiment to verify the given hypothesis. Describe the solutions and equipment you would use and the procedure you would follow.

Questions:

1. Several times in this chapter, it was asserted that radioactive radon gas, radon-222, is produced by the radioactive decay of radium-226, a component of granite rocks and concrete building materials. Devise an experiment to verify that the radon produced is indeed radioactive.
2. Several times throughout this textbook, it has been asserted that the equilibrium between gaseous carbon dioxide gas, $CO_2(g)$, and aqueous carbonic acid, $H_2CO_3(aq)$, is dynamic and may be shifted according to Le Chatelier's principle depending on the partial pressure of $CO_2(g)$ above the solution.

   $$H_2CO_3(aq) \rightleftharpoons H_2O(l) + CO_2(g)$$

   This dynamic equilibrium has been used to account for the behavior of carbonated beverages and the formation of stalactites and stalagmites. Devise an experiment using radioactive labeling to verify that this dynamic equilibrium does indeed exist.
3. Storing a solution of a strong base (such as sodium or calcium hydroxide) of known concentration is very difficult under ordinary laboratory conditions because it is hypothesized that carbon dioxide gas from the air may react with the base to lower its concentration. Carbonic acid, formed in a dynamic equilibrium when the carbon dioxide gas dissolves in water, is responsible for the reaction with hydroxide. The hydroxide, a base, undergoes a neutralization reaction with the carbonic acid that has formed. Consider a dilute solution of calcium hydroxide, $Ca(OH)_2$, which is hypothesized to react with carbon dioxide from the air. Write a balanced chemical equation to describe the neutralization reaction. Devise and describe an experiment with radioactive labeling to verify directly that gaseous carbon dioxide participates in the neutralization reaction.

17.11 nuclear energy

Radioisotopes and nuclear reactions have many other practical applications besides medical uses. For example, they provide an important energy source. All nuclear power plants in operation today employ **nuclear fission**, the splitting of one isotope into smaller nuclei. Nuclear fission produces a great deal of energy, which can be transferred as heat by a closed loop of liquid sodium or liquid water

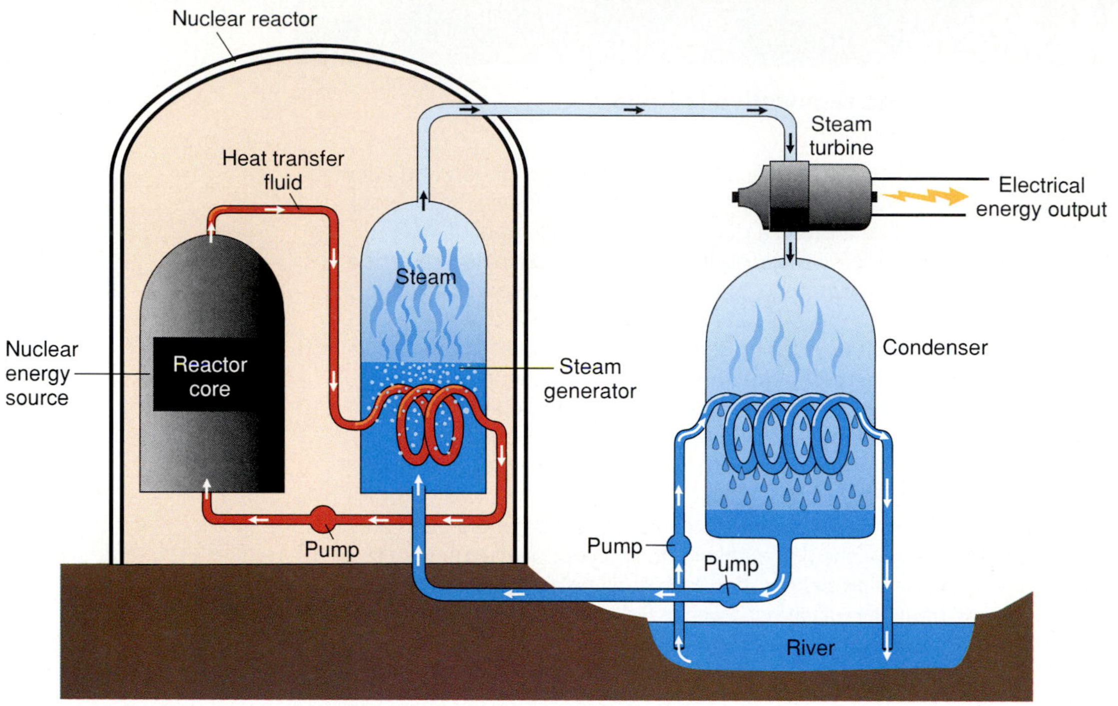

Figure 17.17

The energy produced by nuclear reactions in the core of a nuclear reactor is transferred by a closed loop of liquid sodium or liquid water under pressure to a steam generator. The steam runs a steam turbine, which produces electricity. Steam from the turbine is cooled by water from a nearby source, such as a river, and is pumped back into the steam generator.

under pressure to a steam generator. Steam is then used to drive turbines and produce electricity in the same way that heat energy from fossil fuels is used (Figure 17.17).

Nuclear fission of uranium-235 is the splitting of an atom upon bombardment by a neutron to yield barium-141 and krypton-92 according to the balanced equation

$$^{235}_{92}U + ^{1}_{0}n \rightarrow ^{141}_{56}Ba + ^{92}_{36}Kr + 3\ ^{1}_{0}n$$

Because more neutrons (three) are produced than are required for the fission to occur, a self-sustaining **chain reaction** will occur as long as there is a sufficient mass (**critical mass**) of uranium-235 with which the new neutrons can interact (Figure 17.18). The chain reaction is self-sustaining because the product neutrons can react with more uranium-235. After the first fission, no additional energy need be added and a great deal of energy is released. The amount of energy released is controlled by the presence of rods composed of boron or cadmium, elements that absorb some of the neutrons produced. If neutrons are absorbed, they cannot participate further in the chain reaction. By varying the depth of the control rods within the reactor core, the extent of the chain reaction is carefully controlled or stopped.

A major disadvantage of the use of nuclear fission in power plants is the fact that many of the products of fission are themselves radioactive. For example, the barium-141 and krypton-92 shown in the fission equation are both radioactive and decay to other radioactive species. Another waste product is plutonium-239, which has a half-life of 24,400 years. By the year 2000, it is estimated that 25,000 m^3 of high-level nuclear waste will be produced by commercial reactors each year. That volume of waste corresponds approximately to the volume of 25 three-story houses. Such radioactive waste presents a formidable problem of waste storage (disposal) for hundreds of thousands of years and poses enormous expense and potential health hazards for future generations throughout the world. In addition to the potential future health hazards from storage, the leakage of radiation following the major accident in 1986 at the Chernobyl nuclear power plant in the former USSR (CIS) and the smaller nuclear accident in 1979 at Three Mile Island in Pennsylvania dramatize the need for stringent safety precautions and regulation in using nuclear energy for

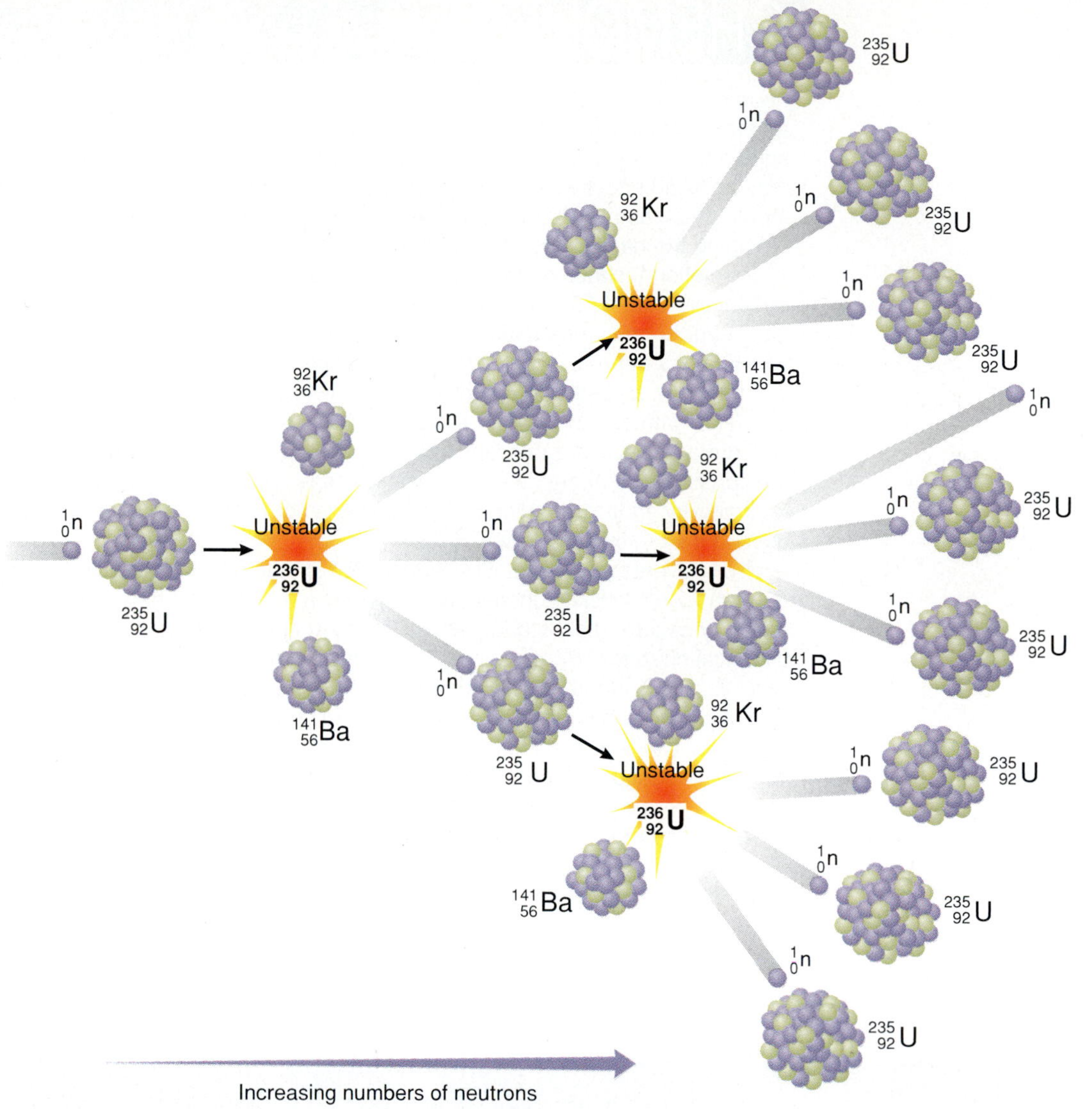

Figure 17.18

The fission of one uranium-235 nucleus by one neutron can set off a chain reaction if there is a sufficient number (critical mass) of the uranium-235 nuclei present.

commercial purposes. Handling plutonium waste is particularly hazardous because it flames upon contact with air.

Much more energy can be generated without the production of radioactive waste through **nuclear fusion**, which is the combination of two or more small nuclei into one larger nucleus. For example, the energy of the sun comes about through a series of nuclear fusion reactions in which four protons combine to form a helium nucleus, two positrons, and energy according to the balanced equation

$$4\,^{1}_{1}H \rightarrow {}^{4}_{2}He + 2\,^{0}_{1}e + \text{Energy}$$

where $^{0}_{1}e$, a *positively charged particle with the mass of an electron,* is called a **positron**.

Fusion is preferable to fission as an energy source because (1) fusion reactions produce more energy per amount of starting material than fission reactions, (2) the starting materials for fusion are more abundant than fissionable isotopes, and (3) the products of fusion are safe, i.e., not radioactive. However, there are extreme technological problems to be overcome before nuclear fusion can be a practical means of energy production. Most notably, fusion reactions require extremely high temperatures (1 to 5 million °C) for initiation, and no known substance can withstand these temperatures and act as a container for the reaction. However, research continues with the hope that fusion will meet the energy needs of the twenty-first century.

CHEMQUEST

Food may be sterilized by passing it under a beam of radiation to kill insects, parasites, and toxic microorganisms. Pathogenic bacteria such as vibrio in seafood, salmonella in eggs and poultry, and staphylococcus, as well as parasites such as trichina in pork, may be killed effectively. As a consequence of sterilization, the shelf life of irradiated cut meats, fresh fish, fresh fruits, and vegetables may be prolonged without significantly changing the natural character of the food.

Irradiation technology involves exposing food to β-rays, X-rays, or gamma rays from radioactive cobalt-60 or cesium-137. In 1985, it was approved for use on pork, but it has never been commercially used. The only extensive use of food irradiation in America has been to kill insects or bacteria on herbs and spices, such as those that go into hot dogs and other processed meats. In 1992, strawberries that had been irradiated with gamma rays at a plant in Mulberry, Florida, were first marketed.

Contrary to popular misconception, irradiated food presents no radiation threat to the consumer. Gamma-, X-, or β-radiation may penetrate the food without inducing radioactivity. Inducing radioactivity in foods is not a concern as long as neutrons are not used in the sterilizing beam.

A disadvantage of irradiation sterilization is that ionizing radiation that passes through food provokes chemical changes as a result of the oxidation and reduction of naturally occurring substances within the food itself. One consequence of these chemical changes is a lowering of the nutrient content of the food, especially vitamins A, C, E, and K. Another consequence is the possible formation of new compounds (radiolytic products), which may include carcinogens such as formaldehyde and benzene. The U.S. Food and Drug Administration estimates that all radiolytic products in irradiated food total no more than 30 parts per million. In some instances, the radiolytic products found in irradiated foods are less than those found in foods preserved by the more traditional methods of canning, pasteurization, and freezing.

Research which foods are suitable for irradiation. Is there any relationship between the water content of food and its suitability for irradiation? Reports on the relative amounts of radiolytic products in irradiated food and food preserved by pasteurization, canning, freezing, and freeze drying.

chapter accomplishments

After completing this chapter, you should be able to define all key terms and do the following.

17.1 Introduction

- ❑ Distinguish between "ordinary" chemical reactions and nuclear reactions.

17.2 The nucleus and radioactive decay

- ❑ Given a periodic table and the symbol for a nucleus that includes the mass number, indicate the numbers of protons and neutrons in the nucleus.

17.3 Radioactivity

- ❑ Name the three types of nuclear radiation.

17.4 Properties of α-, β-, and γ-radiation

- ❑ Write symbols that indicate the composition of α-, β-, and γ-radiation.
- ❑ State the charge and relative mass number of each of the three types of radiation.
- ❑ State the relative penetrating powers of α-, β-, and γ-radiation.

17.5 Nuclear reactions

- ❑ Write a nuclear equation representing the emission of an α-particle from some specified radioisotope.
- ❑ Write a nuclear equation representing the emission of a β-particle from some specified radioisotope.
- ❑ Complete nuclear equations representing bombardment reactions.

17.6 Radioactive decay series

- ❑ Explain why radioisotopes decay in a series of steps.
- ❑ State the event that terminates a radioactive decay series.

17.7 Ionizing radiation

- ❑ State the effect on matter of interaction with α-, β-, and γ-, or X-radiation.
- ❑ State the relative ionizing abilities of α-, β-, and γ-radiation.

- ❑ Name three ways to detect radiation.
- ❑ State the property of radiation on which the design of the cloud chamber and Geiger counter are based.

17.8 Half-life, $t_{1/2}$

- ❑ Given the amount of a radioactive sample and the half-life of the radioisotope, calculate the amount of sample remaining after some specified time interval.
- ❑ Explain how carbon-14 dating works.

17.9 Physiological effects of radiation

- ❑ Describe some of the physiological effects of exposure to nuclear radiation.

17.10 Uses of radioisotopes

- ❑ State at least one each of the medical, chemical, and industrial uses of radioisotopes.

17.11 Nuclear energy

- ❑ State one advantage and one disadvantage of using nuclear fission in electric power plants.
- ❑ State three advantages of nuclear fusion over fission as an energy source.
- ❑ State the technological problem that prohibits the practical use of fusion today.

key terms

radioactivity	transmutation	half-life	chain reaction
radioisotopes	trans-uranium elements	labeled compound	critical mass
alpha (α) radiation	disintegrate (decay)	radioisotope tracing	nuclear fusion
beta (β) radiation	ionizing radiation	nuclear fission	positron
gamma (γ) radiation	inverse square law		

problems

17.1 Introduction

7. What are the differences between "ordinary" chemical reactions and nuclear reactions in terms of the following?
   a. Subatomic particles participating
   b. Amount of energy change
   c. Effect of the state of chemical combination of an element

17.2 The nucleus and radioactive decay

8. Tell the number of protons and the number of neutrons in each of the following.
   a. ^{241}Am c. ^{3}H
   b. ^{63}Cu d. ^{40}Ar
9. Write nuclear symbols corresponding to the following.
   a. Magnesium-25 c. Iodine-131
   b. Gold-197 d. Lead-210
10. Write nuclear symbols for two isotopes of carbon.

17.3 Radioactivity

11. What is radioactivity?
12. How many types of nuclear radiation are there? Name them.
13. Describe the difference between the terms *radioisotope* and *isotope*.

17.4 Properties of α-, β-, and γ-radiation

14. Describe the mass and charge characteristics of α-, β-, and γ-radiation.
15. Write the proper nuclear symbols for α- and β-particles.
16. If only a heavy cloth curtain stood between you and a radioactive source, would you be endangered more by α- or γ-radiation? Explain.
17. Describe the type of shielding necessary for protection against the following types of radiation.
   a. α-particles
   b. β-particles
   c. γ-rays
   d. x-rays

17.5 Nuclear reactions

18. Give an example of transmutation.
19. Why does α-emission lead to transmutation?
20. Why does β-emission lead to transmutation?
21. Write nuclear equations for α-emission from the following isotopes.
   a. $^{222}_{86}Rn$
   b. $^{227}_{89}Ac$
   c. $^{235}_{92}U$

22. Write nuclear equations for β-emission from the following isotopes.

a. ${}^{40}_{19}K$

b. ${}^{131}_{53}I$

c. ${}^{14}_{6}C$

23. Based on the following partial equations, decide whether the emitting isotopes are α- or β-emitters.

a. ${}^{3}_{1}H \rightarrow {}^{3}_{2}He + ?$

b. ${}^{218}_{84}Po \rightarrow {}^{214}_{82}Pb + ?$

c. ${}^{211}_{83}Bi \rightarrow {}^{207}_{81}Tl + ?$

d. ${}^{35}_{16}S \rightarrow {}^{35}_{17}Cl + ?$

24. Identify X in each of the following nuclear reactions.

a. ${}^{15}_{8}O \rightarrow {}^{15}_{7}N + X$

b. ${}^{219}_{86}Rn \rightarrow {}^{4}_{2}He + X$

c. $X \rightarrow {}^{41}_{20}Ca + {}^{0}_{-1}e$

25. Why is the emission of γ-radiation often ignored in writing nuclear equations?

26. What is the nuclear "bullet" in each of the following bombardment reactions?

a. ${}^{242}_{96}Cm + ? \rightarrow {}^{245}_{98}Cf + {}^{1}_{0}n$

b. ${}^{6}_{3}Li + ? \rightarrow {}^{4}_{2}He + {}^{3}_{2}He$

c. ${}^{27}_{13}Al + ? \rightarrow {}^{28}_{14}Si$

d. ${}^{113}_{48}Cd + ? \rightarrow {}^{0}_{0}\gamma + {}^{114}_{48}Cd$

27. Complete the following bombardment reactions.

a. ${}^{197}_{79}Au + {}^{1}_{1}H \rightarrow ? + {}^{1}_{0}n$

b. ${}^{27}_{13}Al + {}^{1}_{0}n \rightarrow {}^{4}_{2}He + ?$

c. ${}^{12}_{6}C + {}^{2}_{1}H \rightarrow {}^{13}_{7}N + ?$

28. Complete the following nuclear equations.

a. ${}^{90}_{37}Rb \rightarrow {}^{90}_{38}Sr + ?$

b. ${}^{98}_{42}Mo + {}^{2}_{1}H \rightarrow {}^{99}_{43}Tc + ?$

c. ${}^{216}_{85}At \rightarrow {}^{212}_{83}Bi + ?$

d. ${}^{39}_{19}K + ? \rightarrow {}^{36}_{18}Ar + {}^{4}_{2}He$

e. ${}^{23}_{11}Na + {}^{2}_{1}H \rightarrow ? + {}^{4}_{2}He$

f. ${}^{3}_{1}H \rightarrow {}^{0}_{-1}e + ?$

g. ${}^{228}_{90}Th \rightarrow {}^{4}_{2}He + ?$

29. Which of the equations in Problem 28 represent the following?

a. α-emissions

b. β-emissions

c. Bombardment reactions

30. Give the symbols of all trans-uranium elements shown in your periodic table.

17.6 Radioactive decay series

31. Fill in the blanks in the following partial radioactive decay series.

$${}^{224}_{88}Ra \rightarrow {}^{4}_{2}He + \_\_\_\_\_ \rightarrow {}^{4}_{2}He + \_\_\_\_\_ \rightarrow$$
$${}^{4}_{2}He + {}^{212}_{82}Pb \rightarrow {}^{0}_{-1}e + \_\_\_\_\_ \rightarrow \_\_\_\_\_ + {}^{0}_{-1}e \rightarrow {}^{208}_{82}Pb + {}^{4}_{2}He$$

32. The series shown in Problem 31 terminates with the formation of ${}^{208}_{82}Pb$. Would you conclude that this isotope is radioactive? Explain.

17.7 Ionizing radiation

33. What is the effect of nuclear radiation on matter?

34. a. Which type of radiation is most penetrating?

b. Which type of radiation produces the greatest amount of ionization?

c. Which radiation measurement unit gives an amount of radiation weighted for different types of radiation?

35. Explain how radiation "tracks" form in a cloud chamber.

36. Explain how radiation causes a Geiger counter to click.

37. What actually exposes the film badges worn by those who work with radioactive materials?

17.8 Half-life, $t_{1/2}$

38. Compare sodium-24 and carbon-14 in Table 17.4. Why is sodium-24 a better choice than carbon-14 for use in blood circulation studies?

39. In an experiment, 1.6 g of ${}^{90}Sr$($t_{1/2}$ = 28 years) was buried in 1952. How much of this sample will be left in 2036?

40. A nuclear chemist needs 12 mg of ${}^{68}_{29}Cu$ ($t_{1/2}$ = 32 seconds) to do a particular experiment. At 10 A.M. a colleague brings 750 mg of ${}^{68}Cu$, which has just been synthesized. The chemist is distracted and does not begin to work until 10:10 A.M. Is there still sufficient ${}^{68}Cu$ for the experiment?

41. A hospital has a 24 g supply of ${}^{131}I$ ($t_{1/2}$ = 8 days) on January 1. Even if none is used by the staff, by what date will the sample have diminished to only 3 g?

42. Two ancient scraps of paper are found. Scrap A has a ${}^{14}C/{}^{12}C$ ratio, twice as large as that for scrap B. Which piece of paper is older?

43. The ${}^{14}C/{}^{12}C$ ratio in a deeply buried fossil is only 0.125 times that in a living plant. How old is the fossil?

44. Phosphorus-32 ($t_{1/2}$ = 14.3 days) can be used to locate brain tumors because, when injected into the body, it is preferentially absorbed by diseased brain tissue. Technetium-99 ($t_{1/2}$ = 6 hours) works just as well for this purpose. Why is the newer isotope technetium-99 used more prevalently?

45. The half-life of a particular radioisotope is 50 years.

a. Will all of the radioisotope be gone in 100 years?

b. If your answer to part a is no, what fraction of the original sample will be left after 100 years?

46. The element selenium occurs in crops in certain regions and is toxic to animals. The radioisotope selenium-75 can be used to label plants, and the uptake of selenium by animals can then be measured readily. The half-life of selenium-75 is 120 days. If an animal ingested 0.1 mg on January 1, 1985, in what year will there be less than 1 μg (1×10^{-6} g) in the animal (assuming that none is excreted)?

17.9 Physiological effects of radiation

47. What part of a living cell is most susceptible to radiation damage?
48. What is the first clinical symptom of exposure to higher levels of radiation?
49. Besides immediate clinical symptoms, what are the possible long-term effects of radiation exposure?

17.10 Uses of radioisotopes

50. Describe one way in which radioisotopes are used in a medical diagnostic procedure.
51. Describe one way in which radioisotopes are used in a medical treatment.
52. What is the function of radiation in the treatment of cancer?
53. How might ^3HCl (i.e., HCl in which the hydrogen is the radioactive isotope tritium, ^3_1H) be used to demonstrate the following equilibria in water?

$$HCl + H_2O \rightleftharpoons H_3O^+ + Cl^-$$

$$H_3O^+ \rightleftharpoons H^+ + H_2O$$

17.11 Nuclear energy

54. Identify the following as fission or fusion reactions.
    a. $^{235}_{92}U + ^1_0n \rightarrow ^{94}_{38}Sr + ^{139}_{54}Xe + 3\ ^1_0n$
    b. $^3_1H + ^1_1H \rightarrow ^4_2He$
    c. $^{13}_6C + ^1_0n \rightarrow ^{10}_4Be + ^4_2He$
55. Give at least two advantages for the fusion process compared to fission.
56. What is one practical problem preventing the commercial use of fusion?

Additional problems

57. When boron-10 is bombarded by α-particles, nitrogen-13 forms and a neutron is released. Write the equation for this reaction.
58. a. A sample of a fallen redwood tree exhibited a $^{14}C/^{12}C$ ratio one-half of that normally present in a living redwood. How long ago did the tree fall?
    b. Assume that the $^{14}C/^{12}C$ ratio is one-fourth of that normally present in a living redwood. How long ago did the tree fall?
59. Fill in the missing symbols in the following equations.
    a. $^{67}_{31}Ga + ^0_{-1}e \rightarrow ?$
    b. $? \rightarrow ^{13}_6C + ^0_{-1}e$
    c. $^{249}_{98}Cf + ^{18}_8O \rightarrow ? + 4\ ^1_0n$
60. a. A sample of a uranium compound contains all of the uranium isotopes in their normal natural distribution. Which of the following samples, each 100 g, will exhibit the greatest amount of radioactivity: UF_4, UCl_4, or UBr_4?
    b. Which of the following samples, each 0.50 mol, will exhibit the greatest radioactivity: UF_4, UCl_4, or UBr_4?
61. The first atomic explosion was detonated in the desert north of Alamogordo, New Mexico, on July 16, 1945. Approximately what fraction of the strontium-90 produced by the explosion will remain on July 16, 2001?
62. Suppose you have three radioactive isotopes, *A, B,* and *C,* which decay in the following series to a stable isotope, *D.*

$$A \xrightarrow{\text{4 seconds}} B \xrightarrow{\text{15 days}} C \xrightarrow{\text{1 second}} D$$

The half-life of *A* is 4 seconds, that of *B* is 15 days, and that of *C* is 1 second. Suppose you start with 1 mol of *A* and no *B, C,* or *D.* After 15 days, how many moles of *A, B, C,* and *D* would you have?

appendix a

basic arithmetic review

A.1 place values

The decimal system uses the ten digits, 0, 1, 2, 3, 4, 5, 6 7, 8, 9, to indicate the value of a particular place and thereby give the magnitude of a number. For example, a number indicating several common places is shown below.

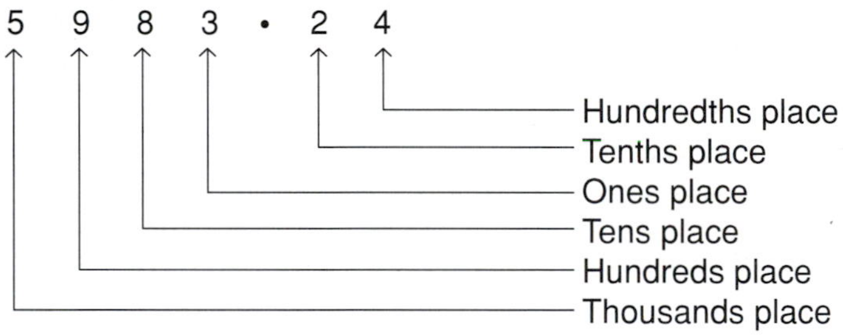

The digits 0 through 9 are used to indicate the value of each place. Thus, the number above is made up of 5 thousands, 9 hundreds, 8 tens, 3 ones, 2 tenths, and 4 hundredths.

Notice that the place values are always one-digit numbers. Two digits in a place indicate that one digit should be carried over to the next higher place. For example, 17 in the ones column should be written as 1 in the tens column and 7 in the ones column.

A.2 addition of positive decimal numbers

The key to the successful addition of decimal numbers is to line up the numbers so that digits of the same place value are found in the same vertical column.

sample exercise 1

Add 12.25 + 5.14 + 6.3

solution

1. Write the numbers one under another so that the digits of the same place value fall into the same vertical column. Note that 6.3 has no digit in the hundredths column.

$$\begin{array}{r} 12.25 \\ 5.14 \\ 6.3 \\ \hline \end{array}$$

2. Add the numbers in each column, "carrying over" any value of ten or greater in any column as one or more in the next column to the left.

$$\begin{array}{r} 12.25 \\ 5.14 \\ 6.3 \\ \hline 23.69 \end{array}$$

Note that the decimal point is brought straight down.

The value in the ones column was 13; 10 ones were carried to the tens column as 1 ten.

sample exercise 2

Add 8.93 + 0.0014 + 117

solution

1. Line up the numbers.

$$\begin{array}{r} 8.93 \\ 0.0014 \\ 117 \\ \hline \end{array}$$

2. Add.

$$\begin{array}{r} 8.93 \\ 0.0014 \\ 117 \\ \hline 125.9314 \end{array}$$

problem 1

Add the following

a. 1.79 + 5.32

b. 7.301 + 11.3 + 2.0903

A.3 subtraction of positive decimal numbers

The key again in subtraction is to line up the place values properly. Decimal point should lie under decimal point. In any place value, if you are trying to subtract a larger digit from a smaller digit, you must borrow ten units from the next place value to the left. This is illustrated in the exercise following.

sample exercise 3

Do the following subtraction

635.49 – 173.52

solution

1. Line up the numbers by place value, putting decimal point under decimal point.

$$\begin{array}{r} 635.49 \\ -173.52 \\ \hline \end{array}$$

2. Subtract, borrowing where necessary.

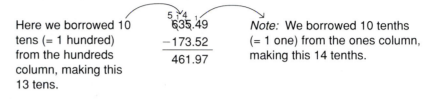

Here we borrowed 10 tens (= 1 hundred) from the hundreds column, making this 13 tens.

$$\begin{array}{r} {}^{5}\not{6}\,{}^{1}\not{3}\,{}^{4}\not{5}.{}^{1}49 \\ -173.52 \\ \hline 461.97 \end{array}$$

Note: We borrowed 10 tenths (= 1 one) from the ones column, making this 14 tenths.

• • • problem 2

Subtract the following

a. 18.93 – 11.76

b. 109.40 – 52.87

A.4 multiplication of positive decimal numbers

The decimal numbers are first multiplied using the same procedure as is used with whole numbers. The significant difference is that you must now locate the decimal point in the product. Add the number of decimal places (digits to the right of the decimal point) in the numbers being multiplied, and place the decimal point in the product so that the number of decimal places in the product will be equal to this sum. Zeros are added to the left of the product answer when there are not enough digits for the required number of decimal places.

sample exercise 4

Multiply 18.4 × 0.9

solution

1. Multiply the numbers as if they were ordinary whole numbers.

$$\begin{array}{r} 18.4 \\ \times 0.9 \\ \hline 16\ 56 \end{array}$$

2. Add the decimal places in the numbers that were multiplied and place a decimal point in the product so that the number of decimal places in the product equals this sum.

$$\begin{array}{rl} 18.4 & \leftarrow \text{One decimal place} \\ \times 0.9 & \leftarrow \text{One decimal place} \\ \hline 16.56 & \leftarrow \text{Requires two decimal places in the product} \end{array}$$

sample exercise 5

Multiply 1.7×0.04

solution

$$\begin{array}{r} 1.7 \\ \times\,0.04 \\ \hline 68 \end{array}$$

Simple multiplication gives → 68

← One decimal place (1.7)

← Two decimal places (0.04)

← Product requires three decimal places

A zero must be added to the left of 68 to hold the tenths place.

$$\begin{array}{r} 1.7 \\ \times\,0.04 \\ \hline .068 \end{array}$$

← Final answer

••• problem 3

Multiply the following

a. 78.45×3.2

b. 19.01×0.05

c. 0.14×0.25

A.5 division of positive decimal numbers

The key step in the division of decimal numbers is to move the decimal point in the *divisor* (denominator of a fraction) as many places to the right as necessary to make the divisor a whole number. The decimal point in the *dividend* (numerator of a fraction) is then moved to the right an equal number of places, with zeros being added if necessary to hold a decimal place.

$$\frac{148.2}{11.1} = 148.2 \div 11.1$$

$11.1\overline{)148.2}$ Move decimal in dividend one place to right.

Move decimal one place to right.

sample exercise 6

Divide the following

$$\frac{17.5}{.35}$$

$$\frac{17.5}{.35} = 17.5 \div .35$$

1. Move the decimal point in the divisor to obtain a whole number.

$.35\overline{)17.5}$

2. Move the decimal point in the dividend an equal number of places to the right, filling in zeros to hold places.

$35.\overline{)17.50}$ Decimal point was moved two places to right with a zero filled in.

3. Locate the decimal point in the *quotient* (answer) immediately above its current position in the dividend.

```
35.)1750.
```

4. Zeros may be placed to the right of the decimal in the dividend and the division carried out to as many decimal places as desired.

```
      50.0
35.)1750.0
    175
    ----
      000
```

sample exercise 7

Divide 1.482 by .111

solution

1. .111)1.482

2. 111.)1.482

3. 111.)1482.

4.

```
        13.35
111.)1482.00
     111
     ---
      372
      333
      ---
       390
       333
       ---
        570
        555
        ---
         15
```

••• problem 4

Divide the following

a. $78.3 \div 1.1$

b. $12.90 \div .638$

c. $0.384 \div .15$

A.6 direct and inverse proportionality

direct proportionality

There are many experiments in chemistry in which one measures two variable properties of a sample and finds that an increase or decrease in one variable leads to an increase or decrease in the other variable. For example, an increase in the volume of a liquid leads to an increase in the mass of the liquid. (Volume and mass are properties of matter that are defined and discussed in Section 2.4.)

If an increase (or decrease) by a given multiplying factor in one variable produces an increase (or decrease) by the same factor in the other variable, then the two variables are said to be directly proportional. Experiments show that if we double the volume of a liquid, we find that the mass doubles. If we triple the volume, we triple the mass. If we cut the volume in half, we find that the mass is also cut in half. Because this is true, volume and mass are said to be directly proportional. The symbol $\propto$ is shorthand for the words *is proportional to.* We can write

$$M \propto V$$

The direct proportionality can be converted into an equation by inserting a proportionality constant k. Thus,

$$M \propto V \qquad \text{can also be expressed} \qquad M = kV$$

---

sample exercise 8

Solve for k in $M = kV$

solution

Divide both sides by V.

$$\frac{M}{V} = \frac{k\cancel{V}}{\cancel{V}} = k$$

$$k = \frac{M}{V}$$

In Section 3.4, the proportionality constant in this problem is given the name density.

If any two variables x and y are directly proportional, then the equation relating them will have the form $x = ky$, where k is a constant. As we did in Sample Exercise 8, we can rearrange this equation to read $k = x/y$. This tells us that if two variables are directly proportional, their ratio is a constant.

---

sample exercise 9

Given the following data, are x and y directly proportional?

x	y
3	6
6	9
9	12

solution

If two variables are directly proportional, then their ratio is a constant. When we examine the ratio y/x, we find $6/3 = 2/1$, $9/6 = 3/2$, and $12/9 = 4/3$. The ratio y/x is not constant, and therefore x and y are *not* directly proportional. Another way of identifying directly proportional variables is to graph them (Section 2.6). Such a graph is always a straight line through the origin.

The concept of direct proportionality is particularly important in understanding Boyle's law (Section 11.4).

inverse proportionality

If two variables are related such that an increase (decrease) in one by a given factor produces a decrease (increase) in the other by the same factor, then the two variables are said to be inversely proportional. For example, average exam score and difficulty of a test may be inversely proportional. If we make the test twice as difficult (increase by a factor of 2), the average exam score will be cut in half (decrease by a factor of 2).

The inverse proportionality between two variables *x* and *y* can be represented as follows.

$$x \propto \frac{1}{y}$$

We can convert the inverse proportionality into an equation by inserting a proportionality constant. Thus,

$$x \propto \frac{1}{y} \qquad \text{can also be expressed} \qquad x = \frac{k}{y}$$

sample exercise 10

Solve for *k* in $x = \dfrac{k}{y}$

solution

Multiply each side of the equation by *y*.

$$xy = k$$

Sample Exercise 10 shows that if two variables *x* and *y* are inversely proportional, then their product must be equal to a constant; that is, $xy = k$. This means that if *x* is doubled, *y* must be cut in half in order for the product *xy* to remain a constant. The equation of two variables *x* and *y* that are inversely proportional is always of the form $xy = k$.

sample exercise 11

Given the following data, are *x* and *y* inversely proportional?

x	*y*
16	12
8	24
4	48

solution

If two variables are inversely proportional, then their product is a constant. When we examine the product *xy*, we find (16)(12) = 192, (8)(24) = 192, and (4)(48) = 192. The product *xy* is a constant, and therefore *x* and *y* are inversely proportional.

The concept of inverse proportionality is particularly important in understanding Charles' law (Section 11.5).

appendix b

chemical arithmetic using a hand calculator

B.1 introduction

Using an electronic hand calculator can speed up solving numerical problems. In this review, we summarize those operations that apply to problems in this text. Calculators use an operating system based on either algebraic notation or reverse polish notation. Most beginning students have the first type, and we limit our discussion to that type.

B.2 addition/subtraction

Before doing a new calculation, press the CLEAR key to delete any previous data or operation. The + key is used for addition and – for subtraction. The ± key changes the sign of a number in the display. Pressing the = key provides the result of the operations in the order in which they were entered. Operations are entered by reading calculations from left to right. Let's look at a few examples.

sample exercise 1

Calculate $8.94 + 0.39 - 3$

1. Enter the number 8.94.
2. Press + for addition.
3. Enter the number 0.39. [At this point, the calculator will have added 0.39 to 8.94. You could check this by pressing =.]
4. Press – for subtraction.
5. Enter 3.
6. Press =. The calculator will display the result, 6.33.

sample exercise 2

Calculate $-4.95 - (-2.1) + 0.93$

1. Enter the number 4.95.
2. Press ± to change the sign of 4.95 to –4.95.
3. Press – for subtraction.
4. Enter 2.1.

5. Press ± to change the sign of 2.1 to –2.1.
6. Press + for addition.
7. Enter 0.93.
8. Press = to display the result, –1.92.

B.3 multiplication/division

The × key is pressed for multiplication, and the ÷ key is pressed for division.

sample exercise 3

Calculate $7.21 \times 4.01 \div 6.35$

1. Enter the number 7.21.
2. Press ×.
3. Enter 4.01.
4. Press ÷.
5. Enter 6.35.
6. Press = to display the result, 4.55.

sample exercise 4

Calculate $-8.50 \div 2.30 \times -1.75$

1. Enter the number 8.50.
2. Press ± to change the sign of 8.50 to –8.50.
3. Press ÷.
4. Enter 2.30.
5. Press ×.
6. Enter 1.75.
7. Press ± to change the sign of 1.75 to –1.75.
8. Press = to display the result, 6.47.

Note that in multiplication and division, it is not necessary to change signs when numbers are entered. The appropriate sign in the answer can be determined as described in Appendix C.2.

B.4 combined addition/subtraction and multiplication/division

In combining addition/subtraction with multiplication/division operations, you need to be careful about the order in which the operations are carried out. For example, a chain of operations such as $3 \times 5 + 2 \times 4$, shown without parentheses, could be carried out as

$$
\begin{aligned}
&(3 \times 5) + (2 \times 4) = 23 \\
\text{or}\quad &((3 \times 5) + 2) \times 4 = 68 \\
\text{or}\quad &3 \times (5 + 2) \times 4 = 84 \\
\text{or}\quad &3 \times (5 + (2 \times 4)) = 39
\end{aligned}
$$

Some calculators using the algebraic operating system use a priority system of operations. Multiplications and divisions are carried out before additions and subtractions. With such a calculator, the result 23 is obtained.

Other calculators simply apply the operations in order from left to right. Such a calculator would yield 68 by first multiplying $3 \times 5 = 15$, then adding 2 to obtain 17, and then multiplying 17×4 to obtain 68.

If your calculator has parenthesis keys, you can have it carry out a chain operation in any order you wish. For example, by using parentheses before 5 and after 2, the example reads $3 \times (5 + 2) \times 4$. The calculator will first carry out the operations within parentheses, in this case 5 + 2 yielding 7, then multiply 3×7 to obtain 21, and then multiply by 4 to display 84.

B.5 scientific notation

Many calculators have an EE or EXP (exponential) key that allows decimal numbers to be entered in scientific notation. Numbers in scientific notation can be used in arithmetic operations with other numbers in scientific notation or with ordinary decimal numbers.

---

sample exercise 5

Enter 6.02×10^{23}

1. Enter the number 6.02.
2. Press EE or EXP.
3. Enter 23.

---

sample exercise 6

Enter 1.67×10^{-24}

1. Enter the number 1.67.
2. Press EE or EXP.
3. Enter 24.
4. Press ± to change the exponent 24 to −24.

---

sample exercise 7

Multiply the following numbers in scientific notation

$$(6.02 \times 10^{23}) \times (1.67 \times 10^{-24})$$

1. Enter 6.02×10^{23} as in Sample Exercise 5.
2. Press ×.
3. Enter 1.67×10^{-24} as in Sample Exercise 6.
4. Press = to display the result, 1.01.

In using results from your calculator, remember that the calculator does not necessarily yield an answer to the correct number of significant figures. In fact, it rarely does. You must consider the measured quantities and the arithmetic operations and round to the appropriate number of sig figs as described in Sections 2.8 and 2.9.

appendix c

arithmetic with signed numbers

C.1 signed numbers

What is a signed number? Every number has a magnitude and a sign. The sign can be either positive or negative. If the sign is omitted, it is assumed to be positive.

+17.8	−13.2	4.9
Magnitude is 17.8.	Magnitude is 13.2.	Magnitude is 4.9.
Sign is positive.	Sign is negative.	Sign is omitted, so we assume it is positive (+).

C.2 multiplication of signed numbers

The product of the multiplication of two factors is positive if both numbers have the same sign and negative if they have opposite signs.

In the following examples, the factors have *like signs,* so the products are *positive.*

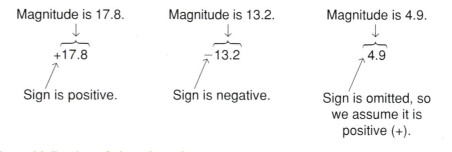

$(3)(6) = 18$ — *Both* positive → Positive

$(-3)(-6) = 18$ — *Both* negative → Positive

In the following examples, the factors have *opposite signs,* so the products are *negative.*

$(-3)(6) = -18$ — Negative, Positive → Negative

$(3)(-6) = -18$ — Positive, Negative → Negative

The sign of the product of more than two numbers can always be worked out two numbers at a time as Sample Exercise 1 shows.

sample exercise 1

What is the product of the following multiplication?

$$\begin{array}{lccccc} & (-3) & (6) & (4) & (5) & (-2) \\ \text{Factor} & a & b & c & d & e \end{array}$$

solution

1. Work out the magnitude of the product of all of the factors, which is | 720 |. The two vertical parallel lines indicate that no sign is implied.
2. Work out the sign of the product, taking two signs at a time. Taking the first two factors,

$$\begin{array}{ccc} a\ b & & a \times b \\ (-)(+) & = & (-) \end{array}$$

Multiplying times the third factor,

$$\begin{array}{ccc} a \times b\ c & & a \times b \times c \\ (-)\ (+) & = & (-) \end{array}$$

Then times the fourth factor,

$$\begin{array}{ccc} a \times b \times c\ d & & a \times b \times c \times d \\ (-)\quad (+) & = & (-) \end{array}$$

Finally times the fifth factor,

$$\begin{array}{ccc} a \times b \times c \times d\ e & & a \times b \times c \times d \times e \\ (-)\quad (-) & = & (+) \end{array}$$

So the product is +720.

It turns out that if there is an *even number* of minus signs in a series of factors to be multiplied, the product will always be positive. This is the case in Sample Exercise 1, in which there are two minus signs. An *odd number* of minus signs leads to a negative product.

$$(-3)(-6)(4)(-5) = -360$$

$$\begin{array}{cc} + & + \\ + & - \\ & - \end{array}$$

• • • problem 1

Indicate the sign of the product for the following multiplications.

a. (3)(–6)(2)(4)
b. (–4)(–5)(–3)(5)

C.3 division of signed numbers

The sign of a quotient is determined in the same way that we determine the sign of a product. That is, the quotient of two numbers is positive if both numbers have the same sign and negative if they have opposite signs.

$$18 \div 6 = 3 \qquad -18 \div -6 = 3$$
$$-18 \div 6 = -3 \qquad 18 \div -6 = -3$$

Also as in multiplication, even numbers of minus signs lead to positive quotients and odd numbers of minus signs lead to negative quotients.

$-18 \div 6 \div -3 = +1$	$-18 \div 6 \div 3 = -1$
Two negative signs	One negative sign

Where more than one division is indicated, proceed to divide from left to right two terms at a time.

C.4 addition of two signed numbers

1. In the addition of two numbers of like sign, simply add the magnitudes and attach the sign involved.

$$3 + 6 = 9$$
$$-3 + (-6) = -9$$
$$+41 + 13 = 54$$

2. In the addition of two numbers of opposite sign, subtract the one of smaller magnitude from the one of larger magnitude and apply to the answer the sign of the number of larger magnitude.

sample exercise 4

Add −16 + 3.

solution

1. Recognize that the number of smaller magnitude is 3.
2. Subtract 3 from 16: 16 − 3 = 13.
3. The sign of the number of larger magnitude is negative, so the answer is −13.

C.5 subtraction of two signed numbers

The subtraction problem is converted into an addition problem by the following steps.

1. Change the sign for the subtraction process (−) into the sign for the addition process (+).
2. Change the sign of the subtrahend (the second number) into the opposite sign.

14 −	+ 6	=	14 +	(−6)
↑	↑		↑	↑
Sign indicating subtraction	Sign of subtrahend is plus		Sign indicating addition	Sign is now minus

Subtraction	Addition
$-38 - (-14)$	$= -38 + (+14)$
$14 - (-15)$	$= 14 + (+15)$
$-18 - 13$	$= -18 + (-13)$

This method works because subtraction is defined as addition of the negative (opposite sign) subtrahend.

Now that we have converted the problem into an addition problem, we simply follow the rules for the addition of two signed numbers.

$$14 + (-6) = 8$$
$$-38 + 14 = -24$$
$$14 + 15 = 29$$
$$-18 + (-13) = -31$$

••• problem 2

Work out the following.

a. (–1.4)(1.9)
b. (–3.1)(–2.4)
c. (6)(3)(2)
d. (–6)(–3)(–2)

••• problem 3

Do the following divisions.

a. 3.9 ÷ –1.3
b. –3.9 ÷ –1.3
c. –3.9 ÷ 1.3

••• problem 4

Calculate the following.

a. 7.4 + 3.8
b. 7.4 + (–3.8)
c. –7.4 + (–3.8)
d. 7.4 – 3.8
e. –7.4 – 3.8
f. –7.4 – (–3.8)

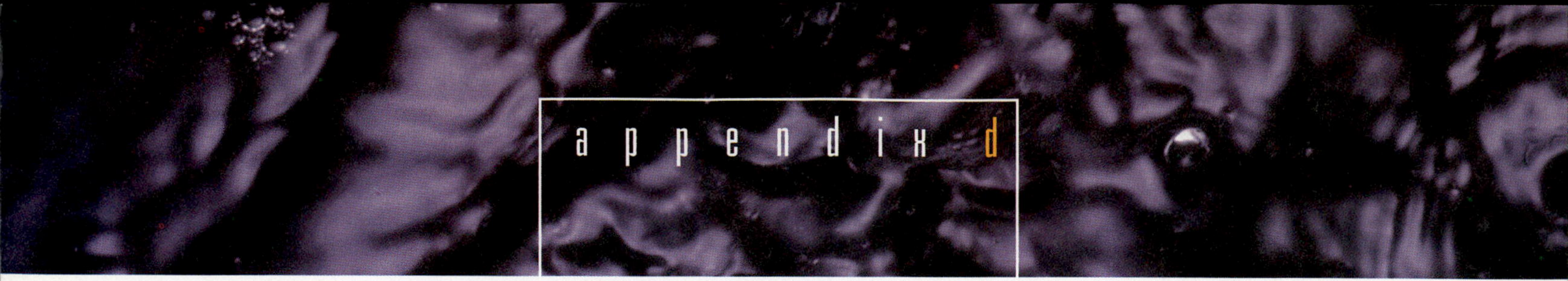

appendix d

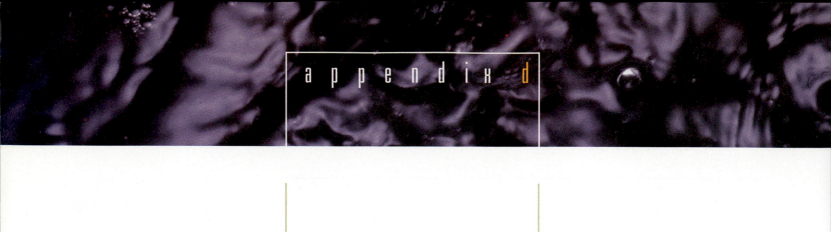

vapor pressure of water at various temperatures

Temperature (°C)	Vapor pressure (mmHg)	Temperature (°C)	Vapor pressure (mmHg)
5	6.5	34	40.0
10	9.2	35	42.2
15	12.8	36	44.6
16	13.6	37	47.1
17	14.5	38	49.7
18	15.5	39	52.4
19	16.5	40	55.3
20	17.5	45	71.9
21	18.6	50	92.5
22	19.8	55	118.0
23	21.1	60	149.4
24	22.4	65	187.5
25	23.8	70	233.7
26	25.2	75	289.1
27	26.7	80	355.1
28	28.3	85	433.6
29	30.0	90	525.8
30	31.8	95	633.9
31	33.7	100	760.0
32	35.7		
33	37.7		

appendix e

answers to odd-numbered problems

chapter 1 classification of matter

1. Some possible categories might be:

   Food—fats, proteins, carbohydrates or Chinese, Italian, Mexican, etc.

   Clothes—casual, dress, work, or winter, summer, spring and fall.

   Friends—boy (or girl) friends, close friends, best friend, and casual friends.

   Jobs—part-time, full time, white collar or blue collar.

   Courses—required, elective, hard, easy, majors or non-majors.

3. a. A page in this book would be a mixture because paper is made up of a combination of different fibers and chemicals.
   b. Skin is a mixture because there are several layers to the skin, each made up of its own special combination of substances.
   c. Pure iron metal is a pure substance because it is composed of only iron atoms and nothing else.
   d. A rusty nail is a mixture because it is composed of the original iron atoms, but also contains rust, a compound which is made up of iron and oxygen.

5. Some physical properties of laboratory glassware would be it is odorless, colorless, transparent, brittle, solid at room temperature, and a poor conductor of electricity.

7. You might distinguish a test tube full of methane from a test tube full of oxygen by attempting to ignite each. Methane will burn and oxygen will not.

9. Of the materials listed in Sample Exercise 6 and Problem 8, only the elements and compounds are pure substances. They would be gold, iron filings and iodine combined into a new material, chalk, magnesium and vitamin C.

11. Matter is defined as anything that occupies space and has mass.

13. The statement "a gas has indefinite volume", means that a gas will expand to fill any size container.

15. Gases and liquids can be poured from one container to another container. Solids in a powdered form can also be poured.

17. To convert a liquid into a solid, you would lower its temperature to its freezing point. To convert a liquid into a gas, you would raise its temperature to its boiling point.

19. The names of the following processes are:
    a. Ice cubes to water is melting.
    b. Steam forming water droplets on a mirror is condensation.
    c. Milk made into ice milk is freezing.
    d. Frozen laundry "drying" is sublimation.
    e. Perspiration "drying" is evaporation.

21. Three physical states are represented inside the sealed bulb; gas by the air, liquid by the water; and solid by the ice and wood.

23. Classifying each of the materials in Problem 18 as either a pure substance or a mixture would give the following results:
    a. Ice cream is a mixture.
    b. Iodized table salt is a mixture.
    c. Cola is a mixture.
    d. A nickel is a mixture (an alloy).
    e. Dry ice is a pure substance.
    f. Snow is a pure substance (water). (Dirty snow could be considered a mixture.)
    g. Methane is a pure substance.
25. a. An experiment to prove that a cola is a mixture could include observing escaping gas and the consequent "flatness" acquired by the liquid. You could heat the cola to evaporate the solvent. The syrup remaining is thicker than the original cola at room temperature.
    b. An experiment to prove that dry ice is a pure substance could involve letting the dry ice sublime in a closed, evacuated container. Then decrease the temperature to the freezing point and the solid will be the same as the original solid.
27. Identifying the chemical and physical changes in each of the following examples:
    a. A lump of sugar is ground to a powder (physical change) and then heated in air. It melts (physical change), then darkens (chemical change), and finally bursts into flames and burns (chemical change).
    b. Gasoline is sprayed into the carburetor, mixed with air (physical change), converted to a vapor (physical change), burned (chemical change), and the combustion products expand (physical change) in the cylinder.
29. Identifying each of the following as a chemical or physical change:
    a. Melting butter is a physical change.
    b. Iron rusting is a chemical change.
    c. A flash cube going off is a chemical change.
    d. Banging on a drum is a physical change.
31. Any three of the following properties might be used to identify specimens as pure substances: color, odor, boiling point, melting point, and flammability.
33. Properties that distinguish white solids such as table salt and table sugar from one another are melting point, tendency to burn in air, and the ability to dissolve in cold and hot water.
35. The chemical and physical properties in this example are: chlorine is a yellow-green gas (physical property) which is toxic to humans (chemical property). It combines with sodium to form sodium chloride (chemical property), a solid that melts at 808°C (physical property).
37. The original blue powder is a compound because it formed two new substances when heated. An element might have changed physical state, but would have been retrievable in the end.
39. Pure substances can be classified into two categories, compounds and elements.
41. There are more compounds in the world than elements. The periodic table lists only 109 elements (depending on which table you use). The are millions of compounds formed by combining those elements in various proportions and configurations.
43. If elements *A* and *B* combine in a fixed proportion of 3:1 by weight, trying to combine 4 g of *A* with 1 g of *B* will result in 3 g of *A* combining with 1 g of *B* to form 4 g of *AB*. One gram of *A* will remain unreacted.
45. Classifying each of the following as elements, compounds or mixtures:
    a. Sodium is an element.
    b. Spaghetti with meatballs is a mixture.
    c. Carbon particles in a hydrogen gas atmosphere is a mixture.
    d. A cup of tea is mixture.
    e. Laughing gas is a compound.
    f. Methane is a compound.
47. The names of the elements represented by the following symbols are: H = hydrogen, Ca = calcium, Si = silicon, C = carbon, Cl = chlorine, P = phosphorus, and K = potassium.
49. Classifying the elements in Problems 47 and 48 as metals or nonmetals:

In Problem 47,	hydrogen	nonmetal
	calcium	metal
	silicon	nonmetal
	carbon	nonmetal
	chlorine	nonmetal
	phosphorus	nonmetal
	potassium	metal
In Problem 48,	bromine	nonmetal
	nitrogen	nonmetal
	mercury	metal
	silver	metal
	gold	metal

51. Classifying each of the following as elements, compounds or mixtures:
   a. C_2H_2 is a compound.
   b. O_2 and N_2 taken together is a mixture.
   c. Mg is an element.
   d. K_2SO_4 is a compound.
   e. H_2O and NaCl taken together is a mixture.
53. Classifying each of the following as either a chemical or physical property:
   a. Aluminum melting at 660°C is a physical property.
   b. Aluminum forming a thin oxide coating in air is a chemical property.
   c. Naphthalene (moth balls) subliming at room temperature is a physical property.
   d. Mercury (II) oxide forming mercury and oxygen when heated is a chemical property.
   e. Butter turning rancid when left unrefrigerated in open air is a chemical property.
   f. Water evaporating from an open container is a physical property.
   g. Sugar dissolving in water is a physical property.
   h. Oil burning and giving off carbon dioxide, water and heat is a chemical property.
55. The names of the elements represented by the following symbols are: Cu = copper, Hg = mercury, Pb = lead, Sn = tin, S = sulfur, and P = phosphorus.
57. In the following symbols:
   a. CH_4 is correctly written.
   b. AL is incorrect, the second letter must be lowercase—Al.
   c. cL is incorrect, the first letter should be capital, second lower case—Cl.
   d. Zn is correctly written.
   e. h is incorrect, the letter must be capital—H or diatomic H_2.
59. A substance that is soft, shiny, and conducts electricity describes properties usually found in a metal.
61. Identifying each of the following as a chemical or physical change:
   a. Tea being sweetened by adding sugar describes a physical change because the sugar just dissolves.
   b. Smoking tobacco in a pipe describes a chemical change because the tobacco is burning in the presence of oxygen.
   c. Frozen lemonade being reconstituted is a physical change because the lemonade is being melted and diluted.
   d. A pond's surface freezing in the winter is a physical change because the water just changes states from liquid to solid.
   e. A candle's light slowly diminishing and going out is a chemical change of the wax burning in the presence of oxygen. The light is a by-product of this reaction.

chapter 2 measurement

1. a. 1.73×10^2
   b. 2.9×10^{-3}
   c. 1.31982×10^5
   d. 4.01×10^{-6}
   e. 1.64×10^1
3. a. 2.30×10^7
   b. 1.65×10^2
   c. 3.84×10^{-14}
   d. 3.16×10^2
5. 68°F
7. Using the graph of the data given to determine the mass corresponding to a volume of 6.5 mL would give approximately 8.4 g.

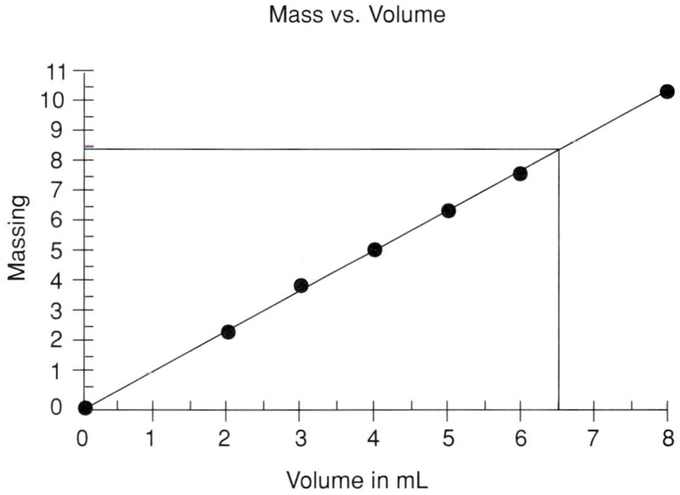

9. 23
11. Measurement is the comparison of some quantity to a reference standard.
13. a. 5.87×10^1
   b. 8.2×10^{-2}
   c. 6.31×10^8
   d. 3.012×10^3
   e. 7.398×10^1
   f. 7.18×10^{-10}

15. a. 2.0×10^4
 b. 5.0×10^3
 c. 1.9×10^0 or 1.9
 d. 4.5×10^{-2}
 e. 3.3×10^{-3}
 f. 2.0×10^{-32}
17. kilo means 1000
 deci means 1/10 or 0.1
 centi means 1/100 or 0.01
 milli means 1/1,000 or 0.001
 micro means 1/1,000,000 or 0.000001
19. a. mm stands for millimeter
 b. mL stands for milliliter
 c. g stands for gram
 d. mg stands for milligram
 e. km stands for kilometer
21. a. Centi- would be 10^{-2}
 b. Kilo- would be 10^3
 c. Milli- would be 10^{-3}
23. –25°C is higher than –15°F
25. 58°C
27. a.

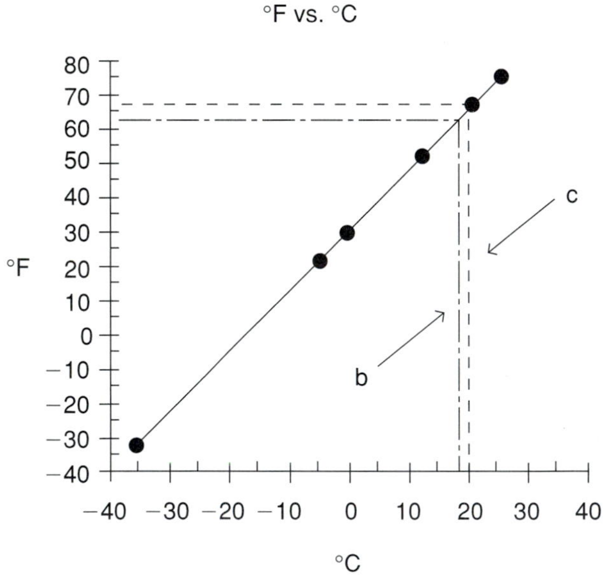

 b. 18°C is approximately 64°F
 c. 68°F is approximately 20°C

29. a.

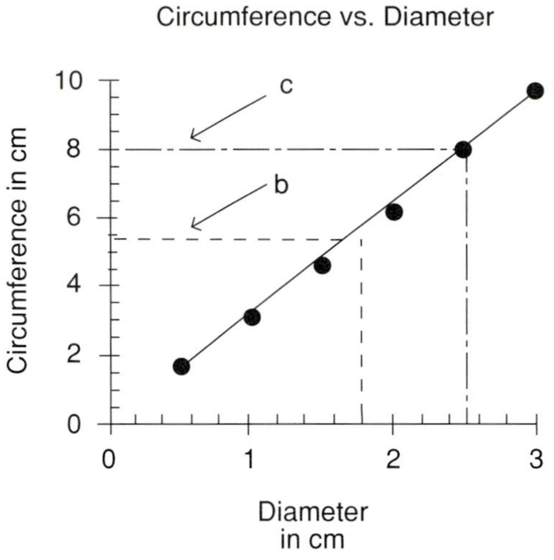

 b. 5.4 cm
 c. 2.6 cm
31. a. three significant figures
 b. one significant figure
 c. four significant figures
 d. two significant figures
 e. one significant figure
 f. two significant figures
 g. three significant figures
 h. three significant figures
33. a. 9.4
 b. 9
 c. 0.002
 d. 1.91×10^{-2}
 e. 5.1×10^{-3}
35. a. 7.015 lb
 b. 7.01 lb

chapter 3 applying measurements to chemical calculations

1. a. 1 kg = 2. 2 lb, are $\frac{1\text{ kg}}{2.2\text{ lb}}$ and $\frac{2.2\text{ lb}}{1\text{ kg}}$
 b. 1 L = 1,000 mL, are $\frac{1\text{ L}}{1{,}000\text{ mL}}$ and $\frac{1{,}000\text{ ml}}{1\text{L}}$
3. 96. g Cu

5. 53 mL carbon tetrachloride
7. a. 0.045
   b. 0.33
   c. 0.00092
9. Percent = 56%
11. 253 J
13. a. 12
    b. −3.8
    c. 56.6
    d. 96
15. $\frac{3}{19} \neq \frac{8}{57}$
17. a. $\frac{1}{2}$
    b. $\frac{-5}{48}$
    c. $\frac{3 \text{ feet}}{100 \text{ centimeters}}$
    d. $\frac{8{,}760 \text{ hours}}{1 \text{ year}}$
    e. $\frac{-64}{27}$
19. a. 0.0189 km
    b. 0.375 m
    c. 145. mL
    d. 0.452 L
    e. 645 mL
    f. 89. g
    g. 1.9×10^3
    h. 3.84×10^4
21. a. 5.3×10^3
    b. 0.000385 km
    c. 0.003910 kg
    d. 8.2 cm
23. a. 8.4×10^4 cm
    b. 0.106 gal
    c. 42.6 g
25. 109 yd
27. 212 mi
29. 41 cm
31. 18.4 m
33. a. 1.5×10^3 m
    b. 0.93 mi
35. 30.2 cents per liter
37. $\frac{11 \text{ km}}{\text{L}}$
39. a. $\frac{88 \text{ km}}{1 \text{ hr}}$
    b. $\frac{9.6 \text{ km}}{1 \text{ L}}$
41. a. $\frac{55 \text{ mi}}{1 \text{ hr}}$
    b. $\frac{23 \text{ mi}}{1 \text{ gal}}$
43. a. 3.85×10^8 m
    b. 1.28 s
45. $\frac{0.7016 \text{ g}}{1 \text{ mL}}$
47. a. 23 g of gasoline
    b. 999 g of vinegar
    c. 17 g of olive oil
49. 1 lb
51. a. 2.2 lbs of blood
    b. 1.4 lbs of gasoline
53. $\frac{11.3 \text{ g}}{1 \text{ cm}^3}$
55. 1.9×10^5 g
57. $C = \frac{B}{A}$
59. $C = \frac{B}{A - 9}$
61. x = 65
63. $x = 2 \times 10^1$
65. $H = 4.5 \times 10^3$ J
67. Percentage = 76%
69. Percentage = 82.5%
71. Percentage = 16.66% = 17%
73. $\frac{3}{8}$ is greater
75. The two major classifications of energy are potential energy and kinetic energy.

77. a. Examples of potential energy as a result of position would be the energy stored in a compressed spring or the energy stored in a book held above the floor.

    b. An example of potential energy as a result of composition would be the chemical energy stored in a compound, e.g., gasoline, TNT, or nitroglycerine.

    c. Yes. An object with potential energy due to its composition could also posses potential energy because of its position.

79. The law of conservation of energy states that although energy can change from one form to another, the total amount of energy in the universe is constant.

81. Humans that take in more energy than they use up store the excess chemical energy as fat.

83. There are 4.35×10^4 joules in 43.5 kilojoules.

85. $13{,}167\ J \equiv 1.3 \times 10^4\ J$

87. 2.3×10^2 grams of water

89. $\text{Specific Heat} = 0.444 \dfrac{J}{g°C}$

91. a. X°C = 20°C

    b. X°C = 27°C

93. a. 1.6×10^4 cm

    b. 5.7 cm

    c. 3.340×10^{-3} kg

    d. 0.914 m

    e. 1.68×10^{-3} lb

    f. 3.2×10^2 ml

95. 0.23 cm^3

97. a. 103.1°F

    b. 86.4 kg

    c. 4.3×10^3 mg

    d. 1.7×10^2 mL

99. a. $\dfrac{9.46 \times 10^{12}\ km}{1\ year}$, $\dfrac{5.88 \times 10^{12}\ mi}{1\ year}$

    b. $\dfrac{3.00 \times 10^5\ km}{1\ s}$, $\dfrac{1.86 \times 10^5\ mi}{1\ s}$

chapter 4 elements and their invisible structures

1. Compounds are pure substances with a constant composition because atoms are combined in simple whole-number ratios. These ratios lead to a fixed proportion by weight. Chemical changes can disrupt the combination of atoms.

3. a. The atomic number of magnesium is 12, sulfur is 16, and silver is 47.

   b. Magnesium has 12 protons, sulfur has 16 protons, and silver has 47 protons.

5. Aluminum has 13 electrons, potassium has 19 electrons and copper has 29 electrons.

7. All isotopes of magnesium contain the same number of protons, which is 12.
9. See the following table for the answers to this question.

Atom	a. # protons, neutrons and electrons	b. Identity	c. Diagram	d. Isotopes
$^{32}_{16}X$	$16p^+$, $16\,^1_0n$, $16e^-$	Sulfur	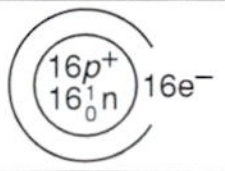	Not an isotope
$^{32}_{15}X$	$15p^+$, $17\,^1_0n$, $15e^-$	Phosphorus	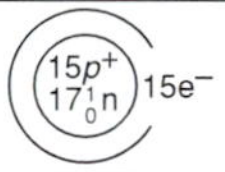	Isotope of phosphorus
$^{127}_{53}X$	$53p^+$, $74\,^1_0n$, $53e^-$	Iodine	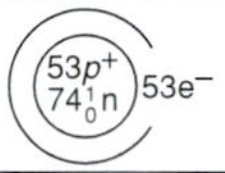	Isotope of iodine
$^{31}_{15}X$	$15p^+$, $16\,^1_0n$, $15e^-$	Phosphorus	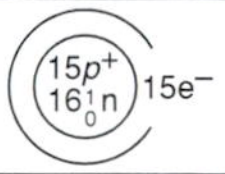	Isotope of phosphorus
$^{130}_{53}X$	$53p^+$, $77\,^1_0n$, $53e^-$	Iodine	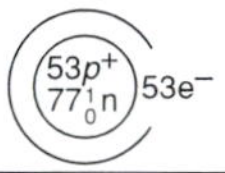	Isotope of iodine

11. The atomic mass of Q is 16.
13. Properties common to nonmetals would be: they lack luster, do not conduct electricity, and may be gases at room temperature or they are brittle if they are solid.
15. Atoms of different elements are similar because all atoms have protons and neutrons in the nucleus and electrons outside the nucleus.
17. A picture of an atom might look like a dense center region surrounded by a loosely packed electron cloud.

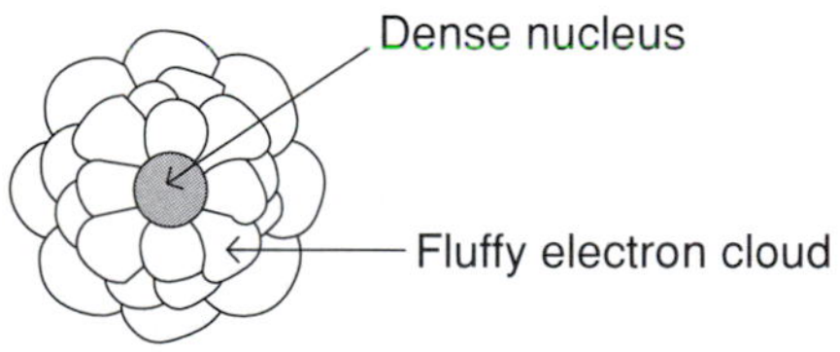

19. The subatomic particle with zero mass is the electron.
21. The mass of a proton approximately equals the mass of a neutron. Both are considered to be 1 amu. An electron is considered to have a relative mass of zero.
23. The protons account for the charge on the nucleus, each proton having one positive charge.
25. Protons and neutrons are in the nucleus of the atom. Electrons move in a region of space around the nucleus.
27. The atomic number of an atom equals the number of protons. The number of electrons may or may not be equal to this number. If the atom is neutral, then they are equal.
29. a. The atom with 7 electrons is nitrogen.
    b. The atom with 11 electrons is sodium.
    c. The atom with 18 electrons is argon.
    d. The atom with 2 electrons is helium.
31. Two isotopes of the same element contain equal numbers of protons. They differ in the number of neutrons they contain.

33. The number of protons for two isotopes of the same element must be the same.

35. The answers to complete the following table are in bold. The number of protons plus the number of neutrons will equal the mass number. The number of electrons is the same as the number of protons.

	Mass Number	Number of Neutrons	Number of Protons	Number of Electrons
a.	51	28	**23**	**23**
b.	28	14	**14**	**14**
c.	19	10	**9**	**9**

37. Complete the table using the following facts. The atomic number equals the number of protons. The atomic mass is the total number of protons and neutrons. The number of electrons equals the number of protons.

	Symbol	Number of Protons	Number of Neutrons	Number of Electrons
a.	${}^{9}_{4}Be$	4	5	4
b.	${}^{127}_{53}I$	53	74	53
c.	${}^{31}_{15}P$	15	16	15
d.	${}^{40}_{18}Ar$	18	22	18

39. Atoms with the same atomic number are isotopes. ${}^{28}_{14}X$ and ${}^{30}_{14}X$ are isotopes and ${}^{14}_{7}X$ and ${}^{15}_{7}X$ are isotopes.

41. a. The symbol for potassium would be ${}^{39}_{19}K$.

    b. A possible isotope of potassium is ${}^{40}_{19}K$.

43. The symbol for an isotope of nitrogen with a mass number of 15 would be ${}^{15}_{7}N$.

45. Yes, there is a difference between carbon-12 and ${}^{13}_{6}C$. Carbon-12 is ${}^{12}_{6}C$; it has six neutrons, while ${}^{13}_{6}C$ has seven neutrons. Carbon-13 and ${}^{13}_{6}C$ both represent atoms with seven neutrons.

47. The number of protons, neutrons and electrons for each of the following ions is listed in the table. Remember to use the charge on the ion to adjust the number of electrons.

	Ion	Number of Protons	Number of Neutrons	Number of Electrons
a.	${}^{88}_{38}Sr^{2+}$	38	50	36
b.	${}^{23}_{11}Na^{+}$	11	12	10
c.	${}^{14}_{7}N^{3-}$	7	7	10
d.	${}^{127}_{53}I^{-}$	53	74	54

49. The symbols for the ions containing the given number of protons, neutrons and electrons are in the following table.

	Symbol	Number of Protons	Number of Neutrons	Number of Electrons
a.	$^{31}_{15}P^{3-}$	15	16	18
b.	$^{39}_{19}K^{+}$	19	20	18
c.	$^{1}_{1}H^{+}$	1	0	0
d.	$^{19}_{9}F^{-}$	9	10	10
e.	$^{85}_{37}Rb^{+}$	37	48	36

51. Anions are formed when a neutral atom gains electrons.
53. a. An ion having 10 electrons and 9 protons would be F^{-}.
    b. An ion having 10 electrons and 12 protons would be Mg^{2+}.
    c. An ion having 10 electrons and 11 protons would be Na^{+}.
    d. An ion having 10 electrons and 8 protons would be O^{2-}.
    e. An ion having 10 electrons and 7 protons would be N^{3-}.
55. The atomic mass of an isotope of cobalt that is five times as heavy as carbon with an atomic mass of 12, would be 5×12 or 60 amu.
57. The mass ratio of magnesium-24 to carbon-12 is 24:12 or 2:1.
59. Sulfur-32 = 0.950, sulfur-33 = 0.0076, and sulfur-34 = 0.422.
61. Average atomic mass = 20.2 amu.
63. Using the periodic table, sodium, potassium, and lithium belong to Group IA. Oxygen and sulfur belong to Group VIA. Phosphorus and arsenic belong to Group VA.
65. a. Period 2, Group IIA is beryllium, Be.
    b. Period 5, Group VIA is tellurium, Te.
    c. Period 2, Group VIIA is fluorine, F.
    d. Period 4, Group VIIIA is krypton, Kr.
67. The elements with atomic numbers of 11, 37, and 55 all belong to Group IA and should have similar properties.
69. Boron at the top of the Group IIIA is a nonmetal. Thallium, being at the bottom of the group would have the most metallic properties.
71. Hydrogen is not a noble gas.
73. The answers in the following table are shown in the unshaded boxes.

Symbol	Atomic Number	Number of Neutrons	Mass Number
$^{39}_{19}K$	19	20	39
$^{48}_{20}Ca$	20	28	48
$^{23}_{11}Na$	11	12	23
$^{84}_{36}Kr$	36	48	84

75. Graphite and diamond are made up of identical carbon atoms. The differences in their appearances is due to the fact that the atoms are arranged and bonded differently.

77. a. The maximum positive charge that can exist on a hydrogen ion is +1, because hydrogen only has one electron to lose.
    b. The maximum positive charge that can occur on a beryllium ion is +4, because beryllium has four electrons that can be removed.

chapter 5 electronic structure of the atom

1. Food is less stable than its metabolic products. The energy diagram to show this:

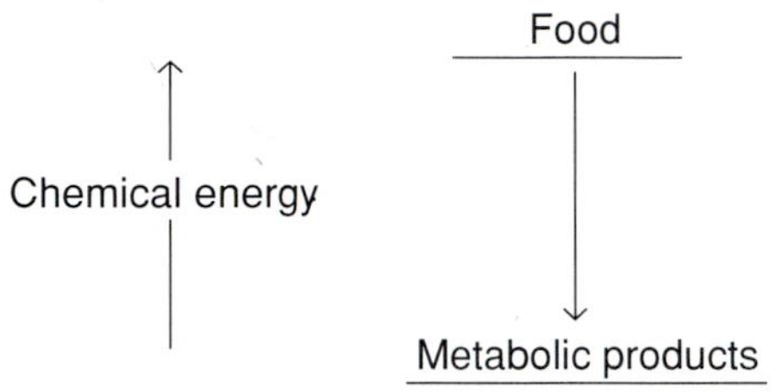

3. Describing the electron arrangement of each of the following:
    a. Sulfur with sixteen electrons would have two electrons in the 1*s*, two in the 2*s*, six in the 2*p*, two in the 3*s*, and four in the 3*p*.
    b. Phosphorus with 15 electrons would have two electrons in the 1*s*, two in the 2*s*, six in the 2*p*, two in the 3*s*, and three in the 3*p*.
    c. Magnesium with twelve electrons would have two electrons in the 1*s*, two in the 2*s*, six in the 2*p*, and two in the 3*s*.

5. Using the Aufbau principle and Hund's rule, phosphorus would have three single (unpaired) electrons.

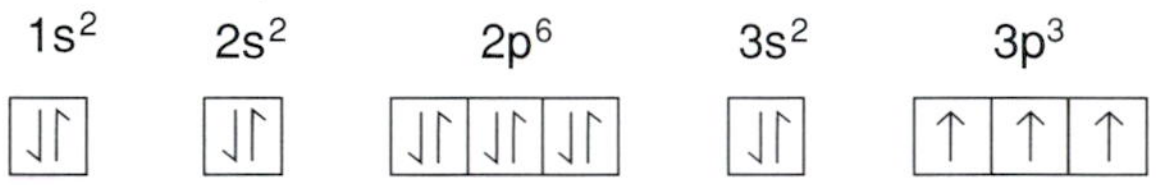

7. Since oxygen is in the second period and in Group VI, there are six electrons to account for in the second level. This means the configuration for the valence electrons of oxygen would be $2s^2\ 2p^4$.

9. a. The Group IA elements have the lowest ionization energy because they are farthest to the left of any of the Group A elements. In any period, the elements with fewer protons and hence a smaller positive charge to attract the electrons, occur on the left.
    b. The Group IA elements are alkali metals.

11. a. We would predict helium to have the highest ionization energy since it is at the top of the periodic table and on the right side.
    b. We would predict francium to have the lowest ionization energy since it is at the bottom of the periodic table and on the left side.

13. Explaining each of the following in terms of energy change:
    a. The apple on the ground is in a lower potential energy state than the apple in the tree.
    b. Water seeks a lower energy state which would be at a lower elevation.
    c. Close approach of negatively charged electrons to the positively charged nucleus is a low energy condition.

15. a. To remove an electron from an atom, The force of attraction between the electrons and protons must be overcome.
    b. The forces that would be relieved by removing an electron from an atom would be the force of repulsion between electrons.
    c. Energy is required to remove an electron from an atom.
    d. The type of force in part a of this problem is greater than the one in part b. According to the experimental results in part c, The force in part a must be greater than in part b.

17. a. The attractive force in a hydrogen atom is between the electron and proton.
    b. The attractive forces that could exist within and between two hydrogen atoms would be the force between the electron of one atom and the proton of the same atom, and the attraction of the electron of the first atom and the proton of the second atom.
    c. The repulsive forces that could exist between two hydrogen atoms would be between the electron of the first atom with the electron of the second atom and between the proton of the first atom and the proton of the second atom.
    d. As the distance between the atoms is made smaller, the attractive and repulsive forces will become larger.
    e. A minimum-energy structure could and does exist; it is a hydrogen molecule.

19. The shape of the s orbital is spherical and the p orbital is like a long balloon tied in the middle.

21. Any two electrons in the same orbital must be of opposite spin to be in the lowest energy state.

23. Within a particular energy level, the sublevel with the lowest energy is the s sublevel.

25. a. Calcium $1s^2\ 2s^2\ 2p^6\ 3s^2\ 3p^6\ 4s^2$
    b. Arsenic $1s^2\ 2s^2\ 2p^6\ 3s^2\ 3p^6\ 4s^2\ 3d^{10}\ 4p^3$
    c. Potassium $1s^2\ 2s^2\ 2p^6\ 3s^2\ 3p^6\ 4s^1$
    d. Bromine $1s^2\ 2s^2\ 2p^6\ 3s^2\ 3p^6\ 4s^2\ 3d^{10}\ 4p^5$
    e. Indium $1s^2\ 2s^2\ 2p^6\ 3s^2\ 3p^6\ 4s^2\ 3d^{10}\ 4p^6\ 5s^2\ 4d^{10}\ 5p^1$

27. The symbols of all sublevels within the third energy level are *s*, *p*, and *d*.

29. a. Sodium has one unpaired electron.
    b. Silicon has two unpaired electrons.
    c. Beryllium has no unpaired electrons.
    d. Carbon has two unpaired electrons.
    e. Fluorine has one unpaired electron.
    f. Cobalt has 3 unpaired electrons.

31. a. Four criteria for identifying an electron in an atom are the principal level, sublevel, orbital, and spin.
    b. The criteria in part a that relate to the energy of an electron in an atom are principal level and sublevel.

33. Absorption spectra are obtained by measuring, quantitatively, the amount of *energy absorbed* by a sample in promoting its electrons to higher levels, while emission spectra are obtained by measuring the intensity of *light emitted* when the excited electrons of a sample fall to a lower energy state.

35. We can attribute the cherry red color of a heated crucible to light energy emitted as excited electrons fall back to the ground state. The wavelength of the light energy corresponds to red light.

37. Elements with atomic numbers 7, 15, 33, 51, and 83 are all in Group VA with an electron configuration of $ns^2\ np^3$.

39. a. $1s^2\ 2s^2\ 2p^1$ is B
    b. $1s^2\ 2s^2\ 2p^6\ 3s^2\ 3p^6\ 4s^1$ is K
    c. $1s^2\ 2s^2\ 2p^6\ 3s^2\ 3p^6\ 4s^2\ 3d^7$ is Co
    d. $1s^2\ 2s^2\ 2p^6\ 3s^2\ 3p^6\ 4s^2\ 3d^{10}\ 4p^1$ is Ga
    e. $1s^2\ 2s^2\ 2p^6\ 3s^2\ 3p^6\ 4s^2\ 3d^{10}\ 4p^6$ is Kr

41. a. The configuration of sulfur is $3s^2\ 3p^4$.
    b. The configuration of phosphorus is $3s^2\ 3p^3$.
    c. The configuration of iodine is $5s^2\ 5p^5$.
    d. The configuration of krypton is $4s^2\ 4p^6$.

43. a. $3s^2\ 3p^4$ describes sulfur; similar elements would be O, Se, Te, or Po.
    b. $5s^2$ describes strontium; similar elements would be Be, Mg, Ca, Ba, or Ra.
    c. $2s^2\ 2p^5$ describes fluorine; similar elements would be Cl, Br, I, or At.
    d. $3s^1$ describes sodium; similar elements would be Li, K, Rb, Cs, or Fr.
    e. $1s^2$ describes helium; similar elements would be Ne, Ar, Kr, Xe, or Rn.

45.

47. a. Elements with low ionization energies are found near the bottom and to the left of the periodic table.
    b. Elements with high ionization energies are found near the top and to the right of the periodic table. These are nonmetals.
    c. The inert gases make up the group with the highest ionization energies. This indicates that these elements have very stable electron arrangements.

49. a. The element from the list with the highest ionization energy is C, carbon.
    b. The element from the list with the smallest size is F, fluorine.
    c. The element from the list with the largest size is Cs, cesium.
    d. The element from the list with the smallest ionization energy is Cs, cesium.

51. Transition metals are those elements found in the B Group of the periodic table. The transition metals tend not to show the periodic trends of the A Group elements.

53. Some of the characteristics you can glean from the information on the periodic table would be atomic number, atomic weight, relative size, relative ionization energy, electron configuration, nature of the element (metal or nonmetal), physical or chemical properties.

55. a. C has four valence electrons.
    b. Sn has four valence electrons.
    c. F has seven valence electrons.
    d. P has five valence electrons.

57. a. The configuration of silicon is $3s^2\ 3p^2$.
    b. The configuration of chlorine is $3s^2\ 3p^5$.
    c. The configuration of potassium is $4s^1$.
    d. The configuration of calcium is $4s^2$.

59.

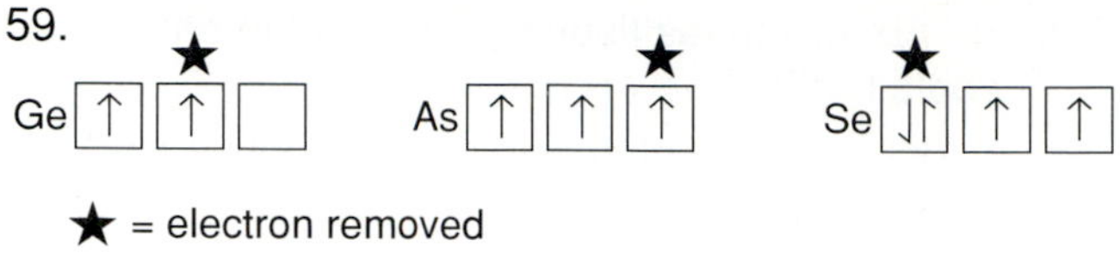

Taking a paired electron reduces the electron-electron repulsion and is easier than taking an unpaired electron.

61. Elements produce a characteristic pattern of colors when there is an emission of energy from an excited state to a ground state because only certain electron levels are allowed and therefore only certain electron transitions. These allowed transitions correspond to certain colors which we see as the spectrum.

63. The maximum number of electrons that can be accommodated in a 4*p* sublevel is six, the 2*s* level is two, and the 3*d* level is 10.

65. a. $3s^1$ is in Group IA (first).
    b. $3s^2\ 3p^2$ is in Group IVA (fourth).
    c. $3s^2\ 3p^5$ is in Group VIIA (seventh).
    d. $4s^1$ is in Group IA (first).

chapter 6 chemical bonding

1. a. Aluminum is in Group IIIA, which means it will lose three electrons to be isoelectronic with Neon.
   b. Sulfur is in Group VIA, which means it must gain two electrons to be isoelectronic with Argon.
   c. Aluminum has a magnitude three positive charge, making it a cation. Sulfur has a magnitude two negative charge, making it an anion.

3. Since phosphorus is in Group VA, its ion will have a negative charge of magnitude three.

5. a. The ratio of Ca:O is 1:1 and there are no polyatomic ions.
   b. The ratio of Mg:OH is 1:2 and OH^- is a polyatomic ion.
   c. The ratio of $Na:PO_4$ is 3:1 and PO_4^{3-} is a polyatomic ion.
   d. The ratio of $K:C_2H_3O_2$ is 1:1 and $C_2H_3O_2^-$ is a polyatomic ion.
   e. The ratio of $Ca:PO_4$ is 3:2 and PO_4^{3-} is a polyatomic ion.
   f. The ratio of $K:ClO_3$ is 1:1 and ClO_3^- is a polyatomic ion.

7. Using the shortcut method to determine the correct formula.

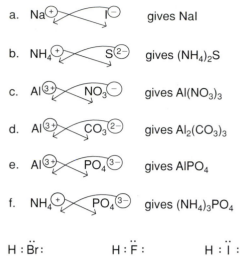

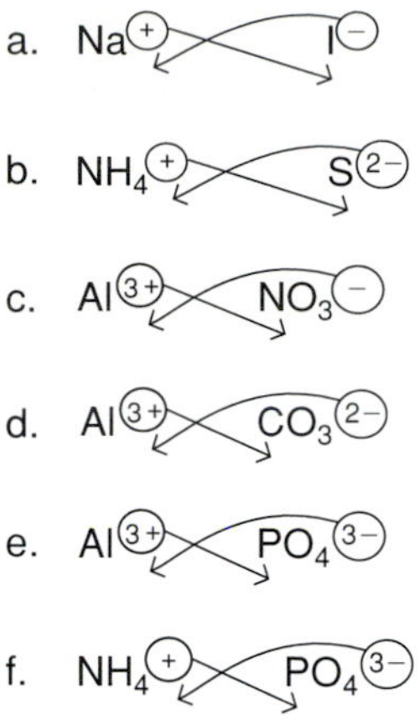

   a. gives NaI
   b. gives $(NH_4)_2S$
   c. gives $Al(NO_3)_3$
   d. gives $Al_2(CO_3)_3$
   e. gives $AlPO_4$
   f. gives $(NH_4)_3PO_4$

9.

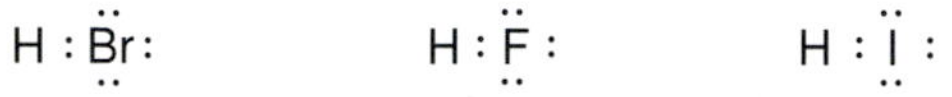

11. The H_2O and H^+ can form H_3O^+ with the following covalent bond formation.

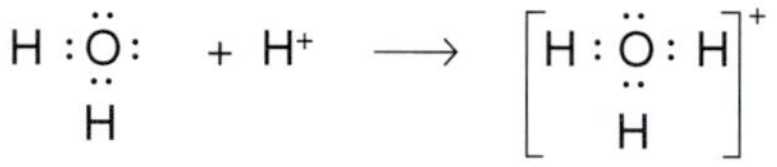

13. In the compound, H_2S, hydrogen has the relative weight of 1 and sulfur has the relative weight of 32. Therefore, the ratio of the weights of two hydrogen to one sulfur would be 2:32 or 1:16.

15. A chemical bond is the force that holds elements together in compounds.

17. Noble-gas electron configurations have filled valence electron levels, i.e., $1s^2$ duet for helium, and octet (ns^2np^6) for other noble gases.

19. Atoms can obtain a noble-gas electron configuration by transferring electrons or sharing of electrons.

21. The loss of two electrons from calcium produces the argon noble-gas configuration. The loss of a third electron would destroy that stable configuration.

23. a. Barium must lose two electrons to form a noble-gas configuration.
    b. Barium will become isoelectronic with xenon

25. a. Krypton is isoelectronic with Sr^{2+}.
    b. Helium is isoelectronic with Li^+.
    c. Krypton is isoelectronic with Rb^+.

27. Group VIIA nonmetals need only to gain one electron to become isoelectronic with a noble gas.

29. a. The electron configuration for the S^{2-} ion is $1s^2\ 2s^2\ 2p^6\ 3s^2\ 3p^6$.
    b. Argon is isoelectronic with this configuration.

31. a. Krypton is isoelectronic with Se^{2-}.
    b. Argon is isoelectronic with P^{3-}.
    c. Xenon is isoelectronic with I^-.

33. a. A pictorial representation of the formation of the ionic compound from sodium and fluorine.

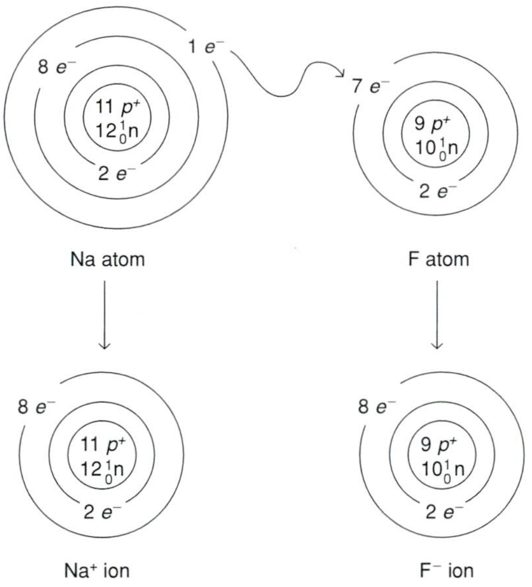

b. Sodium needs to lose only one electron and fluorine needs only to gain one electron in order for both to attain the neon noble-gas configuration. Then the single positive charge on Na^+ is exactly neutralized by the single negative charge on F^-.

35. a. A pictorial representation of the formation of the ionic compound from aluminum and oxygen.

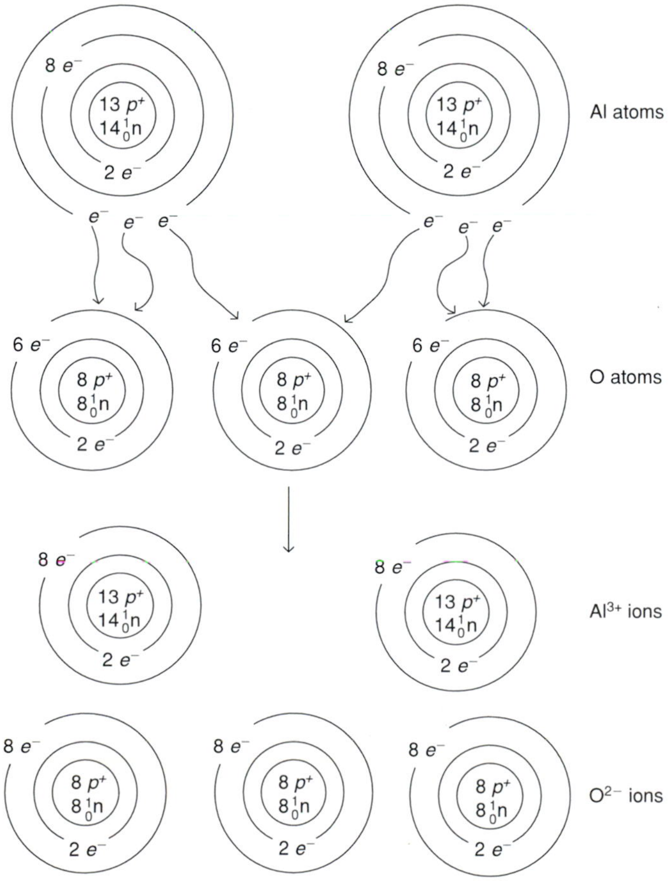

b. Aluminum atoms need to lose three electrons to attain a neon noble-gas configuration. Each oxygen atom needs to gain two electrons to achieve the neon configuration. In order for equal numbers of electrons to be lost and gained, two aluminum atoms lose three electrons each (total of six electrons) and three oxygen atoms gain two electrons each (total of six electrons). The total positive charge of +6 (2 Al^{3+}) is exactly neutralized by the negative charge of –6 (3 O^{2-}).

37. a. The formula and charge for carbonate is CO_3^{2-}.
b. The formula and charge for hydrogen carbonate is HCO_3^-.
c. The formula and charge for phosphate is PO_4^{3-}.
d. The formula and charge for ammonium is NH_4^+.
e. The formula and charge for sulfate is SO_3^{2-}.

39. The correct formulas for the compounds formed by combining the following.
a. Lithium and oxygen form Li_2O.
b. Barium and chlorine form $BaCl_2$.
c. Sodium and sulfur form Na_2S.
d. Nickel and nitrogen form Ni_3N_2.
e. Silver and bromine form AgBr.
f. Iron (the +2 ion) and oxygen form FeO.
g. Iron (the +3 ion) and oxygen form Fe_2O_3.
h. Tin (the +2 ion) and chlorine form $SnCl_2$.
i. Tin (the +4 ion) and chlorine form $SnCl_4$.
j. Gallium and sulfur form Ga_2S_3.
k. Gallium and iodine form CaI_2.

41. If X has one valence electron and Y has six valence electrons, the compound they form would be X_2Y.

43. The step in ionic bond formation that releases the greatest amount of energy is the attraction between cations and anions.

45. The total number of ions in one formula unit of each of the following is given.
a. $CaCl_2$ has 3 ions. (1 Ca^{2+} and 2 Cl^-)
b. $Ca_3(PO_4)_2$ has 5 ions.
c. $Al(HCO_3)_3$ has 4 ions.
d. $(NH_4)_2SO_4$ has 3 ions.

47. Ionic bonds are formed through the transfer of electrons from a metal to a nonmetal. Covalent bonds are formed through the sharing of electrons between nonmetals.

49. Helium has a filled outer level, therefore, it has no need to share electrons. Each hydrogen atom has one valence electron and requires one more to obtain a noble-gas configuration. If two hydrogen atoms share their electrons, they both obtain the noble-gas configurations.

51. a. Six nonbonding electrons surround each chlorine atom in a chlorine molecule.
    b. Two nonbonding electrons surround each nitrogen atom in a nitrogen molecule.

53. a. The energy content of a covalent bond when separation distances are greater than bond length would be higher.
    b. The energy content of a covalent bond when separation distances are less than the bond length would be higher.
    c. The energy content of a covalent bond when separation distances are equal to the bond length would be minimum, so most stable.

55. The chlorine molecule is more stable than two separated chlorine atoms because energy must be added to the molecule in order to separate the atoms. The atoms in a chlorine molecule are at the noble-gas configuration, the separate chlorine atoms are not.

57. a. S is the central atom of H_2S.
    b. C is the central atom of $CHCl_3$.
    c. S is the central atom of H_2SO_4.
    d. N is the central atom of NO_2^-.

59. See Problem 9 in this chapter for help in drawing Lewis structures.

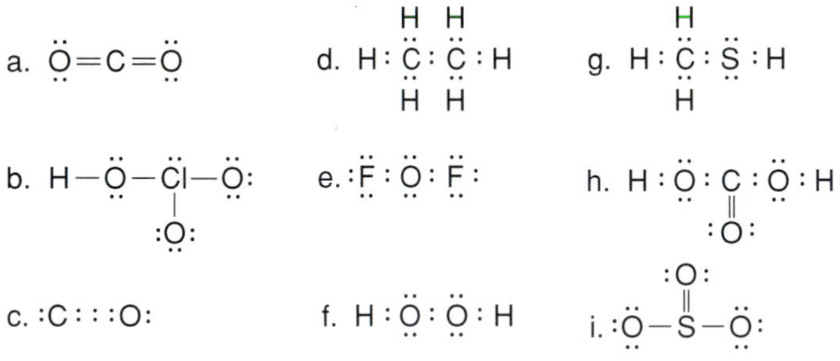

61. The Lewis structures for SO_2 and O_3.
    a. 
    b. Similarities in their structures are they have the same numbers of valence electrons and each central atom forms a single bond and a double bond.
    c. Yes, because the alternate Lewis structure can be written:

which means that each bond can be regarded as "1 1/2" i.e., intermediate between a single and a double bond.

63. a. The Lewis structure for BF_3.

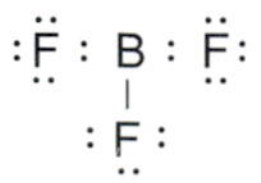

, where the line (I) represents two electron dots.

    b. BF_3 could form a coordinate covalent bond with either of the following.

$\dot{N}H_3$ or $H_2\ddot{O}$:

    c. The Lewis structure for the complex formed from BF_3 and NH_3.

, where the line (I) represents two electron dots.

65. Nonbonding electrons can participate in forming a coordinate covalent bond.

67. In the periodic table, electronegativity increases across a period from left to right. Electronegativity also increases as you move from bottom to top of a group in the periodic table.

69. a. P—Br. The electronegativity difference is 0.7, making it a polar covalent bond.
    b. H—O. The electronegativity difference is 1.4, making it a polar covalent bond.
    c. I—Br. The electronegativity difference is 0.3, making it a slightly polar covalent bond.
    d. N—H. The electronegativity difference is 0.9, making it a polar covalent bond.

71. The dipole bonds from Problem 69.

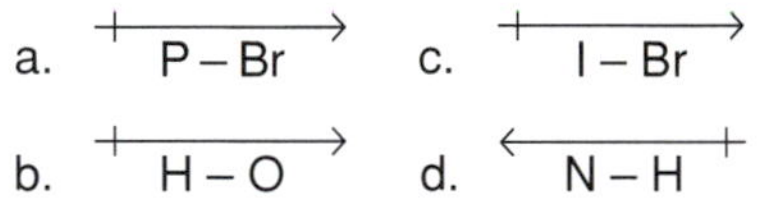

73. The bond polarities for Problem 58.

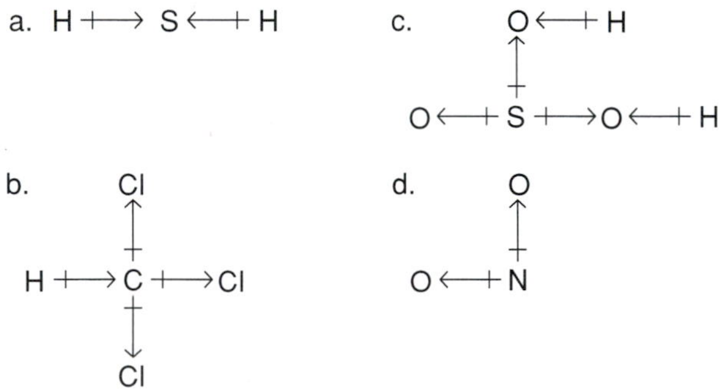

75. Identifying the following as ionic or molecular bonds and the smallest basic unit as a molecule or a formula unit.
    a. H_2O is molecular and a molecule.
    b. NaBr is ionic and a formula unit.
    c. Al_2O_3 is ionic and a formula unit.
    d. CH_4O is molecular and a molecule.
    e. $PbClO_4$ is ionic and a formula unit.

f. HBr is molecular and a molecule.
g. K_2CO_3 is ionic and a formula unit.
h. CS_2 is molecular and a molecule.
i. $CaCl_2$ is ionic and a formula unit.

77. a. C and H, with a difference of 0.4, would form a covalent bond.
b. K and Br, with a difference of 2.0, would form a ionic bond.
c. N and H, with a difference of 0.9, would form a polar covalent bond.
d. O and S, with a difference of 1.0, would form a polar covalent bond.
e. Ca and Cl, with a difference of 2.0, would form a ionic bond.
f. Al and F, with a difference of 2.5, would form a ionic bond.

79. Carbon has a relative mass of 12 and hydrogen 1. In the molecule, C_3H_8, three carbons and eight hydrogen would give the ratio 36:8 or 9:2.

81. Sulfur has a relative mass of 16 and oxygen 8. For the compound SO_2, one sulfur combining with two oxygen gives a 16:16 or 1:1 ratio. Therefore, if the decomposition yields 7.50 g of oxygen, it will also yield 7.50 g of sulfur.

83. The ratio of nitrogen to hydrogen is 20.6:4.41. Dividing both sides by 4.41 to make hydrogen equal to one, yields the ratio 4.67:1.

85. Ionic solids have very strong bonding, requiring a large amount of energy to break the bonds. This results in a higher melting point.

87. Na^- is not a stable ion because it has two outer electrons and it needs to lose two electrons to reach the noble-gas configuration.

89. The structural formulas follow.

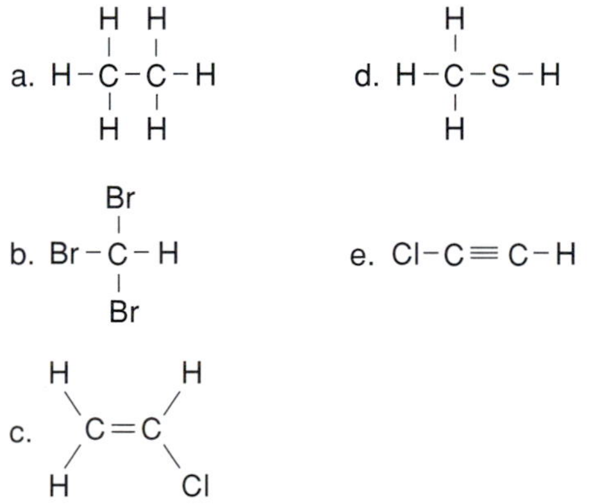

91. The Lewis structures for a, propane and b, butane.

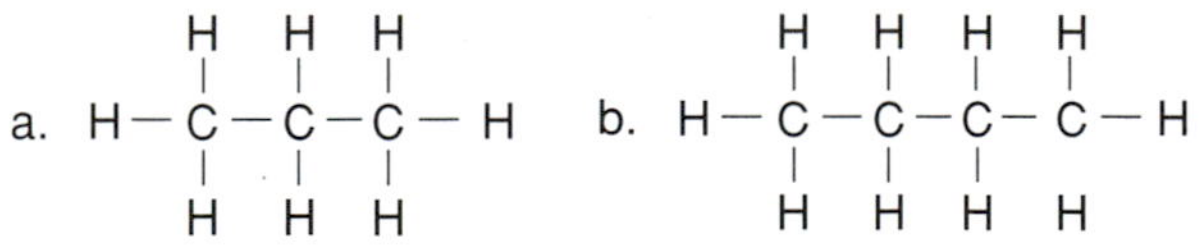

93. 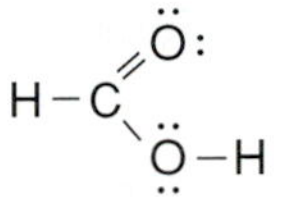

95. The ionic compound formation between A and B can be an energy-lowering process because the attraction of the ions for each other releases a large amount of energy, (lattice energy) and more than compensates for the ionization energy and electron affinity difference.

97. a. :C≡O: :N≡O:$^+$:C≡N:$^{1-}$:N≡N:
b. Characteristics common to all four of these structures are they are isoelectronic, all have a triple bond, and each atom has one lone pair of electrons.

chapter 7 nomenclature, weight, and percent composition

1. a. NaCl is sodium chloride.
b. CaF_2 is calcium fluoride.
c. BaO is barium oxide.

3. a. K_2CO_3 is potassium carbonate.
b. NH_4NO_3 is ammonium nitrate.
c. $Pb_3(PO_4)_2$ is lead(II) phosphate.

5. a. CO_2 is carbon dioxide.
b. CCl_4 is carbon tetrachloride.
c. PCl_3 is phosphorus trichloride.
d. N_2O_4 is dinitrogen tetroxide.
e. OF_2 is oxygen difluoride.
f. HF is hydrogen fluoride.

7. a. H_2CO_3 is carbonic acid.
b. $HC_2H_3O_2$ is acetic acid.
c. H_2SO_3 is sulfurous acid.
d. $HClO_3$ is chloric acid.
e. $HClO_4$ is perchloric acid.

9. a. $MgCl_2$ is magnesium chloride.
b. HBr is hydrobromic acid in water solution. Hydrogen bromide if not in a water solution.
c. H_2SO_4 is sulfuric acid.
d. H_2SO_3 is sulfurous acid.
e. $Ca_3(PO_4)_2$ is calcium phosphate.
f. OF_2 is oxygen difluoride.

11. NaOH = 40.00 amu to the nearest hundredth (Na = 22.99, O = 16.00 and H = 1.008).
$CaCl_2$ = 110.98 amu (Ca = 40.08, 2 Cl = 70.90).
$Al_2(CO_3)_3$ = 233.99 amu (2 Al = 53.96, 3 C = 36.03, and 9 O = 144.00).

13. a. Table salt is sodium chloride.
    b. Lye is sodium hydroxide.
    c. Baking soda is sodium hydrogen carbonate or sodium bicarbonate.
15. a. $AlCl_3$ (aluminum and chlorine).
    d. Fe_2O_3 (iron and oxygen).
    f. Li_3N (lithium and nitrogen).
17. a. Potassium and oxygen form K_2O or potassium oxide.
    b. Aluminum and chlorine form $AlCl_3$ or aluminum chloride.
    c. Calcium and bromine form $CaBr_2$ or calcium bromide.
    d. Silver and oxygen form Ag_2O or silver oxide.
    e. Copper (+2 ion) and iodine form CuI_2 or copper(II) iodide.
    f. Copper (+1 ion) and chlorine form CuCl or copper(I) chloride.
    g. Tin (+2 ion) and bromine form $SnBr_2$ or tin(II) bromide.
    h. Tin (+4 ion) and bromine form $SnBr_4$ or tin(IV) bromide.
    i. Gallium and sulfur form Ga_2S_3 or gallium sulfide.
    j. Barium and iodine form BaI_2 or barium iodide.
    k. Nickel and nitrogen form Ni_3N_2 or nickel nitride.
19. a. Ca^{2+} and HCO_3^- form $Ca(HCO_3)_2$ or calcium bicarbonate.
    b. Li^+ and PO_4^{3-} form Li_3PO_4 or lithium phosphate.
    c. Sn^{2+} and Cl^- form $SnCl_2$ or tin(II) chloride.
    d. Fe^{3+} and OH^- form $Fe(OH)_3$ or iron(III) hydroxide.
    e. Mg^{2+} and ClO_3^- form $Mg(ClO_3)_2$ or magnesium chlorate.
21. IO_4^- is periodate.
    IO_3^- is iodate.
    IO_2^- is iodite.
    IO^- is hypoiodite.
23. The names of the compounds formed from the anions and cations are listed in the table.

	OH^-	SO_4^{2-}	PO_4^{3-}
NH_4^+	Ammonium hydroxide	Ammonium sulfate	Ammonium phosphate
Fe^{2+}	Iron(II) hydroxide	Iron(II) sulfate	Iron(II) phosphate
Al^{3+}	Aluminum hydroxide	Aluminum sulfate	Aluminum phosphate

25. a. PCl_5 is phosphorus pentachloride.
    b. SO_2 is sulfur dioxide.
    c. NO is nitrogen(II) oxide (or nitrogen monoxide or nitrogen oxide).
    d. HF is hydrogen fluoride.
    e. CBr_4 is carbon tetrabromide.
    f. Cl_2O_7 is dichlorine heptoxide.

27. N_2O is dinitrogen monoxide.
    NO is nitrogen monoxide.
    N_2O_3 is dinitrogen trioxide.
    NO_2 is nitrogen dioxide.
    N_2O_4 is dinitrogen tetroxide.
    N_2O_5 is dinitrogen pentoxide.
29. The completed table of molecular compounds. The answers are in bold print.

Name	Formula
a. Sulfur tetrafluoride	**SF_4**
b. **Nitrogen trichloride**	NCl_3
c. Boron trichloride	**BCl_3**
d. Dinitrogen trioxide	**N_2O_3**
e. **Carbon tetrabromide**	CBr_4

31. a. H_3PO_4 is an oxoacid.
    b. HF is a nonoxoacid.
    c. H_2SO_3 is an oxoacid.
    d. H_2S is a nonoxoacid.
    e. HBr is a nonoxoacid.
    f. HNO_3 is an oxoacid.
33. a. H_2SO_4 is sulfuric acid.
    b. H_2SO_3 is sulfurous acid.
    c. HNO_3 is nitric acid.
    d. H_3PO_4 is phosphoric acid.
    e. $HClO_3$ is chloric acid.
    f. $HClO_4$ is perchloric acid.
    g. H_2CO_3 is carbonic acid.
    h. $HC_2H_3O_2$ is acetic acid.
35. a. K_2CrO_4 is potassium chromate.
    b. NaF is sodium fluoride.
    c. FeO is iron(II) oxide.
    d. Cu_2CO_3 is copper(I) carbonate.
    e. LiOH is lithium hydroxide.
    f. CO is carbon monoxide.
    g. HCl is hydrogen chloride.
    h. $HgCl_2$ is mercury(II) chloride.
    i. $HClO_2$ is chlorous acid.
    j. Cl_2O is dichlorine oxide.

37. a. I_2 is 253.8 amu (2 I = 2 × 126.9)
    b. NH_3 is 17.0 amu (1 N = 14.0, 3 H = 3.0)
    c. $(NH_4)_2CO_3$ is 96.0 amu (2 N = 28.0, 8 H = 8.0, 1 C = 12.0, 3 O = 48.0)
    d. K_2SO_4 is 174.3 amu (2 K = 78.2, 1 S = 32.1, 4 O = 64.0)
    e. Cl_2O_7 is 182.9 amu (2 Cl = 70.9, 7 O = 112.0)
    f. $(NH_4)_3PO_4$ is 149.0 amu (3 N = 42.0, 12 H = 12.0, 1 P = 31.0, 4 O = 64.0)
    g. H_2SO_3 is 82.1 amu (2 H = 2.0, 1 S = 32.1, 3 O = 48.0)
    h. $HClO_4$ is 100.5 amu (1 H = 1.0, 1 Cl = 35.5, 4 O = 64.0)
    i. $KMnO_4$ is 158.0 amu (1 K = 39.1, 1 Mn = 54.9, 4 O = 64.0)
    j. $Ca(HCO_3)_2$ is 162.1 amu (1 Ca = 40.1, 2 H = 2.0, 2 C = 24.0, 6 O = 96.0)
39. % Ag = 63.51%
41. % Sn = 88.1%, % O = 11.9%
43. % Cu = 79.9%, % O = 20.1%
45. a. Ammonium sulfate is $(NH_4)_2SO_4$
    b. Potassium sulfide is K_2S
    c. Bromic acid is $HBrO_3$
    d. Perbromic acid is $HBrO_4$
    e. Iron(III) chromate is $Fe_2(CrO_4)_3$
    f. Mercuric oxide is HgO
47. a. H_2O_2 = 34.0 amu (2 H + 2 O)
    b. $C_6H_8O_6$ = 176.0 amu (6 C + 8 H + 6 O)
    c. $K_2Cr_2O_7$ = 294.2 amu (2 K + 2 Cr + 7 O)
    d. Al_2Cl_6 = 267.0 amu (2 Al + 6 Cl)
    e. C_8H_{18} = 114.0 amu (8 C + 18 H)
    f. $C_{56}H_{88}O_2$ = 792.0 amu (56 C's + 88 H's + 2 O's)
49. % C = 42.1%, % H = 6.5%, and % O = 51.4%
51. % C = 92.3 and % H = 7.8%

chapter 8 the mole concept

1. a. The GAW of aluminum is 26.98 g/mole.
   b. The GAW of phosphorus is 30.97 g/mole.
   c. The GAW of silicon is 28.09 g/mole.
3. 1 Ca = 40.08 g/mol
   2 Cl = 70.90 g/mol

   $CaCl_2$ = 110.98 g/mol
5. 8 PO_4^{3-} ions.
7. a. The empirical formula is KCl.
   b. The empirical formula is C_2H_5Cl.
9. Some other examples of "packages" might be a pair, gross, or ream.
11. 20.2:1

13. a. No, the number of sodium atoms will not be equal to the number of sulfur atoms because 22.99 grams of sodium (atomic weight = 22.99) is equivalent to 1 mole of sodium and therefore contains Avogadro's number of atoms or 6.022×10^{23} atoms. For sulfur (atomic weight = 32.06), 22.99 grams is less than 1 mole and therefore will contain less than 6.022×10^{23} atoms.

    b. Sodium contains the greater number of atoms.

15. a. He = 4.003 g
    b. Se = 78.96 g
    c. F = 19.00 g
    d. Au = 197.0 g
    e. Li = 6.941 g
    f. Ge = 72.59 g

17. One mole equals 6.022×10^{23} atoms which equals one GAW. Using the periodic table, the mass of Avogadro's number of atoms of magnesium would be 24.31 grams.

19. 127 g S

21. 0.213 mol Mg

23. 0.0392 mol Fe

25. One mole of any element has the same number of atoms (Avogadro's number), 1.75 moles of chromium will contain the same number of atoms as 1.75 moles of iron.

27. 9.99 g Ar

29. Answers to complete the table are in bold print. The calculation to arrive at each answer is shown to the right of the table.

Element	Number of Moles	Mass in grams	Calculation
Lithium	1.50	**10.4**	$1.50 \text{ mol Li} \times \frac{6.941 \text{ g Li}}{1 \text{ mol Li}}$
Calcium	1.00	**40.1**	$1.00 \text{ mol Ca} \times \frac{40.08 \text{ g Ca}}{1 \text{ mol Ca}}$
Phosphorus	**0.417**	12.9	$12.9 \text{ g P} \times \frac{1 \text{ mol P}}{30.97 \text{ g P}}$
Silicon	**3.16**	88.8	$88.8 \text{ g Si} \times \frac{1 \text{ mol Si}}{28.09 \text{ g Si}}$
Zinc	3.90	**255**	$3.90 \text{ mol Zn} \times \frac{65.38 \text{ g Zn}}{1 \text{ mol Zn}}$

31. a. 74.55 g
    b. 80.06 g
    c. 98.08 g
    d. 68.14 g
    e. 84.01 g
    f. 262.87 g

33. a. One mole of any molecule contains the same number of molecules (Avogadro's number). In this case, 1.50 moles of I_2 would have the same number of molecules as 1.50 moles of H_2.

    b. We need the same number of moles of both H_2 and I_2. Calculate the number of moles in 4.04 grams of H_2.

$$4.04 \text{ g } H_2 \times \frac{1 \text{ mol } H_2}{2.016 \text{ g } H_2} = 2.003 \text{ or } 2.00 \text{ mol } H_2$$

We will need an equal number of moles of I_2 or 2.00 moles. Calculate the number of grams of I_2 needed.

$$2.00 \text{ mol } I_2 \times \frac{253.8 \text{ g } I_2}{1 \text{ mol } I_2} = 507.6 \text{ or } 508 \text{ g } I_2$$

35. 4 P = 123.88
10 O = 160.00
283.88 g or 283.9 g

37. a. 19.0 g F_2
b. 19.7 g H_2
c. 0.426 g NH_3
d. 2.64×10^3 g $BaSO_4$
e. 12.97 g LiF
f. 125 g S^{2-}

39. 96.4 mL CCl_4

41. $\frac{3 \text{ } Ca^{2+} \text{ ions}}{1 \text{ } Ca_3(PO_4)_2 \text{ formula unit}}$ and $\frac{2 \text{ } PO_4^{3-} \text{ ions}}{1 \text{ } Ca_3(PO_4)_2 \text{ formula unit}}$

43. There is a total of three ions in one formula unit of $(NH_4)_2CO_3$. There are two NH_4^+ ions and one CO_3^{2-} ion.

45. A conversion factor to relate the number of oxygen atoms in one molecule of oxygen would be

$$\frac{2 \text{ O atoms}}{1 \text{ } O_2 \text{ molecule}}$$

47. a. 5.25 moles K^+; 5.25 moles I^-
b. 4.00 moles Al^{3+}; 12.0 moles Cl^-
c. 1.10×10^{-3} moles Ca^{2+}; 2.20×10^{-3} moles of NO_3^-
d. 0.750 moles Ba^{2+}; 0.500 moles PO_4^{3-}

49. a. 64.06 g
b. 64.06 grams of SO_2 will contain Avogadro's number of molecules or 6.022×10^{23} molecules.
c. The mass in grams of one SO_2 molecule $= 1.064 \times 10^{-22}$ g.

51. The molecular formula gives the actual number of atoms in one molecule. The empirical formula gives the smallest whole-number ratio of the atoms in one molecule.

53. a. NO
b. CH
c. CH_4
d. NCl_3
e. C_3H_8O
f. $PbSO_4$
g. KNO_2
h. Na_3PO_4

55. The empirical formula is FeS.

57. C_2H_6

59. $C_2H_4O_2$

61. $C_6H_9N_3O_2$

63. $C_6H_6N_2$.

65. a. 263.0 g "starwarsane".
b. One mole equals one GAW, the GAW of "starwarsane" would be 263.0 grams.

67. The mass of Avogadro's number of adult males would be (6.022×10^{23} males) times (70 kg/male) or 4.2×10^{25} kilograms. Comparing this mass to the mass of the earth (6.0×10^{24} kg), it is seven times greater than the mass of the earth.

69. $CuCl_2$.

71. $PtCl_2$ and $PtCl_4$.

chapter 9 chemical reactions

1. $Pb + O_2 \rightarrow PbO_2$

3. Follow the steps to write a balanced equation for the reaction between hydrogen and bromine, which yields hydrogen bromide.

Step 1. Write the equation with the correct chemical formulas

$$H_2 + Br_2 \rightarrow HBr$$

Step 2. Determine if the equation is balanced.

Reactants (Coefficient × subscript)	Products (Coefficient × subscript)
$1 \times 2 = 2$ H	$1 \times 1 = 1$ H
$1 \times 2 = 2$ Br	$1 \times 1 = 1$ Br

Step 3. Begin balancing by placing a 2 in front of HBr to balance the number of bromine. Now check the number of hydrogen. They now balance.

$$H_2 + Br_2 \rightarrow 2\ HBr$$

Step 4. Verify the equation is balanced.

Reactants (Coefficient × subscript)	Products (Coefficient × subscript)
$1 \times 2 = 2$ H	$2 \times 1 = 2$ H
$1 \times 2 = 2$ Br	$2 \times 1 = 2$ Br

Use this example and the hints in Section 9.6 to help write balanced equations where called for in the remaining problems.

5. The balanced equation for the reaction between calcium bromide and silver nitrate which yields calcium nitrate and silver bromide.

Step 1. $CaBr_2 + AgNO_3 \rightarrow Ca(NO_3)_2 + AgBr$

Step 2.

Reactants (Coefficient × subscript)	Products (Coefficient × subscript)
$1 \times 1 = 1$ Ca	$1 \times 1 = 1$ Ca
$1 \times 2 = 2$ Br	$1 \times 1 = 1$ Br
$1 \times 1 = 1$ Ag	$1 \times 1 = 1$ Ag
$1 \times 1 = 1$ NO_3	$1 \times 2 = 2$ NO_3

Step 3. $CaBr_2 + 2\ AgNO_3 \rightarrow Ca(NO_3)_2 + 2\ AgBr$

Step 4.

Reactants (Coefficient × subscript)	Products (Coefficient × subscript)
$1 \times 1 = 1$ Ca	$1 \times 1 = 1$ Ca
$1 \times 2 = 2$ Br	$2 \times 1 = 2$ Br
$2 \times 1 = 2$ Ag	$2 \times 1 = 2$ Ag
$2 \times 1 = 2\ NO_3$	$1 \times 2 = 2\ NO_3$

7. a. There is no reaction in this case, because hydrogen is more reactive than gold.
   b. There is a reaction in this case, because aluminum is more reactive than hydrogen.
9. Chemists use an arrow to represent the word *yields.*
11. a. The reactants are left of the arrow. They are zinc and hydrogen chloride (hydrochloric acid).
    b. The products are right of the arrow. They are zinc chloride and hydrogen.
13. a. Barium chloride reacts with sodium sulfate to yield sodium chloride and barium sulfate.
    b. Silicon reacts with oxygen to yield silicon dioxide.
    c. Potassium reacts with iodine to yield potassium iodide.
    d. Sulfuric acid reacts with calcium hydroxide to yield calcium sulfate and water.
15. a. Balanced (2 K, 2 I, and 2 Br on both sides).
    b. Not balanced (2 P on the left, 4 P on the right).
    c. Not balanced (3 H on the left, 6 H on the right).
    d. Balanced (2 N, 8 H, 1 S, and 4 O on both sides).
    e. Not balanced (2 O on the left, 3 O on the right).
    f. Not balanced (2 H and 2 Cl on the left, 1 H and 1 Cl on the right).
17. The words to correctly fill in the blanks in the following statements are shown. To balance chemical equations, one should use *coefficients.* In balancing equations, *subscripts* should never be altered.
19. a. $2\ KI + Cl_2 \rightarrow 2\ KCl + I_2$
    b. $2\ Cu + O_2 \rightarrow 2\ CuO$
    c. $Li_2O + H_2O \rightarrow 2\ LiOH$
    d. $2\ SO_2 + O_2 \rightarrow 2\ SO_3$
    e. $3\ H_2 + N_2 \rightarrow 2\ NH_3$
    f. $2\ Na + ZnSO_4 \rightarrow Na_2SO_4 + Zn$
21. a. $CH_4 + 2\ O_2 \rightarrow CO_2 + 2\ H_2O$
    b. $2\ FeS + 3\ O_2 \rightarrow 2\ FeO + 2\ SO_2$
    c. $P_4O_{10} + 6\ H_2O \rightarrow 4\ H_3PO_4$
    d. $Cl_2 + H_2O \rightarrow HCl + HClO$
    e. $(NH_4)_2Cr_2O_7 \rightarrow Cr_2O_3 + N_2 + 4\ H_2O$
    f. $2\ Al + 3\ CuSO_4 \rightarrow Al_2(SO_4)_3 + 3\ Cu$
23. a. $C + 2\ Cl_2 \rightarrow CCl_4$
    b. $6\ K + N_2 \rightarrow 2\ K_3N$
    c. $Ba(NO_3)_2 + H_2SO_4 \rightarrow BaSO_4 + 2\ HNO_3$
    d. $Ca(OH)_2 \rightarrow CaO + H_2O$
    e. $4\ P + 5\ O_2 \rightarrow 2\ P_2O_5$
25. a. The correct formula for sodium iodide is NaI.
    $2\ Na + I_2 \rightarrow 2\ NaI$
    b. The correct formula for oxygen is O_2.
    $2\ Ag_2O \rightarrow 4\ Ag + O_2$
    c. The correct formulas for lead(II) chloride and potassium nitrate are $PbCl_2$ and KNO_3.
    $Pb(NO_3)_2 + 2\ KCl \rightarrow PbCl_2 + 2\ KNO_3$
    d. The correct symbol for silicon is Si.
    $Si + O_2 \rightarrow SiO_2$
27. a. $AlSO_4$ should be $Al_2(SO_4)_3$.
    $2\ Al + 3\ H_2SO_4 \rightarrow Al_2(SO_4)_3 + 3\ H_2$
    b. NH_4 should be NH_3.
    $N_2 + 3\ H_2 \rightarrow 2\ NH_3$
    c. Na_2Cl_2 should be NaCl.
    $BaCl_2 + Na_2SO_4 \rightarrow 2\ NaCl + BaSO_4$
    d. MgO_2 should be MgO.
    $2\ Mg + O_2 \rightarrow 2\ MgO$
29. a. $Cl_2O_7(l) + H_2O(l) \rightarrow 2\ HClO_4(aq)$
    b. $NH_4NO_3(s) \rightarrow N_2O(g) + 2\ H_2O(l)$
    c. $BaCl_2(aq) + (NH_4)_2CO_3(aq) \rightarrow BaCO_3 \downarrow + 2\ NH_4Cl(aq)$
    d. $Ca(s) + 2\ HCl(g) \rightarrow CaCl_2(s) + H_2 \uparrow$
31. a. $I_2(s) + 3\ Cl_2(g) \rightarrow 2\ ICl_3(s)$
    b. $Zn(s) + 2\ HCl(aq) \rightarrow ZnCl_2(aq) + H_2 \uparrow$
    c. $2\ Ag_2O(s) \xrightarrow{\Delta} 4\ Ag(s) + O_2 \uparrow$
    d. $2\ Al(s) + 3\ CuSO_4(aq) \rightarrow Al_2(SO_4)_3(aq) + 3\ Cu(s)$
33. a. Combination: $A + B \rightarrow AB$.
    b. Decomposition: $XY \rightarrow X + Y$.
    c. Single replacement: $A + BX \rightarrow AX + B$.
    d. Double replacement: $AX + BY \rightarrow AY + BX$.
35. Combination: 12 a, c; 13 b, c; 15 b, d, f; 19 b, c, d, e; 20 a, c, e, f, j; 21 c; 23 a, b, e; 24 d; 25 a, d; 26 b, c, d; 27 b, d; 28 d, e; 29 a; 31 a.
    Decomposition: 15 e; 20 i; 21 e; 23 d; 25 b; 29 b; 30 b; 31 c.
    Single replacement: 15 a, c; 19 a, f; 20 b, h; 21 f; 22 e; 24 b; 27 a; 28 h; 29 d; 30 a; 31 b, d.
    Double replacement: 12 b, d; 13 a, d; 20 d, g; 21 a, b, d; 22 a, b, c, d, f; 23 c; 24 a, c; 25 c; 26 a; 27 c; 28 a, b, c, f; 29 c; 30 c.

37. The answer to correctly fill in the blank in the following statement is shown.

    Elements react in combination reactions to form *a single product* (*a compound*).

39. a. $Ca + Br_2 \rightarrow CaBr_2$ (calcium bromide)
    b. $SO_2 + H_2O \rightarrow H_2SO_3$ (sulfurous acid)
    c. $3\ Mg + N_2 \rightarrow Mg_3N_2$ (magnesium nitride)
    d. $CaO + H_2O \rightarrow Ca(OH)_2$ (calcium hydroxide)

41. a. Yes, magnesium is more reactive than copper.
    b. No, magnesium is less reactive than calcium.
    c. Yes, barium is more reactive than hydrogen.
    d. No, silver is less reactive than hydrogen.
    e. Yes, aluminum is more reactive than nickel.
    f. Yes, calcium is more reactive than hydrogen.
    g. Yes, aluminum is more reactive than tin.
    h. No, gold is less reactive than hydrogen.

43. a. $2\ Li + CuO \rightarrow Li_2O + Cu$
    b. $2\ Li + 2\ HNO_3 \rightarrow 2\ LiNO_3 + H_2\uparrow$

45. a. $K_2SO_4(aq) + Ba(NO_3)_2(aq) \rightarrow 2\ KNO_3(aq) + BaSO_4$
    b. $(NH_4)_2CO_3(aq) + MgCl_2(aq) \rightarrow 2\ NH_4Cl(aq) + MgCO_3(s)$
    c. $2\ (NH_4)_3PO_4(aq) + 3\ Ca(NO_3)_2(aq) \rightarrow 6\ NH_4NO_3 + Ca_3(PO_4)_2$
    d. $3\ FeCl_2(aq) + 2\ K_3PO_4(aq) \rightarrow Fe_3(PO_4)_2 + 6\ KCl(aq)$
    e. $Na_2S(aq) + Ni(NO_3)_2(aq) \rightarrow 2\ NaNO_3(aq) + NiS$

47. a. Single replacement.

       $3\ Li(s) + AuCl_3(aq) \rightarrow 3\ LiCl + Au$

    b. Double replacement.

       $2\ Al(NO_3)_3(aq) + 3\ K_2CO_3(aq) \rightarrow Al_2(CO_3)_3 + 6\ KNO_3$

    c. Combination.

       $Ba(s) + F_2(aq) \rightarrow BaF_2$

    d. Single replacement.

       $Ba(s) + SnF_2(aq) \rightarrow BaF_2 + Sn$

    e. Combination.

       $3\ Mg(s) + 2\ P(s) \rightarrow Mg_3P_2$

    f. Double replacement.

       $3\ SnCl_2(aq) + 2\ Na_3PO_4(aq) \rightarrow Sn_3(PO_4)_2 + 6\ NaCl$

49. $2\ NaHCO_3(s) \xrightarrow{\Delta} CO_2(g) + H_2O(g) + Na_2CO_3(s)$

51. $ZnO + H_2 \rightarrow Zn + H_2O$

53. $2\ C_8H_{18} + 25\ O_2 \rightarrow 16\ CO_2 + 18\ H_2O$

55. The balanced equation for sulfur dioxide reacting with oxygen.

    $2\ SO_2 + O_2 \rightarrow 2\ SO_3$

    The balanced equation for sulfur trioxide reacting with water.

    $SO_3 + H_2O \rightarrow H_2SO_4$

    The balanced equation for sulfuric acid reacting with calcium carbonate.

    $H_2SO_4 + CaCO_3 \rightarrow CaSO_4 + CO_2 + H_2O$

57. a. $2\ Ca(s) + O_2(g) \rightarrow 2\ CaO(s)$
    b. $Li_2O(s) + H_2O(l) \rightarrow 2\ LiOH(s)$
    c. $2\ NaClO_3(s) \xrightarrow{\Delta} 2\ NaCl(s) + 3\ O_2(g)$
    d. $SO_3(g) + H_2O(l) \rightarrow H_2SO_4(aq)$
    e. $MgO(s) + CO_2(g) \rightarrow MgCO_3(s)$

59. a. $Pb(NO_3)_2 + 2\ KOH \rightarrow Pb(OH)_2 + 2\ KNO_3$
    b. $H_2SO_4 + Ca(OH)_2 \rightarrow CaSO_4 + 2\ H_2O$
    c. $3\ HCl + Al(OH)_3 \rightarrow AlCl_3 + 3\ H_2O$
    d. $2\ LiBr + Pb(ClO_4)_2 \rightarrow PbBr_2 + 2\ LiClO_4$

61. $C_{12}H_{22}O_{11} + 12\ O_2 \rightarrow 12\ CO_2 + 11\ H_2O$

63. If 4 grams of hydrogen are formed from a total of 36 grams of water, then the maximum amount of oxygen that can be formed is 36 grams minus 4 grams or 32 grams of oxygen.

65. The balanced equation for the burning of octane in a limited amount of oxygen.

    $2\ C_8H_{18} + 17\ O_2 \rightarrow 16\ CO + 18\ H_2O$

67. a. $Br_2 + 2\ NaI \rightarrow 2\ NaBr + I_2$
    b. This reaction is an example of a single replacement.
    c. The displacement activity of chlorine is greater than that of bromine, and the displacement activity of bromine is greater than iodine. The displacement activity of nonmetals increases as you move towards the top of the periodic table. The displacement activity of metals is just the opposite, it increases as you move towards the bottom of the periodic table.

chapter 10 stoichiometry

1. a. Two moles of $Al(OH)_3$ plus three moles of H_2SO_4 yields one mole of $Al_2(SO_4)_3$ plus six moles of H_2O.
   b. Four moles of Li plus one mole of O_2 yields two moles of Li_2O.

3. $\dfrac{1\ \text{mol SnO}}{2\ \text{mol HF}}$

5. 224 g KCl

7. To calculate the mass of SO_3 produced, use molar masses to determine the moles of each reactant.

$$3.60 \text{ g } SO_2 \times \frac{1 \text{ mol } SO_2}{64.06 \text{ g } SO_2} = 0.0562 \text{ mol } SO_2$$

$$2.40 \text{ g } O_2 \times \frac{1 \text{ mol } O_2}{32.00 \text{ g } O_2} = 0.0750 \text{ mol } O_2$$

Determine the molar amount of SO_3 that could be obtained from each reactant.

$$0.0562 \text{ mol } SO_2 \times \frac{2 \text{ mol } SO_3}{2 \text{ mol } SO_2} = 0.0562 \text{ mol } SO_3$$

$$0.0750 \text{ mol } O_2 \times \frac{2 \text{ mol } SO_3}{1 \text{ mol } O_2} = 0.150 \text{ mol } SO_3$$

The mass of SO_3 is determined by using the molar mass of SO_3 and the number of moles derived from the limiting reactant.

$$0.0562 \text{ mol } SO_3 \times \frac{80.06 \text{ g } SO_3}{1 \text{ mol } SO_3} = 4.50 \text{ g } SO_3$$

9. 244 kJ
11. Any equation used in a stoichiometric calculation must be balanced.
13. a. Two moles of P plus three moles of H_2 yields two moles of PH_3.
    b. One mole of HBr plus one mole of KOH yields one mole of KBr plus one mole of H_2O.
    c. Two moles of CO plus one mole of O_2 yields two moles of CO_2.
    d. One mole of C plus two moles of Cl_2 yields one mole of CCl_4.
    e. One mole of Mg plus two moles of HCl yields one mole of $MgCl_2$ plus one mole of H_2.
15. No, equations cannot be read directly in units of mass. The amount of matter must be conserved in a chemical reaction. In the example cited, one gram of matter would have been lost. Also, equal weights of substances do not ensure equal numbers of particles.
17. a. $\frac{2 \text{ mol P}}{3 \text{ mol } H_2}$, $\frac{2 \text{ mol P}}{2 \text{ mol } PH_3}$, and $\frac{3 \text{ mol } H_2}{2 \text{ mol } PH_3}$.
    b. $\frac{1 \text{ mol Mg}}{2 \text{ mol HCl}}$, $\frac{1 \text{ mol Mg}}{1 \text{ mol } MgCl_2}$, $\frac{1 \text{ mol Mg}}{1 \text{ mol } H_2}$, $\frac{2 \text{ mol HCl}}{1 \text{ mol } MgCl_2}$, $\frac{2 \text{ mol HCl}}{1 \text{ mol } H_2}$, and $\frac{1 \text{ mol } MgCl_2}{1 \text{mol } H_2}$.
19. The law of conservation of mass is satisfied by the use of mole ratios because the molar ratios are equal to the particle ratios. The total number of atoms of each element is equal regardless of how the balanced equation is interpreted.
21. Using the balanced equation given:
    a. 0.600 mol CS_2
    b. 1.20 mol SO_2
    c. 28.0 mol CO
    d. 7.50 mol C
23. $4 \text{ Al} + 3 O_2 \rightarrow 2 Al_2O_3$
    a. 3.0 mol Al_2O_3
    b. 4.5 mol O_2
25. a. 0.45 mol H_2SO_4
    b. 0.345 mol H_2O
    c. 6.94 mol KOH
27. a. 8 g N_2H_4
    b. 7 g N_2
29. a. 10.2 g Cl_2
    b. 3.48 g HCl
    c. 4.44 g H_2O
31. a. 466 g SiO_2
    b. 2.58 mol CO_2
    c. 0.720 mol $CaCO_3$
33. $2 \text{ Fe} + 3 Br_2 \rightarrow 2 FeBr_3$
    a. 12.3 g Fe
    b. 0.330 mol Br_2
35. a. 298 g O_2
    b. 168 g H_2O
37. a. 2.43 kg NH_3
    b. 1.51 kg H_2
39. a. 3.00 moles of Cl_2
    b. 1.50 moles of CH_4
    c. 0.50 mole of Cl_2
41. 1.00 mol $CHCl_3$
43. 20. g HF
45. 17 g N_2
47. 101 g CO_2
49. The limiting reactant is 2.0 grams of Pb.
    The yield is 2.3 g PbS.
51. a. The actual yield is the amount of product which is actually collected in the lab. This is always less than the theoretical yield.
    b. No, the actual yield can only result from carrying out the reaction in the lab.
53. 64%
55. a. The limiting reactant is 145 g Fe_2O_3.
    The theoretical yield is 101 g Fe.
    b. 92%

57. a. An exothermic process is any in which heat is a product. Wood burning is an example.
    b. An endothermic process is any in which heat is a reactant. Ice melting is an example.
59. a. 1.60×10^3 kJ
    b. 240. kJ
61. a. 29 kJ
    b. 27 g $CaCO_3$
63. a. 40. kJ
    b. 88°C
65. a. $2\ NH_3 + 3\ CuO \rightarrow 3\ Cu + N_2 + 3\ H_2O$
    b. 185 g Cu
    c. 81.6%
67. a. $C_6H_{12}O_6 \rightarrow 2\ C_2H_5OH + 2\ CO_2$
    b. 29.2 g C_2H_5OH
    c. 37.0 mL C_2H_5OH
69. The limiting reactant is 30.0 g NaBr.
    The yield of bromine is 23.3 g Br_2.

chapter 11 gases

1. Original density = 1.9 g/L. Final density = 0.080 g/L.
3. The total pressure of a mixture of gases is equal to the sum of all the partial pressures of each gas in the mixture. In this case the total is 3.0 atm.
5. a. 273 K
   b. 298 K
   c. 373 K
7. 283 K.
9. Hydrogen = 0.0902 g/L, helium = 0.179 g/L, and air = 1.29 g/L
11. $n = 0.0499$ mol
13. $Mg + Cl_2 \rightarrow MgCl_2$
    $V = 15.7$ L Cl_2
15. 1. Gases can be easily compressed.
    2. Gases expand to fill the entire volume of their container.
    3. Gases have indefinite densities.
    4. Gases have low densities.
    5. Gases can diffuse rapidly through each other.
    6. Gases exert a pressure on the walls of any container or surface that they touch.
    7. Gases expand when heated and contract when cooled.
17. A container must be evacuated before it can be used in determining the mass of a gas because any gas not removed from the container would affect the measurement.
19. Atmospheric pressure is the result of the atmosphere being composed of gases, it exerts pressure on any surface it touches.
21. Mercury does not run out of a mercury barometer because the atmospheric pressure acting on the surface of the reservoir of mercury is sufficient to support the column of mercury in the barometer.
23. a. 580. mmHg
    b. 0.763 atm
25. 887 mmHg, or 1.167 atm
27. Gases flow spontaneously from a region of higher pressure to a region of lower pressure. Gas 1 will flow into mixture A and gas 2 will flow into mixture B. Eventually the partial pressures of each gas will be equal on both sides of the barrier.
29. a. $V \propto 1/P$. The volume of a gas is inversely proportional to the pressure when the temperature and the number of moles are held constant.
    b. PV = k. For a fixed number of moles at a given temperature, the product of the pressure and volume of a gas is equal to a constant.
    c. Breathing is an example that demonstrates Boyle's law. Another example is increasing pressure on a piston, decreases the volume in a cylinder.
31. a. The gas will occupy a volume of 3 L at 2 atm.
    b. The value of *PV* for the volume in part a is 6 L • atm.
    c. The value of *PV* at a volume of 1 L is 6 L • atm.
    d. The value of *PV* at a volume of 2 L is 6 L • atm.
33. You could measure the volume of trapped air in Figure 11.19 with a thermometer because the volume increase is proportional to the height or length increase of trapped air. The temperature markings in °C are equally spaced and can be used as a measure of length (height).
35. Warm air rises above cold air because warm air molecules are moving faster and therefore are better able to overcome gravitational attraction and thus rise up over colder air. Warm air molecules are moving faster and therefore more spread out and less dense than cooler air.
37. a. No, the volume of a gas is not directly proportional to Celsius temperature.
    b. The volume of gas is directly related to the Celsius temperature, but not directly proportional. An increase in Celsius temperature leads to an increase in volume but not by the same factor.
39. $V_2 = 1.13$ L

41. No, the answers in Problem 40 would not be the same if the pressure varied because the volume of a gas changes with pressure.

43. $T_2 = 754$ K or 481°C

45. $P_2 = 2.77$ atm

47. $V_2 = 2.43$ L

49. Use Avogadro's law, which states that equal volumes of all gases at the same temperature and pressure contain the same number of molecules (moles). The SO_2 would occupy the same volume as the CO_2 gas or 36 L.

51. $V_2 = 69.5$ mL

53. 67 L N_2

55. One mole of N_2 equals 28.02 grams. One mole of any gas at STP is 22.4 liters. The density of one mole of nitrogen gas at STP is 28.02 g/22.4 L or 1.25 g/L.

57. Mass = 26.0 g/mol.

59. a. No, one mole of a gas will not always occupy 22.4 liters.
    b. For one mole of gas to equal 22.4 liters, the gas must be at STP (0°C and 1 atm).

61. $n = 1.52$ mol C_3H_8

63. 93. 76 mol He, 375 g He

65. 9.0 mol CH_4, 144 g CH_4

67. Density = 0.610 g/L

69. 1. Gases are made up of small particles which are constantly moving in random, straight-line motion.
    2. The distance between particles is very large compared with the size of the particles. A gas is mostly empty space.
    3. There are no attractive forces between particles. The particles move independently of each other.
    4. The particles collide with each other and with the walls of the container without incurring a loss of energy.
    5. The average kinetic energy of the particles is directly proportional to the Kelvin temperature.

71. The pressure in an automobile tire increases after the tire has been driven at high speed because friction with the pavement causes the tire to heat up which in turn causes the increased motion of the gas inside the tire increasing the rate and force at which the gas particles collide with the inside wall of the tire, thus increasing the pressure.

73. A given gas decreases in density as its temperature increases because at a constant pressure, a given mass of a gas increases in volume as its temperature rises. Since density is inversely proportional to volume, the density decreases.

75. The kinetic energy is directly proportional to the Kelvin temperature. To double the kinetic energy of nitrogen gas, you must double its temperature in degrees Kelvin. Since 0°C equals 273 Kelvin, the temperature would be 546 Kelvin. Convert back to °C and you get 273°C.

77. 1.5 L O_2

79. $V = 23.7$ L CO_2

81. The physical process that allows a student ten feet away from an open bottle of perfume to smell it is that some of the liquid in the bottle evaporates to a vapor. Then the vapor diffuses through the air to reach the student.

83. The gas pressure would be much lower in outer space. In outer space, there are considerably fewer molecules at a significantly lower temperature, resulting in few collisions at less force per collision.

85. 0.973 atm.

87. Yes, the pressure of air can be increased without increasing the oxygen pressure by increasing the amount of any other gas besides oxygen.

89. Heating the air in a hot air balloon increases the temperature at a constant pressure. This causes the volume of the air to increase. The same amount of gas in a larger volume results in decreased density. The volume of air of lower density rises, so the balloon rises.

91. $V_2 = 7.23$ L

93. $n = 6.8 \times 10^{-9}$ mol

95. Density = 0.588 g/L
    Water is about 1,630 times more dense.

97. $V_2 = 0.0363$ L

99. $V = 84.4$ L

101. The kinetic molecular theory of gases would explain that the space above the boiling liquid in the heated can is full of water vapor, a hot gas. After the can is sealed and the gas cools, the volume of the gas decreases. The decreased volume lowers the pressure inside the can. Since the can is sealed (air can not get in to equalize the pressure), the sides collapse in.

103. Dalton's law of partial pressures would be more accurate for ideal gases because the behavior of ideal gases doesn't take into account the attraction of molecules for each other and variations due to pressure and temperature.

105. a. 0.50 L O_2
     b. 1.0 L H_2O

107. Molar mass = 16.0 g/mol

109. $P = 1.67$ atm

111. The kinetic molecular theory of gases would explain that as the temperature rises, pressure inside the table-tennis ball increases at a constant volume. As the pressure builds up, it can pop the dent out, slightly increasing the volume and relieving a little of the increased pressure.

chapter 12 liquids and solids

1. a. Two sets of bonding electron pairs and no nonbonding pairs gives us linear geometry and linear shape.
   b. Two sets of bonding electron pairs and one set of nonbonding electrons gives a total of three sets. This gives the planar triangular geometry. One set of nonbonding electrons gives the bent shape.
   c. Four sets of bonding electron pairs and no nonbonding electron pairs gives us tetrahedral geometry and a tetrahedral shape.
3. Molecules with N—H, O—H or F—H bonds will show hydrogen bonding. Both b, acetamide, and c, methylamine, have N—H bonds.
5. 50.8 kJ
7. a. Four sets of electron pairs gives tetrahedral geometry.
   b. Three sets of electron pairs gives planar triangular geometry.
   c. Four sets of electron pairs gives tetrahedral geometry.
   d. Four sets of electron pairs gives tetrahedral geometry.
9. a. Moving electron pairs farther apart lowers their energy because the greater the distance between electron pairs, the lower the repulsive force between the electron pairs is.
   b. An angle of 200° would bring the electron pairs closer together than an angle of 180°. The electron repulsion would prevent their moving to 200°.
11. a.

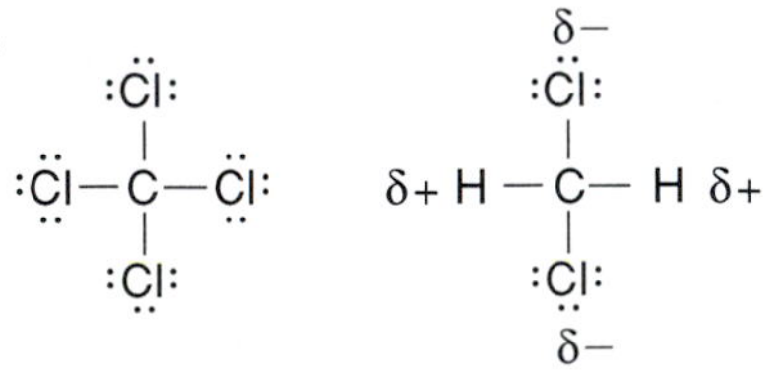

   b. CCl_4 is nonpolar because bond dipoles cancel. CH_2Cl_2 is polar because the bond dipoles and geometry lead to – and + ends of the molecule.
13. a.

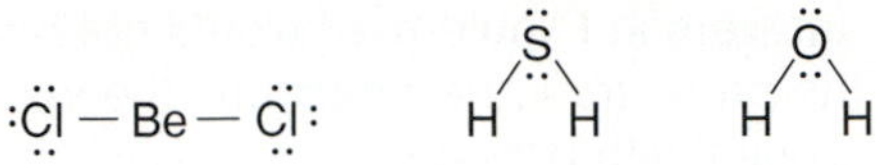

   b. $BeCl_2$ probably has the smallest molecular polarity because it is linear and symmetrical, therefore the bond dipoles cancel.
   c. H_2O has the largest molecular polarity.
15. To measure the molecular polarity of a molecule, place the compound between two charged plates. An observable change in the voltage between the plates indicates that polar molecules have lined themselves up with the electric field.
17. Problem 10 part b, SO_3, has a zero dipole moment.
19. NH_3 would have a larger dipole moment than PH_3 because nitrogen is more electronegative than phosphorus.
21. a. The three intermolecular attractive forces listed in order of increasing strength are dispersion forces, dipole-dipole interactions and hydrogen bonds.
   b. All molecules display dispersion forces. Dipole-dipole interactions occur between molecules with permanent dipole moments, which require polar bonds and noncanceling geometry. Hydrogen bonds exist when a hydrogen atom bonded to an N, O, or F in a molecule is attracted to a nonbonding electron pair on N, O, or F in another molecule.
23. a. F_2 is dispersion forces.
   b. H_2O is hydrogen bonding.
   c. CH_2F_2 is dipole-dipole interactions.
   d. HF is hydrogen bonding.
25. a. NH_3 is hydrogen bonding.
   b. HCN is dipole-dipole interactions.
   c. Ne is dispersion forces.
27. The boiling point of the inert gases decreases with decreasing molar mass because the number of electrons decrease resulting in weaker dispersion forces.
29. Polar molecules exhibit dipole-dipole attractive forces. The molecules that would show this kind of attractive forces between their molecules are b (HCl), d (HCN), f (PF_3), and h (SO_2).
31. $C_{10}H_{22}$ would have stronger intermolecular forces than C_5H_{12} because the greater number of atoms means there are more electrons.

33. An increase in the intermolecular attractions leads to an increase in the boiling point of a liquid because more heat is required to overcome stronger intermolecular attractions.

35. He< CH_4 < CH_3F < CH_2F_2 < $CH_3CH_2CH_2OH$

37. Molecules in a liquid will flow and those in a solid will not because in a solid, the molecules are very close and the intermolecular forces are very strong. The molecules in a liquid are not as close together and the forces between them are weak enough to allow them to move past one another.

39. Liquids and solids have low compressibility when compared to gases because the molecules of gases are much farther apart than those of a liquid or solid. This permits gases to be compressed more easily.

41. a. The kinetic energy of an isolated liquid decreases as the liquid evaporates because the molecules with higher kinetic energy escape into the gas phase leaving behind the molecules with lesser kinetic energy on the average in the liquid phase.
    b. No, the remaining liquid will not evaporate as fast as the initial liquid because the remaining liquid has a lower overall kinetic energy.

43. a. CH_4 would show the higher vapor pressure because H_2O has hydrogen bonding.
    b. CH_3OH would show the higher vapor pressure because of hydrogen bond attraction.
    c. C_3H_8 would show the higher vapor pressure because the dipole-dipole attraction is stronger in C_2H_4O.

45. Increasing the temperature of a liquid increases the average kinetic energy of molecules so that more can escape into the gas phase. This increases its vapor pressure.

47. Differences between evaporation and boiling.

    Evaporation

    Occurs only at the liquid's surface.
    Occurs independently of the external pressure.

    Boiling

    Bubbles of gas form throughout the liquid.
    Occurs only when vapor pressure = external pressure.

49. a. The statement that water has a normal boiling point of 100°C means that the vapor pressure of water equals one atmosphere at 100°C.
    b. Yes, you can make water boil at lower temperatures by reducing the external pressure, or it can boil at higher temperatures by increasing the external pressure.

51. a. The name given to the quantity of heat required to convert a mole of liquid to a gas at its boiling point is the molar heat of vaporization.
    b. The gas will have the greater amount of energy over the liquid at the boiling point. The increased energy comes form the heat of vaporization.
    c. The liquid state has the stronger attractive forces at the boiling point.
    d. On the molecular level, the molar heat of vaporization is the energy required to overcome the stronger intermolecular attractive forces in the liquid.

53. Calculate the energy necessary to vaporize 31.6 grams of H_2O at 100°C.

$$31.1\text{ g H}_2\text{O} \times \frac{2{,}260\text{ J}}{1\text{ g H}_2\text{O}} = 70{,}286\text{ J} = 7.03 \times 10^4\text{ J}$$

55. Increasing the temperature would decreased surface tension of a liquid.

57. Cheese should be kept covered to prevent the escape of volatile components of the cheese which gives it its aroma and flavor. The cover also retards loss of water vapor, which would dry out the cheese.

59. a. Use the graph in Figure 12.13 to determine that the typical boiling point of water in Mexico City is 93°C.
    b. It would take longer to hard-boil an egg in Mexico City compared to sea level because water boils at a lower temperature in Mexico City.

61. One way to determine if an unknown white, powdery solid is crystalline or amorphous is to heat it to determine if it has a definite melting point.

63. a. The lattice points of an ionic crystalline solid are occupied by opposite charged ions.
    b. The lattice points of a molecular crystalline solid are occupied by molecules.
    c. The lattice points of a covalent crystalline solid are occupied by atoms.
    d. The lattice points of a metallic crystalline solid are occupied by metallic cations.

65. a. The four types of crystalline solids and their attractive forces.

Ionic	electrostatic forces
Molecular	dipole-dipole interactions, dispersion forces, and hydrogen bonds
Covalent	covalent bonds
Metallic electrostatic	electrostatic attraction between metal cation and mobile electrons

b. Molecular crystalline solids generally have the weakest attractive forces.

67. The melting point of sodium chloride is so much higher than pure hydrogen iodide because hydrogen iodide is a molecular compound in which the strongest intermolecular force is dipole-dipole interaction. This is much weaker than the electrostatic forces in the ionic crystal, sodium chloride.

69. a. Fullerene resembles a covalent network because the carbon atoms in its structure are covalently bonded to one another and the network closes in upon itself with a definite number of atoms.

b. Fullerene resembles a molecular crystal because the structures described in part a arrange themselves into a lattice structure of indefinite size.

71. 1.37×10^4 J = 13.7 kJ

73. The heat of sublimation would be larger than the heat of vaporization because the heat of sublimation is a conversion from a tightly packed solid directly to a gas. This change is similar to two changes in state, changing a solid to a liquid and the liquid to a gas.

75. 65.872 kJ = 6.59×10^4 J = 65.9 kJ

77. 106 kJ

79. 96.0 kJ

81. a. Pyramidal.

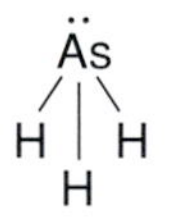

b. Bent.

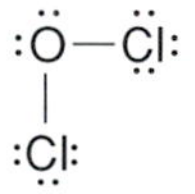

c. Pyramidal.

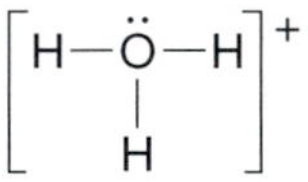

d. Planar triangular.

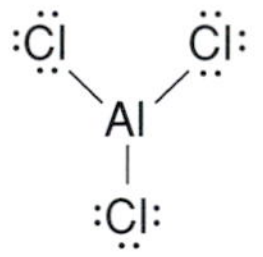

e. Tetrahedral.

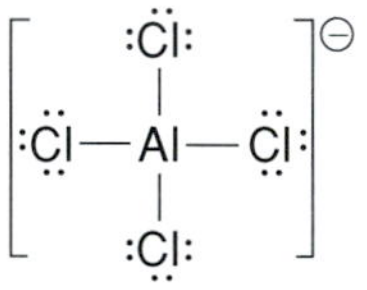

83. The molecule, OCS, has a higher molecular dipole than CO_2. Carbon dioxide has two C—O bonds, the polarity of which offset each other; however in OCS, the polarity of the carbon to sulfur bond cannot offset the polarity of the carbon to oxygen bond.

85. Use the graph in Figure 12.13 to determine the boiling point of water at 400 mmHg; about 82°C.

87. Even though they have similar molar masses, the boiling point of ethyl alcohol is much higher than that of propane because ethyl alcohol has hydrogen bonding and propane does not.
89. a. Magnesium is a metallic crystal.
    b. Carbon dioxide is a molecular crystal.
    c. Silicon dioxide is a covalent crystal.
    d. Sodium is a metallic crystal.
    e. Sodium chloride is an ionic crystal.
    f. Water is a molecular crystal.
91. A burn from steam at 100°C can cause more damage than a burn from water at 100°C because steam can also condense to a liquid releasing heat of evaporation as well as the heat due to the rise in temperature.
93. Obtaining water by ingesting snow or ice would cause the body to expend energy because the body must supply the heat of fusion to melt ice or snow.

chapter 13 solutions

1. a. 15 grams of KBr at 20°C is unsaturated.
   b. 15 grams of $KMnO_4$ at 60°C is unsaturated.
   c. 115 grams of $AgNO_3$ at 20°C is unsaturated.
3. 31.6% w/w
5. 700 mL solution =7.0×10^2 ml solution
7. 4.22 L solution
9. $Ba^{2+}(aq) + SO_4^{2-}(aq) \rightarrow BaSO_4\downarrow$
11. A solution is a homogeneous mixture of two or more substances. To be truly homogeneous, the intermixed substances (solute and solvent) must be ionic or molecular in size.
13. Examples of the three types of solutions in which the solvent is a liquid.
    A gas in a liquid is oxygen gas in water.
    A liquid in a liquid is oil in gasoline.
    A solid in a liquid is sugar in water.
15. a. Helium gas dispersed with air.
    b. Scotch whiskey
    c. Urine
    e. Soda water
    f. Bronze (tin homogeneously dispersed in copper).
17. Dusty air is not considered a solution because it is not homogeneous; dust particles settle out.
19. a. No, a saturated solution does not have to be a concentrated solution. If solubility is low, a solution can be saturated but dilute.
    b. An example of a solution that is saturated yet dilute would be silver acetate. Silver acetate has a solubility limit of 1.04 grams per 100 grams of water. This solution would be saturated but dilute.
21. a. 17 grams of KBr in 25 grams of H_2O at 60°C is unsaturated.
    b. 7.5 grams of $KMnO_4$ in 50 grams of H_2O at 60°C is unsaturated.
    c. 50. grams of NaCl in 150 grams of H_2O at 60°C is unsaturated.
    d. 41 grams KNO_3 in 125 grams H_2O at 20°C is saturated.

23. To prepare a supersaturated solution of potassium nitrate in water, dissolve 50. grams of KNO_3 in 100. grams of H_2O at 60°C. Carefully cool the solution to 20°C. The solution will now be supersaturated.

25. a. The terms "miscible" and "immiscible" apply to mixtures of liquids with liquids.
    b. An example of two miscible substances is ethylene glycol and water. This is the antifreeze mixture used in automobiles.
    c. An example of two immiscible substances is turpentine and water.

27. a. Solutes tend to dissolve in solvents which have similar intermolecular forces. Polar solutes dissolve in polar solvents; nonpolar solutes in nonpolar solvents.
    b. Polar solutes will best dissolve in water, a polar solvent.
    c. Nonpolar solutes will best dissolve in gasoline, a nonpolar solvent.

29. a. A bottle of soda goes "flat" if left open to the air because opening the bottle relieves the pressure on the solution. The solubility of CO_2 in water decreases as pressure decreases, releasing the gas to the atmosphere.
    b. Keeping the bottle of soda warm would not help keep it fizzier longer. The solubility of a gas in water decreases with increasing temperatures.

31. a. Stirring enables a solute to dissolve more quickly because stirring establishes better contact between the solute and solvent particles.
    b. No, stirring does not affect the solubility of a solute in a solvent.

33. Solubility refers to the maximum amount of solute that can dissolve in a given amount of solvent at a given temperature. For example, a maximum of 36 grams of NaCl can dissolve in 100. grams of H_2O at 20°C. Concentration is a statement of the amount of solute dissolved in a given amount of solvent or solution. 10 grams of NaCl in 90 grams of H_2O produces a 10% solution. Concentration is variable.

35. a. 22.1% w/w.
    b. The difference between the mass of the solution and the mass of the solute will be the mass of the solvent. In this case, 139. grams of water.

37. To prepare 250.g of a 5.00% w/w solution of $BaCl_2$, dissolve 12.5 grams of $BaCl_2$ in 237.5 grams of water.

39. 24.0% w/w

41. 13.5 g sugar, 76.5 g water

43. 38.7% w/v

45. 31 mL solution

47. 0.01% w/v solution is more concentrated.

49. Percent mass/volume relates grams of solute to milliliters of solution. Molarity relates moles of solute to liters of solution.

51. 0.0125 mol NaCl.

53. Place 26.3 grams of NaCl in a 500 milliliter volumetric flask. Add some water and swirl the flask to dissolve the NaCl. Add water to the etched mark which gives 0.500 liters of solution.

55. 4.25 g $AgNO_3$

57. 0.568 *M*

59. 1.81×10^3 mL

61. 0.200 *M*

63. 4.8×10^2 mL

65. Evidence for the existence of ions in a solution of NaCl would be that the solution conducts electricity.

67. One covalent substance can be a strong electrolyte and another a weak electrolyte because some covalent materials react with water to form many ions in solution (e.g., hydrogen bromide). Other covalent materials form only a few ions (e.g., $HC_2H_3O_2$).

69. HCl reacts with water to form ions and therefore conducts electricity. In benzene no ions are produced.

71.

Strong Acids	*Weak Acids*
Hydrochloric acid; HCl	Acetic acid; $HC_2H_3O_2$
Hydrobromic acid; HBr	Boric acid; H_3BO_4
Hydroiodic acid; HI	Carbonic acid; H_2CO_3
Nitric acid; HNO_3	Hydrofluoric acid; HF
Sulfuric acid; H_2SO_4	Hydrocyanic acid; HCN
Perchloric acid; $HClO_4$	

73. a. $Ba^{2+}(aq) + SO_4^{2-}(aq) \rightarrow BaSO_4 \downarrow$
    b. $Zn(s) + 2H^+(aq) \rightarrow Zn^{2+}(aq) + H_2 \uparrow$
    c. $Ag^+(aq) + Cl^-(aq) \rightarrow AgCl \downarrow$
    d. $Pb^{2+}(aq) + S^{2-}(aq) \rightarrow PbS \downarrow$
    e. $Fe^{3+}(aq) + 3\ OH^-(aq) \rightarrow Fe(OH)_3 \downarrow$
    f. $2\ H^+(aq) + CO_3^{2-}(aq) \rightarrow H_2O + CO_2 \uparrow$

75. a. Na^+ and SO_4^{2-} ions
    b. FeS $\downarrow$
    c. K^+ and OH^- ions
    d. $Ca_3(PO_4)_2 \downarrow$
    e. Mg^{2+} and NO_3^- ions
    f. NH_4^+ and CO_3^{2-} ions

77. 0.600 *M*
79. 7.60 g
81. 1.78 g
83. a. Water has a higher vapor pressure.
    b. The glucose solution has a higher boiling point.
    c. Water has the higher freezing point.
    d. The glucose solution has the greater osmotic pressure.
85. a. Net water flow from A to B.
    b. Net water flow from A to B.
    c. Net water flow from B to A.
    d. No net flow.
87. a. 2% w/v NaCl is hypertonic.
    b. 0.15 *M* glucose is hypotonic.
    c. 0.35 *M* glucose is hypertonic.
    d. 0.12 *M* $CaCl_2$ is hypertonic.
    e. 0.30 *M* glucose is isotonic.
89. A 10% NaCl solution would have the higher osmotic pressure because of the higher ion content due to the lower molar mass of NaCl compared to KCl.
91. 0.00918 *M*
93. The higher concentration of ions in the brine creates an osmotic pressure that draws the water out of the pickle.
95. M = 1.33 *M*
97. 5.05 *M* of each
99. Methanol, CH_3OH, is polar molecule and dissolves readily in polar solvent, water. Methane,CH_4, is nonpolar and does not mix or dissolve in water.
101. 3.8% $MgSO_4 \cdot 7\ H_2O$.
103. a. Salt is added to ice in an ice cream machine because the salt lowers the freezing point of water and makes it colder so the ice cream freezes faster.
    b. Sea water has a lower freezing point than fresh water because sea water contains salt which lowers the freezing point.
105. Adding HCl to the solution of sodium nitrate and silver nitrate will cause AgCl to precipitate out. NaCl is very soluble.
107. a. 1.5×10^{-6} % w/v
    b. 1.9×10^{-7} *M*

chapter 14 chemical equilibrium

1. a. $K_{eq} = \dfrac{[HI]^2}{[H_2]\,[I_2]}$

    The purple color due to the I_2 would decrease and become constant when the reaction reached equilibrium. The acrid smell of HI would also be apparent.

    b. $K_{eq} = \dfrac{[NH_3]^2}{[N_2]\,[H_2]^3}$

    The pungent odor of NH_3 would become apparent after its formation.
3. a. $K_{sp} = [Ca^{2+}]\,[CO_3^{2-}]$
    b. $K_{sp} = [Pb^{2+}]\,[Cl^-]^2$
    c. $K_{sp} = [Ba^{2+}]^3\,[PO_4^{3-}]^2$
5. a. Increasing the concentration of H_2 would cause the equilibrium to shift right (→).
    b. Adding C_2H_4 would cause the equilibrium to shift left (←).
    c. Increasing the total pressure would cause the equilibrium to shift right (→).
    d. Decreasing the reaction temperature would cause the equilibrium to shift right (→).
7. Two nonchemical examples of a dynamic equilibrium could be: (1) at the melting point, melting and freezing occur at the same rate, solid and liquid exist in dynamic equilibrium; (2) in a closed soda bottle, CO_2 gas enters and leaves the liquid at the same rate. See also section 14.2.
9. a. The evaporation of water in a closed container leads to a dynamic equilibrium because the water molecules leave the liquid state and enter the gaseous state; after a while some gaseous water molecules are recaptured into the liquid state. Eventually the rate of gas changing to a liquid is equal to the rate of liquid changing to a gas and an equilibrium condition is present.
    b. Evaporation of water in an open container would not lead to a dynamic equilibrium because water molecules can leave the surroundings of the liquid and not be recaptured into the liquid.
11. Equal concentrations require [R] = [P]. Constant concentration requires [R] remains the same, e.g. 0.5, and [P] remains the same, e.g. 0.36, but [R] and [P] do not have to be equal.
13. At the equilibrium condition the rate of the forward reaction equals the rate of the reverse reaction.
15. $K_{eq} = \dfrac{k}{k'} = \dfrac{[B]}{[A]^2}$

17. a. $K_{eq} = \frac{[H_2]\,[I_2]}{[HI]^2}$

b. You would know equilibrium has been reached when the color of the system remains constant.

c. $K_{eq} = \frac{[H_2]\,[I_2]}{[HI]^2} = \frac{[0.86]\,[0.86]}{[0.27]^2} = \frac{0.7396}{0.0729} = 10$

19. a. $K_{eq} = \frac{[CO]\,[H_2]}{[H_2O]}$

b. $K_{eq} = \frac{[H_2]^4}{[H_2O]^4}$

c. $K_{eq} = [H_2]^2\,[O_2]$

d. $K_{eq} = \frac{[MnCl_2]\,[Cl_2]}{[HCl]^4}$

e. $K_{eq} = \frac{[CO_2]}{[H_2CO_3]}$

21. a. $K_{eq} = \frac{[\beta\text{-D-glucose}]}{[\alpha\text{-D-glucose}]}$

b. $K_{eq} = \frac{[CH_3COOCH_3]}{[CH_3OH]\,[CH_3COOH]}$

c. $K_{eq} = \frac{[\text{oxyhemoglobin}]}{[\text{hemoglobin}]\,[O_2]}$

d. $K_{eq} = \frac{[\text{single stranded DNA}]^2}{[\text{double stranded DNA}]}$

23. a. $K_{sp} = [Ag^+]\,[Br^-]$

b. $K_{sp} = [Cu^{2+}]\,[OH^-]^2$

c. $K_{sp} = [Al^{3+}]\,[OH^-]^3$

d. $K_{sp} = [Mg^{2+}]^3\,[PO_4^{3-}]^2$

e. $K_{sp} = [Bi^{3+}]^2\,[S^{2-}]^3$

25. a. In this reaction, essentially all products are at equilibrium.

b. In this reaction, more reactants than products are at equilibrium.

c. In this reaction, significant amounts of reactants and products are at equilibrium.

d. In this reaction, essentially all reactants are at equilibrium.

27. a. Equation 2 is the reaction that is most complete from left to right as written.

b. Equation 3 is the reaction that is least complete from left to right.

29. a. Increasing the concentration of oxygen will shift the reaction to the right (→).

b. Adding N_2 to the equilibrium mixture will shift the reaction to the left (←).

c. Removing H_2O as it is formed will shift the reaction to the right (→).

d. Increasing the reaction temperature will shift the reaction to the left (←).

e. Decreasing the total pressure will shift the reaction to the right (→).

31. a. The total pressure should be decreased to maximize the equilibrium concentration of H_2.

b. The total pressure should be increased to maximize the equilibrium concentration of H_2O.

33. a. Adding Ca^{2+} ions will shift the reaction toward $Ca_3(PO_4)_2\,(s)$.

b. Adding PO_4^{3-} ions will shift the reaction toward $Ca_3(PO_4)_2\,(s)$.

c. Removing PO_4^{3-} will shift the reaction toward the product ions.

35. $BaSO_4$ will be more soluble in pure water. In sulfuric acid, the sulfate ion shifts the equilibrium toward the solid (common ion effect).

37. The addition of H^+ and the consequent shift causes the bicarbonate $[HCO_3^-]$ concentration to be lowered.

39. a. $Ag^+(aq) + Br^-(aq) \rightarrow AgBr\downarrow$

b. $H^+(aq) + OH^-\,(aq) \rightarrow H_2O(l)$

c. No reaction.

d. $Ba^{2+}(aq) + 2\,OH^-(aq) + 2\,H^+(aq) + SO_4^{2-}(aq) \rightarrow BaSO_4\downarrow + 2\,H_2O$

e. $3\,Ca^{2+}(aq) + 2PO_4^{3-}(aq) \rightarrow Ca_3(PO_4)_2\downarrow$

41. The formation of a gas drives a reaction to completion because the product gas is being removed from the system. This causes the shift in equilibrium toward the products.

43. Using Le Chatelier's principle to maximize the yield of CH_3OH, use high pressure, as low a temperature as possible, and high concentrations of CO and H_2.

45. The equilibrium constant for the decomposition of phosgene at 527°C.

$K_{eq} = 0.00463$

47. Calculate the equilibrium concentration of NO_2.

$K_{eq} = \frac{[NO_2]^2}{0.630}$

$[NO_2]^2 = 0.126$

$[NO_2] = .35\,M$

49. a. Increasing the reaction temperature will cause the reaction to shift toward the left (←).

b. Increasing the concentration of hydrogen will cause the reaction to shift toward the right (→).

c. Decreasing the concentration of CH_4 will cause the reaction to shift toward the right (→).

d. Increasing the volume of the reaction vessel will cause the reaction to shift toward the left (←).

51. $K_{eq} = [0.242]^2 [0.121] = 0.00709$

chapter 15 acids and bases

1. a. $KOH(s) \rightleftharpoons K^+(aq) + OH^-(aq)$
   b. $Ca(OH)_2(s) \rightleftharpoons Ca^{2+}(aq) + 2OH^-(aq)$
3. $HNO_3 > HF > H_2CO_3 > H_3BO_3 > H_2O$.
5. $[OH^-] = 3.1 \times 10^{-6}$ *M*; Basic.
7. a. pH = 5.17.
   b. pH = 13.85.
9. pH = 7.41.
11. Refer to Table 15.2 to determine two acids and two bases found in commercial, household products. Some examples are acetic acid (vinegar), acetylsalicylic acid (aspirin), sodium hydroxide (lye), and ammonia.
13. The equation to show the formation of the hydronium ion from the ionization of HNO_3.
    $$HNO_3 + H_2O \rightleftharpoons H_3O^+ + NO_3^-$$
15. a. A Brønsted-Lowry acid is an H^+ donor; a Brønsted-Lowry base is an H^+ acceptor.
    b. The definitions of a Brønsted-Lowry acid and an Arrhenius acid are essentially identical.
    c. A Brønsted-Lowry base is an H^+ acceptor (a broad, general definition). An Arrhenius base produces OH^- in solution (a more limited definition).
17. $NH_4^+ + H_2O \rightleftharpoons H_3O^+ + NH_3$
19. a. H_2O is OH^-
    b. HF is F^-
    c. $H_2PO_4^-$ is HPO_4^{2-}
    d. NH_4^+ is NH_3
    e. H_2SO_4 is HSO_4^-
    f. NH_3 is NH_2^-
21. HCO_3^- is an amphoteric substance since it is capable of both losing and accepting protons as shown in the following equations.
    $$HCO_3^- \rightarrow H^+ + CO_3^{2-}$$
    $$HCO_3^- + H^+ \rightarrow H_2CO_3$$
23. a. $NH_4^+ + NH_3 \rightleftharpoons NH_3 + NH_4^+$
       acid base
    b. $H_2SO_4 + H_2O \rightleftharpoons HSO_4^- + H_3O^+$
       acid base
    c. $HCl + CN^- \rightleftharpoons HCN + Cl^-$
       acid base
    d. $H_2CO_3 + NO_3^- \rightleftharpoons HCO_3^- + HNO_3$
       acid base
    e. $HSO_4^- + OH^- \rightleftharpoons SO_4^{2-} + H_2O$
       acid base
    f. $HSO_4^- + NO_3^- \rightleftharpoons SO_4^{2-} + HNO_3$
       acid base
25. $ClO_4^- < SO_4^{2-} < H_2BO_3^- < CN^-, OH^-$
27. a. $K_{eq} = \dfrac{[H_2SO_4]\,[OH^-]}{[HSO_4^-]}$
    b. The concentrations at equilibrium would have to be determined experimentally to evaluate this equilibrium constant.
29. a. All steps showing the complete ionization of H_2CO_3.
       $H_2CO_3 \rightleftharpoons H^+ + HCO_3^-$
       $HCO_3^- \rightleftharpoons H^+ + CO_3^{2-}$
    b. All steps showing the complete ionization of H_3PO_4.
       $H_3PO_4 \rightleftharpoons H^+ + H_2PO_4^-$
       $H_2PO_4^- \rightleftharpoons H^+ + HPO_4^{2-}$
       $HPO_4^{2-} \rightleftharpoons H^+ + PO_4^{3-}$
31. a. $[H^+] = 1.1 \times 10^{-3}$ *M* is acidic.
    b. $[H^+] = 6.0 \times 10^{-11}$ *M* is basic.
    c. $[OH^-] = 2.3 \times 10^{-11}$ *M* is acidic.
    d. $[OH^-] = 5.0 \times 10^{-7}$ *M* is basic.
    e. $[H^+] = 0.00013$ *M* is acidic.
    f. $[H^+] = 0.0000001$ *M* is neutral.
33. Compare the concentrations from Problem 32 to 1.0×10^{-7} to determine if they are acidic, basic, or neutral.
    a. $[OH^-] = 2.5 \times 10^{-9}$ *M* is acidic.
    b. $[OH^-] = 3.0 \times 10^{-6}$ *M* is basic.
    c. $[OH^-] = 0.0015$ *M* is basic.
35. a. $[H^+] = 5.0 \times 10^{-7}$ *M* is acidic.
    b. $[H^+] = 2.4 \times 10^{-3}$ *M* is acidic.
    c. $[H^+] = 0.000095$ *M* is acidic.
37. a. pH = 2.30; acidic.
    b. pH = 7.01; basic-neutral.
    c. pH = 1.33; acidic.
    d. pH = 0.041; acidic.
    e. pH = 8.54; basic.
    f. pH = 5.08; acidic.
    g. pH = 12.49; basic.
    h. pH = 9.30; basic.
39. $[OH^-] = 0.625$ *M*
    pH = 13.80
41. The pH meter will read 7 when the titration has reached the equivalence point and $[OH^-] = [H^+]$.
43. a. $KOH + HCl \rightarrow KCl + H_2O$
    b. $Ba(OH)_2 + 2\ HNO_3 \rightarrow Ba(NO_3)_2 + 2\ H_2O$
    c. $3\ NaOH + H_3PO_4 \rightarrow Na_3PO_4 + 3\ H_2O$
    d. $Ca(OH)_2 + H_2SO_4 \rightarrow CaSO_4 + 2\ H_2O$
    e. $3\ Ca(OH)_2 + 2\ H_3PO_4 \rightarrow Ca_3(PO_4)_2 + 6\ H_2O$

45. a. $HNO_3 + KHCO_3 \rightarrow KNO_3 + H_2O + CO_2$
b. $2\ HCl + Li_2CO_3 \rightarrow 2\ LiCl + H_2O + CO_2$
c. $Zn + 2\ HCl \rightarrow ZnCl_2 + H_2$
d. $Na_2CO_3 + 2\ HC_2H_3O_3 \rightarrow 2\ NaC_2H_3O_2 + H_2O + CO_2$
e. $Mg + H_2SO_4 \rightarrow MgSO_4 + H_2$

47. *M* of HCl = 0.0697 *M*

49. 5.02 mL.

51. A 0.30 *M* H_2SO_4 solution is 0.60 *M* in H^+ because each mole of H_2SO_4 is capable of releasing two moles of H^+.

53. 0.0350 mol HCl will neutralize 0.0350 mol NaOH.

55. M of vinegar = 0.9009 *M.*

57. A buffer can be defined as a weak acid-weak base pair that, by reacting with added amounts of a base or acid, can resist large changes in the solution's pH.

59. The following equilibrium equations explain how a hydrosulfuric acid-sodium hydrogen sulfide mixture can act as a buffer.

$H_2S + OH^- \rightleftharpoons HS^- + H_2O$

$HS^- + H^+ \rightleftharpoons H_2S$

61. When a small amount of hydrochloric acid is added to a carbonic acid-bicarbonate buffer, the $[H^+]$ will increase, the $[H_2CO_3]$ will increase, and the $[HCO_3^-]$ will decrease.

63. pH = 6.4

65. pH = 6.1

67. Even though the concentration of H_2CO_3 in blood plasma is only one-twentieth of the HCO_3^- concentration, the coupling of the equilibria between H_2CO_3 and the unlimited supply of CO_2 in the lungs provides for the high buffering capacity towards base through appropriate shifts in equilibrium.

69. Aqueous solutions with a high OH^- concentration must have a low H^+ concentration because $[H_3O^+][OH^-]$ must equal 1×10^{-14} at 25°C. If $[OH^-]$ rises, then to keep $[H_3O^+][OH^-] = Ka$, $[H^+]$ must go down.

71. The second ionization of a diprotic acid has a smaller equilibrium constant because the first proton comes from a neutral species, but the second proton must come from a –1 ion. It is harder to take a proton from a –1 ion than from a neutral species, so fewer come off and Ka decreases.

73. The hydroxide ion is the strongest base that can be present in water because a stronger base would take a proton from water leaving only OH^- in water.

75. a. HBr is the conjugate acid for Br^-.
b. H_3O^+ is the conjugate acid for H_2O.
c. HCO_3^- is the conjugate acid for CO_3^{2-}.

77. Beer with a pH of 5.0 would be acidic, any pH less than 7 is acidic.

79. a. $pH = -\log(1.0 \times 10^{-8}) = 8.00$
b. $pH = -\log(3.2 \times 10^{-4}) = 3.49$
c. $[H^+] = \dfrac{1 \times 10^{-14}}{3.2 \times 10^{-4}} = 3.1 \times 10^{-11}$

$pH = -\log(3.1 \times 10^{-11}) = 10.51$

81. pH = 4.81.

83. a. The conjugate acid-base pairs for this equilibrium are CH_3NH_2 (base)-$CH_3NH_3^+$ (conjugate acid) and H_2O (acid)-OH^- (conjugate base).
b. The conjugate acid-base pairs for this equilibrium are H_2O (base)-H_3O^+(conjugate acid) and H_3PO_4 (acid)-$H_2PO_4^{-1}$ (conjugate base).

85. 0.128 mol H_3PO_4

87. pH (HCl) = 1.52

pH (H_2SO_4) =1.22

89. a. 3×10^{-12} *M*
b. $\dfrac{[H^+]}{[OH^-]} = 1 \times 10^9$

chapter 16 oxidation - reduction

1. a. This reaction is an oxidation because Cr loses three electrons.
b. This reaction is a reduction because Cl_2 gains two electrons.
c. This reaction is an oxidation because S^{2-} loses two electrons.
d. This reaction is a reduction because Hg^{2+} gains two electrons.

3. a. The oxidation number of N_2O_5 is +5.
b. The oxidation number of NO_3^- is +5.
c. The oxidation number of NO_2 is +4.

5. $Pb(s) + PbO_2 + 2\ H_2SO_4(aq) \rightarrow 2\ PbSO_4 + 2\ H_2O(l)$

7. $10\ HNO_3 + I_2 \rightarrow 10\ NO_2 + 2\ HIO_3 + 4\ H_2O$

9. Oxidation and reduction occur simultaneously in a chemical reaction because when electrons are lost by one species (oxidation), they must be gained by another species (reduction).

11. a. The electrons lost will not equal the electrons gained if we simply add the given two half-reactions.
b. In order to add the two half-reactions so that the electron change will cancel out in the sum, we must multiply the first half-reaction by 2 and the second by 3.

$2\ Fe + 3\ Br_2 + \cancel{6\ e^-} \rightarrow 2\ Fe^{3+} + 6\ Br^- + \cancel{6\ e^-}$

13. a. The oxidizing agent can be defined as the material reduced. The reducing agent is the material oxidized.
    b. In the half-reactions in Problem 12, K^+ is an oxidizing agent in a, Zn is a reducing agent in b, MnO_4^- is an oxidizing agent in c, and SO_2 is a reducing agent in d.
15. a. Oxidation: $Sn^{2+}(aq) \rightarrow Sn^{4+}(aq) + 2e^-$
    The reducing agent is Sn^{2+}.
    Reduction: $Fe^{3+}(aq) + 1e^- \rightarrow Fe^{2+}(aq)$
    The oxidizing agent is Fe^{3+}.
    b. Oxidation: $Zn(s) \rightarrow Zn^{2+}(aq) + 2e^-$
    The reducing agent is Zn.
    Reduction: $2H^+(aq) + 2e^- \rightarrow H_2(g)$
    The oxidizing agent is H^+.
    c. Oxidation: $Zn(s) \rightarrow Zn^{2+}(aq) + 2e^-$
    The reducing agent is Zn.
    Reduction: $Fe^{3+}(aq) + 3e^- \rightarrow Fe(s)$
    The oxidizing agent is Fe^{3+}.
17. a. The oxidation number of each element in K_2SO_4 is K = +1, S = +6, and O = –2.
    b. The oxidation number of Cl in ClO_3^- is +5.
    c. The oxidation number of Cl in Cl_2 is 0.
    d. The oxidation number of N in NH_4^+ is –3.
    e. The oxidation number of N in NH_3 is –3.
    f. The oxidation number of P in H_3PO_4 is +5.
    g. The oxidation number of C in CH_4 is –4.
19. a. The oxidation number of Cl in Cl_2 is 0.
    b. The oxidation number of Cl in Cl_2O is +1.
    c. The oxidation number of Cl in Cl_2O_3 is +3.
    d. The oxidation number of Cl in Cl_2O_7 is +7.
    e. The oxidation number of Cl in Cl_2O_5 is +5.
21. a. In H_2O, hydrogen has an oxidation number of +1.
    b. In NaH, hydrogen has an oxidation number of –1.
    c. In H_2, hydrogen has an oxidation number of 0.
23. a. NO_3^- is reduced and is the oxidizing agent.
    I^- is oxidized and is the reducing agent.
    b. Bi^{3+} is reduced and is the oxidizing agent.
    SnO_2^- is oxidized and is the reducing agent.
    c. I_2O_5 is reduced and is the oxidizing agent.
    CO is oxidized and is the reducing agent.
    d. H^+ is reduced and is the oxidizing agent.
    Mg is oxidized and is the reducing agent.
    e. O_2 is reduced and is the oxidizing agent.
    H_2 is oxidized and is the reducing agent.
25. a. $2H_2O + NO \rightarrow NO_3^- + 4H^+ + 3e^-$
    b. $5e^- + MnO_4^- + 8H^+ \rightarrow Mn^{2+} + 4H_2O$
    c. $6e^- + Cr_2O_7^{2-} + 14H^+ \rightarrow 2Cr^{3+} + 7H_2O$
    d. $C_2O_4^{2-} \rightarrow 2CO_2 + 2e^-$
27. a. $3Ag + 4H^+ + NO_3^- \rightarrow 3Ag^+ + NO + 2H_2O$
    b. $4H^+ + 2NO_2^- + 2I^- \rightarrow 2NO + I_2 + 2H_2O$
    c. $2Pb^{2+} + S + 3H_2O \rightarrow 2Pb + 6H^+ + SO_3^{2-}$
    d. $H_2O_2 + 2H^+ + 2I^- \rightarrow I_2 + 2H_2O$
29. a. $3Zn + 4H_2O + 2MnO_4^- \rightarrow 3Zn^{2+} + 2MnO_2 + 8OH^-$
    b. $Cd + 2H_2O + NiO_2 \rightarrow Cd^{2+} + Ni^{2+} + 4OH^-$
    c. $8OH^- + S^{2-} + 4I_2 \rightarrow 8I^- + SO_4^{2-} + 4H_2O$
    d. $6OH^- + 3Br_2 \rightarrow 5Br^- + BrO_3^- + 3H_2O$
    e. $2Bi^{3+} + 6OH^- + 3SnO_2^{2-} \rightarrow 2Bi + 3SnO_3^{2-} + 3H_2O$
    f. $5OH^- + 8Al + 2H_2O + 3NO_3^- \rightarrow 8AlO_2^- + 3NH_3$
31. a. $3Cu + 8HNO_3 \rightarrow 3Cu(NO_3)_2 + 2NO + 4H_2O$
    b. $2K_2Cr_2O_7 + 2H_2O + 3S \rightarrow 3SO_2 + 4KOH + 2Cr_2O_3$
    c. $6H_2O + P_4 + 10HOCl \rightarrow 4H_3PO_4 + 10HCl$
    d. $PbO_2 + 4HI \rightarrow PbI_2 + I_2 + 2H_2O$
    e. $2HNO_3 + 3H_2S \rightarrow 2NO + 3S + 4H_2O$
33. Referring to Table 16.1, reactions a and d will occur spontaneously because aluminum appears above copper and iron appears above silver in the table.
35. a. Any metal appearing above Mg in the table is a better reducing agent than Mg.
    b. Any metal appearing below H_2 in the table will not spontaneously react with HCl.
37. Voltaic cells could be constructed with reactions a and d from Problem 33. In reaction a, place an Al electrode in $Al_2(SO_4)_3$ solution and a Cu electrode in $CuSO_4$ solution and construct a salt bridge between the two solutions. In reaction d, use Fe in $Fe(NO_3)_2$ and Ag in $AgNO_3$ solution.
39. a. In the battery described, Al acts as the anode and Sn acts as the cathode.
    b. The half-reactions and total cell reaction for the battery described.
    $2(Al \rightarrow Al^{3+} + 3e^-)$
    $3(Sn^{2+} + 2e^- \rightarrow Sn)$
    $2Al + 3Sn^{2+} \rightarrow 2Al^{3+} + 3Sn$
    c. The cathode, or Sn electrode will become heavier as the battery runs.
41. Reactions b and c from Problem 33 could be induced to occur in an electrolytic cell.

43. To plate the metal bracelet with gold, make the bracelet the cathode and the gold strip the anode in the circuit of an electrolytic cell. Submerge both electrodes in the $Au(NO_3)_3$ solution. The gold strip is oxidized to Au^{3+}. Au^{3+} ions plate out of solution as $Au(s)$ on the cathode bracelet.

45. a. The Pb plate is the anode and the PbO_2 plate is the cathode.
    b. The half-reactions that occur as a car's generator recharges a battery.
       $Pb^{2+}(aq) + 2\ e^- \rightarrow Pb(s)$
       $Pb^{2+}(aq) + 2\ H_2O \rightarrow PbO_2(s) + 4\ H^+(aq) + 2\ e^-$
    c. During recharging, the anode and cathode are reversed, making the Pb plate the cathode and the PbO_2 plate the anode.

47. a. Oxidizing agent is Fe^{3+}. Reducing agent is $CoQH_2$.
    b. $CoQH_2 + 2\ Fe^{3+} \rightarrow CoQ + 2\ Fe^{2+} + 2\ H^+$.

49. a. Reduction
    b. Oxidation
    c. Oxidation

51. Reducing agents supply the electrons to reduce other species, while the reducing agent is oxidized. Most metals want to lose electrons and form the positive ions, so they make good reducing agents.

53. $16\ HCl(aq) + 2\ K_2Cr_2O_7(aq) + C_2H_5OH(l) \rightarrow 4\ CrCl_3(aq) + 2\ CO_2(g) + 11\ H_2O(l) + 4\ KCl(aq)$

55. The balanced dry-cell reaction in a basic medium.
    $Zn(s) + 2\ NH_4Cl(aq) + 2\ MnO_2(s) \rightarrow ZnCl_2(aq) + 2\ NH_3(aq) + Mn_2O_3(s) + H_2O(l)$

57. a. A reducing agent that will convert Cd^{2+} to Cd, but not Mg^{2+} to Mg could be Al, Zn, or Fe.
    b. A metal that will react with $CaCl_2$ to form calcium metal could be Li, K, or Ba.

59. a. The balanced equation to determine the amount of sulfurous acid present in rainwater.
       $5\ H_2SO_3 + 2\ MnO_4^- \rightarrow 5\ SO_4^{2-} + 2\ MnO + H_2O + 8\ H^+$
       $5\ H_2SO_3 + 2\ KMnO_4 + 8\ K^+ \rightarrow 5\ K_2SO_4 + 2\ MnO + H_2O + 8\ H^+$
    b. Calculate the number of moles of H_2SO_3 from the given information.
       $0.00850\ M\ KMnO_4 \times 0.00812\ L = 0.0000690\ mol\ KMnO_4$
       $$0.0000690\ mol\ KMnO_4 \times \frac{5\ mol\ H_2SO_3}{2\ mol\ KMnO_4} \doteq 0.000173\ mol\ H_2SO_3$$
    c. Calculate the number of grams of SO_2 present from the given information.
       $$0.000173\ mol\ H_2SO_3 \times \frac{1\ mol\ SO_2}{1\ mol\ H_2SO_3} = 0.000173\ mol\ SO_2$$
       $$0.000173\ mol\ SO_2 \times \frac{64.06\ g\ SO_2}{1\ mol\ SO_2} = 0.0111\ g\ SO_2 \text{ per } 50.0\ mL$$

chapter 17 nuclear chemistry

1. $^{21}_{11}Na$ has 11 protons and 10 neutrons.
   $^{27}_{13}Al$ has 13 protons and 14 neutrons.
   ^{23}Na has 11 protons and 12 neutrons.
   ^{79}Br has 35 protons and 44 neutrons.
   ^{206}Pb has 82 protons and 124 neutrons.
   ^{241}Am has 95 protons and 146 neutrons.

3. $^{227}_{89}Ac \rightarrow ^{0}_{-1}e + ^{227}_{90}Th$
5. 0.12 g.
7. Comparison of "ordinary" chemical reactions and nuclear reactions.

	"Ordinary"	Nuclear
Particles participating	Only electrons	Protons and neutrons
Energy change	Relatively small	Very large
Effect of state	Different state of combination leads to different reactions.	No effect

9. a. $^{25}_{12}Mg$
   b. $^{197}_{79}Au$
   c. $^{131}_{53}I$
   d. $^{210}_{82}Pb$
11. Radioactivity is the spontaneous emission of radiation.
13. Isotopes are atoms with the same atomic number but different mass numbers. Radioisotopes are radioactive isotopes.
15. α-particle = $^{4}_{2}He$
    β-particle = $^{0}_{-1}e$
17. a. A few sheets of paper will stop α-particles.
    b. A sheet of aluminum, a block of wood, or heavy protective clothing will stop β-particles.
    c. Several layers of lead will stop most γ-rays.
    d. Several layers of lead will stop most x-rays.
19. α-emission leads to transmutation because the number of protons in the emitting nucleus is changed, yielding a new element.
21. a. $^{222}_{86}Rn \rightarrow ^{4}_{2}He + ^{218}_{84}Po$
    b. $^{227}_{89}Ac \rightarrow ^{4}_{2}He + ^{223}_{87}Fr$
    c. $^{235}_{92}U \rightarrow ^{4}_{2}He + ^{231}_{90}Th$
23. a. The atomic mass is unchanged; this isotope is a β-emitter.
    b. The atomic mass changes by 4; this isotope is an α-emitter.
    c. The atomic mass changes by 4; this isotope is an α-emitter.
    d. The atomic mass is unchanged; this isotope is a β-emitter.
25. The emission of γ-radiation is often ignored when writing nuclear equations because gamma emission affects neither nuclear charge nor mass.
27. a. $^{197}_{80}Hg$
    b. $^{24}_{11}Na$
    c. $^{1}_{0}n$
29. a. Equations c and g are α-emissions.
    b. Equations a and f are β-emissions.
    c. Equations b, d, and e are bombardment reactions.
31. $^{220}_{86}Rn$, $^{216}_{84}Po$, $^{212}_{83}Bi$, and $^{212}_{84}Po$.
33. The effect of nuclear radiation on matter is it often ionizes atoms and molecules.

35. Radiation "tracks" form in a cloud chamber because the radiation produces ions. Water molecules cluster about the ions forming clouds or "tracks."

37. Radiation exposes the film badges worn by those who work with radioactive materials.

39. 0.20 g ^{90}Sr

41. 24 days or January 25

43. 17,160 years = 1.72×10^4 years

45. a. No, all of a radioisotope with a half-life of 50 years will not be gone in 100 years.

    b. One hundred years is equivalent to 2 half-lives. One quarter of the sample will be left.

47. The part of a living cell that is most susceptible to radiation damage is the nucleus of the cell.

49. Beside immediate clinical symptoms, the possible long-term effects of radiation exposure include sterility and birth defects in offspring.

51. Radioisotopes are used in the treatment of cancer cells. Malignant tissues are irradiated with radiation from a Cobalt-60 source. See section 17.10 for other examples.

53. Radioactive ^3HCl could be used to demonstrate the given equilibria because if the equilibria are taking place, then radioactive ^3_1H will be found in the H_3O^+ ions and in H_2O molecules.

55. (1) Fusion reactions produce more energy than fission reactions.

    (2) The starting materials for fusion are more abundant than fissionable isotopes.

    (3) The products of fusion reactions are not radioactive.

57. $^{10}_{5}B + ^{4}_{2}He \rightarrow ^{13}_{7}N + ^{1}_{0}n$

59. a. $^{67}_{30}Zn$

    b. $^{13}_{5}B$

    c. $^{263}_{106}Sg$

61. Approximately one-fourth of the Strontium-90 will be left.

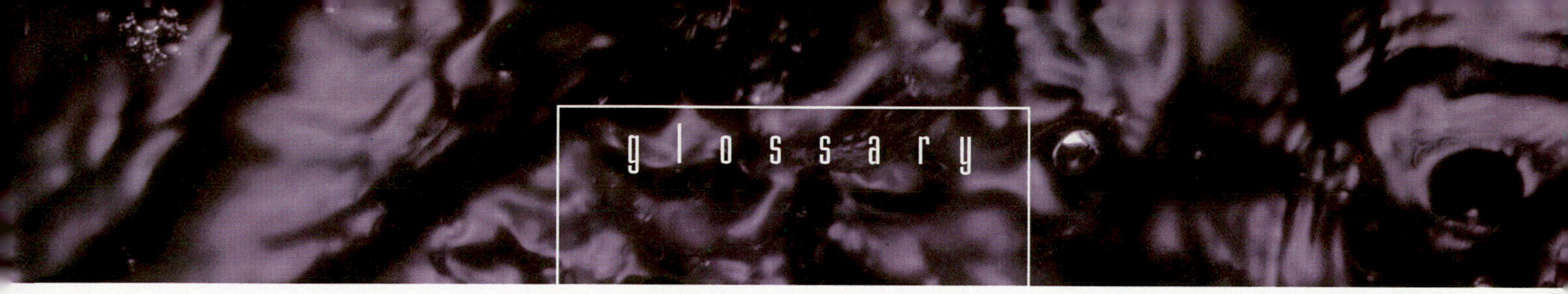

glossary

A

absolute temperature scale
See Kelvin temperature scale

acid
a compound that provides H+ ions to water solutions; generally the hydrogen is bound to a more electronegative element

acidity
the excess concentration of hydrogen ions relative to hydroxide ions in an aqueous solution

actual yield
the quantity of a product actually obtained in a reaction

alkali metal
a metal in group IA of the periodic table—Li, Na, K, Rb, Cs, or Fr

alkaline earth metal
a metal in group IIA of the periodic table—Be, Mg, Ca, Sr, Ba, or Ra

alloy
a homogeneous mixture of two or more metals

alpha (α) radiation
involves the emission of particles from a radioactive nucleus, consists of particles identical to the helium nuclei

amorphous solid
a solid with an internal structure where the particles are packed in an irregular manner

amphoteric
a substance that can act as either an acid or a base

anion
a negatively charged ion

aqueous
a water solution

Arrhenius acid
a compound that provides H^+ ions in aqueous solution

atmosphere (unit)
a unit of pressure equal to 760 Torr or 760 mm Hg

atmospheric pressure
the pressure of the atmosphere

atom
the smallest particle that retains the characteristic composition of an element

atomic mass (unit)
a mass equal to 1/12 of the mass of a carbon-12 atom; abbreviated amu

atomic number
the number of protons in the nucleus of an atom; an identifying characteristic of an atom

atomic symbol
letter(s) that represent(s) an element

Aufbau principle
Aufbau means “buildup” in German; a principle that shows the orderly placement of electrons from lower to higher energy levels

average atomic mass
the weighted average of all the different masses of isotopes of the same atom

Avogadro’s law
a law stating that equal volumes of all gases at the same temperature and pressure contain equal numbers of molecules

B

balanced chemical equations
an equation that has the same number of atoms (or ions) of each element on both sides of the arrow

base
a compound that provides hydroxide ions in aqueous solution or accepts hydrogen ions (Bronsted base)

bases
the number that is multiplied by a coefficient in an exponential number; for example, the ten in 1.5×10^2

basicity (alkalinity)
the excess concentration of hydroxide ions relative to hydrogen ions in an aqueous solution

beta (β) radiation
involves the emission of particles from a radioactive nucleus and consists of particles that are identical to electrons

binary compound
a compound composed of only two elements

Bohr’s model
a model where electrons have a certain definite energy and thus definite orbitals located at a specific distance from the nucleus

boiling point
the temperature at which the vapor pressure of a liquid equals the external pressure

bond
the force that holds elements together in compounds

bond dipole
the separation of partial charges (δ^+ and δ^-) between atoms

bond-dissociation energy
the energy required to break a covalent bond

bonding pair of electrons
the shared pair of electrons in a covalent bond

bond length
the distance between the nuclei of two atoms joined by a chemical bond

Boyle’s law
at constant temperature, the volume of a given sample of gas is inversely proportional to its pressure

Bronsted base
a substance that accepts H^+ ions

buffer
a weak acid-weak base pair which, by reacting with added amounts of a base or acid, can resist large changes in the solution's pH

buffering capacity
the capacity of a particular buffer to resist changes in pH; depends on the ratio of anion to acid

C

calorie
a unit of energy equal to the quantity of heat necessary to raise the temperature of 1 g of water by 1 degree C; equivalent to 4.184 J

capillary action
the ability of a substance to climb upward in a narrow tube

catalyst
a material that affects the speed of a chemical reaction without any permanent change in its own composition

cation
a positively charged ion

chain reaction
a nuclear reaction whose products cause the same reaction to occur again

Charles' law
at constant pressure, the volume of a given sample of gas is directly proportional to its absolute temperature

chemical change
a chemical reaction whereby one or more new materials are formed, each with distinctly different properties.

chemical energy
the potential energy stored in substances that can be released during a chemical reaction

chemical equation
a record of chemical change showing the conversion of reactants to products

chemical equilibrium
the state of a chemical reaction in which the rate of the forward reaction equals the rate of the reverse reaction

chemical formulas
the representation of a compound or ion in which elemental symbols represent types of atoms and subscripts show the relative numbers of atoms; in chemical language they are the words that represent compound

chemical properties
properties of matter that are observable when the substance undergoes a chemical change

chemical reaction
a change in which the composition (or structure) of one or more substances is altered

chemistry
the study of matter and the changes that matter undergoes

colligative properties
properties of solutions that are dependent only on the number of dissolved particles in solution

combination reaction
a reaction in which two substances combine to form one product

common name
an arbitrarily assigned name

compound
a chemical combination of elements that has a definite composition, a simple whole number ratio of atoms, and its own set of properties

concentrated solution
a solution which contains a relatively large quantity of a solute in a given quantity of solvent

concentration
the quantity of solute per unit volume of solution or per unit mass of solvent

concentration gradient
a difference in concentration between solutions

condensation
a change of state from a gas to a liquid; reverse process for evaporation

condensation point
temperature at which gas begins to condense into a liquid

conjugate acid
the cation (or molecule) that results from the reaction of a base with a proton

conjugate base
the anion (or molecule) that results from the loss of a proton by an acid

constancy of composition
a characteristic of elements and compounds that results in their having fixed properties

conversion factor
a ratio equal to 1, which can be multiplied by a quantity to change its units without changing its value

coordinate covalent bond
a covalent bond in which the bonding electron pair is contributed by only one of the bonded atoms

covalent bond
a bond resulting from sharing of two electrons

covalent compounds
compounds held together by covalent bonds

covalent (network) crystal
crystalline solid in which the lattice points are occupied by atoms covalently bonded to one another

critical mass
the smallest mass capable of sustaining a nuclear chain reaction

crystal lattice
unit of a solid characterized by a regular arrangement of components

crystalline solid
a solid with a regular internal structure of repeating units having a characteristic angle between their faces

D

Dalton's law of partial pressures
the total pressure of a gas mixture is equal to the sum of the partial pressures of its components

decomposition reaction
a reaction in which one compound is broken down into two or more substances

degrees Celsius
a point found on the Celsius temperature scale which is based on the melting point of ice at 0°C and the boiling point of water at 100°C

degrees Fahrenheit
a point found on the Fahrenheit temperature scale which is based on the melting point of ice at 32°F and on the boiling point of water at 212°F

density
the mass per unit volume of a sample of matter, a characteristic property that can be used to identify any material

diatomic
two atoms

dilute solution
a solution which contains a relatively small quantity of a solute in a given quantity of solvent

dilution
involves adding solvent (usually water) to a concentrated solution

dipole
difference in charge between atoms in a molecule

dipole-dipole force
an attractive force between polar molecules

dipole moment
an unequal distribution of charge in a molecule resulting from unsymmetrical orientation of polar bonds; also referred to as a dipole

diprotic acid
an acid that can produce two moles of protons per mole of acid

disintegrate (decay)
when radioactive elements disappear as they emit radiation and are changed into other elements

dispersion forces (London forces)
instantaneous attractive forces arising from the unsymmetrical distribution of electrons; the only attractive force between nonpolar molecules

dissociation
separation of ions from their close proximity in a crystal lattice to a distance

double bond
the sharing of two pairs of electrons between atoms

double-replacement reaction
reaction where two compounds react with each other to form two different compounds

duet
a pair of electrons associated with a hydrogen, helium, lithium, or beryllium atom, resulting in the stable configuration of a noble gas (helium); two valence electrons

dynamic (active) equilibrium
a state in which two opposite processes occur at equal rates

E

electrolyte solutions
solutions that conduct electricity

electrolytic (electrolysis) cell
a cell where a nonspontaneous reaction is made to occur by the application of electric energy

electron
a negatively charged subatomic particle; a fundamental particle of nature

electron affinity
the energy released when an electron is added to an isolated atom

electron configuration
the arrangement of the electrons in an atom, ion, or molecule

electronegativity
the attractive force that an atom exerts on shared electrons

electron spin
the spin of an electron around its axis

electron structure
the arrangement of electrons around the nucleus of an atom

electroplating
common application of electrolysis, an example is the silver plating of a fork

electrostatics
concept dealing with the interacting of charged particles; objects with opposite charges attract while objects with like charges repel each other

element
a substance that cannot be broken down into simpler substances by chemical means; one of the basic building blocks of which all matter is composed. Composed of one kind of atom.

empirical formula
the simplest formula for a compound that shows the atomic ratio of elements in the compound

empirical weight
mass of the empirical formula in grams

endothermic reaction
a reaction that absorbs heat

endpoint
the point in a titration when the indicator signals that the reaction is complete

energy
the capacity to do work

energy levels
the set of the discrete energy values that an electron may have in an atom

energy states
a description of the electrons in their energy levels around an atom

equilibrium concentrations
the concentrations of reactants and products at equilibrium

equilibrium constant
a constant that tells how far a reaction will proceed until it reaches equilibrium; at equilibrium at a given temperature, the ratio of product to reactant is always the same

equivalence point
the point at which a given quantity of a reactant is totally consumed in a reaction with an equivalent quantity of another reactant; point of neutralization

evaporation
a change of state from liquid to gaseous

exothermic reaction
a reaction that releases heat

F

formula mass
See formula weight

formula unit
the smallest grouping of ions corresponding to the correct fixed ratio in a formula

formula weight
the mass of a formula unit of a compound in amu

freezing
the process of change from a liquid to a solid state

freezing point
the temperature at which a liquid converts to a solid; note the solid and liquid phases of a substance are in equilibrium

G

gamma (γ) radiation
high-energy emission from nuclear processes, traveling at the speed of light; gamma rays are not particles

gas
a state of matter; a sample of matter that has indefinite volume and shape, where volume and shape are determined by a container

Gay-Lussac's law
at constant volume and mass, the pressure of a given sample is directly related to the Kelvin temperature

gram
the basic unit of mass in the metric system; one-thousandth of the SI standard mass—the kilogram

gram atomic weight
the atomic mass of an element expressed in grams; represents one mole of an element

gram formula weight
the formula weight expressed in grams, represents one mole of an ionic compound

gram molecular weight
the molecular weight expressed in grams represents one mole of the compound

graph
a visual display to view the change in one quantity compared to a change in another quantity

ground state
the lowest energy state of the set of electrons in an atom

group
a column in the periodic table, which includes elements with similar chemical properties; a family

H

half-life
the period of time it takes for half of a radioactive sample to disintegrate naturally

half-reaction
the oxidation or reduction half of a redox reaction

half-reaction method
the method of balancing oxidation-reduction equations that involves the completion and balancing of the oxidation and reduction half-reactions separately, followed by the combining of the two

halogen
an element of periodic group VIIA—fluorine, chlorine, bromine, iodine, or astatine

heat of condensation
heat given off when a gas condenses to the liquid state

heat of fusion
the heat required to change a one gram of solid to the liquid state

heat of solidification
heat released for solidification of one mole of liquid

heat of vaporization
the heat required to change a fixed mass of liquid to the gas state at a fixed temperature

Henderson-Hasselbach equation
the relationship between pH, pK_a, and the concentrations of the acid and base components in a buffer solution

heterogeneous equilibrium
equilibrium involving pure solids and liquids

heterogeneous mixture
a physical combination of substances that are neither uniform nor homogeneous throughout

homogeneous equilibrium
equilibrium not involving pure solids and pure liquids

homogeneous mixture
a mixture that is uniform throughout

Hund's rule
the electrons in a partially filled subshell in an atom occupy the orbitals singly before pairing

hydrate
an ionic compound that has water molecules attached to each formula unit

hydration
a process in which a dissolved ion bonds water molecules

hydration energy
energy released by the formation of new bonds when ions dissolve in water

hydrogen bond
the force resulting from the attraction of a hydrogen atom to a small, highly electronegative atom (F, O, or N) on another molecule (or the same molecule)

hydronium ion
the H_3O+ ion

hypertonic
a concentrated solution with a high osmotic pressure

hypotonic
a dilute solution with a low osmotic pressure

I

ideal gas
a hypothetical gas which obeys the ideal gas laws perfectly

ideal gas law
the pressure, volume, number of moles, and temperature of a sample of gas are related by the equation $PV = nRT$, where $R = 0.0821$ L • atm/mol • K

immiscible liquids
liquids that are completely insoluble in one another regardless of the proportions involved

indicator
a compound that has different colors in solutions of different pH, used to signal the end of a titration

insoluble
a solute that doesn't dissolve in the solvent

intermolecular force
a force between molecules

inverse square law
the law that states that intensity is inversely proportional to the square of the distance from the source

ion
a charged atom or group of atoms wherein the number of protons and electrons are no longer equal

ionic bond
the attractive force between oppositely charged ions

ionic compound
compound that consists of ions held together by ionic bonds

ionic crystal
a crystalline solid in which the lattice consists of oppositely charged ions

ionic equation
a chemical equation in which soluble, ionic substances are written with their ions separated

ionization
the formation of ions from neutral atoms or molecules

ionization constant
the equilibrium constant for the reaction of a weak acid or base with water to form ions in solution

ionization energy
the energy required to remove the least tightly bound electron from a gaseous atom or ion

ionizing radiation
radiation that is sufficiently high in energy to cause ion formation upon impact; nuclear radiation is an example

ion product of water
the product of the molar concentrations of hydronium and hydroxide ions in pure water

isoelectronic
two or more atoms or ions having the same electron configuration

isotonic
solutions with exactly the same particle concentration having identical osmotic pressure

isotope
a form of an element whose atoms all have the same number of protons but different number of neutrons

J

joule
the SI unit of energy; it takes 4.184 J to raise the temperature of 1.000 g of water 1.000°C

K

Kelvin temperature scale
the temperature scale with 273 as the freezing point of water and 373 as the normal boiling point of water; the scale required for gas law and certain other scientific calculations

kilogram
1,000 grams; the standard mass in SI

kinetic energy
energy of motion

L

labeled compound
a compound containing a radioactive isotope; can be detected and traced throughout the body

lattice energy
the energy released from the attraction between a gaseous cation and a gaseous anion to form an ionic compound

lattice structure
regular arrangement of ions in a solid

law of conservation of matter
since matter can neither be created nor destroyed, atoms must be conserved in a chemical equation

law of definite composition
a compound always contains its component elements in a fixed ratio by weight

Le Chatelier's principle
when a stress is applied to a system at equilibrium, the equilibrium shifts so as to tend to reduce the stress

length
the distance between two points

Lewis electron dot structure
a structure where symbols (dots) represent the valence electrons around an atom

Lewis structure
a combination of Lewis symbols to represent a molecule or a polyatomic ion. Shared electron pairs are represented as dashes and unshared electron pairs are shown as dot pairs.

limiting reactant (limiting reagent)
the reactant in a chemical reaction that limits the quantity of a product obtained

liquid
a state of matter; a sample of matter that has a definite volume and indefinite shape; shape is determined by a container

liquid crystal
liquid phase that retains some degree of fluidity but also exhibits some long-range ordering of crystals

litmus
a commonly used acid-base indicator

lone pair
an unshared pair of electrons on a bonded atom; also called nonbonding electrons

M

mass
the amount of matter in an object

mass number
the sum of the number of protons and the number of neutrons in an atom

matter
anything that has mass and occupies space

melting
a change of state from solid to liquid

melting point (temperature of fusion)
the temperature where solid and liquid states are both present; temperature where a solid turns to a liquid

metallic crystals
crystalline solids in which the lattice points are occupied by metallic cations surrounded by a sea of freely moving valence electrons

metalloid
an element near the dividing line between metals and nonmetals on the periodic table; has properties of both metals and nonmetals; generally, they are B, Si, As, Sb, Te, and At

metals
an element on the left in the periodic table, or a mixture of such elements; they are shiny, conduct electricity and are malleable

metric system
a system of measurement whose units are based on multiples of ten

miscible liquids
liquids that will dissolve in each other in all proportions

mixture
a physical combination of two or more pure substances that has an arbitrary composition and properties characteristic of its components, in no fixed proportions

molar gas volume
the volume occupied by one mole of any gas is 22.4 L at STP

molar heat of fusion
the heat needed for liquidification of one mole of solid

molar heat of vaporization
gives a measure of the strength of the intermolecular forces in a liquid expressed in kilojoules per mole

molarity
a measure of concentration defined as the number of moles of solute per liter of solution; symbolized *M*

molar mass
the mass in grams of 1 mol of a substance

mole
the chemical unit of matter equal to 6.02×10^{23} individual atoms, molecules, or formula units of the substance; abbreviated mol

molecular compound
a compound that consists of molecules

molecular crystal
crystalline solid in which molecules occupy the lattice points

molecular dipole
charge difference located over the whole molecule

molecular shape
the shape of the bonded atoms in a molecule

molecular weight
the mass of a molecule in amu

molecule
an uncharged, covalently bonded group of nonmetal atoms

mole ratios
conversion factors that relate molar amounts of one reactant to molar amounts of another reactant or the molar amounts of a product

monatomic ion
an ion consisting of one atom only

N

net ionic equation
an ionic equation in which the ions that do not change (the spectator ions) are omitted

neutralization reaction
a reaction of an acid and a base

neutron
a subatomic particle that has no charge and a mass of 1 amu; located in the nucleus

noble (inert) gas
an element of periodic group VIIIA—He, Ne, Ar, Kr, Xe, or Rn

nomenclature
a system for naming chemical compounds

nonbonding electrons
the valence electrons not involved in bonding

nonmetals
hydrogen or any element on the right in the periodic table; generally they lack luster, do not conduct electricity, are brittle, and many are gases at room temperature

nonoxoacids
molecular compounds composed of hydrogen and some nonmetal other than oxygen (or carbon)

nonpolar bond
a covalent bond in which the bonding electron pair is shared equally between the bonded atoms

nonpolar molecule
a molecule that does not contain a molecular dipole

nuclear fission
the splitting of a nucleus into two more or less equally sized smaller nuclei, releasing energy

nuclear fusion
combination of nuclei in a nuclear reaction to form a larger nucleus

nuclear reactions
reactions in which the nucleus undergoes changes

nucleus
the center of an atom; the protons and neutrons are located in this tiny area

O

octet
a set of eight electrons in the outermost shell of an atom or ion

octet rule
atoms or ions with an octet (8 electrons) are stable

orbital
the volume element within which an electron is likely to be found

osmosis
the diffusion of a solvent through a semipermeable membrane

osmotic pressure
measure of pressure forcing the solvent, water, to flow across a membrane

oxidation
an increase in oxidation number caused by the loss of electron(s)

oxidation number
for a monatomic ion, the charge on the ion; for an atom in a polyatomic ion or a molecule, the charge the atom would have if the pairs of electrons in each bond belonged to the more electronegative atom

oxidation-number method
the method of balancing redox reactions by equating the changes in oxidation number of the oxidizing and reducing agents

oxidation-reduction reaction
a reaction in which the oxidation number of one element is raised and the oxidation number of another element (or the same element) is lowered by the transfer of electron(s)

oxidizing agent
a species that can cause the oxidation of another reactant; the material that is reduced in a redox reaction

oxoacid
an acid containing an oxygen atom as a central atom with a hydrogen and a nonmetal attached to it

P

paramagnetic
a condition where unpaired electrons are present in the atoms

partial pressure
the pressure of each gas in a mixture of gases

Pauli exclusion principle
the rule that requires that electrons minimize their energy in the same orbital by having different spins

percentage composition (percentage by weight)
the percentages by mass of all elements in a compound

percentage concentration
a measure of the concentration defined as unit of solute per unit of solvent giving rise to three possible percentage expressions

percentage yield
the ratio of actual yield to theoretical yield, expressed as a percentage by multiplying by 100

period
one of the seven horizontal rows of the periodic table

physical change
a process in which no change in composition occurs

physical property
a property of matter that can be directly observed or measured without changing the composition of the substance being observed

physical state
the form in which matter is present; that is, solid, liquid, or gas

polar bond
a covalent bond in which there is unequal sharing of electrons

polarity
the measure of the inequality in the distribution of bonding electrons

polar molecule
a molecule that has a permanent dipole

polyatomic ion
an ion composed of two or more atoms

position of equilibrium
property of a system at equilibrium

positron
a subatomic particle created in a nuclear reaction that has the same properties as an electron except for being *positively* charged

potential energy
the energy possessed by an object by virtue of its position or composition

precipitate
an insoluble solid which separates out of a liquid reaction mixture

pressure
the force exerted per unit area

products
any substance produced in a reaction and appearing on the right-hand side of a chemical equation

properties
characteristics that are distinct for each material

proton
a subatomic particle with a mass slightly greater than 1 amu and a relative charge of 1 +, located in the nucleus

pure substance
an element or compound with a constant composition; always having the same properties for each physical state regardless of its origin

Q

quantization
a characteristic that energy can occur only in discrete energy states

quantum mechanics
a mathematical description of the electron states of atoms

R

radioactivity
a spontaneous process of unstable nuclei emitting energy

radioisotopes
isotopes formed by the spontaneous reaction of nuclei

radioisotope tracing
the medical diagnostic technique of using radioactive labeled compounds

rate constant
the proportionality constant in the rate law for a chemical reaction

reactant
the starting material that undergoes a reaction and appears on the left-hand side of a chemical equation

reaction rate
measure of the change in concentration of either a reactant or product in a given amount of time

reducing agent
a species that reduces the oxidation number of another reactant; the material that is oxidized in a redox reaction

reduction
the lowering of oxidation number by the gain of electron(s)

reversible reaction
reaction in which products can react with themselves to reform reactants

round off
upon completion of a calculation, the answer must be adjusted to reflect the correct significant figures, if the first figure to be dropped is less then five, the preceding digit is not changed. If the first figure to be dropped is five or more, the preceding digit is increased by one.

S

salt
an ionic compound formed from the reaction of an acid with a base

salt bridge
a bridge that allows positive and negative ions to pass between half-cells to maintain an electrical neutrality

saturated solution
a solution in which no more solute can dissolve in the given amount of solvent at the given temperature

scientific method
a method of problem-solving based on experimentation and reasoning that leads from observations to the construction of unifying principles for explanation and prediction

scientific notation
a format for writing large and small numbers using a coefficient with one (nonzero) integer digit times 10 to an integral exponent

significant figures
the reported number of figures that can be read accurately from a measuring device plus one more figure that must be estimated

single-replacement reaction
a reaction in which one element replaces one of the existing elements in a compound

SI system of units
a system of measurement whose units are meter for length, kilogram for mass, and liter for volume

solid
a state of matter; a sample of matter that has a definite shape and volume

solubility
the measure of how much solute can dissolve in some given amount of solvent

solubility product constant
the equilibrium constant for the equilibrium between a solid and its dissolved ion

soluble
when the solute dissolves in the solvent

solute
the component of a solution that is dissolved in another component—the solvent

solution (homogeneous mixture)
a homogeneous mixture composed of two or more pure substances whose proportions may vary in different mixtures

solvation
the process in which solvent molecules surround solute particles

solvent
the component of a solution that does the dissolving

specific heat
the quantity of heat required to raise the temperature of 1 g of a substance by 1°C

spectator ion
an ion that is present but does not change during a chemical reaction

spectroscope
an instrument which separates the components of visible light into distinct bands

spectrum
an array of radiation arranged in the order of some varying characteristic such as the wavelength

sphere of hydration
shell of water molecules surrounding each ion in solution

spontaneously
something that will occur by itself without external assistance

spontaneous reaction
a reaction that will occur by itself without external assistance

stable
a condition where an atom is in its ground state; generally something is considered stable if it has low potential energy

standard pressure
defined as 1 atmosphere

standard temperature
defined as 273 K

stoichiometric coefficients
the numerical coefficients found in a chemical equation

stoichiometry
a calculation relating the amounts of products and products in chemical reaction

strong acid
an acid that ionizes completely in water to form ions in solution

strong electrolyte
an electrolyte that substantially ionizes or dissociates in an aqueous solution

sublimation
a phase change in which a solid goes directly into the gas phase

sublimation point
the temperature at which a solid converts directly into a gas

supersaturated solution
a solution holding more solute than it can hold stably at a given temperature

surface tension
the energy required to increase or stretch a unit area of a liquid's surface

T

temperature
a measure of the hotness or coldness of an object

temperature of fusion (melting point)
point where solid and liquid states are both present; point where solid melts to become a liquid

temperature of sublimation
(*see* sublimation point)

ternary compound
a compound consisting of three elements

theoretical yield
the calculated quantity of product that would result from a chemical reaction based on the amount of limiting reactant given and the balanced chemical equation

titration
an experimental technique used to determine the concentration of a solution of unknown concentration or the number of moles in an unknown sample of a substance

torr
a unit of pressure equal to 1 mm Hg or *1/760* atm

transmutation
the conversion of one element into another by a nuclear reaction

trans-uranium element
an element with an atomic number greater than that of uranium

triple bond
a covalent bond consisting of three pairs of electrons shared between two atoms

triprotic acid
an acid that can produce three moles of protons per mole of the acid

U

unit cell
the smallest characteristic structural unit of a crystalline substance

universal gas constant
a proportionality constant relating temperature, volume, pressure, and number of moles of a gas

unsaturated solution
a solution in which the amount of solute dissolved is less than the solubility limit

unstable
: a condition in which an object has high potential energy; it will fall to a lower, more stable state

V

valence electron
: an electron in the outermost energy level of an atom

valence shell electron-pair repulsion (VSEPR) theory
: the theory that the valence shell electron pairs of the central atom of a molecule keep as far as possible from one another

vaporization
: a phase change from liquid to gas (vapor)

vapor pressure
: the pressure of the vapor over a liquid in equilibrium with its vapor

voltaic cell
: an electrochemical cell which uses a spontaneous redox reaction to generate electric current

volume
: the extent of space occupied by a sample of matter; the basic unit of volume in the metric system and SI system is the liter

W

weak electrolyte
: an electrolyte that ionizes or dissociates to only a small degree in aqueous solution

weight
: the force that gravity exerts on an object

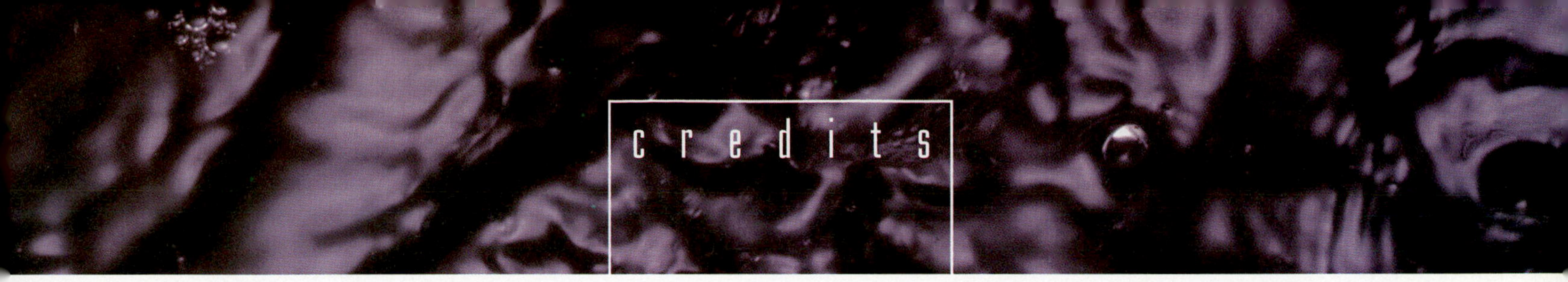

credits

Photographs

Chapter 1

Opener: Courtesy of Chemical Abstracts Service; **1.1:** © Matt Bradley/Southern Stock Photos; **1.2, 1.5, 1.6, 1.12, 1.14a–c, 1.20b, 1.21, 1.25:** © Yoav Levy/Phototake; **1.4, 1.16, 1.17:** © Len Lessin/Peter Arnold, Inc.; **1.7:** © Susan Kaprov; **1.8:** © Carol Rasegg/Martha Swope Associates; **1.9:** © John Kelly/The Image Bank; **1.10a:** © Bonnie Rauch/Photo Researchers, Inc.; **1.10b:** © ColorDay Productions/The Image Bank; **1.13:** © Nowitz/Phototake; **1.15:** Longcore Maciel Studios © Wm. C. Brown Communications, Inc.; **1.20a:** American Museum of Natural History

Chapter 2

Opener, 2.1, 2.13: © Yoav Levy/Phototake; **2.2, 2.3:** © Susan Kaprov; **2.6:** © NASA/Tom Stack & Associates; **2.8a and b:** © Wm. C. Brown Communications, Inc.

Chapter 3

Opener, 3.4a and b, 3.6, 3.7: © Yoav Levy/Phototake; **3.8:** © Michael Collier/Discover Magazine; **3.9:** © Walter Iooss/The Image Bank; **3.10:** © Paolo Koch/Photo Researchers, Inc.

Chapter 4

Opener, 4.12, 4.13, 4.14, 4.15, 4.17f: © Yoav Levy/Phototake; **4.1:** © Michael Ross/IBM; **4.11a:** © B. Seitz/Photo Researchers, Inc.; **4.11b:** © Susan Leavines/Science Source/Photo Researchers, Inc.; **4.16:** © Nimrod/Phototake; **4.17d:** © Len Lessin/Peter Arnold, Inc.; **4.17e:** © V. Fleming/Photo Researchers, Inc.

Chapter 5

Opener, 5.13(all), 5.16: © Yoav Levy/Phototake

Chapter 6

Opener(top): © J. & L. Weber/Peter Arnold, Inc.; **(bottom), 6.8, 6.11a and b, 6.12, 6.13, 6.15, 6.16a, 6.21a–d:** © Yoav Levy/Phototake; **6.1:** © M. Romanelli/The Image Bank; **6.16b:** © Dr. Dennis Kunkel/Phototake; **6.17:** American Museum of Natural History; **6.20, 6.24:** The Bettmann Archives; **6.23:** © Wm. C. Brown Communications, Inc./Jim Shaffer, photographer

Chapter 7

Opener, 7.1, 7.2, 7.4, 7.6, 7.8, 7.9, 7.10, 7.11: © Yoav Levy/Phototake; **7.3:** © Susan Kaprov

Chapter 8

Opener, 8.4, 8.5, 8.9, 8.10: © Yoav Levy/Phototake; **8.2:** NASA; **8.3:** © Susan Kaprov; **8.6:** © Tom McHugh/Photo Researchers, Inc.; **8.7:** © G. Hammell/The Stock Market; **8.8:** © Ed Wheeler/The Stock Market

Chapter 9

Opener: © D. Lawrence/The Stock Market; **9.1:** © D. Stoecklein/The Stock Market; **9.2:** © Susan Kaprov; **9.3a and b:** © Wm. C. Brown Communications, Inc./Jim Shaffer, photographer; **9.4, 9.6, 9.9, 9.10, 9.11, 9.12, 9.13, 9.14, 9.15, 9.16, 9.17:** © Yoav Levy/Phototake; **9.5:** © Mascardi/The Image Bank; **9.8:** Courtesy of Carbonaire Co. Inc., Palmerton, PA; **9.18:** © Ed Degginger

Chapter 10

Opener, 10.1a and b, 10.3, 10.5, 10.6, 10.14, 10.16, 10.17, 10.18, 10.19: © Yoav Levy/Phototake; **10.2:** © Wm. C. Brown Communications, Inc./Bob Coyle, photographer; **10.10:** The Bettmann Archives; **10.12, 10.13:** © Ed Degginger

Chapter 11

Opener: Courtesy of General Motors; **11.1:** © Susan Kaprov; **11.2, 11.8a and b, 11.20, 11.21:** © Yoav Levy/Phototake; **11.3:** © G. V. Faint/The Image Bank; **11.4:** © M. F. Kahl/Photo Researchers, Inc.; **11.9:** NASA

Chapter 12

Opener, 12.1, 12.2, 12.3, 12.4, 12.5, 12.18b: © Yoav Levy/Phototake; **12.15:** © Ed Degginger; **12.16:** © Karl Hartmann/Phototake; **12.20:** © Susan Kaprov

Chapter 13

Opener, 13.2, 13.4a and b, 13.10a–d, 13.13, 13.14, 13.15: © Yoav Levy/Phototake

Chapter 14

Opener, 14.3, 14.4a–c, 14.6, 14.7, 14.8a and b: © Yoav Levy/Phototake; **14.1a:** © R. Bennett/FPG; **14.1b:** © Cogolini/The Image Bank; **14.1c, 14.2:** © Ed Degginger

Chapter 15

Opener, 15.1, 15.2, 15.3, 15.4, 15.7a–c, 15.10, 15.11a–d: © Yoav Levy/Phototake; **15.6:** © Wm. C. Brown Communications, Inc./Jim Shaffer, photographer; **15.8:** © Explorer/Photo Researchers, Inc.

Chapter 16

Opener: © A. Jenick/The Image Bank; **16,1, 16.4, 16.5, 16.14:** © Yoav Levy/Phototake; **16.2:** © Richard Weiss/Peter Arnold, Inc.; **16.3:** © Ed Degginger; **16.10:** © Gabe Palmer/The Stock Market; **16.11:** © SOHM/Chromosohm/The Stock Market; **16.13:** © Phil Degginger; **16.15:** © D. Stocklein/The Stock Market

Chapter 17

Opener: © U.S. Dept. of Energy/Photo Researchers, Inc.; **17.1a:** © Wm. Rivelli/The Image Bank; **17.1b:** © Will McIntyre/Photo Researchers, Inc.; **17.2:** © James L. Ruhle & Associates; **17.3a, 17.4:** The Bettmann Archives; **17.3b:** Emilio Segre Visual Archives/American Institute of Physics; **17.11b:** © Lawrence Berkeley Lab/Photo Researchers, Inc.; **17.14** Art Resource, NY.; **17.15:** © Novosti/The Gamma Liaison Network; **17.16:** © Yoav Levy/Phototake

Illustrator

All art by Wilderness Graphics

index

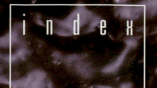

C

D

E

F

G

H

N

O

P

Q

R

U

V

W

table of atomic weights (based on carbon-12)

Name	Symbol	Atomic number	Atomic weight
actinium	Ac	89	(227)
aluminum	Al	13	26.9815
americium	Am	95	(243)
antimony	Sb	51	121.75
argon	Ar	18	39.948
arsenic	As	33	74.922
astatine	At	85	(210)
barium	Ba	56	137.34
berkelium	Bk	97	(247)
beryllium	Be	4	9.0122
bismuth	Bi	83	208.980
boron	B	5	10.811
bromine	Br	35	79.904
cadmium	Cd	48	112.40
calcium	Ca	20	40.08
californium	Cf	98	242.058
carbon	C	6	12.0112
cerium	Ce	58	140.12
cesium	Cs	55	132.905
chlorine	Cl	17	35.453
chromium	Cr	24	51.996
cobalt	Co	27	58.933
copper	Cu	29	63.546
curium	Cm	96	(247)
dysprosium	Dy	66	162.50
einsteinium	Es	99	(254)
erbium	Er	68	167.26
europium	Eu	63	151.96
fermium	Fm	100	257.095
fluorine	F	9	18.9984
francium	Fr	87	(223)
gadolinium	Gd	64	157.25
gallium	Ga	31	69.723
germanium	Ge	32	72.59
gold	Au	79	196.967
hafnium	Hf	72	178.49
hahnium	Ha	105	(262)
hassium	Hs	108	(265)
helium	He	2	4.0026
holmium	Ho	67	164.930
hydrogen	H	1	1.0079
indium	In	49	114.82
iodine	I	53	126.904
iridium	Ir	77	192.2
iron	Fe	26	55.847
krypton	Kr	36	83.80
lanthanum	La	57	138.91
lawrencium	Lr	103	260.105
lead	Pb	82	207.19
lithium	Li	3	6.941
lutetium	Lu	71	174.97
magnesium	Mg	12	24.305
manganese	Mn	25	54.938
meitnerium	Mt	109	(266)
mendelevium	Md	101	258.10

*A value given in parentheses denotes the mass number of the longest-lived or best-known isotope.